The Fascination of Groups

F. J. BUDDEN

CAMBRIDGE
AT THE UNIVERSITY PRESS
1972

Published by the Syndics of the Cambridge University Press
Bentley House, 200 Euston Road, London N.W.1
American Branch: 32 East 57th Street, New York, N.Y.10022

Library of Congress Catalogue Card Number: 77–142126

ISBN: 0 521 08016 9

Text set in 10/12 pt. Monotype Times, printed by photolithography, and bound in Great Britain at The Pitman Press, Bath

Contents

Preface xi

Notation xvii

1 **Mathematical structure** 1

2 **Putting things together: binary operations on a set** 5

Laws of composition: illustrations of binary operations on numbers, on sets, on people, on matrices, on ordered pairs and triples, in geometry. Unary operations. Matrices: a brief summary. Commutative and non-commutative operations. Exercises.

3 **Self-contained systems: closure** 14

Binary operations within a set. Illustrations; counter-examples – failure for closure. Closure by including extra elements, or by omitting elements. Exercises.

4 **Combined Ops: composition of operations** 21

Composition of geometrical transformations, of functions by successive substitution, of permutations, of matrices. Games with operations. Notation – first operation on the right: juxtaposition – the multiplicative notation. Matrices describing symmetries; permutation matrices. Successive transformations of the points of a circle. Exercises.

5 **Associativity** 36

Punctuation in mathematics. Testing for associativity. Examples of operations which are associative, and counter-examples where associativity fails. Associativity of mappings: examples. Exercises.

6 ***Status Quo*: identity elements** 45

Examples of identity elements. Formal definition. Finite arithmetics. Cases when identity elements are obscure. Geometrical examples. Left and right identities. Exercises.

7 **As you were! – inverses** 55

The idea of inverse – inverse permutations: the dual aspect of permutations. Inverses in finite arithmetics. Geometrical examples. General definition. The operation Δ on sets. Solution of linear equations – singular matrices. $(xy)^{-1} = y^{-1}x^{-1}$. Exercises.

8 **Group structure** 73

Requirements of a group – definition. Examples of finite and infinite groups. Structure tables for finite groups. A group of six permutations. Introduction to groups of symmetries; the two-group; further groups of symmetries. Abelian groups. Groups of two-dimensional transformations. Systems which fail to be groups. Exercises.

9 **Properties of groups** 89

Cancellation law. Latin-square property; proof. Solution of $ax = b$, etc. Use of Latin-square property to complete a group table. Cayley's theorem – verification. Regular representation of finite groups by matrices. Doing 'algebra' in a group. Examples discussed. Exercises.

10 **Period (or order) of an element: permutations and cycles** 106

Definition. Finding the period of an element from the group table. Subgroups generated by single elements. Period of a permutation – cycles; cyclic permutations. Period of product of disjoint cycles. Overlapping cycles. Period of a function; of a matrix. Matrices with complex terms. Infinite period; the infinite cyclic group. Exercises.

11 **Carbon copy groups: abstract groups: isomorphism** 131

A group of order 4 (the Klein four-group) in thirteen different situations. Abstract groups. Isomorphism of finite groups. Two essentially different groups of order 4; of order 6. Setting up an isomorphism; definition of isomorphic groups. Isomorphism between infinite groups. Isomorphism between additive and multiplicative groups in finite arithmetics. Slide rules for modular arithmetics. Automorphisms, of C_4, of D_2, of C_6. Automorphisms of non-Abelian groups; inner automorphisms. Exercises.

12 **Cyclic groups** 157

Nine realisations of the cyclic group of order 6. Recurring decimals. Cyclic groups of prime order. Regular polygons. Cyclic groups of composite order. *n*th roots of unity and the cyclotomic equation. Some problems. The infinite cyclic group. Exercises.

13 **The dihedral group** 187

Direct and opposite symmetries; mirror reflections. The full group of the regular hexagon (D_6). Other realisations of D_6, and of D_3 and D_4. Reflections in two intersecting mirrors – the kaleidoscope. The group D_3 of the equilateral triangle – isomorphic to S_3. Permutations of vertices – fixed and moving axes. Illustration of Cayley's theorem by rotations of triangular prism. 'Dividing up' the group D_6. The general dihedral group of order $2n$: two generators; defining relations. The infinite dihedral group. Groups generated by two elements of period 2. Exercises.

14 **Groups within groups: subgroups** 214

Definition; examples; Lagrange's theorem. Cyclic subgroups generated by single elements. Centre of a group. Subgroups of S_4. Invariant properties associated with subgroups. Normalisers and centralisers: use in finding subgroups. Non-cyclic subgroups generated by two or more elements. Subgroups of infinite groups: e.g. $\mathbf{C},+$; $\mathbf{C},\times$; $\mathscr{M}$, $\times$. Geometric transformations in two dimensions. Finite groups of plane transformations, and their subgroups. Exercises.

15 **Group specifications: generators and defining relations** 241

Groups requiring two generators. Defining relations. How many generators? How many relations? Form of defining relations; independence of generators. Groups of quaternions – dicyclic groups. Cayley diagrams; Cayley graph for Q_4, the quaternion group. Further note on Cayley diagrams. Exercises

16 **Bigger and better groups: direct product groups** 255

Definition and illustration. Building new groups, e.g. $C_6 \times C_2$. Associativity and commutativity of direct product; non-Abelian cases. Periods of elements in direct product groups. The 'packing-case' group; some groups of order 24. The groups of order 8. The group $C_2 \times C_2 \times C_2$: realisations: automorphisms: subgroups. An excursion into finite geometry. The groups $C_4 \times C_2$; $C_3 \times C_3$. Direct product of infinite groups; of finite with infinite groups. Exercises.

17 **Catalogue of groups: symmetry groups** 290

Table of all groups of order $n \leqslant 12$. Symmetry groups of two- and three-dimensional figures. The rotation group; enantiomorphs and the full group. Group of regular tetrahedron; representation by matrices. Cube and octahedron; representation by matrices and by permutations. Icosahedral group A_5; the full group $A_5 \times C_2$. Exercises.

18 **Permutations** 311

Odd and even permutation – inversions of order. Change of parity due to a single transposition. Permutations as product of transpositions. Pairing odd and even permutations – alternating groups. A set of permutations to represent S_4; two generators only needed. Resolution of permutation into disjoint cycles. Period of permutation is the L.C.M. of the lengths of its disjoint cycles. Manipulation of cycles; overlapping cycles. The group of a polynomial. Cross-ratio; the six cross-ratios of four numbers. Exercises.

19 **Cosets in finite and infinite groups: equivalence classes** 333

Cosets in the group D_6. Reading off cosets from the group table. Interpretation of cosets; illustrations. Preview of normal subgroups. Properties of cosets. Cosets in infinite groups: groups of vectors; of real and of complex numbers; of plane transformations. Lagrange's theorem – proof. Construction of structure tables. Equivalence relations and equivalence classes; partitioning into disjoint subsets. Cosets as equivalence classes. A binary operation on subsets – the product set. Product of subsets from infinite groups and from finite groups. Product of cosets. Exercises.

20 **Conjugate elements: normal subgroups (1)** 363

The transform of an element. Conjugate elements have the same period. Similar permutations. Reflection in a moved axis; successive reflections in moved axes. Successive rotations about moved vertices. Transformed operations in general; further illustrations. Conjugacy classes; finding conjugacy classes – the 'snap' method. The transforming element. Conjugacy classes and cosets; centralisers. The transform of a given subgroup; conjugate subgroups. Normal subgroups; to discover whether a given subgroup is normal or not – seven methods. Normaliser of a subset. Postscript. Exercises.

21 **Homomorphism: quotient groups: normal subgroups (2)** 396

Homomorphic mappings. Many-to-one mappings in general. Homomorphisms of finite groups. Kernel of a homomorphism. Normal subgroups; groups of cosets; quotient groups. Homomorphic images of Abelian groups. Failure when subgroup is not normal. Product of cosets of normal subgroups. Quotient or Factor groups. Quotient groups in direct product groups. Chains of normal subgroups. Simple groups: insolubility of quintic equation. Proof that A_5 is simple, and that A_5 is the only normal subgroup of S_5. Infinite groups with finite and infinite quotient groups. Summary. Exercises.

22 **Automorphisms** 421

Inner and outer automorphisms. Automorphisms of Abelian groups; of non-Abelian groups. Automorphisms of S_4. Proof that inner automorphisms are a normal subgroup of the full group of automorphisms.

23 **Groups and music** 429

Musical pitch; the octave; musical intervals. The harmonic series. The perfect fifth. The pentatonic scale; the Pythagorean scale; just intonation. Equal temperament; tempered intervals – the group C_{12}. Groups and musical form – the round, the canon and the fugue. Imitation, inversion, augmentation, sequences. Exercises.

24 **Ringing the changes: groups and campanology** 451

The campanologist's rules for ringing changes, and the principles of composition of methods. Three and four bells. Methods of producing twenty-four changes on four bells, using group theory. Five bells – Stedman Doubles; leads and plain courses, subgroups and cosets. Six bells; bobs and singles; Plain Bob Minor. Symmetry; other methods.

25 **Groups in geometrical situations** 480

Dihedral groups on a circle and on a parabola generated by two involutions. Poncelet's porism. Relettering of geometrical configurations. The complete quadrilateral; the orthocentric quadrilateral and nine-point circle; Pappus' theorem; Pascal's hexagon; Desargues' perspective triangle theorem. Stabiliser subgroups.

26 **Patterns** 504

Patterns obtained by systematic repetitions of a motif. The point groups C_n and D_n. Analytical and synthetic approaches. Fundamental regions. The seven frieze patterns. Classification of patterns. The two-dimensional (wall-paper) patterns. The twelve plane patterns which contain opposite isometries. Sub-patterns; subgroups of plane patterns; normal subgroups. Exercises.

Appendices 539

Answers 546

Bibliography 585

Index 589

Preface

This is not a text-book on group theory, and it does not pretend either to ideal balance and emphasis, or to a universally acceptable sequence of development and arrangement, nor does it make a fetish of mathematical rigour: we shall feel free to use results before we have proved them, and some of the more tedious proofs are relegated to appendices! The volume aims to interest, to enlighten and to transport the reader, rather than to provide him with the strict discipline of a mathematical education.

It takes 545 pages to cover what would be completed in most text-books in one to two hundred pages. But that is precisely its *raison d'être* – to be expansive, to examine in detail with care and thoroughness, to pause – to savour the delights of the countryside in a leisurely country stroll with ample time to study the wild life, rather than to plunge from definition to theorem to corollary to next theorem in the feverish haste of a cross-country run.... Perhaps some of the alternative titles which I had in mind before *The Fascination of Groups* finally crystallised may help to make its purpose clear: *The Appreciation of Groups*; *The Student's Guide to Groups*; *A Concrete Approach to Groups*; *The Architecture of Groups*; *Insight into Groups*; *Background to the Understanding of Groups*, ... The objective is to provide a wealth of illustration and examples of situations in which groups may be found and to examine their properties in detail, and the development of the elementary theory in the light of these widely ranged examples. The weakness of most texts is that, so often, not enough illustration is given, and the reader is compelled to construct his own examples, or is subjected to the frustration resulting from an undue degree of abstraction.

Now there are many admirable text-books on Groups, and even more less admirable ones!† At the top end of the scale one finds well-established, erudite university texts, from (chronologically) Burnside onwards. At the lower end of the scale, groups are now given more than a brief mention in most mathematical books intended for use in schools. In many of these, the treatment is excellent, but unfortunately they usually only scratch the surface of the subject, and there is generally a lack of *motivation*, a failure

† Some of which do not contain a single *diagram* (though that may possibly be their only defect). All the books listed in the Bibliography are admirable in my opinion for various reasons. The omission of any title is not necessarily to be interpreted as an indication of unworthiness, but more likely betrays my unfamiliarity with it.

to admit further development, and little indication of where the ideas are leading.

It was to bridge the gap between the extremes of the rigorous university text on the one hand, and the pleasant but shallow schoolbooks on the other that was thought to be the principal justification for embarking on the present work. There seemed to be a need for a book which would tackle the extremely important subject of groups in a way which would combine some measure of thoroughness with a wealth of attractively presented detail in a way which would satisfy a wide range of readers. The average student of mathematical courses may well be able to prove Lagrange's theorem, or to enunciate Sylov's theorem, but he may be hard put to point to some of even the simplest illustrations dealt with in the following pages. In short, he would lack a *feeling* for groups, as a musician with a rigorous training in classical harmony and counterpoint may yet not feel the shape or direction of a musical phrase, or appreciate the relative importance of the climaxes in the unfolding of an extended movement... So in the pages ahead, some of the most important groups are set out and analysed in greater detail than may be found at this elementary/intermediate level; while the application of groups to such fields as Music and Bell-ringing are given more than superficial treatment.

The ideas for this book have grown over a number of years out of my own enthusiasm for the subject, which has been catalysed by attempts to teach some of the topics to mathematical students. Indeed it is no exaggeration to claim that the book has grown directly from what has gone on in the classroom, though of course only a small fraction of the material may have actually received classroom treatment. One example of how class discussion affected the expansion of the book may be given:

I was showing a group of students the interesting non-Abelian group of order 27 which appears in the text on pp. 100–104, and to which the reader may care to refer. In listing the elements of this group, it happened that those for which $z = 0$ appeared first, i.e. $e = (0, 0, 0)$; $a = (1, 0, 0)$; $a^2 = (2, 0, 0)$; $b = (0, 1, 0)$; $b^2 = (0, 2, 0)$; $ab = ba = (1, 1, 0)$; $a^2b = (2, 1, 0)$; $ab^2 = (1, 2, 0)$; $a^2b^2 = (2, 2, 0)$. Clearly the y_2z_1 term in the composition law (on page 100) is superfluous, and we have the group $C_3 \times C_3$. Similarly, Gp $(a, c) \cong C_3 \times C_3$. The question arose: What is Gp (b, c)? Someone thought it would be all those elements for which $x = 0$, but this proved not to be the case. Now Gp (b, c) is undoubtedly of order 3, 9, or 27. Obviously it is not of order 3. But neither can it be of order 9, since $bc \neq cb$, and there are no non-Abelian groups of order 9. The unlikely conclusion remained that b and c *generate the whole group*. I say 'unlikely' since the group had given the appearance of being a three-generator group, and this I had previously supposed it to be. One was thus forced to admit that it must be possible to express a in terms of

b and c. As it turned out, $a = cbc^{-1}b^{-1}$, and this led to the example following in the text, wherein the same group is studied from the starting-point of an abstract definition in which $a = cbc^{-1}b^{-1}$ is, in effect, one of the defining relations. The reader may possibly be wondering why this group was introduced at all. The answer is that it is the simplest counter-example to refute the conjecture (from a scholar) 'Are two groups isomorphic if they contain equal numbers of elements of the same periods?' In this sort of way, the book grew (and still continues to grow, though new material is too late to be printed!), and explains how it came about that the size inflated to 545 pages when, at the outset, I had supposed that all I knew about groups could be contained within a very slim paperback!

Certain passages which may be safely by-passed without detriment are marked by red stars in the left margin. There are certain classes of reader who may well wish to omit much of the more mathematical portions of the book. I am thinking for example of musicians and campanologists. I would suggest that musicians might have read most of the first eight chapters, and possibly also Chapter 12, in order to get the most out of Chapter 23. Bell-ringers would need a little more theory, particularly on permutations (Chapters 10 and 18), as well as a little on subgroups and cosets (Chapter 19) and the beginning of Chapter 20. The more general reader, not wishing to become too deeply immersed in the theoretical sections, should not allow himself to be deterred by those more difficult sections from further reading, but would do well to press on to resume the text at the next point where it engages his interest.

There are some who may complain about the way the subject is developed and the sequence used; some who may say that undue emphasis is given to some aspects out of proportion to their true importance, as for example that there is a preoccupation with the display of group tables, which purists may well frown upon; or that lines of development are pursued which are unprofitable; and that the reader is encouraged to tackle problems which would have been best left till he had proceeded further in the theory and acquired more sophisticated techniques. For such imperfections I express my regrets, and hope that whatever may be lacking in mathematical orthodoxy, may be compensated by a measure of clarity and an ample degree of motivation in many diverse contexts.

Another kind of criticism may well be that there is so much wealth of detail that the thorough treatment leaves the student with little to do. Against this possible complaint, let it be pointed out that there is a very large number of questions and exercises, amounting perhaps to a unique collection at this level. The reader is urged to dig deeply into these examples, since as much of the fascination of groups lies there as in the text. Those questions which are woven into the fabric of the text are useful in helping along the development of the subject matter, which more often than not grows out of them. Many of these questions, and of the exercises at the end of each chapter are open ended and will launch the student upon a programme of study which may rivet his interest and curiosity for many hours. Quite a number of the questions are difficult, and some attempt has been made to classify and to grade the sets of exercises.

It is hoped that *The Fascination of Groups* will cater for the needs of the scholar who wishes to go beyond the superficial treatment to be found in the odd chapter or two of his text-book, or of the books which he may now find in his mathematical library; and will enable him to discover the endless fascination of a study which is now growing rapidly into school syllabuses. Indeed, it is hoped that this book will go some way towards demonstrating the suitability of elementary group theory as a subject for school study for seventeen and eighteen year-olds. And it could well be, in the years ahead, that a course in Matrices and Group theory may prove to be the very core of the Mathematics Syllabus, and this book seeks to clear a path, or rather to widen the paths already cleared, and to extend them to lead to the more interesting country beyond.

But the classes of reader to whom it is expected the book will make the strongest appeal are, first, teachers – particularly those who want to intensify and enrich a superficial knowledge of Groups, or who want to study the subject with very little previous knowledge behind them. To those, it is hoped that this will be a valuable source book. Second, university students, to whom it may be a companion volume to their main text-books, and one to which they may turn for enlightenment and for the provision of a new dimension to their studies. The student working on his own (as for example one pursuing a mathematical course of the Open University in Britain) should find the book especially helpful. Finally I hope it may be to senior pupils in schools and non-specialist students, a volume which they may either dip into, or make a detailed study of, according to their needs, their previous knowledge, or their mathematical ability.

The collection and processing of ideas, the writing of the text, the designing of the diagrams, and the amassing and devising of the repertoire of examples has been an absorbing and rewarding task, occupying over 1100 hours, and the work has been spurred on, as noted, by interchanges with classes. A very great deal of extra work has been done by a number of persons whose contributions I should like to acknowledge. First Mr D. Wanless, Scholar of King's College, Cambridge, who read the typescript and made many valuable suggestions as well as eradicating a large number of errors. Mr G. K. Rockett and Mr C. M. Turk, Scholars of University College, Oxford and of Trinity College, Cambridge, respectively, spent an enormous time working out answers to the questions and exercises. To these, my ex-pupils, I am particularly grateful and can only hope that they may have derived some benefit from the labour. I am also indebted to Mr J. Wolstenholme, Director of Music at the Royal Grammar School, Newcastle, for his study of Chapter 23 which led to many valuable additions and alterations and corrections of errors. The influence of Mr Wolstenholme's scholarship has added much to the authenticity and to the value of that chapter. A similar task was undertaken in respect of Chapter 24 by Mr A. Craddock, one of the campanological experts of the north-east region, and I am grateful for the great deal of trouble which he took. Finally I must record the tremendous amount I owe to Dr T. J. Fletcher, HMI, who inspired much of the material and who kindly read the whole work in typescript, offering a great deal of new material and valuable ideas, most of which became incorporated thereby providing much enrichment of the substance of the work.

F. J. Budden

Royal Grammar School
Newcastle upon Tyne

15 Westfield Avenue
Gosforth
Newcastle upon Tyne

March 1971

Notation

Throughout this book, we shall use the following notation:

$\mathbf{Z}$ = the set of integers; $\mathbf{Z}^+$ (or $\mathbf{N}$) the set of positive integers; $\mathbf{Q}$ = the set of rationals; $\mathbf{Z}\backslash 0$ the set of non-zero integers; $\mathbf{R}$ = the set of real numbers; $\mathbf{C}$ = the set of complex numbers.

The above will enable us to make the statements 'a and b are complex numbers' in the abbreviated form: '$a, b \in \mathbf{C}$'; 'p is a non-zero rational number' in the form '$p \in \mathbf{Q}\backslash 0$'; '$n$ is an even positive integer' in the form '$\frac{1}{2}n \in \mathbf{Z}^+$'. The latter statement may also be written '$n \in 2\mathbf{Z}^+$', for we shall use $2\mathbf{Z}$ to denote the set of even integers, a notation which may evidently be extended thus:

$5\mathbf{Z}$ denotes the set $\{\ldots -10, -5, 0, 5, 10, \ldots\}$
$5\mathbf{Z} + 2$ denotes the set $\{\ldots -8, -3, 2, 7, 12, \ldots\}$, and so on.

We may also use the notation $\mathbf{Z}_n$ to denote the set of residues modulo n, e.g. $\mathbf{Z}_5 = \{0, 1, 2, 3, 4\}$. Each member of such a set is really an equivalence class; indeed, the members are respectively the sets $5\mathbf{Z}$, $5\mathbf{Z} + 1$, $5\mathbf{Z} + 2$, $5\mathbf{Z} + 3$ and $5\mathbf{Z} + 4$.

$\mathbf{V}_2$ will denote the set of two-dimensional vectors, or ordered pairs of reals.
$\mathbf{V}_3$ will denote the set of three-dimensional vectors, or ordered triples of reals.
$\mathcal{M}_2$ will denote the set of 2×2 matrices $\Big\}$ with terms in $\mathbf{R}$, unless
$\mathcal{M}_3$ will denote the set of 3×3 matrices $\Big\}$ otherwise stated.

Groups will be denoted by specifying the set, followed by the operation, e.g. ($\mathbf{Z}_7$, + mod. 7), or $\{0, 1, 2, 3, 4, 5, 6\}$ + mod. 7, is the additive group of the residues, modulo 7.

When a group is multiplicative, it will be understood that the 'zero' element(s) has been excluded, thus:

$(\mathbf{Q}, \times)$ refers to the multiplicative group of the set $\mathbf{Q}\backslash 0$
$(\mathcal{M}_n, \times)$ refers to the multiplicative group of *non-singular* matrices $(n \times n)$
$(\mathbf{Z}_7, \times)$ refers to the multiplicative group of $\{1, 2, 3, 4, 5, 6\}$ or $\{\pm 1, \pm 2, \pm 3\}$ under multiplication modulo 7.
$\mathbf{S}$ refers to the set $\{x : x \in \mathbf{R}, 0 \leqslant x < 1\}$.

Elements of groups will generally be denoted by small letters, a multiplicative (juxtaposition) notation being used, with 1 as the identity element, though there may be exceptions (see pp. 30 and 147).

When x and y are operations or mappings of any sort, xy will denote that the operation y is applied first (see pp. 28–30).

Operands will follow the operators, and will generally be enclosed in parentheses, thus:
$R(\triangle)$ might mean 'apply the rotation R to the triangle';
$TR(\triangle)$ would mean 'first rotate the triangle, then apply the translation T to it'.

Bold letters in tables throughout the text indicate non-commutative products.

The symbol $\cong$ is used to denote isomorphism between groups (see p. 142). Thus $\{1, i, -1, -i\}, \times \cong C_4$.

Acknowledgments

The substance of Chapter 23 first appeared in part as an article in the *Mathematical Gazette*. The author and the publishers are grateful to the editor for permission to reproduce it in this book. Acknowledgment is also due to Dr T. J. Fletcher for Appendix 3 which also appeared first in the *Mathematical Gazette*.

1 *Mathematical structure*

Introduction: structure

'Structure' is an overworked, modish word. We hear of social structure, political structure, class structure, incomes structure, tax structure, the structure of the Church, the structure of the trades unions, of the transport system, of education, of an industry, of a language, of the universe, of an atom ... A geographer will see structure in something so amorphous to the eye of the layman as the rocky mass of a mountain like Tryfan, and the geologist will consider the crystalline structure of the rocks in the layers of earth beneath. In civil engineering, in architecture and in music, 'structure' retains its original meaning, but the word has extended beyond its architectural significance in the other contexts quoted. In mathematics, the word has acquired a special meaning again, and we need to look into this straight away. Indeed, of several words which might have been used in the mathematical context 'structure' is probably the most appropriate that could have been chosen to convey the meaning. Consider the following sets:

(a) {Napoleon's left ear, the weather, the Greek language, an electric razor};
(b) {the components of a TV set which has been dismantled};
(c) {A, B, C, ..., X, Y, Z} (the letters of the alphabet);
(d) {1, 2, 3, 4, ...} (the counting numbers);
(e) the set of positive real numbers;
(f) the set of students in the chess club;
(g) {{the even integers}, {the odd integers}} (a set consisting of two sets);
(h) {violet, indigo, blue, green, yellow, orange, red};
(i) the set of paints in a box;
(j) the set of points in a plane.

Some of these sets contain a finite number of objects (i.e. they correspond to one of the elements of set (d)!). Set (a) contains four elements, (c) contains twenty-six, (b) might contain several thousand. Some, on the other hand, contain an unlimited number, such as (d), (e) and (j), though you might instinctively feel that there are far more in (j) than there are in (e), and again, far more in (e) than in (d) (but see App. I).

Varying degrees of mathematical structure

In the case of (a), there is absolutely no sense in which this set can be said to have structure. Indeed, it is arguable that the set is not very well defined, and there are some people who would doubt whether there is any sense in which the set can be said to exist at all! In the case of (b), the components, consisting of transistors, resistors, capacitors, switches, cathode-ray tube, chassis, wires, etc., could be put together according to the designer's plans to build a complete TV set which would function correctly. Thus, what was an amorphous set of objects without structure could be fashioned into a unity, a single structure. In a similar way, a cathedral is a structure built from a set of stones, wooden beams, glass, etc.; a symphony is built from a set of thousands of notes to which the composer's art gave form, shape and unity. But, though it is illuminating to recall the basic meaning of the word, this is not quite the mathematician's idea of structure. Sets (c), (d) and (e) do have mathematical structure in varying degrees. The letters of the alphabet, however, have only a faint semblance of mathematical structure: they are 'ordered': A < B < C < ... < P < ... < Z. The order is, however, an arbitrary order, and is in no sense a natural, inevitable order as in the case of (d). There, $5 < 6$ precisely because 6 is the next integer after 5; in fact that is what we mean by 6 ! But F is the letter after E because men accepted by agreement and custom, that it should be so. If an order had not been agreed for the letters of the alphabet, everyday life would be made much harder. Think how often, apart from dictionaries, one relies on alphabetical lists, and how tedious life would be if man had not imposed upon the alphabet the rudimentary structure of an 'order'. Not only do we have A < B < C < ..., etc., but the relation '<' between letters is extended to be useful as a relation between words, so that we may say: ... ANGER < ANVIL < AORTA < ...; while in the telephone directory, ... LEE, M. R. < LEE, P. H. < LEE, R. < ..., and so we are able, very quickly to locate a particular name.

Operations on the natural numbers

The sense in which set (d) is a superior structure – mathematically – to set (c) lies in the fact that we may carry out 'operations' on the set of natural numbers. We may add two or more of them together; we may multiply two or more of them; we may subtract a number from a larger number; sometimes a natural number may be divided by another, sometimes not. Two or more natural numbers will always have a Highest Common Factor and a Lowest Common Multiple. Sometimes a natural number will have a square root in the set, sometimes (as in the case of all the natural numbers except 1, 4, 9, 16, 25, ...) a square root does not exist within the set. For example, given the numbers 18 and 12, one may derive from them the following new numbers: 30, by addition; 216 by multiplication; 6 by subtraction; 18^{12} (a very large number), by 'exponentiation'; 36, the L.C.M., 6, the H.C.F.; but 18 ÷ 12 does not have a solution *so long as we confine our attention to the set of natural numbers.*

Lower structures

But in the set {A, B, C, . . ., Y, Z}, there is no analogous process; there is no way in which one can combine two letters such as H and X to produce a new letter, F. One *could* impose artificially such operations, but it would be a pointless exercise. Set (i) does have the ghost of a mathematical structure about it: yellow and red paints when mixed produce orange, but that is about as far as we can go! The set (f) of students in the chess club has a slight mathematical interest: they can be ordered alphabetically, or according to order of merit in an examination; they can be placed into subsets: {those who take chemistry}, {those in the orchestra}, {those who have no sister}, {those whose surnames are ordinary English words}, and so on.

Higher structures

When we consider set (e) this will include the counting numbers, but will also include $1\frac{1}{2}$, so that here, the division, 18 ÷ 12 is possible *within the set*. Indeed, any positive real number may be divided by another to produce a third positive real number. Thus we begin to see that (e) is a 'higher' structure than (d), because it has *less imperfections* than (d). But (e) still possesses one of (d)'s imperfections – that it may not always be possible to find $a - b$ when a and b belong to the set (e). Thus, $\pi - 2{\cdot}7$ is possible, while $2{\cdot}7 - \pi$ is not. The set of positive reals is not a *perfect* structure. What do we mean by a perfect structure in the mathematical sense? It would be better to consider the question: 'How perfect can the structure of a set be made?' It may surprise you to learn that, of the sets listed under (a) to (j), it is set (g) which qualifies for first prize on grounds of structural completeness, for we have in this case what mathematicians call a 'field'. Briefly, this means that addition, subtraction, multiplication and division are always possible within the set. Here there are only two members of the set, the set of even integers, and the set of odds. A more familiar example of a field is the set of real numbers, in which it is possible to carry out the 'four rules' (or 'operations') of addition, subtraction, multiplication and division, always provided that, in the latter case, one does not attempt to divide by zero.
Examples of operations on these sets are as follows:

addition: odd + odd = even, etc.
subtraction: odd − even = odd, etc.
multiplication: odd × odd = odd, etc.
division: odd ÷ odd = odd; even ÷ odd = even.

However, we are unable to predict the result of dividing an even number by an even factor – the result may be odd or even, so even this system has an imperfection.

Set (j) – the set of points in a plane – can be given a mathematical structure. Thus, given any two points A and B, there is always a third point C lying on AB produced so that $AB = BC$, i.e. such that B is the mid-point of AC. Again there are *two* possible points C such that the triangle ABC is equilateral. Further, two distinct points determine a unique line, and two non-parallel lines determine a unique point; three non-collinear points determine a unique circle. And so on. Moreover, the points in a plane (like the students in the chess club) have a fascinating variety of subsets – curves, regions, shapes, all of which have their own intrinsic mathematical interest.

'*What is electricity?*' 'Electricity is what electricity does', someone replied. In the same way, though it is scarcely possible to *define* structure, a perfectly reasonable meaning can be ascribed to the relation '. . . has the same structure as . . .'

If you are still hazy about the concept of mathematical structure, do not be deterred from reading on. Many illustrations will be recorded on the following pages which it is hoped will help to clarify ideas and illuminate the study of the subject.

2 *Putting things together: binary operations on a set*

Words with alternative meanings

It is unfortunate that in mathematics, as in life in general, there are words which have two entirely different meanings. Such, for example, is the word 'identity', which may be used in the sense which we shall meet (see p. 45) or may refer to a mathematical expression which may be written in two alternative ways, such as:

$$\frac{3}{x-1}+\frac{2}{x}=\frac{5x-2}{x(x-1)}.$$

Again, the word 'inverse' may be used in mathematics in the sense in which it is chiefly used in this book (see Ch. 7), and also to refer to geometrical inversion with respect to a circle or sphere, as well as to the entirely different process of reflection in a point, usually in three dimensions (see pp. 271, 302). The word 'inverse' is used in yet another sense in logic (see p. 138). Other words which have distinct meanings within the realm of mathematics are 'involution', 'congruent', 'function', 'centre', 'linear', 'closed', 'model', 'field', 'order', 'complex', etc., as well as a substantial number of symbols. However, the context usually enables us to ascertain in what sense a particular word or symbol is being used.

Combining two objects to produce a third

One other such word is 'binary'. Frequently this is used, particularly in these days of computers, to refer to the use of 2 as the base of a number scale.† In this chapter, we refer to binary operations,‡ and this has nothing whatever to do with base 2. A binary operation is an operation on *two* objects from a set which produces from them a *unique* third object, as for example, the operation of multiplication on the set of integers, which takes the *two* integers -5 and $+3$ and produces from them the unique integer -15. Here we write $(-5)\times(+3)=(-15)$, the familiar sign '$\times$' being used to denote the binary operation multiplication. More generally, a binary operation may be indicated by a symbol, such as '$*$',

† See F. J. Budden, *Number Scales and Computers* (Longmans), 1966.
‡ A binary operation, which combines an ordered pair of elements of a set to produce a unique element is sometimes called a 'law of composition'.

so we shall denote by $x * y$ that object which results from the application of the operation $*$ to the two objects x and y.

Such an operation, as we have seen, endows the set with mathematical structure. No such structure is possible in the set of students in the chess club, because there is no binary operation which takes 'Marshall' and 'Graham' and from them produces the answer 'Sherwood'. It is possible to manufacture a binary operation on a subset of N human beings by (say) assigning to each individual a number from 0 to $N - 1$, and then adding modulo N; but there would be enormous practical difficulties! On the whole set of human beings living and dead, one might invent a binary operation as follows: for any two given persons, there may be traced a 'latest common forefather'. In the case of two brothers, this would be their father; for two cousins, their grandfather; a person and his natural uncle (or aunt) would have that uncle's father as their l.c.f.; while for any pair of human beings we may, in theory determine their l.c.f. by tracing back enough generations, provided the precaution is taken of eliminating ambiguity by selecting from all possible people obeying the above definition the *one* who was born *latest* in time.

Examples of binary operations

However, it is possible to do binary operations on *two sets*, and one of these could be the set of students in the chess club! Suppose the second set is the set of students whose names end with '-man'. Call these sets respectively **A** and **B**. Then you may be familiar with the operations $\cap$ and $\cup$:

$\mathbf{A} \cap \mathbf{B} =$ the set of students in both **A** and **B**, i.e. who are in the chess club and whose names end with '-man'.

Q. 2.01. What is the meaning of $\mathbf{A} \cup \mathbf{B}$?

Unary operations

It is also possible to have a *unary* operation. For example, in the set of natural (counting) numbers, any number may be replaced by its square, $3 \to 9$, $7 \to 49$, and so on. Again, for a given real number x, the unary operation $x \to \sin x$ maps the number x into a real number lying between -1 and $+1$, x having been regarded as the measure of an angle in radians. You may invent other mathematical examples of unary operations within other sets. In the set of human beings, every member has a unique father: Queen Elizabeth II of Great Britain → King George VI of Great Britain. Again, there is a unary operation on a single *set*, known as 'complementation': if **A** is a set possessing a particular attribute, then the complement, $\mathbf{A}'$, is the set of all those objects under consideration which do not possess that attribute. For example, if the universal set is the set of all students in a college, and **A** is the set of students in a particular class, then **A**$'$ is the set of all students who are not in this class. Evidently $(\mathbf{A}')' = \mathbf{A}$. Differentiation is a unary operation on the set of functions of x.

Q. 2.02. Is integration a unary operation on this set?

Taking the value of the determinant of a square matrix, e.g.

$$\begin{bmatrix} a & b \\ c & d \end{bmatrix} \rightarrow \begin{vmatrix} a & b \\ c & d \end{vmatrix} = ad - bc,$$

is a unary operation on the set of square matrices, and so on.

More examples of binary operations

In this book, we shall be almost exclusively concerned with binary operations, though you should note that any binary operation might be thought of as a *unary operation on ordered pairs*; thus for example:

$$(6, 9) \xrightarrow{\times} 54; \qquad (6, 9) \xrightarrow{-} -3; \qquad (6, 9) \xrightarrow{\div} \tfrac{2}{3}.$$

In Ch. 1 we have considered such examples as addition, subtraction, multiplication and division on various sets of numbers, and of exponentiation, of Highest Common Factor, and of Lowest Common Multiple on the set of positive integers. There is, moreover, no limit to the diversity of binary operations which can be invented. For example, if x and y are two real numbers we might combine them according to the rule:

$$x * y = \frac{xy}{x + y} \quad (x \neq -y),†$$

to give a third real number. The symbol $*$ denotes symbolically the binary operation given explicitly by the formula on the right-hand side of the '$=$' sign. There is evidently no limit to the complexity of the formula which might appear on the right-hand side to define the binary operation. Indeed, it is not even necessary for the operation to be expressible by a formula at all, *so long as* $x * y$ *may be uniquely determined.* For example, if x and y are in the set of positive integers, and the operation is that of finding the H.C.F. (or L.C.M.), this cannot be expressed by an algebraic, or any other, formula. We could write:

12 H 18 = 6 (where H means 'take the H.C.F.');
12 L 18 = 36 (where L means 'take the L.C.M.').

† To have a binary operation on the set of reals, it would be necessary to supplement the given definition by specifying the meaning of $x * y$ in the case when $x = -y$. Otherwise the 'domain' of the operation would have to be restricted to a subset of the reals.

Again, the operation $*$ might be interpreted:

$x * y =$ *the smaller of x and y*, e.g. $7 * -3 = -3$,

and this could not be expressed by a simple formula. In Fortran computer language, the operation of multiplication is indicated $*$, and exponentiation $**$, so that $4*3$ is 12, while $4**3$ is 64.

All sorts of eccentric methods of combining numbers may be invented. For example:

(a) $0{\cdot}234 \rightsquigarrow 0{\cdot}181 = 0{\cdot}213841$;

(b) $234 \diamond 181 = 244$;

(c) $234 \hat{\uparrow} 181 = 160$.

You will certainly spot how the operation $\rightsquigarrow$ works – the decimal digits of the two given numbers are written down alternately. This is a perfectly good binary operation, and it is not without its uses (see App. I).

The operation $\diamond$ will be more obscure: corresponding digits of the two numbers are multiplied modulo 10; for the tens digits, for example:

$3 \times 8 = 24 \equiv 4 \pmod{10}$.†

Again, $976 \diamond 75 = 090$.

It is unlikely that you will see the rule which gives the operation $\hat{\uparrow}$. The numbers are first converted into the binary scale:

$$\begin{aligned} 234 &= 11101010 \\ 181 &= 10110101 \\ \hline & \quad 10100000 = 160, \end{aligned}$$

and a '1' appears in the answer only when *both* the given numbers have a '1' in that particular column.

Q. 2.03. What is the connection between this and the previous operation $\diamond$? Invent other 'eccentric' operations on integers.

Suppose we have $x, y \in \mathbf{Z}$, and a binary operation $*$ is defined as follows:

$x * y = 2$ when $|x - y| \equiv 0 \pmod{3}$;
$x * y = -1$ when $|x - y| \equiv 1$ or $2 \pmod{3}$.†

Surprisingly enough, the operation *can* be expressed by a formula in this case:

$x * y = 2 \cos \frac{2}{3}\pi(x - y)$,

† For example, $n \equiv 2 \pmod{3}$ means that n is of the form $3k + 2$, where k is an integer, i.e. n leaves a remainder 2 when divided by 3. See pp. 46, 340, 352, 413 ff.

and the use of trigonometry here may be an even greater source of amazement.

With two sets, besides the operations $\cap$ and $\cup$, we could define an operation Δ as follows:

$$\mathbf{A} \,\Delta\, \mathbf{B} = (\mathbf{A} \cap \mathbf{B}') \cup (\mathbf{A}' \cap \mathbf{B}).$$

Q. 2.04. How could this operation be described in words?

Matrices: a brief summary

Matrices will feature in this book. If you have no knowledge of matrices, you need not allow yourself to be held back by those sections of the book that deal with matrices. If, however, you feel you might be missing something,† the chief properties are summarised as follows.

A matrix is a rectangular array of numbers, such as $\begin{bmatrix} a & b & c \\ p & q & r \end{bmatrix}$ which may be regarded as a *store of numerical information*. In the example quoted, there are two rows and three columns. The matrix may in fact be regarded as *two row vectors*, each of three components; or alternatively, as *three column vectors*, each of two components.

Matrix addition

A pair of 2×3 matrices may be 'added' by the rule:

$$\begin{bmatrix} a_1 & b_1 & c_1 \\ p_1 & q_1 & r_1 \end{bmatrix} + \begin{bmatrix} a_2 & b_2 & c_2 \\ p_2 & q_2 & r_2 \end{bmatrix} = \begin{bmatrix} a_1 + a_2 & b_1 + b_2 & c_1 + c_2 \\ p_1 + p_2 & q_1 + q_2 & r_1 + r_2 \end{bmatrix}.$$

This is a binary operation on the set of 2×3 matrices. The rule for matrix addition may be generalised so that any pair of p by q matrices may be combined.

Matrix multiplication

A p by q matrix may be 'multiplied' by a q by r matrix by the rule illustrated in the case $p = 3$, $q = 2$, $r = 4$:

$$\begin{bmatrix} a_1 & b_1 \\ p_1 & q_1 \\ x_1 & y_1 \end{bmatrix} \times \begin{bmatrix} a_2 & p_2 & u_2 & x_2 \\ b_2 & q_2 & v_2 & y_2 \end{bmatrix}$$

$$= \begin{bmatrix} a_1a_2 + b_1b_2 & a_1p_2 + b_1q_2 & a_1u_2 + b_1v_2 & a_1x_2 + b_1y_2 \\ p_1a_2 + q_1b_2 & p_1p_2 + q_1q_2 & p_1u_2 + q_1v_2 & p_1x_2 + q_1y_2 \\ x_1a_2 + y_1b_2 & x_1p_2 + y_1q_2 & x_1u_2 + y_1v_2 & x_1x_2 + y_1y_2 \end{bmatrix}.$$

† See, for example: G. Matthews, *Matrices I* and *II* (Arnold); A. Coulson, *Matrices* (Longmans).

Here we have a second binary operation on matrices, but for this operation to be possible, it is essential that the number of columns in the first matrix shall be equal to the number of rows in the second. In general, we shall mostly be concerned with square ($p = q$) matrices. Thus two 2×2 matrices may be 'multiplied'; and it is worth verifying (if you are not already familiar with the process) that the resulting 'product' matrix, also a 2×2 matrix, would in general be different if the order of the two given matrices were reversed. For example:

$$\begin{bmatrix} 2 & 0 \\ -3 & 1 \end{bmatrix} \times \begin{bmatrix} 0 & -1 \\ 1 & 3 \end{bmatrix} = \begin{bmatrix} 0 & -2 \\ 1 & 6 \end{bmatrix} \quad \text{whereas}$$

$$\begin{bmatrix} 0 & -1 \\ 1 & 3 \end{bmatrix} \times \begin{bmatrix} 2 & 0 \\ -3 & 1 \end{bmatrix} = \begin{bmatrix} 3 & -1 \\ -7 & 3 \end{bmatrix}.$$

On the other hand, $\begin{bmatrix} 5 & -7 \\ -2 & 3 \end{bmatrix} \times \begin{bmatrix} 1 & 14 \\ 4 & 5 \end{bmatrix} = \begin{bmatrix} -23 & 35 \\ 10 & -13 \end{bmatrix}$

and $\begin{bmatrix} 1 & 14 \\ 4 & 5 \end{bmatrix} \times \begin{bmatrix} 5 & -7 \\ -2 & 3 \end{bmatrix} = \begin{bmatrix} -23 & 35 \\ 10 & -13 \end{bmatrix}.$

The latter event is rare. In general, it may be assumed that multiplication of square matrices is *not commutative*, i.e. the matrices may not be commuted, or interchanged – the 'order' of multiplication matters.

Commutative operations

In general, an operation $*$ is said to be commutative when $x * y = y * x$ for any pair of elements x, y of the set. For example, this is true in the case of the binary operation given by the formula $xy/(x + y)$, where $(x \neq -y)$ above, but would not be true if the formula were, say, $(x - 2y)/(2x + y)$.

Q. 2.05. In general, what property does the formula on the right-hand side have to possess in order for the operation to be commutative? Give simple examples of non-commutative operations.

An important special case of matrix addition is the case of 1×2 matrices, or 'ordered pairs', the addition rule for which is:

$$[x_1, y_1] + [x_2, y_2] = [x_1 + x_2, y_1 + y_2].$$

Ordered pairs could also be written as 'column vectors': $\begin{bmatrix} x \\ y \end{bmatrix}$.

Again, 1×3 matrices, or 'ordered triples', may be added in a similar way.

Here is an artificial example of a method of combining ordered pairs:

$$[x_1, y_1] + [x_2, y_2] = [x_1 + x_2 \text{ (mod. 5)}, y_1 \times y_2 \text{ (mod. 3)}],$$

where $x_1, x_2 \in \{0, 1, 2, 3, 4\}$ and $y_1, y_2 \in \{0, 1, 2\}$; e.g. $(4, 2) + (3, 2) = (2, 1)$, (compare pp. 256, 262 and Ex. 3.15).

Geometrical examples

We have already mentioned examples of binary operations on geometrical objects such as points and lines. In *one dimension* – on a line – we could easily invent a binary operation on the set of points on the line, e.g. having selected any two points A and B, we define $A * B$ to be the mid-point C of the line segment AB, and write $A * B = C$.

★Or, if O were a fixed point on the line, C could be defined as the harmonic conjugate of O with respect to A and B, so that the cross ratio $(OACB) = -1$.† Perceptive readers will realise that the first example was merely a special case of the second, with the point O at infinity.

In three dimensions, we may define a binary operation $\cap$ on two planes by the following: $\mathbf{\Pi}_1 \cap \mathbf{\Pi}_2 = l$, where l is the line of intersection of the two planes $\mathbf{\Pi}_1$, $\mathbf{\Pi}_2$. Here, however, the composition of the two planes does not produce another plane, but *a different kind of object*, a line – the operation is not *closed* but yields a resulting object which is outside the set (see Closure, Ch. 3).

An illustration from topology

A different kind of geometric example could be invented by a method of composition of circuits (or regions) in two dimensions. Suppose each circuit in fig. 2.01 (think of Kirchoff's laws for electrical circuits) is denoted by a letter thus: $BDCB = a$; $CDAC = b$; $ADBCA = c$; $ABCA = d$; $BDCAB = p$; $CDABC = q$; $ADBCA = r$. Then we may consider the circuits $ADBA$ and $CDAC$ to be 'combined' to give the single

† The method could be extended into the Argand plane: if O is the origin, and a and b are *complex* numbers, $a * b$ could be defined as $a * b = c$, where:

$$(Oacb) = -1, \text{ i.e. } \frac{(O-a)(c-b)}{(O-b)(c-a)} \quad -1 \Rightarrow c = \frac{2ab}{a+b}.$$

Oddly enough, we get *almost* the relation $x * y$ given as an example on p. 7. Readers may like to consider what happens in the case $a + b = 0$. See F. J. Budden, *Complex Numbers* (Longmans), p. 62.

circuit *BDCAB*. This is a possible *definition* of 'addition' of circuits, and we may write $c + b = p$. Some readers may think that we are merely adding areas, but we may also agree that by our definition, $b + p = p$,

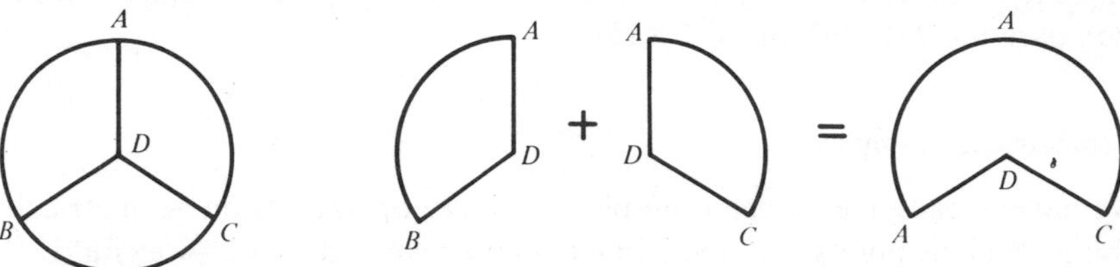

Fig. 2.01.

since a circuit of p 'includes' circuits of both p and b. The way in which all possible circuits combine is shown in table 2.01. Some readers may

Table 2.01

	a	*b*	*c*	*d*	*p*	*q*	*r*
a	*a*	*r*	*q*	*d*	*d*	*q*	*r*
b	*r*	*b*	*p*	*d*	*p*	*d*	*r*
c	*q*	*p*	*c*	*d*	*p*	*q*	*d*
d	*d*	*d*	*d*	*d*	*d*	*d*	*d*
p	*d*	*p*	*p*	*d*	*p*	*d*	*d*
q	*q*	*d*	*q*	*d*	*d*	*q*	*d*
r	*r*	*r*	*d*	*d*	*d*	*d*	*r*

recognise that what we are doing is really taking the binary operation $\cup$ between the sets of points on the perimeters of the regions *a*, *b*, *c*, etc.

You may try to construct the table for the networks shown in fig. 2.02

Fig. 2.02.

but you will come up against a snag in the second of these.

We described this composition of circuits as a geometrical example. It would have been more correct to describe it as a 'topological' example, our concern being not with size or shape, but only with configuration. Indeed, the circuit of fig. 2.01 is merely a distortion of the well-known Wheatstone Bridge circuit, as you may easily verify.

EXERCISES 2

1 What binary operation on the set of integers might the $*$ stand for in the following cases:
(a) $28 * 12 = 4$; (b) $9 * 13 = 4$; (c) $3 * 4 = 81$;
(d) $7 * 9 = 3$, $8 * 5 = 0$; (e) $53 * 11 = 4$, $4 * 7 = 0$.

2 Using the operations $\Diamond$, $\ltimes$ and $\pitchfork$ defined on p. 8 find:
(a) $47 \Diamond 608$; (b) $0{\cdot}047 \ltimes 0{\cdot}608$; (c) $0{\cdot}608 \ltimes 0{\cdot}047$;
(d) $187 \pitchfork 593$.

3 Is the operation 'anagram' an example of a unary operation on the set of words? e.g. LIVE $\xrightarrow{\text{anag.}}$ EVIL; ROSYTH $\xrightarrow{\text{anag.}}$ SHORTY.

4 If 1011001 is a binary numeral, then its complement is 0100110. Invent methods of combining natural numbers, e.g. addition with no carrying; first number + complement of second number, with or without carrying, etc.

5 If the matrices $\begin{bmatrix} a & b \\ c & d \end{bmatrix}$ and $\begin{bmatrix} p & q \\ r & s \end{bmatrix}$ commute under matrix multiplication, prove that $r = cq/b$ and $s = \{p - (q/b)\}(a - d)$.

6 Invent other binary operations on the set of human beings (e.g. given the age in seconds of two persons A and B, find the mean. Then find persons who have this age in seconds, and A $*$ B is that one who lives nearest to the Eiffel Tower. Is this a good binary operation?).

7 Binary operation on the points of a spherical surface. Given two points A and B on the earth's surface, the point $A * B$ might be defined as that point on the earth's surface which is equidistant from A and B and also from (say) the north pole, i.e. the circumcentre of the spherical triangle ABN. Or, if the latitude and longitude of A and B were (l_1, L_1), (l_2, L_2), then $A * B$ might be defined as the point $(-l_1, -L_2)$, or in other ways. Invent other binary operations on the points of a sphere.

8 $\overrightarrow{AB}$ and $\overrightarrow{CD}$ are two directed line segments ('arrows') in a plane. Invent a binary operation in the set of arrows. Discuss whether your definition may be extended to three dimensions.

9 On p. 7 we gave the example: $x * y = xy/(x + y)$ where $(x, y \in \mathbf{R}, x + y \neq 0)$. Supplement this definition to cover the cases when $x + y = 0$, and also when both $x = 0$ and $y = 0$; (a) so that $*$ is commutative, (b) so that $*$ is not commutative.

3 *Self-contained systems: closure*

When we multiply two numbers together, we expect the result to be a number – we should be surprised if the answer was a bunch of parsley.

W. W. Sawyer, *Prelude to Mathematics*, Penguin

In the last chapter we considered the binary operation $\cap$ on the set of planes in three-dimensional space and said that, since the resulting object, a line, is not in the set of planes, the operation is described as being not 'closed' in the set. For those of you familiar with elementary vector analysis, another example is the scalar product of two vectors, which is not a vector but a scalar. 'Vector multiplication' of vectors in three-dimensional space *is*, however, closed, since $\mathbf{a} \times \mathbf{b}$ is another vector. Of the examples of binary operations quoted in the first two chapters, the one about the intersection of two planes was the only example which failed for closure, for in the other cases we were chiefly concerned with binary operations *within a set*. When we mentioned the operation of subtraction on the set of positive integers **N**, we stated that the operation was not always possible, *within the set*. Of course, $3 - 5 = -2$, but this takes us *outside the set*, and the operation is not closed so far as the set **N** is concerned. Addition and multiplication, however, are closed, since $x, y \in \mathbf{N} \Rightarrow x + y \in \mathbf{N}$, and $xy \in \mathbf{N}$.

Some analogies for closure

To see why the term 'closed' is used, we may consider the following analogies. If one starts inscribing a polygon in a given circle so that each side subtends an angle θ at the centre (see fig. 3.01), then if θ were, say, 18°, the polygon would have exactly 20 sides. If θ were 54°, we should still get a *closed* polygon of 20 sides, returning to our starting-point and thus *closing* the polygon after three complete circuits of the circumference. But if the angle θ had an awkward† value, it is conceivable that the starting-point would never be reached again. (You will be familiar with this sort of thing if you have made patterns with *Spirograph*. Nevertheless, however many teeth the wheels may have, one will always return to the starting-point after a sufficiently large number of revolutions have been made.)

† In fact, irrational when expressed in revolutions.

Another case of 'closure' may be found on a mathematical billiard table, with perfectly smooth and perfectly elastic balls. A ball struck from position P (see fig. 3.02), after striking the cushions repeatedly, may never again pass through the point P. But it may easily be shown that, provided the direction of projection is *parallel to a diagonal* of the table, the ball will

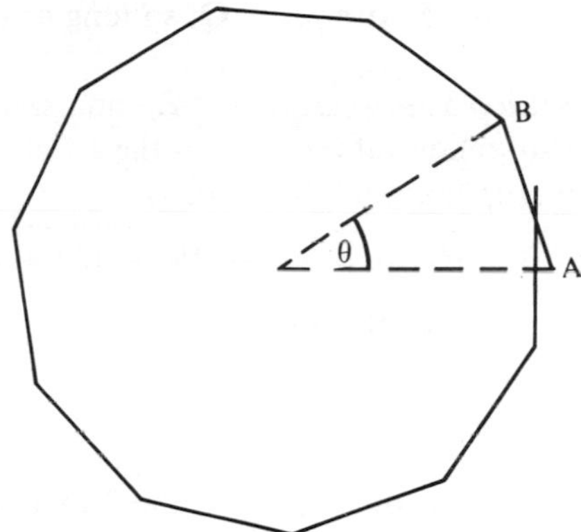

Fig. 3.01. When $\theta/360$ is irrational, the polygon is not 'closed'.

return to P after striking the four cushions, and will thereafter continue tracing the same parallelogram for ever. Thus we have a form of 'closure' for paths requiring a single impact with each of the four cushions, in the case when projection is parallel to a diagonal, and not otherwise.

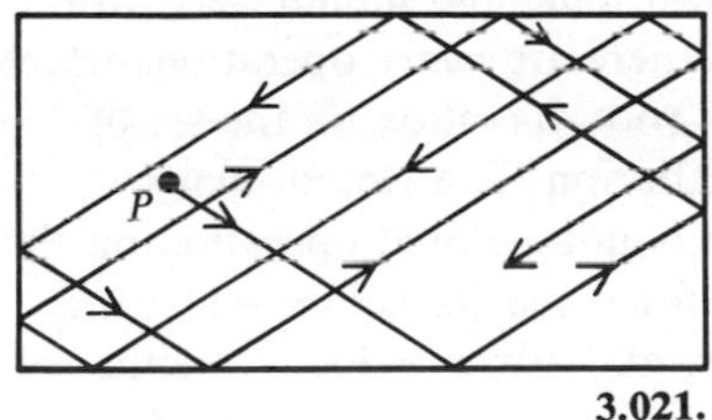

3.021.

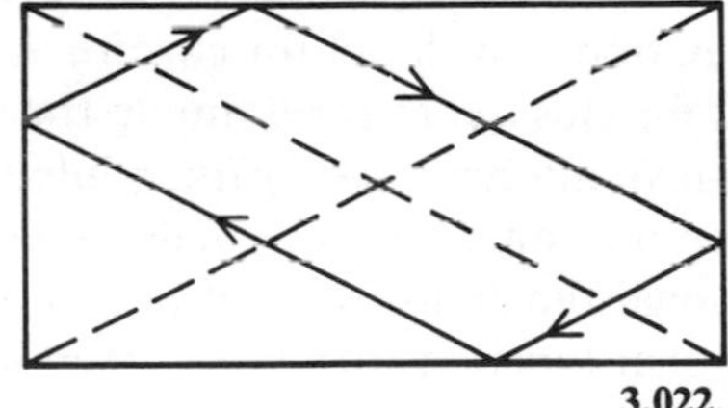

3.022.

Fig. 3.02. Path of billiard ball.
3.021. Path not closed.
3.022. Path closed when direction of projection parallel to diagonal.

Consider now numbers of the form $a + b\sqrt{5}$, where a and b are *rational*. This set is closed under addition, since $(a + b\sqrt{5}) + (c + d\sqrt{5}) = (a + c) + (b + d)\sqrt{5}$, which is of the required form, since $a + c \in \mathbf{Q}$, and also $b + d \in \mathbf{Q}$.

Q. 3.01. Verify that the set is also closed under multiplication.

Q. 3.02. Would closure apply if a and b were drawn from the set of integers?

We could go further in the case $a, b \in \mathbf{Q}$, and show that the set is also closed under division, for,

$$\frac{a + b\sqrt{5}}{c + d\sqrt{5}} = \frac{(a + b\sqrt{5})(c - d\sqrt{5})}{(c + d\sqrt{5})(c - d\sqrt{5})} = \frac{(ac - 5bd) + (bc - ad)\sqrt{5}}{c^2 - 5d^2},$$

which is of the form $p + q\sqrt{5}$ with $p, q \in \mathbf{Q}$, so long as c and d are not both zero.

Consider next the set $a\sqrt{3} + b\sqrt{2}$, $(a, b \in \mathbf{Z})$, and show that it is closed under addition and also under subtraction. Is the set closed under the binary operation of multiplication? No, because

$$(a\sqrt{3} + b\sqrt{2})(c\sqrt{3} + d\sqrt{2}) = 3ac + 2bd + (bc + ad)\sqrt{6},$$

and this number does *not* lie in the given set!

Failing for closure

You may have got the idea from chapters 1 and 2 that closure is almost inevitable, but this is not so – it is easy to construct examples like the one above which *fail for closure*: the odd integers are not closed under addition; the set of all polynomials with integral coefficients is closed under both addition, subtraction and multiplication, but the set of (say) *cubic* polynomials is not closed under multiplication; the binary operation of exponentiation on the set of positive rationals is not closed, as is proved by the single counter-example: $3^{\frac{1}{2}} \notin \mathbf{Q}$. You should have no difficulty in thinking of many examples taken from both finite and infinite sets with operations which fail for closure. Indeed, there are *unary* operations which fail for closure, as for example the square root operation on the set of positive integers; and again, while differentiation† is a closed unary operation on the polynomials, integration is not a closed operation on the rational functions (ratio of polynomials), for while $\int dx/x^2 = -1/x$, is a rational function, $\int dx/(1 + x^2) = \tan^{-1} x$, and $\int dx/x = \log_e x$, neither of which is a rational function.

An interesting example comes from music, which we shall investigate in greater detail in Ch. 23. Musical intervals, measured by the ratio of the frequencies of the two notes, are most pleasing to the ear when the ratio is that of *small* integers, e.g. $\frac{5}{4}$ (major third), $\frac{4}{3}$ (perfect fourth). When simple intervals are combined, the resulting interval may be harsh because the numbers involved in the ratio are no longer small, e.g. $\frac{8}{5}$ represents a minor sixth, $\frac{6}{5}$ a minor third. The interval produced when they are combined has ratio $\frac{8}{5} \times \frac{6}{5} = \frac{48}{25}$, and this cannot be considered to be a

† Integration *can be made* a unary operation by suitable definition. The difficulty arises from the arbitrary constant.

pleasant, consonant interval. It was, indeed, the fact that the simple musical intervals are not closed under composition, that led to the necessity for the invention of the equal-tempered scale (see pp. 436 ff.).

Successive geometrical transformations: the identity transformation

In Ch. 4 we shall be considering the composition of successive *operations*. A glimpse of this in advance may be illuminating. Suppose you take a rectangular box, and consider the three operations:

> X, a half-turn about the axis labelled x (see fig. 3.03),
> Y, a half turn about the axis labelled y,
> Z, a half-turn about the axis labelled z.

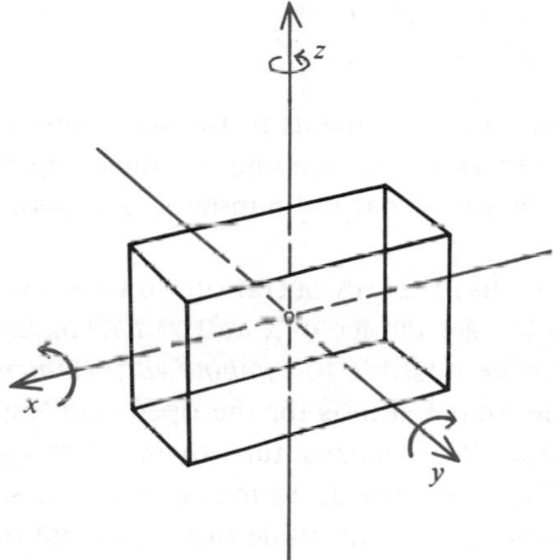

Fig. 3.03. Half-turn symmetries of rectangular box.

What happens if we perform first X and then Y? A practical experiment should help you to see that the result is the same as if we had performed the single operation Z. Thus we may say:

$$Y \text{ following } X = Z.$$

Similarly,

$$Z \text{ following } Y = X \quad \text{and} \quad X \text{ following } Z = Y,$$

and you may verify that the operations *commute*.

Does this set of three movements of the box possess the property of closure under successive applications? The answer is 'No', because if one performs X, and then X again, the result is neither X, Y nor Z. But you may object to *repeating* an operation. All right, if one performs X, then

Y, then Z, the final result is neither X nor Y nor Z. What happens is that the box returns to its original position! Now we can get over the difficulty of the set $\{X, Y, Z\}$ not being closed, by annexing to it what may appear to be a most unlikely candidate for admission, the *identity*, or 'stay-put' operation (see Ch. 6). Calling this I, the set $\{I, X, Y, Z\}$ will now be closed, and we can see how these operations combine by consulting table 3.01, which is *closed*, or *self-contained*.†

Table 3.01

		1st operation			
		I	X	Y	Z
2nd operation	I	I	X	Y	Z
	X	X	I	Z	Y
	Y	Y	Z	I	X
	Z	Z	Y	X	I

The device of adding more elements to the set in order to remedy the fact that it is not closed is a common one: by annexing zero and the negative integers to the set of natural numbers, one obtains a set which is *closed under subtraction.*

It is important to realise that an operation $*$ on a set is only closed if $x*y$ lies in the set for *every* choice of x and y. *This includes the case when* $x = y$, so that it must be possible to *combine each element with itself.* Suppose, for example, that s stands for the operation 'put your socks on', t stands for 'take them off', x means 'turn them inside out', and i means 'make no change'. Then we write sx to mean 'put your socks on inside out'; while $xstx$ means 'turn them inside out, put them on, take them off, and finally turn them inside out again', so evidently we may write $xstx = i$. At first sight it may appear as if this set is closed, till you contemplate the composite operation ss, which means 'put your socks on, and then put your socks on', which is clearly nonsense.

***Summary:* A binary operation $*$ on a set S is closed when, for every possible choice of x and y in S, the element $x*y$ also lies in S.**

EXERCISES 3

1 Consider carefully the question of the closure of all the examples considered so far in this text. Invent examples where closure fails by reducing the set, or by changing the operation. For example, $\{0, 1, 2, 3, 4\}$ is closed under addition modulo 5, but would not be if one of the elements were removed; nor would it be if the operation were (say) multiplication modulo 6.

† This is the first example in this book of a Group table. There will be many more to come!

2 Is the set $\{1, i, -1\}$, where $i^2 = -1$, closed under $\times$? How can it be made so? How can the set $\{-1, +1\}$ be made closed under addition modulo 3?

3 **Consider the closure or otherwise of the following number systems under the operations $+, -, \times, \div$:**
(a) $\mathbf{N}$; (b) $\mathbf{Z}$; (c) $\mathbf{Z}^-$; (d) $\mathbf{Q}$; (e) $\mathbf{Q}^+$; (f) $\mathbf{R}$; (g) $\mathbf{R}^+$;
(h) $\mathbf{C}$; (i) $\cos\theta + i\sin\theta$ $(0 \leqslant \theta < 2\pi)$; (j) $x + iy$ $(x, y \in \mathbf{Z})$;
(k) $x + iy$ $(x, y \in \mathbf{Q})$.

4 Is the set of all functions of x in a particular interval $a \leqslant x \leqslant b$ closed under addition, under multiplication?

5 Can you invent a closed binary operation on circles in the plane, i.e. given any two circles, suggest some construction which will produce a unique third circle for all possible choices of the two circles? (See also the Mid Circle, *Mathematical Gazette*, Feb. 1968.)

6 Show that the set of all positive integers which are prime to a given integer (i.e. have no common factor with it) are closed under multiplication (e.g. 5 and 13 are each prime to 8, and so is 5×13).

7 Examine numbers of the form $a\sqrt{2} + b\sqrt{3} + c\sqrt{6}$ for closure under addition, subtraction, multiplication and division in the cases:
(a) $a, b, c \in \mathbf{Z}$; (b) $a, b, c \in \mathbf{Q}$.

8 Discuss the closure of the following binary operations in cases when x and y are drawn from: (a) $\mathbf{Z}$; (b) $\mathbf{Q}$; (c) $\mathbf{R}$.
(a) $x * y = (x + y)/(x - y)$; (b) $x * y = (x/y) + (y/x)$;
(c) $x * y = (x + y)/(1 + xy)$; (d) $x * y = (x^2 + y^2)^{\frac{1}{2}}$;
(e) $x * y = |x| + |y|$.
Also state in each case whether these operations are commutative.

9 Consider the operations of rotation of a figure about a point O in its plane through angles 40°, 80°, 120°, 160°, 200°, 240°, 280°, 320°. Is this set of rotations closed? What needs to be done to secure closure? Are all the rotations in the plane about a given point closed under successive application? What about the rotations of a solid body in three dimensions which has one point fixed?

10 If $e = \begin{bmatrix} 1 & 0 & 0 \\ 0 & 1 & 0 \\ 0 & 0 & 1 \end{bmatrix}$, $x = \begin{bmatrix} 1 & 0 & 0 \\ 0 & 0 & 1 \\ 0 & 1 & 0 \end{bmatrix}$, $y = \begin{bmatrix} 0 & 0 & 1 \\ 0 & 1 & 0 \\ 1 & 0 & 0 \end{bmatrix}$, find three more matrices to give a set of six which are closed under matrix multiplication.

11 Show that the set of $n \times n$ matrices each containing n 1's, no two in the same row or column, and 0's elsewhere contains $n!$ matrices, and that they are closed under the operation of matrix multiplication.

12 What must we add to the statement 'the set of all matrices is closed under addition' to make it true? Construct subsets of matrices which give closure under (a) addition, (b) multiplication.

13 Under which operations are polynomials with integer coefficients closed: (a) when the degree is specified; (b) when the degree is not specified? Consider the same question when the coefficients are drawn from the rationals; from the reals.

14 Consider the formula $y = a + b/x$. If we had another formula of the same type, $z = c + d/y$, then we may eliminate y to obtain a formula giving z in terms of x, thus:

$$z = c + \frac{d}{a + b/x} = c + \frac{dx}{b + ax}$$

and this is a formula of a *different* type. We say that the functions of the form $a + b/x$ are *not closed* under successive composition (see p. 22). **Show that functions of the form $y = (ax + b)/(cx + d)$ are closed under successive composition.** Find other examples of functions of forms which are closed, and others which are not closed.

15 If (x, y) are ordered pairs with $x \in \{0, 3, 6, 9\}$ and $y \in \{1, 2, 4\}$ and these are combined by the binary operation $*$ defined:

$$(x_1, y_1) * (x_2, y_2) = (x_1 + x_2, \text{mod. } 12, y_1 y_2, \text{mod. } 7)$$

show that these twelve elements form a set which is closed under this operation.

16 On a chessboard, a knight's move followed by a knight's move cannot possibly be a knight's move because the colour of the square changes at each jump. Consider for what x an x's move followed by an x's move can (though need not) be an x's move.

17 If $\mathbf{Z}$ is the set of integers, and $1/\mathbf{Z}$ the set of the reciprocals of the non-zero integers, is the set $\mathbf{Z} \cup 1/\mathbf{Z}$ closed under any of the arithmetic operations?

18 **Two-dimensional vectors, represented by ordered pairs of reals, (x, y) are closed under addition: $(x_1, y_1) + (x_2, y_2) = (x_1 + x_2, y_1 + y_2)$. The subset for which $y = 0$ is itself closed, for $(x_1, 0) + (x_2, 0) = (x_1 + x_2, 0)$ which belongs to the same subset. Find other subsets of the two-dimensional vectors which are closed under vector addition.**

19 Are the plane rotations about a single point together with the plane translations, a closed set of transformations?

20 Two points O and U are fixed on a circle. A binary operation is set up between points on the circle by agreeing that, if A and B are any two points of the circle, then $A * B$ shall be the intersection of AO and BU. Is this closed? Is it commutative? Invent an example of a binary operation on the set of points on the circle which is closed.

21 A piece of plasticine is deformed into a different shape. Such a deformation may be termed a 'topological transformation'. A binary operation on any two topological transformations would be to apply them one after the other. Is the set of topological transformations closed? Do topological transformations commute?

22 Show that matrices of the form $\begin{bmatrix} \cosh u & \sinh u \\ \sinh u & \cosh u \end{bmatrix}$ are closed under matrix multiplication. Are they closed under matrix addition?

4 *Combined ops: composition of operations*

In the previous chapters we have mostly been concerned with operations on sets of objects such as numbers, points, regions, polynomials, sets, etc. We also considered the composition of geometrical movements (see pp. 17 ff), and of matrices. We shall now consider in more detail the question of binary operations on objects which *themselves may be operations*, such as geometrical transformations, algebraic functions, permutations, etc. So that you may not be discouraged from proceeding further at this early stage, let us give some simple examples of what is meant in these cases.

Composition of geometrical transformations

In fig. 4.01 we show how, when the figure F is rotated (R_1) about the point A through an angle α to a new position F', and then 'translated' by a

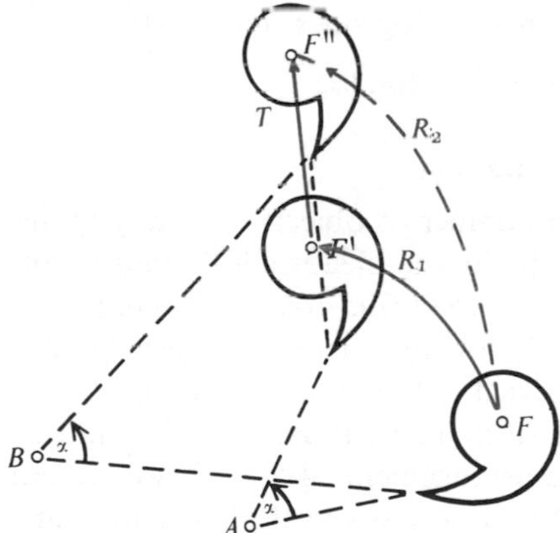

Fig. 4.01. Composition of transformations. Rotation about A through α (R_1) followed by translation T is equivalent to rotation about B through α (R_2): $R_2 = T \circ R_1$.

parallel movement T to its final position F'', the result is the same as if it had been rotated through α about the point B, and we call this single rotation R_2. We may say that a binary operation on the set of movements in the plane may be defined by performing these movements one after the other, and we may write in the present case: $T \circ R_1 = R_2$, where '$\circ$'

would be read 'following', i.e. the translation T *following* the rotation R_1 produces the same effect as the single rotation R_2.

Composition of functions

Function of a function

Suppose $f_1(x) \equiv 2x/(1 + x)$ $(x \neq -1)$ and $f_2(x) \equiv 4 - x$.

Then if $y = 2x/(1 + x)$ and $z = 4 - y$, so that $z = 4 - 2x/(1 + x)$ $= (4 + 2x)/(1 + x)$, and we may say that $z = f_2(y)$ where $y = f_1(x)$, i.e. that $\mathbf{z = f_2[f_1(x)]}$.

This is known as the 'composition' of the two functions $\mathbf{f_1}$ and $\mathbf{f_2}$. We later omit the square brackets and write:

$$z = f_2f_1(x) \equiv \frac{4 + 2x}{1 + x} \quad (x \neq -1).$$

Q. 4.01. Find $f_1f_2(x)$, $f_1f_1(x)$, $f_2f_2(x)$, etc.

Contrariwise, if $z = (3x^2 + 1)^{-\frac{1}{2}}$, we could write $z = u^{-\frac{1}{2}}$, $u = 3x^2 + 1$, and this is the sort of thing we need to do in calculus in order to differentiate such a 'function of a function'. In the same say, if $f_1(x) \equiv \sin x$, $f_2(x) \equiv \log x$, then:

$$f_2f_1(x) \equiv \log(\sin x) = \log \sin x \text{ (for short)}$$

and $f_2f_2(x) \equiv \log(\log x) = \log \log x$.

Composition of permutations

A permutation of a number of objects is usually thought of as a 'rearrangement'. Actually the idea is a little more subtle than this (see D. B. Hunter, *Mathematical Gazette*, Oct. 1966, pp. 290–4). We are not so much concerned with how or where the objects lie after the shuffling process has been carried out, as with *which object takes the place of which.* Thus, if five people sit in a row thus: 1 2 3 4 5, and then reshuffle themselves into this arrangement: 4 5 1 3 2, we are not so much concerned with the fact that No. 4 now sits in the left-hand seat, and No. 2 in the seat nearest the window, and that No. 1 will now be occupying the most comfortable chair, as with the fact that No. 1 has *been replaced by* No. 4; No. 2 by No. 5, and so on.

Suppose the five people were all string players,† each with the rare gift

† The politically minded may prefer, in what follows, to think in terms of Cabinet reshuffles, and who takes over the Ministry of Proliferation from whom. The energetic may prefer to draw their example from Scottish country dancing, in which case they should consult *Discovering Modern Algebra* Ch. 1, by K. L. Gardener (O.U.P., 1966). Motorists may prefer to think in terms of the rearrangement of the five wheels of a car to ensure even wear of the tyres.

of being able to play violin, viola and cello. They play two quintets, and on the first occasion, No. 1 plays 1st violin, No. 2 plays 2nd violin, No. 3 1st viola, No. 4 – 2nd viola, and No. 5 plays cello. On the second occasion, they change over with No. 4 playing 1st violin, No. 5 – 2nd violin, No. 1 1st viola, No. 3 – 2nd viola, and No. 2 cello. We are not concerned with the fact that the players have been re-arranged – where they sit is of no concern. We are interested in the fact that, for example, *No. 4 took over* (the 1st violin part) *from No. 1*, i.e.

1 was replaced by 4
2 by 5
3 by 1
4 by 3
and 5 by 2.

This particular permutation is abbreviated $\begin{pmatrix} 1 & 2 & 3 & 4 & 5 \\ 4 & 5 & 1 & 3 & 2 \end{pmatrix}$ and we shall call this substitution operation 'x'. Note that Nos. 2 and 5 have changed places, while 1, 3 and 4 have 'moved round one place'.

The numbers of the objects (people in this example) are arranged in numerical order in the original line for convenience only: this same permutation could equally well have been written

$$x = \begin{pmatrix} 2 & 1 & 4 & 3 & 5 \\ 5 & 4 & 3 & 1 & 2 \end{pmatrix} \quad \text{or} \quad \begin{pmatrix} 3 & 5 & 4 & 1 & 2 \\ 1 & 2 & 3 & 4 & 5 \end{pmatrix}.$$

How do we combine permutations? Suppose the above permutation is now followed by a different permutation $\begin{pmatrix} 1 & 2 & 3 & 4 & 5 \\ 5 & 3 & 2 & 4 & 1 \end{pmatrix}$, which we call '$y$'; i.e. Nos. 1 and 5 change places, and so do Nos. 2 and 3, while No. 4 plays the same part. What would be the final effect?

x		y		combined effect: x then y	
1 was replaced	by 4	4 was replaced	by 4	1 was replaced	by 4
2	by 5	5	by 1	2	by 1
3	by 1	1	by 5	3	by 5
4	by 3	3	by 2	4	by 2
5	by 2	2	by 3	5	by 3

This may be abbreviated:

$$\text{First, } x = \begin{pmatrix} 1 & 2 & 3 & 4 & 5 \\ 4 & 5 & 1 & 3 & 2 \end{pmatrix}$$

$$\text{Then, } y = \begin{pmatrix} 4 & 5 & 1 & 3 & 2 \\ 4 & 1 & 5 & 2 & 3 \end{pmatrix}$$ (rewriting y, as suggested above, in one of the alternative ways),

$$y \circ x = \begin{pmatrix} 1 & 2 & 3 & 4 & 5 \\ 4 & 1 & 5 & 2 & 3 \end{pmatrix}$$ (y following x).

Note that the composite permutation interchanges 3 and 5, and moves 1, 2 and 4 round (see Cycles, pp. 111–117). Again,

$$\begin{pmatrix}1 & 2 & 3 & 4\\3 & 2 & 4 & 1\end{pmatrix} \text{ followed by } \begin{pmatrix}1 & 2 & 3 & 4\\4 & 3 & 2 & 1\end{pmatrix},$$

i.e.

$$\begin{pmatrix}3 & 2 & 4 & 1\\2 & 3 & 1 & 4\end{pmatrix} \text{ gives } \begin{pmatrix}1 & 2 & 3 & 4\\2 & 3 & 1 & 4\end{pmatrix}.$$

It is well known that the number of possible permutations on a set of n objects is $n!$, i.e. $1\ 2\ 3 \ldots (n-1)n$, and it is evident that this set is *closed* under the operation of successive application.

Note: It is most important to realise that $\begin{pmatrix}1 & 2 & 3 & 4 & 5\\4 & 5 & 1 & 3 & 2\end{pmatrix}$ does *not* mean 1 'replaces' 4, etc. This point is dealt with again in several places in this book (see pp. 55, 199), and arises especially in connection with bell-ringing (Ch. 24).

Remember
$$\begin{pmatrix}\mathbf{1} & \mathbf{2} & \mathbf{3} & \mathbf{4} & \mathbf{5}\\ \downarrow & \downarrow & \downarrow & \downarrow & \downarrow\\ & & \textit{\textbf{is replaced by}} & & \\ & & \textit{\textbf{or `becomes'}} & & \\ \downarrow & \downarrow & \downarrow & \downarrow & \downarrow\\ \mathbf{4} & \mathbf{5} & \mathbf{1} & \mathbf{3} & \mathbf{2}\end{pmatrix}.$$

Games with operations

In order to get used to the idea of combining operations, we may consider a number of simple games suggested by Dr T. J. Fletcher.

The three coins game

Three coins are placed in a row in positions A, B, C, and the following operations may be performed on them:

X: turn over the coin in position A, and interchange coins in positions B and C.
Y: turn over the coin in position C, and interchange coins in positions A and B.

For example, if we started with the three coins showing:

	A	B	C	
	tail	tail	head	
X would give	h	h	t	abbreviate: $tth \xrightarrow{X} hht$
Y would then give	h	h	h	$hht \xrightarrow{Y} hhh$.

$Y \circ X$ is therefore a new operation which takes *tth* into *hhh*, and you may verify that $X \circ Y$ is different again.

We may ask all sorts of questions about the 'algebra' generated by the symbols X and Y. For example, $hhh \xrightarrow{X \circ Y \circ X} ttt$, meaning that the successive application of X, then Y and then X again results in all three coins being turned over. Can you find another succession of moves which will have the same result? Does $X \circ Y \circ X$ *always* equal $Y \circ X \circ Y$ irrespective of the starting position? How do we get from *thh* to *hht*; from *hth* to *ttt*? Which changes are impossible? Is it always possible to restore a previous position – to undo what has been done in a series of moves, and so on.

The angles game

In the second game, we use the configuration of fig. 4.02, with the parallel lines making the angles a, b, c, d equal. We use the following 'instructions'. Starting from a given angle,

V: jump to the vertically opposite angle	(e.g. from b to a);
A: jump to the alternate angle	(e.g. from b to c);
C: jump to the corresponding angle	(e.g. from d to b).

A sequence of operations such as $VCACVCACCCVAAV$ could be interpreted:
'Jump to the vertically opposite angle (b to a), then to the corresponding angle (a to c), then to the alternate angle (c to b), and so on.† One obvious

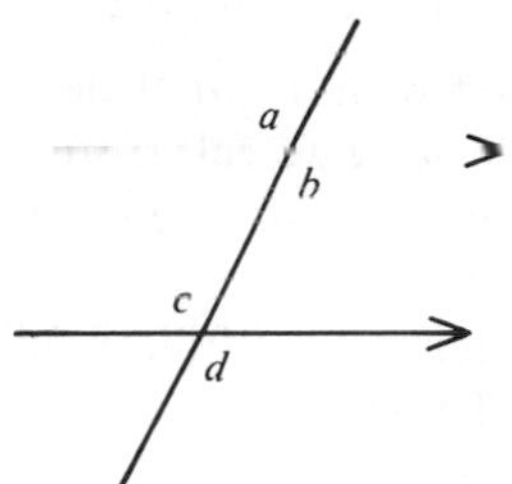

Fig. 4.02. Angles with parallel lines: $a = bV$; $c = bA$; $d = bC$.

thing we can do is to try to shorten such a 'word': are there *rules* for shortening such words? We note that the sequence of operations is not closed (since VV or AA or CC all result in a return to the starting position) till we introduce a further element of the set, the stay-put, or *identity* operation, which we shall call I. Any consecutive pair of repeated letters 'cancel out', and so the word above reduces to $VCACVCACIVIV$

† But see the paragraph on p. 28ff. on the two ways of writing a sequence of operations.

$= VCACVCACVV = VCACVCACI = VCACVCAC$. Again, we note that $a(VA) = c$ (i.e. starting from a and moving to the vertically opposite and then to the alternate angle, we arrive at c), and that $a(C) = c$, so that the operations VA and C are equivalent, and we may write $VA = C$. We may also verify that $AV = C$; $CV = VC = A$ and $AC = CA = V$. But there is an imperfection which you may have noticed: how do we obtain $a(A)$, or $d(A)$? The answer is to *extend the meaning of the word 'alternate'*, so that, since $A = VC = CV$ when starting from b or c, we shall take it that this will also apply when starting from a or d. It will then be seen that $d(A) = d(VC) = c(C) = a$, and that $a(A) = d$. Indeed, angles like a and d are often called 'exterior alternate' angles. Having agreed thus, our system is now complete. The 'word' quoted ($VCACVCAC$) now reduces further: $(VC)A(CV)A(C) = AAAAC = (AA)(AA)C = IIC = C$. The series of instructions $VCACVCACCCVAAV$, moreover, will take us to the *corresponding* angle whether we start from a, b, c or d. (*Note*: the set $\{I, V, A, C\}$ under successive application give the group D_2 – see p. 132.

The three jugs game

Imagine three jugs in a row, some or all of which may be filled with water, and let us consider 'words' formed from the instructions:

X: empty the jug in position A;
Y: pour the contents of jug in position B into the jug in position C;
Z: move the jugs around, that in position B being replaced by that in position A, the jug in position C by the one in position B, and the jug in position A by the one in position C (a 'cyclic' permutation).

(There is one snag connected with Y: what do we do if the jug at C is already full? We may make our own rules: we may either agree that B should in such a case not be emptied, or else that it is emptied all the same. Here we assume the latter.)

Suppose the jugs were originally: A full, B full, C empty. Represent this 'state' symbolically by *ffe*. Thus:

$$ffe \xrightarrow{X} efe; \quad ffe \xrightarrow{Y} fef; \quad ffe \xrightarrow{Z} eff$$

Again, performing successive operations, and using XY to denote† 'do X and then do Y', we see that:

$$ffe \xrightarrow{XY} eef; \quad ffe \xrightarrow{XYZ} fee; \quad ffe \xrightarrow{XYZY} fee.$$

A group of three letters such as (*ffe*) represents a 'state' of the system, and a 'word' such as $XYZY$ represents an operation, which effects a shift from one state to another.

† But see p. 28ff.

We may ask such questions as 'how do we get from one given state to another?' e.g. $ffe \xrightarrow{?} fee$. Is the answer, $XYZY$, unique? By no means! For example, $eff \rightarrow efe$ may be achieved by the use of YZZ, or ZZX, or $ZXZZ$, etc.; for the latter, we note that

$$eff \xrightarrow{Z} fef \xrightarrow{X} eef \xrightarrow{Z} fee \xrightarrow{Z} efe,$$

and, combining these,

$$eff \xrightarrow{ZXZZ} efe.$$

Thus there are different ways of achieving a particular change of state, and some are more complicated than others. We may further ask the question: Will a sequence of operations (represented by a particular 'word'), *always* produce the same change of state *wherever we start from*? The answer to this is seen to be 'No', for $ffe \xrightarrow{YZ} ffe$ (the same state, or $YZ = I$ in this case), whereas $eff \xrightarrow{YZ} fee$ (a change of state). Thus YZ sometimes preserves the *status quo*, and sometimes produces a change of state. Find other instances of this sort of thing.

However, there are some invariable 'rules', such as $ZZZ = I$ (always), $XX = X$ (always), $YY = Y$ (always), $XY = YX$ (always), and you should verify that these are true in all cases, and why.

Sometimes a law may be deduced from ones already established, e.g. $XYXY = (XY)(YX)$† $= X(YY)X$‡ $= XYX = X(YX) = X(XY)$ $= (XX)Y = XY$.

Q. 4.02. Experiment with this algebra on the following lines:

(a) find as many rules as possible for word shortening;

(b) find as many words as possible which are equivalent to I, from various initial states;

(c) find which words result in all the water being emptied (final state *eee*), from various initial states. What is the shortest word which takes us from *fff* to *eee*?

Note that in this game the total amount of water can never be increased, so that on the whole, operations like X and Y cannot be 'undone' – we can never get from *efe* to *fef*, say. However, you may like to invent a game which allows for buckets being *filled*, and it may well be that you are able to produce a more interesting system.

† Note the 'punctuation' brackets.
‡ See Associativity, Ch. 5.

Passing the parcel

Three boys sit in a row of three chairs, and a parcel is passed from one to the other. There are two instructions:

> L: whoever has the parcel, give it to the boy on his left;
> R: whoever has the parcel, give it to the boy on his right.

How do we interpret L (or R) when the boy on the extreme left (right) has the parcel? We may agree to interpret it how we like: either (1) he throws it out and the game stops, or (2) he keeps it, and the game restarts when the opposite instruction is received, or (3) he passes it to the boy at the far end. In the last case, we would, in effect, be arranging the boys in a circle instead of in a line, and this turns out to be the most satisfactory method of play. For using rule (3), we should *always* have $RL = I = LR$; $LLL = RRR = I$; $LL = R$, and $RR = L$. You may think that, for this very reason, (3) gives the *least interesting* situation, and in that case you may well like to investigate the various compound operations and how they may be simplified when (1) or (2) are agreed to as the rules of the game.

Notation: first operation on the right

At this point, we must stop and decide once and for all whether, in this book, we are going to be right-handed or left-handed: when two operations x and y are combined successively, first x and then y, are we going to write the x on the left and y on the right, or the other way round? You have seen several instances of the latter:

> $T \circ R_1 = R_2$ meaning 'T following R_1', i.e. first R_1, then T (p. 21).
> $f_2f_1(x)$ meant 'first form $f_1(x)$ (call it y), then form $f_2(y)$' (p. 22),

and two cases of the former. In the 'angles game' (p. 25), we used AV to mean 'first A, then V', and in the three jugs game, we used ZY to mean 'first Z, then Y'.

You will find that different books use different conventions, but that it is more common to find the *first operation being written on the right*:

> **$X \circ Y$ means 'first Y, then X' or 'X following Y',**

and this is the convention that will be used invariably throughout this book.†

There are mathematicians who would urge us to use the other method, and there are those who do not feel strongly about it, and think that so long as we settle the choice, and stick to one or the other, then all will be

† With the exception that, if X is a row vector of n components, and A and B are square $n \times n$ matrices, then XAB is the row vector obtained from X by applying first the matrix A, then B. However, we shall generally work with column vectors.

well. You may well think that putting the first operation on the right is against all sense, since one naturally moves from left to right in writing across the page. But you might as well accuse the French of being illogical in putting an adjective after a noun! My own view is that there is a slight preference for the convention that the first operation should be written on the right. Table 4.01 gives some instances of both notations.

Table 4.01

Notation to be used in this book: $X \circ Y$ means 'first Y, then X' ('$\circ$' = 'following').	Notation used (?) in some other books: $X \circ Y$ means 'first X, then Y' ('$\circ$' = 'followed by').
$f(x)$	$x{:}f$
$f_2[f_1(x)] = f_2 f_1(x)$	$(x{:}f_1)f_2 = x{:}f_1 f_2$
$\log(\sin x) = \log\sin x$	$(x \sin)\log = x \sin\log$
$\dfrac{dy}{dx} = \dfrac{d}{dx}(y) = Dy$	$y\dfrac{d}{dx} = yD$
$\dfrac{\partial^2 z}{\partial x \partial y} = \dfrac{\partial}{\partial x}\left(\dfrac{\partial z}{\partial y}\right)$	$z\dfrac{\partial}{\partial y \partial x} = \left(z\dfrac{\partial}{\partial y}\right)\dfrac{\partial}{\partial x}$
$\begin{bmatrix} a & b & c \\ d & e & f \end{bmatrix}\begin{bmatrix} x \\ y \\ z \end{bmatrix}$	$[x \quad y \quad z]\begin{bmatrix} a & d \\ b & e \\ c & f \end{bmatrix}$
$\sqrt{53}$	$53\surd$
$T(A)$ = 'apply the transformation T to the point A'	$A{:}T$

It seems to me more natural to put the operator *before* the operand, and the generally accepted notation for differential operators in the calculus seems to provide strong support for this belief. At least it is more natural for the English! Because, surely, the operator is the *verb*, and the operand is the *object* of the verb, and we say 'Phyllis loves Ferdinand',† and not 'Phyllis Ferdinand loves', as they do in some languages (find out which!). Again, 'Take the square root of 53', ($\sqrt{53}$), not '53, square root it' ($53\surd$). (The latter notation is, however, used in some machines.) Again, 'Turn the car round please', not 'Please the car to turn round, so!' If one accepts this, then the order of the operations, ($X \circ Y$ = first Y, then X) must follow naturally: suppose 'Turn the car round' is abbreviated $T(C)$, (T is the operator, C is the operand), and 'accelerate the car' is abbreviated $A(C)$, then 'turn the car round and accelerate' becomes $T(C)$, then $AT(C)$, so this will be written $AT(C)$, i.e. accelerate the turned car!

† This is a *relation* rather than a *function*, since the fickle Phyllis may love another!

Juxtaposition: the multiplicative notation

One further point: while we are now deciding to use $X \circ Y$ to mean 'X following Y', we may as well also agree to drop the '$\circ$' sign and to use *juxtaposition* (as we did in the games on pp. 25–28), i.e.

XY shall mean first Y, then X.

'Juxtaposition' is known as the 'multiplicative notation' since it resembles the method of writing a product in ordinary algebra.

Another controversial point emerges at this stage: whether we shall use lower or upper-case letters on various occasions. It is common practice to use capital letters for sets, for matrices, and for transformations, in which case I is usually used for the identity. When small letters are used (e.g. for numbers) e is often used for the neutral element (see Ch. 6). When the operation is multiplication, or something akin to multiplication, or something which might be described as multiplication, 1 may be used; for an addition process, 0 is more appropriate; when the objects being added are vectors, then we would use $\mathbf{0}$. Some books make a practice of using the *additive* notation for *commutative* operations, in Abelian groups (see, e.g. P. Alexandroff, *Introduction to the Theory of Groups* (Blackie/Hafner). We do not feel bound to adhere to any inflexible convention in these matters, but will allow the context, commonsense, and usual practice to be our guide. (But see also note on Abstract Groups, p. 147 footnote.)

Geometrical: symmetries and matrices

At the beginning of this chapter we remarked that we would be considering binary operations on sets of operations, and mentioned that the latter might be sets of transformations, of substitutions in functions, or of permutations. The set may also be a set of symmetries of a geometrical figure (see p. 78 and Ch. 17), or a set of matrices, but all these entities are closely interwoven. For example, a parallelogram is symmetric by a half-turn about its centre; the geometric transformation which reflects in the centre may be represented by the complex number functional relation $z' = -z$,† and this may also be expressed by the matrix $\begin{bmatrix} -1 & 0 \\ 0 & -1 \end{bmatrix}$.

Again, the three half-turns X, Y, Z of page 17 may be expressed by the matrices

$$\begin{bmatrix} 1 & 0 & 0 \\ 0 & -1 & 0 \\ 0 & 0 & -1 \end{bmatrix}, \quad \begin{bmatrix} -1 & 0 & 0 \\ 0 & 1 & 0 \\ 0 & 0 & -1 \end{bmatrix} \quad \text{and} \quad \begin{bmatrix} -1 & 0 & 0 \\ 0 & -1 & 0 \\ 0 & 0 & 1 \end{bmatrix}$$

† See F. J. Budden, *Complex Numbers and their Applications* (Longmans), 1968, pp. 32, 51, Ch. 4.

and you should verify that matrix multiplication gives $YZ = ZY = X$, etc. Finally, the transformation on a parabola $y^2 = 4ax$, which takes the point P into the point P' where PP' passes through the focus of the parabola, can be represented by the algebraic relation $t' = -1/t$, where t and t' are the parameters of P and P', the point P having coordinates $(at^2, 2at)$.

Permutation matrices

Permutations may also be represented by matrices:

$$\begin{bmatrix} 1 & 0 & 0 \\ 0 & 1 & 0 \\ 0 & 0 & 1 \end{bmatrix}\begin{bmatrix} a \\ b \\ c \end{bmatrix} = \begin{bmatrix} a \\ b \\ c \end{bmatrix}; \quad \begin{bmatrix} 0 & 1 & 0 \\ 0 & 0 & 1 \\ 1 & 0 & 0 \end{bmatrix}\begin{bmatrix} a \\ b \\ c \end{bmatrix} = \begin{bmatrix} b \\ c \\ a \end{bmatrix};$$

$$\begin{bmatrix} 0 & 1 & 0 & 0 \\ 0 & 0 & 1 & 0 \\ 1 & 0 & 0 & 0 \\ 0 & 0 & 0 & 1 \end{bmatrix}\begin{bmatrix} a \\ b \\ c \\ d \end{bmatrix} = \begin{bmatrix} b \\ c \\ a \\ d \end{bmatrix}.$$

These are examples of *permutation matrices* pre-multiplying column vectors. We may also have post-multiplication of row vectors:

$$[a \quad b \quad c \quad d]\begin{bmatrix} 0 & 0 & 0 & 1 \\ 0 & 1 & 0 & 0 \\ 1 & 0 & 0 & 0 \\ 0 & 0 & 1 & 0 \end{bmatrix} = [c \quad b \quad d \quad a];$$

$$\begin{bmatrix} a & b & c \\ p & q & r \end{bmatrix}\begin{bmatrix} 1 & 0 & 0 \\ 0 & 0 & 1 \\ 0 & 1 & 0 \end{bmatrix} = \begin{bmatrix} a & c & b \\ p & r & q \end{bmatrix};$$

$$[a \quad b \quad c \quad d \quad e]\begin{bmatrix} 0 & 0 & 1 & 0 & 0 \\ 0 & 0 & 0 & 0 & 1 \\ 0 & 0 & 0 & 1 & 0 \\ 1 & 0 & 0 & 0 & 0 \\ 0 & 1 & 0 & 0 & 0 \end{bmatrix} = [d \quad e \quad a \quad c \quad b].$$

Q. 4.03. Write down all the 3×3 permutation matrices, and experiment with them by using the same matrix both for pre- and for post-multiplication of column and row vectors respectively, and note the effect in each case. Consider also 4×4 permutation matrices, and find their effect on a vector both by pre- and post-multiplication.

Q. 4.04. The matrix

$$\begin{bmatrix} 0 & 0 & 1 & 0 & 0 \\ 0 & 0 & 0 & 0 & 1 \\ 0 & 0 & 0 & 1 & 0 \\ 1 & 0 & 0 & 0 & 0 \\ 0 & 1 & 0 & 0 & 0 \end{bmatrix}$$

may be used to produce the permutation x referred to on p. 23. What matrix (post-multiplying a row-vector), produces the permutation y in the text? Consider the products xy and yx.

Note that $\begin{bmatrix} 0 & 1 & 0 \\ 0 & 0 & 1 \\ 1 & 0 & 0 \end{bmatrix} \begin{bmatrix} a & p \\ b & q \\ c & r \end{bmatrix} \begin{bmatrix} 0 & 1 \\ 1 & 0 \end{bmatrix} = \begin{bmatrix} b & q \\ c & r \\ a & p \end{bmatrix} \begin{bmatrix} 0 & 1 \\ 1 & 0 \end{bmatrix} = \begin{bmatrix} q & b \\ r & c \\ p & a \end{bmatrix}$.

Investigate the results of applying permutation matrices before and after a $m \times n$ matrix M, thus: $P_m M P_n$, where P_m is a $m \times m$ permutation matrix, and P_n is a $n \times n$ permutation matrix. Note that when $m = 3$, $n = 2$, as in the example above, there are six possible P_3's, and two possible P_2's, so that $P_3 \begin{bmatrix} a & p \\ b & q \\ c & r \end{bmatrix} P_2$ may produce twelve different results. Are they all different? Are these twelve permutations of the elements of the matrix $\begin{bmatrix} a & p \\ b & q \\ c & r \end{bmatrix}$ closed under the composition of permutations?

Q. 4.05. Is it possible to find matrices which *transpose* a given square matrix, i.e. can matrices be applied before and after $\begin{bmatrix} a & p & u \\ b & q & v \\ c & r & w \end{bmatrix}$ to change it into $\begin{bmatrix} a & b & c \\ p & q & r \\ u & v & w \end{bmatrix}$? (Try with 2×2 matrices first.)

Successive transformations on the points of a circle

To complete this chapter a further geometrical example may be illuminating. Suppose we are given a circle and two fixed points A and B, as in fig. 4.031. Let a† be the transformation which takes a point P on the circle and replaces it by the point Q_1 on the circle where PQ_1 passes through A, and let b be the transformation which replaces P by Q_2 where PQ_2 passes through B. Consider the transformation ba (first a, then b) (see fig. 4.032):

$$a(P) = Q; \qquad b(Q) = R = b(aP). \qquad \text{Hence, } ba(P) = R.$$

The transformation ba ($= c$) might be repeated:

$$cc(P) = c[c(P)] = c(R) = ba(R) = b(S) = T.$$

Thus, $\quad R = c(P)$, and $\quad T = c(R) = cc(P) = c^2(P)$.

(When using the multiplicative notation, cc is usually contracted, by analogy with ordinary algebra, to c^2; evidently the notation can be extended to an operation repeated any number of times:

$$cccc = c^4; \qquad cc \ldots c \text{ (n factors)} = c^n.)$$

† Note that a and b themselves are each *unary* operations on the points of the circle. In our example, the *operation* is the *composition* of a and b; these transformations being the *operands*.

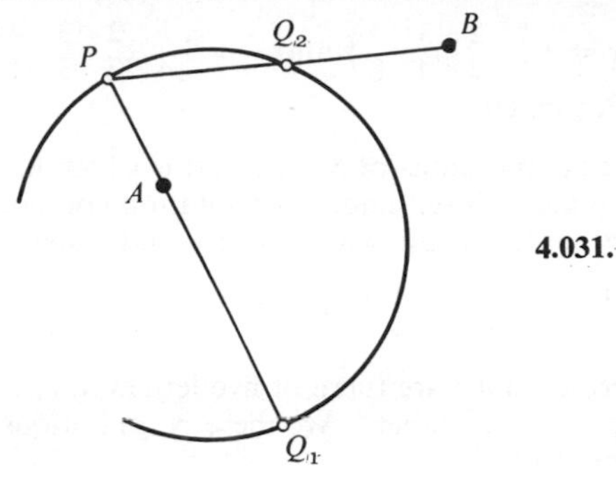

4.031.

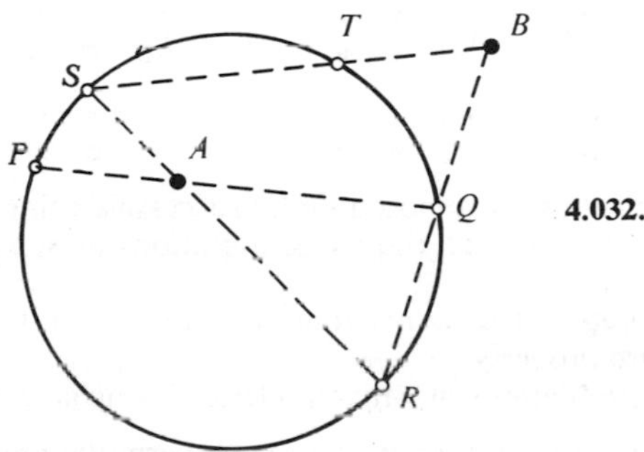

4.032.

Fig. 4.03. Transformations on a circle.

4.031. $Q_1 = a(P)$; $Q_2 = b(P)$.

4.032. $Q = a(P)$; $R = b(Q) = ba(P)$; $S = aba(P)$; $T = baba(P)$.

EXERCISES 4

1 Is the set of permutations of four objects which interchange two pairs, e.g. $\begin{pmatrix} 1 & 2 & 3 & 4 \\ 3 & 4 & 1 & 2 \end{pmatrix}$, closed under successive application of the permutations?

2 Find another permutation which will close the set

$$\begin{pmatrix} 1 & 2 & 3 & 4 \\ 4 & 1 & 2 & 3 \end{pmatrix}, \quad \begin{pmatrix} 1 & 2 & 3 & 4 \\ 3 & 4 & 1 & 2 \end{pmatrix}, \quad \begin{pmatrix} 1 & 2 & 3 & 4 \\ 2 & 3 & 4 & 1 \end{pmatrix}$$

(known as 'cyclic' permutations).

3 Represent the permutations of Exs. 1 and 2 above by 4×4 matrices acting on column vectors.

4 Consider those permutations of five objects which keep at least one object in the same position, e.g. $\begin{pmatrix} 1 & 2 & 3 & 4 & 5 \\ 2 & 1 & 5 & 4 & 3 \end{pmatrix}$, $\begin{pmatrix} 1 & 2 & 3 & 4 & 5 \\ 1 & 5 & 3 & 2 & 4 \end{pmatrix}$. How many of these permutations are there, and are they closed under composition of permutations?

5 Show that if $x = \begin{pmatrix} 1 & 2 & 3 & 4 \\ 1 & 2 & 4 & 3 \end{pmatrix}$ and $y = \begin{pmatrix} 1 & 2 & 3 & 4 \\ 4 & 2 & 3 & 1 \end{pmatrix}$, then $xy \neq yx$. Find x^2, y^2, x^3, yx^2, etc.

6 Consider those permutations of A, B, C and D which keep A and C next to each other. Are these closed under permutation composition? Is the complementary set (i.e. those which keep A and C separated, such as $\begin{pmatrix} A & B & C & D \\ A & D & C & B \end{pmatrix}$, $\begin{pmatrix} A & B & C & D \\ C & B & D & A \end{pmatrix}$ closed?

7 How many permutations are there of five letters A, B, C, D, E such that A and B are next to each other? Are these permutations closed under successive composition?

8 Consider the subset of the twenty-four permutations of ABCD which leave D alone, i.e.

$$\begin{pmatrix} A & B & C & D \\ A & B & C & D \end{pmatrix}, \quad \begin{pmatrix} A & B & C & D \\ C & A & B & D \end{pmatrix}, \quad \begin{pmatrix} A & B & C & D \\ B & C & A & D \end{pmatrix},$$

$$\begin{pmatrix} A & B & C & D \\ A & C & B & D \end{pmatrix}, \quad \begin{pmatrix} A & B & C & D \\ C & B & A & D \end{pmatrix} \text{ and } \begin{pmatrix} A & B & C & D \\ B & A & C & D \end{pmatrix}.$$

It is evident that this set is closed under successive composition because in effect what we have is all the six permutations of A, B and C, with D discarded.
Find other subsets of the twenty-four permutations of four letters which have the closure property.
Is the set of permutations in which no letter is unchanged, e.g. $\begin{pmatrix} A & B & C & D \\ C & A & D & B \end{pmatrix}$, closed (there are nine such permutations)? If these nine permutations are represented by 4×4 matrices, what could you say about the positions of the 1's in the matrices?

9 Consider the set of six permutations

$$\begin{matrix} 1 & 2 & 3 & 4 & 5 & 6 \\ 2 & 3 & 1 & 5 & 6 & 4 \\ 3 & 1 & 2 & 6 & 4 & 5 \\ 4 & 6 & 5 & 1 & 3 & 2 \\ 5 & 4 & 6 & 2 & 1 & 3 \\ 6 & 5 & 4 & 3 & 2 & 1 \end{matrix}$$

in which the pairs (1, 4), (2, 5) and (3, 6) are always separated by two numbers between them. Show that this set is closed under successive composition. Find other subsets of the 720 permutations of six digits which have the closure property. Give also examples of subsets which are constructed according to some simple rule, and yet are not closed.

10 Show that those permutations of six people sitting at a round table which are such that each person always has the same two people sitting next to him, are closed.
(*Note*: we are not concerned with the question as to who sits on the left and who on the right side.)

11 Discover how many times the permutation $\begin{pmatrix} 1 & 2 & 3 & 4 & 5 & 6 & 7 \\ 3 & 1 & 4 & 2 & 5 & 7 & 6 \end{pmatrix}$ has to be repeated in order to give the identity. (See Period, Ch. 10.) Repeat with $\begin{pmatrix} 1 & 2 & 3 & 4 & 5 & 6 & 7 \\ 6 & 5 & 7 & 1 & 3 & 4 & 2 \end{pmatrix}$

12 The permutations $\begin{pmatrix}1 & 2 & 3 & 4\\ 2 & 1 & 4 & 3\end{pmatrix}$, $\begin{pmatrix}1 & 2 & 3 & 4\\ 3 & 4 & 1 & 2\end{pmatrix}$, $\begin{pmatrix}1 & 2 & 3 & 4\\ 4 & 3 & 2 & 1\end{pmatrix}$ (in which two pairs of digits are interchanged in each case), together with the identity, are closed under successive composition (see Ex. 2 above). Consider a similar set of permutations of six digits in which three pairs are interchanged, e.g. $\begin{pmatrix}1 & 2 & 3 & 4 & 5 & 6\\ 3 & 5 & 1 & 6 & 2 & 4\end{pmatrix}$. There are fifteen such permutations – is this set together with the identity closed? If not, can you find a subset of it which is closed? Or can you form the smallest set of permutations of six digits which contains all fifteen and is closed? Repeat for permutations of eight digits.

13 Show that the permutations:

1	2	3	4	5	6	7	8
4	3	2	1	8	7	6	5
5	6	7	8	1	2	3	4
2	1	4	3	6	5	8	7
6	5	8	7	2	1	4	3
3	4	1	2	7	8	5	6
8	7	6	5	4	3	2	1
7	8	5	6	3	4	1	2

are a closed set under successive composition.

14 **Show that the set of linear functions, $x \to ax + b$, where a and b are constants, is closed under successive function composition, and that if $f(x) = ax + b$, and $g(x) = px + q$, then $fg(x) = apx + aq + b$. Find also $gf(x)$.**

15 **Repeat the previous question for the general bilinear function, $x \to (ax + b)/(cx + d)$.**

16 If $f_1(x) = 1/x$, $f_2(x) = x/(x - 1)$, find $f_1f_2(x)$, $f_2f_1(x)$, $f_1f_1(x)$, etc., forming all possible composite functions. If $f(x) = 1/(1 - x)$, find $ff(x)$, $fff(x)$, $f^4(x)$, and so on.

17 A toy consists of six cubes each of side 5 cm which fit into a box 15 cm × 10 cm. Each face of each cube has a portion of a picture on it, there being six pictures altogether. When the cubes are in the right position, a complete picture shows on the top surface, and each of the other five complete pictures may be obtained by performing the same operation on each cube (e.g. by rotating each cube clockwise through 90° about an axis pointing from left to right). Some operations will give a jumbled picture. Obtain (or make!) one of these toys, and examine the mathematical structure of the operations, finding which combinations of operations do, and which do not, give a proper picture.

5 *Associativity*

'I have thirty odd boys in my class', said the high school teacher.

Punctuation in mathematics

When I was a small boy I remember hearing the sentence from Genesis: '... and the earth was without form and void', and reflecting that there were two things which the earth hadn't got in those early days, (1) form, and (2) void. I had some idea what form was, but 'void' remained a mysterious object to me for some time. The confusion, of course, lies in the punctuation. I thought of it as 'the earth was without (form and void)', whereas I should have read it as '(the earth was without form), and void'.

Such confusion due to lack of punctuation is common in everyday experience: 'Going down hill on a bicycle I saw a cow', for example. And one which appeared recently in a newspaper's advertisement column: 'LOST: a black man's umbrella'.

Many boys, when doing calculations with the angles of a triangle will write:

$x = 180 - 50 + 70 = 60$, when of course they mean:
$x = 180 - (50 + 70) = 180 - 50 - 70 = 60$.

They have left out the 'punctuation marks' of mathematics, as the S.M.P. text-books so aptly call brackets (see *Book* 1, pp. 51–3 and *Additional Mathematics*, *Book* 1, pp. 2–3, C.U.P.). And in this instance the brackets are essential, for $180 - 50 + 70 = 200$, not 60. Brackets may be needed to indicate *which operation shall be done first*. Normally we take $a - b \times c$ to mean $a - (b \times c) = a - bc$, but there is no logical reason why it should not mean $(a - b) \times c$. In ordinary algebra, we conventionally accept a 'hierarchy of operations', whereby, in the absence of brackets, multiplication and division take precedence over addition and subtraction. This priority is recognised and followed in computer languages such as Algol and Fortran.

The above examples were complicated by the fact that the operations ($+$, $-$ and $\times$) are 'mixed up'. Let us confine our attention to cases of three numbers which are combined by two applications of a single operation. Which of the following need brackets to avoid a double meaning?

(a) $a + b + c$; (b) $a - b - c$; (c) $a \times b \times c$;
(d) $a \div b \div c$; (e) a^{b^c}; (f) H.C.F. (a, b, c).

Evidently, brackets may be completely omitted with no fear of ambiguity in cases (a) and (c). On the other hand, (b) may mean $(a - b) - c$, or $a - (b - c)$; (d) may mean $(a/b) \div c$ $(= a/bc)$, or else $a \div (b/c)$ $(= ac/b)$; (e) may mean $(a^b)^c$ or $a^{(b^c)}$, which are different. In (f), suppose a, b and c are natural numbers, and H is the binary operation of taking the H.C.F. on the set of natural numbers, so that a H b is the H.C.F. of a and b. We want to know whether (a H b) H c is equal or is not equal to a H (b H c)? A little thought will convince you that they are in fact equal, and so the brackets may be safely omitted, and we may speak of 'the H.C.F. of a, b and c' without being misunderstood.

Q. 5.01. Investigate whether the operation L, 'take the L.C.M.' possesses the same property.

Definition: An operation$*$ is said to be *Associative* if $(a*b)*c = a*(b*c)$ for all a, b and c in the set of operands.

Thus we have seen that $+$, $\times$ are associative operations on the set of real numbers; H and L are associative on the set of natural numbers. And we have also found counter-examples of non-associative operations.

Q. 5.02. Verify that the operations $\cap$ and $\cup$ are each associative binary operations on sets, i.e. that if A, B and C are sets, then

$$(A \cap B) \cap C = A \cap (B \cap C), \quad \text{and} \quad (A \cup B) \cup C = A \cup (B \cup C).$$

You should verify these by means of a Venn diagram in each case, and also by a logical argument.

Q. 5.03. Discover whether the operation $A \Delta B = (A \cap B') \cup (A' \cap B)$ (see p. 9) is associative – use Venn diagrams, or set algebra, or logical argument.

Q. 5.04. Is the 'logical connective' *or* associative when used between statements? That is, if a, b and c are statements, can we say (a or b) or c is the same as a or (b or c)? Repeat for the conjunction 'and', and find a conjunction which is not associative.

Q. 5.05. Investigate whether addition of vectors: u + v is associative, and illustrate by diagrams.

Testing for associativity

Sometimes it is a very laborious matter to discover whether a particular operation is associative or not. For an example of this, involving heavy manipulation, see F. J. Budden, *Complex Numbers* (Longmans), 1968, p. 15. For an example involving much deep thought, consider the operation 'latest common forefather' (see p. 6). Is it associative or not . . . ? Nor is it at all obvious in many cases by merely 'looking at it' whether a particular operation is associative or not. Take for example, the operation

$*$ on the set of real numbers defined thus (as on p. 7): $x*y = xy/(x + y)$.† Is it associative or not? We can only tell the hard way:

$$(x*y)*z = \frac{xy}{x+y}*z = \frac{\{xy/(x+y)\}z}{xy/(x+y)+z} = \frac{xyz}{xy+xz+yz},$$

$$x*(y*z) = x*\frac{yz}{y+z} = \frac{x+(yz)/(y+z)}{x+(yz)/(y+z)} = \frac{xyz}{xy+xz+yz}$$

$$= (x*y)*z,$$

so $*$ is associative. But you may claim that you could have predicted this: it is *bound* to be, you might say – and perhaps you might be swayed by the fact that the formula $xy/(x + y)$ is *symmetrical* in x and y. All right, try it in the case of the operation $\square$, where:

$$x \square y = \frac{2xy}{x+y} \quad \text{(we met this one on p. 11).}$$

Indeed, if you write down a simple formula in x and y at random in place of the $2xy/(x + y)$ above, it is *most probable* that it would turn out to be a non-associative operation. For we already know that $-$ and $\div$ are not associative, so that $x - y$ or x/y would be simple counter-examples of a non-associative formula.

As an example of a laborious verification of associativity, let us consider the following binary operation on sets of ordered pairs of real numbers:

$$(x_1, y_1) \Diamond (x_2, y_2) = (x_1x_2 + x_1y_2 + y_1x_2, y_1y_2).$$

$$[(x_1, y_1) \Diamond (x_2, y_2)] \Diamond (x_3, y_3)$$

$$= (x_1x_2 + x_1y_2 + y_1x_2, y_1y_2) \Diamond (x_3, y_3)$$

$$= [(x_1x_2 + x_1y_2 + y_1x_2)x_3 + (x_1x_2 + x_1y_2 + y_1x_2)y_3 + y_1y_2x_3, y_1y_2y_3]$$

$$= (x_1x_2x_3 + x_1x_3y_2 + x_2x_3y_1 + x_1x_2y_3 + x_1y_2y_3 + x_2y_1y_3 + x_3y_1y_2, y_1y_2y_3)$$

$$(x_1, y_1) \Diamond [(x_2, y_2) \Diamond (x_3, y_3)]$$

$$= (x_1, y_1) \Diamond (x_2x_3 + x_2y_3 + y_2x_3, y_2y_3)$$

$$= [x_1(x_2x_3 + x_2y_3 + y_2x_3) + x_1y_2y_3 + y_1(x_2x_3 + x_2y_3 + y_2x_3) y_1y_2y_3]$$

$$= (x_1x_2x_3 + x_1x_2y_3 + x_1x_3y_2 + x_1y_2y_3 + x_2x_3y_1 + x_2y_1y_3 + x_3y_1y_2, y_1y_2y_3)$$

† See footnote p. 7. We have deliberately avoided the difficulty which arises from the case $x + y = 0$. In fact it is impossible to invent a meaning for $x * (-x)$ in such a way that the operation $*$, now defined on the domain of the complete set of reals, is associative.

and since these agree, we conclude that ◇ is associative on these ordered pairs.

Q. 5.06. Is this operation ◇ commutative?

Q. 5.07. Discover for what values of the constants a and b, the operation ○ on the set of reals defined thus $x \bigcirc y = ax + by$ is associative ($x, y \in \mathbf{R}$).

Repeat in the cases where the formula on the right is replaced by:

$$ax + by + c; \qquad \frac{ax + by}{cx + dy}; \qquad ax^2 + by^2; \qquad \sqrt{(ax^2 + by^2)}$$

In table 5.01 we show how five objects of a set (never mind what they

Table 5.01

*	a	b	c	d	e
a	a	b	c	d	e
b	b	c	e	a	d
c	c	a	d	e	b
d	d	e	a	b	c
e	e	d	b	c	a

(for example, $c * d = e$).

are) combine according to a binary operation $*$ (never mind what it is). Is the operation associative? We might start by trying $a*c*e$.

$$\left.\begin{array}{l}(a*c)*e = c*e = b\\ a*(c*e) = a*b = b\end{array}\right\} \text{ agreement!}$$

Again, $(b*b)*c = b*(b*c)$, as you may verify. This is encouraging! But to verify associatively for *every* possible choice of three elements, each selection in every possible order (and there are 105 such permutations of three of the letters, allowing for repetitions as in the second case above) would be extremely tedious.† However, we may quickly find *one* selection that does not work: $(b*c)*d = e*d = c$; while $b*(c*d) = b*e = d$. No need to go any further! A *single counter-example* such as we have found, is sufficient for disproof. The operation as defined by the table is NOT associative.

When is associativity guaranteed?

We mentioned that usually associativity cannot be checked by a quick inspection. But, mercifully, there are certain situations where associativity can instantly be recognised.

(1) **Multiplication and addition in any field of numbers, such as the rationals, the reals, the complex numbers, finite fields; as well as in systems**

† Refer to an article by Bruckheimer, 'Checking for Associativity', *Mathematics Teaching*, Nov. 1968.

which are not fields. (It is not necessary at this stage to know precisely what is meant by a 'field'.)

(2)† **Composition of *successive operations* such as we considered in the last chapter: (a) permutations; (b) geometrical transformations and symmetries; (c) successive substitutions in functions; (d) matrix addition and multiplication.**

You may regard it as axiomatic that operations are associative under successive composition. For it is a matter of common experience that if you shave and then wash, and (pause) then dress, the final result is the same as if you had shaved (long pause), then washed and dressed. Mathematicians are not so easily satisfied as to accept such intuitively evident truths on trust, and if you would feel happier with a proof, you may find one‡.

It is most important to realise that the *order of the operands* is not altered in all these cases: the shaving, washing and dressing take place in that order in both cases; if the order of these were changed the final result could be different. When we say $a - (b - c) \neq (a - b) - c$, the order of the a, b and c is the same in both cases. It is the two *operations*, the two minuses, whose order of coming into action, is different.

In the games of the previous chapter, we had instances of associative systems, and these are sometimes called 'semi-groups'. In having introduced the concept of associativity, we are moving nearer to the goal – groups. In point of fact, associativity is central to the whole theory of the subject of groups. Why is associativity of such vital importance? The answer lies in the fact that in the proof of practically every theorem and result in the study of group theory, the associative property has to be assumed. Instances may be found on pp. 65, 90, 100, 102, 106, 206–7, 355, 364, 368–9, 380, App. II, etc.

A geometrical example

★Consider now a geometrical example (see fig. 5.01). O is a fixed point on a circle. A binary operation$*$ on pairs of points on the circle is defined as follows: $A*B = P$, where P is obtained by the following construction. Produce AB to D so that $AB = BD$. Join OD meeting the circle at P. We ask ourselves, is$*$ associative?

The answer (as in some of the algebraic examples considered earlier) is neither obviously 'Yes' nor obviously 'No'. We have to decide whether, for *all* possible choices of A, B and C, $A*(B*C) = (A*B)*C$ or not. What you would probably do here would be to draw a few

† These are all special cases of the single result that *mappings* are associative.
‡ See for example A. Bell, *Algebraic Structures* (Allen and Unwin), p. 60; E. M. Patterson and D. E. Rutherford, *Elementary Abstract Algebra* (Oliver & Boyd), p. 15.

diagrams and to construct the position of $(A*B)*C$ and of $A*(B*C)$ for three selected points A, B and C. If there was a wild disagreement, you would assume associativity not to be satisfied, and would search for a single counter-example. If there were agreement, you would try again

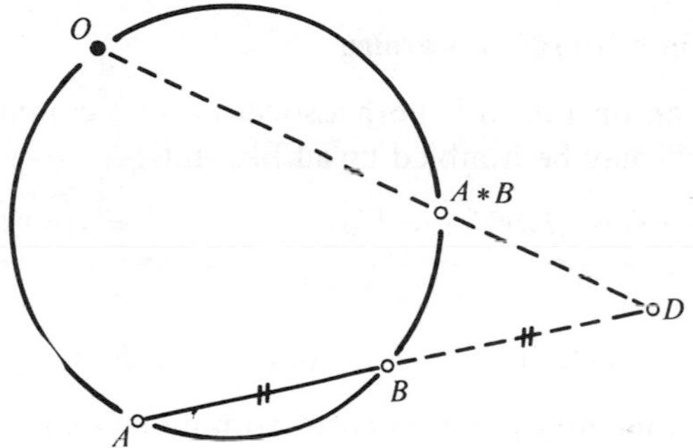

Fig. 5.01. A binary operation on the set of points on a circle – is it associative?

with other positions of A, B, C, and if your experiments led you strongly to believe in associativity, you would then consider it worth while seeking a positive proof. In actual fact, we can find a counter-example which disproves associativity, as follows (see fig. 5.02). Suppose

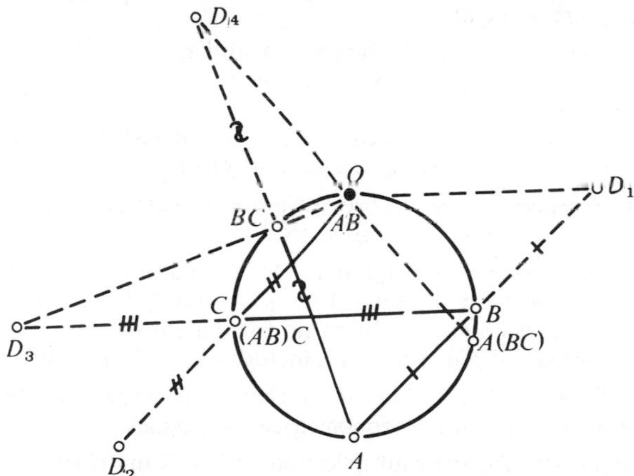

Fig. 5.02. Counter-example: when $OABC$ is a square, $(AB)C \neq A(BC)$.

A, B, C are selected so that $OBAC$ is a square. Then OD will be a tangent to the circle, so that $A*B$ (the asterisk is omitted in the diagram) coincides with O. In constructing $(AB)C$, D_2 lies on OC produced, and so $(AB)C$ actually coincides with C. The position of BC is also marked on the diagram, and the construction for $A(BC)$ leads to a point in the neighbourhood of B – about as far from $(AB)C$ as it could conceivably be.

★**Q. 5.08.** Is this operation commutative? What happens when A and B coincide? What happens when A (or B) coincides with O?

Order of letters in a 'word': a warning

Note that when an operation is both associative and commutative, the letters in a 'word' may be jumbled up ad lib, thus:

$$abcd = (ab)(cd) = (ba)(dc) = b(ad)c = b(da)c = (bd)(ac) = bdac, \text{ etc.}$$

Again,

$$abaa = aba^2 = a(ba^2) = a(a^2b) = (aa^2)b = a^3b$$

and so on. But when an operation is not commutative, *even though it be associative*, the order of the letters in a 'word' must be strictly unaltered, so that:

aba may not be written a^2b or ba^2.

EXERCISES 5

1 Examine the following binary operations on the set of real numbers and say whether they are associative:
(a) $x * y = \max(x, y)$ (i.e. the larger of x and y);
(b) $x * y = x^3 + y^3$; (c) $x * y = x(x + y)$;
(d) $x * y = x + y + xy$;
(e) $x * y = x$ (the first mentioned), $x * y = y$ (the last mentioned);
(f) $x \sim y = |x - y|$; (g) $x \circ y = (x + y)/xy$.

2 Show that the operation $*$ on ordered pairs of reals defined $(a, b) * (c, d) = (ac, bc + d)$ is associative.

3 Is the set of functions z, $-z$, $1/z$ and $-1/z$ closed under successive substitution? Repeat for the set z, $1 - z$, $1/z$ and $1/(1 - z)$. Find other functions which, together with these, do form a closed set.
Find another set of *four* functions, including z and $1/z$, which are closed.

4 Show that if $x * y = (xy - 1)/(x + y)$, then $*$ is associative. Invent other associative operations with trigonometrical analogues.

5 Verify associativity for the multiplication of 2×2 matrices.

6 Show that the operation '$\circ$' on ordered pairs of reals, defined $(x_1, y_1) \circ (x_2, y_2) = (x_1x_2, 2y_1 + 2y_2 - y_1x_2)$ fails for associativity. (Find a single counter-example.)

7 Show that the operation $*$ on ordered triples, defined $(p_1, q_1, r_1) * (p_2, q_2, r_2) = (p_1 + p_2 + q_2r_1, q_1 + q_2, r_1 + r_2)$ is associative (cf. Ch. 9, pp. 100 ff). Find $(p, q, r) * (p, q, r) * (p, q, r)$.

8 Verify associativity for the composition of functions in various cases when $f_1(x) \equiv -1/x$; $f_2(x) \equiv x + 3$; $f_3(x) \equiv x^2$.

9 If A and B are points of a plane, define $A * B$ as the mid-point of AB. Show that $*$ is not associative. Invent a binary operation on the points of a plane which is associative.

10 Suppose A and B are two points on a line through O, and a binary operation $*$ is set up as follows: let $(OACB)$ be a harmonic range, i.e. $(OA \, . \, CB)/(OB \, . \, CA) = -1$, then $C = A * B$ (see fig. 5.03). Show that if the

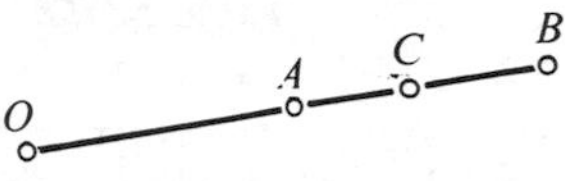

Fig. 5.03. Harmonic range $\dfrac{OA}{AC} = \dfrac{-OB}{BC} \Leftrightarrow \dfrac{2}{OC} = \dfrac{1}{OA} + \dfrac{1}{OB}$.

directed lengths OA, OB, OC are denoted a, b, c then $2/c = 1/a + 1/b$. (Note that OC is the Harmonic Mean between OA and OB, hence the term 'harmonic' range.)

Show that $*$ is not associative, but that an associative operation would have been achieved if the length of OC had been doubled. (Cf. p. 38, where $x * y = xy/(x + y)$ was shown to be associative.)

11 Consider the associativity or otherwise of the operations defined below, in which A and B are points on a line through O; $C = A * B$, where
(a) OC is the arithmetic mean between OA and OB;
(b) OC is the geometric mean between OA and OB, with A, B and C all on the same side of O;
(c) for the case when OC is the harmonic mean, see Q. 5.10 above.

12 Given a line l and a fixed point O not on l, for any two points A and B in the plane of O and l, we define $C = A * B$ by the following construction: Let AO meet l in D. Then C is the fourth vertex of the parallelogram $ADBC$ (see fig. 5.05). Consider special positions of A and B (e.g. AB passing through O; A or B lying on l; A coinciding with O; A and B coincident, etc.). Is $*$ associative?

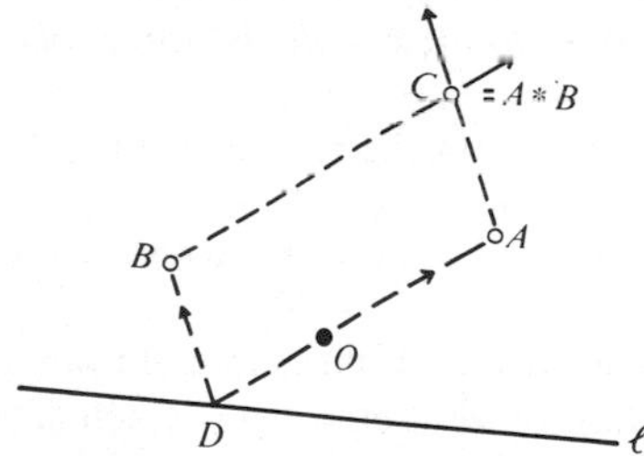

Fig. 5.05. A binary operation on the points of the plane: O is a fixed point; l is a fixed line; C is obtained by drawing the parallelogram $ADBC$.

13 Assuming that free vectors may be added by the rule shown in fig. 5.04, consider the geometrical aspect of commutativity and of associativity. Draw the various diagrams from which $\mathbf{a} + \mathbf{b} + \mathbf{c} + \mathbf{d}$ may be constructed.

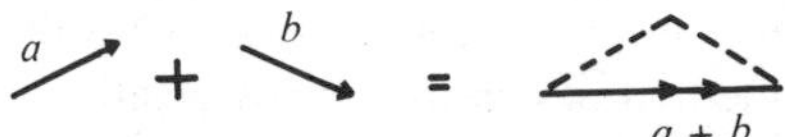

Fig. 5.04.

14 If **a**, **b** and **c** are vectors in three dimensions, is the operation of *vector product* associative, i.e. may be say $(\mathbf{a} \times \mathbf{b}) \times \mathbf{c} = \mathbf{a} \times (\mathbf{b} \times \mathbf{c})$?

15 Check triple 'products' for associativity in the following 'multiplication' tables.

Table 5.021

×	0	1	2	3	4	5
0	0	1	2	3	4	5
1	1	2	3	0	5	4
2	2	3	4	5	0	1
3	3	0	5	4	1	2
4	4	5	0	1	2	3
5	5	4	1	2	3	0

Table 5.022

×	I	A	B	C	D
I	I	A	B	C	D
A	A	I	C	D	B
B	B	D	I	A	C
C	C	B	D	I	A
D	D	C	A	B	I

16 Draw flow diagrams illustrating a sequence of operations in computing expressions like $a - b + c$; $a/b/c$; $a \times b - c$; $a - b \times c$, etc. when brackets are inserted in various positions.

17 If $*$ is an operation which is commutative and associative, then all six expressions: $a*b*c$, $a*c*b$, $c*b*a$, $b*a*c$, $c*a*b$ and $b*c*a$ are equal. How many of these triple products are equal:
(a) when $*$ is commutative but not associative (e.g. $a*b = |a - b|$);
(b) when $*$ is neither commutative nor associative? (Give an example.)

18 Show that $x*y = (xy + k)/(x + y)$ is associative for all k.

19 Show that $x*y = (x + y)/(a + kxy)$ is associative for all k provided $a = 1$.
Give interpretations in the above for the cases $k = 0, -1$ and $+1$, and invent other associative operations on the reals.

20 Prove that $x*y = (ax + by)/(cx + dy)$ is not associative for general a, b, c or d.

21 Show that the operation $\square$ on the reals defined

$$x, y \in R \qquad \begin{aligned} x \square y &= xy \quad (x > 0) \\ x \square y &= x/y \quad (x < 0) \quad \text{when } y \neq 0, \\ x \square y &= 0 \quad (x = 0) \end{aligned}$$

is associative. (Consider triple products in eight possible cases.)

22 Find a binary operation on the set of reals which is (a) commutative but not associative; (b) associative but not commutative.

23 If $f(x)$ is a function for which an inverse function $f^{-1}(x)$ can be found, show that the operations $\oplus$ and $\otimes$ on the reals, defined

$x \oplus y = f^{-1}[f(x) + f(y)]$; $\quad x \otimes y = f^{-1}[f(x) \, . \, f(y)]$ are associative

Obtain formulae giving $x \oplus y$ and $x \otimes y$ where appropriate, when $f(x)$ takes the forms:
(a) $1/x$; (b) $1 + 1/x$; (c) $1/(x - 1)$; (d) e^x; (e) $k + x^2$;
(f) $\sin^{-1} x$; (g) $\tan^{-1} x$; (h) $\cosh^{-1} x$; (i) $\coth^{-1} x$.

24 Show how the above method of constructing associative operations on the reals may be still further generalised.

6 *Status quo: identity elements*

We have already referred to 'identity' (p. 18), and in the case of transformations and permutations, we called it the 'stay-put' operation. This may lead you to hope that this chapter may be a nice short one – after all, you will say, surely there is not much to be said about leaving things unaltered. That, you may think, is that, and here is the end of the chapter!

Unfortunately there is more in it than meets the eye. Some systems possess an Identity operation (alternatively known as the *Neutral* or *Null* element), some do not. In the latter event, one may invent, or introduce one, as we found necessary for closure in the example about the set of half-turn symmetries of the cuboid, on p. 18). Sometimes the identity is easy to spot; sometimes it is obscure.

Examples of identity elements

The set N of natural numbers does not possess an identity element for addition, so the number *zero* is introduced. And if that seems a perfectly obvious thing to do, remember that it was centuries after the Greek civilisation that the necessity for the invention of a symbol for zero was realised. For multiplication within a set, the identity is *Unity*, 1, (provided the set includes this number). The identity element for multiplication is therefore called the 'unit' of the number system. For matrix addition, the identity is a matrix, of the appropriate number of rows and columns every entry of which is zero (the 'null' matrix). While for matrix multiplication, the identity is a square matrix with zeros everywhere except on the leading (NW.–SE.) diagonal, which contains 1's, e.g. $\begin{bmatrix}1 & 0 & 0\\ 0 & 1 & 0\\ 0 & 0 & 1\end{bmatrix}$ is the identity for multiplication of 3×3 square matrices, and is called the 'unit' 3×3 matrix, usually shown as I_3. It also acts as an identity for the post-multiplication of $p \times 3$ matrices (including row-vectors), and for the pre-multiplication of $3 \times q$ matrices, including column vectors. For example,

$$\begin{bmatrix}1 & 0 & 0\\ 0 & 1 & 0\\ 0 & 0 & 1\end{bmatrix}\begin{bmatrix}a & p\\ b & q\\ c & r\end{bmatrix} = \begin{bmatrix}a & p\\ b & q\\ c & r\end{bmatrix}.$$

In general, the $n \times n$ identity matrix would be denoted I_n.

Again, for the set of real numbers under subtraction, we have, for any real number x, $x - 0 = x$, so that 0 is a *right identity*. But 0 is *not* a *left* identity, since $0 - x$ is not equal to x.

Q. 6.01. Discuss the identity element in the set of reals for (a) division, (b) exponentiation, and in the set of natural numbers for (a) H.C.F., (b) L.C.M.

Q. 6.02. In the set of, say, cubic polynomials, under addition, the identity polynomial is $0x^3 + 0x^2 + 0x + 0$. Call this set $P_3(x)$. What is the identity element in the set $P_n(x)$ under multiplication?

Formal definition

The above are examples giving the general idea of identity. We now give a formal definition.

> **If a set has a binary operation $*$, and there is a fixed element e of the set such that $x * e = x$ for *every* element x of the set, then e is called a *right-identity* of the set under the operation $*$.**

The definition for left identity is similar. It may happen that a set has a right but not a left identity (or vice versa), as the example of subtraction shows. If, however, a set with an operation possesses an element e which is both a left and a right identity, then we should normally refer to it as 'the identity', or 'neutral element' of the set. We shall see that it is *by no means necessary* for the operation to be commutative in order that an identity element may exist which commutes with every element.

Q. 6.03. Can you find a set and an operation with a left identity but no right identity? Can you find a case where there is a left and a right identity, but which are different?

Q. 6.04. What is the identity element in the set of functions of x under successive substitution? (See p. 22.)

Q. 6.05. For the operations $\cap$ and $\cup$ on sets, what are the identity elements respectively? What is the identity for Δ (see p. 9)?

Some examples from finite arithmetics

Sometimes we are in for a surprise. Consider the set of residue classes {2, 4, 8} under multiplication modulo 14 (e.g. $11 \times 3 = 33 \equiv 5$ (mod. 14)). When working modulo 14, we shall regard the number 2, for example, as being the representative of the class {... −12, 2, 16, 30, ...} of integers which leave remainder 2 when divided by 14. '2' is called the 'least positive residue' of this residue class. When we say that $8 \times 4 = 4$ (mod. 14), we really mean: 'Any integer from the 8 class (such as −6) multiplied by any integer from the 4 class (such as 18) will produce a number from the 4 class (here, $-6 \times 18 = -108 \equiv 4$ (mod. 14)). Table 6.01† shows how each pair of elements combine under multiplication. Which is the identity element, if any?‡

Table 6.01

× mod. 14	2	4	8
2	4	8	2
4	8	2	4
8	2	4	8

We see that $2 \times 8 = 2$ (mod. 14)
$4 \times 8 = 4$ (mod. 14)
$8 \times 8 = 8$ (mod. 14).

Therefore 8 is the identity element (or 'unit') in this type of multiplication for this particular set.

Q. 6.07. Find another set for which 8 is the identity element for multiplication mod. 14.

Q. 6.08. What are the other elements in a set containing 3 and 11 which is closed under multiplication mod. 14? What is the identity element? Can you find a smaller subset of odd integers which are closed under multiplication mod. 14?

In case you were unable to do the above questions, we now proceed to answer a question similar to the above, but this time in the 'finite arithmetic' of multiplication modulo 15. One could write out the whole multiplication (mod. 15) table showing all possible products of pairs from the set $\{1, 2, 3, \ldots, 13, 14\}$ (evidently there is no point in including the number zero), and then analyse the table. Or one could proceed as follows. Suppose we want a set containing 7. Then it must, (to be closed) also contain $7 \times 7 = 4$ (mod. 15). Similarly it must contain $4 \times 7 = 13$ (mod. 15). Then again we must have $4 \times 13 = 7$ (already listed); $4 \times 4 = 1$ (mod. 15) (a new one), as well as $7 \times 13 = 1 \ldots$, and we find that the set is complete: $\{1, 4, 7, 13\}$, with 1 as the unit element, of course. The multiplication is shown in table 6.02.§ It will be seen that an even smaller set exists, $\{1, 4\}$ which is closed under multiplication modulo 15.

† In writing a multiplication table like this, it is invariable practice to enter the unit element *first* both across and down, so that the numbers would normally appear in the order 8, 4, 2, or 8, 2, 4.

‡ **Q. 606.** Is it necessary to say 'if any'? – is it possible to find a set closed under multiplication modulo 14, which does *not* possess an identity? Can you answer the same question for multiplication modulo any integer?

§ This, and table 6.01 are our second and third instances of a group table, the first having appeared on p. 18.

Suppose we next look for a set containing 6. $6 \times 6 = 36 = 6$ (mod. 15), so nothing new here! Now 9×9 also $= 6$ (mod. 15), so let us try $6 \times 9 = 9$ (mod. 15). Therefore $\{6, 9\}$ is a set closed under multiplication modulo 15, and its unit element is 6.

Table 6.02

× mod. 15	1	4	7	13
1	1	4	7	13
4	4	1	13	7
7	7	13	4	1
13	13	7	1	4

Q. 6.09. Find a larger set which has 6 as its identity Find also a set of two numbers for which the identity is neither 1 nor 6. Find other self-contained sets.

Q. 6.10. Experiment with multiplication modulo 6, mod. 10, mod. 12, mod. 18, mod. 20, mod. 21, mod. 22, mod. 24, mod. 26. (You could of course experiment with multiplication modulo any number, but those mentioned are specially interesting because they contain subsets which have an identity other than 1.)

Q. 6.11. A set of six numbers under multiplication modulo n have 10 as the unit. Find n and the other five numbers.

Keep a list of the multiplication tables obtained in the questions above for future reference.

Cases when identity elements are obscure

Consider next the binary operation$*$ on the set of real numbers as defined on p. 7, i.e. $x*y = xy/(x + y)$. What is the identity element? Perhaps you would guess it to be 0: $x*0 = x \,.\, 0/(x + 0) = 0$, but we require $x*0 = x$. Try 1: $x*1 = x \,.\, 1/(x + 1) = x/(x + 1)$ (no good). We had better give up guessing! We require e such that $x*e = e*x = x$ for all x. Then $x = xe/(x + e)$, so assuming $x \neq 0$, we conclude that $x + e = e$, which means that no such value of e exists.

Again, suppose we consider the operation $x*y = (x + y)/xy$ $(x, y \neq 0)$. An identity element would have to satisfy $x = (x + e)/xe$ for all x (other than $x = 0$) so that $e = x/(x^2 - 1)$. But this *will not do*, for e must be a *fixed* element of the set, whereas here it depends on x. We conclude that *there does not exist* an identity element for this particular operation. You may now begin to see what was meant by saying that there is more than meets the eye to the question of leaving things as they were!

Consider next the operation $\oplus$ on the set of real numbers defined: $x \oplus y = x + y - 1$. If e is to be an identity, we must have $x = x + e - 1$, and this is true for all x when $e = 1$. Moreover, $e = 1$ is a left identity, since in any case the operation is commutative.

Some geometrical examples

Considerable interest may be aroused by hunting for identity elements in the case of some geometrical examples. Can we find a null element for the example on p. 40? Yes, because when B is at O, we see from fig. 6.01

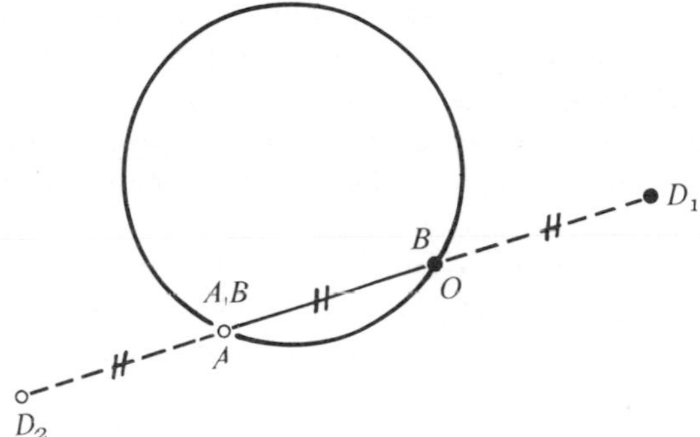

Fig. 6.01. O is the identity point for the transformation of fig. 5.01.

that the construction for the point $A*B$ brings us back to the point A: $A*O = A$. You should check that $O*A = A$ also, so that the point O is both a left and a right identity for this operation between points of a circle.

In fig. 6.02 we show how another binary operation may be set up on the points of a circle. O and U are fixed points on the circumference. $A*B$ is found by allowing AB to meet the tangent at U in D and taking the

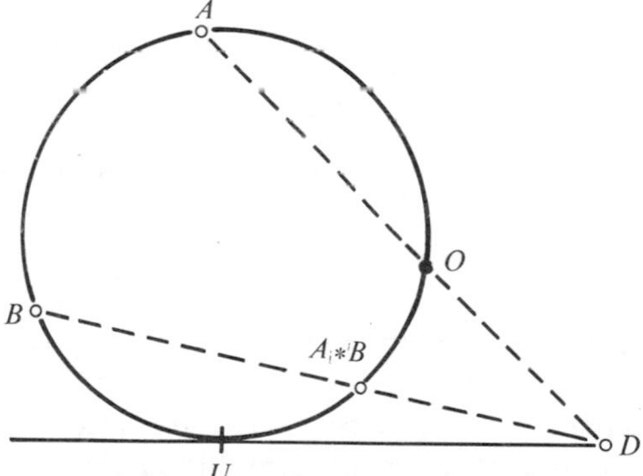

Fig. 6.02. Another binary operation on a circle.

remaining intersection of DO with the circle. If AB happened to be parallel to the tangent at U, then the line through O would also be parallel, for D would then be regarded as being at infinity. If B were at U, then D would also be at U, and $A*B$ would then coincide with U. Thus $A*U = U$.

But we require $A*?=A$ for an identity, and you may easily discover the answer to the question.

Is the operation associative? It is easily seen to be commutative, for A and B are interchangeable without affecting $A*B$. But this is no help in deciding whether it is associative, and this is a very difficult question to settle. In point of fact, the operation, unlike that of fig. 5.01, is associative. You may 'verify' this experimentally by taking various positions of A, B and C (including special positions, such as when A and B coincide and the line AB becomes a tangent), and follow this by attempting a geometrical proof. Failing this, the following coordinate method is recommended: take the circle to be $x^2+y^2=a^2$, and the point U to be (a, O), so that the tangent is $x=a$. If A and B are $(a\cos\theta_1, a\sin\theta_1)$ and $(a\cos\theta_2, a\sin\theta_2)$, and O is $(a\cos\alpha, a\sin\alpha)$, and if $t_1=\tan\frac{1}{2}\theta_1$; $t_2=\tan\frac{1}{2}\theta_2$ and $T=\tan\frac{1}{2}\alpha$, show that:

$$t=\frac{t_1t_2T}{T(t_1+t_2)-t_1t_2},$$

where $A*B$ is the point $(a\cos\theta, a\sin\theta)$ with $t=\tan\frac{1}{2}\theta$. Associativity may then be checked. Consider also that in the special case when $T\to\infty$ (OU is a diameter), the formula reduces to:

$$t=t_1*t_2=\frac{t_1t_2}{t_1+t_2},$$

which is familiar!

Left and right identities

Earlier, we noticed systems which have a right identity but not a left, and it is easy to invent ones which have a left but not a right identity. We may easily show that, if a system possesses a left identity (i.e. an element e such that $e*x=x$ for all x in the set), and also a right identity e', then these must be the same. For consider the element ee'. Since e is a left identity, $ee'=e'$; and since e' is a right identity, $ee'=e$, so it follows that $e=e'$. In the case of rectangular $p\times q$ matrices ($p\neq q$), which have I_p as left identity under matrix multiplication, whereas the right identity is I_q, it would appear that we have a contradiction of the above assertion. But it must be remembered that I_p and I_q, being square, lie outside the set $\mathcal{M}_{p,q}$ of $p\times q$ matrices, and the product ee' is now meaningless.

Can one find an operation on the real numbers which is not associative, and yet which has an element which is both a right and a left identity? In view of the fact that associativity plays such a vital part in the proofs of the relevant theorems (see App. II), it might be thought that the answer would be 'No'. Consider, however;

$$x \sim y = |x - y| \quad (x,\ y \text{ real and non-negative}).$$

Then you will have no difficulty in showing that O is a left and right identity, while failure for associativity could be proved by finding a single counter-example.

Q. 6.12. What would happen if x and y were drawn from the set of all the reals?

In App. II, we prove the following results.

If an associative system is provided with *left* identity and *left* inverses (see Ch. 7), then the left identity must be a right identity, and left inverses are also right inverses (i.e. inverse elements commute); and also that the identity element is unique, and that each element possesses a unique inverse.

Thus in a system which possesses a right identity which is not also a left identity (or vice versa), one may conclude either that the operation is not associative (as in the cases of subtraction, division, exponentiation, etc.), or else that inverses are not defined, as in the case $x * y = x$ on a set containing two elements only. Here y is a right identity, but neither element is a left identity, yet the operation is associative.

Again, if one were to find, say, two or more identities in a system, this would mean that one of the conditions requisite for the above conclusions must break down, and the usual cause of failure would be in associativity.

In spite of various digressions, this chapter is not, after all, unduly long, but long enough, perhaps, to convince you that, though on the surface nothing should be easier than 'staying put', there is in fact much to be said about the concept of 'identity'.

EXERCISES 6

1 Find the identity elements in the operations $*$ on the real numbers defined: (a) $x * y = (x^2 + y^2)^{\frac{1}{2}}$; (b) $x * y = (x + y)/(1 + xy)$; and also for the operations given as examples in the text and in the exercises of the previous chapters, particularly those given in Exs. 3.08, 3.15, 5.01, 5.06, 5.07, 5.18 and 5.19.

2 Find the identity elements for the composition of ordered pairs of reals:

(a) $(x_1, y_1) * (x_2, y_2) = (x_1x_2 + x_1y_2 + y_1x_2, y_1y_2)$;
(b) $(x_1, y_1) * (x_2, y_2) = (x_1x_2 - y_1y_2, x_1y_2 + x_2y_1)$.

3 For the operations 'L' (find the L.C.M.) and 'H' (find the H.C.F.) on the set of natural numbers, show that in one case there is an identity element, but not in the other case.

4 What is the identity element in the set of musical intervals (see Ch. 23)?

5 Does the operation of 'adding circuits' (p. 12) possess an identity element?

6 Identify where possible the identity element in the set of vectors in three dimensions, when the composition is by: (a) vector addition, $\mathbf{a} + \mathbf{b}$; (b) scalar product, $\mathbf{a} \cdot \mathbf{b}$; (c) vector (cross) product, $\mathbf{a} \times \mathbf{b}$.

7 Show that the set of residues {2, 4, 8, 10, 14, 16} is closed under multiplication modulo 18. What is the identity? Find other sets of integer residues not containing 1 which are closed under multiplication modulo 10, 12, 20, 24, and pick out the unit element in each case.

8 Given a fixed point O in a plane, a binary operation is defined thus: $A * B$ shall be the foot of the perpendicular from O to AB when $A \neq B$, and $A * A = A$. Show that no identity point exists. Make up a similar operation on the points on the earth's surface, taking O to be the North Pole.

9 Consider the geometrical example of fig. 5.05. Can you find a point in the plane which plays the part of an identity?

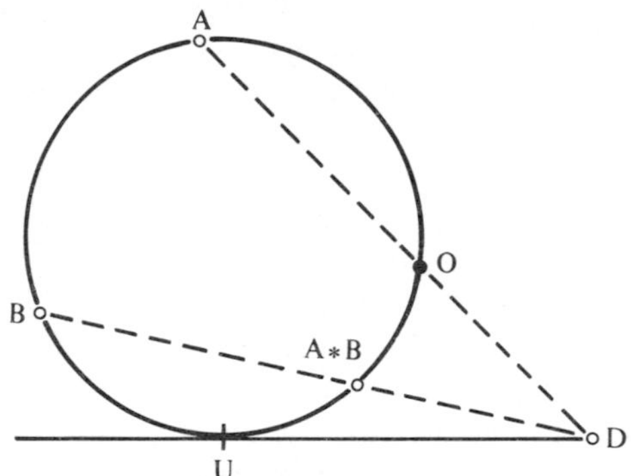

Fig. 6.03. Binary operation on a circle (not to be confused with fig. 6.05).

10 If the construction for $A * B$ (see p. 49, fig. 6.02) were modified to that shown in fig. 6.03 (i.e. AO meets the tangent at U in D, then $A * B$ is the remaining intersection of BD with the circle), show that $A * O = A$, and find $O * A$. What do you conclude?

11 U and V are fixed points, and a binary operation is set up between the points of a plane containing U and V by defining $A * B$ to be the intersection of AU and BV (see fig. 6.04). Is $*$ associative? Can you find an identity point?

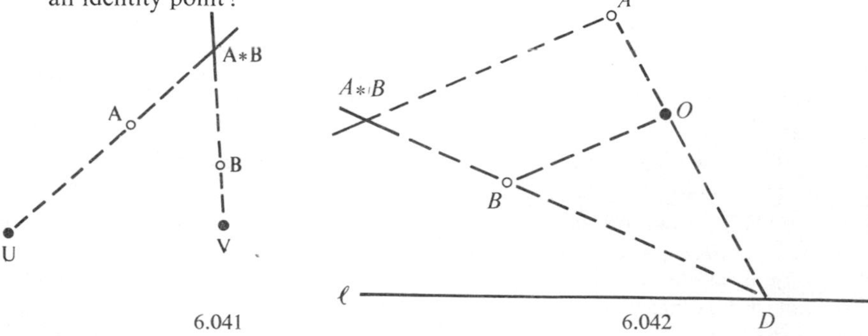

Fig. 6.04. Binary operations on the points of a plane.

12 O is a fixed point and l is a fixed line (see fig. 6.042). A and B are any two points in the plane of O and l. $A * B$ is defined by the following construction: AO meets l in D, DB meets a line through A parallel to OB in the point $C = A * B$. Consider this in detail: is it commutative; what happens when A coincides with O; what happens when A and B coincide; what happens when A or B lies on l; etc.? Show that O is an identity point, and that the operation is associative.

13 O is a fixed point on a conic (see fig. 6.05). A binary operation is defined between points of the conic as follows: given A and B, let the tangents at A and B meet at T. Then the remaining intersection C of OT with the conic is $A * B$. Discover whether this operation is associative, and find the identity point, if any. (Note that T is the pole of AB, and that the points O, A, C and B form a harmonic set of points on the conic.)

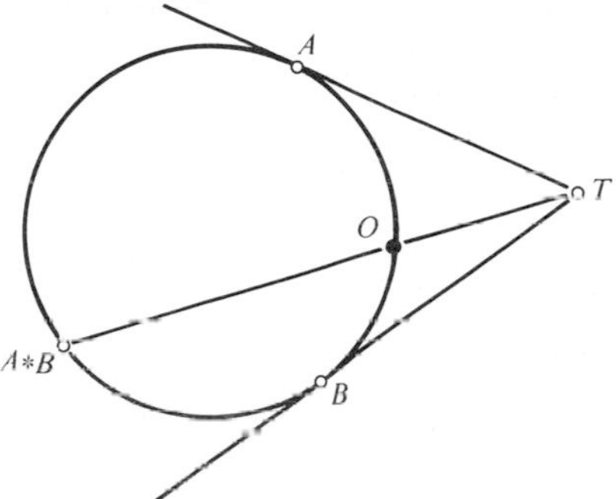

Fig. 6.05. Binary operation on a circle. Is $*$ associative? Where is the identity point?

14 Given a point O in a plane, and selecting two points A and B, the point $A * B$ is defined as the orthocentre of the triangle OAB when $A \neq B$. Invent a suitable definition for $A * A$. Is $*$ associative; can you find a point which is a neutral element for $*$?

15 In the examples above circles considered previously in the text, and the examples above, in how many cases would it be possible to replace 'circle' by 'conic'?

16 A straight line meets a (polynomial) cubic curve in three points. Therefore, given two points A and B on a cubic curve, a binary operation may easily be set up by allowing AB to meet the cubic again at C, and letting $C = A + B$. Has this operation an identity element? Experiment to discover whether it is associative. *Note*: if the inflexion of the cubic is taken as origin, O, then the equation of the cubic is of the form $y = ax^3 + bx$. We may then easily show that the abscissae of collinear points satisfy $x_1 + x_2 + x_3 = O$. Thus $x_1 * x_2 = -x_1 - x_2$.

For a full discussion of this, see D. S. MacNab, 'The cubic curve and associated structure', *Mathematical Gazette*, pp. 105–110, May 1966. Note also that a binary operation may be set up between the points on a parabola by allowing the normals at A and B to meet at N, and calling $A * B$ the foot of the third normal from N to the parabola. In what way does this example resemble the example above about the cubic curve?

17 Consider the following operation $\circ$ on the set of integers $\{0, 1, 2, 3, \ldots 98, 99\}$: $(10x_1 + y_1) \circ (10x_2 + y_2) = (10x_1 + y_2)$, where $x_1, y_1, x_2, y_2 \in \{0, 1, 2, \ldots 9,\}$ e.g. $43 \circ 75 = 45$. Show that $\circ$ is associative though not commutative. Show, however, that there is no element which is an identity element for every member of the set, though any particular element has ten right identities and ten different left identities. Compare with the operation on 'arrows' (see Ex. 2.08; Ex. 7.01).

18 Suppose an operation $\circ$ on 2×2 matrices is defined $A \circ B = AB - BA$, e.g.

$$\begin{pmatrix} 2 & 1 \\ 4 & -1 \end{pmatrix} \circ \begin{pmatrix} 0 & 5 \\ -3 & 1 \end{pmatrix} = \begin{pmatrix} 2 & 1 \\ 4 & -1 \end{pmatrix}\begin{pmatrix} 0 & 5 \\ -3 & 1 \end{pmatrix} - \begin{pmatrix} 0 & 5 \\ -3 & 1 \end{pmatrix}\begin{pmatrix} 2 & 1 \\ 4 & -1 \end{pmatrix}$$

$$= \begin{pmatrix} -3 & 11 \\ 3 & 19 \end{pmatrix} - \begin{pmatrix} 20 & -5 \\ -2 & -4 \end{pmatrix} = \begin{pmatrix} -3 & 11 \\ 3 & 19 \end{pmatrix} + \begin{pmatrix} -20 & 5 \\ 2 & 4 \end{pmatrix} = \begin{pmatrix} -23 & 16 \\ 5 & 23 \end{pmatrix}$$

Show that $\circ$ is associative. Do we have right or left identities?

19 Show that the set $\{2, 4, 6, 8, 12, 14, 16, 18\}$ is closed under multiplication modulo 20, but has no identity.

7 As you were! —inverses

We now come to the final requirement for a group – the existence of an *inverse* for each element of the set. This means that the operation can be 'undone' so as to restore the *status quo*. For example, if x represents the permutation $\begin{pmatrix}1 & 2 & 3 & 4 & 5\\ 4 & 5 & 1 & 3 & 2\end{pmatrix}$ (discussed on p. 23), what is the inverse permutation? The permutation x replaces 1 by 4, 2 by 5, 3 by 1, 4 by 3 and 5 by 2. The inverse permutation will have to replace 4 by 1, 5 by 2, 1 by 3, 3 by 4 and 2 by 5, that is

$$x^{-1} = \begin{pmatrix}4 & 5 & 1 & 3 & 2\\ 1 & 2 & 3 & 4 & 5\end{pmatrix} = \begin{pmatrix}1 & 2 & 3 & 4 & 5\\ 3 & 5 & 4 & 1 & 2\end{pmatrix}^{\dagger}$$

You will recall that (see p. 23) we actually wrote x in the alternative form $\begin{pmatrix}3 & 5 & 4 & 1 & 2\\ 1 & 2 & 3 & 4 & 5\end{pmatrix}$. The relationship between x and its inverse stands out, and we see immediately that the 'product' of x with its inverse:

$$x^{-1}x = \begin{pmatrix}1 & 2 & 3 & 4 & 5\\ 3 & 5 & 4 & 1 & 2\end{pmatrix} \circ \begin{pmatrix}1 & 2 & 3 & 4 & 5\\ 4 & 5 & 1 & 3 & 2\end{pmatrix}$$

$$= \begin{pmatrix}1 & 2 & 3 & 4 & 5\\ 3 & 5 & 4 & 1 & 2\end{pmatrix} \circ \begin{pmatrix}3 & 5 & 4 & 1 & 2\\ 1 & 2 & 3 & 4 & 5\end{pmatrix} = \begin{pmatrix}1 & 2 & 3 & 4 & 5\\ 1 & 2 & 3 & 4 & 5\end{pmatrix},$$

is the identity permutation. In other words, the permutation which we have called x^{-1} has 'unshuffled the pack', and we are back where we started. You should verify that $x \circ x^{-1}$ is also equivalent to the identity permutation.

Q. 7.01. Find the inverses of the following permutations:

$$\begin{pmatrix}1 & 2 & 3 & 4\\ 2 & 4 & 3 & 1\end{pmatrix}; \quad \begin{pmatrix}1 & 2 & 3 & 4 & 5 & 6\\ 4 & 5 & 1 & 3 & 6 & 2\end{pmatrix}; \quad \begin{pmatrix}5 & 3 & 1 & 4 & 2\\ 1 & 2 & 3 & 4 & 5\end{pmatrix}.$$

Verify that each permutation and its inverse satisfy $x \circ x^{-1} = x^{-1} \circ x = e$, where e is the identity permutation.

Q. 7.02. If $x = \begin{pmatrix}1 & 2 & 3 & 4 & 5\\ 1 & 4 & 5 & 3 & 2\end{pmatrix}$ and $y = \begin{pmatrix}1 & 2 & 3 & 4 & 5\\ 3 & 4 & 5 & 1 & 2\end{pmatrix}$, find $x \circ y$, and the inverse of $(x \circ y)$. Find also x^{-1} and y^{-1}, and also $y^{-1} \circ x^{-1}$. Can you see any reason for the result?

† The reason for the notation x^{-1} will appear later.

The dual aspect of permutations

If we may take up again the case of the quintet reshuffle introduced in Ch. 4, p. 22, we may look at this in two ways. Calling the instruments 1 (1st violin), 2 (2nd violin), 3 (1st viola), 4 (2nd viola), and 5 (cello), we argued in Ch. 4 as shown in table 7.01. Here we were thinking on these

Table 7.01

Instrument	V1 1	V2 2	Va1 3	Va2 4	Vc 5	
Originally played by person No.	1	2	3	4	5	
Next played by person No.	4	5	1	3	2	(permutation x)

lines: 'Taking each *instrument* in turn, consider the change of *person*'; i.e. for instrument No. 4, person No. 4 is replaced by person No. 3. This is a *permutation of the persons.*

But we may equally well *think the other way round*: 'Taking each *person* in turn, consider the change of instrument'; e.g. person No. 1 changes from 1st violin to 1st viola, i.e. from instrument No. 1 to instrument No. 3.

Table 7.02

Person No.	1	2	3	4	5	
Originally played instrument No.	1	2	3	4	5	
Next played instrument No.	3	5	4	1	2	(permutation x^{-1})

These changes may be tabulated (table 7.02). This may be summarised by the permutation $\begin{pmatrix} 1 & 2 & 3 & 4 & 5 \\ 3 & 5 & 4 & 1 & 2 \end{pmatrix}$ which is the *permutation of the instruments*, and we observe that it is in fact the inverse permutation, x^{-1}.

The above might have been clearer had we used, say, letters for people (Arthur, Bill, Charlie, Don and Ed), and numbers for instruments. But the point of using numbers for both was that it enabled us more easily to see that the two permutations, of the people and of the instruments, were *inverse*.

This idea of looking at permutations from two dual points of view is important, and is closely bound up with the way in which we can take rotation axes for the transformation of bodies to be either fixed in space, or else fixed relative to the body. See pp. 192 and 200–1. See also the chapter on bell-ringing, Ch. 24. For a fuller discussion of the whole problem of permutations and arrangements, see D. B. Hunter, *Mathematical Gazette*, No. 373, p. 290.

Q. 7.03. Consider the problem of the two consecutive permutations $x = \begin{pmatrix} 1 & 2 & 3 & 4 & 5 \\ 4 & 5 & 1 & 3 & 2 \end{pmatrix}$ followed by $y = \begin{pmatrix} 1 & 2 & 3 & 4 & 5 \\ 5 & 3 & 2 & 4 & 1 \end{pmatrix}$ in Ch. 4 from the other point of view, i.e. as permutations of *instruments*.

Notation

The notation x^{-1} for the inverse of x is used whenever a multiplicative notation (such as $x.y$ or xy) is used for the composition of the two operations. This derives from the fact that, in the set of real† numbers under multiplication, the neutral element is 1, and if x and y are inverses, then $xy = yx = 1$, so that $y = 1/x = x^{-1}$ ($x \neq 0$). The notation is imitated,‡ by analogy, whenever the composition of elements of a set is indicated by juxtaposition. For example, in the game about emptying jugs (see pp. 26 ff), we used X for the operation 'empty the first jug'. Then X^{-1} would mean 'fill the first jug', though we did not in fact include this in our set of rules.

Q. 7.04. In that game, what would Y^{-1} and Z^{-1} mean?

When the operation (on a number system) is *addition*, this notation, i.e. juxtaposition, is not used. For, the neutral element being 0, if y is the additive inverse of x, we have $x + y = y + x = 0 \Rightarrow y = -x$. The notation $-x$ ('negative x') is commonly used when the operation is described, or written, as addition. For the binary operation $\oplus$ of p. 48, $x \oplus y = x + y - 1$, the identity was found to be 1. So if x and y are inverses under $\oplus$, we have;

$$x \oplus y = 1 \Rightarrow x + y - 1 = 1 \Rightarrow y = 2 - x.$$

Check: $x \oplus (2 - x) = x + (2 - x) - 1 = 1.$

Inverse functions

Let us now consider the question of the inverse of a function. Suppose $f(x) = 5 + 2x$, how do we determine the inverse function which undoes the work of this function? In the diagram (fig. 7.01), the function $f(x)$ maps

† Can you find a subset of **R** for which all the remarks of this paragraph are true?
‡ But the terminology is not imitated: x^{-1} is commonly called the *reciprocal* of x, rather than the multiplicative inverse.

the length OL $(= x)$ into the length OM $(= y)$ given by $y = 2x + 5$. We want a function which will take us *in the reverse direction* from OM back to OL, i.e. we want to obtain x in terms of y. Symbolically:

$$x \xrightarrow{\mathrm{f}} y; \qquad y \xrightarrow{\mathrm{g}(?)} x.$$

Now $y = 2x + 5 \Rightarrow x = \frac{1}{2}(y - 5)$. If we let $\mathrm{g}(y) \equiv \frac{1}{2}(y - 5)$, or $\mathrm{g}(x) \equiv \frac{1}{2}(x - 5)$, we can see that $\mathrm{gf}(x)$ will give the identity function. For $\mathrm{gf}(x) = \mathrm{g}(2x + 5) \equiv \frac{1}{2}[(2x + 5) - 5] \equiv x$, i.e. the identity function which replaces x by itself. You should check that $\mathrm{fg}(x)$ also gives the function x.

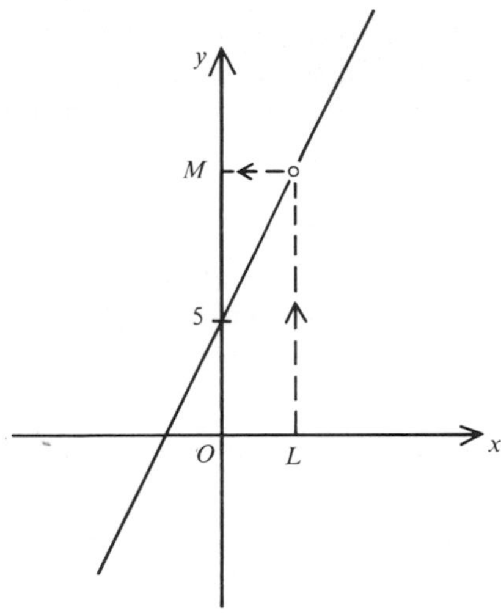

Fig. 7.01. The mapping $x \rightarrow 2x + 5$.

All we have done to obtain the inverse function is to 'make y the subject of the formula'. By analogy with the foregoing, we shall denote the inverse function by $\mathrm{f}^{-1}(x)$, so that $\mathrm{ff}^{-1}(x) \equiv \mathrm{f}^{-1}\mathrm{f}(x) \equiv x$.† The above may be summarised:

$$x \xrightarrow{\mathrm{f}} 2x + 5$$

$$\tfrac{1}{2}(x - 5) \xleftarrow[\mathrm{f}^{-1}]{} x$$

In a second example, suppose

$$\mathrm{f}(x) \equiv \frac{x + 1}{1 - 2x} \quad (x \neq \tfrac{1}{2}).$$

† See App. II.

Then if $y = \dfrac{x+1}{1-2x}$, we have $y(1 - 2x) = x + 1$, so that

$$x = \frac{y-1}{2y+1} \equiv \mathrm{f}^{-1}(y).$$

The inverse function is therefore

$$\mathrm{f}^{-1}(x) \equiv \frac{x-1}{2x+1},$$

and you should verify that the composition of this function with the given function produces the identity function.

Q. 7.05. Obtain the inverse functions in the following cases: $\mathrm{f}(x) \equiv$

(a) $\dfrac{1}{x-1}$; (b) $\dfrac{x}{1-x}$; (c) $\dfrac{4x-7}{3x-5}$; (d) $\dfrac{3x-6}{2x-3}$;

(e) $\dfrac{x-1}{2x-1}$; (f) $\log_e x$; (g) $\sin x$; (h) $\log \log x$; (i) x^3;

(j) a^x; (k) $\dfrac{ax+b}{cx+d}$.

In each case, find also $\mathrm{ff}(x)$, and $(\mathrm{ff})^{-1}(x)$. Discover some functions for which $\mathrm{f}(x) = \mathrm{f}^{-1}(x)$, and find the condition for this to be true in case (k).

Inverses in finite arithmetics

Next, let us consider the question of inverses in a finite arithmetic. For example, in the arithmetic modulo 10 (i.e. 'final digits arithmetic'), we have $7 + 9 = 6$ (mod. 10); $8 \times 9 = 2$ (mod. 10), since we are discarding all superfluous tens. The neutral elements for + and ×† are 0 and 1. Let us set up the addition and multiplication tables (see tables 7.03 and 7.04). Inverses can readily be found by inspection; for the inverse of 3 under + mod. 10, we have $3 + ? = 0$ (mod. 10), and by consulting the row starting '3', and looking along until we find 0, we see that $3 + 7 = 0$ (mod. 10), so that the additive inverse of 3 is 7.

† A dot is commonly favoured to indicate multiplication, thus: $8 . 9 \equiv 2$ (mod. 10). In this book, we prefer a cross (×) on grounds of (a) visibility, (b) less likelihood of confusion, though we realise that the dot, with a symmetry group D_∞ may be deemed a more beautiful object than a cross, whose symmetry group is merely D_4. Thus, we shall refer, later, to the group (**C**, ×), rather than to (**C**, .) to mean 'the Complex numbers under multiplication'. Furthermore, while working in finite arithmetics, we shall not feel bound to use the official sign '≡' to indicate a congruence, but we shall be deliberately careless by making such daring statements as $8 \times 9 = 2$ (mod. 10), confident in the presumption that the context makes clear the sense in which the '=' sign is being used. See also the footnote on p. 414.

Table 7.03

+ mod. 10	0	1	2	3	4	5	6	7	8	9
0	0	1	2	3	4	5	6	7	8	9
1	1	2	3	4	5	6	7	8	9	0
2	2	3	4	5	6	7	8	9	0	1
3	3	4	5	6	7	8	9	0	1	2
4	4	5	6	7	8	9	0	1	2	3
5	5	6	7	8	9	0	1	2	3	4
6	6	7	8	9	0	1	2	3	4	5
7	7	8	9	0	1	2	3	4	5	6
8	8	9	0	1	2	3	4	5	6	7
9	9	0	1	2	3	4	5	6	7	8

Table 7.04

× mod. 10	0	1	2	3	4	5	6	7	8	9
0	0	0	0	0	0	0	0	0	0	0
1	0	1	2	3	4	5	6	7	8	9
2	0	2	4	6	8	0	2	4	6	8
3	0	3	6	9	2	5	8	1	4	7
4	0	4	8	2	6	0	4	8	2	6
5	0	5	0	5	0	5	0	5	0	5
6	0	6	2	8	4	0	6	2	8	4
7	0	7	4	1	8	5	2	9	6	3
8	0	8	6	4	2	0	8	6	4	2
9	0	9	8	7	6	5	4	3	2	1

With × mod. 10, the situation is more complicated. What is the inverse of 3? We require 3 × ? = 1 (mod. 10), and by consulting the table, we see that 3 × 7 = 1 (mod. 10), so that 3 and 7 are 'reciprocals' in this finite arithmetic; 9 is its own inverse, since $9^2 = 1$ (mod. 10).

What is the multiplicative inverse of 6? Evidently it does not have one, for 1 does not appear in row 6. Nor indeed, can any *even* number be equivalent to 1 (mod. 10) – obviously. But even numbers are not the only numbers to be disqualified, as you may discover for yourself.

Table 7.04 is 'spoiled' so to speak by the presence of zeros. If we excise all those numbers which cause zero to appear, including the useless border of zeros along the first row and column, we shall obtain table 7.05.† The

Table 7.05

× mod. 10	1	3	7	9
1	1	3	7	9
3	3	9	1	7
7	7	1	9	3
9	9	7	3	1

imperfections have been expunged: now *every* element in the set has an inverse.

You should now write out the table for multiplication modulo 10 of the set {2, 4, 6, 8} and make a full investigation of it. Compare with the work on p. 47. You may be puzzled by the fact that for the set {1, 3, 7, 9} the identity for × mod. 10 is 1, whereas for the set {2, 4, 6, 8} the identity is 6, because on p. 51 we stated that the identity element is unique. Here there would appear to be two distinct identities. What is the explanation? The answer is that, *in the set* {1, 3, 7, 9}, the identity is 1, whereas *in the set* {2, 4, 6, 8} it is 6. We are working in *two different sets*. In the set

† This is our fourth instance of a group table, the others having appeared on pp. 18, 47, 48.

{1, 2, 3, 4, 5, 6, 7, 8, 9} there are not two neutral elements, but only one: 6 is *not* an identity for this set, since $3 \times 6 \neq 3$. Thus the mystery is cleared up.

Geometrical examples

In some of the geometrical examples we met in the last chapter, we were unable to find an identity element. In such a case, the question of inverses does not arise. But in the example on p. 49 (fig. 6.02), we found that the point O was the neutral point on the circle, for $A*O = O*A = A$ for all positions of A. We now require to find B (given A) so that $A*B = B*A = O$.

Now if $A*B$ is to be at O, the point D must lie on the tangent at O, and so A^{-1} is located as the point where AD meets the circle again (see fig. 7.02).

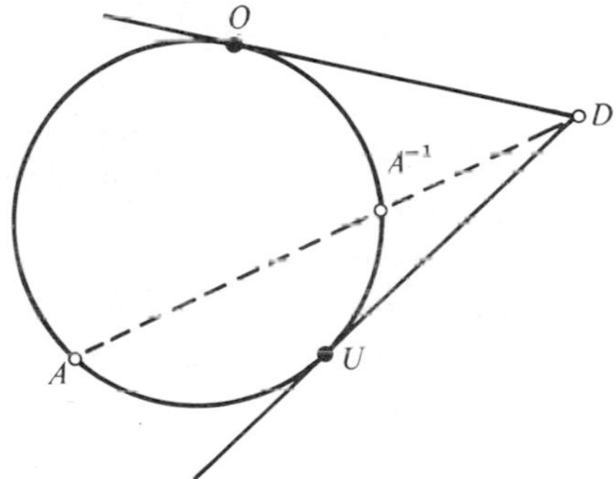

Fig. 7.02. Construction of A^{-1}.

Q. 7.06. What happens when A is at O or at U?
Find the inverse of A in other geometrical examples in the text where an identity point has been established (e.g. figs. 6.01, 6.042, and Ex. 7.27).

General definition

We are now in a position to give a general definition of Inverse:

> **If a set has an operation $*$, and an identity element e, and if, for each and every given element x there can be found an element y† such that $x*y = y*x = e$, then y is called the Inverse of x,** (and will be denoted x^{-1} when multiplicative notation is being employed).

† Thus y *depends on* x and it is important to realise that in some cases y may be x, that is, an element may be its own inverse.

Note: There are cases when only a *right* inverse can be found, i.e. $x*y = e \neq y*x$; and other cases when only a left inverse exists. But such exceptions do not occur in well-behaved systems – that is to say, when the operation is associative. In App. II, we shall prove that, in an associative system, a right inverse is also a left inverse.

Q. 7.07. Consider the questions of identity and inverse for the real numbers under the operation of subtraction.

Reversible and irreversible operations in general

We have spoken of the inverse as 'undoing' an operation. This presumes, of course, that the operation is 'undoable', or *reversible*. This may depend on circumstances: emptying a bucket is reversible, but not in the middle of a desert! Some actions are always irreversible – striking a match, sticking on a stamp, diving into the sea, shaving, etc.

Q. 7.08. Name the inverse operations: dressing; getting out of bed; switching on a light; driving a car forwards; an aircraft taking off. Invent other examples, and also of irreversible operations.

Note: $y = \mathrm{f}(x) \Rightarrow x = \mathrm{f}^{-1}(y)$ (if f is reversible); similarly: John *loves* Mary ⇒ Mary *is loved by* John. Thus, in a sense, 'active' and 'passive' are inverse aspects of a transitive verb. A better illustration might be:

$$\text{water} \underset{\text{melt}}{\overset{\text{freeze}}{\rightleftharpoons}} \text{ice}$$

showing that melting and freezing may be regarded as inverses.
Again, in a mathematical context, we have *squaring* as a unary operation on numbers, whose inverse is *taking the square root*; while 'reciprocal' is its own inverse. In yet another context, we have inverse relations between, say, male human beings, such as 'is the son of' and 'is the father of'. To a clarinettist playing a B flat instrument, all the notes played will sound a tone lower than the notes as written – the player would have to *transpose down a tone* to ascertain the true pitch of his part. If, however, he wishes to produce, notes of a given pitch, he must perform the inverse operation – *transpose up a tone* in order to describe the notes as played on his transposing instrument. Similarly, a navigator using a magnetic compass and making allowance for the deviation of the compass due to the ship's magnetism will have to apply the deviation in one sense in reading a bearing off the chart, and in the inverse sense in drawing a line on the chart from a compass reading.

Operations on sets

In Ch. 6 you discovered the identity elements for the operations $\cup$ and $\cap$ on sets:

$$A \cup \phi = A \quad (\phi = \text{the empty set}),$$
$$A \cap \mathscr{E} = A \quad (\mathscr{E} = \text{the universal set}).$$

Can we have inverses in this system? We shall answer the question for $\cap$, and you may do it for $\cup$. We require B such that $A \cap B = \mathscr{E}$, the neutral element for $\cap$. Evidently there can be no solution to this problem, for any element which is not in A cannot possibly be in $A \cap B$ whichever set B may be, even if it is the largest possible set, $\mathscr{E}$. In fact, $A \cap \mathscr{E} = A$ (see fig. 7.03).

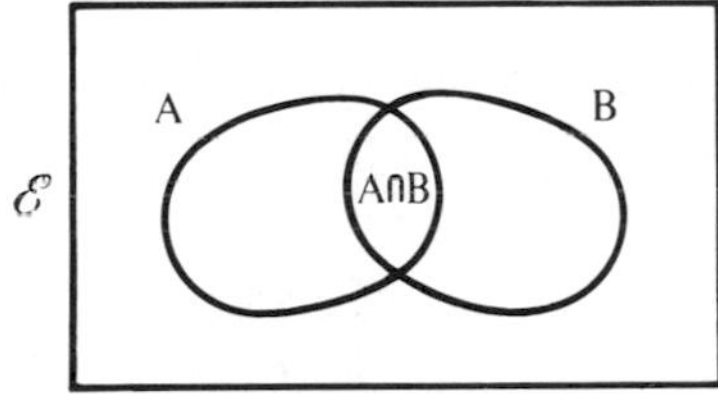

Fig. 7.03. For a given set A, it is impossible to solve $A \cap B = \mathscr{E}$ for B.

Is there an operation on sets which admits an inverse? Yes! It is the operation Δ introduced on p. 9, and defined:

$$A \,\Delta\, B = (A' \cap B) \cup (A \cap B'),$$

and is known as the 'symmetric difference' of the two sets. The region corresponding to $A \,\Delta\, B$ is shaded in the Venn diagram in fig. 7.041. We may describe $A \,\Delta\, B$ in words as the set of objects which are *either* in A, *or* in B, but not both.

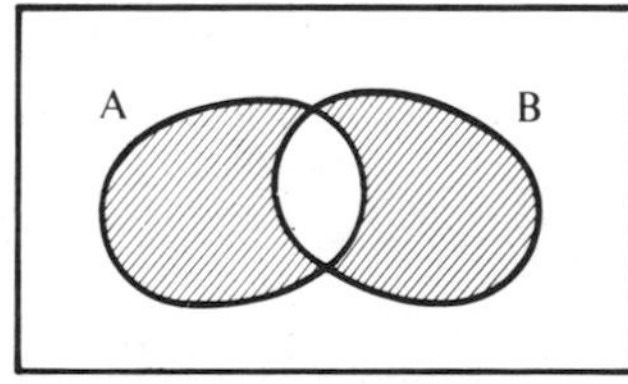

Fig. 7.041. $A \,\Delta\, B$: Symmetric difference.

The identity is $\emptyset$, since $A \,\Delta\, \emptyset = A$. Associativity was referred to in Ch. 5 (see Q. 5.03), and we now consider the problem of discovering an inverse: Given a set A, can we find B such that $A \,\Delta\, B = \emptyset$? A little thought will convince you that the set which satisfies this requirement is

A itself. For a set B which is *nearly* the same as A (see fig. 7.042) there would be very few elements which were in A or B but not both . . . Finally, when B = A, there are no such elements, and we see that $A \Delta A = \emptyset$.

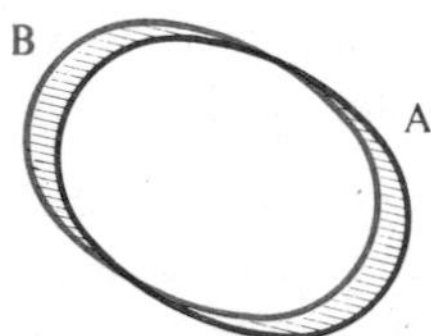

Fig. 7.042. $A \Delta B$ when $A \simeq B$.

Q. 7.09. Find $A \Delta (A \cap B)$; $A \Delta (A \cup B)$; $A \Delta B \Delta C$ (illustrate in Venn diagrams).
What is $A \Delta B$ when $A \cap B = \emptyset$? What is $A \Delta B$ when $B \subset A$?

Matrices

Additive inverses of matrices are simple enough:

$$\begin{pmatrix} a & b & c \\ p & q & r \end{pmatrix} + \begin{pmatrix} ? & ? & ? \\ ? & ? & ? \end{pmatrix} = \begin{pmatrix} 0 & 0 & 0 \\ 0 & 0 & 0 \end{pmatrix} \quad \text{(the neutral element),}$$

and it is evident that $\begin{pmatrix} -a & -b & -c \\ -p & -q & -r \end{pmatrix}$ is what we are looking for.

For multiplicative inverses, the problem is not so simple. Taking the case of a 2×2 matrix, we require:

$$\begin{pmatrix} a & b \\ c & d \end{pmatrix} \times \begin{pmatrix} ? & ? \\ ? & ? \end{pmatrix} = \begin{pmatrix} 1 & 0 \\ 0 & 1 \end{pmatrix}.$$

$$M \quad \times \quad M^{-1} \quad = \quad I_2$$

Suppose that $M^{-1} = \begin{pmatrix} a' & b' \\ c' & d' \end{pmatrix}$. Then $\begin{pmatrix} a & b \\ c & d \end{pmatrix} \times \begin{pmatrix} a' & b' \\ c' & d' \end{pmatrix} = \begin{pmatrix} 1 & 0 \\ 0 & 1 \end{pmatrix}$.

It follows from the rule for matrix multiplication that:

(i) $aa' + bc' = 1$
(ii) $ab' + bd' = 0$
(iii) $ca' + dc' = 0$
(iv) $cb' + dd' = 1$

Solving these equations (details omitted), we obtain, from (i) and (iii): $a'(ad - bc) = d$, and $c'(ad - bc) = -c$; and from (ii) and (iv):

$b'(ad - bc) = -b$ and $d'(ad - bc) = a$. Thus the inverse of $\begin{pmatrix} a & b \\ c & d \end{pmatrix}$

is $\dfrac{1}{ad - bc}\begin{pmatrix} d & -b \\ -c & a \end{pmatrix} = M^{-1}$.

Evidently the inverse does not exist whenever $ad - bc = 0$; a matrix for which this is true is called a '*singular*' matrix.

The inverse matrix is useful in the solution of a pair of simultaneous linear equations.

$$\left.\begin{matrix} ax + by = p \\ cx + dy = q \end{matrix}\right\} \Rightarrow \begin{pmatrix} a & b \\ c & d \end{pmatrix} \begin{pmatrix} x \\ y \end{pmatrix} = \begin{pmatrix} p \\ q \end{pmatrix}.$$

Letting $\mathbf{X}$ denote the unknown column vector $\begin{pmatrix} x \\ y \end{pmatrix}$, and $\mathbf{P}$ to denote the given vector $\begin{pmatrix} p \\ q \end{pmatrix}$, this now reads: $M\mathbf{X} = \mathbf{P}$. Pre-multiplying by the inverse matrix, M^{-1} on both sides, we obtain: $M^{-1}(M\mathbf{X}) = M^{-1}\mathbf{P}$. But matrix multiplication is associative,† so $M^{-1}(M\mathbf{X}) = (M^{-1}M)\mathbf{X}$, and since $M^{-1}M = I$ (the unit matrix), and $I\mathbf{X} = \mathbf{X}$, we find that:

$$\mathbf{X} = M^{-1}\mathbf{P},$$

and so the unknown vector $\mathbf{X}$ has been found.

Solution of linear equations

We may make a brief diversion here to note that there are three ways of interpreting the solution of the equations: $ax + by = p$; $cx + dy = q$.

(1) Which vector $\begin{pmatrix} x \\ y \end{pmatrix}$, when transformed by the matrix $\begin{pmatrix} a & b \\ c & d \end{pmatrix}$, will become the vector $\begin{pmatrix} p \\ q \end{pmatrix}$? This is the interpretation considered above.

OR (2) Which point (x, y) lies on each of the lines whose Cartesian equations are those given? This is the most usual interpretation.

OR (3) $x \begin{pmatrix} a \\ c \end{pmatrix} + y \begin{pmatrix} b \\ d \end{pmatrix} = \begin{pmatrix} p \\ q \end{pmatrix} \Rightarrow x\mathbf{A} + y\mathbf{B} = \mathbf{P}$; i.e. given vectors $\mathbf{A}$, $\mathbf{B}$ and $\mathbf{P}$, what linear combination of $\mathbf{A}$ and $\mathbf{B}$ will produce $\mathbf{P}$?

Q. 7.10. For three linear equations in three unknowns, give three separate interpretations on the lines followed above.

Singular matrices

In the preceding work of this chapter, we have made one glaring oversight. In the example above about the solution of two simultaneous equations, you will observe that unique finite solutions do not exist when $ad - bc = 0$.

† See the remark on p. 40.

Q. 7.11. Interpret this exceptional case in terms of the three situations (1), (2) and (3) in the previous paragraph.

The inverse of a matrix $\begin{pmatrix} a & b \\ c & d \end{pmatrix}$ does not exist in this circumstance. You are reminded that such a matrix, which has $ad - bc = 0$, is called 'singular'.

Q. 7.12. What is the condition for a 3×3 matrix to be singular?

When the matrix operation is addition, there are no exceptions to the existence of an additive inverse. Moreover, a number system may always be provided with *additive* inverses. But we have already noted difficulties in the case of multiplication (mod. 10) – and the difficulty centred around the number *zero*. In a number system with the operation $\times$, the number zero does not possess an inverse (reciprocal). Nor is it possible to invent a number (∞) to place in the set to remedy the deficiency.† The only possible remedy is to *remove zero* from the set, and to say, for example, that:

In the set of rational numbers under multiplication, every number has a reciprocal *provided zero has been omitted from the set*. This is a prerequisite to the endowment of the rationals with group structure under $\times$. We write the set of rationals with zero removed thus: $\mathbf{Q}\backslash 0$.

Inverse of the product of two or more elements

There is an important theorem on inverses, which we must next consider. If x and y are any two elements of a set which combine under an associative operation $*$, then the inverse of $x*y$ is $y^{-1}*x^{-1}$; i.e. using juxtaposition and dropping the $*$: $(\boldsymbol{xy})^{-1} = \boldsymbol{y}^{-1}\boldsymbol{x}^{-1}$. The proof is straightforward:

$$\begin{aligned}(xy)(y^{-1}x^{-1}) &= xyy^{-1}x^{-1} \\ &= x(yy^{-1})x^{-1} \quad \text{(associative property)} \\ &= xex^{-1} \quad \text{(letting } e \text{ be the neutral element)} \\ &= (xe)x^{-1} = xx^{-1} = e.\end{aligned}$$

Similarly, $(y^{-1}x^{-1})(xy) = e$, so that xy and $y^{-1}x^{-1}$ are inverse elements.

Q. 7.13. What is the inverse of the continued 'product' xyz; $xyzw$, etc.?

† See F. J. Budden, *Complex Numbers* (Longmans), p. 9.

As an illustration of the result $(xy)^{-1} = y^{-1}x^{-1}$, we may note that the inverse of the combined operation of putting on one's socks (y) and then one's shoes (x), (i.e. the combined operation of removing one's shoes and socks) is made up of, first the operation of removing one's shoes (x^{-1}) and then one's socks (y^{-1}). As a mathematical illustration of $(xyzw\ldots)^{-1} = \ldots w^{-1}z^{-1}y^{-1}x^{-1}$, consider the step by step process of 'changing the subject of a formula' in ordinary manipulative algebra:

Make x the subject	$y = \sqrt[3]{\left(\dfrac{a(x^2+p^2)}{a-x^2}\right)}$	
c (cube both sides) ↓		↑ c^{-1} (cube root)
	$y^3 = \dfrac{a(x^2+p^2)}{a-x^2}$	
m (mult. b.s.) ↓		↑ m^{-1} (= d)(divide b.s)
	$y^3(a-x^2) = a(x^2+p^2)$	
b (remove brackets) ↓		↑ b^{-1} (= f) (factorise)
	$y^3a - x^2y^3 = ax^2 + ap^2$	
r (rearrange) ↓		↑ r^{-1} (= r) (rearrange)
	$y^3a - ap^2 = ax^2 + y^3x^2$	
f (factorise) ↓		↑ f^{-1} (= b) (remove brackets)
	$x^2(a^2+y^3) = ay^3 - ap^2$	
d (divide b.s.) ↓		↑ d^{-1} (= m) (multiply b.s.)
	$x^2 = \dfrac{ay^3 - ap^2}{a+y^3}$	
s (square root) ↓		↑ s^{-1} (square)
	$x = \sqrt{\left(\dfrac{ay^3-ap^2}{a+y^3}\right)}$	*Make y the subject*

The letters c, m, b, r, f, d, s stand for the successive operations which are to be performed to both sides of the formula, and these letters are chosen so that there is no chance of confusion with the letters y, a, x, p, which, of course, stand for numbers. Now when one transforms the formula:

$$x = \sqrt{\left(\frac{ay^3-ap^2}{a+y^3}\right)}$$

so as to give y in terms of x, the whole process would be reversed, and the sequence of operations would be first s^{-1} (square both sides; squaring being the inverse of taking the square root), then d^{-1}, then f^{-1}, and so on, the final step being c^{-1} (taking the cube root). These inverse processes are shown on the right, and we see that $(sdfrbmc)^{-1} = c^{-1}m^{-1}b^{-1}r^{-1}f^{-1}d^{-1}s^{-1}$. The whole combined operation – that of transforming the formula to obtain y in terms of x – is really that of *finding the inverse function* which gives x in terms of y. It resembles the simpler cases that we considered earlier: such as;

$$\mathrm{f}(x) \equiv \frac{1}{1-x} \Leftrightarrow \mathrm{f}^{-1}(x) \equiv \frac{x-1}{x},$$

only here the formula is deliberately a good deal more complicated, and involves the constants a and p.

EXERCISES 7

1 $\overrightarrow{AB}$, $\overrightarrow{CD}$ are 'arrows' (see Ex. 2.08) in two or three dimensions. Suppose arrows are 'added' thus: $\overrightarrow{AB} + \overrightarrow{CD} = \overrightarrow{AD}$. Discuss associativity, identity and inverse. Repeat for other methods of composition. Is there a way of combining arrows which is associative and provides identity and inverses?

2 Consider the operation 'L' on the set of natural numbers: $a \mathrm{L} b$ = L.C.M. of a and b. We have seen that the neutral element is 1 since $a \mathrm{L} 1 = 1 \mathrm{L} a = a$ ($\forall a \in \mathbf{N}$). Is it possible to have inverses?

3 Find the reciprocals of $3 - 2\sqrt{5}$; $4\sqrt{3} - \sqrt{2}$; $3i - 2$; $2 + i$; $a\sqrt{2} + b\sqrt{3}$ ($a, b \in \mathbf{Q}$).

4 Consider the operations of subtraction and division on the set of reals from the point of view of inverses.

5 Find the inverse of $3x^3 - 5x^2 + 2$ in the set of cubic polynomials under addition. Discuss identities and inverses in the set of all polynomials under multiplication under various conditions.

6 Find the inverse of x for the operation $x \sim y = |x - y|$ (x, y real and non-negative).

7 Turn to table 5.01, p. 39 showing the combination of the five elements a, b, c, d and e. Which element is the identity? Find inverses for each element where possible.

8 Repeat the above question for the set {0, 1, 2, 3, 4, 5} under the combination table 5.021, p. 44.

9 What is the inverse of a given vector under vector addition?

10 Describe inverses of the elements in the examples of multiplication in finite arithmetics quoted on pp. 46 ff.

11 Consider the addition of binary numerals without carrying (see Ex. 2.04, pp. 74, 269). Show that it is associative, find the identity, and show how the inverse of a given numeral can be written down.

12 Suppose X means 'translate a passage from English to French'.
Y means 'translate a passage from French to German'.

Then X^{-1} means ...? Y^{-1} means ...?

How many of the composite operations XY, YX, $X^{-1}Y^{-1}$, $Y^{-1}X^{-1}$, $X^{-1}Y$, etc. are meaningful? What assumptions are we making in saying $XX^{-1} = X^{-1}X = I$? Check whether, in this context, $(YX)^{-1} = X^{-1}Y^{-1}$.

13 For the binary operation $(x_1, y_1) \diamondsuit (x_2, y_2) = (x_1x_2 + x_1y_2 + x_2y_1, y_1y_2)$ which we have already shown to be associative, and have identified the neutral element as $(0, 1)$, find the inverse of (x, y), and the conditions under which it exists.

14 Ordered pairs of reals are combined thus: $(a, b) * (c, d) = (ac, bc + d)$. Show that $*$ is associative but not commutative, find the neutral element, and inverse of (a, b).

15 We have seen that $(xy)^{-1} = y^{-1}x^{-1}$. Find the inverses of x^py^q, $x^py^qz^r$, and of $y^qx^py^r$, where p, q and r are integers.

16 In Q. 7.05 we found some functions such that $\mathrm{f}^{-1}(x) \equiv \mathrm{f}(x)$, such as $\mathrm{f}(x) \equiv 1/x$; $\mathrm{f}(x) \equiv (ax + b)/(cx - a)$. Such functions are described as 'involutory'. Is $(ax + b)/(cx - a)$ the most general form for an involutory function? (Consider any symmetric relation between x and y.)

17 Consider the multiplication of polynomials modulo $(x^2 + 1)$, e.g.

$$2x^2 + 5x - 3 \equiv 2(x^2 + 1) + 3x - 3 \equiv 3x - 3 \quad (\text{mod. } x^2 + 1)$$
$$2x^3 - 3x^2 - 3 \equiv (2x - 3)(x^2 + 1) - 2x \equiv -2x \quad (\text{mod. } x^2 + 1).$$

We are working with *remainders* mod. $x^2 + 1$, in a similar way as with remainders modulo n in a finite arithmetic:

$$(2x^2 + 5x - 3) \qquad \times (\text{mod. } x^2 + 1) \qquad (2x^3 - 3x^2 - 3)$$
$$\equiv (3x - 3)(-2x) \equiv -6x^2 + 6x \equiv -6(x^2 + 1) + 6x + 6$$
$$\equiv 6x + 6 \quad (\text{mod. } x^2 + 1).$$

Find the identity for multiplication modulo $x^2 + 1$, and the inverses of $2x - 1$; of x^3; and of $2x(x - 3)$. Show that any polynomial has an inverse unless it reduces to $0x + 0$. (This question should be compared with Ex. 21.19.)

18 Consider the set of linear polynomials with rational coefficients. Define a binary operation thus: $(a + bx) * (c + dx) = ac + bdk + (bc + ad)x$, where, in the ordinary product, x^2 has been replaced by a given rational number k. Find the neutral element, and show that the inverse of $ax + b$ is

$$\frac{a}{a^2 - b^2k} + \frac{-b}{a^2 - b^2k}\, x.$$

Discuss the question as to when inverses break down, and find for what values of k an inverse is guaranteed for all rational a and b. What value might k have if a and b were drawn from the reals? Compare with the previous question in the case when $k = -1$.

19 Note that $[a\ b\ c]\begin{bmatrix} 0 & 0 & 1 \\ 1 & 0 & 0 \\ 0 & 1 & 0 \end{bmatrix} = [b\ c\ a]$ and that $\begin{bmatrix} 0 & 0 & 1 \\ 1 & 0 & 0 \\ 0 & 1 & 0 \end{bmatrix}\begin{bmatrix} a \\ b \\ c \end{bmatrix} = \begin{bmatrix} c \\ a \\ b \end{bmatrix}$.

Find permutation matrices A and B such that:

$$[a\ b\ c]\quad A = [c\ a\ b] \quad \text{and} \quad B\begin{bmatrix} a \\ b \\ c \end{bmatrix} = \begin{bmatrix} c \\ a \\ b \end{bmatrix}.$$

What are the inverses of $\begin{bmatrix} 0 & 0 & 1 \\ 1 & 0 & 0 \\ 0 & 1 & 0 \end{bmatrix}$, of $\begin{bmatrix} 0 & 0 & 1 \\ 0 & 1 & 0 \\ 1 & 0 & 0 \end{bmatrix}$ and $\begin{bmatrix} 0 & 1 & 0 & 0 \\ 0 & 0 & 1 & 0 \\ 1 & 0 & 0 & 0 \\ 0 & 0 & 0 & 1 \end{bmatrix}$?

20 Consider the permutation matrix $\begin{bmatrix} 0 & 0 & 1 & 0 & 0 \\ 0 & 0 & 0 & 0 & 1 \\ 0 & 0 & 0 & 1 & 0 \\ 1 & 0 & 0 & 0 & 0 \\ 0 & 1 & 0 & 0 & 0 \end{bmatrix}$.

Post multiplying the row-vector $[a\ b\ c\ d\ e]$ changes it into $[d\ e\ a\ c\ b]$. Consider the effect of the same matrix in pre-multiplying the column vector $\begin{bmatrix} a \\ b \\ c \\ d \\ e \end{bmatrix}$.

What matrix would be needed to change the column vector $\begin{bmatrix} a \\ b \\ c \\ d \\ e \end{bmatrix}$ into $\begin{bmatrix} d \\ e \\ a \\ c \\ b \end{bmatrix}$?

How can you write down *at sight* the inverse of a permutation matrix?

21 Three simultaneous linear equations are expressed compactly in the form $MX = P$, where M is the general 3×3 matrix $\begin{bmatrix} a_1 & b_1 & c_1 \\ a_2 & b_2 & c_2 \\ a_3 & b_3 & c_3 \end{bmatrix}$, X is the column vector $\begin{bmatrix} x \\ y \\ z \end{bmatrix}$ and P is a given column vector.

Show how the inverse matrix M^{-1} may be used to solve the equations, and consider the threefold interpretation of the solution of the three linear equations on the lines of the discussion on p. 65. (*Note*: a knowledge of determinants and cofactors is desirable here.)

22 Show that, if a finite set of n complex numbers is to be closed under multiplication, then the modulus of each complex number must be 1, and deduce that the numbers must be cis $2\pi k/n$ ($k = 0, 1, 2, \ldots, n - 1$). Does this exhaust the finite sets of complex numbers which are closed under multiplication?

23 If $R(z) \equiv iz$ and $T(z) \equiv 2z + 1$ are transformations on the Argand diagram, find formulae for $R^{-1}(z)$, $T^{-1}(z)$, $TRT^{-1}(z)$, $T^{-1}RT(z)$, $RTR^{-1}(z)$, $R^{-1}TR(z)$ and interpret geometrically.

24 O is a fixed point, and a binary operation is set up on the points of a plane through O such that $C = A * B$ where (a) $OACB$ is a parallelogram; (b) $OABC$ is a parallelogram. Show that $*$ is associative in case (a) but not in case (b). Investigate identities and inverses in each case.

25 O and U are fixed points, and $C = A * B$ is obtained by making triangles OUA, OBC similar, the vertices being named in corresponding order. Show that $*$ is associative. Find the identity point, and show how to construct the inverse of a given point.

26 Discuss inverses for the examples considered of operations on the points of a conic (see Ch. 4).

27 Given a circle and a point O on it, and a fixed line cutting the circle in U and V (see fig. 7.05), a binary operation $*$ on the points of the circle is set up by defining $A * B$ by the following construction: AB meets UV in D; then OD meets the circle again in $A * B$. Show that $*$ is commutative and associative. Find the point which plays the part of identity, and describe the construction for obtaining the inverse of a given point A.

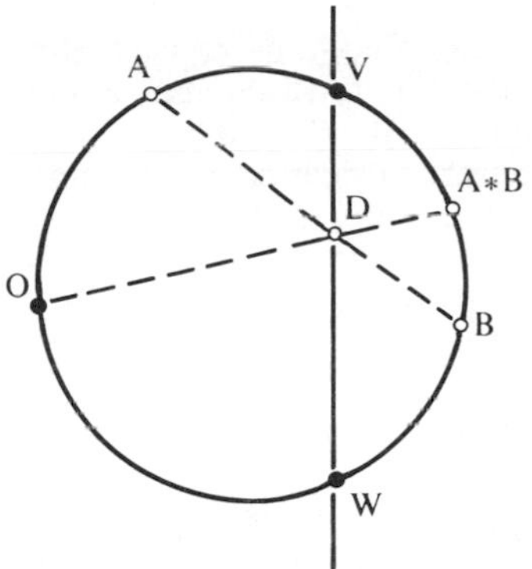

Fig. 7.05. Binary operation between points on a circle. For any choice of A and B, $A * B$ is uniquely determined.

28 O and U are fixed points (see fig. 7.06). A binary operation $*$ on the points

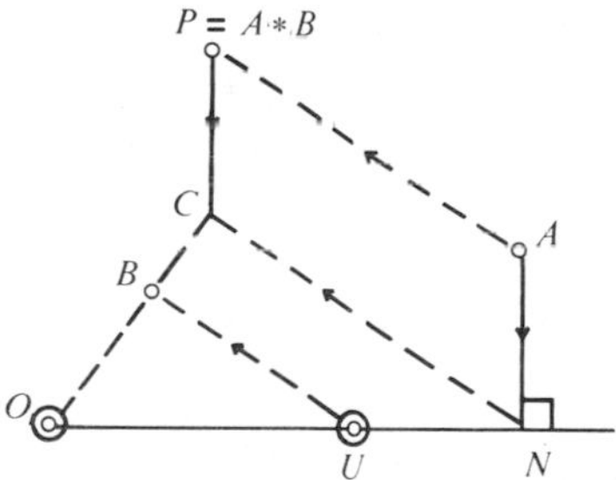

Fig. 7.06.

of a plane containing O and U is defined by the construction: Draw AN perpendicular to OU. Through N draw NC parallel to UB to meet OB at C. Complete the parallelogram $ANCP$. Then $P = A * B$. Establish that $*$ is associative. Find the neutral point, and obtain a construction for the inverse of a given point A. (This should be compared with Exs. 4.14, 5.02, 7.14, 8.30 and 14.25.)

29 The hand calculating machine group. Let r indicate a clockwise rotation of the crank in a hand calculating machine, which increases the number in the

multiplying register by 1, and let m denote a movement of the carriage one place to the right. Then to multiply by 10, we may either perform the operation r^{10} followed by m (i.e. mr^{10}), *or* we may first move the carriage to the right and crank the handle once (= rm). Hence $mr^{10} = rm$ (note that the operations m and r do not commute). This relation may be rewritten: $r^{-1}mr^{10} = m$, or $r^{10} = m^{-1}rm$.†

To multiply by 8, we can either do r^8, or else $r^{-2}m^{-1}rm$ (interpret). Again, to multiply by 86, we could carry out the operations r^8mr^6, or $mr^{-4}m^{-1}r^9m$, or $mr^{-4}m^{-1}r^{-1}m^{-1}rm^2$. Give various methods of registering the multipliers: 37; 218; 803; 996; etc.

Which numbers are recorded in the register when the following sequence is performed; $r^2mr^{-3}m^{-1}rm$? Give alternative equivalent 'words' in r and m which produce the same result. Compose other problems, e.g. about how to multiply by a given number using the smallest possible number of operations. Using a 'reverse notation', we may write $8 = 1\bar{2}$; $86 = 9\bar{4} = 1\bar{1}\bar{4} = 1\bar{2}6$, etc., in which the bar above the digit denotes that the value of that particular digit is to be negative, e.g. $1\bar{2}6 = 100 - 20 + 6$. Relate this notation to the present problem.

† It is supposed that the final position of the carriage is the same after equivalent operations.

8 *Group structure*

Throughout the first seven chapters we have been working towards a more and more highly organised mathematical structure – the Group, and we have now reached a climax in the development of our studies, for after all these preliminaries, we are now in a position to define a group. After this definition has been stated, we shall give a variety of examples of systems which are groups, and also a number of counter-examples of systems which are not groups for one reason or another, and in several cases we shall meet systems which are very nearly groups.

Definition

A group is a set G† provided with a binary operation $*$‡ with the following properties.

(1) **The set is closed under $*$** *C.*
(2) **$*$ is associative** *A.*
(3) **There is an identity element** *N.*§
(4) **Each element has an inverse** *I.*

We shall now redefine a group, using set language.
A group is an ordered pair (G, $*$), with the following properties:

C $\forall x, y \in \mathbf{G} \Rightarrow x*y \subset \mathbf{G}$
A $\forall x, y, z \in \mathbf{G}\ (x*y)*z = x*(y*z)$
N $\forall x \in \mathbf{G}\ \exists\ e \in \mathbf{G}$ **such that** $x*e = e*x = x$
I $\forall x \in \mathbf{G},\ \exists y$ **such that** $xy = yx = e$.

You should study this alternative expression of the definition. *Note*: It can be shown that the above set of 'axioms' contains slight redundancies – there is more information there than we *need*: it does not constitute a *minimum* set of axioms. (Compare, for example, App. II; also see Ex. 9.02 where an alternative set of axioms is given.) It is as if we were to define a square as a rhombus with four right-angles, when it is only necessary to require a single right-angle, from which the three remaining right-angles may be deduced.

For a concise description *in words* of a group, we cannot improve on that given by A. Bell and T. J. Fletcher *Symmetry Groups*, p. 2 (A.T.M.

† A set of what? – *objects*, never mind what!
‡ What sort of a binary operation? – never mind!
§ *N* for Null, or Neutral element.

pamphlet): 'A *group* is a mathematical system consisting of elements, with *inverses*, which can be combined by some operation without going outside the system.'

Examples of groups

First, let us give our attention to groups which have already been met with in the first seven chapters. You should check that C, A, N, I are satisfied in each case. We give first the set, then the operation.

(1) Sets, under the operation Δ (symmetric difference). See pp. 9, 37, 63.
(2) $p \times q$ matrices, under matrix addition. See pp. 9, 64.
(3) $n \times n$ non-singular matrices, under matrix multiplication. See pp. 9, 64.
(4) Integers with a specified number n of binary digits, under addition modulo 2 of corresponding binary digits. See pp. 133, 269 and Ex. 8.12.
(5) $\mathbf{Z}$, $+$.
(6) $a + b\sqrt{p}$ ($a, b \in \mathbf{Q}$, p prime), under $+$, and also under $\times$. See pp. 15–16.
(7) The set $\{I, X, Y, Z\}$ (half-turns about three perpendicular axes) under successive composition. See p. 17.
(8) Shapes of a piece of plasticine, under topological deformations. See Ex. 3.21.
(9) Isometries in the plane,† under successive composition. See pp. 231 ff.
(10) Certain sets of functions, under successive substitution. See pp. 22, 160 and Ex. 4.16.
(11) Sets of permutations on n objects, under successive application. See Ch. 18.
(12) The set $\{I, V, C, A\}$; performing the operation in sequence. See p. 25.
(13) The set $\{I, L, R\}$; performing the operation in sequence. See p. 28.
(14) Vectors in 3-space; vector addition. See p. 283.
(15) Ordered pairs of real numbers, $\diamond$ as defined on p. 38.
(16) Polynomials in x, under addition. See Ex. 7.05 Q. 6.02, Q. 19.23.
(17) $\{0, 1, 2, 3, \ldots, n-1\}$, $+$ mod. n. See Ex. 8.11, pp. 93, 162, 170, etc.
(18) Certain subsets of $\{1, 2, 3, \ldots, n-1\}$, $\times$ mod. n. See pp. 47 ff, 60.
(19) $\mathbf{R}$, $\oplus$, as defined on p. 48.
(20) $\mathbf{R}$, $\times$; also $\mathbf{Q}$, $\times$.
(21) The set of points on a circle; the operation $*$ defined on p. 49, Ex. 7.27.
(22) The operations a and b on a conic, under successive composition. See figs. 4.03, 11.02, 11.03, 24.01–24.05.

† I.e. Transformations which preserve distance; see p. 83, and p. 231.

We have *finite groups* in cases (4) order 2^n; (7) order 4; (10), (11) order $n!$; (12) order 4; (13) order 3; (17) order n; (18) order less than n (actually $\phi(n)$, see Ex. 12.26, and App. IV).
(The *order* of a finite group is simply the number of elements in the group.)

Structure tables for finite groups

The Group table (showing all possible 'products' of pairs of elements) has been exhibited in cases (7), (17) and (18). You should construct group tables where possible in the remaining cases. (See Exs. 8.02, 8.04, 8.05 etc.)

In all the remaining cases, we have *infinite groups*. A structure table cannot of course be set out in full in the case of an infinite group.

When a Group table is being drawn up, we always adopt the following convention:

(a) The identity always appears first, both across and down.

(b) The order in which the elements appear shall be the same across as down.

(c) To find the product $x*y$ (or xy), we look in *row* x and *column* y, i.e. the first operation is given across, the second operation is given down.

Other methods of showing the same information as the group table 3.01 p. 18, might be as shown in table 8.01. Although there is nothing wrong

Table 8.01

	X	Y	Z	I
X	X	Y	Z	I
Y	Z	I	X	Y
Z	Y	X	I	Z
I	X	Y	Z	I

or

	I	X	Y	Z
I	I	X	Y	Z
Z	Z	Y	X	I
X	X	I	Z	Y
Y	Y	Z	I	X

or

	I	Z	X	Y
I	I	Z	X	Y
Z	Z	I	Y	X
X	X	Y	I	Z
Y	Y	X	Z	I

with the first two, they do contravene the above conventions. The last one is a perfectly good alternative to the version on p. 18.

Q. 8.01. Regarding a group table as a matrix, show how permutation matrices may be used to transform one version of a group table into another. Take the group table in its original version in the text, and use matrix operations to transform into each of the three 'metamorphoses' above. If rules (i) and (ii) are to be adhered to, what are the corresponding rules for the use of these matrices? (Compare Ex. 10.37.)

A group of six permutations

We now illustrate these conventions in the case of the composition of the permutations:

$$e = \begin{pmatrix} 1 & 2 & 3 & 4 & 5 & 6 \\ 1 & 2 & 3 & 4 & 5 & 6 \end{pmatrix}; \quad p = \begin{pmatrix} 1 & 2 & 3 & 4 & 5 & 6 \\ 2 & 3 & 1 & 5 & 6 & 4 \end{pmatrix};$$

$$q = \begin{pmatrix} 1 & 2 & 3 & 4 & 5 & 6 \\ 3 & 1 & 2 & 6 & 4 & 5 \end{pmatrix}; \quad a = \begin{pmatrix} 1 & 2 & 3 & 4 & 5 & 6 \\ 4 & 6 & 5 & 1 & 3 & 2 \end{pmatrix};$$

$$b = \begin{pmatrix} 1 & 2 & 3 & 4 & 5 & 6 \\ 5 & 4 & 6 & 2 & 1 & 3 \end{pmatrix}; \quad c = \begin{pmatrix} 1 & 2 & 3 & 4 & 5 & 6 \\ 6 & 5 & 4 & 3 & 2 & 1 \end{pmatrix}.$$

There are thirty-six possible 'products', only twenty-five of which need to be worked out since e makes no change. We note that:

$$\begin{aligned} p^2 &= \begin{pmatrix} 1 & 2 & 3 & 4 & 5 & 6 \\ 2 & 3 & 1 & 5 & 6 & 4 \end{pmatrix} \circ \begin{pmatrix} 1 & 2 & 3 & 4 & 5 & 6 \\ 2 & 3 & 1 & 5 & 6 & 4 \end{pmatrix} \\ &= \begin{pmatrix} 2 & 3 & 1 & 5 & 6 & 4 \\ 3 & 1 & 2 & 6 & 4 & 5 \end{pmatrix} \circ \begin{pmatrix} 1 & 2 & 3 & 4 & 5 & 6 \\ 2 & 3 & 1 & 5 & 6 & 4 \end{pmatrix} \\ &= \begin{pmatrix} 1 & 2 & 3 & 4 & 5 & 6 \\ 3 & 1 & 2 & 6 & 4 & 5 \end{pmatrix} = q \end{aligned}$$

and $pq = pp^2 = p^3 = e = p^2p = qp$. (This should be verified the 'hard way'.) Also that $a^2 = b^2 = c^2 = e$. Suppose now we want aq. We have:

$$\begin{aligned} aq &= \begin{pmatrix} 1 & 2 & 3 & 4 & 5 & 6 \\ 4 & 6 & 5 & 1 & 3 & 2 \end{pmatrix} \circ \begin{pmatrix} 1 & 2 & 3 & 4 & 5 & 6 \\ 3 & 1 & 2 & 6 & 4 & 5 \end{pmatrix} \\ &= \begin{pmatrix} 3 & 1 & 2 & 6 & 4 & 5 \\ 5 & 4 & 6 & 2 & 1 & 3 \end{pmatrix} \circ \begin{pmatrix} 1 & 2 & 3 & 4 & 5 & 6 \\ 3 & 1 & 2 & 6 & 4 & 5 \end{pmatrix} \\ &= \begin{pmatrix} 1 & 2 & 3 & 4 & 5 & 6 \\ 5 & 4 & 6 & 2 & 1 & 3 \end{pmatrix} = b. \end{aligned}$$

Again, $$\begin{aligned} qa &= \begin{pmatrix} 1 & 2 & 3 & 4 & 5 & 6 \\ 3 & 1 & 2 & 6 & 4 & 5 \end{pmatrix} \circ \begin{pmatrix} 1 & 2 & 3 & 4 & 5 & 6 \\ 4 & 6 & 5 & 1 & 3 & 2 \end{pmatrix} \\ &= \begin{pmatrix} 1 & 2 & 3 & 4 & 5 & 6 \\ 6 & 5 & 4 & 3 & 2 & 1 \end{pmatrix} = c. \end{aligned}$$

It would be tedious to work through all possible combinations. Fortunately there are various short ways of filling in the table, as we shall see later (see pp. 77, 78, 92, 348). For the present, we note that the composition of the permutations may be contracted and simplified (see table 8.021). If we want bp, we ask ourselves 'What is 1 replaced by?' First

Table 8.021

e	1	2	3	4	5	6
p	2	3	1	5	6	4
q	3	1	2	6	4	5
a	4	6	5	1	3	2
b	5	4	6	2	1	3
c	6	5	4	3	2	1

p replaces 1 by 2, then b replaces 2 by 4 (follow the arrows), so we conclude that bp replaces 1 by 4. The question 'What is 6 replaced by?' is also answered by following the other arrows in the table of permutations above. Again, if we wanted ab, we should say (following the arrows in the table 8.022), '1 becomes 5, 5 becomes 3, so ab replaces 1 by 3', and since

Table 8.022

e	1	2	3	4	5	6
a	4	6	5	1	3	2
b	5	4	6	2	1	3

we know this set of permutations is closed, we can only arrive at the permutation (3 1 2 6 4 5), i.e. $ab = q$. The final example showing the structure of this group is to be found in table 8.03. (Results already

Table 8.03

2nd operation \ 1st operation	e	p	q	a	b	c
e	e	p	q	a	b	c
p	p	q	e	b	c	a
q	q	e	p	c	a	b
a	a	c	b	e	q	p
b	b	a	c	p	e	q
c	c	b	a	q	p	e

$(D_3 \cong S_3)$

obtained, $p^2 = q$, $pq = qp = e$, $aq = b$, $qa = c$, $ab = q$ are shown underlined.)

Note that one way in which we could be spared the labour of combining all the pairs of permutations would be as follows: Suppose we want ab, and that we have already found that $aq = b$. Then $ab = a(aq) = (a^2)q$. But $a^2 = e$, and $eq = q$, hence $q = ab$. Again, if we required bp, then we could proceed as follows:

$aq = b$ (already found) $\Rightarrow aqp = bp$. But $qp = e$ and $ae = a$, so $bp = a$.

There is a remarkable property of this table. You will observe that, for example, the row labelled b, i.e. $b\ a\ c\ p\ e\ q$, is a permutation of the first row, i.e. $e\ p\ q\ a\ b\ c$. The extraordinary thing is that it is the very same permutation which b represents, namely $\begin{pmatrix}1 & 2 & 3 & 4 & 5 & 6\\ 5 & 4 & 6 & 2 & 1 & 3\end{pmatrix}$, only with the figures 1 2 3 4 5 6 replaced respectively by the letters $e\ p\ q\ a\ b\ c$. You should check that the same property is possessed by the other rows. The consequence of the foregoing, moreover, is that the tables 8.02 and 8.03 are exact replicas under the interchange $\begin{pmatrix}e & p & q & a & b & c\\ 1 & 2 & 3 & 4 & 5 & 6\end{pmatrix}$. Check, for example, that the element c occupies the same relative positions in the group table as does the digit 6 in the table of permutations. This is an instance of Cayley's Theorem, which we shall look at later on (see Ch. 9, p. 92). Of course, there are 6! (= 720) permutations of six objects, and these form a group of order 720 – the table for this would be huge. The above subset $\{e, p, q, a, b, c\}$ of these 720 permutations is a group in its own right, and qualifies to be called a *Subgroup* of the full group of permutations of order 720. (See Ch. 14 on Subgroups.)

Introduction to groups of symmetries

A two-dimensional object which has an axis of symmetry, such as fig. 8.01, gives rise to a group in the following way. Denote by M the operation of reflecting† in the axis of symmetry, so that P $\xrightarrow{M}$ ꟼ, and ꟼ $\xrightarrow{M}$ P. Then this operation M when applied to the symmetrical figure *brings it into coincidence with itself*: if you turned your head away while I performed M, and then looked round, you wouldn't know the difference – you might think that I had performed I! This idea of a *symmetry operation* – that is, a geometrical transformation which brings a figure (in two or in three dimensions) into coincidence with itself has far wider applications, as we shall presently see (Ch. 17). In the case of 'bilateral symmetry', it is evident that the operation M, when repeated, restores the original position: $MM = M^2 = I$. Thus we obtain table 8.04. This is a group of order 2 – the smallest possible group, except for the trivial case of a group containing I alone.

$\circ$	I	M
I	I	M
M	M	I

Table 8.04

You may wonder why we did not start with this easy 2-group, and gradually introduce bigger and bigger groups, instead of the larger ones we have so far considered. The answer is that the larger groups are more

† See 'flips and reflections', p. 187.

likely to arouse interest: there is nothing much to be said about a 2-group! Or is there? Certainly not much can be said about its structure, but the interest lies in *the large number of situations in which it may arise.*

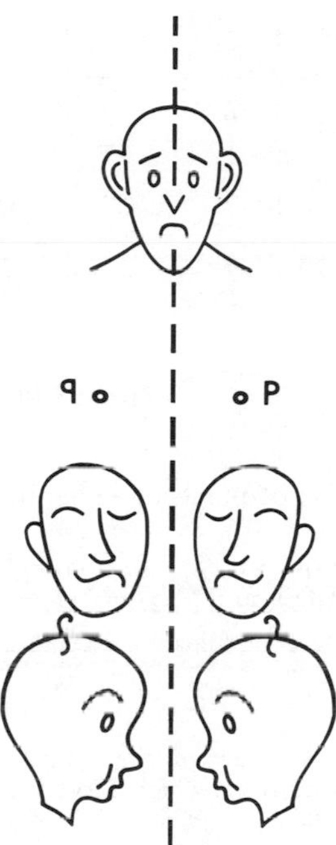

Fig. 8.01. Bilateral symmetry.

The two-group

Besides the bilateral symmetry of a two-dimensional object, there is the symmetry of a three-dimensional body about a plane. For example, a car, a ship, or a pair of semi-detached houses have a vertical plane of symmetry running 'fore-and-aft', and a reflection in this plane would map the whole figure into itself, apart from minor details such as the steering-wheel of the car, etc.

Q. 8.02. What sort of symmetry has a playing card? (the Queen of Hearts.) Here, the symmetry operation is a *half-turn* (H) about an axis through its centre perpendicular to the plane of the card.

Q. 8.03. What kinds of plane quadrilateral have (a) bilateral symmetry, (b) half-turn symmetry (or symmetry about a point)?

In the latter case, the group table is:

Table 8.05

∘	I	H
I	I	H
H	H	I

Q. 8.04. Consider the following group tables, and see if you can discover the (?) operation in each case:

?	0	1
0	0	1
1	1	0

?	1	2
1	1	2
2	2	1

?	+	−
+	+	−
−	−	+

?	1	3
1	1	3
3	3	1

?	1	7
1	1	7
7	7	1

?	+1	−1
+1	+1	−1
−1	−1	+1

($C_2 \cong D_1 \cong S_1$)

Interpret table 8.06.

Table 8.06

+	E	O
E	E	O
O	O	E

Find as many as possible pairs of numbers which form a 2-group under multiplication modulo n (e.g. {4, 16}, × mod. 20).
Next consider the set of functions $\mathrm{i}(x) \equiv x$; $\mathrm{f}(x) \equiv 1 - x$. These form a 2-group, for $\mathrm{if}(x) \equiv 1 - x \equiv \mathrm{f}(x) \equiv \mathrm{fi}(x)$, and $\mathrm{f}^2(x) = \mathrm{f}(1 - x) = 1 - (1 - x) = x \equiv \mathrm{i}(x)$.
So the two functions combine according to the table:

∘	i	f
i	i	f
f	f	i

Q. 8.05. Find a number of functions which have the same property (compare p. 117).

Again, think of the permutations of two objects: $e \begin{pmatrix} 1 & 2 \\ 1 & 2 \end{pmatrix}$ and $a \begin{pmatrix} 2 & 1 \\ 1 & 2 \end{pmatrix}$.
You can easily see that $a^2 = e$, and that these permutations combine thus:

	e	a
e	e	a
a	a	e.

Another way in which the 2-group can arise may be seen by looking again at the group of order 6 in table 8.03.

Table 8.07

	e	p	q	a	b	c
e	e	p	q	a	b	c
p	p	q	e	b	c	a
q	q	e	p	c	a	b
a	a	c	b	e	q	p
b	b	a	c	p	e	q
c	c	b	a	q	p	e

⟶

	E	A
E	E	A
A	A	E

You will see that the top left-hand and the bottom right-hand corners contain only the elements of the subgroup $\{e, p, q\}$, whereas the other two quarters contain $\{a, b, c\}$. Calling these subsets E and A respectively, we see that they give the pattern of a 2-group, as shown on the right: what this table tells us is that, for example, any element of subset A when combined with any element of A is bound to give an element of E.

Q. 8.06. Is A a subgroup?

This is an example of a 'homomorphic mapping' (see Ch. 21), and E and A are called 'cosets' (see Ch. 19). Both the ideas of cosets and of homomorphic mappings are of great importance. Note that, if the elements in the table had been set down in some other order, such as e, a, p, c, b, q, then the overall 2-group pattern which stood out clearly above would now be obscured (though of course, it would still be there!)

Two-groups which occur as subgroups of larger groups

Finally, 2-groups always arise as subgroups of larger groups whenever there is an 'involutory' element present, that is, an element whose square is equal to the identity. For example, in the group of permutations of five symbols, the permutation $x = \begin{pmatrix} 1 & 2 & 3 & 4 & 5 \\ 3 & 2 & 1 & 5 & 4 \end{pmatrix}$ will generate a subgroup, since $x^2 = e$, and so we have

∘	e	x
e	e	x
x	x	e.

Further groups of symmetries

As a further example of a group of symmetries, consider, for example, fig. 16.13. It possesses four symmetries, including the identity, since rotations about its centre through angles 90°, 180°, 270°† will bring it into coincidence with itself. Calling these rotations R, H and S respectively, they may be combined as shown in table 8.08.

Table 8.08

(a)

∘	I	R	H	S
I	I	R	H	S
R	R	H	S	I
H	H	S	I	R
S	S	I	R	H (C_4)

and this may be written

(b)

∘	I	R	R^2	R^3
I	I	R	R^2	R^3
R	R	R^2	R^3	I
R^2	R^2	R^3	I	R
R^3	R^3	I	R	R^2

or

(c)

∘	R^0	R^1	R^2	R^3
R^0	R^0	R^1	R^2	R^3
R^1	R^1	R^2	R^3	R^0
R^2	R^2	R^3	R^0	R^1
R^3	R^3	R^0	R^1	R^2.

† Positive anticlockwise, by general agreement.

We might draw up a group table showing the 'indices' only, which, since $R^4 = I = R^0$, are being added *modulo* 4:

(d)

	0	1	2	3
0	0	1	2	3
1	1	2	3	0
2	2	3	0	1
3	3	0	1	2 (C_4)

Another group of symmetries we might consider is that consisting of the rotations of the rectangular box (e.g. a brick) through 180° about the joins of the centres of opposite faces. This is precisely what we did when we considered the half-turns about three mutually perpendicular axes on p. 17, and the group table is reproduced here, for *contrast* with the table for the 4-group above:

Table 8.09

$\circ$	I	X	Y	Z
I	I	X	Y	Z
X	X	I	Z	Y
Y	Y	Z	I	X
Z	Z	Y	X	I (D_2)

Q. 8.07. Find (a) a two-dimensional, (b) a three-dimensional figure which has a group of symmetries of order 3.

Abelian groups

A group in which every element *commutes* with every other element is called an *Abelian* group, after the famous Norwegian mathematician, Abel, (1802–29), or simply a commutative group. More formally;

$$\textbf{(G},*\textbf{) is an Abelian group} \Leftrightarrow \forall x, y \in \textbf{G}, x*y = y*x.$$

All the groups of order four in the previous paragraph are Abelian, but not the group of order 6 on p. 77, since $aq \neq qa$.

Q. 8.08. If a group is Abelian, what can you say about the group table (provided the conventions of p. 75 are adhered to)?

Q. 8.09. Can you say a firm 'Yes' or 'No' to the questions: are the following Abelian:

(i) groups of permutations under successive application;
(ii) finite arithmetic groups, under + mod. n. or × mod. n (see pp. 46–8);
(iii) symmetry groups of figures in two and three dimensions (see Ch. 17);
(iv) groups obtained by the composition of functions (see p. 22, 160)?

Groups of two-dimensional transformations

To qualify as a group, the requirements *C*, *A*, *N*, *I* of p. 73 must all be satisfied. Consider the set of rotations in the plane through any angle about any point of the plane. Such transformations are undoubtedly associative (see p. 40). The set contains the null element, since the angle of rotation may be zero; while to any rotation, say $R_{A,\alpha}$ (about the point A through angle α) there is an inverse rotation $R_{A,-\alpha}$. Is the composition of two rotations always a rotation – is the set *closed*? At first sight you may be inclined to think that it is. Certainly rotations *about a particular point* are closed, but not when all points may be used. $R_{B,\beta} \circ R_{A,\alpha}$ causes a resultant rotation through an angle $\alpha + \beta$ about some point C_1, i.e.

$R_{B,\beta} \circ R_{A,\alpha} = R_{C_1,\alpha+\beta}$. These are not commutative.

$R_{A,\alpha} \circ R_{B,\beta} = R_{C_2,\alpha+\beta}$,

the centre of rotation for the composite rotation, C_2, in this case being different from C_1. (See F. J. Budden, *Complex Numbers and their Applications*, p. 141.)

What happens when $\alpha + \beta = 0$, or $\alpha + \beta = 360°$? You may think that the *identity* is then obtained, but in fact this only happens when $A = B$, i.e. when we rotate through angles α and $-\alpha$ about the *same point*. When

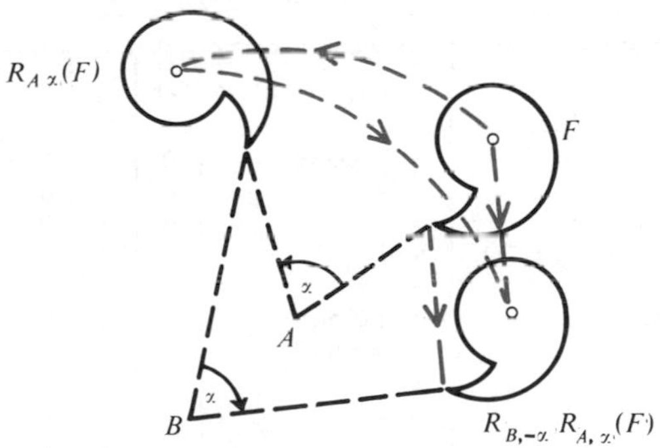

Fig. 8.02. Product of two rotations through equal and opposite angles is a translation.

we rotate about different points, the body is restored to its original *orientation* (i.e. 'facing the same way'), but is moved to a new position. The final resultant of the two opposite rotations is a *translation*:

$R_{B,-\alpha} \circ R_{A,\alpha} = T$. (See fig. 8.02.)

We might have seen this another way. On p. 21, we saw that

$T \circ R_1 = R_2$ (where we now shall write $R_1 = R_{A,-\alpha}$, $R_2 = R_{B,-\alpha}$ for short).

So, $T \circ R_1 \circ R_1^{-1} = R_2 \circ R_1^{-1}$.

But $R_1 \circ R_1^{-1} = I$ (= R_0, the null rotation).

Hence, $R_2 \circ R_1^{-1} = T$. But $R_1^{-1} = R_{A,\alpha}$, and so we have $R_{B,-\alpha}\, R_{A,\alpha} = T$, just as before.

Composition of transformations in two dimensions may be described by multiplication of 2 × 2 matrices, but it is simpler by using complex number methods (see F. J. Budden, op. cit. p. 137). The above discussion shows that the set of rotations in the plane is *not a group* because it *fails for closure*, even though it satisfies the requirements *A*, *N* and *I*. Note that it *only just* fails to be a group: we could make it a group by throwing into the set all the translations in the plane; then we should have the group of the 'direct isometries' in the plane.

Systems which fail to be groups

In the table 8.10, are listed examples of sets with operations which fail to be groups for various reasons. In the columns headed *C*, *A*, *N*, *I* are inserted a 0 or 1 rather than × or √ to indicate whether or not the system possesses that characteristic.

Table 8.10

Set	Operation	*C*	*A*	*N*	*I*	Score
All people living or dead	l.c.f. (see p. 6)	0	0	0	0	0
Odd integers	+	0	1	0	0	4
$a\sqrt{2} + b\sqrt{3}\ (a, b \in \mathrm{Q})$	×	0	1	0	1	5
$a\sqrt{2} + b\sqrt{3} + c\ (a, b, c \in \mathrm{Q})$	×	0	1	1	0	6
$\{\frac{1}{2}, \frac{1}{3}, 1, 2, 3\}$	×	0	1	1	1	7
Vectors in 3 dimensions	Vector product	1	0	0	0	8
R	~ (see p. 51)	1	0	1	1	11
Even integers *(2Z)*	×	1	1	0	0	12
Sets	∩ (or ∪)	1	1	1	0	14
Any Group		1	1	1	1	15
R	−	1	0	0*(R)*	0*(R)*	—
R	$x * y = 2x + y$	1	0	0*(L)*	0*(L)*	—

Q. 8.10. Interpret the last two rows.

The column headed 'score' has summed up the information of the four columns by regarding them as constituting a binary numeral (e.g. 0110 = 0.8 + 1.4 + 1.2 + 0.1 = 6). The numbers, running from 0 (junk) to 15 (group) therefore show all shades of 'groupworthiness'. Certain combinations (represented by 1, 2, 3, 9, 10, 13) are missing. Some of these might possibly be artificially constructed by an arbitrary table such as we met on p. 39. You may find you can succeed in filling some of the gaps.†

Q. 8.11. Draw up a table to indicate the characteristic binary numeral in the following cases: $(\mathbf{Z}, \times)$; $(\mathbf{Z}|0, \times)$; $(\mathbf{N}, \text{L.C.M.})$; (odd integers, $\times$); (all 2×2 matrices, $+$); (all 2×2 matrices, $\times$); $(\mathbf{Z}^+, -)$; $(\{\frac{1}{3}, \frac{1}{2}, 2, 3\}, \times)$ (primes, $+$); (primes, $\times$); all isometries in the plane under composition; {vectors in 3-D}, scalar product; $(\cos\theta + i\sin\theta, \times)$; $(\cos\theta + i\sin\theta, +)$; the permutations

$$\begin{pmatrix}1 & 2 & 3 & 4\\ 2 & 1 & 4 & 3\end{pmatrix}, \quad \begin{pmatrix}1 & 2 & 3 & 4\\ 4 & 3 & 2 & 1\end{pmatrix}, \quad \begin{pmatrix}1 & 2 & 3 & 4\\ 1 & 2 & 3 & 4\end{pmatrix}.$$

In each case, state how group structure may be gained by either inserting into, or removing elements from the set.

Q. 8.12. Repeat the above question for

(a) the permutations $\begin{pmatrix}1 & 2 & 3\\ 1 & 2 & 3\end{pmatrix}, \begin{pmatrix}1 & 2 & 3\\ 1 & 3 & 2\end{pmatrix}, \begin{pmatrix}1 & 2 & 3\\ 3 & 2 & 1\end{pmatrix}, \begin{pmatrix}1 & 2 & 3\\ 2 & 1 & 3\end{pmatrix}$;

(b) the matrices $\begin{bmatrix}1 & 0\\ 0 & -1\end{bmatrix}, \begin{bmatrix}-1 & 0\\ 0 & -1\end{bmatrix}, \begin{bmatrix}-1 & 0\\ 0 & 1\end{bmatrix}, \begin{bmatrix}0 & 1\\ 1 & 0\end{bmatrix}$;

(c) the functions (i) $\left\{x, \dfrac{1}{1-x}, 1 - \dfrac{1}{x}, \dfrac{1}{x}, 1 - x\right\}$,

(ii) $\left\{2 - x, \dfrac{x-2}{x-1}, \dfrac{x}{x-1}\right\}$.

EXERCISES 8

1 Consider the following and determine whether or not they are groups:‡

(a) $(\mathbf{Z}^+, +)$; (b) $(\mathbf{Z}, +)$; (c) $(\mathbf{Z}^-, +)$; (d) $(\mathbf{Z}, \times)$; (e) $(\mathbf{Q}, +)$; (f) $(\mathbf{Q}^+, +)$; (g) $(\mathbf{Q}^+, \times)$; (h) $(\mathbf{Q}, \times)$; (i) $(\mathbf{R}, +)$; (j) $(\mathbf{R}^-, +)$; (k) $(\mathbf{R}^+, \times)$; (l) $(\mathbf{R}^-, \times)$; (m) $(\mathbf{R}, \times)$; (n) $(\mathbf{C}, +)$; (o) $(\mathbf{C}, \times)$; (p) $(\cos\theta + i\sin\theta, \times)$; (q) $(a + b\sqrt{2} + c\sqrt{3}$ where $a, b, c \in \mathbf{Z}, +)$; (r) $(a + b\sqrt{2} + c\sqrt{3}$ where $a, b, c \in \mathbf{Q}, \times)$; (s) $(2\mathbf{Z}, +)$; (t) $(2\mathbf{Z} + 1, +)$; (u) $(2\mathbf{Z} + 1, \times)$; (v) $(5\mathbf{Z}, +)$; (w) $(5\mathbf{Z}, \times)$; (x) $(\mathbf{Z}_6, +)$; (y) $(\mathbf{Z}_6, \times)$; (z) $(\mathbf{Z}_7, \times)$.

2 Consider the set of permutations

$$e = \begin{pmatrix}1 & 2 & 3 & 4\\ 1 & 2 & 3 & 4\end{pmatrix}, \quad x = \begin{pmatrix}1 & 2 & 3 & 4\\ 2 & 1 & 4 & 3\end{pmatrix}, \quad y = \begin{pmatrix}1 & 2 & 3 & 4\\ 3 & 4 & 1 & 2\end{pmatrix}$$

and $z = \begin{pmatrix}1 & 2 & 3 & 4\\ 4 & 3 & 2 & 1\end{pmatrix}$.

Show that they form a group under composition of permutations, and write out the group table.

† The complexity of the definitions for N and I and their interdependence means that the above is really an over-simplification.

‡ For the notation, see pp. xvii, xviii.

3 If $p = \begin{pmatrix} 1 & 2 & 3 & 4 \\ 2 & 1 & 3 & 4 \end{pmatrix}$ and $q = \begin{pmatrix} 1 & 2 & 3 & 4 \\ 1 & 2 & 4 & 3 \end{pmatrix}$, find pq, qp, p^2, q^2, etc., and discover a group of permutations which contains p and q.

4 Draw up the multiplication table of the set G{1, 2, 4, 5, 7, 8} under multiplication modulo 9. From it, solve the following equations, multiplications being modulo 9 in every case: $4x = 7$; $7x = 4$ $(x \in \text{G})$; $2^n = 1$, $7^n = 1$ $(n \in \mathbf{Z})$. Find also subsets of G which are themselves groups (i.e. subgroups of G).

5 Obtain the table for multiplication mod. 11 of the set G{1, 3, 4, 5, 9}. Show that $\forall\, x \in \text{G}$, $x^5 = 1$ (mod. 11). Find the value of $\frac{5}{9}$ in this system (i.e. find x such that $9x = 5$).

6 Show that the matrices

$$\begin{bmatrix} 1 & 0 \\ 0 & 1 \end{bmatrix}, \quad \begin{bmatrix} -1 & 0 \\ 0 & 1 \end{bmatrix}, \quad \begin{bmatrix} 1 & 0 \\ 0 & -1 \end{bmatrix}, \quad \begin{bmatrix} -1 & 0 \\ 0 & -1 \end{bmatrix}$$

form a group under matrix multiplication, and construct the group table.

7 Show that the matrices

$$A = \begin{bmatrix} 0 & 1 \\ 1 & 0 \end{bmatrix} \quad \text{and} \quad B = \begin{bmatrix} 0 & -1 \\ -1 & 0 \end{bmatrix}$$

and two other matrices form a group of order 4 under matrix multiplication. Specify the other two matrices, and draw the group table.

8 Prove that a group of order 4 is necessarily Abelian.

9 If $\omega = \cos \frac{1}{3}\pi + i \sin \frac{1}{3}\pi$ $(i^2 = -1)$, show that $\{1, \omega, \omega^2, \omega^3, \omega^4, \omega^5\}$ is a group under multiplication, and construct the group table. What connection is there between this and the group {0, 1, 2, 3, 4, 5} under addition modulo 6?

10 Show that the eight permutations of Ex. 4.13 form a group, and write out its table (call them 1, p, q, r, x, y, z and i).

11 Draw up the group tables in the following cases (keep your tables for further use):

(a) {0, 1, 2, 3, 4, 5, 6, 7, 8, 9}, + mod. 10 ('final digits addition').

(b) Find as many group tables as you can which resulted from your experiments with multiplication mod. n of subsets of $\{1, 2, 3, \ldots, n-1\}$ for various n. (See Q.6.10 etc.)

(c) Construct the group table for the six permutations

$$e = \begin{pmatrix} 1 & 2 & 3 \\ 1 & 2 & 3 \end{pmatrix}, \quad p = \begin{pmatrix} 1 & 2 & 3 \\ 3 & 1 & 2 \end{pmatrix}, \quad q = \begin{pmatrix} 1 & 2 & 3 \\ 2 & 3 & 1 \end{pmatrix},$$

$$a = \begin{pmatrix} 1 & 2 & 3 \\ 1 & 3 & 2 \end{pmatrix}, \quad b = \begin{pmatrix} 1 & 2 & 3 \\ 3 & 2 & 1 \end{pmatrix}, \quad c = \begin{pmatrix} 1 & 2 & 3 \\ 2 & 1 & 3 \end{pmatrix}.$$

12 A group table of order 8 resulting from addition (mod. 2) of corresponding digits of the numbers {0, 1, 2, 3, 4, 5, 6, 7} expressed in binary, with no carrying, e.g. $6 \oplus 3 = 110 \oplus 011 = 101 = 5$. Repeat to give a group of order 9 using the following method of 'addition': express the numbers {0, 1, 2, 3, 4, 5, 6, 7, 8} in the scale of 3, and add corresponding digits (mod. 3), again with no carrying, e.g. $4 \oplus 5 = 11 \oplus 12 = 20 = 6$.

13 If R is a rotation in two dimensions through an angle 60°, R^2 is a rotation through 120°, and so on. Write out the group table for the composition of the rotations $\{I, R, R^2, R^3, R^4, R^5\}$.

14 What are the symmetry groups of MUM, bud, pod, dip, ᴍow, hoʮ, doop.

15 Highway Code Geometry (see fig. 8.03). Give the symmetry groups of (a) the whole board; (b) the sign itself, in the case of each of the signs.

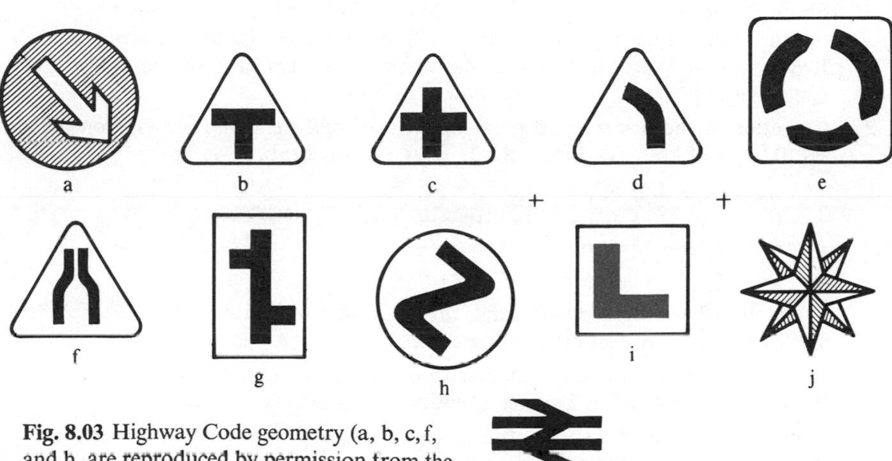

Fig. 8.03 Highway Code geometry (a, b, c, f, and h, are reproduced by permission from the Highway Code)

16 In the group table 8.03, p. 77 (D_3), solve the following for the unknown element x: $xp = b$; $px = b$; $x^2 = p$; $x^2 = e$; $xax = c$; $px = xq$. In the same table, show that $ap^2 = pa$, $bq^2 = qb$, and obtain similar relations between other pairs of elements.

17 Show that all matrices $\begin{bmatrix} a & b \\ c & 0 \end{bmatrix}$ (for which $d = 0$) are a group under matrix addition. Construct other infinite subgroups of $(\mathcal{M}_2, +)$.

18 What is the inverse of the matrix $\begin{bmatrix} \cosh u & \sinh u \\ \sinh u & \cosh u \end{bmatrix}$ under matrix multiplication? Do matrices of this form, where $u \in R$, form a subgroup of $(\mathcal{M}_2, \times)$?

19 Show that the fractions whose numerators and denominators are both odd and coprime are a group under multiplication.

20 Consider all rational numbers whose denominators do not contain the factor 7 (e.g. we include $\frac{28}{135}$ but exclude $\frac{48}{35}$). Are they a group under addition? under multiplication?

21 Is the set of irrational numbers together with the number zero a group under addition? Is the set of irrational numbers together with the number unity a group under multiplication?

22 Show that the integral powers of 3 form a group under multiplication. Generalise.

23 Show that the functions x, $1/x$, $-x$ and another function of x form a group. Find the fourth function and construct the table.

24 Show that the functions $i(x) = x$; $f(x) = 1/(2 - 4x)$; $g(x) = \frac{1}{2} - 1/4x$ form a group under successive substitution, and construct the group structure table.

25 If $f(x) = (2x + 1)/4(1 - x)$, find $ff(x)$ (abbreviate $f^2(x)$), $f^3(x)$, etc., and finally obtain a group of functions, displaying the group table.

26 Show that the set $\{x, 1, -x, -1\}$ of linear polynomials in x, when multiplied modulo $x^2 + 1$ (see Ex. 7.17) are a group. Set up the group table. Investigate how group structure may arise in the operation of multiplication of polynomials modulo $x^2 - 2$, and in other cases.

27 Consider the addition of polynomials whose coefficients are drawn from the set $\{0, 1\}$ and are added modulo 2. Thus, for quadratics, we have:

$$\begin{array}{ll} 0x^2 + 0x + 0 = e & 1x^2 + 0x + 0 = d \\ 0x^2 + 0x + 1 = a & 1x^2 + 0x + 1 = f \\ 0x^2 + 1x + 0 = b & 1x^2 + 1x + 0 = g \\ 0x^2 + 1x + 1 = c & 1x^2 + 1x + 1 = h \end{array}$$

Show that these eight polynomials are a group, and draw up the table showing how they combine, using the notation suggested.

28 Repeat the above question for linear polynomials $Ax + B$, where $A, B \in \{0, 1, 2\}$ and coefficients are added mod. 3. Try to generalise.

29 Note that each of the six permutations e, p, q, a, b, c discussed in the text (see p. 76) may be thought of as a cyclic permutation of 1, 2, 3 and the corresponding cyclic permutation of 4, 5, 6, these triads being interchanged in a, b and c; also that the latter three permutations each contains three pairs of interchanges: e.g. a has (1 4), (2 6) and (3 5). Compare with Ch. 4 and Ch. 13.

30 Consider $(a, b) * (c, d) = (ac, ad + b)$ $(a, b, c, d \in \mathbf{R})$. Find a left and right identity, and the inverse of (a, b).
Show that these ordered pairs with the operation $*$ form a group. Compare with Exs. 4.14, 5.02, 7.14, 14.25.
Suppose now that (a, b) and (c, d) are points in the Cartesian plane. Devise a construction for the point $(a, b) * (c, d)$, and compare with Ex. 7.28. (*Hint*: Select the point U to be (1, 0).)

31 Why is the set of polynomials with real coefficients not a group under multiplication?
From which number system may the coefficients of *rational functions* (i.e. the ratio of two polynomials) be drawn, so that they will be a group under multiplication?

32 Consider the non-negative integers under the operation + mod. 10, so that for example, $53 * 249 = 2$. Why does this fail to be a group?

33 Discuss the following operation on $\mathbf{R}$. First express each number as the sum of an integer (positive, negative or zero) and a non-negative decimal, thus: $a = \alpha + 0 \cdot a_1 a_2 a_3 a_4 \ldots$ where $\alpha \in \mathbf{Z}$ and $a_1, a_2, \ldots \in \{0, 1, 2, \ldots, 9\}$. If $b = \beta + 0 \cdot b_1 b_2 b_3 b_4 \ldots$, define $a * b = c$ by $c = \gamma + 0 \cdot c_1 c_2 c_3 c_4 \ldots$ where $\gamma = \alpha + \beta$, and $\begin{pmatrix} c_1 & c_2 \\ c_3 & c_4 \end{pmatrix} = \begin{pmatrix} a_1 & a_2 \\ a_3 & a_4 \end{pmatrix} \begin{pmatrix} b_1 & b_2 \\ b_3 & b_4 \end{pmatrix}$, all the operations in the matrix product being modulo 10. How far short does this system fall from being a group?

9 *Properties of groups*

At the British Mathematical Association's Annual Conference in 1966 at Keele University in England, a Brains Trust was asked the question, 'What answer would you give to a pupil who asked to be shown instances of the uses of group theory, and how it can be applied to the solutions of problems ...' No very satisfactory answer was given by the team. Dr T. J. Fletcher has subsequently, in a letter to the *Mathematical Gazette* (Dec. 1966, pp. 398–401), given a large list of applications, and these are reproduced with his permission in Appendix III.

The short answer can be boiled down to the following:

(1) because groups crop up *everywhere* in mathematics, in the most diverse and unexpected situations. A wide variety of (not only mathematical†) situations can be described, and problems solved, by group-theoretical methods.

(2) Once a system is recognised to exhibit group structure, then any rules that have been established for groups (e.g. the cancellation law – see below) can be applied to that system.

The cancellation law

We shall consider some of the properties of groups which can be deduced from the definition. First, the 'cancellation law':

If a, x and y are elements in a group, and if

$xa = ya$, then $x = y$; or if $ax = ay$, then $x = y$.
(right cancellation) (left cancellation)

† Thus, for example, groups are involved even in the simple act of putting on one's trousers. I was reminded of this in dressing my small son: one can put the trousers on correctly, back-to-front, inside-out, or back-to-front-and-inside-out. This gives rise to the group D_2 in much the same way as we got the group $\{I, V, C, A\}$ on p. 25. It is a great pity that one cannot put the trousers on upside-down, for then we might get a group of order eight. However, this may be remedied by the simple process of having twins and getting their trousers mixed up – see p. 269. Again, when drying dishes one holds a pile of say four plates in the left hand, and having wiped the top and bottom surfaces, transfers the top plate to the bottom (operation r), and repeats the drying operation. When r has been performed four times, the plates are in their original order, and the process is complete: $r^4 = e$, and the permutations form the cyclic group C_4 (see pp. 81, 140).

Proof (right cancellation)

$$xa = ya \Rightarrow (xa)a^{-1} = (ya)a^{-1} \quad (a^{-1} \text{ certainly exists, by I, p. 73})$$

$$\Rightarrow \quad x(aa^{-1}) = y(aa^{-1}) \quad \text{(by } A\text{)}$$

$$\Rightarrow \quad xe = ye \quad \text{(by } I\text{)}$$

$$\Rightarrow \quad x = y \quad \text{(by } N\text{)}$$

Q. 9.01. Write out the proof for left cancellation.

The final result is that we may 'cancel on the right': $x\not{a} = y\not{a}$; or 'cancel on the left': $\not{a}x = \not{a}y$. But ***beware:*** if $ax = ya$ (or if $xa = ay$), we cannot deduce that $x = y$. For $ax = ya \Rightarrow a^{-1}ax = a^{-1}ya \Rightarrow ex = a^{-1}ya \Rightarrow x = a^{-1}ya$ and $a^{-1}ya$ is not equal to y in general. (See also Ch. 20.)

Q. 9.02. $a^{-1}ya$ would certainly be equal to y if the group . . . ?

The Latin square property

The structure table for any finite group is always a Latin square, that is to say, each row and each column contains all the elements of the set with no repetitions. Table 5.01, p. 39 is a Latin square, but is not a group table. (Why?)

Q. 9.03. Some (Latin) squares are shown in table 9.01. How many of them are group tables? (Assume e is an identity in each case.) Make up other examples.

Table 9.01

(a)

e	*a*	*b*	c
a	c	*e*	*b*
b	*e*	c	*a*
c	*b*	*a*	*e*

(b)

e	*a*	*b*	c	*d*
a	c	*d*	*b*	*e*
b	*d*	*a*	*e*	c
c	*b*	*e*	*d*	*a*
d	*e*	c	*a*	*b*

(c)

e	*a*	*b*	c	*d*
a	*b*	*d*	*e*	c
b	c	*a*	*d*	*e*
c	*d*	*e*	*a*	*b*
d	*e*	c	*b*	*a*

(d)

e	*a*	*b*	c	*d*	*f*
a	*e*	*f*	*b*	c	*d*
b	*d*	*a*	*e*	*f*	c
c	*f*	*e*	*d*	*b*	*a*
d	*b*	c	*f*	*a*	*e*
f	c	*d*	*a*	*e*	*b*

Show how the order of the letters in the second example can be rearranged so as to show the identical pattern to the group table for the set {0, 1, 2, 3, 4} under addition mod. 5:

Table 9.02

0	1	2	3	4
1	2	3	4	0
2	3	4	0	1
3	4	0	1	2
4	0	1	2	3.

These examples serve as a warning that the form of a Latin square is by no means sufficient to guarantee a group table – associativity will usually fail somewhere along the line.

Proof of the Latin square property

Suppose in row p of a finite group of order n, there is a repetition q in columns r and s. Then $pr = q = ps$. By the cancellation law, $r = s$, which is a contradiction, since r and s are two different elements. Therefore all the elements in row p are different. But row p contains n elements, and since they are all different, they must consist of a permutation of all the n elements of the group. Prove the similar property for columns, and the Latin square property is then established.

Note that in table 7.03 for multiplication of the set $\{1, 2, 3, \ldots, 9\}$ mod. 10 (p. 60) those rows and columns which had to be deleted to give a group were just those which did not contain a permutation of the whole set. In fact, none of those rows or columns deleted contained a '1'. Can you generalise about this?

Q. 9.04. In forming a group under multiplication modulo n, how does one determine the subset of $\{1, 2, 3, \ldots, n-1\}$ to give a group?

For example, when $n = 24$, the largest group we can get is $\{1, 5, 7, 11, 13, 17, 19, 23\}$; there are subsets of this which are groups (subgroups); and there is also the set $\{9, 3, 15, 21\}$ which is a group under multiplication mod. 24, with identity 9; this latter itself possesses three subgroups in its turn.

An immediate consequence of the Latin square property is that, if a and b are given elements of a group, then an x can always be found to satisfy the 'equation' $ax = b$ (and also $xa = b$). Another way of saying this is that 'in row a, we shall somewhere (under column x) find b'.

In fact $\quad ax = b \Rightarrow a^{-1}ax = a^{-1}b \Rightarrow ex = a^{-1}b \Rightarrow x = a^{-1}b.$

Similarly, $\quad xa = b \Rightarrow x = ba^{-1}.$

Q. 9.05. Interpret the above in the cases of groups of permutations under successive composition; finite arithmetics, under + and × mod. n; sets, under symmetric difference (see p. 9).

The last question may be very puzzling to many people: $A \Delta X = B$, where A and B are given sets and X is a set to be found; or 'find an unknown set such that any object which is either in A or in X but not in both is necessarily in B, and vice versa'. Even with the help of Venn diagrams, it is still extremely difficult to visualise. However, we must remember that we are *working with a group*, and we have already noted (see p. 63) that the identity is the empty set, ϕ, and that $A \Delta A = \phi$, so that $A^{-1} = A$, or each set is its own inverse. Thus we may solve $A \Delta X = B$ by premultiplying by $A^{-1} (= A)$:

$$A \Delta A \Delta X = A \Delta B \Rightarrow \phi \Delta X = A \Delta B \Rightarrow X = A \Delta B.$$

Q. 9.06. Check this by reference to a Venn diagram.

Use of the Latin square property to fill up a Group table

We mentioned earlier that the labour of constructing a group table can be greatly reduced. Indeed, in the case considered on p. 77, once one has filled in the subgroup $\{e, p, q\}$ in the top left-hand corner (Table 8.07), the whole group table may be completed by working out only *three* further products. Referring to table 9.03, where the obvious results have already

Table 9.03

	·	first e	p	q	a	b	c
	e	e	p	q	a	b	c
	p	p	q	e	b	.	.
second	q	q	e	p	.	.	.
	a	a	.	.	e	.	.
	b	b	.	.	.	e	.
	c	c	.	.	.	.	e

been filled in, as well as $pa = b$, which we suppose we have worked out. Then the remaining positions can easily be filled in by the use of the Latin square property: in row p, the remaining two elements are a and c. But c cannot come again in column c, so we conclude $pb = c$, $pc = a$. We are now able to fill in the remaining blanks in row q, which we know must be a permutation of the remaining elements a, b and c: each of columns a, b and c contains already a pair of these, so this settles that $qa = c$, $qb = a$ and $qc = b$. A similar procedure may be used for the bottom two 3×3 squares once one has filled in *one* of the gaps in each. Later we shall see how properties of cosets may be used to cut down the work even further.

Cayley's theorem

Just as Pythagoras' theorem is one of the dominant results of Euclidean geometry, there are in Group Theory theorems of central importance. One of these (already referred to, see p. 78) is **Cayley's theorem.**

> **Any finite group (of order n) is isomorphic to a subgroup of the group S_n of permutations of n objects.**

We are familiar with groups of permutations, and have met some subsets of the complete set of $n!$ permutations of n objects, which themselves form a group. For example, in Q. 8.11 we found that the permutations (1234) 2143, 3412 and 4321 form a group on their own, and this must be a subgroup of the group of all the permutations of four digits, which is of order 24. **The full group of all $n!$ permutations of n symbols is called the *Symmetric Group*** and is denoted S_n or P_n, so that the full group of permutations of four letters, is denoted S_4 or P_4, and is of order 24. Again, on p. 78 we found six permutations of six letters which gave the group table which we there worked out. As we pointed out, this group is a subgroup of the group of order 720 of all the permutations of six objects which form the group S_6 or P_6.

The word 'isomorphic' will be discussed more thoroughly in Ch. 11. Briefly, it means 'having the same structure', or 'cast in the same mould' if you like. We can find, says Cayley's theorem, a set of n permutations which give the identical group table as that of any given group of order n. Thus, for example, the theorem says that the group of order 10 for the addition of the set {0, 1, 2, 3, ..., 9}, mod. 10 must be a subgroup of the group of permutations of ten objects. The latter, S_{10}, is an enormous group of order 10! – well over a million – so Cayley's theorem may not seem to be a very strong result! It is as if, in elementary plane geometry, there were a theorem saying 'In triangles ABC, PQR, if $AB > PQ$, $BC > QR$, $CA > RP$, then the area ABC > area PQR'. A rather timid result, you may well think.

Verification of Cayley's theorem

However, Cayley's theorem though not a powerful result, is an elegant and interesting one. Let us see how it can be verified. Consider the group {0, 1, 2, 3, 4}, + (mod. 5) with table 9.04. Now consider the *rows* of the

Table 9.04

+ (mod. 5)	0	1	2	3	4	perms.
0	0	1	2	3	4	e
1	1	2	3	4	0	x
2	2	3	4	0	1	y
3	3	4	0	1	2	z
4	4	0	1	2	3	w

table to be permutations of the set {0, 1, 2, 3, 4} (here we are using the Latin square property), and label these permutations as shown on the right. We shall show that the five permutations $\{e, x, y, z, w\}$ form a group with the same table as the group table which led to them.

For example, xz is 40123, i.e. w; z^2 is 12340, i.e. x, and so on. You may verify that the set of five permutations is closed, and combine to give table 9.05; this is the 'same' as the original table, with 0, 1, 2, 3, 4

Table 9.05

	e	x	y	z	w
e	e	x	y	z	w
x	x	y	z	w	e
y	y	z	w	e	x
z	z	w	e	x	y
w	w	e	x	y	z

(C_5)

substituted respectively by e, x, y, z, w. The identical *patterns* of the two tables can be seen instantly by observing the 'stripes' running from NE. to SW. This group of permutations is a subgroup of the full group of the permutations of five digits which is of order $5! = 120$.

Lest the above should seem a rather obvious example (seeing that the permutations were simply cyclic permutations, see pp. 111, 157), we would refer you again to the similar verification of Cayley's theorem in the case of the group table D_3 (of order 6) on p. 78. However, we take our illustration of the theorem one stage further by considering an even larger group, as defined by table 9.06. Think of each row as a permutation of the

Table 9.06

		first								
		e	*a*	*b*	*c*	*d*	*f*	*g*	*h*	perms.
	e	*e*	*a*	*b*	*c*	*d*	*f*	*g*	*h*	*e′*
	a	*a*	*e*	*h*	*g*	*f*	*d*	*c*	*b*	*a′*
	b	*b*	*f*	*e*	*h*	*g*	*a*	*d*	*c*	*b′*
second	*c*	*c*	*g*	*f*	*e*	*h*	*b*	*a*	*d*	*c′*
	d	*d*	*h*	*g*	*f*	*e*	*c*	*b*	*a*	*d′*
	f	*f*	*b*	*c*	*d*	*a*	*g*	*h*	*e*	*f′*
	g	*g*	*c*	*d*	*a*	*b*	*h*	*e*	*f*	*g′*
	h	*h*	*d*	*a*	*b*	*c*	*e*	*f*	*g*	*h′*

first row, and denote the permutations which change the first row into the second, third, ... eighth rows by e', a', b', c', d', f', g', h' (shown on the right); the identity permutation is e'.

We want to show that these eight permutations are a group which is isomorphic to the given group. This amounts to showing that any product in the original table (e.g. $cf = b$) is exactly imitated by the composition of the permutations, i.e. we expect $c'f' = b'$, and similar agreement in every possible case. Let us begin by verifying that in fact $c'f'$ does $= b'$. We reproduce only the rows of the table which concerns us. (See table 9.07.)

Table 9.07

perms.								
e′	*e*	*a*	*b*	*c*	*d*	*f*	*g*	*h*
f′	*f*	*b*	*c*	*d*	*a*	*g*	*h*	*e*
c′	*c*	*g*	*f*	*e*	*h*	*b*	*a*	*d*
c′f′ = *b′*	*b*	*f*	*e*	*h*	*g*	*a*	*d*	*c*

You should trace through (remember: *e* becomes *f*, *f* becomes *b*; *a* becomes *b*, *b* becomes *f*, etc.) and check that the permutation b' is in fact obtained.

Finally, after checking all such products, we observe that the eight perms. e', a', b', c', d', f', g', h' form a group which has the identical structure table as the original, and so is isomorphic to it. It is a subgroup of the full group of permutations of eight objects, of order 8! (= 40,320).

Rather than giving a general proof of Cayley's theorem, which is rather difficult,† we should prefer to satisfy you of the underlying reasons for its truth. It is not enough to *notice* that $b = cf$ and that $b' = c'f'$. Could this not be a fluke? The real question is *why* do we get this imitation of the behaviour of the elements of the original group by the respective permutations?

Let us again write out the rows we are concerned with (table 9.08). Now

Table 9.08

e'	e	a	b	c	d	f	g	h
f'	f	b	c	d	a	g	h	e
c'	c	g	f	e	h	b	a	d
b'	b	f	e	h	g	a	d	c

we single out a typical element, say g, and consider the effect on it of the permutations first f', then c'.

f' changes g into h, and we may express this: $f'(g) = h$ (here f' is a *function*).

Also, $h = fg$ in the group itself.

Hence, $h = f'(g) = fg$.

Similarly, $c'(h) = d = ch = cfg$,

i.e. $c'[f'(g)] = c'f'(g) = cfg = bg = b'(g)$.

Thus the effect of the permutation $c'f'$ upon the element g is the same as that of the permutation b'. By similar reasoning we can show that the effect of the composite permutation $c'f'$ upon *any* element of the group is the same as that of the single permutation b', so we may say $b' = c'f'$. The above argument may be illustrated diagrammatically with arrows representing the functions as shown in table 9.09. Similar considerations would apply to any pair of elements whatsoever.

For a general proof of Cayley's theorem, the above would have to be

† For a more straightforward, though less detailed description of Cayley's theorem, see Bell *Algebraic Structures* (Allen and Unwin), pp. 106–7. For rigorous proofs, see Birkoff and Maclane, *A Survey of Modern Algebra*, p. 123; and E. Patterson and D. Rutherford, *Elementary Abstract Algebra* (Oliver and Boyd), p. 53.

extended to apply to any pair of elements in any finite group. The difficulties in the general proof are difficulties of notation rather than of concept. If you have grasped the idea of the above demonstration, you will at least be able to see the lines on which a general proof would run.

Table 9.09

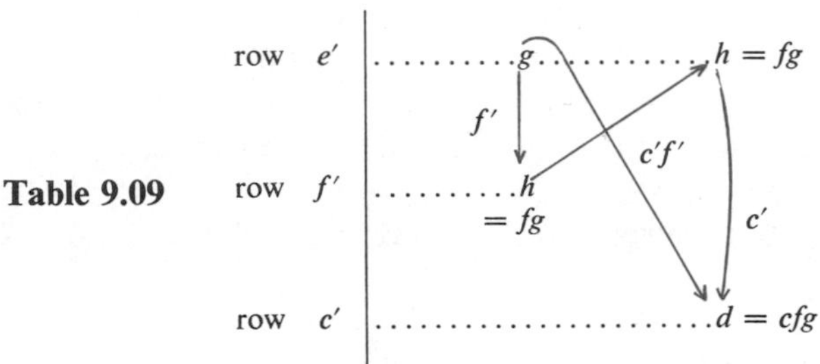

In table 9.01 we showed some Latin squares and asked whether or not they could be group tables. The answer centred around the problem of associativity. Strictly speaking, to show that a Latin square is a group table, we should have to check every possible triple product for associativity. A single failure would of course be sufficient to prove that it was not a group, but in either case, the labour would be enormous. Cayley's theorem gives us a short cut.

In the case of table (c), reprinted here as table 9.10, we label the

Table 9.10

Perm.					
e'	e	a	b	c	d
a'	a	b	d	e	c
b'	b	c	a	d	e
c'	c	d	e	a	b
d'	d	e	c	b	a

permutations of the rows e', a', b', c', d', as previously. Now $ab = d$, so we consider the permutation $a'b'$ In fact $a'b' = \begin{pmatrix} e & a & b & c & d \\ d & e & b & c & a \end{pmatrix}$ whereas $d' = \begin{pmatrix} e & a & b & c & d \\ d & e & c & b & a \end{pmatrix}$. So, the set of permutations $\{e', a', b', c', d'\}$ is not even *closed*, and certainly therefore does not form a group. Thus the Latin square table does not represent a group table. (When more familiar with groups you will know that a group of order 5 is bound to be cyclic, and that a cyclic group is always Abelian. Here, the fact that $ab = d$, $ba = c$ immediately dismisses the possibility of a group.)

Q. 9.07. Use the same method for investigating the other Latin squares shown on p. 90 and also on p. 44. Discover whether or not table 9.11 is a group table.

Table 9.11

e	a	b	c	d	f	g	h
a	g	h	f	e	b	d	c
b	f	g	e	h	d	c	a
c	h	e	g	f	a	b	d
d	e	f	h	g	c	a	b
f	c	a	d	b	g	h	e
g	d	c	b	a	h	e	f
h	b	d	a	c	e	f	g

★The 'Regular' representation of any finite groups by matrices

Since any finite group may be represented by a group of permutations (Cayley's theorem), and since permutations may be achieved by the use of permutation matrices (see p. 31), it is to be expected that any group of order n may be represented by a set of $n \times n$ permutation matrices. These matrices may very easily be constructed, and we illustrate by reference to the group D_3 whose table is displayed on p. 77, and below.

The first thing to do is to contravene the convention on p. 75 relating to the order in which the elements of the group are displayed along the border of the table! Suppose the elements across the top border are 1, a, b, c, ..., then instead of arranging them in the same order down the left border, we rearrange them in the order of corresponding inverses, thus: 1, a^{-1}, b^{-1}, c^{-1}, ... The effect of this is that the leading diagonal of the group table will now consist entirely of identity elements, as shown in the accompanying skeleton table.

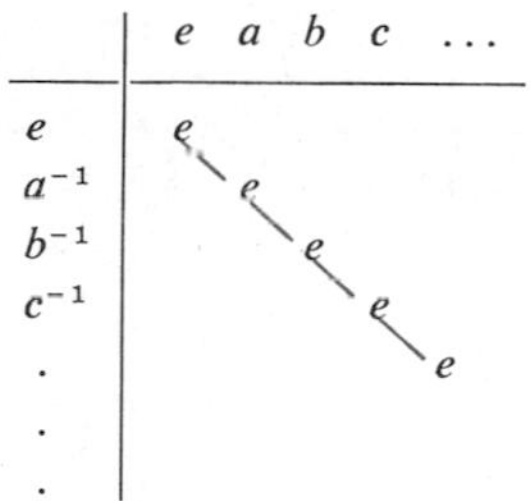

The table for D_3 becomes transformed as shown:

Table 9.12

(a)

	e	p	q	a	b	c
e	e	p	q	a	b	c
p	p	q	e	b	c	a
q	q	e	p	c	a	b
a	a	c	b	e	q	p
b	b	a	c	p	e	q
c	c	b	a	q	p	e

→

(b)

	e	p	q	a	b	c
e	e	p	q	a	b	c
q	q	e	p	c	a	b
p	p	q	e	b	c	a
a	a	c	b	e	q	p
b	b	a	c	p	e	q
c	c	b	a	q	p	e

The matrices are now constructed as follows:

$$e' = \begin{bmatrix} 1&0&0&0&0&0\\ 0&1&0&0&0&0\\ 0&0&1&0&0&0\\ 0&0&0&1&0&0\\ 0&0&0&0&1&0\\ 0&0&0&0&0&1 \end{bmatrix}; \qquad p' = \begin{bmatrix} \cdot&1&\cdot&\cdot&\cdot&\cdot\\ \cdot&\cdot&1&\cdot&\cdot&\cdot\\ 1&\cdot&\cdot&\cdot&\cdot&\cdot\\ \cdot&\cdot&\cdot&\cdot&\cdot&1\\ \cdot&\cdot&\cdot&1&\cdot&\cdot\\ \cdot&\cdot&\cdot&\cdot&1&\cdot \end{bmatrix};$$

$$q' = \begin{bmatrix} \cdot&\cdot&1&\cdot&\cdot&\cdot\\ 1&\cdot&\cdot&\cdot&\cdot&\cdot\\ \cdot&1&\cdot&\cdot&\cdot&\cdot\\ \cdot&\cdot&\cdot&\cdot&1&\cdot\\ \cdot&\cdot&\cdot&\cdot&\cdot&1\\ \cdot&\cdot&\cdot&1&\cdot&\cdot \end{bmatrix}; \qquad a' = \begin{bmatrix} \cdot&\cdot&\cdot&1&\cdot&\cdot\\ \cdot&\cdot&\cdot&\cdot&1&\cdot\\ \cdot&\cdot&\cdot&\cdot&\cdot&1\\ 1&\cdot&\cdot&\cdot&\cdot&\cdot\\ \cdot&1&\cdot&\cdot&\cdot&\cdot\\ \cdot&\cdot&1&\cdot&\cdot&\cdot \end{bmatrix};$$

$$b' = \begin{bmatrix} \cdot&\cdot&\cdot&\cdot&1&\cdot\\ \cdot&\cdot&\cdot&\cdot&\cdot&1\\ \cdot&\cdot&\cdot&1&\cdot&\cdot\\ \cdot&\cdot&1&\cdot&\cdot&\cdot\\ 1&\cdot&\cdot&\cdot&\cdot&\cdot\\ \cdot&1&\cdot&\cdot&\cdot&\cdot \end{bmatrix}; \qquad c' = \begin{bmatrix} \cdot&\cdot&\cdot&\cdot&\cdot&1\\ \cdot&\cdot&\cdot&1&\cdot&\cdot\\ \cdot&\cdot&\cdot&\cdot&1&\cdot\\ \cdot&1&\cdot&\cdot&\cdot&\cdot\\ \cdot&\cdot&1&\cdot&\cdot&\cdot\\ 1&\cdot&\cdot&\cdot&\cdot&\cdot \end{bmatrix}.$$

You will see that the matrix for q' (for example) is built up by putting 1's† wherever q occurs in the transformed table (b) above; and 0's

$$p' = \begin{bmatrix} \cdot&p&\cdot&\cdot&\cdot&\cdot\\ \cdot&\cdot&p&\cdot&\cdot&\cdot\\ p&\cdot&\cdot&\cdot&\cdot&\cdot\\ \cdot&\cdot&\cdot&\cdot&\cdot&p\\ \cdot&\cdot&\cdot&p&\cdot&\cdot\\ \cdot&\cdot&\cdot&\cdot&p&\cdot \end{bmatrix}.$$

We should then have

$$b'q' = \begin{bmatrix} \cdot&\cdot&\cdot&\cdot&b&\cdot\\ \cdot&\cdot&\cdot&\cdot&\cdot&b\\ \cdot&\cdot&\cdot&b&\cdot&\cdot\\ \cdot&\cdot&b&\cdot&\cdot&\cdot\\ b&\cdot&\cdot&\cdot&\cdot&\cdot\\ \cdot&b&\cdot&\cdot&\cdot&\cdot \end{bmatrix} \begin{bmatrix} \cdot&\cdot&q&\cdot&\cdot&\cdot\\ q&\cdot&\cdot&\cdot&\cdot&\cdot\\ \cdot&q&\cdot&\cdot&\cdot&\cdot\\ \cdot&\cdot&\cdot&\cdot&q&\cdot\\ \cdot&\cdot&\cdot&\cdot&\cdot&q\\ \cdot&\cdot&\cdot&q&\cdot&\cdot \end{bmatrix}$$

$$= \begin{bmatrix} \cdot&\cdot&\cdot&\cdot&\cdot&bq\\ \cdot&\cdot&\cdot&bq&\cdot&\cdot\\ \cdot&\cdot&\cdot&\cdot&bq&\cdot\\ \cdot&bq&\cdot&\cdot&\cdot&\cdot\\ \cdot&\cdot&bq&\cdot&\cdot&\cdot\\ bq&\cdot&\cdot&\cdot&\cdot&\cdot \end{bmatrix} = \begin{bmatrix} \cdot&\cdot&\cdot&\cdot&\cdot&c\\ \cdot&\cdot&\cdot&c&\cdot&\cdot\\ \cdot&\cdot&\cdot&\cdot&c&\cdot\\ \cdot&c&\cdot&\cdot&\cdot&\cdot\\ \cdot&\cdot&c&\cdot&\cdot&\cdot\\ c&\cdot&\cdot&\cdot&\cdot&\cdot \end{bmatrix} = c'.$$

† Instead of 1's, these matrices could have the appropriate element of the original group, e.g.

elsewhere (we have put dots instead of 0's so that the 1's stand out more).

We may now demonstrate the isomorphism between the original group and the set of six matrices. For example, the product $bq = c$ is imitated by the matrices, for you will see that matrix multiplication gives:

$$b'q' = \begin{bmatrix} \cdot & \cdot & \cdot & \cdot & 1 & \cdot \\ \cdot & \cdot & \cdot & \cdot & \cdot & 1 \\ \cdot & \cdot & \cdot & 1 & \cdot & \cdot \\ \cdot & \cdot & 1 & \cdot & \cdot & \cdot \\ 1 & \cdot & \cdot & \cdot & \cdot & \cdot \\ \cdot & 1 & \cdot & \cdot & \cdot & \cdot \end{bmatrix} \begin{bmatrix} \cdot & \cdot & 1 & \cdot & \cdot & \cdot \\ 1 & \cdot & \cdot & \cdot & \cdot & \cdot \\ \cdot & 1 & \cdot & \cdot & \cdot & \cdot \\ \cdot & \cdot & \cdot & \cdot & 1 & \cdot \\ \cdot & \cdot & \cdot & \cdot & \cdot & 1 \\ \cdot & \cdot & \cdot & 1 & \cdot & \cdot \end{bmatrix} = \begin{bmatrix} \cdot & \cdot & \cdot & \cdot & \cdot & 1 \\ \cdot & \cdot & \cdot & 1 & \cdot & \cdot \\ \cdot & \cdot & \cdot & \cdot & 1 & \cdot \\ \cdot & 1 & \cdot & \cdot & \cdot & \cdot \\ \cdot & \cdot & 1 & \cdot & \cdot & \cdot \\ 1 & \cdot & \cdot & \cdot & \cdot & \cdot \end{bmatrix} = c'$$

and you may verify that structure is preserved in respect of other products.

Why does it work? Evidently the answer to this question is closely bound up with Cayley's theorem, and we may see in the present case that e', p', q', a', b', c' are permutation matrices which *change the first column* of the original structure table 9.12 into the columns headed e, p, q, a, b, c, respectively. For example,

$$c' \begin{bmatrix} e \\ p \\ q \\ a \\ b \\ c \end{bmatrix} = \begin{bmatrix} \cdot & \cdot & \cdot & \cdot & \cdot & 1 \\ \cdot & \cdot & \cdot & 1 & \cdot & \cdot \\ \cdot & \cdot & \cdot & \cdot & 1 & \cdot \\ \cdot & 1 & \cdot & \cdot & \cdot & \cdot \\ \cdot & \cdot & 1 & \cdot & \cdot & \cdot \\ 1 & \cdot & \cdot & \cdot & \cdot & \cdot \end{bmatrix} \begin{bmatrix} e \\ p \\ q \\ a \\ b \\ c \end{bmatrix} = \begin{bmatrix} c \\ a \\ b \\ p \\ q \\ e \end{bmatrix},$$

and this is indeed the final column in table 9.12 above.

★**Q. 9.08.** What would you have to do to obtain a set of matrices which, operating on the *rows* of the original table would produce its successive *rows*?

★**Q. 9.09.** Obtain matrix representations for finite groups whose tables are given in the text, e.g. C_4, D_2, C_6, etc.

Doing algebra in a group

Certain manipulations on the elements of a group may be performed, and we shall illustrate by examples. While performing such 'algebra', we must bear in mind (see warnings pp. 42, 90, 106), that in general multiplication is not commutative, so that an expression such as *pap* may *not* be reduced to p^2a.

Example 1

In a certain group, r and a are elements which satisfy the relations $a^2 = r^2$; $r^4 = 1$, $ar = r^3a$, where 1 is the identity.

Prove: that $ar^2 = r^2a$ and that $ar^3 = ra$, and that both these elements are of period 4†. How many elements are there in the group? Construct the group table. Is there an element which commutes with all the others?

Solution: $a^2 = r^2 \ldots (1)$; $r^4 = 1 \ldots (2)$; $ar = r^3a \ldots (3)$.

From (1), $ar^2 = aa^2 = a^3 = a^2a = r^2a \ldots (4)$.

Similarly, $a^2r = ra^2 \ldots (5)$.

Now, $ar = r^3a$ (3) $\Rightarrow a^2r = ar^3a \Rightarrow ra^2 = ar^3a$ (by (5))

$\Rightarrow ra = ar^3$ (cancellation on the right)...(6).

Note that we have not used (2); it is however easier to prove (6) when $r^4 = 1$ is used – try this.

Q. 9.10. Show that the relation $r^4 = 1$ may be deduced from $ar = r^3a$ and $a^2 = r^2$ but that neither of these may be deduced from the remaining two.

By (4), $(ar^2)^2 = ar^2.r^2a = ar^4a = a1a$ (by (2)) $= a^2$.

Hence, $(ar^2)^4 = (a^2)^2 = a^4 = 1$, so ar^2 is of period 4.

Again, $(ar)^2 = ar.ar = arr^3a$ (by (3)) $= ar^4a = a1a = a^2 \Rightarrow$
$(ar)^4 = (a^2)^2 = 1$;

and $(ra)^2 = ra.ra = ar^3ra$ (by (6)) $= a^2 = (ra)^4 = 1$.

Moreover, $(r^3)^2 = r^6 = r^2$ (by (2)) so that $(r^3)^4 = 1$.

Hence the elements r, r^3, a, ar $(= r^3a)$, ar^2 $(= r^2a)$, ar^3 $(= ra)$ are all of period 4.

(You are left to show that none of the cubes of these elements is 1.)

In addition, we have the elements 1 and r^2, the latter of period 2. There can be no other elements than these eight, and we can show they form a closed system: any conceivable combination of a's and r's is bound to reduce to one of the eight terms. For example, $r^3ar = r^3(ar) = r^3(r^3a) = r^6a = r^2a = ar^2$. Abbreviating the composite elements as follows: $ar = b$, $ar^3 = c$, $ar^2 = d$, we obtain group table 9.13. The table shows that the element (other than 1) which commutes with all the others is r^2.

Example 2

In Ex. 5.07 we considered the binary operation on ordered triples defined:

$$(x_1, y_1, z_1)*(x_2, y_2, z_2) = (x_1 + x_2 + y_2z_1, y_1 + y_2, z_1 + z_2),$$

and showed that it is associative. We note that (0, 0, 0) is both a left and a right identity, and that the operation is not commutative.

Q. 9.11. What is the inverse of (x, y, z)? Find the cube of (x, y, z).

† See Ch. 10. Briefly, 'x is of period 4' means $x^4 = 1$, while x, x^2 and x^3 are not 1.

Table 9.13

first

	1	r	r^2	r^3	a	ar r^3a	ar^3 ra	ar^2 r^2a	
second 1	1	r	r^2	r^3	a	b	c	d	
r	r	r^2	r^3	1	c	a	d	b	
r^2	r^2	r^3	1	r	d	c	b	a	
r^3	r^3	1	r	r^2	b	d	a	c	
a	a	b	d	c	r^2	r^3	r	1	
$r^3a = ar$	b	d	c	a	r	r^2	1	r^3	
$ra = ar^3$	c	a	b	d	r^3	1	r^2	r	
$r^2a = ar^2$	d	c	a	b	1	r	r^3	r^2	(Q_4)

Now suppose that x, y and z are taken from the set $\{0, 1, 2\}$ (so that there are twenty-seven possible triads), and that addition is modulo 3, so that, for example,

$$(1, 1, 2)*(2, 2, 0) = (2 + 1 + 4, 2 + 1, 0 + 2) \text{ (mod. 3)} = (1, 0, 2).$$

Prove: (1) that every element is of period 3,

(2) letting $a = (1, 0, 0)$, $b = (0, 1, 0)$, $c = (0, 0, 1)$, show that a commutes every element of the group, and that every element may be expressed in terms of a, b and c.†

Solution: (1) Since $(x, y, z)^3 = (3x + 3yz, 3y, 3z) = (0, 0, 0)$ (mod. 3), it follows that (x, y, z) is of period 3 for each element except $(0, 0, 0)$.

$$\text{(2)}\quad (x, y, z)*(1, 0, 0) = (x + 1 + 0.z, y, z) = (x + 1, y, z)$$
$$(1, 0, 0)*(x, y, z) = (1 + x + y.0, y, z) = (x + 1, y, z)$$

so a commutes with each element of the group. Note that if

$$(x_1, y_1, z_1)*(x_2, y_2, z_2) = (x_2, y_2, z_2)*(x_1, y_1, z_1)$$

then $x_1 + x_2 + y_2z_1 = x_2 + x_1 + y_1z_2 \Rightarrow y_2z_1 = y_1z_2$, and this is certainly true when $y_1 = z_1 = 0$, or when $y_2 = z_2 = 0$.

(3) We have $a = (1, 0, 0)$, $a^2 = (2, 0, 0)$, $ab = ba = (1, 1, 0)$
$b = (0, 1, 0)$, $b^2 = (0, 2, 0)$, $cb = (1, 1, 1)$, $bc = (0, 1, 1)$
$c = (0, 0, 1)$, $c^2 = (0, 0, 2)$, $ac = ca = (1, 0, 1)$.

† But see the brief discussion in the Preface, pp. xii–xiii.

As an example of some of the remaining elements, we have

$$(0, 2, 1) = (0, 0, 1)*(1, 2, 0) = (0, 0, 1)*(0, 2, 0)*(1, 0, 0)$$
$$= cb^2a\,(= cab^2 = acb^2 \text{ since } a \text{ commutes}).$$
$$(1, 0, 1) = (0, 2, 0)*(1, 1, 1) = (0, 2, 0)*(0, 0, 1)*(0, 1, 0) = b^2cb.$$

You should complete the work by obtaining the remaining 14 elements in terms of a, b and c, a set of generators of the group (see Ch. 15).

Q. 9.12. What can you say about those elements for which $z = 0$? Are those elements for which $x = 0$ a subgroup?

Q. 9.13. If eight ordered triples are taken with each x, y and z either 0 or 1, and the operation is as defined above, but where addition is modulo 2, investigate the group so obtained.

Q. 9.14. Suppose the law of composition were modified so that terms z_2x_1 and/or x_2y_1 were added respectively to the y and z components, investigate.

★*Example 3*

A group has elements a, b, c which satisfy the relations $a^3 = b^3 = c^3 = 1$; $ab = ba$; $ac = ca$; $cb = bca$. Find out all you can about the group.

Solution: $a^3 = b^3 = c^3 = 1$ (1)

$ab = ba,\ ac = ca$ (2)

$cb = bca$ (3)

Equation (1) says that a, b and c are each of period 3. Equation (2) says that a commutes with b and c, and so with any element built from them. Equation (3) enables any 'word' or product in which c precedes b to be rewritten with c following b. Any conceivable combination of a's, b's and c's may therefore be expressed in the form $a^xb^yc^z$ (which we shall call 'standard form'), for the a's may be 'brought to the front' by (2), and the b's may be brought to the left of the c's by repeated applications of (3).

Now $cb = bca$ (by (3)) $= abc$ (by (2))

$$c^2b = c(cb) = c(bca) = acbc \text{ (by (2))} = a(cb)c = a(bca)c = a^2bc^2$$
$$cb^2 = (cb)b = (bca)b = abcb = ab(cb) = ab(bca) = a^2b^2c$$
$$c^2b^2 = (c^2b)b = (a^2bc^2)b \text{ (proved)} = a^2b(c^2b) = a^2b(a^2bc^2) \text{ (proved)}$$
$$= a^4bbc^2 = ab^2c^2 \text{ (by (1))}.$$

These results may be summarised by the formula $c^zb^y = a^pb^yc^z$,

where $p = 1$ when $y = 1, z = 1$, or when $y = 2, z = 2$

$p = 2$ when $y = 1, z = 2$, or when $y = 2, z = 1$.

Thus, $p = yz$ (mod. 3), so that $a^p = a^{yz}$ (since $a^3 = 1$).

Hence, $c^z b^y = a^{yz} b^y c^z$ (4).

Now consider the product of two general terms. First express them in standard form: $a^{x_1} b^{y_1} c^{z_1}$ and $a^{x_2} b^{y_2} c^{z_2}$, where the indices are drawn from {0, 1, 2}.

Then the product:

$$= a^{x_1+x_2} b^{y_1} c^{z_1} b^{y_2} c^{z_2} \quad \text{(by 2)}$$

$$= a^{x_1+x_2} b^{y_1}(c^{z_1} b^{y_2}) c^{z_2} = a^{x_1+x_2} b^{y_1}(a^{y_2 z_1} b^{y_2} c^{z_1}) c^{z_2} \quad \text{(by 4)}$$

$$= a^{x_1+x_2+y_2 z_1} b^{y_1+y_2} c^{z_1+z_2} \quad \text{(by 2).}$$

We see, therefore, that the indices are combined exactly as for the group of example 2, and an isomorphism† is established between these groups of order 27.

On p. 143 you will read that we shall claim that for $n < 27$, groups which agree on the number of elements of each period are isomorphic.† But when we get to $n = 27$, this rule for isomorphism breaks down. The above group has one element of period 1 and 26 of period 3, and unfortunately there is a totally different group, $C_3 \times C_3 \times C_3$ (see Ch. 16) of order 27 which also contains 26 elements of period 3.

Example 4

Prove, by arriving at a contradiction, that there does not exist‡ a group in which $r^5 = 1 = a^2$ and $ar^3 = ra$.

Solution: $ra = ar^3 \Rightarrow ra^2 = ar^3a \Rightarrow r = ar^3a$ (since $a^2 = 1$)

$$\Rightarrow r^2 = (ar^3a)(ar^3a) = ar^3a^2r^3a = ar^3 1 r^3 a = ar^6a = ara$$

$$\Rightarrow ar^2 = a(ara) = a^2ra = 1ra = ra \quad (\text{since } r^5 = 1).$$

But $ra = ar^3$, and so $ar^2 = ar^3 \Rightarrow 1 = r$ (by right cancellation), and this is a contradiction, since r must be an element distinct from the identity. Alternatively, we might proceed thus:

$$ar^3 = ra \Rightarrow ar^3r^2 = rar^2 \Rightarrow ar^5 = rar^2 \Rightarrow a = rar^2$$

$$a = rar^2 \Rightarrow a = r(rar^2)r^2 = r^2ar^4 = r^2(rar^2)r^4 = r^3ar^6 = r^3ar.$$

Continue the process and complete the argument.

Example 5

p and q are elements of a group such that $p = qpq$, $q = pqp$.

† See Ch. 11.

‡ Strictly these relations define the trivial group consisting of a single element. We need to specify that a and r are to be *distinct* in order to be able to state that there does not exist a group in these cases. See also Ch. 15, p. 245.

Prove: that $p^4 = 1 = q^4$.

Solution: $p^2 = p(qpq) = (pq)^2$

$$q^2 = (pqp)q = (pq)^2. \quad \text{Hence, } p^2 = q^2.$$

Now, $p = qpq = q(qpq)q = q^2pq^2 = p^2pp^2$ (since $p^2 = q^2$, proved) $= p^5$. Hence, by the cancellation law, $p^4 = 1$, and similarly $q^4 = 1$.

Q. 9.15. Given that $a^2 = b^2 = p^3 = 1$, $ap = pb$ and $bp = pab$, prove that $ab = ba$.
A group has $a^2 = 1$, $p^3 = 1$, $(pa)^3 = 1$. Show that there are elements a, pap^2 and p^2ap of period 2, and six more elements of period 3 other than p and pa.

Q. 9.16. Using equations (1), (2) and (3) on p. 102, express in standard form ($a^xb^yc^z$) the following: bc^2b; b^2cb^2; cb^2c^2b; $b^2cbc^2bc^2b^2$; $b^2cbcb^2c^2b^2c$.

Q. 9.17. Show, by arriving at a contradiction, that a group cannot exist† for which $p^3 = 1$, $r^4 = 1$, $pr^2 = rp^2$.

Q. 9.18. If $bc = cba$, $ab = ba$ and $ac = ca$, prove that $b^rc^s = a^{rs}c^sb^r$ (compare pp. 102–3).

EXERCISES 9

1 *Prove:* that $xy = yx \Rightarrow x^my^n = y^nx^m$ for all integers m, n.

2 **Show that, instead of the requirements N and I for a group (see p. 73), we might have substituted the single requirement: for any given $a, b \in G$ there exists a unique x satisfying $a * x = b$, and also a unique y satisfying $y * a = b$.**

3 'Solve' to obtain x in terms of a, b and c:

(a) $axb = c$; (b) $xab = c$; (c) $ax^2 = b$ and $x^3 = 1$;
(d) $x^{-1} = abc$; (e) $x^5 = 1$ and $x^3 = a$; (f) $xaxaba = xbc$.

4 If $xy^2 = y^3x$, prove that $xy \neq yx$.

5 If $rqp = q$ and $prq = r$, prove that $rqr = qrq$.

6 If $a^2 = b^2 = (ab)^2 = 1$, prove that $ab = ba$.

7 Is it true that a Latin square may be regarded as a group table so long as every triple product is associative?

8 In the Latin square shown (see table 9.14), give three independent reasons why the table is not a group table in spite of being a Latin square. (Use Lagrange's theorem in only one answer.)

Table 9.14

1	a	b	c	d
a	c	d	b	1
b	1	c	d	a
c	d	a	1	b
d	b	1	a	c

9 Obtain a set of matrices to represent various groups whose tables are displayed elsewhere in the text, by using the technique described on pp. 97–9.

† See footnote, p. 103.

10 Given that $a^2 = 1 = r^4$ and also that $ar = r^4a$, prove that $a = rar$, that $ar^3 = ra$ and that $r^nar^n = a$ ($n \in \mathbf{Z}$). Is there an element which commutes with all the others? How many elements has this group? Identify its structure. Repeat by deducing other relations from $a^2 = 1 = r^5$, $ar = r^4a$.

11 If $r^4 = 1$, $a^2 = 1$, and $r^3a = (ar)^2$, prove that ra is of period 3.
Interpret the above when a is the permutation $\begin{pmatrix} 1 & 2 & 3 & 4 \\ 2 & 1 & 3 & 4 \end{pmatrix}$ and r is $\begin{pmatrix} 1 & 2 & 3 & 4 \\ 2 & 3 & 4 & 1 \end{pmatrix}$.

12 Show that those permutations of the nine letters ABC, PQR, UVW which move each triad round cyclically, but keep the three triads separate, e.g. $\begin{pmatrix} \text{ABC} & \text{PQR} & \text{UVW} \\ \text{BCA} & \text{PQR} & \text{WUV} \end{pmatrix}$ number 26 besides the identity, and form an Abelian group of order 27.

13 If $a^2 = 1$, $axa = x^3$, prove that $x^8 = 1$, and generalise.

14 If $p^3 = 1$, $pxp^{-1} = x^4$, prove x has finite period, and find it.

15 If $rs^2 = s^3r$ (1) and $sr^2 = r^3s$ (2) show from (1) that $r^2 = s^3rsrs^{-2}$ and $r^3 = s^3rsrsrs^{-2}$; and by substituting in (2) go on to prove that $(sr)^3 = (rs)^3$. By using the interchangeability of r and s, deduce that $rs^2r = 1$, and proceed to establish that $rs = 1$, and finally that $r = s = 1$!

16 If $a^2 = 1$, $r^6 = 1$, $ar = r^4a$, find the period of ar, and express its inverse in the form a^mr^n and also in the form r^pa^q, where m, n, p, q are integers.

17 If $p^3 = 1 = q^4$, $qp = pq^3$, prove that $q^2p = pq^2$ and that $q^3p = pq$. Also that $qpq = p$ and that $q^{-1}p^{-1}q^{-1} = p^2$. Find the period of pq, of qp and of pq^2.

18 In a certain group, $r^5 = p^4 = 1$, and $rp = pr^3$. Prove that $pr = r^2p$ and $rp^3 = p^3r^2$.

19 Given that $p^3 = 1$ and $pxp^{-1} = x^2$, prove that $x^7 = 1$. (*Hint*: begin by proving that $x^4 = p^2xp$, and then that $x^8 = x$.)

10 *Period (or order) of an element: permutations and cycles*

In the group of order 6 studies on pages 76 and 77, we observed that $a^2 = e$, $b^2 = e$, $c^2 = e$, $p^3 = e$, $q^3 = e$. We say that a, b and c are of *period* (or *order*) 2, and p and q are of *period* (or *order*) 3. The word 'order' is also used, of course, to give the *number* of elements in a finite group. This is not entirely a separate concept since, as we shall see, an element of order 4 (say) does in fact generate a subgroup of that order. However, on the whole we shall prefer to use the word 'period', and so we should describe p as being 'of period 3'. Of course, $p^6 = p^3p^3 = ee = e$, but we do not say p has period 6; **the period of an element x is in fact the *smallest* positive integer m such that $x^m = e$; when no such integer exists, x is said to have 'infinite period'.**

Q. 10.01. Prove that in this case, $1, x, x^2, \ldots x^{m-2}, x^{m-1}$ are all distinct. (Use the cancellation law.)

We may use 'negative indices' when multiplicative notation is employed: for example, x^{-3} will be taken to mean $(x^{-1})^3$ (by definition), and in general, $x^{-n} = (x^{-1})^n$.
The usual Laws of indices follow immediately, for example,

$$\begin{aligned} x^4x^{-7} &= x^4(x^{-1})^7 = x^3(xx^{-1})(x^{-1})^6 \quad \text{(associative law)} \\ &= x^3(x^{-1})^6 \quad (\text{since } x.x^{-1} = e) \\ &= x^2(xx^{-1})(x^{-1})^5 = x^2(x^{-1})^5 = x(xx^{-1})(x^{-1})^4 \\ &= x(x^{-1})^4 = x.x^{-1}(x^{-1})^3 = e(x^{-1})^3 = (x^{-1})^3 = x^{-3}. \end{aligned}$$

In general, for any integers k, l, positive or negative,

$$x^kx^l = x^{k+l}, \quad \text{also} \quad x^0 = e.$$

Q. 10.02. Prove that $(x^k)^l = x^{kl}$ when k and l are integers. But note that $(xy)^k$ is not in general equal to x^ky^k. State under what conditions $(xy)^k$ does equal x^ky^k.

An immediate consequence of the foregoing is that the inverse of x^k is x^{-k} (k being an integer), for we have $x^kx^{-k} = x^0 = e$, and it follows from this that:

if x is an element of period m, then x^{-1} is also of period m,

since $x^m = e \Rightarrow x^{-m} = e \Rightarrow (x^{-1})^m = e$, and there is no smaller integer than m for which this is true. Furthermore, if x is of period m, then $x^m = e$, and it follows that $(x^m)^k = x^{mk} = e$ for all integral values of k. No other 'powers' of x will give the identity save positive and negative integral multiples of m. For if $x^r = e$, where r is some integer, we know that, for given m, r may be expressed in the form $r = km + s$, where the integers k and s are uniquely determined, with $0 \leqslant s < m$. Then:

$$x^{km+s} = e \Rightarrow x^{km}x^s = e \Rightarrow (x^m)^k x^s = e \Rightarrow x^s = e, \text{ since } x^m = e.$$

But m is the smallest integer to satisfy $x^m = e$, so that $s = 0$. Thus an integer k may be found so that $r = km$.

An important theorem may be deduced from the fact that in any group x and x^{-1} have the same period: in any group

the number of elements of period 3, 4, 5 and upwards is always even.

For they occur in inverse pairs, and can only coincide ($x = x^{-1}$) in the case when x is of period 2.

Q. 10.03. Interpret this result in terms of the structure table of the group.

Finding the period of an element from the group table

If we wish to find the period of an element x of a group, it is simply a matter of finding x^2, x^3, x^4, ... and going on till we get the identity. When the group table is available, this is an easy task, and we illustrate the method in the case of a rather larger group, of order 12.

Suppose in the top row, the elements are displayed in the order: $e, a, b, c, d, f, g, h, j, k, p, q$ and we require the period of g. We need to show below only the g row:

Table 10.01

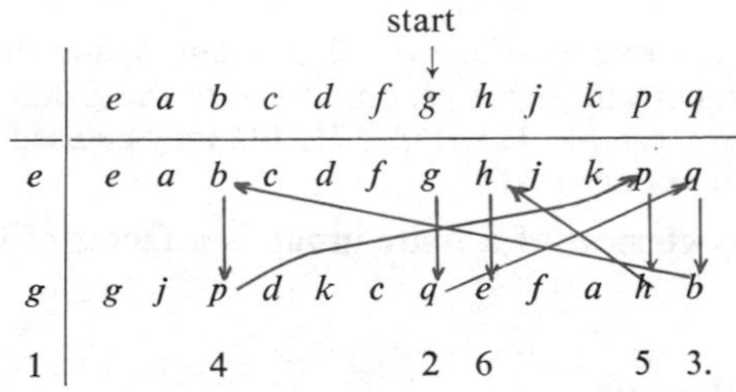

	e	a	b	c	d	f	g (start)	h	j	k	p	q
e	e	a	b	c	d	f	g	h	j	k	p	q
g	g	j	p	d	k	c	q	e	f	a	h	b
1			4				2	6			5	3.

We follow the arrows, which point to successive values of g^2, g^3, g^4, etc.:

$$g^2 = q; \quad g^3 = gq = b; \quad g^4 = gb = p; \quad g^5 = gp = h; \quad g^6 = gh = e.$$

Hence the period of g is 6. You do not need to write all these down, or even to bother about the intermediate values q, b, p, h. All you need to do is to *count* as you follow the arrows, and the count is indicated at each point below the row.

We proved that the period of an element is equal to that of its inverse. Note in the row given that $gh = e$, so that the inverse of g is h. Check from row h, given below, that the period of h is also 6:

e	e	a	b	c	d	f	g	h	j	k	p	q
h	h	k	q	f	c	j	e	p	a	d	b	g.

The values of h^2, h^3, h^4 and h^5 are p, b, q, g. What is the reason that they are the same as in the case of g, only in the reverse order?

Q. 10.04. Find the period of each element in group table 10.02. In this question,

Table 10.02

	e	a	b	c	d	f	g	h	j	k	l	m
e	e	a	b	c	d	f	g	h	j	k	l	m
a	a	k	l	d	f	g	h	j	c	b	m	e
b	b	l	e	g	h	j	c	d	f	m	a	k
c	c	j	g	b	k	a	e	m	l	h	f	d
d	d	c	h	l	b	k	a	e	m	j	g	f
f	f	d	j	m	l	b	k	a	e	c	h	g
g	g	f	c	e	m	l	b	k	a	d	j	h
h	h	g	d	a	e	m	l	b	k	f	c	j
j	j	h	f	k	a	e	m	l	b	g	d	c
k	k	b	m	f	g	h	j	c	d	l	e	a
l	l	m	a	h	j	c	d	f	g	e	k	b
m	m	e	k	j	c	d	f	g	h	a	b	l (Q_6)

you will have noticed that there are elements of periods 2, 3, 4 and 6, and that these are all factors of 12. Experiment with other groups (some of the group tables which appear later in the text, e.g. pp. 123, 158, 171, 192 may be used) to confirm a result which we shall prove later that:

The order (or period) of any element of a finite group is a factor of the order of the whole group.

Subgroups generated by single elements

Now take an element of period 6 from the above group – you should have found, for example, that a is one of the elements of period 6. In fact we have:

$$a^2 = k; \quad a^3 = b; \quad a^4 = l; \quad a^5 = m; \quad a^6 = e.$$

Now the elements $\{e, a, k, b, l, m\}$ evidently satisfy all the requirements of a group in their own right, for the multiplication table may be written

Table 10.03

	e	a	a^2	a^3	a^4	a^5
e	e	a	a^2	a^3	a^4	a^5
a	a	a^2	a^3	a^4	a^5	e
a^2	a^2	a^3	a^4	a^5	e	a
a^3	a^3	a^4	a^5	e	a	a^2
a^4	a^4	a^5	e	a	a^2	a^3
a^5	a^5	e	a	a^2	a^3	a^4

(C_6)

or as

	e	a	k	b	l	m
e	e	a	k	b	l	m
a	a	k	b	l	m	e
k	k	b	l	m	e	a
b	b	l	m	e	a	k
l	l	m	e	a	k	b
m	m	e	a	k	b	l

We have therefore a *subgroup* of the group of order 12. A group of this type, every element of which can be written in terms of a single element a, is called a *cyclic group*. The cyclic group of order n is denoted C_n. Both subgroups and cyclic groups are important enough to deserve chapters to themselves (see Chs. 14 and 12).

It is sometimes convenient to redraw the 12×12 group table so as to make this subgroup stand out, by rearranging the order of the elements in the first row so that the subgroup appears in the top left-hand corner, as in table 10.04.

Table 10.04

	e	a	k	b	l	m						
	e	a	a^2	a^3	a^4	a^5	c	d	f	g	h	j
e	e	a	a^2	a^3	a^4	a^5	c	d	f	g	h	j
a	a	a^2	a^3	a^4	a^5	e	d	f	g	h	j	c
a^2	a^2	a^3	a^4	a^5	e	a	f	g	h	j	c	d
a^3	a^3	a^4	a^5	e	a	a^2	g	h	j	c	d	f
a^4	a^4	a^5	e	a	a^2	a^3	h	j	c	d	f	g
a^5	a^5	e	a	a^2	a^3	a^4	j	c	d	f	g	h
c	c	j	h	g	f	d	a^3	a^2	a	e	a^5	a^4
d	d	c	j	h	g	f	a^4	a^3	a^2	a	e	a^5
f	f	d	c	j	h	g	a^5	a^4	a^3	a^2	a	e
g	g	f	d	c	j	h	e	a^5	a^4	a^3	a^2	a
h	h	g	f	d	c	j	a	e	a^5	a^4	a^3	a^2
j	j	h	g	f	d	c	a^2	a	e	a^5	a^4	a^3

(Q_6)

Again, you will have found that there were several elements of period 4, e.g. h; for $h^2 = b$, $h^3 = d$, $h^4 = e$. Therefore the elements $\{e, h, b, d\}$ form a cyclic subgroup of order 4 (see table 10.05).

Table 10.05

	e	h	h^2	h^3
e	e	h	h^2	h^3
h	h	h^2	h^3	e
h^2	h^2	h^3	e	h
h^3	h^3	e	h	h^2

or

	e	h	b	d
e	e	h	b	d
h	h	b	d	e
b	b	d	e	h
d	d	e	h	b

(C_4).

Q. 10.05. You should write out the group tables of the subgroups 'generated' by each of the other elements in turn. Can you find subgroups of the group of order 6 generated by a? If not, these will appear in the course of your work in the above exercise.

We shall see later that a group may be determined uniquely by what are called 'defining relations', these being certain essential relations between a few of the elements. Thus, in the case of the above group, it is only necessary to know that a and b satisfy the defining relations:

$$a^6 = e = b^4; \qquad a^3 = (ab)^2 = b^2$$

to be able to build up the whole table. The other elements of the group will consist of various combinations of a's and b's, among which will be:

$$\{a, a, a^2, a^3, a^4, a^5, b, ab, \ldots\}.$$

Another element would be, for example, a^2b. This cannot be one of those already listed, for:

$$a^2b = e \Rightarrow a^4a^2b = a^4 \Rightarrow a^6b = a^4 \Rightarrow b = a^4$$

$$\left.\begin{aligned} a^2b &= a \Rightarrow ab = e \\ a^2b &= a^2 \Rightarrow b = e \\ a^2b &= a^3 \Rightarrow b = a \\ a^2b &= a^4 \Rightarrow b = a^2 \\ a^2b &= a^5 \Rightarrow b = a^3 \\ a^2b &= b \Rightarrow a^2 = e \\ a^2b &= ab \Rightarrow a = e \end{aligned}\right\} \quad \text{(cancellation law each time.)}$$

In each case we arrive at a contradiction.

Suppose we wish to know the period of a^2b. This can be found *from the defining relations*, without the labour of constructing the whole table:

The defining relations were:

$$\begin{aligned} a^6 &= e & (1)\\ b^4 &= e & (2)\\ a^3 &= (ab)^2 & (3)\\ b^2 &= (ab)^2 & (4). \end{aligned}$$

From (3), $a^3 = abab$ (NOT a^2b^2 – see warning in Q.10.02, p. 106)
$\Rightarrow a^2 = bab$ (5) (by cancellation law).

From (4) $b^2 = abab \Rightarrow b = aba$ (6) (cancellation law).

Now, $(a^2b)^2 = a^2ba^2b = a(aba)ab = abab$ (from (6)) $= a(bab) = a^3$ (from (3)).

Hence, $(a^2b)^3 = (a^2b)^2(a^2b) = a^3a^2b = a^5b$ $(\neq e)$.

However, $(a^2b)^4 = [(a^2b^2)]^2 = (a^3)^2 = a^6 = e$, so a^2b is of *period 4*.

Q. 10.06. Find the other elements in terms of a and b, and the period of each by similar manipulations. (We did not use (5), but you may find you need it in answering these questions.)

Period of a permutation – cycles

Consider the permutation $a = \begin{pmatrix} 1 & 2 & 3 & 4 & 5 \\ 1 & 3 & 2 & 5 & 4 \end{pmatrix}$. Evidently $a^2 = e$, so this permutation is of period 2.

Q. 10.07. Make a list of all those permutations of 5 numbers which have period 2. (There are 15 of the type quoted, and 15 others.)

This permutation $\begin{pmatrix} 1 & 2 & 3 & 4 & 5 \\ 1 & 3 & 2 & 5 & 4 \end{pmatrix}$ consists essentially of *two interchanges*, or *transpositions* (or switches, or swaps), and is abbreviated to (23)(45) to indicate that these are the two pairs that have been switched.

Q. 10.08. Write out the effect on the letters CRALIPTE of the transpositions (CP)(AR)(TL); and on the letters AICLTUN of the transpositions (AL)(UI)(CN).

Cyclic permutations

But the permutation $p = \begin{pmatrix} 1 & 2 & 3 & 4 & 5 \\ 4 & 2 & 1 & 3 & 5 \end{pmatrix}$ does not have period 2. We observe that $p^2 = \begin{pmatrix} 1 & 2 & 3 & 4 & 5 \\ 3 & 2 & 4 & 1 & 5 \end{pmatrix}$, and finally that $p^3 = e$, so p is a permutation of period 3. What has happened here is that 2 and 5 have remained unchanged throughout, so we can ignore them and think only of the effect on the set $\{1, 3, 4\}$:

$$\underset{(e)}{1\,3\,4} \xrightarrow{p} \underset{(p)}{4\,1\,3} \xrightarrow{p} \underset{(p^2)}{3\,4\,1} \xrightarrow{p} \underset{(p^3)}{1\,3\,4}.$$

Thus we have a *permutation of period 3*. The effect of each permutation has been to move the figures 1, 3, 4 'round in a circle' (see fig. 10.01), and

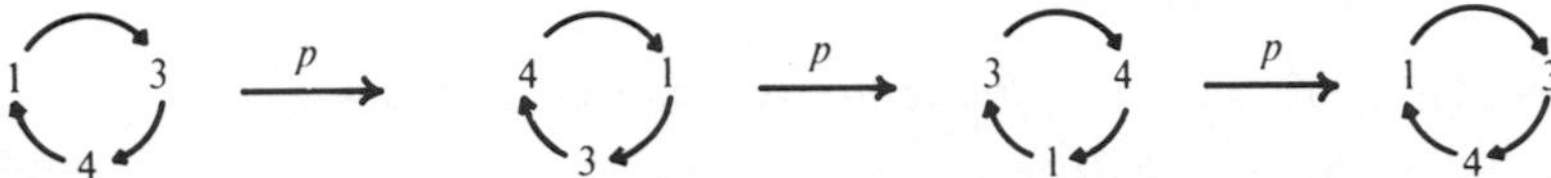

Fig. 10.01.

this particular kind of permutation is known as a *cyclic permutation*, because e, p and p^2 form a cyclic group on their own. Here are other examples:

$$x = \begin{pmatrix} 1 & 2 & 3 & 4 & 5 & 6 & 7 \\ 2 & 3 & 4 & 5 & 6 & 7 & 1 \end{pmatrix}; \qquad y = \begin{pmatrix} 1 & 2 & 3 & 4 & 5 & 6 & 7 \\ 7 & 4 & 2 & 6 & 1 & 5 & 3 \end{pmatrix}.$$

It is immediately evident that x is a cyclic permutation, and that it would have to be repeated seven times in order to restore the original order, so $x^7 = e$, and x is of period 7. This should be compared with the note on drying dishes on p. 89. It is not so obvious in the case of y, until one sees it as a cycle. (See fig. 10.02.) The purpose of this figure is to show

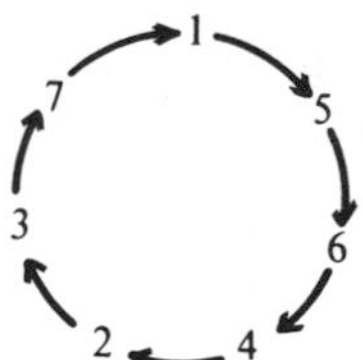

Fig. 10.02.

diagrammatically that the effect of the permutation is that 1 is replaced by 7, 7 is replaced by 3, 3 by 2, 2 by 4, 4 by 6, 6 by 5, and 5 by 1, and it now becomes apparent that $y^7 = e$, and that y is also a cyclic permutation of period 7.

The period of this permutation can be decided without having to draw the circular diagram each time:

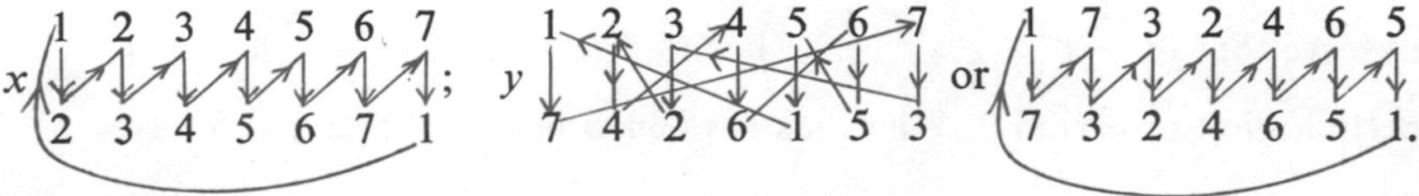

Simply start from 1 (or any number), and follow the continuous line in the direction of the arrows until the starting point is reached again, and it is found that seven steps are required, so the permutations are of period 7.

This suggests a new notation for a cycle: we shall enclose in a bracket the objects belonging to the cycle in the correct order, as shown by the example:

(1 7 3 2 4 6 5) means the cycle by which

1 becomes 7,
7 becomes 3,
3 becomes 2,
2 becomes 4,
4 becomes 6,
6 becomes 5,
and 5 becomes 1,

i.e. it denotes the permutation

$$\begin{pmatrix} 1 & 7 & 3 & 2 & 4 & 6 & 5 \\ 7 & 3 & 2 & 4 & 6 & 5 & 1 \end{pmatrix} \quad \text{or} \quad \begin{pmatrix} 1 & 2 & 3 & 4 & 5 & 6 & 7 \\ 7 & 4 & 2 & 6 & 1 & 5 & 3 \end{pmatrix}.$$

Evidently the cycle could just as well be indicated by (7 3 2 4 6 5 1), or by (3 2 4 6 5 1 7), etc.

In the simpler case of three objects, we have the permutation

$\begin{pmatrix} 1 & 2 & 3 \\ 2 & 3 & 1 \end{pmatrix}$ represented by the notation (1 2 3) or (2 3 1) or (3 1 2)

$\begin{pmatrix} 1 & 2 & 3 \\ 3 & 1 & 2 \end{pmatrix}$ represented by the notation (1 3 2) or (3 2 1) or (2 1 3).

Obviously it would be better if these were arranged in a circular pattern rather than in a line, thus:

(1 2 3) → 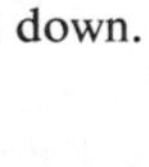

,

but the linear arrangement is more convenient.

Q. 10.09. Show that if $x = (1\ 2\ 3\ 4\ 5)$, then $x^2 = (1\ 3\ 5\ 2\ 4)$. Find x^3, x^4, x^5. Find all the 'powers' of the cycles (1 2 3 4); (1 2 3 4 5 6), etc., and obtain a rule for writing them down.

Permutations containing more than one cycle

Matters may be complicated by a permutation containing several cycles. For example, if $p = \begin{pmatrix} 1 & 2 & 3 & 4 & 5 \\ 4 & 2 & 1 & 3 & 5 \end{pmatrix}$ were followed by

$q = \begin{pmatrix} 1 & 2 & 3 & 4 & 5 \\ 1 & 5 & 3 & 4 & 2 \end{pmatrix}$ (i.e. the transposition (25)), we should get

$qp = \begin{pmatrix} 1 & 2 & 3 & 4 & 5 \\ 4 & 5 & 1 & 3 & 2 \end{pmatrix}$, and this 'composite' permutation, being the product of the cycles (25), of period 2, and (4 3 1) of period 3, is written (2 5)(4 3 1), meaning the permutation $\begin{pmatrix} 2 & 5 & & 4 & 3 & 1 \\ 5 & 2 & & 3 & 1 & 4 \end{pmatrix}$.

Q. 10.10. What is pq?

Again, for the permutation

$$\begin{pmatrix} 1 & 2 & 3 & 4 & 5 & 6 & 7 & 8 & 9 & 10 & 11 & 12 & 13 & 14 & 15 & 16 \\ 11 & 12 & 6 & 3 & 8 & 16 & 1 & 14 & 15 & 2 & 7 & 5 & 13 & 10 & 9 & 4 \end{pmatrix},$$

we can trace through the following cycles, and here the order of the figures has been altered for convenience:

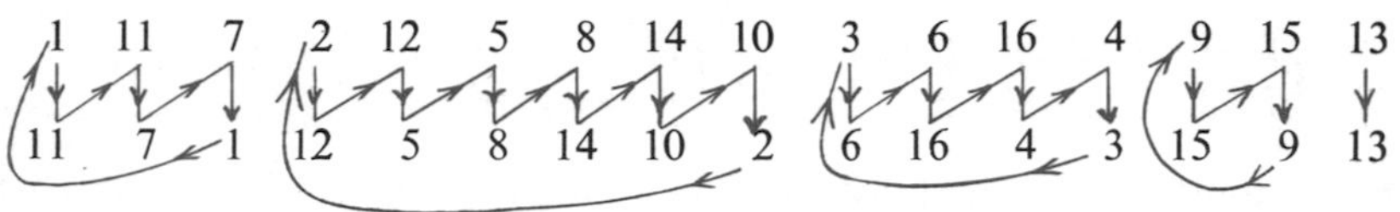

so the cycles are (1 11 7), (2 12 5 8 14 10), (3 6 16 4), (9 15) and (13).

Period of product of disjoint cycles

In the examples above the cycles were entirely independent of each other – they affected disjoint, or non-overlapping sets of the objects – so it is not surprising that they do in fact *commute*. In the case of the permutations labelled p and q, for example, we found that $pq = qp$. Let us now find the period of qp ($= pq$). We write down the successive powers of qp:

e	1	2	3	4	5	(using the method of Q. 10.09)
qp	4	5	1	3	2	
$(qp)^2$	3	2	4	1	5	
$(qp)^3$	1	5	3	4	2	
$(qp)^4$	4	2	1	3	5	
$(qp)^5$	3	5	4	1	2	
$(qp)^6$	1	2	3	4	5,	so qp is of period 6.

This could have been found 'algebraically' as follows. Knowing that $pq = qp$, and that $q^2 = e = p^3$, it follows that:

$$(qp)^2 = qp.qp = pq.qp = pq^2p = pep = p^2$$

$$(qp)^3 = qp(qp)^2 = qp.p^2 = qp^3 = qe = q$$

$$(qp)^4 = (qp)^3qp = q.qp = q^2p = ep = p$$

$$(qp)^5 = (qp)(qp)^4 = qp.p = qp^2$$

$$(qp)^6 = (qp)(qp)^5 = (pq)(qp^2) = pq^2p^2 = pep^2 = p^3 = e.$$

Or, we could argue that, since one cycle is of period 3 and the other is of period 2, we shall only restore the original order after 3 × 2 applications of the permutation.

If p has been of period 3 and q of period 5, then pq would be of period 15. But if p were of period 3 and q of period 6, it is obvious that the *status quo* would be restored after six permutations, so that pq (= qp) would be of period 6 in this case (see also pp. 259 ff and p. 321).

Q. 10.11. What is the period of pq or qp when p and q are independent cycles of periods (a) 4 and 5; (b) 3 and 9; (c) 6 and 9; (d) 4 and 6; (e) 18 and 24; (f) 24 and 30?

Q. 10.12. Show that the cyclic group of order 12 may be realised as a group of permutations of seven objects by partitioning them into disjoint sets containing three and four.

In general the period of *pq* or *qp*, where *p* and *q* are *independent* cycles (i.e. affecting *disjoint* sets of objects so that there is no 'interaction' between them) **is the L.C.M. of the periods of p and q.**

Here is another illustration:

$$p = \begin{pmatrix} 1 & 2 & 3 & 4 & 5 & 6 & 7 & 8 & 9 & 10 & 11 & 12 & 13 & 14 & 15 \\ 11 & 3 & 5 & 9 & 14 & 6 & 2 & 15 & 1 & 7 & 4 & 8 & 13 & 10 & 12 \end{pmatrix}$$

To what cycle does 1 belong? We can trace through.

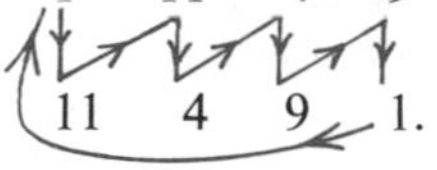

$$\begin{array}{cccc} 1 & 11 & 4 & 9 \\ \downarrow & \downarrow & \downarrow & \downarrow \\ 11 & 4 & 9 & 1. \end{array}$$

Here is a cycle, a (11 4 9 1) of period 4. Again, starting with 2, we can trace the members of that cycle which contains 2:

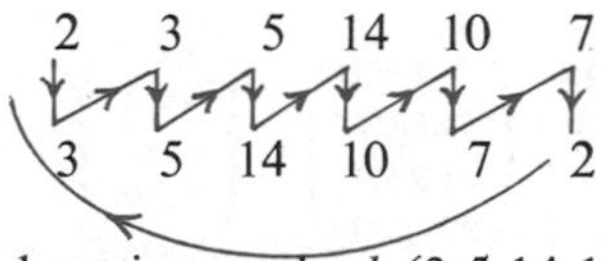

$$\begin{array}{cccccc} 2 & 3 & 5 & 14 & 10 & 7 \\ \downarrow & \downarrow & \downarrow & \downarrow & \downarrow & \downarrow \\ 3 & 5 & 14 & 10 & 7 & 2 \end{array}$$

so here is a cycle, b (3 5 14 10 7 2), of period 6. 6 and 13 are unaltered by the permutation, so they are on their own as trivial cycles of period 1, and may be indicated (6), (13) respectively, though such elements are usually omitted altogether. Finally we have a cycle of period 3: c (8 15 12). *Note:*

The order† of the objects in a cycle is important (unless it is a mere transposition).

$$(8\ 15\ 12) \text{ means } \begin{pmatrix} 8 & 12 & 15 \\ 15 & 8 & 12 \end{pmatrix},$$

and could be written (15 12 8) or (12 8 15)

$$(8\ 12\ 15) \text{ means } \begin{pmatrix} 8 & 12 & 15 \\ 12 & 15 & 8 \end{pmatrix},$$

and could be written (12 15 8) or (15 8 12),

i.e. the figures move forward → in the bracket, as it were. Hence that permutation which we labelled p above could be written in terms of its cycles as:

$$(11\ 4\ 9\ 1)\quad (3\ 5\ 14\ 10\ 7\ 2)\quad (8\ 15\ 12)\quad (6)\quad (13),$$

and we may say $p = abc$ ($= acb = bca = bac = cab = cba$, since these independent cycles are commutative). The *period* of p is 12, the L.C.M. of 4, 6 and 3.

Q. 10.13. Apply the cycles (IC)(NH) to CHIN; (AHPNM) to HAMPTON; (DPRIY)(AM) to DIP MARY; (LETYU)(AQ) to LAY QUIET; (GFDT) to GIFTED.

Q. 10.14. Find the cycles in the permutations $\begin{pmatrix}\text{pre-suction}\\ \text{sup ertonic}\end{pmatrix}$; $\begin{pmatrix}\text{I DANCE OUT}\\ \text{E DUCAT ION}\end{pmatrix}$. What are their periods?

Q. 10.15. Here are two words giving an immediate cyclic permutation $\begin{pmatrix}\text{ETHOS}\\ \text{THOSE}\end{pmatrix}$. Find other examples.

Q. 10.16. Find the cycles which perform the permutation $\begin{pmatrix}\text{TRIANGLE}\\ \text{RELATING}\end{pmatrix}$, and also the inverse of this permutation. What is the period of each? Repeat for $\begin{pmatrix}\text{REPLICA}\\ \text{CALIPER}\end{pmatrix}$, $\begin{pmatrix}\text{OF STREAM}\\ \text{FO REMAST}\end{pmatrix}$, $\begin{pmatrix}\text{HAMILTON}\\ \text{LIONMATH}\end{pmatrix}$, $\begin{pmatrix}\text{HAMILTON}\\ \text{THINLOAM}\end{pmatrix}$.

Q. 10.17. Find the cycles which make the changes from each one of the following sets of words into another:

$$\begin{pmatrix}\text{GARDEN}\\ \text{RANGED}\\ \text{GANDER}\\ \text{DANGER}\end{pmatrix} \begin{pmatrix}\text{TRACE}\\ \text{CATER}\\ \text{CRATE}\end{pmatrix} \begin{pmatrix}\text{ESPIRT}\\ \text{PRIEST}\\ \text{STRIPE}\\ \text{SPRITE}\end{pmatrix} \begin{pmatrix}\text{STAPLE}\\ \text{PETALS}\\ \text{PALEST}\\ \text{PASTEL}\\ \text{PLATES}\end{pmatrix} \begin{pmatrix}\text{PRATED}\\ \text{DEPART}\\ \text{PARTED}\\ \text{PETARD}\end{pmatrix} \begin{pmatrix}\text{SATE}\\ \text{SEAT}\\ \text{EAST}\\ \text{EATS}\\ \text{TEAS}\end{pmatrix}.$$

† One reason for preferring the word 'period' to 'order' (see p. 106). A further reason is that the latter word is also used in the entirely different mathematical sense when referring to an 'order relation' on a set, such as $<$ or $>$ on the set **R**.

Overlapping cycles

When cycles are not independent, i.e. when the sets in the cycles *overlap*, we have a different story. For example, to take the simplest possible case of two cycles of period 2 (two transpositions), if a is the permutation $\begin{pmatrix}1 & 2 & 3 & 4\\ 2 & 1 & 3 & 4\end{pmatrix}$, or the cycle (1 2), and b is permutation $\begin{pmatrix}1 & 2 & 3 & 4\\ 4 & 2 & 3 & 1\end{pmatrix}$ or the cycle (1 4). Then, when these are combined we get $ab = \begin{pmatrix}1 & 2 & 3 & 4\\ 4 & 1 & 3 & 2\end{pmatrix}$, i.e. the cycle (1 4 2) and this is of period 3, not of period 2, or 4 as might possibly have been expected. Thus the product of the transpositions (12) and (1 4) is the cycle (1 4 2) of period 3.

Q. 10.18. Find ba. What is its period? Experiment further with the composition of transpositions on a set of four objects, and extend your work to transpositions on a set of five objects.

Q. 10.19. Apply the cycles first (AMDC) then (ELI) to the word DECLAIM, and finally the cycle (CDLEA). Apply the cycle (STAEUD) to SEA DUTY without writing anything down.

The question of permutations and cycles will be resumed in more detail in Ch. 18.

Period of a function

We have already come across functions of period 2:

$$f(x) = 1 - x \Rightarrow f^{-1}(x) = 1 - x, \text{ and } ff(x) = x = i(x),$$

where i is the identity function; $f(x) = 1/x$ has the same property. You may invent other examples. We may check that:

$$f(x) = \frac{2x+5}{3x-2} \text{ is of period 2 by forming } f^2(x) = f[f(x)] = f\left(\frac{2x+5}{3x-2}\right)$$

$$= \frac{2\{(2x+5)/(3x-2)\}+5}{3\{(2x+5)/(3x-2)\}-2} = \frac{2(2x+5)+5(3x-2)}{3(2x+5)-2(3x-2)} = x.$$

Q. 10.20. Check that the inverse function is identical to the given function.

To find under what circumstances the function $(ax+b)/(cx+d)$ is of period 2, we note that

$$f^2(x) = ff(x) = \frac{a(ax+b)+b(cx+d)}{c(ax+b)+d(cx+d)} = \frac{(a^2+bc)x+(ab+bd)}{(ac+cd)x+(bc+d^2)}.$$

If this is to be equal to x, we must have: $\left.\begin{aligned} a^2+bc &= bc+d^2\\ ab+bd &= ac+cd\end{aligned}\right\}$

and these will certainly be true when $a + d = 0$.

Consider the function $f(x) = 1 - 1/x$. What is its period? We find that:

$$f^2(x) = ff(x) = 1 - \frac{1}{1-(1/x)} = 1 - \frac{1}{x-1} = \frac{1}{1-x};$$

$$f^3(x) = ff^2(x) = 1 - \frac{1}{1/(1-x)} = 1 - (1-x) = x,$$

so this function has period 3. We may now set up a group table as follows:

$$f(x) = 1 - \frac{1}{x}; \qquad g(x) = f^2(x) = \frac{1}{1-x}; \qquad i(x) = x.$$

Table 10.06

∘	i	f	g
i	i	f	g
f	f	g	i
g	g	i	f

showing how these three functions form the C_3 group structure under successive substitution.

Q. 10.21. Find the periods of the following functions (where possible):

(a) $\dfrac{x+1}{2x-1}$; (b) $5 - x$; (c) $\dfrac{1+x}{1-x}$; (d) $\dfrac{x+1}{x-1}$;

(e) $-ix$ $(i^2 = -1)$; (f) $\dfrac{2}{2-x}$; (g) $2 - \dfrac{2}{x}$; (h) $x \operatorname{cis} \dfrac{2\pi}{5}$;

(i) $\dfrac{x-3}{2x-1}$; (j) $\dfrac{1}{2-4x}$; (k) $\dfrac{2x+1}{4-4x}$; (l) $x \operatorname{cis} 40°$;

(m) $x \operatorname{cis} 160°$; (n) $\dfrac{x+1}{2-x}$; (o) $1 + x$; (p) $3x$.

The question of the period of the bilinear transformation $(ax + b)/(cx + d)$ is also dealt with elsewhere (see Exs. 4.15, 10.28, 14.35 and 15.43).

Period of a matrix

We have already met examples of finite sets of 2 × 2 matrices which form a group under matrix multiplication. Moreover (see p. 30), there are the four 3 × 3 matrices,

$$I_3 = \begin{bmatrix} 1 & 0 & 0 \\ 0 & 1 & 0 \\ 0 & 0 & 1 \end{bmatrix}, \qquad X = \begin{bmatrix} 1 & 0 & 0 \\ 0 & -1 & 0 \\ 0 & 0 & -1 \end{bmatrix},$$

$$Y = \begin{bmatrix} -1 & 0 & 0 \\ 0 & 1 & 0 \\ 0 & 0 & -1 \end{bmatrix}, \qquad Z = \begin{bmatrix} -1 & 0 & 0 \\ 0 & -1 & 0 \\ 0 & 0 & 1 \end{bmatrix}$$

which form a group of order 4. Indeed, either X, Y or Z, with I, will give a 2-group, a subgroup of that group of order 4. We also found finite sets of permutation matrices which formed groups.

Consider the matrix $R = \begin{bmatrix} 0 & 1 \\ -1 & 0 \end{bmatrix}$. Then $R^2 = \begin{bmatrix} -1 & 0 \\ 0 & -1 \end{bmatrix}$; $R^3 = \begin{bmatrix} 0 & -1 \\ 1 & 0 \end{bmatrix}$; and $R^4 = \begin{bmatrix} 1 & 0 \\ 0 & 1 \end{bmatrix} = I_2$. So the matrix $\begin{bmatrix} 0 & 1 \\ -1 & 0 \end{bmatrix}$ is of period 4 under matrix multiplication. We may now observe the geometrical aspect:

$$\begin{bmatrix} 0 & 1 \\ -1 & 0 \end{bmatrix} \begin{bmatrix} x \\ y \end{bmatrix} = \begin{bmatrix} y \\ -x \end{bmatrix},$$

and the effect of this matrix R is to rotate the vector $\begin{bmatrix} x \\ y \end{bmatrix}$ to a new position $\begin{bmatrix} y \\ -x \end{bmatrix}$, i.e. through a right-angle, and this of course quickly explains why the period of the matrix is 4 (see fig. 10.03).

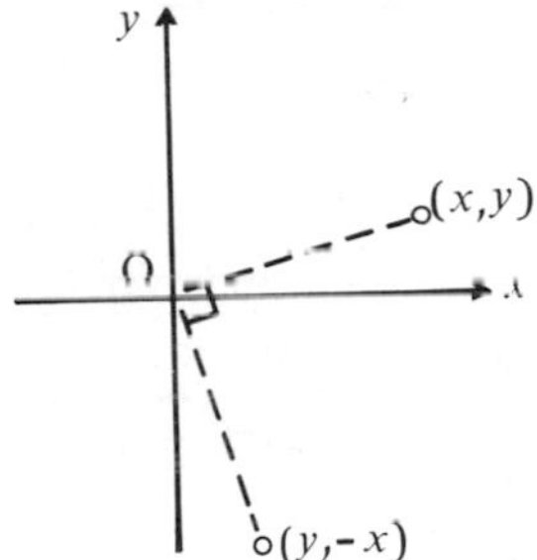

Fig. 10.03. The matrix $\begin{bmatrix} 0 & 1 \\ -1 & 0 \end{bmatrix}$.

Q. 10.22. Draw the group table in the above cases.

Q. 10.23. Consider the matrices

$$I_2, \quad P = \begin{bmatrix} 1 & 0 \\ 0 & -1 \end{bmatrix}, \quad Q = \begin{bmatrix} -1 & 0 \\ 0 & 1 \end{bmatrix}, \quad \text{and} \quad H = \begin{bmatrix} -1 & 0 \\ 0 & -1 \end{bmatrix},$$

and show that they form a group. Draw up the group table, and give a geometrical explanation.

Q. 10.24. Find some 2×2 matrices of period 2, and also of period 4. Find a rule for $\begin{bmatrix} a & b \\ c & d \end{bmatrix}$ to be of period 2, or period 4.

Those of you who are more familiar with matrices will know that the matrix which causes a rotation through angle θ is $\begin{bmatrix} \cos\theta & -\sin\theta \\ \sin\theta & \cos\theta \end{bmatrix}$†.

† See F. J. Budden, *Complex Numbers* (Longmans, 1968), p. 133.

Evidently we can get a matrix of *any* period we please by choosing θ suitably. For example, when $\theta = 30°$ we get $\begin{bmatrix} \frac{1}{2}\sqrt{3} & -\frac{1}{2} \\ \frac{1}{2} & \frac{1}{2}\sqrt{3} \end{bmatrix} = \frac{1}{2}\begin{bmatrix} \sqrt{3} & -1 \\ 1 & \sqrt{3} \end{bmatrix}$ of period 12.

Q. 10.25. Find the inverse of $\begin{bmatrix} \cos\theta & -\sin\theta \\ \sin\theta & \cos\theta \end{bmatrix}$, and also three other matrices of period 12.
Find matrices of periods 3, 5, 8 and 9.

Q. 10.26. What can be said about θ if the matrix $\begin{bmatrix} \cos\theta & -\sin\theta \\ \sin\theta & \cos\theta \end{bmatrix}$ has finite period?
Give a geometrical interpretation.

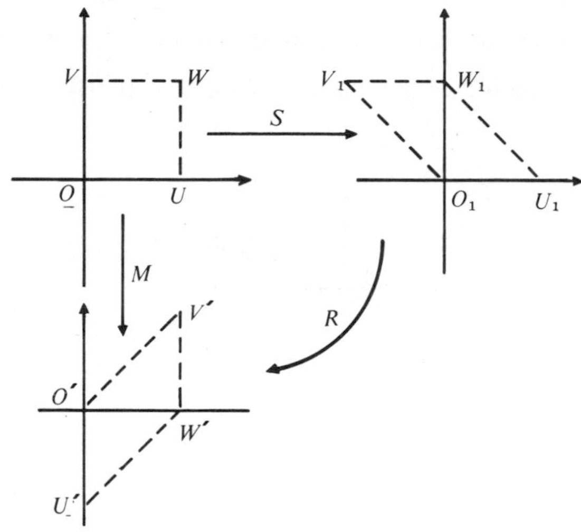

Fig. 10.04. Product of shear and quarter-turn: $M = RS$.

Next let M denote the matrix $\begin{bmatrix} 0 & 1 \\ -1 & 1 \end{bmatrix}$ and we find

$$M^2 = \begin{bmatrix} 0 & 1 \\ -1 & 1 \end{bmatrix}\begin{bmatrix} 0 & 1 \\ -1 & 1 \end{bmatrix} = \begin{bmatrix} -1 & 1 \\ -1 & 0 \end{bmatrix};$$

$$M^3 = \begin{bmatrix} 0 & 1 \\ -1 & 1 \end{bmatrix}\begin{bmatrix} -1 & 1 \\ -1 & 0 \end{bmatrix} = \begin{bmatrix} -1 & 0 \\ 0 & -1 \end{bmatrix};$$

$$M^6 = \begin{bmatrix} -1 & 0 \\ 0 & -1 \end{bmatrix}^2 = \begin{bmatrix} 1 & 0 \\ 0 & 1 \end{bmatrix}$$

and this is the lowest power of M which is equal to the identity. Hence this matrix has period 6. It is worthwhile to consider the geometrical significance of this transformation.

The effect of this matrix on the unit square $OUWV$ (see fig. 10.04) is to transform it into the parallelogram $O'U'W'V'$. The transformation is not immediately recognisable as a standard elementary one, but it would appear to combine a shear with a quarter-turn as shown on the right-hand side of the diagram. Indeed, we may verify that

$$\begin{bmatrix} 0 & 1 \\ -1 & 1 \end{bmatrix} = \begin{bmatrix} 0 & 1 \\ -1 & 0 \end{bmatrix}\begin{bmatrix} 1 & -1 \\ 0 & 1 \end{bmatrix}, \quad \text{or} \quad M = RS,$$

so that the transformation is compounded of first the shear S, represented by the matrix $\begin{bmatrix} 1 & -1 \\ 0 & 1 \end{bmatrix}$, and then the clockwise quarter-turn R. Check that $M \neq SR$. It may be thought strange that a transformation which includes a *quarter*-turn should be of period 6, but it must be remembered that $M^6 = RSRSRSRSRSRS$, and the R's cannot be 'collected together' since $RS \neq SR$.

Q. 10.27. What is the geometrical interpretation of SR?

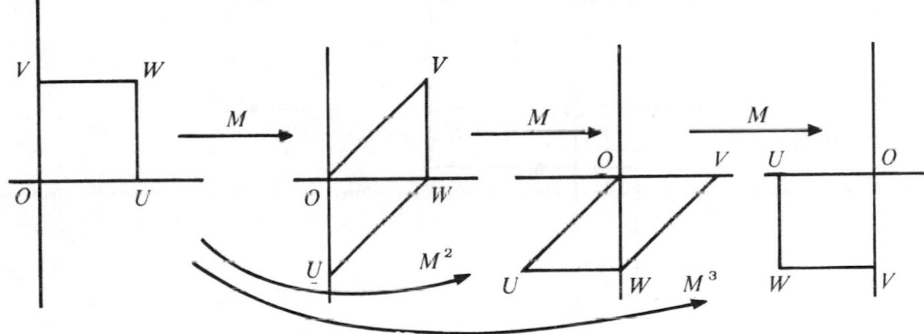

Fig. 10.05. $M = \begin{bmatrix} 0 & 1 \\ -1 & 1 \end{bmatrix}$; M^3 is a half-turn.

Figure 10.05 shows the effect of the transformations M, M^2, M^3 on the unit square, and that M^3 is a half-turn. The next diagram, fig. 10.061,

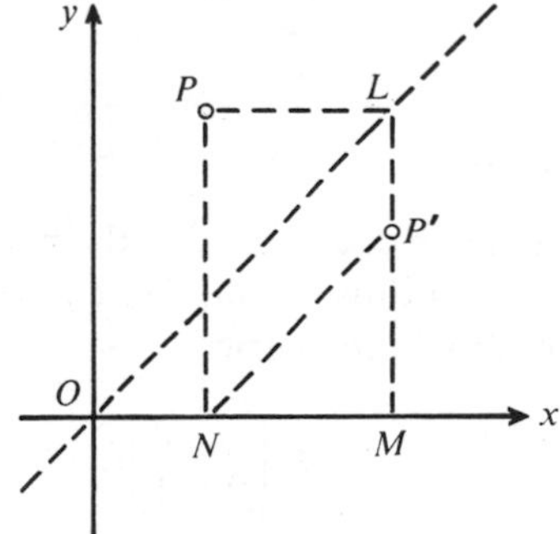

Fig. 10.061. The matrix $M = \begin{bmatrix} 0 & 1 \\ -1 & 1 \end{bmatrix}$ – geometrical construction.

gives a geometrical construction for the image of a given point P under the transformation M:

$$\begin{bmatrix} x' \\ y' \end{bmatrix} = \begin{bmatrix} 0 & 1 \\ -1 & 1 \end{bmatrix} \begin{bmatrix} x \\ y \end{bmatrix} \Rightarrow \begin{array}{l} x' = y \\ y' = y - x. \end{array}$$

We begin by drawing PN perpendicular to Ox, and let a horizontal line through P meet the line $y = x$ in L. Then P' is the intersection of a line through N parallel to OL with the perpendicular LM from L to Ox.

Q. 10.28. Verify this construction.

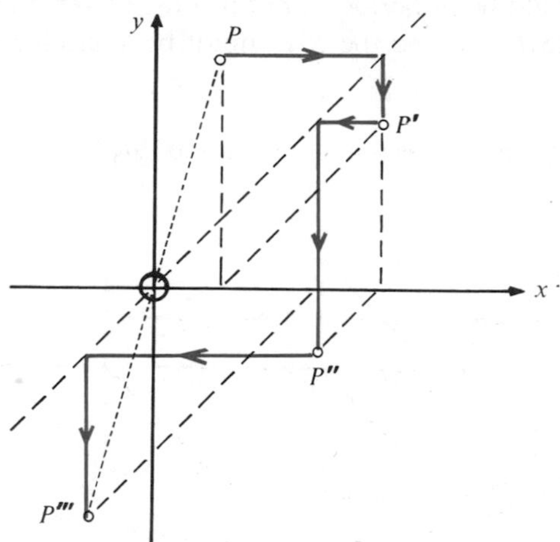

Fig. 10.062. M^3 is a half-turn.

In fig. 10.062 we see three successive applications of the transformation: $P \xrightarrow{M} P' \xrightarrow{M} P'' \xrightarrow{M} P'''$, and the figure shows that M^3 represents a half-turn about O.

Matrices with complex terms

Complex numbers may arise as coefficients in the matrices, and the following example provides an illustration. Here $\omega = \text{cis } \frac{2}{3}\pi = -\frac{1}{2} + \frac{1}{2}\sqrt{3}i$ $(i = \sqrt{-1})$. Show that each of the matrices:

$$P\begin{bmatrix} 1 & 0 \\ 0 & \omega \end{bmatrix}, \quad Q\begin{bmatrix} 1 & 0 \\ 0 & \omega^2 \end{bmatrix}, \quad R\begin{bmatrix} \omega & 0 \\ 0 & 1 \end{bmatrix}, \quad S\begin{bmatrix} \omega^2 & 0 \\ 0 & 1 \end{bmatrix},$$

$$T\begin{bmatrix} \omega & 0 \\ 0 & \omega \end{bmatrix}, \quad U\begin{bmatrix} \omega^2 & 0 \\ 0 & \omega^2 \end{bmatrix}, \quad V\begin{bmatrix} \omega & 0 \\ 0 & \omega^2 \end{bmatrix}, \quad W\begin{bmatrix} \omega^2 & 0 \\ 0 & \omega \end{bmatrix},$$

is of period 3, and together with the unit matrix I_2 form a group of order 9 which is not cyclic.

$$\text{Now } P^2 = \begin{bmatrix} 1 & 0 \\ 0 & \omega \end{bmatrix} \begin{bmatrix} 1 & 0 \\ 0 & \omega \end{bmatrix} = \begin{bmatrix} 1 & 0 \\ 0 & \omega^2 \end{bmatrix} = Q;$$

$$P^3 = PQ = \begin{bmatrix} 1 & 0 \\ 0 & \omega \end{bmatrix} \begin{bmatrix} 1 & 0 \\ 0 & \omega^2 \end{bmatrix} = \begin{bmatrix} 1 & 0 \\ 0 & 1 \end{bmatrix}$$

(since $\omega^3 = 1) = I_2$, so P is of period 3.
Moreover, $Q^2 = (P^2)^2 = P^4 = P$, so that $Q^3 = (Q^2)Q = PP^2 = P^3 = I_2$. Similarly we may show that

$$\begin{array}{lll} R^2 = S, & R = S^2, & R^3 = S^3 = I_2 \\ T^2 = U, & U^2 = T, & T^3 = U^3 = I_2 \\ V^2 = W, & W^2 = V, & V^3 = W^3 = I_2. \end{array}$$

Closure (the only condition needed to ensure group structure, since all inverses have been accounted for, e.g. $P^{-1} = Q$, etc.) is easy to establish. For example,

$$QU = \begin{bmatrix} 1 & 0 \\ 0 & \omega^2 \end{bmatrix} \begin{bmatrix} \omega^2 & 0 \\ 0 & \omega^2 \end{bmatrix} = \begin{bmatrix} \omega^2 & 0 \\ 0 & \omega \end{bmatrix} = W,$$

and the table for the group is shown in table 10.07. It will be noted (1) that

	I	P	P^2	R	R^2	T	T^2	V	V^2
I	I	P	Q	R	S	T	U	V	W
P	P	Q	I	T	W	V	S	R	U
P^2	Q	I	P	V	U	R	W	T	S
R	R	T	V	S	I	W	Q	U	P
R^2	S	W	U	I	R	P	V	Q	T
T	T	V	R	W	P	U	I	S	Q
T^2	U	S	W	Q	V	I	T	P	R
V	V	R	T	U	Q	S	P	W	I
V^2	W	U	S	P	T	Q	R	I	V

($C_3 \times C_3$)

Table 10.07

the group is Abelian, and (2) that it is *not* the cyclic group of order 9, for the latter is generated by a single element, say x, and is of the form $\{e, x, x^2, x^3, x^4, x^5, x^6, x^7, x^8\}$ where $x^9 = e$. The present group does not contain an element of period 9, so cannot be the cycle group (see pp. 142 ff, Isomorphism).

It may be of interest to note that we could regard these matrices as ordered pairs, $\begin{bmatrix} a & 0 \\ 0 & b \end{bmatrix} \leftrightarrow (a, b)$, with multiplication defined $(a, b) \times (c, d)$

$= (ac, bd)$. Or, if each non-zero number is expressed as a power of ω, we may take the ordered pairs formed by the *indices* of ω, which are then added modulo 3:

$$\text{e.g.} \quad S = \begin{bmatrix} \omega^2 & 0 \\ 0 & 1 \end{bmatrix} \leftrightarrow (\omega^2, 1) = (\omega^2, \omega^0) \leftrightarrow (2, 0),$$

$$W = \begin{bmatrix} \omega^2 & 0 \\ 0 & \omega \end{bmatrix} \leftrightarrow (\omega^2, \omega) = (\omega^2, \omega^1) \leftrightarrow (2, 1),$$

$$SW \leftrightarrow (2, 0) \underset{\text{(mod. 3)}}{+} (2, 1) = (1, 1) \leftrightarrow \begin{bmatrix} \omega & 0 \\ 0 & \omega \end{bmatrix} = T.$$

Q. 10.29. Show that the matrix $M = \begin{bmatrix} -\omega & 0 \\ 0 & -\omega^2 \end{bmatrix}$ ($\omega = \text{cis } \frac{2}{3}\pi$) is of period 6. Construct five other matrices which with M together form the cyclic group of order 6 under matrix multiplication.

Q. 10.30. Find the period of the matrices $A = \begin{bmatrix} 0 & \omega^2 \\ \omega & 0 \end{bmatrix}$ and $P = \begin{bmatrix} \omega^2 & 0 \\ 0 & \omega \end{bmatrix}$. Form all possible products containing A and P, and so obtain a group generated by these two matrices. (*Hint*: the group is of order 6.)

Infinite period

When there does not exist a smallest positive integer m such that, for a typical element x of the group, $x^m = e$, we are bound to have an *infinite group* for it will contain all integral 'powers' of x:

$$\{\ldots x^{-3}, x^{-2}, x^{-1}, e, x, x^2, x^3, \ldots\}$$

and these are all distinct, and there are an infinity of them.

For example, if x were the matrix $\begin{bmatrix} a & b \\ c & d \end{bmatrix}$, it is unlikely in general that one would ever obtain the unit matrix by forming integral powers of x; e.g. if $x = \begin{bmatrix} -1 & 1 \\ 2 & 0 \end{bmatrix}$, then $x^2 = \begin{bmatrix} 3 & -1 \\ -1 & 2 \end{bmatrix}$, $x^3 = \begin{bmatrix} -5 & 3 \\ 2 & 2 \end{bmatrix}$, ..., and the unit matrix is never obtained. Again, if $f(x)$ is a function of x, it is most unlikely that $f^m(x)$ will ever become identically equal to the function $i(x) \equiv x$. In the simple case $f(x) = 2x$, we have $f[f(x)] = f^2(x) = 2(2x) = 4x$, and in general, $f^m(x) = 2^m x$. Again, if $f(x) = x + 4$, then $f^2(x) = (x + 4) + 4 = x + 8$; $f^m(x) = x + 4m$, and in neither of these cases shall we ever reach $f^m(x) \equiv x$.

Q. 10.31. If $f(x) \equiv ax + b$, find for what possible values of a and b the function may have finite period.

In the case $f(x) = x + c$, for any c, this represents, geometrically, a shift through a distance c along the x-axis. Evidently, however many times this shift is repeated, we can never return to the starting point! More generally, in the plane, the transformation $z' = z + a$, where z', z and a are complex numbers (or vectors) represents a *translation* through a vector represented by a, and so, when reproduced successively, will generate an infinite linear pattern, as in fig. 10.07.

Fig. 10.07. Infinite strip pattern generated by a single translation t; the group C_∞.

In the same way, plane rotations about a fixed point through angle $\theta°$ are *closed* only when θ is a rational fraction of a revolution. When $\theta/360$ is irrational, we get an infinite group. (Compare the remarks on closure, p. 14.)

Consider a modular arithmetic with a very large modulus (e.g. mod. 100 000, which is the modulus for readings on a car milometer). The set would consist of {0, 1, 2, 3, 4, . . .99 999} and the period of, say, 2000 would be 5, since $2000 + 2000 + 2000 + 2000 + 2000 = 0$ (mod. 10^5). But the period of any number *prime to* 100 000 would be 100 000; e.g. $63 + 63 + 63 + \ldots + 63$ (100 000 terms) $= 0$ (mod. 10^5). Thus, as long as we persevere and go on far enough we shall eventually reach the identity. (Of course, it is simpler to generate this (cyclic) group by using the element 1 rather than 63.)

The infinite cyclic group, Z,+

Now if we consider the set **Z**, under addition, the zero is never reached in the process $1 + 1 + 1 + 1 + \ldots$, and here the period of the element 1 (and of every element) is *infinite*. It is as if our milometer were not to stop at 99 999, but had facilities for an unlimited number of digits. Here we have perhaps the simplest example of the generation of the 'infinite cyclic group' by the operation (+1).

Q. 10.32. Has the group (**Z**, +) any other generators?

One final remark, the truth of which must be self-evident: if a finite group is of order n, then the period of every element is *at most n*. In point of fact, we have already made a much more powerful statement than this: the period of any element of a finite group is a factor of the order of the group (see p. 108). For example, for a group of order 30, the only possible periods that its elements can have is 1, 2, 3, 5, 6, 10, 15 and 30. This is a corollary of the result of great importance known as Lagrange's theorem (see Chs. 14 and 19).

EXERCISES 10

1 What is another way of saying that an element of a group may be its own inverse?

2 What is the period of a rotation in the plane about a fixed point through angles:
(a) 90°, (b) 40°, (c) 18°, (d) 0.2°, (e) 160°, (f) 144°, (g) 73°, (h) π°?

3 Give various values of θ so that the rotation $R_{0,\theta}$ may have period 3, 5, 6, 8, 9, 10, 12, etc.

4 **Prove that the period of xy is equal to the period of yx, where x and y are any two elements of a finite group. Find elements with period equal to that of xyz. Prove that, in general, the period of xyz is not necessarily equal to that of yxz. Find a counter-example to show that the period of xy is not necessarily equal to that of $xyx^{-1}y^{-1}$, where x and y are independent generators in a non-Abelian group.**

5 **If every element of a group is of period 2, prove that the group is necessarily Abelian.**

6 **Prove that every group of even order must contain at least one element of period 2.**

7 **Prove that a cyclic group of even order contains exactly one element of period 2.**

8 Prove that a finite group cannot have exactly two elements of period 2.

9 Prove further that a finite group cannot have an even number of elements of period 2. Can an infinite group contain an even number of elements of period 2?

10 **If y is an element of period k, and if x is any other element of the group, prove that $x^{-1}yx$ and xyx^{-1} also have period k.**

11 Find the period of all the elements in the group tables on pp. 204, 337.

12 Discover how many of the twenty-four permutations of four letters have period 2, how many have period 3, period 4, 5, and so on. Classify the permutations of five letters in the same way, according to their periods.

13 Show that the product of the cycles (1 4 5 6)(2 1 5) is (1 6)(2 4 5)(3).

14 Show that the set of functions $\{x, 1/x, -x, -(1/x)\}$ form a group under successive substitution. Show how a group may be set up on the parabola $y^2 = 4ax$ by applying to the general point $(at^2, 2at)$ the transformations $t' = t$; $t' = 1/t$, $t' = -t$ and $t' = -1/t$, and interpret the result geometrically. Using the same transformations, give a geometrical interpretation in the case of the rectangular hyperbola $x = ct$, $y = c/t$ (cf. Q. 25.02, p. 481).

15 What can you say about the order of a group if it contains elements of periods 3 and 4?

16 If p has period 4 and $rqp = q$, show that r also has period 4.

17 If a finite group has only one element a of period 2, show that a commutes with every other element of the group; i.e. $a^2 = 1, x^2 \neq 1, (\forall x) \Rightarrow ax = xa$ (compare Ex. 20.01).

18 If the period of x is equal to the period of x^2 $(= m)$, what can you say about m?
If the period of x is equal to the period of x^3 $(= m)$, what can you say about m?

19 If the period of x^2 is equal to the period of x^3, what may be said about the period of x? Generalise your result (period of x^p = period of x^q).

20 If four elements of a group, 1, x, y, z, where 1 is the identity, are situated in a group table so as to form the corners of a rectangle with x in the same row as 1 and y in the same column as 1, prove that $yx = z$.
What can be said about xy?

21 If $xy = yx$, prove that the period of xy is a factor of the product of the periods of x and of y.

22 What are the periods of the permutations: DEVIL $\xrightarrow{x}$ LIVED, AGNOSTIC $\xrightarrow{w}$ COASTING, DIRTY $\xrightarrow{y}$ DRY IT, MOTHER BY LAW $\xrightarrow{z}$ BLAMEWORTHY.

23 Show that the cycles $a = (1\ 2)$ and $r = (4\ 5\ 3)$ on five digits generate a group of order 6, and draw up its group table. (Prove that $ra = ar$ is of period 6.)
Which other permutation derived from a and r is also of period 6?

24 Show that the permutation $x = \begin{pmatrix} A & B & C & D & E & F \\ C & D & F & A & E & B \end{pmatrix}$ can be obtained as the product of a number of transpositions (for example, start with (AC) to get C at the front). Do this by several different combinations of transpositions. Also start from x and obtain the identity by using transpositions. What is the least possible number of transpositions you can find which will change 7426153 into 1234567? Obtain a series of transpositions which combine to give the permutation

$$\begin{pmatrix} 1 & 2 & 3 & 4 & 5 & 6 & 7 & 8 & 9 & 10 & 11 & 12 & 13 & 14 & 15 \\ 11 & 3 & 5 & 9 & 14 & 6 & 2 & 15 & 1 & 7 & 4 & 8 & 13 & 10 & 12 \end{pmatrix}.$$

25 A given permutation may be brought about as a succession of transpositions in an infinite number of ways, e.g.

$$1234 \xrightarrow[(12)]{} 2134 \xrightarrow[(14)]{} 2431 \xrightarrow[(12)]{} 1432 \xrightarrow[(34)]{} 1342 \xrightarrow[(14)]{} 4312 \xrightarrow[(12)]{} 4321$$

(six transpositions)

or $1234 \xrightarrow[(14)]{} 4231 \xrightarrow[(23)]{} 4321$ (two transpositions).

What can be said about the *number* of transpositions needed, irrespective of how many or how few?

26 Investigate the question: under what circumstances do two permutations commute in the case of permutations of four objects and then of five

objects? Note that 'powers' of the same permutation commute (e.g. if $p = (1\ 2\ 3\ 4\ 5)$ and $p^3 = (1\ 4\ 2\ 5\ 3)$, then $pp^3 = p^3p$), as also do disjoint cycles (e.g. if $a = (6\ 7)$, then $pa = ap$), but these are not the only cases, for example, if $r = (1\ 2\ 3\ 4)$, then $r^2 = (1\ 3)(2\ 4)$ commutes with $(1\ 2)(3\ 4)$, with $(1\ 4)(2\ 3)$, with $(1\ 3)$ and with $(2\ 4)$ as well as with r and r^3.

27 We saw in the text that the group of order 9 (see p. 123) could be produced by adding (mod. 3) the ordered pairs

$I = (0, 0)$, $P = (0, 1)$, $Q = (0, 2)$, $R = (1, 0)$, $S = (2, 0)$,
$T = (1, 1)$, $U = (2, 2)$, $V = (1, 2)$ and $W = (2, 1)$.

Show that we may find nine permutations of six letters which give the same group by starting as follows:

e	$ABC\ PQR$	Find the remaining five permutations, and check
p	$ABC\ RPQ$	products of pairs of permutations by reference to the
q	$ABC\ QRP$	group table. Note that we could think of these as
r	$CAB\ PQR$	permutations of the names of two triangles, e.g.

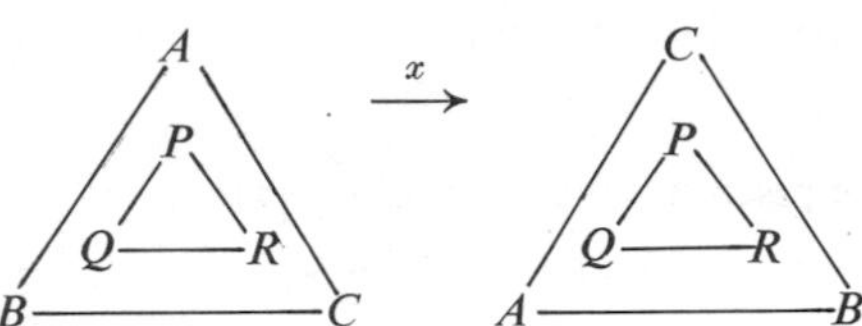

by which the cyclic order of the names of each triangle is preserved as well as its anticlockwise sense. Evidently the group is a subgroup of the group of order 6! (= 720) of all the permutations of six letters, but as we shall see later, a subgroup may generally be associated with some special property: here, that property is that each set of three letters is used for each separate triangle, and that each set rotates cyclically. Thus we do not include

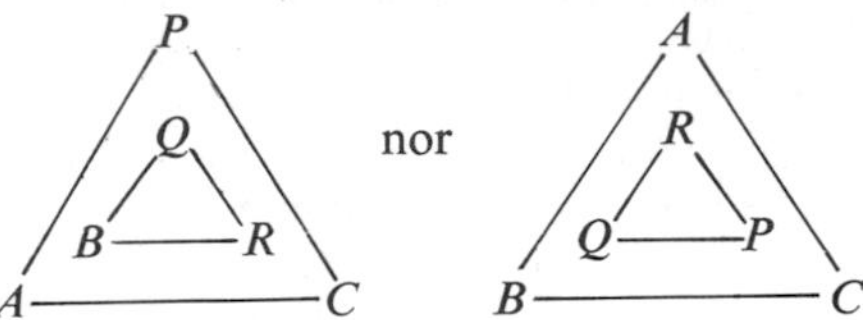

28 Show that, if $f(x) \equiv \dfrac{ax+b}{cx+d}$ and $g(x) \equiv \dfrac{px+q}{rx+s}$, then

$$fg(x) \equiv \frac{(ap+br)x+(aq+bs)}{(cp+dr)x+(cq+ds)}.$$

Note that $\begin{bmatrix} a & b \\ c & d \end{bmatrix} \begin{bmatrix} p & q \\ r & s \end{bmatrix} = \begin{bmatrix} ap+br & aq+bs \\ cp+dr & cq+ds \end{bmatrix}$

so that the coefficients in the composite function may be found by *matrix multiplication.*

Now since $\dfrac{ax+b}{cx+d} = \dfrac{kax+kb}{kcx+kd}$,

we should regard the matrices $\begin{bmatrix} a & b \\ c & d \end{bmatrix}$ and $\begin{bmatrix} ka & kb \\ kc & kd \end{bmatrix}$ as being 'equivalent' in the context of this problem. (See p. 352 'Equivalence classes'.)

The appropriate matrix for the identity function, $i(x) \equiv x$ is $\begin{bmatrix} 1 & 0 \\ 0 & 1 \end{bmatrix}$ or $\begin{bmatrix} k & 0 \\ 0 & k \end{bmatrix}$. Now let $M = \begin{bmatrix} a & b \\ c & d \end{bmatrix}$. By considering M^2, find the condition for $f(x)$ to be of period 2.

Show that the condition for $f(x)$ to be of period 3 (find M^3) is $a^2 + ad + d^2 + bc = 0$; and to be of period 4 is $a^2 + d^2 + 2bc = 0$; and to be of period 6 is $a^2 + d^2 - ad + 3bc = 0$.

How do you account for the fact that b and c do not appear separately in either of these conditions, but that only their *product* is important? What is the significance of the fact that every one of these relations is homogeneous of degree 2 in a, b, c and d?

29 What is the connection between the matrix $\begin{pmatrix} 0 & 1 \\ -1 & 1 \end{pmatrix}$ of period 6 and the function $f(x) \equiv 1/(1 - x)$ of period 3?

30 If $f(x) \equiv -1/x$, $g(x) \equiv (x + b)/(ax - 1)$ (both of period 2), prove that fg and gf cannot have period 3 for any real values of a and b.

31 Show that the matrix $\begin{bmatrix} 0 & e^{i\alpha} \\ -e^{-i\alpha} & 0 \end{bmatrix}$ is of multiplicative period 4.

32 Find the periods (under multiplication) of the matrices:

$$P = \begin{bmatrix} 1 & 0 & 0 & 0 \\ 0 & 0 & 1 & 0 \\ 0 & 1 & 0 & 0 \\ 0 & 0 & 0 & 1 \end{bmatrix} \qquad M = \tfrac{1}{2}\begin{bmatrix} -1 & 1 & 1 & 1 \\ 1 & -1 & 1 & 1 \\ 1 & 1 & -1 & 1 \\ 1 & 1 & 1 & -1 \end{bmatrix}$$

$$J = \begin{bmatrix} -1 & 0 & 0 & 0 \\ 0 & -1 & 0 & 0 \\ 0 & 0 & -1 & 0 \\ 0 & 0 & 0 & -1 \end{bmatrix} \qquad N = \tfrac{1}{2}\begin{bmatrix} -1 & 1 & 1 & -1 \\ 1 & -1 & 1 & -1 \\ 1 & 1 & -1 & -1 \\ -1 & -1 & -1 & -1 \end{bmatrix}.$$

33 Show that the matrix $\begin{bmatrix} 1 & -1 \\ 1 & 0 \end{bmatrix}$ is of period 6, and give a geometrical interpretation.

34 If $M = \begin{bmatrix} 1 & 1 \\ 1 & 0 \end{bmatrix}$, find M^2, M^3, M^4, ... and show that M is of infinite period. Illustrate the connection with the Fibonnacci series, 0, 1, 1, 2, 3, 5, 8, 13, ...

35 Consider the transformation T which changes the ordered pair (x, y) into the ordered pair $[(y, (1 + y)/x]$, where $x, y \in \mathbf{R}$, i.e.

$$(x, y) \xrightarrow{T} \left(y, \frac{1 + y}{x}\right).$$

Show that T is of period 5, and attempt a geometrical illustration.

36 Show that the operation p which transforms the matrix,

$$\begin{pmatrix} A & B & C & D & E \\ F & G & H & I & J \\ K & L & M & N & O \\ P & Q & R & S & T \\ U & V & W & X & Y \end{pmatrix} \text{ into the matrix } \begin{pmatrix} A & I & L & T & W \\ S & V & E & H & K \\ G & O & R & U & D \\ Y & C & F & N & Q \\ M & P & X & B & J \end{pmatrix} \text{ is of period 6.}$$

37 The operation x transforms the 3×3 matrix as follows:

$$\begin{bmatrix} a & b & c \\ d & e & f \\ g & h & i \end{bmatrix} \xrightarrow{x} \begin{bmatrix} i & g & h \\ c & a & b \\ f & d & e \end{bmatrix}$$

What is the period of this transformation?
Obtain the transform of the given matrix under the inverse of x.

If $\begin{bmatrix} a & b & c \\ d & e & f \\ g & h & i \end{bmatrix} \xrightarrow{p} \begin{bmatrix} c & g & e \\ i & d & b \\ f & a & h \end{bmatrix}$, show that $\begin{bmatrix} a & b & c \\ d & e & f \\ g & h & i \end{bmatrix} \xrightarrow{p^3} \begin{bmatrix} d & b & i \\ a & h & f \\ g & e & c \end{bmatrix}$

and find the period of p, and also the effect of the inverse of p.

38 Find the largest possible period of a permutation of fifty-two objects. (If a particular method of shuffling a pack of cards is applied successively, this will give the longest possible process to restore the original order.)

39 If every element of a group (other than 1) is of period 3, and x and y are *any* two elements of the group, prove that $xyx^{-1}y = yxyx^{-1}$.
Apply this result to the case of the group of order 27 discussed on pp. 100ff.
Hint for first part: use $(xy^2)^3 = 1$ and $(yx)^3 = 1$.

40 Find a 2×2 matrix with coefficients in $\mathbf{Z}$, which has period 3.

41 Show that the permutations $x = (1\ 2\ 3\ 4)(5\ 6\ 7\ 8)$, $y = (1\ 5\ 3\ 7)(2\ 8\ 4\ 6)$, $z = (1\ 8\ 3\ 6)(2\ 7\ 4\ 5)$, x^{-1}, y^{-1}, z^{-1}, x^2 and identity form the group Q_4 (table 9.13, p. 101).

42 In the Fibonacci series, $u_n = u_{n-1} + u_{n-2}$, we start with $u_1 = 0$, $u_2 = 1$. Suppose addition is modulo N, then when $N = 3$, we obtain the sequence 0, 1, 1, 2, 0, 2, 2, 1, 0, ... of period 8. Show that when $N = 10$, the period is 60, but that u_1 and u_2 may be selected to obtain different periods. Investigate for other values of N. Write a computer program to print the terms of the series, and the period for values of N running from 2 to 500.

11 *Carbon copy groups: abstract groups: isomorphism*

'Daddy, I know a good joke: "Why do white sheep eat more than black sheep?" ' 'I've no idea'. 'Because there are fewer black sheep than white sheep!' If Daddy had wished to 'repeat' this joke in his club, he would probably have altered it: 'Any of you chaps know why Americans drink more whisky than Scotsmen . . . ?' Here we have an example of what we might call *isomorphic jokes:* they are really the *same* joke, but placed into a different context. So it is with groups . . .

A group of order four in many different situations

One of the fascinating things about groups is, as we have observed before, that they may arise from widely diverse situations. Indeed, the 'same' group may appear in two totally different guises. For example, the group table

Table 11.01

	I	X	Y	Z
I	I	X	Y	Z
X	X	I	Z	Y
Y	Y	Z	I	X
Z	Z	Y	X	I

(D_2)

on p. 18 was found by considering the rotations through half a revolution about three perpendicular axes in space. If X, Y and Z had stood for the operations†

X: turn the trousers back to front,
Y: turn the trousers inside out,
Z: turn the trousers inside out and back to front‡

the very same table would express the results of combining these operations, which belong to an entirely different context!

† See p. 89.
‡ It is important to realise that X is not to be the operation '*put the trousers on* back to front', but merely '*turn* the trousers'. If we had an operation 'put the trousers on' (call this P), we should also have the inverse operation, 'take the trousers off' (P^{-1}). 'Put the trousers on back to front' would be PX (actually XP is physically impossible!). We should have eight elements in the set instead of four. But the set of operations would not have been closed, since P^2 is meaningless!

Again, the group table arising from the game† with vertically opposite, corresponding and alternate angles (see p. 25) is as follows:

Table 11.02

	I	V	A	C
I	I	V	A	C
V	V	I	C	A
A	A	C	I	V
C	C	A	V	I (D_2)

Compare these two tables, and note that their *structures* are identical. Again, we have the following multiplication tables in finite arithmetics:

Table 11.03

(a)

× (mod. 8)	1	3	5	7
1	1	3	5	7
3	3	1	7	5
5	5	7	1	3
7	7	5	3	1 (D_2)

(b)

× (mod. 12)	1	5	7	11
1	1	5	7	11
5	5	1	11	7
7	7	11	1	5
11	11	7	5	1

(c)

× (mod. 15)	1	4	11	14
1	1	4	11	14
4	4	1	14	11
11	11	14	1	4
14	14	11	4	1

(d)

× (mod. 16)	1	7	9	15
1	1	7	9	15
7	7	1	15	9
9	9	15	1	7
15	15	9	7	1

Q. 11.01. How many solutions has the quadratic equation $x^2 = 1$ in these systems?

Q. 11.02. Fin‿ sets of four numbers which give a multiplication table with the same pattern under (a) × mod. 20, and (b) × mod. 24.

Consider next the permutations

$$\begin{matrix} e \\ a \\ b \\ c \end{matrix} \begin{pmatrix} A & B & C & D \\ B & A & C & D \\ A & B & D & C \\ B & A & D & C \end{pmatrix}$$

This set of permutations is closed under composition of permutations, and each of a, b and c is of period 2. You will have observed this property – that every element except the identity is of period 2 – in all the groups mentioned so far in this chapter, and the consequence of it is that the

† See also the game 'Groupo' in the S.M.P. Textbook, *Additional Mathematics*, Book I (C.U.P.), pp. 110–2, 121.

'leading diagonal' of the table consists entirely of identities. In the case of the four permutations, we obtain table 11.04.

Table 11.04

	e	a	b	c	
e	e	a	b	c	
a	a	e	c	b	
b	b	c	e	a	
c	c	b	a	e	(D_2)

Q. 11.03. Find another set of permutations of four letters, one of which is $\begin{pmatrix} A & B & C & D \\ B & A & D & C \end{pmatrix}$, which give the same group table.

Addition of ordered pairs, modulo 2, (denote this operation $\oplus$) on the set e (0, 0), a (1, 0), b (0, 1) and c (1, 1) produces precisely the same table, e.g. $b \oplus c = (1, 0) = a$. Note that this amounts to the same thing as expressing the numbers 0, 1, 2 and 3 in *binary*, and adding them *without carrying* (compare Ex. 8.12).

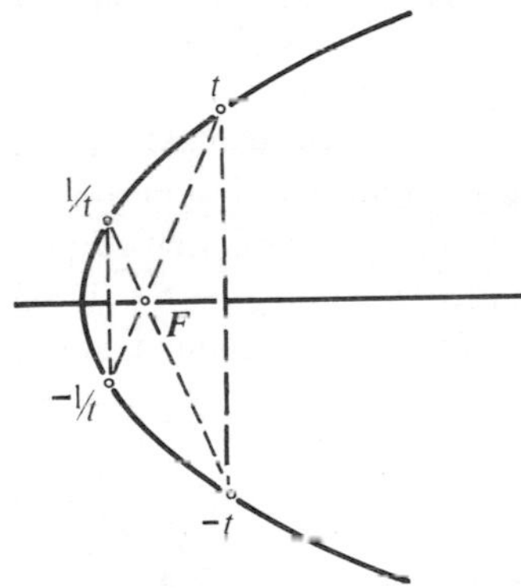

Fig. 11.01. Group structure on a parabola.

Your working in Exs. 5.03, 8.23 and 10.14 should have revealed that the functions $i(x) \equiv x$, $f(x) \equiv 1/x$, $g(x) \equiv -x$, $h(x) \equiv -1/x$ lead to the same group, with the table shown (11.05); and in Ex. 10.14, this same

Table 11.05

o	i	f	g	h	
i	i	f	g	h	
f	f	i	h	g	
g	g	h	i	f	
h	h	g	f	i	(D_2)

set of functions was used to transform any point of a parabola into three remaining points, and these sets of four points were related by the same basic group structure (see fig. 11.01, also p. 480 ff).

★Indeed, this example of group structure on a parabola is merely a special case of group structure on any conic (see fig. 11.02): ABC is a self-polar triangle (see for example, R. Walker, *Cartesian and Projective Geometry* (Arnold), p. 126) with respect to a given conic. Select any point P on the

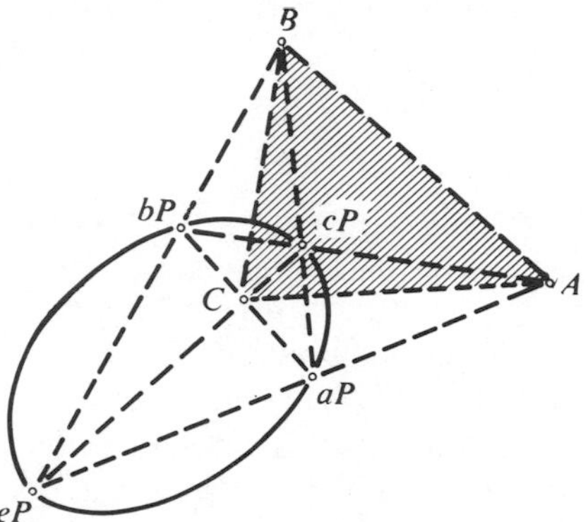

Fig. 11.02. Group structure on any conic: $c = ab = ba$ when ABC is a self-polar triangle.

conic. Then let a be the operation which changes P into the point where PA cuts the conic again; similarly for b and c. Those of you familiar with work on pole and polar may be able to prove that $ab = ba$ *only when* A and B are conjugate points, see fig. 11.031, and that we only get a *closed* system when ABC is a self-polar triangle.

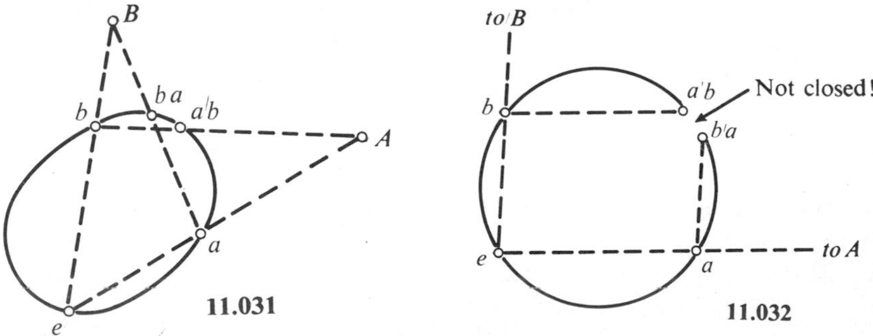

Fig. 11.03. $ab = ba$ only when A and B are conjugate points.

★**Q. 11.04.** Consider the special case when the conic is a circle (see fig. 11.032), and A and B are points at infinity – then we have opposite sides of the quadrilateral parallel. What can you now say about the lines ea and eb if ab and ba are to coincide? Where is C?

Q. 11.05. Another set of functions which give the same group table is z, $a - z$, $\bar{z}$, $a - \bar{z}$, where z is a complex number $(x + iy)$, and $\bar{z}$ represents its conjugate $(x - iy)$, and a is a given *real* number. Give a geometrical interpretation of these transformations on the Argand Diagram. (See F. J. Budden, *Complex Numbers*, Longmans p. 157.) What happens if a is complex? (Compare Ch. 13, fig. 13.16.)

Q. 11.06. Show that the functions

$$\text{(a)}\ z,\quad \frac{z-2}{z-1},\quad 2-z,\quad \frac{z}{z-1};\qquad \text{(b)}\ z,\quad \frac{z-2}{z-1},\quad \frac{2}{z}$$

and another function which should be found, have the same group.

The transformations represented by the complex functions z, $-z$, $\bar{z}$, $-\bar{z}$ take the point with coordinates (x, y) into the point (x, y), $(-x, -y)$, $(x, -y)$, $(-x, y)$ respectively. These transformations are effected by the matrices:

$$\begin{bmatrix}1 & 0\\0 & 1\end{bmatrix},\quad \begin{bmatrix}-1 & 0\\0 & -1\end{bmatrix},\quad \begin{bmatrix}1 & 0\\0 & -1\end{bmatrix},\quad \begin{bmatrix}-1 & 0\\0 & 1\end{bmatrix},$$

and these combine under matrix multiplication to give the same group. These transformations which take the point (x, y) into the points $(\pm x, \pm y)$ may be interperted as:

$(x, y) \to (x, -y)$ reflection in the x-axis,
$(x, y) \to (-x, y)$ reflection in the y-axis,
$(x, y) \to (-x, -y)$ half-turn about the origin.

Inasmuch as reflection in the x- and y- axes may be thought of as half-turns† about those axes, we have once again the situation of p. 17, i.e. rotations about three mutually perpendicular axes. However, looking at them again as mirror reflections,† we may visualise the situation by thinking of two perpendicular mirrors m_1, m_2. There are three images in these mirrors, one by reflection in m_1, one by reflection in m_2, and one by successive reflections in m_1 and m_2. The latter is, in fact, a half-turn about the common line of the planes of the two mirrors, and you may see this image by looking into this line separating the two mirrors. The plan view (fig. 11.04) show the position of object and its three images. The reflections in the two mirrors *commute*, but would not do so if the mirrors were not perpendicular, and this is shown in fig. 11.05. Figure 11.06 is an attempt to show what the observer actually sees when he looks into the two perpendicular mirrors.

Q. 11.07. Find another set of four 2×2 matrices which have the same property of forming the group described in this chapter under matrix multiplication.

† See p. 187.

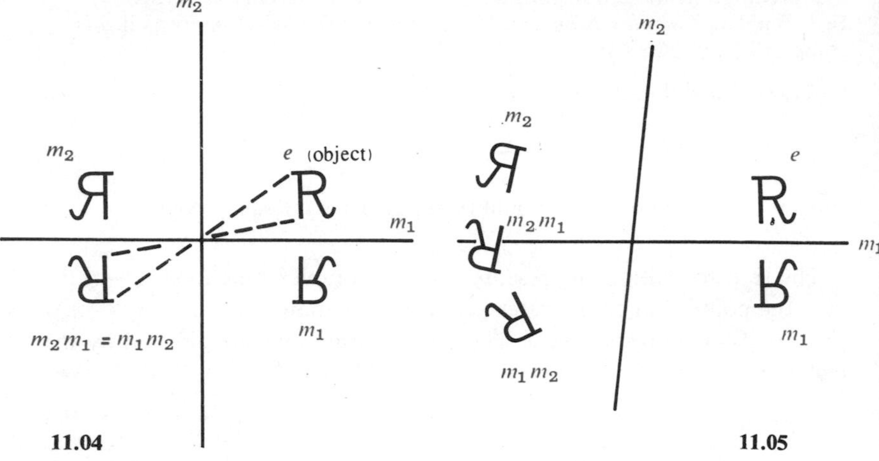

Fig. 11.04. Perpendicular mirrors $m_1m_2 = m_2m_1 =$ half-turn.
Fig. 11.05. Non-perpendicular mirrors reflections do not commute.

In Ch. 8 we saw that the operation Δ on sets possesses all the requirements of a group. Consider a very simple case, where there are only two objects in the universal set, say {apple, banana}. Then we have the

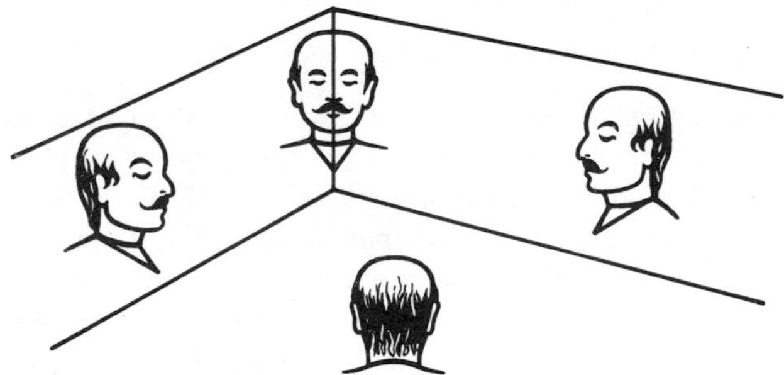

Fig. 11.06. Reflections in perpendicular mirrors.

following possible sets: $\mathscr{E}$ {apple, banana}, A {apple}, B {banana}, ϕ { }. Remembering that $A \Delta B$ yields the set containing the objects in A or B but not both, we see that here, $A \Delta B = \mathscr{E}$, $A \Delta A = \phi$, $A \Delta \mathscr{E} = \phi$, ... and you should check that the group table is:

Table 11.06

Δ	ϕ	A	B	$\mathscr{E}$	
ϕ	ϕ	A	B	$\mathscr{E}$	
A	A	ϕ	$\mathscr{E}$	B	
B	B	$\mathscr{E}$	ϕ	A	
$\mathscr{E}$	$\mathscr{E}$	B	A	ϕ	(D_2)

Next, consider three coins on a table, all showing heads. Suppose we are allowed to turn two of them over, and let the possible operations be labelled

$$\left.\begin{array}{l} x \text{ means `turn } B \text{ and } C \text{ over'} \\ y \text{ means `turn } C \text{ and } A \text{ over'} \\ z \text{ means `turn } A \text{ and } B \text{ over'} \end{array}\right\} \quad \text{Evidently, } x^2 = y^2 = z^2 = e, \text{ the identity.}$$

Then we have

$$h\,h\,h \xrightarrow{x} h\,t\,t \xrightarrow{y} t\,t\,h;$$

$$h\,h\,h \xrightarrow{yx} t\,t\,h.$$

Hence $yx = z$. You should verify that $yx = z = xy$ irrespective of the initial state (compare pp. 24–5). The complete group table will be found to agree with those so far displayed in this chapter.

Q. 11.08. Can one move from $h\ h\ h$ to $h\ h\ t$? Which moves are impossible? Consider the same game, only where one is allowed to turn only one coin over, say, a means 'turn coin A over', etc. Then obviously $x = cb = bc$. How many possible operations will there be? Draw up the group table.

In the next example, we consider a network of four points joined by six lines (a quadrangle), where the lines are coloured red, black and dashed 'opposite' sides being coloured alike. Starting from any point, which we label e (see fig. 11.07), we may take either the red route to reach the

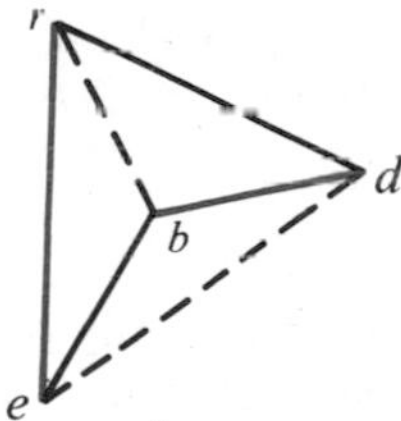

Fig. 11.07.

point r, or the black route to bring us to b, or the dashed to arrive at d. But y could have been reached by first taking the red and then the black route, and also by taking first the black and then the red route. So there is a sense in which we may combine routes, and write the result symbolically $d = br = rb$. It is evident that the symbols $\{e, r, b, d\}$ combine in the same way as in the other examples we have met. We have here an example of what is known as a Cayley diagram (see pp. 247 ff) for the group D_2. One does not need to think of a Cayley diagram as being two-dimensional – even this very simple configuration may be interpreted as a tetrahedron.

It is possible to represent even very complicated groups by means of Cayley diagrams, the elements being shown as vertices, and the operations by links connecting them, and we shall go into this in more detail in Ch. 15. An excellent account of this is given in Grossman and Magnus, *Groups and their Graphs* (Random House).

An instance of the same group occurs in Logic. If p and q are statements, then the sign $\rightarrow$ is used to denote 'implication':

$p \rightarrow q$ means 'If p is true, then q will also be true'.

Denote by a the operation of changing this to

$q \rightarrow p$ (the *converse*).

Using the symbol $\sim$ to denote 'not', we have two other possible propositions:

$\sim p \rightarrow \sim q$ (the inverse, b, of the original)
and $\sim q \rightarrow \sim p$ (the contrapositive, c, of the original).
We are *not* concerned with the *truth* of the propositions,† but merely with the changing of the original statement ($p \rightarrow q$) into the other three, these transformations being labelled a, b, c.

Suppose we form the inverse of a certain proposition, and then the converse of the result, i.e. we perform ab. We require the converse of $\sim p \rightarrow \sim q$, and this is $\sim q \rightarrow \sim p$, the contrapositive. Thus $ab = c$, and we may say that the converse of the inverse is the contrapositive.‡ We can also show that $ba = c$, $ac = ca = b$, $bc = cb = a$, $a^2 = b^2 = c^2 = e$, so the group table is identical with the others in the foregoing sections.

Finally,§ we may note that this same group crops up over and over again as a subgroup of larger groups.

Abstract group

In the foregoing, we have spoken repeatedly of 'the same' group, when what we really mean is that in all these widely differing situations, a group table results which has an *identical pattern*, except that perhaps different

† For example, if p is the statement '$ABCD$ is a cyclic quadrilateral', and q is '$AB.CD + AD.BC = AC.BD$', then e, a, b and c are all true. If p is the statement 'Σu_n is a convergent series', and q is the statement 'u_n tends to zero as n tends to infinity', then only e and b are true. If p is the statement '$AB^2 + BC^2 > AC^2$' and q is 'angle $ABC > \frac{1}{2}\pi$', then none of them is true.

‡ A pupil of mine once asked, apropos of Descargue's and Pappus' and other incidence theorems in projective geometry: 'Is the converse of the dual the same as the dual of the converse?' Those familiar with the principle of duality may care to think over this teaser!

§ See also *Mathematics Teaching*, No. 43, p. 35 in which T. J. Fletcher shows a most unusual realisation of the group D_2 in connection with the marriage rules of the Kariera, a tribe of Australian aboriginals.

letters have been used, and they have had different meanings in the different cases. In every case, the structure of the table has had 1's

Table 11.07

	1	a	b	c
1	1	a	b	c
a	a	1	c	b
b	b	c	1	a
c	c	b	a	1

along the leading diagonal (i.e. $a^2 = b^2 = c^2 = 1$), and $bc = cb = a$, $ca = ac = b$, $ab = ba = c$. (Here we have used 1 rather than e, since a multiplicative notation has been used, and 1 stands out better than e.) When we are not particularly concerned with what $a, b, c, \ldots$ represent, but only with the ways in which they combine, we have then *abstracted* superfluous considerations, and the table shows the *abstract group* which describes any one of the various *concrete* situations. You may like to compare this with the idea of abstraction in the concept of 'number'. We may speak of thirteen men, thirteen peas, thirteen baboons, thirteen chapters, thirteen years, ... and these are merely concrete instances of the abstract number thirteen. The word 'realisation' is the one chiefly used of the reverse process, e.g. the half-turns about three mutually perpendicular axes are one realisation of the abstract group displayed in the above table, and alternatively described by the defining relations, $a^2 = b^2 = 1$, $ab = ba$. Thus the word 'realisation' is, in the group sense, the *inverse* of the word 'abstraction'.

Isomorphic finite groups

Groups which have the same structure are called *isomorphic*; the above group is called the **Klein four-group**, and one may say, for example, that the permutations 1 2 3 4, 2 1 4 3, 3 4 1 2 and 4 3 2 1 combine to give *the* Klein four-group. Some groups have a name (e.g. the cyclic group of order 5, the alternating group of order 60, the quaternion group of order 8, and so on), but we prefer to identify groups by an abbreviation. Some of these abbreviations, which have appeared alongside the tables in previous chapters may have mystified you. Their meaning will unfold in due course. The code name for the Klein four-group is either D_2 or $C_2 \times C_2$, or $D_1 \times D_1$.

Two essentially different groups of order four

How many groups exist of order four? You may try re-arranging the order of the elements (according to the rules on p. 75) in D_2, but you will find that the *pattern* of the table is thereby unchanged. Does this mean that there is only one group of order four? By no means, because we have the following simple counter-examples drawn from finite arithmetics:

Tables 11.08

(a)

+ mod. 4	0	1	2	3
0	**0**	1	2	3
1	1	2	3	**0**
2	2	3	**0**	1
3	3	**0**	1	2

(b)

× mod. 5	1	2	3	4
1	**1**	2	3	4
2	2	4	**1**	3
3	3	**1**	4	2
4	4	3	2	**1**.

Now it looks as if we have produced two further different groups of order 4. But a little consideration may convince you that, in spite of the *apparent* dissimilarity (observe the different pattern of the identity element in each case), there are encouraging similarities.

In the first table there is one element (2) of period 2, and two (1, 3) of period 4. In the second table there is one element (4) of period 2, and two (2, 3) of period 4. This suggests that we should rearrange the elements in the second table so that the two elements of period 4 are separated, either:

Tables 11.09

(a)

× mod. 5	1	2	4	3
1	1	2	4	3
2	2	4	3	1
4	4	3	1	2
3	3	1	2	4

or

(b)

× mod. 5	1	3	4	2
1	1	3	4	2
3	3	4	2	1
4	4	2	1	3
2	2	1	3	4

and the identity of structure with the table for {0, 1, 2, 3}, + mod. 4 is immediately revealed – observe the 'stripes' running diagonally from NE. to SW. This group is the Cyclic Group of order 4, C_4, and we have met it already in numerous situations.

Q. 11.09. Collect all the instances so far found of the group C_4, both in the text and in the exercises.

It is easily possible to show that D_2 and C_4 are the only possible abstract groups of order 4. You should experiment with constructing group tables of order 4 in order to convince yourself that only two essentially different tables exist, and if possible prove this conclusively. In the next paragraph, we outline the rather more difficult proof that there are only *two* possible groups of order 6 (C_6 and D_3).

The groups of order six

To investigate the possible structures of groups of order 6, we shall assume the truth of Lagrange's theorem, from which it follows that the periods of the elements are 2, 3 or 6. If there is an element of period 6, the group is C_6. If not, let us assume there is an element p of period 3, so that $p^3 = e$, the identity element. Letting $p^2 = q$, we arrange the six elements in the order e, p, q, a, b, c along the borders, so that the subgroup $\{e, p, q\}$ appears in the top left-hand corner, and we suppose that b is the name of the element pa. The Latin-square property enables the whole of the first three rows to be filled up, as on p. 92.

Table 11.10

	e	*p*	*q*	*a*	*b*	*c*
e	*e*	*p*	*q*	*a*	*b*	*c*
p	*p*	*q*	*e*	*b*	*c*	*a*
q	*q*	*e*	*p*	*c*	*a*	*b*
a	*a*	*b*	*c*	.	.	.
b	*b*	*c*	*a*	.	.	.
c	*c*	*a*	*b*	.	.	.

(a) $ap = b$

	e	*p*	*q*	*a*	*b*	*c*
e	*e*	*p*	*q*	*a*	*b*	*c*
p	*p*	*q*	*e*	***b***	***c***	***a***
q	*q*	*e*	*p*	***c***	***a***	***b***
a	*a*	***c***	***b***	.	.	.
b	*b*	***a***	***c***	.	.	.
c	*c*	***b***	***a***	.	.	.

(b) $ap = c$

Application of the Latin-square property to the first three *columns* shows that the bottom left-hand 3 × 3 block must be filled entirely by a's, b's and c's, so that $ap =$ either b or c, and this leaves us with no choice for the remaining products in that block other than those shown in tables (a) and (b) above.

We now consider how the block in the bottom right-hand corner may be filled.

Suppose that $a^2 = p$. Then $ap = aa^2 = a^3 = a^2a = pa$, and this is only possible in table (a), the Abelian case. Similarly $a^2 = q$ is possible only in case (a). Hence in case (b) – the non-Abelian case – we must have $a^2 = e$, and similarly $b^2 = e$, $c^2 = e$. Then $ab = aaq = eq = q$, and the rest may be filled in by the Latin-square property:

Table 11.11

	a	*b*	*c*
a	*e*	***q***	***p***
b	***p***	*e*	***q***
c	***q***	***p***	*e*

Returning to the Abelian case (a), if $a^2 = p$, then $a^3 = ap = b \neq e$, so a can only be of period 6, and we have the group C_6. On the other hand, if $a^2 = e$ in the Abelian case, then $b = ap = pa$, so that $b^2 = paap = p^2$, and $b^3 = bb^2 = app^2 = ap^3 = a$, and this time b is of period 6 and the group is again C_6.

Q. 11.10. Show that the group cannot have five elements of period 2.

Setting up an isomorphism

It is one thing to *observe* that the group $\{0, 1, 2, 3\}$, + mod. 4 and $\{1, 2, 3, 4\}$,† × mod. 5 are isomorphic to C_4, and to each other. It is more interesting to see if we can discover the underlying *reason* behind it.

Before doing this, we shall first consider an isomorphism between two *infinite* groups, and this may give us a clue to the proposed question.

Now it is not possible, of course, to give a complete group *table* for an infinite group, so how are we to recognise isomorphism, or identity of structure? This is the point at which it is necessary to give a formal definition of isomorphism rather than to rely upon intuition:

Definition

Two groups $(\mathbf{X}, *)$ and $(\mathbf{Y}, \circ)$ are said to be isomorphic if:

(a) **it is possible to set up a 1,1 correspondence between the elements of X and those of Y,** i.e. such that, for each element x of X there corresponds a ***unique*** element y of Y, and vice versa, and also;

(b) **If x_1 and x_2 are two elements of X, and y_1, y_2 are the two corresponding elements of Y, and if:**

$$x_1 * x_2 = x_3 \quad \textbf{and} \quad y_1 \circ y_2 = y_3$$

then x_3 and y_3 are also corresponding elements.

Also (b) might have been re-stated: if x_1 and y_1 are a pair of mates, and x_2 and y_2 are a pair of mates, then $x_1 * x_2$ and $y_1 \circ y_2$ are also a pair of mates.

Briefly, therefore, we may describe isomorphism as *a 1,1 correspondence which preserves structure*, or which *preserves products*. *Note:* when two groups A and B are isomorphic, we write $A \cong B$. (It would hardly be right to say $A = B$, since A and B may be completely different realisations of the same abstract group, and it is complete nonsense to say that 'turning the trousers inside out' is *equal* to the matrix $\begin{bmatrix} 1 & 0 \\ 0 & -1 \end{bmatrix}$!)

Note that the above definition applies equally well to finite or to infinite sets, but that, in the case of finite groups, the first condition imposes the (obvious) requirement that isomorphic groups must have the *same order*.

† Or $\{\pm 1, \pm 2\}$.

(The corresponding requirement for isomorphic infinite groups is that they must have the same transfinite number (see D. Pedoe, *The Gentle Art of Mathematics*, Ch. 3 (C.U.P., 1957) for a brief discussion). Thus we could not possibly have an isomorphism between the rationals and the points of a line because they have different transfinite numbers, and so it is impossible even to set up a 1,1 correspondence.) But note that, while it may be easy to set up a 1,1 correspondence, it is usually much more difficult to prove that products are preserved – compare the proof of Cayley's theorem, see pp. 93–96.

But 1,1 correspondence (condition (a) above) is *not sufficient on its own.* For consider two groups of order 4:

Tables 11.12

	1	a	b	c
1	1	a	b	c
a	a	1	c	b
b	b	c	1	a
c	c	b	a	1

(D_2) and

	1	x	y	z
1	1	x	y	z
x	x	y	z	1
y	y	z	1	x
z	z	1	x	y

(C_4).

One may set up a 1,1 correspondence: $1 \leftrightarrow 1$
$a \leftrightarrow x$
$b \leftrightarrow y$
$c \leftrightarrow z$

but we do *not* have isomorphism, for though $ab = c$ and $xy = z$, (as required), nevertheless $ac = b$ while $xz = 1$ (not y). Indeed, $a^2 = 1$ whereas $x^2 = y$. The latter disagreement might have been expressed by the statement 'whereas in the first table a is of period 2, in the second table, x is of period 4'. Any alternative attempt to set up an isomorphism is doomed to failure.

A useful working rule

An important result (whose truth is self-evident) is that, if two finite groups are isomorphic, then not only have they the same number of elements, but also they each have the same number of elements of any specified period. The *converse*, which might be thought to be true is, *in fact true only of order less than 27*; i.e. suppose we have two groups of order 24, and each contains an identity, seven elements of period 2, two of period 3, eight of period 4, two of period 6 and four of period 12, then we may be sure that they are isomorphic, but only because $24 < 27$. On the other hand, if two groups *disagree* in the matter of numbers of elements of a given period, then they are *certainly not* isomorphic. Thus

C_4 has elements of period 1, 2, 4 and 4; whereas D_2 has elements of periods 1, 2, 2 and 2. This last result seems almost too obvious to deserve mentioning, yet it provides a most useful *negative test* to disprove isomorphism between two groups of any order (not only for orders less than 27), and we shall have occasion to refer to it in future work. As a further example, the group on p. 94 has two elements of period 4, whereas that on p. 101 has six of period 4, so though both groups are of order 8, they cannot possibly be isomorphic.

Again for infinite groups, the rotations in the plane about a fixed point form a group which is not isomorphic to the group (**R**, +) of the real numbers under addition. For though every rotation can be described by a real number (the angle rotated through, in radians), and these real numbers are added when rotations are combined, the difference is that for the rotations, the real numbers have been placed in equivalence classes (see pp. 352 ff), so that we regard ... $-\frac{1}{2}\pi$, $\frac{1}{2}\pi$, $2\frac{1}{2}\pi$, $4\frac{1}{2}\pi$, $6\frac{1}{2}\pi$, ... as being the same rotation. This means that we are adding the reals *modulo* 2π, and the group *contains elements of finite period*, e.g. $\frac{3}{5}\pi$ is of period 10, since $\frac{3}{5}\pi + \frac{3}{5}\pi + \frac{3}{5}\pi + \frac{3}{5}\pi + \ldots$ (10 terms) $+ \frac{3}{5}\pi = 6\pi = 0$ (mod. 2π), *for any multiple of* 2π *is the identity.*

Q. 11.11. Give examples of elements of period 2, 3, 4, ... in the group (**R**, + mod. 2π).
Describe the group (**R**, + mod. 1). What does this mean?

Isomorphism between infinite groups

To return to the question of isomorphism between infinite groups. Suppose we consider the groups ($\mathbf{R}^+$, ×) and (**R**, +). Then these can be shown to be isomorphic by the simple process of mapping the first set onto the second by the logarithm function: $x \to \log x$.

Then, if $x' = \log x$, $y' = \log y$, $z' = \log z$,
we have, $x' + y' = \log x + \log y = \log xy = \log z = z'$,

so that structure is preserved, and the isomorphism property established.

Q. 11.12. Show that the group (**Z**, +) is isomorphic to the group $\{\ldots -6, -3, 0, 3, 6, 9, \ldots\}$ under addition, and generalise the result. (We denote the latter group by (3**Z**, +).)

Q. 11.13. Show that (**Z**, +) is isomorphic to the infinite group generated by the function $f(x) \equiv x + a$ ($a \in \mathbf{R}$), and also to the group generated by a single vector under vector addition (i.e. if the vector is **u**, the group will contain **u** + **u**, **u** + **u** + **u**, etc.).

Q. 11.14. Show that the group $\{10^m, m \in \mathbf{Z}\}$, × is isomorphic to (**Z**, +).

Q. 11.15. Show that $\{\ldots -x^{-3}, x^{-2}, -x^{-1}, 1, -x, x^2, -x^3, \ldots\}$, $\times$ is isomorphic with the ordered pairs $\{\ldots (-3, 1), (-2, 0), (-1, 1), (0, 0), (1, 1), (2, 0), (3, 1), \ldots\}$ under $*$, where $(x_1, y_1) * (x_2, y_2) = (x_1 + x_2, y_1 + y_2 \text{ (mod. 2)})$.

Isomorphism between additive and multiplicative groups in finite arithmetics

Considering again the groups $\{0, 1, 2, 3\}$, + mod. 4 and $\{1, 2, 3, 4\}$, $\times$ mod. 5, we are now presented with the clue to the isomorphism: that the first set might be thought of as *logs*, or *indices*. Therefore, one tries to express each element in the second set in the form $\{a^0, a^1, a^2, a^3\}$ with a suitable choice of 'base' a, and in such a way that the indices are to be added modulo 4, i.e. so that $a^4 = 1$ (mod. 5). Since $2^4 = 1$ (mod. 5), we may evidently select the value 2 for a.

Q. 11.16. Can you find another value for a?

So we have (mod. 5): $1 = 2^0$, $2 = 2^1$, $3 = 2^3$, $4 = 2^2$, and the reason for the isomorphism now becomes apparent:

$$+ \text{ mod. } 4 \left\{\begin{matrix} 0 \leftrightarrow 2^0 = 1 \\ 1 \leftrightarrow 2^1 = 2 \\ 2 \leftrightarrow 2^2 = 4 \\ 3 \leftrightarrow 2^3 = 3 \end{matrix}\right\} \times \text{ mod. } 5$$

OR

$$+ \text{ mod. } 4 \left\{\begin{matrix} \log_2 1 = 0 \leftrightarrow 1 \\ \log_2 2 = 1 \leftrightarrow 2 \\ \log_2 3 = 3 \leftrightarrow 3 \\ \log_2 4 = 2 \leftrightarrow 4 \end{matrix}\right\} \times \text{ mod. } 5.$$

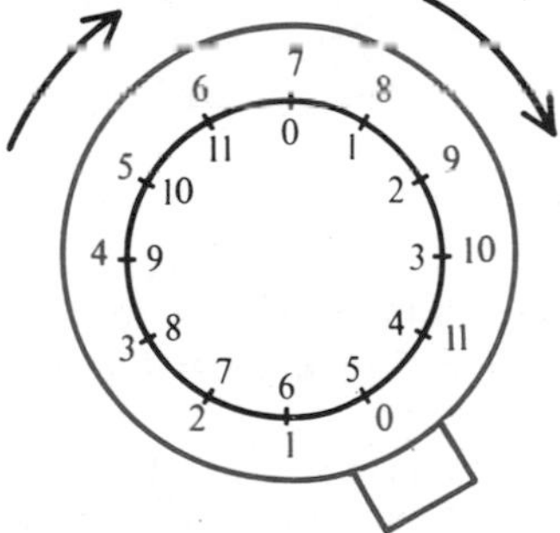

Fig. 11.081. Circular slide rule for addition mod. 12.

Slide rules for modular arithmetic

It is a simple matter to construct a circular slide rule for, say, addition modulo 12 ('clockface arithmetic'). Imagine that the inside scale in fig. 11.081 is fixed, and the outer ring is capable of rotation. In the position shown, the zero of the 'slide' is set against 5 on the fixed scale. Thus we can read off, for example, $5 + 4 = 9$; $5 + 9 = 2$ (mod. 12).

How does one construct a slide rule for *multiplication* in a finite arithmetic? It is merely a matter of regarding the readings on the slide rule for addition as being *indices* of some suitable base. The selection of a base for multiplication modulo n requires the discovery of a of period $n - 1$ such that $a^{n-1} = 1$ (mod. n), e.g. $2^4 = 1$ (mod. 5), as we saw in the previous section. Here we require $a^{12} = 1$ (mod. 13), and $a = 2$ fulfils this requirement. We have

$$\left.\begin{array}{lll} 2^0 = 1 & 2^4 = 3 & 2^8 = 9 \\ 2^1 = 2 & 2^5 = 6 & 2^9 = 5 \\ 2^2 = 4 & 2^6 = 12 & 2^{10} = 10 \\ 2^3 = 8 & 2^7 = 11 & 2^{11} = 7 \\ & & (2^{12} = 1) \end{array}\right\} \text{ modulo 13.}$$

Hence, to construct a circular slide rule for multiplication mod. 13, it is only necessary to replace the numbers 0, 1, 2, 3, 4, 5, 6, 7, 8, 9, 10, 11 on the slide rule for addition modulo 12 by 1, 2, 4, 8, 3, 6, 12, 11, 9, 5, 10, 7

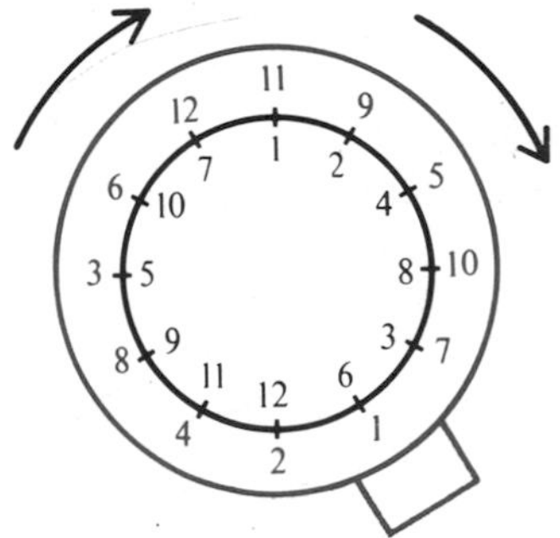

Fig. 11.082. Circular slide rule for multiplication mod. 13.

respectively, and this is shown in fig. 11.082, where you may read off from the given setting that $6 \times 3 = 5$; $6 \times 10 = 8$ (mod. 13).

Q. 11.17. Which other numbers serve as a suitable base for constructing a slide rule for multiplication mod. 13? (Note that, for example, 3 would not do, since $3^3 = 1$ (mod. 13).

Q. 11.18. Construct a circular scale for multiplication modulo 5 and modulo 7. Find out how many essentially different scales are possible.

We could also construct a circular slide rule for the following:

$\{1, 3, 7, 9\}$, $\times$ mod. 10 and $\{1, 5, 7, 11, 13, 17\}$, $\times$ mod. 18.

Find a correct cyclic order for labelling these slide-rules, and construct more examples by reference to pp. 48, 162–3.

Q. 11.19. What condition on n has to be satisfied so that a slide-rule may be successfully constructed for the multiplication of the *whole* set $\{1, 2, 3, \ldots, n-1\}$, mod. n?

In the case of multiplication modulo 5, we observed that either 2 or 3 would serve as a base, but not 4, since 4 was of period 2 ($4^2 = 1$). But the equation $a^4 = 1$ can be satisfied not only in a modular arithmetic, but also in the field of complex numbers: $i^4 = 1$ or $(-i)^4 = 1$. Thus we get the set $\{i^0, i^1, i^2, i^3\}$ or $\{1, i, -1, -i\}$ which gives the group C_4 under multiplication (table 11.13). Now in the Argand diagram, multiplication by

Table 11.13

$\times$	1	i	-1	$-i$
1	1	i	-1	$-i$
i	i	-1	$-i$	1
-1	-1	$-i$	1	i
$-i$	$-i$	1	i	-1

(C_4)

i represents an anticlockwise rotation through a right-angle; multiplication by -1 is a half-turn, and multiplication through $-i$ a clockwise turn through a right-angle. Thus we see the correspondence between the rotations about an axis through integral numbers of right-angles (see pp. 81–2, and the aspect of C_4 just described. Indeed, the table showing $\{0, 1, 2, 3\}$, + mod. 4 simply expresses the *number of right-angles turned through.*

The group $\{1, i, -1, -i\}$, $\times$ is simply a special case of the group $\{1, \omega, \omega^2, \omega^3, \ldots \omega^{n-1}\}$, $\times$, where ω is a (primitive) nth root of unity (see Ch. 12, pp. 150, 162, 167).

Automorphisms of C_4

Here again is the table for C_4, the cyclic group of order 4:

Table 11.14

	1	p	a	q	period
1	1	p	a	q	1
p	p	a	q	1	4
a	a	q	1	p	2
q	q	1	p	a	4

(Note the diagonal 'stripes'.)

(C_4).†

† From now on, we shall, *when convenient*, use the following notation for abstract groups, based on the multiplicative notation using juxtaposition: 1 for the identity (this will enable it to 'stand out'; the letter e has several defects.)
Small letters for other elements: $a, b, c, \ldots$ for elements of period 2, later letters of the alphabet for elements of higher period (e.g. p, p^2, q, q^2 for elements of period 3). The letters $x, y, z, \ldots$ will usually be used either for variable elements of a group, or for elements which are undetermined.

There will, however, be numerous departures from this nomenclature, which we use only, as stated, when convenient.

Now it must have occurred to you that, of the several metamorphoses of the table (see p. 140) above, the following method of redrafting it bears a remarkable similarity to the original:

Table 11.15

	1	q	a	p
1	1	q	a	p
q	q	a	p	1
a	a	p	1	q
p	p	1	q	a

(C_4)

because it is merely the original table *with p and q* interchanged, i.e. the table has been subjected to the transposition $(p\,q)$ (see p. 111). Another way of looking at it is from the matrix point of view:

$$\begin{bmatrix}1&0&0&0\\0&0&0&1\\0&0&1&0\\0&1&0&0\end{bmatrix}\begin{bmatrix}1&p&a&q\\p&a&q&1\\a&q&p&1\\q&1&p&a\end{bmatrix}\begin{bmatrix}1&0&0&0\\0&0&0&1\\0&0&1&0\\0&1&0&0\end{bmatrix}$$

$$=\begin{bmatrix}1&0&0&0\\0&0&0&1\\0&0&1&0\\0&1&0&0\end{bmatrix}\begin{bmatrix}1&q&a&p\\p&1&q&a\\a&p&1&q\\q&a&p&1\end{bmatrix}=\begin{bmatrix}1&q&a&p\\q&a&p&1\\a&p&1&q\\p&1&q&a\end{bmatrix}$$

(compare p. 130).

Obviously p and q are the only elements of this group which could possibly be considered to be interchangeable, because each is of period 4. 1 and a both stand alone, being of periods 1 and 2 respectively. If the group were realised as the rotations through multiples of a right-angle as in fig. 16.13, then p and q would both be quarter-turns, whereas a would be a 'different kind' of transformation, a half-turn. So p and q are interchangeable in the sense that clockwise and anticlockwise are interchangeable by the simple process of *looking at the figure from the other side.*

Thus we have an isomorphism of the group $\{1, p, a, q\}$ on to *itself* $\{1, q, a, p\}$ by the correspondence:

$$1 \leftrightarrow 1$$
$$p \leftrightarrow q$$
$$a \leftrightarrow a$$
$$q \leftrightarrow p.$$

Such an isomorphism of a group onto itself is, for obvious reasons, called an *automorphism*, and you may appreciate an automorphism as a *re-labelling* of the elements of the group. Concrete examples appear later in the text which will make the concept clearer, and the subject is expanded in Ch. 22.

Automorphisms of the Klein four-group

In the case of the group D_2, we have more scope for automorphism, since there are three elements of period 2. Using the notation 1, a, b, c, with $a^2 = b^2 = c^2 = 1$, $a = bc$, $b = ca$, $c = ab$, it is evident that any permutation may be applied to the elements a, b and c to produce an automorphism. For example, consider the permutation $\begin{pmatrix} a & b & c \\ c & b & a \end{pmatrix}$ which takes the original table (11.16)

Table 11.16

	1	a	b	c
1	1	a	b	c
a	a	1	c	b
b	b	c	1	a
c	c	b	a	1

(D_2)

into the form

	1	c	b	a
1	1	c	b	a
c	c	1	a	b
b	b	a	1	c
a	a	b	c	1

.

It is evident that any one of the six cycles: (bc), (ca), (ab), (abc), (acb) and the identity will produce an automorphism of this group. These six permutations themselves form a group (see p. 199 ff). In fact there is a theorem that *the automorphisms of any group themselves form a group* (See p. 422.)

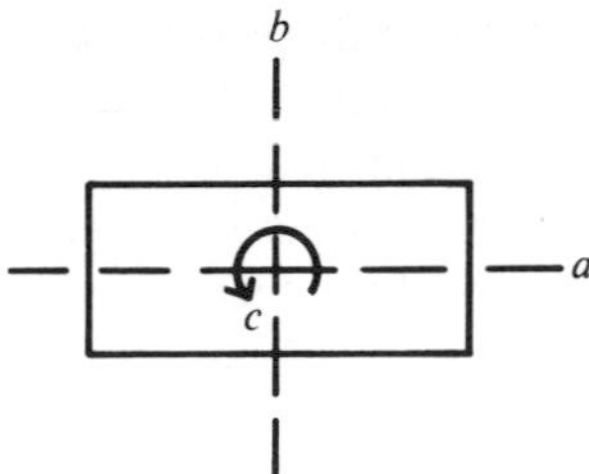

Fig. 11.09. Symmetries of rectangle.

The group D_2 may be realised as the group of the symmetries of the rectangle (or rhombus, or ellipse, etc., see Exs. 11.08 and 17.01). For example, a and b would be half-turns about the axes of symmetry lying in the plane (see fig. 11.09), while c would be the half-turn which interchanges opposite vertices of the rectangle. You may be perfectly satisfied about the automorphism $a \to b$, $b \to a$, but you may feel intuitively that there is 'something different' about the half-turn c about an axis perpendicular to the plane. This is an illusion, and it should be dispelled by the thought that the rectangle could just as well be a rectangular box (brick), and we should then have the situation of p. 17. The rectangle, indeed, may be thought of as a brick of zero thickness!

Automorphisms of C_6

Next consider the cyclic group of order 6 (labelled C_6 (table 11.17)), and in this case, we shall give a realisation of the group in terms of the

Table 11.17

$\times$	1	ω	ω^2	ω^3	ω^4	ω^5	period
1	1	ω	ω^2	ω^3	ω^4	ω^5	1
ω	ω	ω^2	ω^3	ω^4	ω^5	1	6
ω^2	ω^2	ω^3	ω^4	ω^5	1	ω	3
ω^3	ω^3	ω^4	ω^5	1	ω	ω^2	2
ω^4	ω^4	ω^5	1	ω	ω^2	ω^3	3
ω^5	ω^5	1	ω	ω^2	ω^3	ω^4	6

(C_6)

multiplication of the set of complex numbers $1, \omega, \omega^2, \omega^3, \omega^4, \omega^5$, where $\omega = \cos \frac{1}{3}\pi + i \sin \frac{1}{3}\pi$, $\omega^6 = 1$. Which, if any, automorphisms are feasible now? We may consider the possibility of the interchange of the two elements of period 6 (ω and ω^5), and this would automatically have the effect of interchanging ω^2 and ω^4, the elements of period 3, since $\omega^4 = (\omega^5)^2$. This is, indeed, the only possible automorphism of this group. You should complete the table and verify that the transposition (ω, ω^5) has the desired result. We say that ω and ω^5 are 'primitive' elements of the cyclic group, for either of them, of period 6, generates the whole group: the group may be written entirely in terms not only of ω, but also of ω^5:

$$\omega^5 = \omega^5 \qquad (\omega^5)^3 = \omega^3 \qquad (\omega^5)^5 = \omega$$

$$(\omega^5)^2 = \omega^4 \qquad (\omega^5)^4 = \omega^2 \qquad (\omega^5)^6 = 1$$

In the next chapter, where we consider cyclic groups in more detail, we shall deal with the question of automorphisms of cyclic groups of higher order.

Q. 11.20. Show that the group $\{0, 1, 2, 3, 4\}$, + mod. 5 has four automorphisms, including the identity. Investigate the automorphisms of the cyclic groups of order 3, 7 and 8. How many automorphisms are there for a group of order p (p prime)?

Automorphisms of non-Abelian groups

When we come to non-Abelian groups, we find that they are usually much more interesting from the point of view of automorphisms. The non-Abelian group of lowest possible order is D_3 (see pp. 76, 197–202). The table is reproduced below (table 11.18), with the periods of each element listed.

Table 11.18

	1	p	q	a	b	c	period
1	1	p	q	a	b	c	1
p	p	q	1	b	c	a	3
q	q	1	p	c	a	b	3
a	a	c	b	1	q	p	2
b	b	a	c	p	1	q	2
c	c	b	a	q	p	1	2

(D_3)

The first possibility for an automorphism that may occur to us is the interchange of the elements p and q of period 3. (This indeed is the only possible automorphism for the *subgroup* $\{1, p, q\}$. We should find that, in order to preserve structure, the interchange of p and q would also need to be accompanied by an interchange of *one pair* of a, b and c. For example in the case when a and c are interchanged, we obtain the automorphism:

$1 \leftrightarrow 1$
$p \leftrightarrow q$
$q \leftrightarrow p$
$a \leftrightarrow c$
$b \leftrightarrow b$
$c \leftrightarrow a$

and the table

Table 11.19

	1	q	p	c	b	a
1	1	q	p	c	b	a
q	q	p	1	b	a	c
p	p	1	q	a	c	b
c	c	a	b	1	p	q
b	b	c	a	q	1	p
a	a	b	c	p	q	1

(D_3).

You should study the other two automorphisms of this group, and draw up the group tables.

But there is another type of automorphism, and you may have guessed that this would result from a *cyclic* permutation of a, b and c. In this case, we do not have to interchange p and q, for in the original table we have $pa = b$ and since the automorphism replaces a by b and b by c, and since it is true that $pb = c$ (whereas $qb \neq c$), it follows that p and q must evidently be retained in their original positions. Hence the automorphism in this case is

$$\begin{aligned} 1 &\leftrightarrow 1 \\ p &\leftrightarrow p \\ q &\leftrightarrow q \\ a &\leftrightarrow b \\ b &\leftrightarrow c \\ c &\leftrightarrow a \end{aligned}$$

and you would be well advised to write out the table so obtained.

Hence there are *six* (including the identity) automorphisms of the group D_3, and they correspond to all the permutations of $\{a, b, c\}$. You should interpret these automorphisms in terms of the various realisations of the group D_3 discussed in Ch. 13, in particular for the group of the symmetries of the equilateral triangle.

Inner automorphisms

In the case of non-Abelian groups, there is a systematic way of obtaining a certain class of automorphisms (known as 'inner' automorphisms). Suppose x and y are any two elements of a group. Then the element xyx^{-1} is known as the 'transform of y by x', and is of enormous importance. This importance is recognised by the space given to this concept in later chapters (see Ch. 20).

If x remains fixed, while y runs through all the elements of the group, i.e. we consider the elements $xy_1x^{-1}, xy_2x^{-1}, xy_3x^{-1}, \ldots, xy_nx^{-1}$, then we can show that the correspondence $y_r \leftrightarrow xy_rx^{-1}$ constitutes an automorphism. Let us illustrate this in the case of D_3 by taking $x = p$, so that $x^{-1} = q$. Then letting y be equal to 1, p, q, a, b, c in turn:

$$\left.\begin{aligned} p1p^{-1} &= p1q = pq &&= 1 \\ ppp^{-1} & &&= p \\ pqp^{-1} &= pqq = pp &&= q \\ pap^{-1} &= paq = bq &&= c \\ pbp^{-1} &= pbq = cq &&= a \\ pcp^{-1} &= pcq = aqq &&= b \end{aligned}\right\}$$

Thus we derive the automorphism:

$$\begin{aligned} 1 &\leftrightarrow 1 \\ p &\leftrightarrow p \\ q &\leftrightarrow q \\ a &\leftrightarrow c \\ b &\leftrightarrow a \\ c &\leftrightarrow b \end{aligned}$$

Q. 11.21. Obtain the automorphisms by taking $x = q$; $x = a$; $x = b$; $x = c$ in turn.

Note that in the case of an *Abelian* group, $xyx^{-1} = xx^{-1}y = 1y = y$, and so the only 'inner' automorphism is the trivial one which associates each element with itself. Hence our earlier statement that non-Abelian groups are usually more interesting in automorphisms.

Q. 11.22. Find a cyclic group whose automorphism group is not cyclic.

Q. 11.23. All cyclic groups have no inner automorphisms, save the identity. Is there a group which has inner but no outer automorphisms?

Q. 11.24. Show that the order of the group of inner automorphisms is less than or equal to the order of the group. Find a group for which these orders are equal.

EXERCISES 11

1 A student working an examination paper was asked to say whether two given groups A and B were isomorphic. He answered: 'A is isomorphic, but B is not'. Construct a joke isomorphic to this joke.

2 Show that the group $\{1, i, -1, -i\}, \times$ is isomorphic to $\{2, 4, 6, 8\}, \times$ mod. 10

3 Show that an isomorphism may be set up between the permutations:

1 2 3 4 5 6		1 2 3 4 5 6 7 8	
3 1 2 6 4 5		3 1 2 6 4 5 8 7	
2 3 1 5 6 4	on six objects,	3 1 2 6 4 5 7 8	on eight objects.
4 5 6 1 2 3	and the perms.	2 3 1 5 6 4 7 8	
6 4 5 3 1 2		2 3 1 5 6 4 8 7	
5 6 4 2 3 1		1 2 3 4 5 6 8 7	

4 Why cannot an isomorphism be found between:

(a) $(\mathbf{Z}, +)$ and $(\mathbf{Z}_5, +)$?

(b) the symmetries of the equilateral triangle, and the direct symmetries (rotations in its own plane) of the regular hexagon?

(c) the symmetries of the rectangular box and the symmetries of the square?

(d) the complex numbers under multiplication, and ordered pairs of reals (x, y) under addition?

5 Establish isomorphisms between $(\mathbf{Z}, +)$ and $(10^{\mathbf{Z}}, \times)$
and between $(\mathbf{Z}, +)$ and $(2\mathbf{Z}, +)$.
Generalise both results.

6 Show that if $x, y \in \mathbf{R}$, and $x * y$ is defined $(x + y)/(1 - xy)$, then $(\mathbf{R}, *)$ is a group which is isomorphic to the group of rotations about a fixed point. (Put $x = \tan \theta$.)

7 Establish the isomorphism between the tables for Q_4 on p. 97 and p. 101.

8 The group of the rectangle and the group of the rhombus are isomorphic from purely geometrical considerations. Consider the symmetries from the alternative point of view as permutations of the vertices A, B, C, D of the rectangle, or of the sides a, b, c, d of the rhombus.

9 Show that the ordered pairs (0, 0), (0, 1), (1, 0) and (1, 1) combined by the rule

$$(x_1, y_1) \oplus (x_2, y_2) = (x_1 + x_2 - 1, y_1 + y_2 - 1), + \text{ and } - \text{ mod. } 2$$

form the group D_2, with (1, 1) as the identity. Interpret this in terms of binary numerals for the integers 0, 1, 2 and 3. Extend to (a) ordered triples from the set {0, 1}, and (b) ordered pairs from {0, 1, 2}, with addition modulo 3 instead of modulo 2.

10 Consider the statement: 'If you eat less you will lose weight.' Form the converse, the inverse and the contrapositive, i.e. perform the operations a, b and c of p. 138. Consider which of the resulting statements might be true!

11 Suppose a firm makes two shapes of car, a saloon and a shooting brake, and in two colours, black and white. A man always gets this make of car. Denote by a the operation 'he changes his car for a new one with the same shape but a different colour'; and by b the operation 'he changes for a car with the same colour but a different shape'. Show how the group D_2 is involved here. Extend to the case where there are (a) three styles, two colours; (b) three styles, three colours; (c) two styles, two colours and two different makes. In what way is this problem isomorphic with the illustration in the text of $A \Delta B$ from a universal set containing two members? (p. 136).

12 Invent another example about soft and hard-centred chocolates which may be plain or milk; and another about two electric-light switches.

13 Consider the binary operation $*$ on the lines through the origin, defined thus:
If l_1 is the line $y = m_1x$ and l_2 is the line $y = m_2x$, the $l_1 * l_2$ is the line $y = (m_1 + m_2)x$. Show that the lines through the origin form a group under $*$. Is this group isomorphic to the rotations about 0? (i.e. $R_{0,\alpha} * R_{0,\beta} = R_{0,\alpha+\beta}$).

14 Consider the group $\{0, 1, 2, 3, 4\}$, $+$ mod. 5 (the group C_5). Find multiplicative groups to which it is isomorphic.

15 For the group D_2, the equation $x^2 = 1$ (x an unknown element) has four solutions: ($x = 1$, a, b or c in table 11.16). For example, in multiplication mod. 12, the solutions are $x = 1, 5, 7$ or 11. How many solutions has $x^2 = 1$ in the group C_4? Using a finite arithmetic, find other examples of 'quadratic equations' which have more than two roots. Note that if you want a quadratic equation of the form $ax^2 + bx + c = 0$, you must have *two operations*, addition and multiplication, so you must consider a system in which these operations are defined, preferably a *field*, such as $\{0, 1\}$, $+$, $\times$ (mod. 2), or $\{0, 1, 2\}$, $+$, $\times$, mod. 3. In the group $C_2 \times C_2 \times C_2$ (see table 16.10, p. 267 ff), how many solutions of $x^2 = 1$ are there?

16 In Ex. 8.10, the eight permutations

1	1	2	3	4	5	6	7	8
p	4	3	2	1	8	7	6	5
q	5	6	7	8	1	2	3	4
r	2	1	4	3	6	5	8	7
x	6	5	8	7	2	1	4	3
y	3	4	1	2	7	8	5	6
z	8	7	6	5	4	3	2	1
i	7	8	5	6	3	4	1	2

of Ex. 4.13 were shown to form a group. Verify Cayley's theorem in this case, showing that the permutations which take the first row of the group table $\{1, p, q, r, x, y, z, i\}$ into successive rows are precisely these eight permutations themselves, provided they are taken in a suitable order. How many automorphisms are there of this group?

17 Show that $x \to x^{-1}$ is an automorphism of a group G if and only if G is Abelian.

18 Show that $x \to -x$ is an automorphism of $(\mathbf{R}, +)$ and that $z \to \bar{z}$ is an automorphism of $(\mathbf{C}, +)$.
Are these also automorphisms of the respective multiplicative groups?
Find other automorphisms of $(\mathbf{R}, +)$, $(\mathbf{R}, \times)$, $(\mathbf{C}, +)$ and $(\mathbf{C}, \times)$.

19 We saw that the group of automorphisms of D_2 is D_3. Give other instances of Abelian groups whose group of automorphisms is non-Abelian.

20 Can the group of automorphisms of an infinite group be finite?

21 How many automorphisms has C_p; C_{pq} (p and q prime)?

22 Find inner automorphisms of other non-Abelian groups (D_4, Q_4, etc.) by finding the transform by an element of the group of each element in turn, as on p. 152.

23 In the table D_2

	1	a	b	c
1	1	a	b	c
a	a	1	c	b
b	b	c	1	a
c	c	b	a	1

the body of the table is unchanged when the bordering elements are subject to the permutation $\begin{pmatrix} 1 & a & b & c \\ b & c & 1 & a \end{pmatrix}$ giving:

Table 11.20

	b	c	1	a
b	1	a	b	c
c	a	1	c	b
1	b	c	1	a
a	c	b	a	1

Similarly in the group $\{1, p, p^2, p^3\}$, $p^4 = 1$, the permutation $\begin{pmatrix} 1 & p & p^2 & p^3 \\ p^2 & p^3 & 1 & p \end{pmatrix}$ leaves the body of the table unchanged, though in this case no other permutation of the four elements does so. Investigate such permutations in the case of: (a) Abelian groups; (b) D_4 (table, p. 94), and other non-Abelian groups; (c) in general.

24 Given $x, y \in G$, and a is a fixed element of G. A binary operation $*$ is defined $x * y = xay$. What is the identity, and the inverse of x? Prove that $(G, *)$ is a group which is isomorphic to the given group under the transformation $x \to a^{-1}x$.
Can you find another way of establishing the isomorphism? Is this an automorphism of the group? Illustrate in the case of well-known groups such as D_4. Interpret in the case when the original group is a group of sets with the operation Δ (see p. 9).

25 Prove that the automorphisms of a given group themselves form a group. (This means that each automorphism should be regarded as a permutation of the elements of the group. But in the proof you need only show that the mapping $\phi(a)$ which takes x into axa^{-1} satisfies the requirements C, N, I of Ch. 8.)

26 Consider the following two systems:
(a) ordered pairs under the operation $(x_1, y_1) * (x_2, y_2) = (x_1, y_2)$;
(b) points of the plane under the operation: $A * B$ is the intersection of the lines AU, BV, where U and V are two fixed, distinct points of the plane.

Give a geometrical interpretation of (a), and hence establish an isomorphism between the two systems. Show that $*$ is associative but not commutative. Would $*$ be associative if we had (x_2, y_1) on the right-hand side instead of (x_1, y_2)? What can be said about identities and inverses in the above systems?

What is the connection between this example and that of Exs. 2.08, 6.17, and 7.01?

12 *Cyclic groups*

We have already made several references to cyclic groups, and are now going to investigate them more thoroughly. The name 'cyclic' derives from the fact that such groups appear whenever we have a set of *cyclic permutations* of a number of objects. For example, the set

$$\begin{pmatrix} 0 & 1 & 2 & 3 & 4 \\ 1 & 2 & 3 & 4 & 0 \\ 2 & 3 & 4 & 0 & 1 \\ 3 & 4 & 0 & 1 & 2 \\ 4 & 0 & 1 & 2 & 3 \end{pmatrix}$$

of permutations form a cyclic group of order 5. If the first change $\begin{pmatrix} 0 & 1 & 2 & 3 & 4 \\ 1 & 2 & 3 & 4 & 0 \end{pmatrix}$ is called p, then the others are p^2, p^3, p^4, with p^5 equal to the identity permutation. The cycles are $p = (01234)$, $p^2 = (02413)$, $p^3 = (03142)$ and $p^4 = (04321)$. The group table is

Table 12.01

	1	p	p^2	p^3	p^4	period
1	1	p	p^2	p^3	p^4	1
p	p	p^2	p^3	p^4	1	5
p^2	p^2	p^3	p^4	1	p	5
p^3	p^3	p^4	1	p	p^2	5
p^4	p^4	1	p	p^2	p^3	5 .

and if you write p^0 and p^1 for 1 and p respectively, you should see Cayley's theorem (see p. 92) staring you in the face!

In this cyclic group, the elements were all expressed in terms of a single element p, namely p^0, p^1, p^2, p^3, p^4, i.e. all the possible distinct 'powers' of p. This is true of all cyclic groups:

The cyclic group of order n (denoted C_n) has elements $\{1, x, x^2, x^3, \ldots, x^{n-1}\}$ with x satisfying the relation $x^n = 1$ ($x^m \neq 1$ when $1 \leqslant m < n$).

Cyclic groups may be divided into two kinds: those for which the order is prime, and those for which it is composite. The former, like C_5 above, are in most respects less interesting, chiefly because they have no subgroups. Groups of composite order are more rewarding to study. You have already collected instances of the group C_4 in the previous chapter. In Ch. 23 we shall meet two examples of C_4 to be found in *music*, one in respect of harmony, and the other in connection with musical form, or the analysis of musical structure.

Nine realisations of the cyclic group of order six

We shall now consider a number of realisations of the group C_6, the abstract table for which is as follows:

Table 12.02

	1	x	x^2	x^3	x^4	x^5	period
1	1	x	x^2	x^3	x^4	x^5	1
x	x	x^2	x^3	x^4	x^5	1	6
x^2	x^2	x^3	x^4	x^5	1	x	3
x^3	x^3	x^4	x^5	1	x	x^2	2
x^4	x^4	x^5	1	x	x^2	x^3	3
x^5	x^5	1	x	x^2	x^3	x^4	6

(C_6)

Here we have listed the elements in their 'natural order', $1, x, x^2, x^3, x^4, x^5$, the result of which is to produce the NE.–SE. 'stripes', to which we have referred in previous chapters.

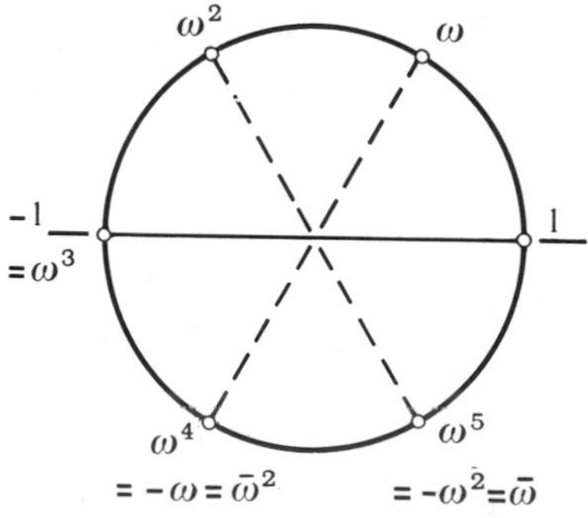

Fig. 12.01. The complex sixth roots of unity.

This group will arise whenever we have a situation represented by $x^6 = 1$.

(1) If $x = \cos 60° + i \sin 60° = \frac{1}{2}(1 + \sqrt{3}i) = \omega$, we obtain the sixth complex roots of unity, shown at the vertices of the regular hexagon in fig. 12.01. These form the group C_6 under multiplication.

(2) If x represents the permutation $\begin{pmatrix} 1 & 2 & 3 & 4 & 5 & 6 \\ 2 & 3 & 4 & 5 & 6 & 1 \end{pmatrix}$, then x^6 is clearly the identity permutation, and the permutations x, x^2, x^3, x^4, x^5 again combine according to the patterns of the table above.

Q. 12.01. Write down the permutations x, x^2, x^3, x^4, x^5 as *cycles*.

(3) Suppose x denotes an anticlockwise rotation through 60° about the point O.Then if O coincides with the centre of a regular hexagon (see fig. 12.02), each of the rotations x^0, x^1, x^2, x^3, x^4, x^5 will be a rotational

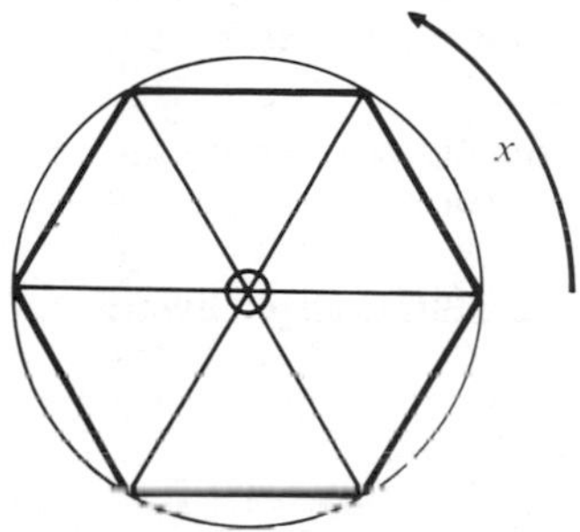

Fig. 12.02. The rotation of the regular hexagon.

symmetry of the hexagon (see pp. 78, 81), i.e. one could rotate the regular hexagon through 0, 60°, 120°, 180°, 240°, or 300°, and nobody would be the wiser if the change had been made in their absence. Note that the regular hexagon has also *reflective* symmetries (see pp. 78, 187 ff).

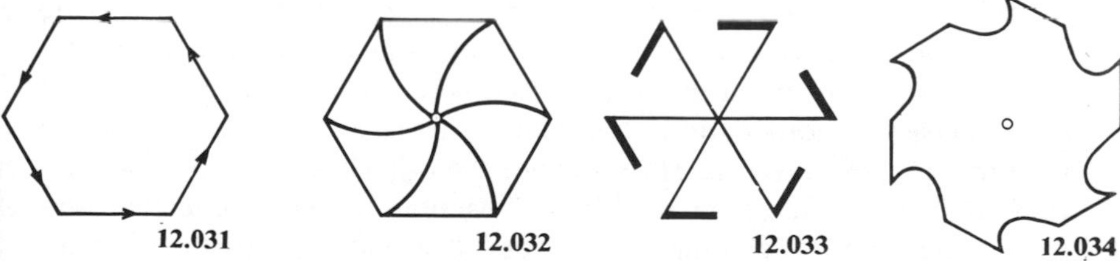

Fig. 12.03. The symmetry group C_6.

We may do better than just ignoring these. We can eliminate them altogether by modifying the figure in some such way as is shown in fig. 12.03.

Q. 12.02. Can you name at least two *solids* which have the symmetry group C_6?

(4) If $f(x) \equiv \dfrac{x+1}{2-x}$,

then $$f^2(x) \equiv \left(\frac{x+1}{2-x}+1\right)\Big/\left(2-\frac{x+1}{2-x}\right) \equiv \frac{1}{1-x};$$

$$f^4(x) = 1\Big/\left(1-\frac{1}{1-x}\right) \equiv \frac{x-1}{x};$$

$$f^6(x) \equiv 1\Big/\left(1-\frac{x-1}{x}\right) \equiv x \equiv i(x),$$

the identity function. Thus, $f^6 = i$, and we have C_6 again; the table above merely needs to have 1 and x replaced by i and f.

Q. 12.03. Find the other functions of this group, and indicate the subgroups. Find a different function of period 6.

(5) Consider the six permutations of five letters:

e	A	B	C	D	E
p	C	A	B	D	E
q	B	C	A	D	E
r	A	B	C	E	D
s	C	A	B	E	D
t	B	C	A	E	D

These are a subset of the 120 permutations of five letters, and you may easily verify that this set is closed under combination. Now we have here the *product of two cycles*, (CBA), of period 3, and the transposition (DE), of period 2. In the first three permutations e, p, q, the letters D and E are undisturbed, while in the case of the permutations e and r, the letters A, B and C are undisturbed. However, when both cycles have 'moved round', as in the case of both s and t, it is evident that such a permutation would have to be repeated six times to produce the original order, and you may verify that $s^6 = e$ and $t^6 = e$. This is all we need to convince us that we have the group C_6. To complete the work, it would be instructive if you were to express every element in terms of s and t; for example, $s^2 = q = t^4$.

(6) Consider a prism whose cross section is the shape shown in fig. 12.04. The triangle ABC is equilateral, and each of the triangles BCX, CAY, ABZ is congruent to half the triangle ABC. Evidently this plane figure has the symmetry group C_3, the symmetry operations being rotations through 0, 120° and 240° about the centre. But the whole prism not only has these rotational symmetries, but also has a reflective symmetry about a plane

midway between the two ends, as shown in fig. 12.042. A reflection in this plane would interchange the two ends of the prism. Call these two ends D and E. Let us now consider all the possible symmetries of the prism. We might, for example, combine a rotation through 120° about its central axis which replaces the letters A, B, C by C, A, B, with the reflection which interchanges the faces D and E. Call this symmetry $CABED$ .... You should not require much convincing that here we have precisely the same situation as we had in the previous section, since each of the possible

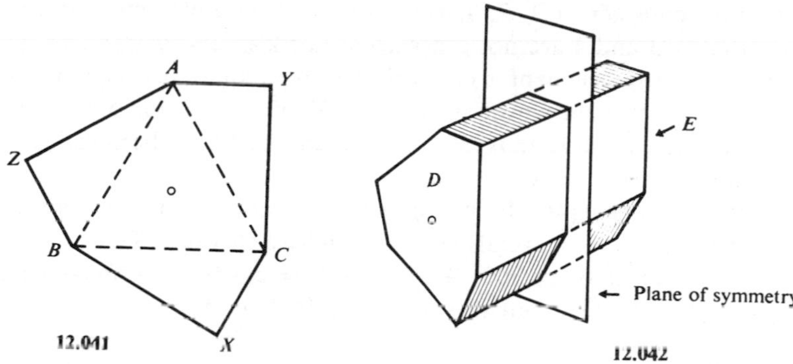

Fig. 12.04. Prism with symmetry group C_6.

symmetries of the prism corresponds to one of the permutations; for example, the rotation which replaces A, B, C by C, A, B, and reflects in the central plane of symmetry corresponds to the permutation s.

Q. 12.04. Consider the geometrical interpretation of the other permutations.

You will have noticed from the table on p. 157 that, in C_6, there is an element of period 2. For example, the multiplicative group of the sixth roots of unity includes the number -1, and $(-1)^2 = 1$. In the case of our prism, the element of period 2 is of course the reflection in the central plane.

Q. 12.05. Which are the two symmetries of period 3? Which is the element of period 2 in the set of functions generated by $f(x) = (x + 1)/(2 - x)$? (See p. 160.)

We used the word 'generated' in the last sentence, and it will appear frequently. A cyclic group of *prime* order is *generated* by any of its elements (except the identity). Suppose you consider the group $\{0, 1, 2, 3, 4\}$, $+$ mod. 5. Start with the element 3. To say that this element will generate the whole group simply means that we can derive *all* the other elements of the group by using it alone:

$$3 + 3 = 1 \text{ (mod. 5)}$$
$$3 + 3 + 3 = 4 \text{ (mod. 5)}$$
$$3 + 3 + 3 + 3 = 2 \text{ (mod. 5)}$$
$$3 + 3 + 3 + 3 + 3 = 0 \text{ (mod. 5)}.$$

Q. 12.06. Check that 2 and 4 also generate the whole group.

If you now consider $\{0, 1, 2, 3, 4, 5\}$ + mod. 6, you will immediately discover that 2, 3 and 4 are not generators. For $2 + 2 = 4$, $2 + 2 + 2 = 0$ (mod. 6), so that the element 2 has period 3. We require, to generate the whole group, an element whose period is 6. Verify that 1 and 5 are the only such elements. (Note that '5' could equally well have been called -1 (mod. 6).)

Elements which generate the whole group C_n (i.e. so that n is the *least* integer such that $x^n = 1$) are also called **primitive elements.** Thus, $\cos 60° + i \sin 60°$ and $\cos 300° + i \sin 300°$ $(= \cos 60° - i \sin 60°)$ are primitive sixth roots of 1, but they are not primitive twelfth, ... roots.

Q. 12.07. What are the primitive fourth roots, primitive eighth roots, of 1?

Q. 12.08. Show that in a group of prime order, p, all the elements $x, x^2, x^3, \ldots x^{p-1}$, are primitive.

(7) It may have struck you that we have wandered from our consideration of representations of the group C_6, so we now resume this quest with an example which you may feel should have appeared earlier: $\{0, 1, 2, 3, 4, 5\}$, + mod. 6.

You will readily write out the group table for this. What is its connection with the table on p. 157? The isomorphism is at once apparent when we realise that the integers 0, 1, 2, 3, 4, 5 are merely the 'indices' of x in the previous table, in which 1 and x have been replaced by x^0 and x^1.

(8) We noted that the group C_6 is bound to appear in any situation where $x^6 = 1$. This happens frequently in finite arithmetics, not only under addition, but also under multiplication. It is, of course, easy to proliferate examples with *addition* to some modulus, e.g. $\{-300, -200, -100, 0, +100, +200\}$, + mod. 600. In order to obtain multiplicative examples, it is only necessary to search for some *base* x, and some modulus n, so that $x^6 = 1$ (mod. n), while none of the smaller powers of x is equal to 1. For example, $2^6 = 64 = 1$ (mod. 21). Now $2^2 = 4$, $2^3 = 8$, $2^4 = 16$, $2^5 = 11$ (mod. 21), and so the set $\{1, 2, 4, 8, 16, 11\}$ is what we were looking for. You may verify, for example, that $4 \times 11 = 2$ (mod. 21).

Now in the above example while the *numbers* are being *multiplied mod.* 21, the *indices* are being *added mod.* 6 (since $2^6 = 1$, mod. 21), and this gives:

$$2^2 \times 2^5 = 2^7 \equiv 2^1 \quad (\text{mod. } 21) \quad (\text{numbers}),$$
$$2 + 5 = 7 \equiv 1 \quad (\text{mod. } 6) \quad (\text{indices}).$$

Another example is provided by noting that $3^6 = 1$ (mod. 14), and we have the set $\{1, 3, 3^2, 3^3, 3^4, 3^5\}$, i.e. {1, 3, 9, 13, 11, 5} with the required structure under multiplication modulo 14. The group table is, of course,

Table 12.03

× mod. 14	1	3^1	3^5	3^2	3^4	3^3
1	1	3^1	3^5	3^2	3^4	3^3
3^1	3^1	3^2	1	3^3	3^5	3^4
3^5	3^5	1	3^4	3^1	3^3	3^2
3^2	3^2	3^3	3^1	3^4	1	3^5
3^4	3^4	3^5	3^3	1	3^2	3^1
3^3	3^3	3^4	3^2	3^5	3^1	1

× mod. 14	1	3	5	9	11	13	period
1	1	3	5	9	11	13	1
3	3	9	1	13	5	11	6
5	5	1	11	3	13	9	6
9	9	13	3	11	1	5	3
11	11	5	13	1	9	3	3
13	13	11	9	5	3	1	2

that on p. 157 with x replaced by 3, but we rearrange in the equivalent version shown in table 12.03. Note that the 'stripes' have disappeared; but the table itself still has some symmetry, *apart from* the symmetry about the leading diagonal due to the commutative property. Look closely again at the table!

Q. 12.09. Use $2^6 = 1$ (mod. 7); $3^6 = 1$ (mod. 13) to construct similar examples. Show that {3, 6, 9, 12, 15, 18} × mod. 21 has the C_6 group structure. Find two sets of six which have the same structure under × mod. 14. Explore other examples.

(9) Miscellaneous examples:

Q. 12.10. Show that the ordered pairs (0, 1), (1, 1), (2, 1), (0, 2), (1, 2), (2, 2) combined thus:

$$(x_1, y_1) * (x_2, y_2) = (x_1 + x_2 - 1, y_1 y_2, \text{mod. } 3) \text{ give the group } C_6.$$

Q. 12.11. 'Addition' of ordered pairs is defined: $(x_1, y_1) \oplus (x_2, y_2) = (x_1 + x_2, \text{mod. } 3, y_1 + y_2, \text{mod. } 2)$, where $x_1, x_2 \in \{0, 1, 2\}$; $y_1, y_2 \in \{0, 1\}$. Show that there are six such ordered pairs, and that $\oplus$ produces C_6 group structure upon this set.

(10) The group C_6 is, of course, obtained by multiplication (mod. 7) in the set {1, 2, 3, 4, 5, 6}. We concentrate on 'row 3' of the group table:

Table 12.04

× mod. 7	1	2	3	4	5	6
.	.	.	.	.	.	.
.	.	.	.	.	.	.
→ 3	3	6	2	5	1	4
.	.	.	.	.	.	.

i.e. rearranging:

× mod. 7	1	3	2	6	4	5
.	.	.	.	.	.	.
.	.	.	.	.	.	.
3	3	2	6	4	5	1
.	.	.	.	.	.	.

This row tells us the successive multiples (mod. 7) of 3. But 3 = 10 (mod. 7), so we may read these as multiples of 10:

$$\left.\begin{aligned} 10 \times 1 &= 3 \\ 10 \times 2 &= 6 \\ 10 \times 3 &= 2 \\ 10 \times 4 &= 5 \\ 10 \times 5 &= 1 \\ 10 \times 6 &= 4 \end{aligned}\right\} \text{mod. } 7$$

Note that row 3 is the cycle (1 3 2 6 4 5) of the numbers of the set.†

Recurring decimals

The above has a hidden connection with *recurring decimals:* for consider the computation of $\frac{1}{7}$ as a decimal:

```
    0.1 4 2 8 5 7 1 ...
  7)1.0 0 0 0 0 0
    7
   ---
    3 0
    2 8
    ---
      2 0
      1 4
      ---
        6 0
        5 6
        ---
          4 0
          3 5
          ---
            5 0
            4 9
            ---
              1 0, etc.
```

† Compare the whole of the preceding section with the discussion of slide rules for multiplication in modular arithmetics, see p. 145 ff.

The numbers in the cycle referred to are the *successive remainders* in the division process, these being shown in bold type. The reason for this is as follows. When 10 is divided by 7, the remainder is 3: this is what we mean when we say that $10 = 3$ (mod. 7). Our next step is to divide 30 by 7, with remainder 2, i.e. $30 = 2$ (mod. 7), and so on, and it will be seen that the remainders follow the cycle exactly as in the row 3 of the multiplication table. The fact that there are *six* (the maximum possible number) of digits in the period of the recurring decimal is due to the fact that *every possible remainder does occur*, and the cycle therefore has its maximum length.

$$\begin{array}{ccccccccccc} 1 & & 3 & & 2 & & 6 & & 4 & & 5 \\ \downarrow & \nearrow & \downarrow & \nearrow & \downarrow & \nearrow & \downarrow & \nearrow & \downarrow & \nearrow & \downarrow \\ 3 & & 2 & & 6 & & 4 & & 5 & & 1 \end{array}$$

All this may become clearer when we consider the case of the fraction $\frac{1}{13}$:

```
       0.0 7 6 9 2 3 0 ...                    0.1 5 3 8 4 6 1 ...
   13)1.0 0 0 0 0 0 0 ...        but      13)2.0 0 0 0 0 0 0 ...
        0                                    1 3
      1 0 0                                    7 0
        9 1                                    6 5
          9 0                                    5 0
          7 8                                    3 9
          1 2 0                                  1 1 0
          1 1 7                                  1 0 4
              3 0                                    6 0
              2 6                                    5 2
                4 0                                    8 0
                3 9                                    7 8
                  1 0, etc.                              2 0, etc.
```

Here we do *not* get a cycle of maximum length (12 digits), but the process 'breaks up into two cycles'. If we had been converting any of the fractions $\frac{10}{13}$, $\frac{9}{13}$, $\frac{12}{13}$, $\frac{3}{13}$ or $\frac{4}{13}$ to decimals, we should evidently have found the same recurring sequence in the decimal ... 0 7 6 9 2 3 0 7 6 9 2 3 0 7 6 9 2 3 ..., because the numerators of those fractions are precisely the remainders which turned up in the first case. The recurring sequence would, of course, start at a different point in the case of each of the fractions. On the other hand, the fractions $\frac{2}{13}$, $\frac{7}{13}$, $\frac{5}{13}$, $\frac{11}{13}$, $\frac{6}{13}$ and $\frac{8}{13}$ all give the second alternative sequence of recurring digits in the quotient.

The underlying reason for all this is bound up with group theory. When we traced through the cycle (1 3 2 6 4 5) in the group $\{1, 2, 3, 4, 5, 6\}$, $\times$ mod, 7, we were using the process described on p. 107, for finding the *period* of an element. Here are the details in the above case:

1		3		2		6		4		5
↓	↗	↓	↗	↓	↗	↓	↗	↓	↗	↓
3		2		6		4		5		1

$$\begin{cases} 1 \times 3 = 3, & \text{i.e.} \quad 3^1 = 3 \\ 3 \times 3 = 2, & 3^2 = 2 \\ 2 \times 3 = 6, & 3^3 = 6 \\ 6 \times 3 = 4, & 3^4 = 4 \\ 4 \times 3 = 5, & 3^5 = 5 \\ 5 \times 3 = 1, & 3^6 = 1, \text{ or } 10^6 = 1 \text{ (mod. 7)}. \end{cases}$$

Thus the *period of* 10 in this group is 6 – i.e. 10 is a *primitive* element in the group.

Now repeat the process using mod. 13:

Table 12.05

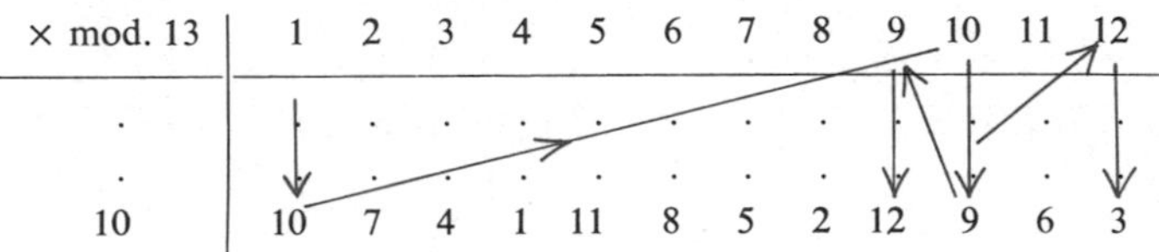

× mod. 13	1	2	3	4	5	6	7	8	9	10	11	12
.												
.												
10	10	7	4	1	11	8	5	2	12	9	6	3

This time we do not get a complete cycle, but the permutation breaks up into two cycles, which you should trace through to find that they are, (1 10 9 12 3 4) and (2 7 5 11 6 8). In other words, while it is true that $10^{12} = 1$ (mod. 13), it is also true that $10^6 = 1$ (mod. 13), so that the period of 10 is 6 and not 12. Note that we do not need to divide one million by 13 to check this: there is a shorter way: $10 = -3$ (mod. 13) $\Rightarrow 10^6 = (-3)^6 = 9^3 = (-4)^3 = -64 = +1$ (mod. 13).

Q. 12.12. Prove that $10^8 = -1$ (mod. 17), and investigate recurring decimals for fractions with denominator 17. Find other prime numbers (p) which give maximum period ($p - 1$) recurring decimals.

★We are here entering the rather specialised branch of mathematics known as the Theory of Numbers. There is an important theorem, attributed to Fermat, which states: **if n is an integer, and p is prime, then $n^{p-1} = 1$ (mod. p).** For example.

if $p = 7$,

$$1^6 = 2^6 = 3^6 = 4^6 = 5^6 = 6^6 = 1 \text{ (mod. 7)}$$

if $p = 13$,

$$1^{12} = 2^{12} = 3^{12} = \ldots = 10^{12} = \ldots = 12^{12} = 1 \text{ (mod. 13)}$$

if $p = 17$,

$$1^{16} = 2^{16} = 3^{16} = \ldots = 10^{16} = \ldots = 16^{16} = 1 \text{ (mod. 17)}.$$

Of course, Fermat's theorem does not enable us to forecast whether a recurring decimal representing the fraction m/p will have its full cycle of $(p - 1)$ digits. For, as we have seen, though $10^{12} = 1$ (mod. 13), it is also true that $10^6 = 1$ (mod. 13), with the result that the cycle of recurring digits has period 6. Fermat's theorem says nothing about this. (Further discussion of the connection between Fermat's theorem and Group Theory may be found in Grossman and Magnus, *Groups and their Graphs* (Random House), p. 88 and F. M. Hall, *Abstract Algebra I* (C.U.P.), p. 287.)

In considering recurring decimals, we were concerned with the period of the number 10, because we were working in *base* 10. Suppose we had been working in some other base, say base 8 (octal). Consider powers of 8 (mod. 11): $8^1 = 8$, $8^2 = 9 = -2$ (mod. 11), $8^3 = 6$ (mod. 11), and so on, and we should find that we do not satisfy the equation $8^k = 1$ (mod. 11) till k reaches the value 10: $8^{10} = (8^2)^5 = (-2)^5 = -32 = +1$ (mod. 11). The consequence of this is that when a fraction with denominator eleven (written 13 in base 8) is expressed in *octal* notation, we get the full period of ten digits recurring:
$(1/11)_{\text{base } 10} = (1/13)_{\text{base } 8} = 0.\dot{0}56{,}427{,}213{,}\dot{5}$, whereas in base ten, $1/11 = 0.\dot{0}\dot{9}$ with only two recurring digits because $10^2 = 1$ (mod. 11).†

Cyclic groups of prime order

Let us return to consider a cyclic group of prime order, ω say $n = 7$. If $\omega = \text{cis } 2\pi/7$, then ω^2, ω^3, ω^4, ω^5, ω^6 are the other complex roots of unity, and these together with 1 itself form the group C_7 under multiplication (see table 12.06). Now every element (save 1) in this group is of period 7, that is to say, cis $2\pi k/7$ is a *primitive* root of 1 for $k = 1, 2, 3, 4, 5, 6$. Another way of saying this is that any element of the group may be used to generate the whole group. Thus, putting $\Omega = \omega^3 = \text{cis } 6\pi/7$, we obtain the revised form in table 12.07, and if we set up a table showing *indices of*

† See F. J. Budden, *Number Scales and Computers*, Longmans, pp. 28–39, 62–74, 173–6. For an easily understood account of recurring decimals which goes into more detail, see an article by R. E. Green, *Mathematical Gazette*, Feb. 1963, pp. 25–33. Also consult *Some Lessons in Mathematics* (ed. T. J. Fletcher), C.U.P.; A. Bell, *Algebraic Structures*, p. 89, Q. 4 (Allen and Unwin), and *The Art of Algebra*, by R. North (Pergamon), where continued fractions are also treated. Other good sources are H. Rademacher and O. Toeplitz, *The Enjoyment of Mathematics*, Ch. 23, (Princeton U.P.), and the *Mathematical Gazette*, Dec. 1964 'Maximum length decimals', by S. N. Collings, pp. 384–6; and May 1956, pp. 137–8 by C. L. Wiseman. In an article 'Algebraic Structure', (*Ideas actuales de la Matematica y su Didactica*, Direccion General de Ensenanza Media, Madrid 1964), T. J. Fletcher traces, in a pack of cards, an unexpected connection between recurring decimals and card shuffles.

Table 12.06

×	1	ω	ω^2	ω^3	ω^4	ω^5	ω^6	period
1	1	ω	ω^2	ω^3	ω^4	ω^5	ω^6	1
ω	ω	ω^2	ω^3	ω^4	ω^5	ω^6	1	7
ω^2	ω^2	ω^3	ω^4	ω^5	ω^6	1	ω	7
ω^3	ω^3	ω^4	ω^5	ω^6	1	ω	ω^2	7
ω^4	ω^4	ω^5	ω^6	1	ω	ω^2	ω^3	7
ω^5	ω^5	ω^6	1	ω	ω^2	ω^3	ω^4	7
ω^6	ω^6	1	ω	ω^2	ω^3	ω^4	ω^5	7. (C_7)

Table 12.07

×	1	Ω	Ω^2	Ω^3	Ω^4	Ω^5	Ω^6
$\omega^0 = 1$	1	Ω	Ω^2	Ω^3	Ω^4	Ω^5	Ω^6
$\omega^3 = \Omega$	Ω	Ω^2	Ω^3	Ω^4	Ω^5	Ω^6	1
$\omega^6 = \Omega^2$	Ω^2	Ω^3	Ω^4	Ω^5	Ω^6	1	Ω
$\omega^2 = \Omega^3$	Ω^3	Ω^4	Ω^5	Ω^6	1	Ω	Ω^2
$\omega^5 = \Omega^4$	Ω^4	Ω^5	Ω^6	1	Ω	Ω^2	Ω^3
$\omega^1 = \Omega^5$	Ω^5	Ω^6	1	Ω	Ω^2	Ω^3	Ω^4
$\omega^4 = \Omega^6$	Ω^6	1	Ω	Ω^2	Ω^3	Ω^4	Ω^5

ω only (these being added modulo 7, of course), table 12.08 appears, and each row is a cyclic permutation of the previous row.

Table 12.08

+ mod. 7	0	3	6	2	5	1	4
0	0	3	6	2	5	1	4
3	3	6	2	5	1	4	0
6	6	2	5	1	4	0	3
2	2	5	1	4	0	3	6
5	5	1	4	0	3	6	2
1	1	4	0	3	6	2	5
4	4	0	3	6	2	5	1

Cyclic groups and regular polygons

Geometrically, this may be interpreted as a tour of the vertices of the star heptagon in fig. 12.05, these vertices being visited in the order of the cycle, each 'leg' of the tour corresponding to a multiplication by cis $6\pi/7$, or a rotation through $\frac{3}{7}$ of a revolution. The fact that every vertex is visited in turn corresponds to the fact that, if you change your socks every (say) three days, then, whichever day you start on, you must make a subsequent change on every possible day of the week. Here we are dealing with the modulo 7 arithmetic, and this statement about changing socks would remain true if 3 were replaced by 1, 2, 4, 5, 6, or for that matter by any number not divisible by 7. If we were dealing with a composite modulus (e.g. modulo 12), it would be a different story. If you pay, or rather receive, your electricity bill once a quarter, and the first account is received in February, then subsequent payments will be due in May, August,

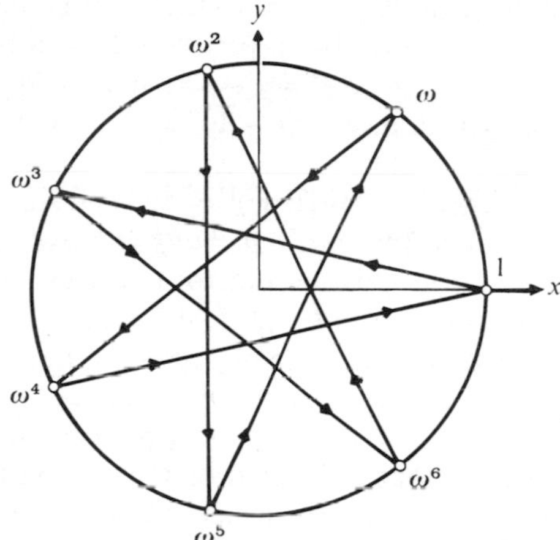

Fig. 12.05. Star heptagon. Vertices in the order 1, ω^3, ω^6, ω^2, ω^5, ω, ω^4.

November, February, . . . and you will never receive the bill in any other month. We shall deal with C_n when n is composite in the second half of this chapter.

Q. 12.13. Repeat the above, only taking $\Omega = \omega^2$, $\Omega = \omega^4$, etc., and consider the geometrical interpretation in each case. Repeat in the case of the eleventh, or the thirteenth, roots of 1.

The fact that every element in C_7 (except 1) is primitive (i.e. of period 7), and may be used as a generator means that here we have a system in which all the elements (apart from the unit) are *of equal status* – every member of this (so to speak) democracy is just as good as every other, except the identity, who is a sort of president. This means that the group C_7 has six automorphisms (see p. 148 ff and Ch. 22).

$\omega \to \omega$
$\omega \to \omega^2$
$\omega \to \omega^3$ (considered on pp. 167–8, Q. 22.07, Exs. 21.06, 12.25
$\omega \to \omega^4$ and 12.35)
$\omega \to \omega^5$
$\omega \to \omega^6$

The members of this 'democracy' may be thought of as being seated at a *round* table, with the Identity in the chair!

Before leaving the subject of cyclic groups *of prime order* (= p), let us prove that in such a group, *every element is of period* p. This stronger result, that there exists only one group of prime order, namely the cyclic group, will be proved later when we have studied Lagrange's Theorem.

If x is one generator, we have $x^p = 1$, and we wish to show that x^k $(k = 2, 3, 4, \ldots, p-1)$ is also of period p. Now it is easy to see that $(x^k)^p = 1$, for $(x^k)^p = x^{kp} = (x^p)^k = 1^k = 1$. What is not so obvious is that $(x^k)^q \neq 1$ for $q = 1, 2, 3, \ldots, p-1$, i.e. that the period of x^k is *not less than* p.

Suppose q is an integer such that $(x^k)^q = 1$. Then since $(x^k)^q = x^{kq}$, it follows that kq must be a multiple of p. *But* p *is prime*, and if it is to be a factor of the product kq, it must be a factor either of k or of q. But p is certainly not a factor of k, since $k \in \{2, 3, 4, \ldots, p-1\}$. Thus p must be a factor of q, or q must be a multiple of p; that is the only powers of x^k which give the identity are $\ldots -3p, -2p, -p, 0, p, 2p, 3p, \ldots$ Clearly then $(x^k)^q \neq 1$ when $q = 1, 2, 3, \ldots, p-1$, but $(x^k)^p$ does $= 1$. This is what we mean by saying that the period of x^k is p, and this completes the proof. But we shall see later an easier method based on Lagrange's Theorem – the above result follows immediately as a consequence of that important theorem – see pp. 215, 346.

Cyclic groups of composite order

We have already considered C_6, and noted that the only elements of period 6 are x and x^5 (where $x^6 = 1$). You will have no difficulty in picking out the following subgroups:

$\{1, x^3\}$ the two-group,

$\{1, x^2, x^4\}$ the group C_3, generated by either of the elements of period 3.

Q. 12.14. (a) Why does the proof at the end of the last section break down when n is composite? (b) Consider the geometrical interpretation of C_6 on the lines used in our consideration of C_7 (see p. 168ff). Notice the difference in the two cases.

In order to discuss cyclic groups of composite order, we shall find it more revealing to use a larger number, say 12, as the order. We shall show the table for $\{0, 1, 2, \ldots, 10, 11\}$, + mod. 12 in several different ways, the purpose of which will become apparent as you read on. (See tables 12.091–096.) The arrangement of the elements in version (b) may be

Table 12.091

+ mod. 12	0	1	2	3	4	5	6	7	8	9	10	11	period	
0	0	1	2	3	4	5	6	7	8	9	10	11	1	
1	1	2	3	4	5	6	7	8	9	10	11	0	12	
2	2	3	4	5	6	7	8	9	10	11	0	1	6	
3	3	4	5	6	7	8	9	10	11	0	1	2	4	
4	4	5	6	7	8	9	10	11	0	1	2	3	3	
5	5	6	7	8	9	10	11	0	1	2	3	4	12	
6	6	7	8	9	10	11	0	1	2	3	4	5	2	
7	7	8	9	10	11	0	1	2	3	4	5	6	12	
8	8	9	10	11	0	1	2	3	4	5	6	7	3	
9	9	10	11	0	1	2	3	4	5	6	7	8	4	
10	10	11	0	1	2	3	4	5	6	7	8	9	6	
11	11	0	1	2	3	4	5	6	7	8	9	10	12	(C_{12}) version (a)

Table 12.092

+ mod. 12	0	1	4	2	9	5	11	3	8	10	7	6	period	
0	0	1	4	2	9	5	11	3	8	10	7	6	1	
1	1	2	5	3	10	6	0	4	9	11	8	7	12	
4	4	5	8	6	1	9	3	7	0	2	11	10	3	
2	2	3	6	4	11	7	1	5	10	0	9	8	6	
9	9	10	1	11	6	2	8	0	5	7	4	3	4	
5	5	6	9	7	2	10	4	8	1	3	0	11	12	
11	11	0	3	1	8	4	10	2	7	9	6	5	12	
3	3	4	7	5	0	8	2	6	11	1	10	9	4	
8	8	9	0	10	5	1	7	11	4	6	3	2	3	
10	10	11	2	0	7	3	9	1	6	8	5	4	6	
7	7	8	11	9	4	0	6	10	3	5	2	1	12	
6	6	7	10	8	3	11	5	9	2	4	1	0	2	(C_{12}) version (b)

puzzling: why the order 0, 1, 4, 2, 9, 5, 11, 3, 8, 10, 7, 6? How was it arrived at? Remember that the set {1, 2, 3, ..., 12}, also gives the group C_{12} under multiplication modulo 13, and suitable generators (of period 12) are 2, 6, 7 and 11. Each element may be written as a *power of 2:*

$2^0 = 1$; $2^1 = 2$; $2^2 = 4$; $2^3 = 8$; $2^4 = 3$; $2^5 = 6$;

$2^6 = 12$; $2^7 = 11$; $2^8 = 9$; $2^9 = 5$; $2^{10} = 10$;

$2^{11} = 7$; $2^{12} = 1$.

When these are arranged in their 'natural order', we get:

1	2	3	4	5	6	7	8	9	10	11	12
2^0	2^1	2^4	2^2	2^9	2^5	2^{11}	2^3	2^8	2^{10}	2^7	2^6 .

The table for this set under multiplication modulo 13 is Table 12.10.

Table 12.093

+ mod. 12	0	6	1	7	2	8	3	9	4	10	5	11	period
0	0	6	1	7	2	8	3	9	4	10	5	11	1
6	6	0	7	1	8	2	9	3	10	4	11	5	2
1	1	7	2	8	3	9	4	10	5	11	6	0	12
7	7	1	8	2	9	3	10	4	11	5	0	6	12
2	2	8	3	9	4	10	5	11	6	0	7	1	6
8	8	2	9	3	10	4	11	5	0	6	1	7	3
3	3	9	4	10	5	11	6	0	7	1	8	2	4
9	9	3	10	4	11	5	0	6	1	7	2	8	4
4	4	10	5	11	6	0	7	1	8	2	9	3	3
10	10	4	11	5	0	6	1	7	2	8	3	9	6
5	5	11	6	0	7	1	8	2	9	3	10	4	12
11	11	5	0	6	1	7	2	8	3	9	4	10	12 (C_{12}) version (c)

Table 12.094

+ mod. 12	0	4	8	1	5	9	2	6	10	3	7	11	period
0	0	4	8	1	5	9	2	6	10	3	7	11	1
4	4	8	0	5	9	1	6	10	2	7	11	3	3
8	8	0	4	9	1	5	10	2	6	11	3	7	3
1	1	5	9	2	6	10	3	7	11	4	8	0	12
5	5	9	1	6	10	2	7	11	3	8	0	4	12
9	9	1	5	10	2	6	11	3	7	0	4	8	4
2	2	6	10	3	7	11	4	8	0	5	9	1	6
6	6	10	2	7	11	3	8	0	4	9	1	5	2
10	10	2	6	11	3	7	0	4	8	1	5	9	6
3	3	7	11	4	8	0	5	9	1	6	10	2	4
7	7	11	3	8	0	4	9	1	5	10	2	6	12
11	11	3	7	0	4	8	1	5	9	2	6	10	12 (C_{12}) version (d)

Table 12.095

+ mod. 12	0	3	6	9	1	4	7	10	2	5	8	11	period	
0	0	3	6	9	1	4	7	10	2	5	8	11	1	
3	3	6	9	0	4	7	10	1	5	8	11	2	4	
6	6	9	0	3	7	10	1	4	8	11	2	5	2	
9	9	0	3	6	10	1	4	7	11	2	5	8	4	
1	1	4	7	10	2	5	8	11	3	6	9	0	12	
4	4	7	10	1	5	8	11	2	6	9	0	3	3	
7	7	10	1	4	8	11	2	5	9	0	3	6	12	
10	10	1	4	7	11	2	5	8	0	3	6	9	6	
2	2	5	8	11	3	6	9	0	4	7	10	1	6	
5	5	8	11	2	6	9	0	3	7	10	1	4	12	
8	8	11	2	5	9	0	3	6	10	1	4	7	3	
11	11	2	5	8	0	3	6	9	1	4	7	10	12	(C_{12}) version (e)

Table 12.096

+ mod. 12	0	2	4	6	8	10	1	3	5	7	9	11	period	
0	0	2	4	6	8	10	1	3	5	7	9	11	1	
2	2	4	6	8	10	0	3	5	7	9	11	1	6	
4	4	6	8	10	0	2	5	7	9	11	1	3	3	
6	6	8	10	0	2	4	7	9	11	1	3	5	2	
8	8	10	0	2	4	6	9	11	1	3	5	7	3	
10	10	0	2	4	6	8	11	1	3	5	7	9	6	
1	1	3	5	7	9	11	2	4	6	8	10	0	12	
3	3	5	7	9	11	1	4	6	8	10	0	2	4	
5	5	7	9	11	1	3	6	8	10	0	2	4	12	
7	7	9	11	1	3	5	8	10	0	2	4	6	12	
9	9	11	1	3	5	7	10	0	2	4	6	8	4	
11	11	1	3	5	7	9	0	2	4	6	8	10	12	(C_{12}) version (f)

Table 12.10

× mod. 13	1	2	3	4	5	6	7	8	9	10	11	12	period	
1	1	2	3	4	5	6	7	8	9	10	11	12	1	
2	2	4	6	8	10	12	1	3	5	7	9	11	12	
3	3	6	9	12	2	5	8	11	1	4	7	10	3	
4	4	8	12	3	7	11	2	6	10	1	5	9	6	
5	5	10	2	7	12	4	9	1	6	11	3	8	4	
6	6	12	5	11	4	10	3	9	2	8	1	7	12	
7	7	1	8	2	9	3	10	4	11	5	12	6	12	
8	8	3	11	6	1	9	4	12	7	2	10	5	4	
9	9	5	1	10	6	2	11	7	3	12	8	4	3	
10	10	7	4	1	11	8	5	2	12	9	6	3	6	
11	11	9	7	5	3	1	12	10	8	6	4	2	12	
12	12	11	10	9	8	7	6	5	4	3	2	1	2	(C_{12})

If one wrote this out showing only the *indices of 2* (or logs to base 2, if you like), under addition, this would be version (b) of the table for the set {0, 1, 2, . . ., 11} under addition modulo 12.

Q. 12.15. Instead of 2 as the base (e.g. $\log_2 9 = 8$ (mod. 13)), use one of the other possible bases 6, 7, or 11 mentioned above. For example, $7^4 = 9$ (mod. 13), or $\log_7 9 = 4$ (mod. 13), and then draw up a further version of the table for {0, 1, 2, . . ., 11} + mod. 12.

Q. 12.16. Examine the symmetries of version (b) of the table. Notice, for example, that row 4 is row 10 backwards; row 1 is row 7 backwards, and so on. Why is this? Compare these pairs of numbers with the pairings in version (c) of the table.

The other four versions (c) to (f) of the table also need some explanation. You may already have observed that, in each case, the top left-hand square is a subgroup.

In (c) we have the subgroup	{0, 6}	(C_2).
In (d)	{0, 4, 8}	(C_3).
In (e)	{0, 3, 6, 9}	(C_4).
In (f)	{0, 2, 4, 6, 8, 10}	(C_6).

In our explanation, it will be easier if we realise the group by considering the rotations through 0, 30°, 60°, 90°, . . ., 330° applied to the regular

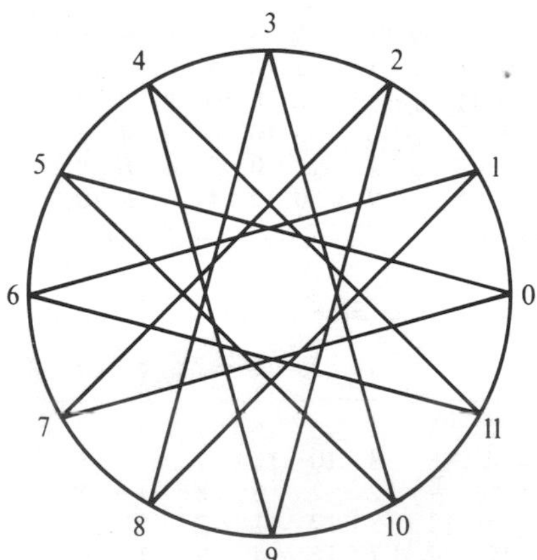

Fig. 12.06. Regular dodecagon: the cycle (0 5 10 3 8 1 6 11 4 9 2 7).

dodecagon (see fig. 12.06). If we start with 0 and keep adding 1 (i.e. rotating through 30°), we cover the perimeter of the dodecagon in the anticlockwise direction. If we had kept adding 11, we should arrive in turn at the vertices 11, 10, 9, . . . 2, 1, 0, and so cover the perimeter in the clockwise direction. If we increased in steps of 5: 0, 5, 10, 3, 8 ... we should also visit each vertex of the dodecagon as shown in the fig. 12.06. This time we have the element 5 generating the whole group.

Q. 12.17. Which element causes the star decagon to be traced out in the reverse order?

But if we go up in steps of 2, thus: 0, 2, 4, 6, 8, 10, 0, we never visit the odd vertices, but get a regular hexagon; this corresponds to the fact that the element 2 is of period 6. What happens in the case of the other element (10) of period 6? These elements form the subgroup C_6.

Elements of period 4 (3 and 9) give us the square 0, 3, 6, 9, and these form the subgroup C_4. Elements of period 3 (4 and 8) give us the equilateral triangle 0, 4, 8, and these form the subgroup C_3. Finally, {0, 6} is the subgroup C_2.

Taking version (e) of the table by way of example, we have shown the C_4 subgroup in the top left-hand corner. The other 'groupings', {1, 4, 7, 10} and {2, 5, 8, 11} are easily seen to correspond to the *two other squares* in fig. 12.06. We shall see later (Ch. 19) that these two sets are called '*cosets*' of the group by the subgroup {0, 3, 6, 9}. In the same way, the four equilateral triangles are the cosets by the subgroup {0, 4, 8} and these are made to stand out in version (d) of the table. In version (c), the cosets are represented by opposite ends of diameters, and these might be regarded as 'regular polygons of two sides' inscribed in the circle. Compare also with the section dealing with C_{12} in the chapter on music (Ch. 23, p. 437).

Q. 12.18. Are these other cosets, such as {1, 5, 9} subgroups?

Useful exercises on the work of this chapter so far might be:

(i) To consider the work on C_{12} from the point of view of various realisations such as were illustrated in the first part of the chapter for C_6.

(ii) To write out alternative versions of the table for {1, 2, 3, ..., 12}, $\times$ mod. 13, corresponding to the tables (a) to (f) for {0, 1, 2, ..., 11}, $+$ mod. 12 in the text.

(iii) To write out alternative versions of the group tables for, say, C_8, C_9 or C_{10}, and to analyse the subgroups as we have done in the case of C_{12}.

Some general results

Having 'seen inside' various cyclic groups, you will now be in a position to appreciate the following theorems:

(i) **Any subgroup of a cyclic group (of composite order) is necessarily cyclic.**

(ii) **If a group of order n has an element of period n, then it is necessarily the cyclic group C_n.**
(iii) **A group of prime order, p, is necessarily cyclic, and every element other than the identity has period p.**
(iv) **If x is a primitive element, then so is x^{-1}.**
(v) **All cyclic groups are Abelian (commutative).**

In connection with (i), we note that subgroups of C_n may be found by taking each element in turn and considering the group which it generates. For example, for the group C_6, $\{1, x, x^2, x^3, x^4, x^5\}$ ($x^6 = 1$), the whole group is generated by x and x^5; but x^2 generates the subgroup $\{1, x^2, x^4\}$, i.e. C_3; and x^3 generates the subgroup $\{1, x^3\}$, i.e. C_2.

Q. 12.19. Find a 'chain' of subgroups of C_{12}, such that each is a subgroup of the next, of the form: $C_1 \subset C_a \subset C_b \ldots \subset C_{12}$.

Q. 12.20. Find the subgroups of C_8, C_9 and C_{10}.

How to identify a cyclic group

(ii) is a useful result, because it means that if we have a set of (say) ten objects which satisfy all the requirements for a group (C, A, N, I, see Ch. 8), then if one element is known to be of period 10, we can be sure that the group is C_{10}. Here is an example illustrating the result. In the finite arithmetic modulo 3, we have $1^2 = 1$ and $2^2 = 1$ (mod. 3). Therefore, while we may say that the square root of 0 is 0, and the square root of 1 is either 1 or 2, there does not exist a square root of 2 ($= -1$, mod. 3). So *invent* a number $j = \sqrt{2} = \sqrt{-1}$ (mod. 3), and construct the following 'numbers':

$$0, 1, 2, j, 1 + j, 2 + j, 2j, 1 + 2j, 2 + 2j.†$$

It may easily be verified that all the group requirements are satisfied for this set under addition modulo 3, e.g. $(2 + j) + (1 + j) = 3 + 2j = 2j$ (mod. 3); $(2 + j) + (1 + 2j) = 0$, and these two elements are inverses. You may easily check that each element is of period 3. The group is therefore *certainly not* C_9, for this would require an element of period 9.

Q. 12.21. Construct the group table.

For multiplication, we must omit the element 0 as usual, and work with the remaining eight elements. As an example of typical products, and remembering that $j^2 = 2$ (or -1), we have:

$$(1 + j)^2 = 1 + 2j + j^2 = 1 + 2j + 2 = 2j \quad \text{(mod. 3)},$$
$$2j(2 + j) = 4j + 4 = j + 1 \quad \text{(mod. 3)}.$$

† They form a finite *vector space* with the basis 1, j over the finite field $\{0, 1, 2\}$.

Now,

$$(1+j)^4 = (2j)^2 = 4j^2 = 8 = 2 \qquad \text{(mod. 3)}.$$

Hence, $(1+j)^8 = 4 = 1$ (mod. 3). Thus $(1+j)$ is of period 8, for there are no smaller powers of $(1+j)$ which are equal to 1 (mod. 3). *Thus the group of order 8 formed by these 8 'numbers' is undoubtedly the group* C_8. (The elements could be written $(1+j)^k$, $k = 0, 1, 2, 3, 4, 5, 6, 7$).

Q. 12.22. Which other elements of this group have period 8?

Theorem (iii) means that there is ***only one group of prime order***. The converse is not true! There is only one group of order 15.

The *n*th roots of unity and the cyclotomic equation

★We have seen that the nth roots of unity form the group C_n under multiplication, and lie in the Argand diagram at the vertices of a regular n-gon. If ω denotes cis $2\pi/n$, then the roots are $1, \omega, \omega^2, \ldots, \omega^{n-1}$, and we may say

$$(x^n - 1) \equiv (x-1)(x-\omega)(x-\omega^2)\ldots(x-\omega^{n-1})$$

Now when n *is prime*, the roots $\omega, \omega^2, \ldots, \omega^{n-1}$ are all primitive. The equation whose roots are the primitive roots is called a cyclotomic equation, and in this case is:

$$(x-\omega)(x-\omega^2)\ldots(x-\omega^{n-1}) = 0,$$

i.e. $$\frac{x^n - 1}{x - 1} = 0,$$

or $x^{n-1} + x^{n-2} + x^{n-3} + \ldots + x^2 + x + 1 = 0$ (n prime).

★**Q. 12.23.** Write down the cyclotomic equation for the third, the fifth and the seventh roots of unity.

★ When n is composite, then not all of the complex roots will be primitive. For example, the roots of $x^4 = 1$ are $1, i, -1, -i$, and although -1 is a root of this equation, it is not a primitive root, for $(-1)^2 = 1$. In fact, -1 is a primitive *square root* of 1, but is not primitive for any higher order of roots. The primitive roots of $x^4 = 1$ are i and $-i$, and so the cyclotomic equation for the 4th roots of 1 is $x^2 + 1 = 0$.

However, though $i^8 = 1$ and $(-i)^8 = 1$, so that i and $-i$ satisfy the equation $x^8 = 1$, they are not primitive 8th roots since, as we have already seen, they are primitive roots of an equation of lower degree,

$x^4 = 1$. The primitive roots of $x^8 = 1$ are cis $\pi/4$, cis $3\pi/4$, cis $5\pi/4$, cis $7\pi/4$, i.e. $\pm 1/\sqrt{2} \pm i/\sqrt{2}$. The cyclotomic equation may be set up using these values, but it is easier to proceed as follows: The roots of $x^8 = 1$ which are *not* primitive roots are $1, i, -1, -i$, i.e. the roots of the equation $x^4 - 1 = 0$. Therefore the cyclotomic equation may be obtained from $x^8 - 1 = 0$ by omitting those factors which give rise to the non-primitive roots, i.e. by dividing by $x^4 - 1$:

$$\frac{x^8 - 1}{x^4 - 1} = x^4 + 1, \text{ so the cyclotomic equation is } x^4 + 1 = 0.$$

★**Q. 12.24.** Form the cyclotomic equation for the primitive sixth roots of unity.

We have covered the cyclotomic equation of the roots up to the eighth degree, and they are as follows:

$n = 1$: $x - 1 = 0$

$n = 2$: $x + 1 = 0$

$n = 3$: $x^2 + x + 1 = 0$

$n = 4$: $x^2 + 1 = 0$

$n = 5$: $x^4 + x^3 + x^2 + x + 1 = 0$

$n = 6$: $x^2 - x + 1 = 0$

$n = 7$: $x^6 + x^5 + x^4 + x^3 + x^2 + x + 1 = 0$

$n = 8$: $x^4 + 1 = 0$

★**Q. 12.25.** You may continue to obtain cyclotomic equations for $n = 9$, 10, 12, 14 . . .

The study of cyclic groups, and the results of the present chapter, enable the theory of cyclotomic equations to be worked out more easily.

★Other problems connected with cyclic groups

In trigonometry, why is it that we can form a cubic equation with integer coefficients whose roots are $+\cos \pi/7$, $-\cos 2\pi/7$, $-\cos 3\pi/7$, but cannot do it when all the signs are positive?

If $\omega = \text{cis } 2\pi/7$, why is it that we can form a quadratic equation with integer coefficients whose roots are $\omega + \omega^2 + \omega^4$ and $\omega^3 + \omega^5 + \omega^6$, whereas we cannot do so for $\omega + \omega^2 + \omega^3$ and $\omega^4 + \omega^5 + \omega^6$, or for $\omega + \omega^3 + \omega^4$ and $\omega^2 + \omega^5 + \omega^6$? Why does the first partition of the complex roots succeed, yet the others fail?

If $\omega^{11} = 1$, why should $(\omega + \omega^3 + \omega^4 + \omega^5 + \omega^9)(\omega^2 + \omega^6 + \omega^7 + \omega^8 + \omega^{10})$ turn out to be $2(\omega + \omega^2 + \omega^3 + \omega^4 + \ldots + \omega^{10}) + 5\omega^{11} = -2 + 5 = 3$, i.e. *real*. Is this a fluke? Why does it work when the indices are partitioned $\{1, 3, 4, 5, 9\}$ and $\{2, 6, 7, 8, 10\}$, and for no other partition?

These are questions which can be answered by the application of the theory of cyclic groups to the cyclotomic equation, and the following section provides an illustration.

Suppose that $\alpha = \text{cis}\,(2\pi/13)$, so that $\alpha^{13} = 1$, and we consider the roots $S = \{\alpha, \alpha^2, \alpha^3, \alpha^4, \ldots, \alpha^{12}\}$ of the cyclotomic equation:

$$x^{12} + x^{11} + \ldots + x^2 + x + 1 = 0 \qquad (0)$$

Now the permutation of the roots in which α is replaced by α^2 (see pp. 167, 171) replaces the indices according to the permutation

$$p = \begin{pmatrix} 1 & 2 & 3 & 4 & 5 & 6 & 7 & 8 & 9 & 10 & 11 & 12 \\ 2 & 4 & 6 & 8 & 10 & 12 & 1 & 3 & 5 & 7 & 9 & 11 \end{pmatrix}$$

i.e. the cycle (1 2 4 8 3 6 12 11 9 5 10 7) of period 12, and we have $p^2 = (1\ 4\ 3\ 12\ 9\ 10)(2\ 8\ 6\ 11\ 5\ 7)$ breaking up into two cycles each of period 6.

Next take the expressions:

$$x_1 = \alpha + \alpha^4 + \alpha^3 + \alpha^{12} + \alpha^9 + \alpha^{10};$$

$$x_2 = \alpha^2 + \alpha^8 + \alpha^6 + \alpha^{11} + \alpha^5 + \alpha^7$$

so that $x_1 + x_2 = -1$ since α satisfies the cyclotomic equation (0). x_1 and x_2 are therefore the roots of a quadratic equation $x^2 + x + a = 0$, where $a = x_1x_2$, and we propose to show, not only that a is integral, but why this is so.

First we note that when x_1 and x_2 are multiplied together we get terms such as $\alpha^3, \alpha^9, \ldots, \alpha^{23}, \ldots, \alpha^{17}$, and all these are in the set S. Thus:

$$x_1x_2 = \lambda_1\alpha + \lambda_2\alpha^2 + \lambda_3\alpha^3 + \ldots + \lambda_{12}\alpha^{12} \qquad (1)$$

where all the λ's are positive integers.

If now α is replaced by α^2 as a generating (primitive) root, we shall have the permutation p acting on both the expressions x_1 and x_2, and on the right-hand side of (1). Clearly p has the effect of interchanging x_1 and x_2, while the right-hand side of (1) will become:

$$\lambda_1\alpha^2 + \lambda_2\alpha^4 + \lambda_3\alpha^6 + \lambda_4\alpha^8 + \lambda_5\alpha^{10} + \lambda_6\alpha^{12} + \lambda_7\alpha + \lambda_8\alpha^3 + \lambda_9\alpha^5 + \lambda_{10}\alpha^7 + \lambda_{11}\alpha^9 + \lambda_{12}\alpha^{11},$$

so we have:

$$x_2x_1 = \lambda_1\alpha^2 + \lambda_2\alpha^4 + \lambda_3\alpha^6 + \ldots + \lambda_{12}\alpha^{11} \qquad (2)$$

Next, suppose the operation p^2 acts on both sides of (1). This time, x_1 and x_2 are each separately unchanged by the permutation – we formed the expressions x_1 and x_2 with this in mind – and we are led to the result:

$$x_1x_2 = \lambda_1\alpha^4 + \lambda_2\alpha^8 + \lambda_3\alpha^{12} + \lambda_4\alpha^3 + \lambda_5\alpha^7 + \lambda_6\alpha^{11} + \lambda_7\alpha^2 + \lambda_8\alpha^6 + \lambda_9\alpha^{10} + \lambda_{10}\alpha + \lambda_{11}\alpha^5 + \lambda_{12}\alpha^9 \qquad (3)$$

We now proceed to apply all the permutations of Gp (p) in turn to (1), and so to obtain equations (1), (2), (3), . . ., (12). It will readily be seen that the permutations e, p^2, p^4, p^6, p^8, p^{10} do not change either x_1 or x_2. The remaining permutations (the coset of the subgroup Gp (p^2)) interchange x_1 and x_2. Meanwhile on the right-hand side, we have $\lambda_1\alpha^2$ in (1), $\lambda_1\alpha^4$ in (2), $\lambda_1\alpha^8$ in (3), and so on, with λ_1 multiplying in turn each member of the set S. Indeed, every one of the λ's will multiply every one of the members of S in the order determined by the permutation p.

Now when the twelve equations are added together, we get:

$$\begin{aligned}12x_1\,x_2 &= \lambda_1(\alpha + \alpha^2 + \alpha^4 + \ldots + \alpha^7)\\ &\quad + \lambda_2(\alpha^2 + \alpha^4 + \alpha^6 + \ldots + \alpha^{10})\\ &\quad + \lambda_3(\alpha^4 + \ldots\\ &\qquad \vdots\\ &\quad + \lambda_{12}(\alpha^7 + \alpha + \alpha^8 + \ldots + \alpha^6)\\ &= (\lambda_1 + \lambda_2 + \lambda_3 + \ldots + \lambda_{12})(-1)\end{aligned}$$

But all the λ's are positive integers, and it follows that x_1x_2 is rational. It can readily be verified that $\lambda_1 = \lambda_2 = \lambda_3 = \ldots = \lambda_{12} = 3$, so that $x_1x_2 = -3$, and x_1, x_2 are the roots of the quadratic equation:

$$x^2 + x - 3 = 0$$

Note that equations (1) to (12) may be written in matrix form:

$$x_1x_2\begin{pmatrix}1\\1\\1\\1\\1\\1\\1\\1\\1\\1\\1\\1\end{pmatrix} = \begin{pmatrix}\alpha & \alpha^2 & \alpha^3 & \alpha^4 & \alpha^5 & \alpha^6 & \alpha^7 & \alpha^8 & \alpha^9 & \alpha^{10} & \alpha^{11} & \alpha^{12}\\ \alpha^2 & \alpha^4 & \alpha^6 & \alpha^8 & \alpha^{10} & \alpha^{12} & \alpha & \alpha^3 & \alpha^5 & \alpha^7 & \alpha^9 & \alpha^{11}\\ \alpha^4 & \alpha^8 & \alpha^{12} & \alpha^3 & \alpha^7 & \alpha^{11} & \alpha^2 & \alpha^6 & \alpha^{10} & \alpha & \alpha^5 & \alpha^9\\ \alpha^8 & \alpha^3 & \alpha^{11} & \alpha^6 & \alpha & \alpha^9 & \alpha^4 & \alpha^{12} & \alpha^7 & \alpha^2 & \alpha^{10} & \alpha^5\\ \alpha^3 & \alpha^6 & \alpha^9 & \alpha^{12} & \alpha^2 & \alpha^5 & \alpha^8 & \alpha^{11} & \alpha^1 & \alpha^4 & \alpha^7 & \alpha^{10}\\ \alpha^6 & \alpha^{12} & \alpha^5 & \alpha^{11} & \alpha^4 & \alpha^{10} & \alpha^3 & \alpha^9 & \alpha^2 & \alpha^8 & \alpha & \alpha^7\\ \alpha^{12} & \alpha^{11} & \alpha^{10} & \alpha^9 & \alpha^8 & \alpha^7 & \alpha^6 & \alpha^5 & \alpha^4 & \alpha^3 & \alpha^2 & \alpha\\ \alpha^{11} & \alpha^9 & \alpha^7 & \alpha^5 & \alpha^3 & \alpha & \alpha^{12} & \alpha^{10} & \alpha^8 & \alpha^6 & \alpha^4 & \alpha^2\\ \alpha^9 & \alpha^5 & \alpha & \alpha^{10} & \alpha^6 & \alpha^2 & \alpha^{11} & \alpha^7 & \alpha^3 & \alpha^{12} & \alpha^8 & \alpha^4\\ \alpha^5 & \alpha^{10} & \alpha^2 & \alpha^7 & \alpha^{12} & \alpha^4 & \alpha^9 & \alpha & \alpha^6 & \alpha^{11} & \alpha^3 & \alpha^8\\ \alpha^{10} & \alpha^7 & \alpha^4 & \alpha & \alpha^{11} & \alpha^8 & \alpha^5 & \alpha^2 & \alpha^{12} & \alpha^9 & \alpha^6 & \alpha^3\\ \alpha^7 & \alpha & \alpha^8 & \alpha^2 & \alpha^9 & \alpha^3 & \alpha^{10} & \alpha^4 & \alpha^{11} & \alpha^5 & \alpha^{12} & \alpha^6\end{pmatrix}\begin{pmatrix}\lambda_1\\ \lambda_2\\ \lambda_3\\ \lambda_4\\ \lambda_5\\ \lambda_6\\ \lambda_7\\ \lambda_8\\ \lambda_9\\ \lambda_{10}\\ \lambda_{11}\\ \lambda_{12}\end{pmatrix}$$

and that, in adding these twelve equations, we are really premultiplying both sides by a 1 × 12 matrix consisting of a row of twelve 1's. Note also that the *indices* of the 12 × 12 matrix form the array:

Table 12.11

1	2	3	4	5	6	7	8	9	10	11	12
2	4	6	8	10	12	1	3	5	7	9	11
4	8	12	3	7	11	2	6	10	1	5	9
8	3	11	6	1	9	4	12	7	2	10	5
3	6	9	12	2	5	8	11	1	4	7	10
6	12	5	11	4	10	3	9	2	8	1	7
12	11	10	9	8	7	6	5	4	3	2	1
11	9	7	5	3	1	12	10	8	6	4	2
9	5	1	10	6	2	11	7	3	12	8	4
5	10	2	7	12	4	9	1	6	11	3	8
10	7	4	1	11	8	5	2	12	9	6	3
7	1	8	2	9	3	10	4	11	5	12	6

This would serve as a table for multiplication modulo 13 (compare table 12.10) with the first column rearranged according to the cycle p, the other columns merely reproducing the order of the cycle, but starting at a different point in it.

★**Q. 12.26.** (a) Show that no other partitions of the indices are possible other than {1, 3, 4, 9, 10, 12} and {2, 5, 6, 7, 8, 11} for selecting the indices to form x_1 and x_2 in such a way that x_1x_2 may be rational.
(b) Repeat the above work in the cases $\alpha^5 = 1$; $\alpha^7 = 1$; $\alpha^{11} = 1$.
(c) Show that if $y_1 = \alpha + \alpha^3 + \alpha^9$ and $y_2 = \alpha^4 + \alpha^{12} + \alpha^{10}$, then y_1, y_2 are the roots of the quadratic equation $y^2 - x_1y + (x_2 + 3) = 0$.
Partition the set {2, 5, 6, 7, 8, 11} of indices to obtain a similar result. Explain in group theoretical terms.

Infinite cyclic group

Finally, we consider an infinite group which has properties closely akin to those of the finite cyclic groups. You are referred back to Ch. 10, p. 125, where we discussed the group (**Z**, +). This group, and isomorphic copies of it, is denoted C_∞, and is termed the 'infinite cyclic group'. The reason for this is that it is generated by a single element, this being either +1 or −1 in the case of (**Z**, +); for every integer can be expressed as the sum of a number of 1's or −1's.

Another way of generating C_∞ is by means of a single *translation*, as in fig. 12.073. The strip pattern is derived from repeated applications of the translation t applied to the letter R.

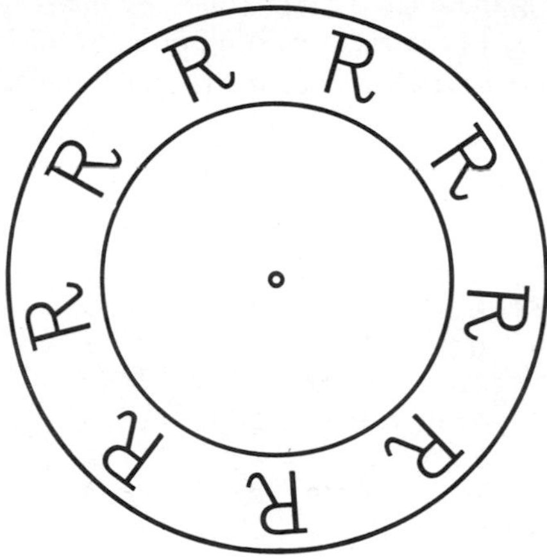

Fig. 12.071. The symmetry group C_9.

To see how this resembles a finite cyclic group, consider fig. 12.071 which shows a pattern of R's round a circle, the sort of pattern one often finds on plates and saucers. Here the cyclic group C_9 is generated by a rotation through an angle of 40°. In the fig. 12.072, we have a small part

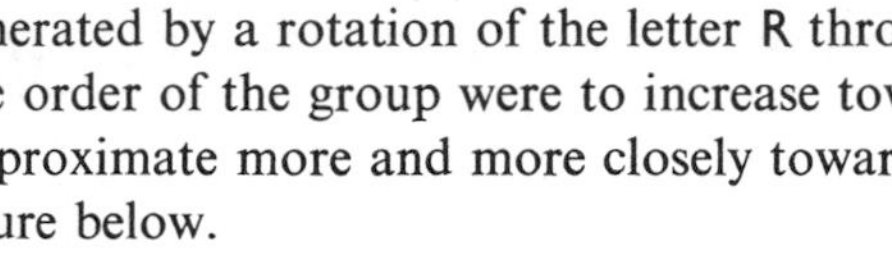

Fig. 12.072. Symmetry group C_{360}.

of the pattern generated by a rotation of the letter R through 1°, giving the group C_{360}. If the order of the group were to increase towards infinity, the pattern would approximate more and more closely towards the strip pattern in the figure below.

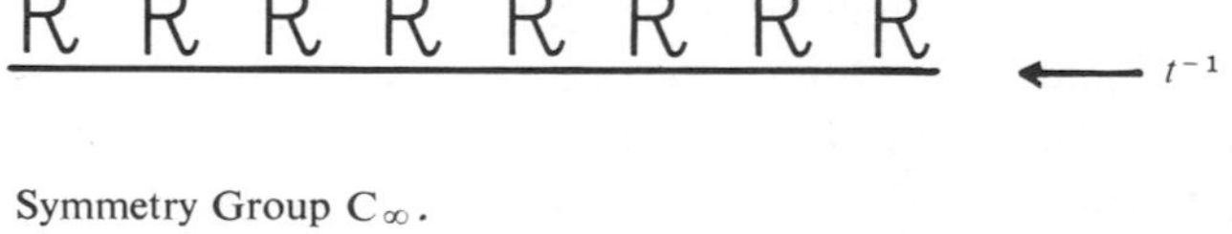

$\cdots t^{-2} \quad t^{-1} \quad 1 \quad t^{1} \quad t^{2} \quad t^{3} \cdots$

R R R R R R R R $\longrightarrow t$ $\longleftarrow t^{-1}$

Fig. 12.073. Symmetry Group C_∞.

The group C_∞ could easily be discovered in many guises, e.g. by repeated substitution in a function such as $f(x) \equiv x + 1$; $f(x) \equiv x^2$; $f(x)\ e^x$, and so on. Also by matrices under addition and also under multiplication: $n \times \begin{pmatrix} a & b \\ c & d \end{pmatrix}$ will never give $\begin{pmatrix} 0 & 0 \\ 0 & 0 \end{pmatrix}$ as long as all of a, b, c, d are not zero, a trivial case; while it is only for very rare matrices that an n will exist such that:

$$\begin{pmatrix} a & b \\ c & d \end{pmatrix}^n = \begin{pmatrix} 1 & 0 \\ 0 & 1 \end{pmatrix} \quad \text{(see p. 118ff).}$$

It is perfectly possible for an infinite group (though not the infinite *cyclic* group) to contain elements of finite period. For example, we might construct a group by starting off with an infinite cyclic group generated by an element x, and then introducing a further element y of period (say) 2. The group generated by these two elements would be infinite, and would contain elements of period 2 (y, and possibly others). For example, if $X = \begin{pmatrix} 1 & 1 \\ 1 & 0 \end{pmatrix}$ and $Y = \begin{pmatrix} 0 & -1 \\ -1 & 0 \end{pmatrix}$, with $Y^2 = I$, then the group obtained by taking all possible products of X with Y (e.g. $XYX^2YX^{-4}Y\ldots$) will be infinite, and contains the matrix Y of period 2.

Q. 12.27. Show that there are no other elements of period 2, but that if $Z = \begin{pmatrix} 1 & 2 \\ 2 & 1 \end{pmatrix}$, then the group generated by Y and Z is Abelian and does contain elements of period 2 other than Y. (The group is $C_\infty \times C_2$ in this case – Ch. 16.)

Q. 12.28. Construct an infinite group of functions containing a function of period 2; repeat with a function of period 3.

Infinite groups can also be generated by elements each of which has a finite period: for example, the reflections in two parallel mirrors (each of period 2) generate an infinite sequence of images. Here is the group D_∞, and this we shall look at in the next chapter (p. 206ff).

Again, the linear pattern shown in fig. 26.167 contains not only a translation (of infinite period), but also two reflections. For the pattern may be brought into coincidence with itself not only by moving an integral number of places to the right or left (the translations), but also by reflecting either in the central axis of the pattern, or in any of the axes perpendicular to it such as those marked in the diagram. Either of these two reflections has period 2.

It may surprise you to know that it is even possible to have an infinite group in which *every* element has finite period! (see Ex. 12.31).

EXERCISES 12

1 Which group is generated by the rotations about a given point through angle:
(a) 15°; (b) 144°; (c) 1°; (d) 85°; (e) 38°; (f) 340°?

2 **Prove that every subgroup of a cyclic group is cyclic.**

3 **Prove that in the group $\{1, 2, 3, \ldots, p-1\}$, $\times$ mod. p (p prime), the only element of period 2 is $p - 1$.**

4 **Prove that every cyclic group of even order has exactly one element of period 2.**

5 If n is a positive integer, and S is the set of complex numbers z such that $z^n = -1$, is $(S, \times)$ a group?

6 Find the period of the function $f(x) \equiv 3/(3 - x)$, and find the other functions of the group which it generates. Repeat for the functions $\frac{1}{2} - x$; $(4x - 1)/(4x + 2)$; $1/(2 - 2x)$; $-ix$.

7 Find a function of period 4 to generate the group C_4 (use the results of Ex. 10.28) and also a function of period 8 to generate C_8. In the latter case, find all the other functions of the group, show the subgroups, and indicate which functions are inverse pairs.

8 Show that the matrix $\frac{1}{2}\begin{pmatrix} 1 & \sqrt{3} \\ -\sqrt{3} & 1 \end{pmatrix}$ has period 6. Write down the other matrices of the group which it generates, and interpret geometrically.

9 Show that the matrix $\frac{1}{\sqrt{3}}\begin{pmatrix} 1 & 1 \\ -1 & 2 \end{pmatrix}$ generates the group C_{12} by matrix multiplication. Find a matrix with integral coefficients which has period 6. (Note that since $M^6 = I$, and since $\det(AB) = \det(A) \times \det(B)$ (see pp. 229, 398), we must have $\det(M) = \pm 1$, i.e. $ad - bc = \pm 1$. Moreover, from the results in Ex. 10.28, p. 129, we must also have $a^2 + d^2 - ad + 3bc = 0$.)

10 Find the period of the permutation $\begin{pmatrix} 1 & 2 & 3 & 4 & 5 & 6 \\ 5 & 4 & 1 & 6 & 3 & 2 \end{pmatrix}$ and obtain the other permutations in the group which it generates.

11 Show that the product of two disjoint cycles of lengths 3 and 4 generate C_{12} (e.g. permutations of the form $x^p y^q$, where $x = \begin{pmatrix} A & B & C & P & Q & R & S \\ C & A & B & P & Q & R & S \end{pmatrix}$ and $y = \begin{pmatrix} A & B & C & P & Q & R & S \\ A & B & C & S & P & Q & R \end{pmatrix}$.
Can you generate C_{12} by using disjoint cycles of lengths 2 and 6?

12 Show that the multiplication of ordered pairs $(x_1, y_1) \times (x_2, y_2) = (x_1x_2, y_1y_2)$ where $x \in \{1, i, -1, -i\}$ and $y \in \{1, \omega, \omega^2\}$ ($\omega = \text{cis } \frac{2}{3}\pi$) produces the group C_{12}. How can this be represented by (a) matrices; (b) *addition* of ordered pairs?

13 Can you arrange four (empty) teacups on a saucer so that the pattern of the symmetry is described by C_4?

14 Use *Spirograph* to draw patterns with cyclic symmetry, and state the order of the cyclic groups involved.

15 Consider the rows of C_5 and then C_6 as permutations of the first row, and interpret in terms of Cayley's theorem.

16 Consider the table for C_7 realised as $\{0, 1, 2, 3, 4, 5, 6\}$, + mod. 7 as an illustration of Cayley's theorem.

17 State the number of elements of periods 2, 3, 4, 5, 6, ... in the groups C_{15}, C_{18}, C_{24}, C_{36}, etc. Draw up a 'catalogue' of cyclic groups on the lines of table 17.01.

18 In the group of residues prime to 100 under multiplication mod. 100, i.e. $\{1, 3, 7, 11, 13, \ldots, 99\}$, find: (a) the order of the group; (b) the period of 73; (c) the inverse of 19. Prove that, if n is a positive integer,

$(2n+1)^{20} = 1$ (mod. 100) and $(2n)^{20} = 76$ (mod. 100), provided neither $2n+1$ nor $2n$ is divisible by 5. Interpret these results.

19 **If $n = mr$, show that there exists a subgroup of C_n with structure C_m, and conversely.**

20 **In the group C_n $\{1, x, x^2, \ldots, x^{n-1}\}$ $x^n = 1$, show that x^r is a generator if and only if r is prime to n (i.e. the H.C.F. of r and n is 1). If r is not prime to n, show that the period of x^r is l/r, where l is the L.C.M. of n and r.**

21 Prove that $\operatorname{cis} 2\pi(k/m)$ where $k, m \in \mathbf{Z}$, and k and m are coprime is a primitive nth root of 1 for $m = n$ and for no other values of m.

22 Consider the ordered pairs (0, 0), (0, 1), (0, 2), (1, 0), (1, 1), (1, 2), (2, 0), (2, 1) and (2, 2). Show that, under addition modulo 3, they form the group isomorphic to the group of order 9 introduced on p. 123. Find a subset of the above set of ordered pairs which form a group under the operation $(a, b) \otimes (c, d) = (ac, bd$, mod. 3), e.g. $(0, 2) \otimes (1, 2) = (0, 1)$.

23 Discuss automorphisms of cyclic groups of composite order, e.g. C_6, C_8, C_{10}. Show that the group of automorphisms of C_{10} is isomorphic with the group $\{1, 3, 7, 9\}$, $\times$ mod. 10.

24 Show that the generators of C_n are the maximum subset $\{0, 1, 2, \ldots, n-1\}$ which are a group under multiplication modulo n (compare the case $n = 10$ in the previous question).

25 Show that a set of residues modulo n forming a group under addition modulo n can only form a group under multiplication modulo n (with zero excluded from the set) provided that n is prime, and also show that the groups are respectively C_n and C_{n-1}.

26 **Prove that, if r and s are both prime to n, then so is rs. Thence prove that those residues less than n which are prime to n form a group under multiplication modulo n. The order of this group is denoted $\phi(n)$, Euler's* function. For example, $\phi(15) = 8$, since there are 8 numbers: 1, 2, 4, 7, 8, 11, 13, 14 less than 15 which are prime to it (see also Ch. 6). Note that 1, 2, 4, 7, 8, 11, 13, 14 are all *primitive* elements, or generators, in the additive group of residues modulo 15.**
Prove generally that, in the group C_n, there are $\phi(n)$ primitive elements, and that these form a group under multiplication modulo n.

27 Prove that, if r and n are coprime integers with $n > r$, then n and $n - r$ are also coprime. Hence prove that Euler's function $\phi(n)$ is even for all n.*

28 **Prove that every infinite cyclic group is isomorphic with $(\mathbf{Z}, +)$.**

29 The transformation shown in fig. 26.25 is known as a *glide reflection* in the line l. Show how a glide reflection may generate an infinite pattern, and that the associated group is isomorphic to C_∞.

30 Show how C_∞ may be generated by a spiral similarity ($z' = z\rho \operatorname{cis} \phi$), and how an equiangular spiral may thus be built up.

31 Consider the complex numbers $\operatorname{cis} \pi t$ where $t \in \mathbf{Q}$, e.g. $\cos 37\pi/83$. Show that these form an infinite group under multiplication, yet every element has a finite period – the period of $\operatorname{cis} 37\pi/83$, for example, is 166, since $(\operatorname{cis} 37\pi/83)^{166} = \operatorname{cis} 74\pi = 1$. Show that this group is isomorphic to the group ($\mathbf{Q}$, + mod. 1).

*See App. IV for a table of values of Euler's function.

32 Show that the group of the direct symmetries of the circle (rotations about its centre) is isomorphic to the group (**S**, + mod. 1) and also to (k**S**, + mod. k) where k is any real number. Give another geometrical description of this group.

33 Discuss an infinite spiral staircase as a geometrical example of C_∞. What is a generator?

34 If $a^k = 1$ mod. $(ab - 1)$ prove that $b^k = 1$ mod. $(ab - 1)$. Interpret in group-theoretical terms.

35 Show that the group of automorphisms of C_n is of order $\phi(n)$, (see Ex. 12.26) and is isomorphic with the group of residues less than n and prime to n under multiplication modulo n.
Consider the group C_7 in the form {0, 1, 2, 3, 4, 5, 6}, + mod. 7. Show that when the generator 1 is replaced by 2, this induces the permutation (1 2 4)(3 6 5) of the non-zero elements, but that when 1 is replaced by 3, we obtain a cycle of period 6. Represent all the automorphisms of C_7 in this way. Repeat in the case of C_{11}, C_{13}, etc.

13 *The dihedral group*

Direct and opposite symmetries

In the previous chapter we considered the rotational symmetries of the regular hexagon, and took the precaution of shutting our eyes to the bilateral, or reflective symmetries which the regular hexagon possesses; the rotational symmetries through multiples of 60° have the group C_6. We are now going to enlarge this group by including the reflective symmetries. There are six axes, or lines of symmetry for the regular hexagon (see fig. 13.01). One may visualise these symmetries in one of two ways: in fig. 8.01

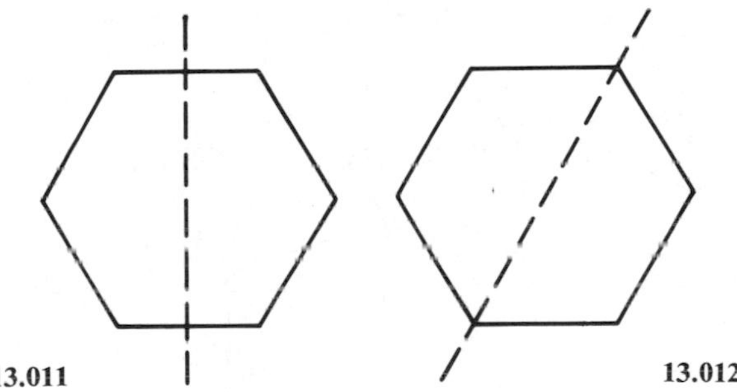

Fig. 13.01. The six bilateral symmetries of the regular hexagon – three like 13.011, three like 13.012.

for example, the line *l* is an axis of symmetry. Either one can think of *l* as an axis of *rotation*, in which case the symmetry movement of the figure consists in performing *a half-turn about l*, or 'flipping it over', an operation which causes the hexagon to coincide with its previous position but with the figure 'turned over'. Or one can think of *l* as being a plane mirror, reflection in which simply reverses the two halves of the figure. With this way of looking at it, there is no 'turning over', but there is a *reversal of sense*, i.e. clockwise ↔ anticlockwise. In either case, we have what is known as an 'opposite' symmetry.† The first way of visualising the symmetry means that, although we are considering a two-dimensional figure, we have to use *the third dimension* in order to perform the operation of flipping it over.

† Direct and opposite symmetries are sometimes known as 'proper and improper', terms which we shall eschew.

Mirror reflections

A difficulty arises when considering a three-dimensional body which has a (bilateral) plane of symmetry, such as a pair of semi-detached houses, an aircraft, the human body, or the prisms in fig. 13.02. Here one has to think of the symmetry (m) as a *mirror reflection*, because it is physically impossible to pick up the prism from position (a) and turn it inside out so as to produce position (b) (unless one first took it into the *fourth dimension*, whatever that may mean!). Opposite symmetries of three-dimensional figures are called *enantiomorphs*: one's left hand is an enantiomorphic image of one's right hand; a car with a right-hand drive

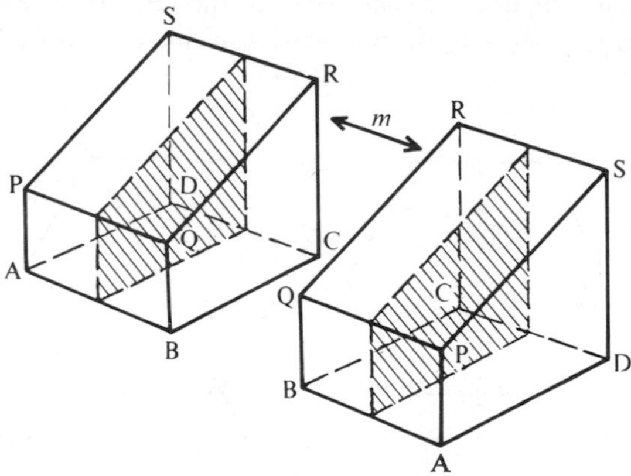

Fig. 13.02. Symmetry of a three-dimensional figure by reflection in a plane.

is an enantiomorphic image of the same model with a left-hand drive; one can imagine two 'spiral' staircases which are mirror images (or enantiomorphic images) of each other – as one ascends the first, one finds oneself always turning left; for the second, the direction of rotation would be reversed. The first is like a right-hand thread on a screw, the second like a left-hand thread. No possible physical movement of one of these can bring it into corresponding position with the second. Merely turning the spiral staircase upside-down will not do it – you may verify this by taking a piece of rope which is laid right-hand (see fig. 13.03), and try the effect of turning it round.

The full group of the regular hexagon: rotations and reflections

Let us now return to the symmetries of the regular hexagon. There are the six rotations through 0, 60°, 120°, 180°, 240° and 300°, which we shall call 1, r, r^2, r^3, r^4 and r^5, where we shall agree that r refers to an anticlockwise

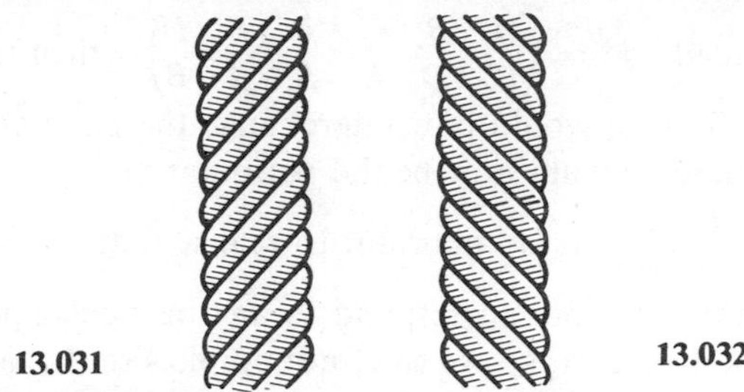

Fig. 13.03. Enantiomorphs.
13.031. Rope layed right-hand.
13.032. Rope layed left-hand.

rotation through 60°. We have also the symmetries about the six axes, and these too are shown in fig. 13.04. We denote the symmetry operations of reflections in these axes by *a*, *b*, *c*, *d*, *f*, *g*, so that, for example, *f* means 'flip the hexagon over about the axes running from N. 60° E. to S 60° W.', or 'reflect the hexagon' in that axis. We could represent such movements by permutations: thus *r* causes the vertices† *A*, *B*, *C*, *D*, *E*, *F* of the

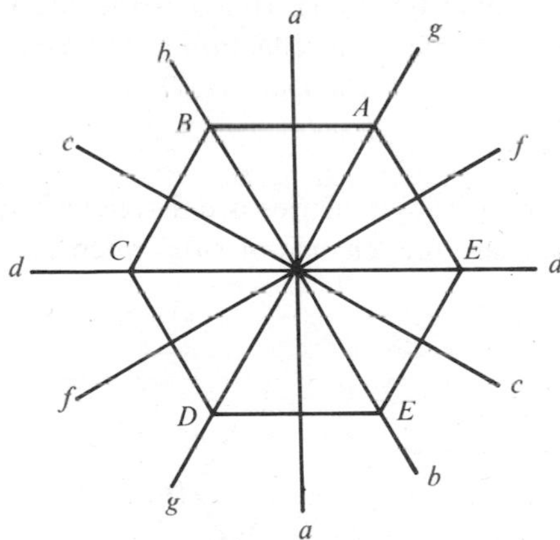

Fig. 13.04. Symmetries of regular hexagon.

hexagon to be replaced by *F*, *A*, *B*, *C*, *D*, *E*, so we may represent *r* by the permutation $\begin{pmatrix} A & B & C & D & E & F \\ F & A & B & C & D & E \end{pmatrix}$, or by the cycle (*FEDCBA*). Again, when the hexagon starts from the position shown in fig. 13.04, the operation *a* corresponds to the permutation $\begin{pmatrix} A & B & C & D & E & F \\ B & A & F & E & D & C \end{pmatrix}$, or the cycles (*AB*)(*CF*)(*DE*) of the vertices; but if the operation r^4 (say) had

† These are labels for the vertices *which move with the hexagon*. See also pp. 24, 197 ff., 296 ff., 365 ff.

first been performed – $r^4 = \begin{pmatrix} A & B & C & D & E & F \\ C & D & E & F & A & B \end{pmatrix}$ – then the operation a (reflection in N.–S. axis) would now interchange the pairs of vertices $(CD)(EB)(AF)$, thus a would now be the permutation $\begin{pmatrix} A & B & C & D & E & F \\ F & E & D & C & B & A \end{pmatrix}$. It is important to realise that the reflection a in the *fixed* N.–S. axis does not correspond to any particular permutation of the *vertices* of the hexagon; so we shall not emphasise the aspect of these

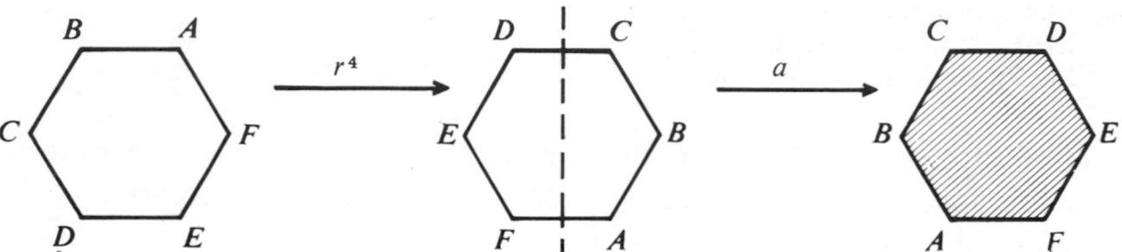

Fig. 13.05. $ar^4 = c$.

symmetries as permutations of the *vertices*. The combined operation ar^4 (shown diagrammatically in fig. 13.05) thus has the effect of interchanging the pairs $(BC)(AD)(EF)$, which is a reflection in the axis labelled c, and we may write:

$$ar^4 = c.$$

(In these diagrams, opposite symmetries are distinguished by the hexagon being shaded to emphasise the 'change of side' when the figure is 'turned over'.)

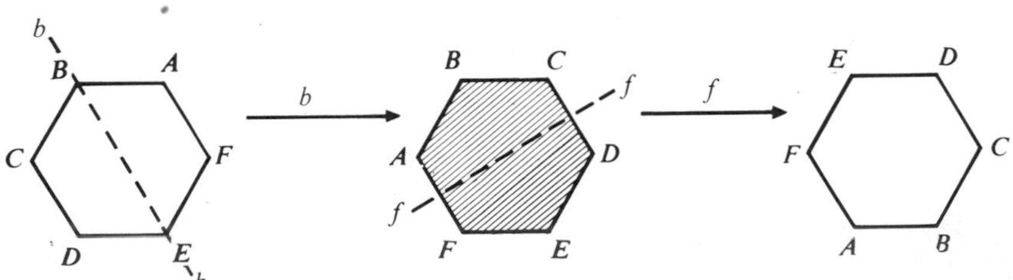

Fig. 13.06. $fb = r^3$.

Again, consider the effect of, first b, then f. Fig. 13.06 shows that $fb = r^3$.

Q. 13.01. Find bf.

If the second operation had been a, we should have $ab = r^5$ (see fig. 13.07). On the other hand, fig. 13.08 shows that $ba = r$.

Thus we have combined several pairs of symmetries of the regular hexagon, and this is a start towards drawing up a complete table of the group of symmetries.

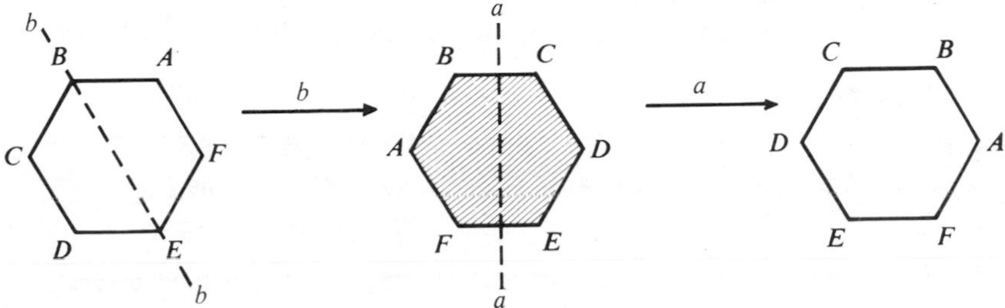

Fig. 13.07. $ab = r^5$.

But it would be laborious to have to draw three diagrams like this to obtain each of the $12 \times 12 = 144$ possible combinations of symmetries. How can we shorten the work? The best way is to consider the effect of the successive movements of the hexagon on *one particular vertex*, say *A*. Thus the above diagram shows that the motion *a* takes the vertex *A* to

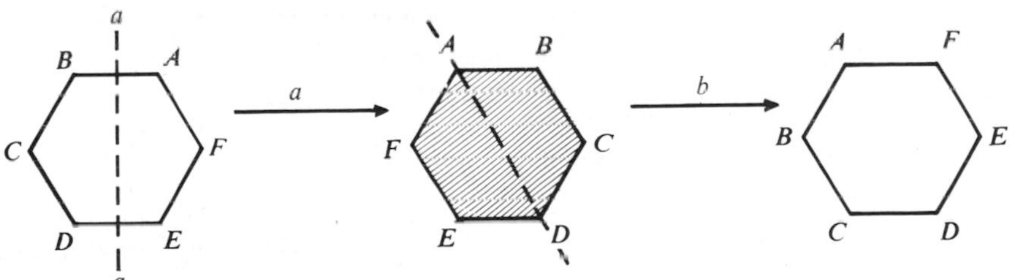

Fig. 13.08. $ba = r$.

position *B*. But then *b* leaves *B* where it was so the resultant effect of the operation *ba* on the vertex *A* is to move it to *B*. But clearly *ba*, being the product of *two opposite* symmetries must be one of the *direct* symmetries,† i.e. one of the rotations about the centre. For *a* changes the direction of the lettering *ABCDEF* from anticlockwise to clockwise, and *b* changes the order back to anticlockwise again; or, if you like, both *a* and *b* turn the hexagon over from heads to tails, and then back from tails to heads again. Which, therefore, of the direct symmetries, 1, r, r^2, r^3, r^4 or r^5 takes *A* to *B*? The answer is *r*, so we have $ba = r$.

† In saying this, we are really partitioning the twelve symmetries into two cosets by the subgroup C_n – see Ch. 19.

Again, consider r^4a. First a takes A to B; then r^4 takes B to F.† Hence r^4a takes A to F. But r^4a is evidently an opposite symmetry, since there is only one 'reversal of side'. Which of a, b, c, d, f, g interchanges A and F? The answer is f, hence $r^4a = f$.

We are now in a position to complete the group table (see table 13.01).

Table 13.01

				first										
									ar^5	ar^4	ar^3	ar^2	ar	
									ra	r^2a	r^3a	r^4a	r^5a	
	1	1	r	r^2	r^3	r^4	r^5	a	b	c	d	f	g	period
	1	1	r	r^2	r^3	r^4	r^5	a	b	c	d	f	g	1
	r	r	r^2	r^3	r^4	r^5	1	b	c	d	f	g	a	6
	r^2	r^2	r^3	r^4	r^5	1	r	c	d	f	g	a	b	3
second	r^3	r^3	r^4	r^5	1	r	r^2	d	f	g	a	b	c	2
	r^4	r^4	r^5	1	r	r^2	r^3	f	g	a	b	c	d	3
	r^5	r^5	1	r	r^2	r^3	r^4	g	a	b	c	d	f	6
	a	a	g	f	d	c	b	1	r^5	r^4	r^3	r^2	r	2
$ra = ar^5 =$	b	b	a	g	f	d	c	r	1	r^5	r^4	r^3	r^2	2
$r^2a = ar^4 =$	c	c	b	a	g	f	d	r^2	r	1	r^5	r^4	r^3	2
$r^3a = ar^3 =$	d	d	c	b	a	g	f	r^3	r^2	r	1	r^5	r^4	2
$r^4a = ar^2 =$	f	f	d	c	b	a	g	r^4	r^3	r^2	r	1	r^5	2
$r^5a = ar =$	g	g	f	d	c	b	a	r^5	r^4	r^3	r^2	r	1	2 (D_6)

The method described, of finding the effect of a composite transformation upon a particular vertex can be somewhat simplified by choosing the vertex wisely. For example, if one requires r^4d, it is easiest to consider the effect on the vertex C (or F), because the first movement keeps that vertex in the same place. Or if we wanted gr^2, consider the effect on F: r^2 moves F to B, and then g moves it back to where it was, so $gr^2 = d$.

† Meaning 'to where F was initially'. The confusion which may arise here can be cleared up by using letters for the vertices of the moving hexagon, and figures for the vertices of the fixed hexagonal 'hole' into which it fits – see pp. 200–1. It is tempting to write $r^4(B) = F$, but this would be misleading, since r^4 is the permutation $\begin{pmatrix} A & B & C & D & E & F \\ C & D & E & F & A & B \end{pmatrix}$ of the vertices only when the hexagon is showing 'heads'. When it is turned over (as here), r^4 is the permutation $\begin{pmatrix} A & B & C & D & E & F \\ E & F & A & B & C & D \end{pmatrix}$ of the vertices.

Parts of the table may be completed by 'algebra'. Suppose we want dg, and we already know that $g = dr^4$ from the part of the table which is already filled in. Then $dg = d(dr^4) = d^2r^4 = 1r^4 = r^4$. And so on.

In point of fact, it is possible to construct the *whole table algebraically* using no more than the following information connecting the 'generators' r and a:

$$r^6 = 1 = a^2; \qquad ar^5 = ra.$$

For since

$$\begin{aligned} ar^5 = ra &\Rightarrow (ar^5)r = rar \Rightarrow ar^6 = rar \Rightarrow a = rar \\ &\Rightarrow r^5a = r^5(rar) = r^6ar = ar, \\ &\quad\; r^4a = r^4(rar) = r^5ar = r^5(rar)r = r^6ar^2 = ar^2, \end{aligned}$$

and similarly, $r^3a = ar^3$. (Note that a and r^3 *commute* precisely because r^3 is a *half-turn* about an axis perpendicular to a, and we know these generate the group D_2, which is Abelian (see pp. 17, 18, 75).

Q. 13.02. What are the elements of this D_2 subgroup?

Furthermore, if we now wanted, say, bd, we could say:

$$b = ra,\ d = r^3a = ar^3 \Rightarrow bd = ra.ar^3 = ra^2r^3 = r1r^3 = r^4.$$

And again, $r^4g = r^4.r^5a = r^3a = d.$

The relations $a^2 = r^6 = 1$; $ar^5 = ra$ are a set of ***defining relations*** for the group. They may be used to describe or specify a group completely, as we shall see in Ch. 15.

The full group of symmetries of the regular hexagon, which includes the bilateral as well as the rotational symmetries is known as the Dihedral Group of order 12, and is denoted D_6.

In so far as the hexagon in the text might be thought of as being a very thin hexagonal *prism*, we can easily see that any regular hexagonal prism, however long, will have the symmetry group D_6, but the axes of the half-turns a, b, c, d, f, g will lie in the plane midway between the two ends (see fig. 13.09).

Q. 13.03. Show that the right regular hexagonal pyramid has the symmetry group C_6 if enantiomorphs are not included, but D_6 when reflections are allowed. Specify the planes of the six reflections.

Q. 13.04. What is the order of the group of symmetries of (1) the regular hexagonal prism when enantiomorphs are admitted, (2) the regular hexagonal bi-pyramid (i.e. two pyramids with a common regular hexagon as base):
(a) direct symmetries only, (b) the full group including enantiomorphs.

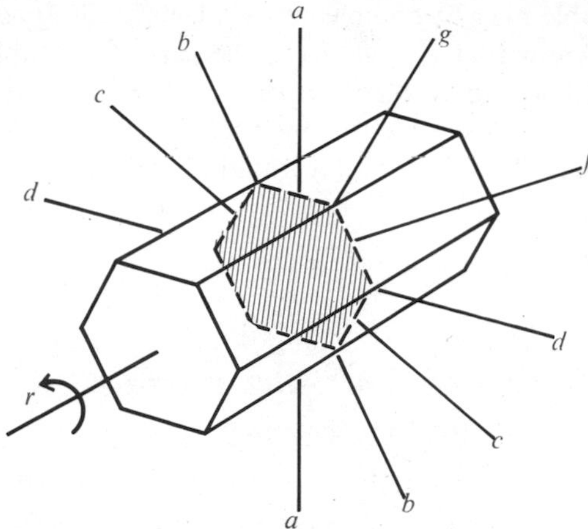

Fig. 13.09. The rotational symmetries of the regular hexagonal prism: a, b, c, d, f, g and r^3 are half-turns.

Q. 13.05. Make a detailed investigation of the group D_5 and also D_4 by considering the symmetries of the regular pentagon, and of the square. Draw up the group tables, in one case by considering the effect of the transformations on a particular vector, and in the other by 'algebraic' methods. Give a set of defining relations for each of these groups.

Other realisations of the dihedral group, D_6

To introduce the group D_6, we considered the symmetries of the regular hexagon. There are many other ways that this particular group may be realised, for example, as a group of permutations:

Q. 13.06. Find a set of 12 permutations of 1, 2, 3, 4, 5, 6, 7, 8, 9, 10, 11, 12 which combine to give D_6. (Use Cayley's theorem: if the elements in our table on p. 192 were numbered 1, 2, ..., 12, then the second row corresponds to the permutation $\begin{pmatrix} 1 & 2 & 3 & 4 & 5 & 6 & 7 & 8 & 9 & 10 & 11 & 12 \\ 2 & 3 & 4 & 5 & 6 & 1 & 8 & 9 & 10 & 11 & 12 & 7 \end{pmatrix} \ldots$)

To find a set of permutations on *six* symbols which give D_6, we might take the cyclic permutations to give the group C_6, and also their *reverses*, e.g.

$$\begin{pmatrix} 1 & 2 & 3 & 4 & 5 & 6 \\ 3 & 4 & 5 & 6 & 1 & 2 \end{pmatrix} \quad \text{and} \quad \begin{pmatrix} 1 & 2 & 3 & 4 & 5 & 6 \\ 2 & 1 & 6 & 5 & 4 & 3 \end{pmatrix}.$$

These are closed under successive composition, and you may discover *why* they describe the symmetries of the regular hexagon.

The group D_6 may also be realised by a set of functions:

Example

If $f(x) \equiv \dfrac{2x-1}{x+1}$ and $g(x) \equiv \dfrac{x}{x-1}$,

find all the functions that can be derived from these by successive substitution. Show that there are exactly twelve of these and find them, specifying which are inverse pairs. Give the period of each function, construct the group table, and find the subgroup C_6 and the subgroup D_3.

Now $\quad f(x) = \dfrac{2x-1}{x+1} \Rightarrow f^2(x) = \dfrac{x-1}{x} \Rightarrow f^3(x) = \dfrac{x-2}{2x-1};$

$$f^6(x) = \frac{(x-2)-2(2x-1)}{2(x-2)-(2x-1)} = \frac{x-2-4x+2}{2x-4-2x+1} = x.$$

Hence, $\quad f(x)$ is of period 6, and generates the subgroup C_6.
Again,

$$g(x) = \frac{x}{x-1} \Rightarrow g^2(x) = \frac{x}{x-(x-1)} = x,$$

so $g(x)$ is of period 2. Possible composite functions are therefore i, f, f^2, f^3, f^4, f^5, fg, f^2g, f^3g, f^4g, f^5g, gf, gf^2, gf^3, gf^4, gf^5, fgf, ..., f^4gf^5g, ..., etc. We must show that only twelve of these are distinct.

Now $fg(x) = (x+1)/(2x-1)$, and you may verify that gf^5 is the same function. Thus $fg = gf^5$. From this, we deduce, as on p. 193, that $g = fgf$, and that

$$f^2g = gf^4, \qquad f^3g = gf^3; \qquad f^4g = gf^2; \qquad f^5g = gf \ \ldots \quad (1).$$

The specification of these functions is left to the reader. Clearly this exhausts the possibilities, for any composite function of the form $g^m f^p g^n f^q g^l f^r \ldots$ can first be reduced to $(gf^p)(gf^q)(gf^r) \ldots$ (since $g^2 = i$), after which we may finally arrive at an expression of the form $f^t g^s$ by a process which we illustrate by an example: suppose we had $gf^3gf^2gfgf^5g$. Then this could be rewritten, using relations (1) as $(f^3g)(gf^2)(f^5g)(gf^5)g$ $= f^3g^2f^7g^2f^5g = f^3ff^5g$ (since $f^6 = i$, $g^2 = i$) $= f^9g = \underline{f^3g}$.
In a similar way, every possible function can be expressed in the form f^tg^s, where $t = 0, 1, 2, 3, 4$ or 5 and $s = 0$, or 1, and so their number is confined to 12. Moreover, each of the functions f^tg has period 2, since:

$(f^tg)(f^tg) = f^tg.gf^{6-t}$ (by (1)) $= f^t i f^{6-t} = f^6 = i$, the identity function.

The group is clearly isomorphic with the group D_6, and the table will correspond to that on p. 192 by substituting f in place of r and g for a.

Dihedral groups D_3 and D_4

To obtain the subgroup D_3, we require two elements of period 3, and these must be $f^2(x) \equiv (x - 1)/x$ and $f^4(x) \equiv 1/(1 - x)$ as well as an independent function of period 2, e.g. $g(x) \equiv x/(x - 1)$. So we get the functions i, f^2, f^4, g, gf^2 and gf^4, and you should verify that they do give the group D_3.

Q. 13.07. Show that $h(x) \equiv 1/x$ will do equally well in place of $g(x) \equiv x/(x - 1)$ to combine with f^2 and f^4 to give D_3. (In fact you should have found that the function $1/x$ is one of those already included in the complete set forming D_6.)

Q. 13.08. Find representations of the groups D_4 by means of functions, and also a different representation of D_3 from the one above.

The dihedral groups may also be represented by sets of matrices under matrix multiplication. For example, D_4 may be generated by $\begin{bmatrix} 0 & 1 \\ 1 & 0 \end{bmatrix}$ (representing a reflection about the line $y = x$), and $\begin{bmatrix} -1 & 0 \\ 0 & 1 \end{bmatrix}$ (representing a reflection about the y-axis).

Q. 13.09. Find all the other matrices of this set, and check that they do form the group D_4 under matrix multiplication.

Reflections in two intersecting mirrors: the kaleidoscope

The above example is interesting because of its connection with the images obtained by reflection in two mirrors. When the two plane mirrors are perpendicular, we saw on pp. 135–6 that the group D_2 is obtained. When the mirrors are inclined at 45°, the images obtained by successive reflection are shown in fig. 13.10. In this diagram, we indicate the operations which lead to the formation of the several images: e.g. that labelled *bab* is obtained from the object labelled 1 by reflecting in mirror *B*, then in mirror *A*, and finally in *B* again.

Q. 13.10. Draw the diagrams showing the images when the mirrors are inclined at (a) 60°, (b) 30°. Which groups are obtained?

The figure obtained above with the mirrors at 45° possesses the same symmetries as the *square*, whose symmetry group is also D_4. When the angle between the mirrors is 30°, you should have obtained a figure with the same symmetries as those of the regular hexagon. *Note:* you will recall having seen pictures of various types of snow crystals. In many cases these are intricate in detail, but they all have the symmetry group D_6.

On p. 193 we mentioned that the group D_6 could be generated by just two elements, r and a, governed by the defining relations $a^2 = r^6 = 1$, $ar = r^5a$. The 'kaleidoscope' illustration above helps to demonstrate that the group could equally well have been *generated by two of the reflections*, and the defining relations (for D_6) would be $a^2 = b^2 = 1$, $(ab)^6 = 1$.

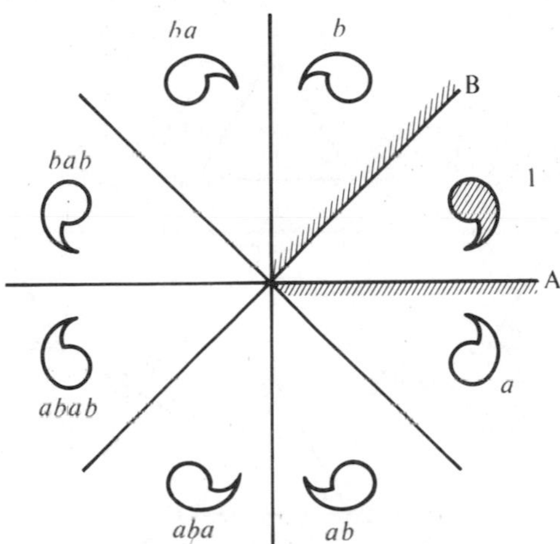

Fig. 13.10. The group D_4 – reflections in mirrors at 45°.

Q. 13.11. Check the truth of the relation $(ab)^4 = 1$ in the group D_4 by reference to the figure 13.10 in which the group is obtained by successive reflections in mirrors inclined at 45°.

Note that every dihedral group D_n requires *two* generators (even the smallest possible dihedral group, for which $n = 2$).

Q. 13.12. Why is this?

The group D_3 of the equilateral triangle

In most books on groups, and in odd chapters of books which deal with groups, you will find that the symmetries of the equilateral triangle comes in the first half-dozen pages. Here it does not appear till p. 197! Perhaps, in leading up to it gradually, we have shown excessive caution, or else a remarkable lack of haste. However, there are good reasons for the delay. In many of the books, there are difficulties in the consideration of the problem which are conveniently slurred over!

Suppose our equilateral triangle has vertices labelled A, B, C, and fits exactly into a hole of the same shape in the plane whose fixed vertices are labelled 1, 2, 3. Then clearly there are six ways in which the triangle may be fitted into the hole, and these correspond to the three direct and the three opposite symmetries of the triangle. Starting from the position in which the vertices A, B, C of the triangle lie in positions 1, 2, 3 respectively of the fixed hole, the six transformations will be denoted as in fig. 13.11. The reverse side ('tails') of the triangle has been shaded to show

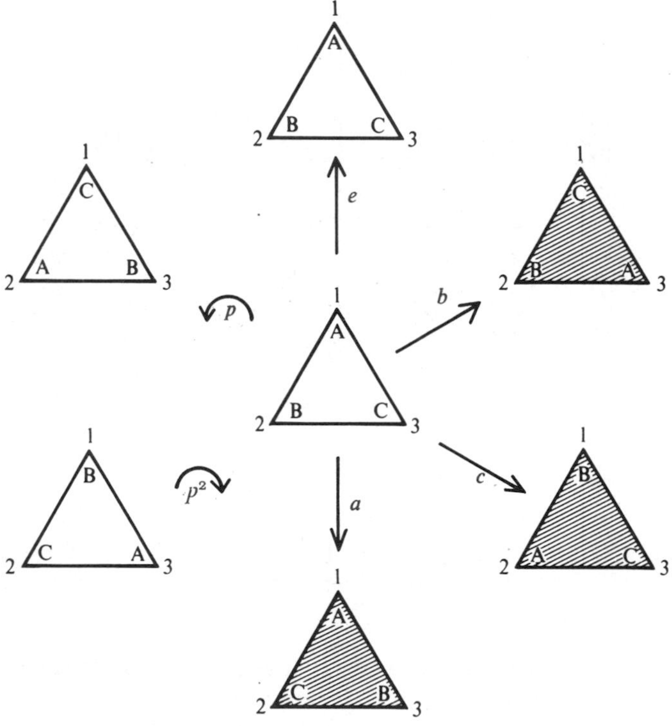

Fig. 13.11. Symmetries of the equilateral triangle. Vertices 1, 2, 3 are fixed, A, B, C move with the triangle, a, b, c are half-turns about axes fixed in space.

clearly that a, b and c are *opposite* symmetries. For example, c represents a rotation about the axis in the direction of the arrow passing through the centre of the triangle.

Now we may consider the composition of the six symmetries in exactly the same way as we did in the case of the regular hexagon on pp. 190 ff., and we obtain table 13.02. By way of verifying one of the entries, say bp^2, fig. 13.12 below shows that $bp^2 = a$.

Group D_3 of the equilateral triangle is isomorphic to the group S_3

It is essential to emphasise here that the axes of the reflections or half-turns a, b, c are *fixed* – the triangle does not carry these axes round with it, but we shall presently consider the situation when it does. Nor, when

Table 13.02

		first				p^2a / ap	pa / ap^2	
		e	p	p^2	a	b	c	period
	e	e	p	p^2	a	b	c	1
	p	p	p^2	e	c	a	b	3
second	p^2	p^2	e	p	b	c	a	3
	a	a	b	c	e	p	p^2	2
	$p^2a = ap = b$	b	c	a	p^2	e	p	2
	$pa = ap^2 = c$	c	a	b	p	p^2	e	2

(D_3)

the triangle is 'turned over', does it carry with it the axis of rotation for p and p^2 – these rotations are reckoned anticlockwise from a *fixed* vantage-point; indeed, the direction of the rotation p is the direction around the *fixed* vertices 1, 2, 3, and not in the sense of the vertices A, B and C of the moving triangle.

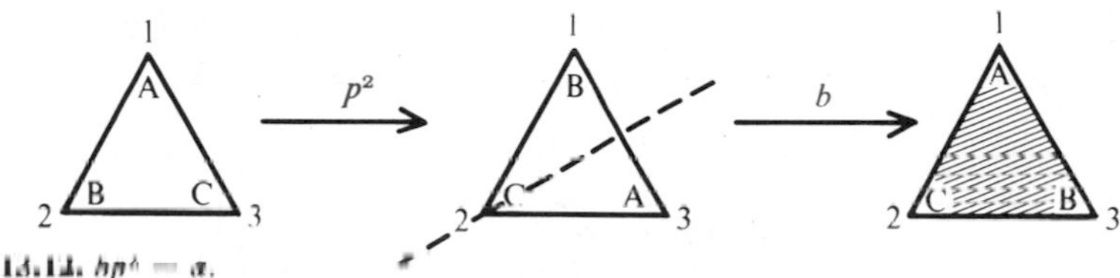

Fig. 13.11. $bp^2 = a$.

Now each of these six movements of the triangle may be shown by a permutation. It is no use considering permutations of the moving vertices A, B, C (see also pp. 189–90), since the flip-over, labelled 'b', does not always correspond to the transposition (AC): in the illustration above, the operation b interchanged A and B, for example. We must instead consider *permutations of the fixed points 1, 2, 3* with respect to the vertices of the moving triangle. This means that we think of p not as the permutation $\begin{pmatrix} A & B & C \\ C & A & B \end{pmatrix}$ of the moving vertices which replaces A by C, B by A, C by B. We think rather of the permutation $\begin{pmatrix} 1 & 2 & 3 \\ 2 & 3 & 1 \end{pmatrix}$, the significance of which is:

> 'One vertex of the triangle moves from position 1 to position 2.'
> 'One vertex of the triangle moves from position 2 to position 3.'
> 'One vertex of the triangle moves from position 3 to position 1.'

We are thinking of *permutations of the positions*† rather than permutations of the vertices of the moving triangle. Every one of the six symmetries may

† Compare what was said about the quintet when we considered permutations in Ch. 4; we could think either of the permutations of the instruments over the persons, or of the permutations of the persons over the instruments (see pp. 23–24).

then be assigned a permutation, and you should verify that these are as follows:

$$e\begin{pmatrix}1&2&3\\1&2&3\end{pmatrix};\quad p\begin{pmatrix}1&2&3\\2&3&1\end{pmatrix};\quad p^2\begin{pmatrix}1&2&3\\3&1&2\end{pmatrix};\quad a\begin{pmatrix}1&2&3\\1&3&2\end{pmatrix};$$

$$b\begin{pmatrix}1&2&3\\3&2&1\end{pmatrix};\quad c\begin{pmatrix}1&2&3\\2&1&3\end{pmatrix}.$$

This should be sufficient to convince us that the group of the equilateral triangle (D_3) is isomorphic to the full group of the six permutations of three symbols (P_3 or S_3). We may say $\mathbf{D}_3 \cong \mathbf{S}_3$. The reason that this particular group has such a rich variety of realisations is that it is masquerading in these two different guises.

Permutations of vertices: moving axes

Now let us think of the symmetries of the equilateral triangle from the point of view of the permutations of its *vertices*. Denote these permutations:

$$e\begin{pmatrix}A&B&C\\A&B&C\end{pmatrix};\quad p^*\begin{pmatrix}A&B&C\\C&A&B\end{pmatrix};\quad q^*\begin{pmatrix}A&B&C\\B&C&A\end{pmatrix};$$

$$a^*\begin{pmatrix}A&B&C\\A&C&B\end{pmatrix};\quad b^*\begin{pmatrix}A&B&C\\C&B&A\end{pmatrix};\quad c^*\begin{pmatrix}A&B&C\\B&A&C\end{pmatrix}.$$

Not a single one of these can be identified with the permutations e, p, p^2, a, b, c above. For while p^* is an anticlockwise rotation when the triangle is showing 'heads', it is a clockwise rotation when the triangle has been turned over and is showing 'tails'. The effect is that *the axis of rotation has been turned over with the triangle* – the triangle carries the axis with it, just as it carries the axes of the bilateral symmetries: a^* means a flip *about the median through A.* Thus we are concerned here with *axes which move with the body.* Whereas a means a flip about the axis which runs north–south – a line fixed in the plane.

In Ch. 20 we shall show that, when the axis a (running N.–S.) is carried by a movement t to a new position, the reflection in the *moved axis,* denoted by a^*, would be given by:

$$\boldsymbol{a^* = tat^{-1}}.$$

For the six permutations of the vertices, A, B, C, we note, for example, that

$$a^*p^* = c^*.$$

The first operation on the triangle, p^*, is the same as p, since the triangle starts from the identity position $\begin{pmatrix} 1 & 2 & 3 \\ A & B & C \end{pmatrix}$. This transformation, however, moves the median through A, and we have:

$$a^* = pap^{-1}.$$

Hence $a^*p^* = (pap^{-1})p = pa = c,$

and this is equivalent to the operation c^* starting from the initial position.

Table 13.03 summarises the two 'dual' aspects of the equilateral triangle symmetries.

Table 13.03

axes fixed in space

	vertices			
	A	B	C	
e	1	2	3	permutations of *positions*
p	2	3	1	
p^2	3	1	2	
a	1	3	2	
b	3	2	1	
c	2	1	3	

axes moving with the triangle

	positions			
	1	2	3	
e	A	B	C	permutations of *vertices*
p^*	C	A	B	
q^*	B	C	A	
a^*	A	C	B	
b^*	C	B	A	
c^*	B	A	C	

Much confusion may arise in the mind if the distinction between axes fixed in space, and axes which move with the body is not faced squarely. Many books are content to steer round this, and some even pretend it does not exist. The present discourse will, it is hoped, clear the matter up once and for all. It is of course, closely bound up with the distinction between permutations of *objects*, on the one hand, and permutations of their *positions* (or *arrangements*) on the other. This is discussed at greater length in an article in the *Mathematical Gazette*, Oct. 1966, D. B. Hunter, pp. 290–4.

Q. 13.13. Go through all the work of the last 4 pages, only with a square instead of an equilateral triangle.

Cayley's theorem illustrated by rotations of triangular prism

Next, let us rearrange the table for D_3 obtained above, by interchanging b and c, and replacing p^2 by q. Now consider the triangular prism in fig. 13.13. Each symmetry of this prism may be represented by a permutation of vertices, e.g. the half-turn about an axis through the centre of the

Table 13.04

	e	*p*	*q*	*a*	*c*	*d*
e	*e*	*p*	*q*	*a*	*c*	*b*
p	*p*	*q*	*e*	*c*	*b*	*a*
q	*q*	*e*	*p*	*b*	*a*	*c*
a	*a*	*b*	*c*	*e*	*q*	*p*
c	*c*	*a*	*b*	*p*	*e*	*q*
b	*b*	*c*	*a*	*q*	*p*	*e*

rectangle $BCPQ$ perpendicular to its plane may be represented by $\begin{pmatrix} AEBP & Q & A & C & B \\ A & B & C & E & Q & P \end{pmatrix}$, and we call this symmetry transformation a. Notice that every permutation keeps the letters A and E separated by two other letters; the same is true of the pairs (B, Q) and (C, P). AE, BQ and

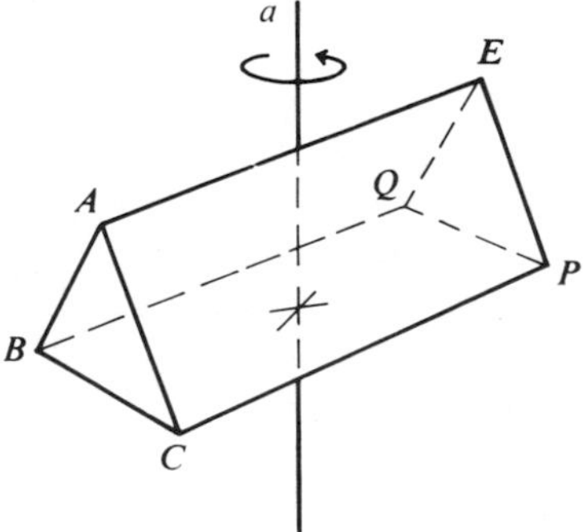

Fig. 13.13. Rotational symmetry of equilateral triangular prism.

CP are, in fact, the parallel edges of the prism. The method of naming the permutations is that the smaller letter is used for the permutation to correspond to the capital letter which E is replaced by in the permutation, so that we have $q = \begin{pmatrix} E & P & Q & A & C & B \\ Q & E & P & B & A & C \end{pmatrix}$. Now when the table of this group of six permutations, which is therefore the table for the direct symmetries of the triangular prism is constructed, and compared with the table above, it will be found that they are identical; and moreover, *each row of the table depicts precisely the permutation which it represents.* Thus Cayley's theorem stands out in all its glory!

Q. 13.14. Show that if the enantiomorphs of the triangular prism are included, the group is D_6. Represent these twelve symmetries by permutations.

Q. 13.15. Construct such a prism and show how it can be used as a 'slide rule' for the group. Compare also with the Cayley diagram (see p. 250) obtained by colouring the edges of the prism.

Subgroups of D_n

When n is prime, as in the case D_5, the only subgroups will be C_5 (of course), and the subgroups of order two which each of the elements of period 2 form with the identity.

When n is composite, the situation is more interesting. Considering the table for D_6 on p. 192, we have not only the subgroups $\{1, r, r^2, r^3, r^4, r^5\}$ (C_6), $\{1, a\}$, $\{1, b\}$, $\{1, c\}$, $\{1, d\}$, $\{1, f\}$, $\{1, g\}$ and $\{1, h\}$ (D_1), but also the subgroups of C_6 itself, namely $\{1, r^2, r^4\}$ and $\{1, r^3\}$.

But this is not all! The equilateral triangle ACE in fig. 13.05 has the symmetries 1, r^2, r^4 and also the bilateral symmetries a, c, f. Hence we have the subgroup $\{1, r^2, r^4, a, c, f\}$ (D_3).

Q. 13.16. Find two other subgroups, one with D_3 structure, the other D_2.

Q. 13.17. Find all the subgroups of the group $\{1, r, r^2, r^3, a, ar, ar^2, ar^3\}$ where $a^2 = r^4 = 1$ and $rar = a$ (D_4).

Q. 13.18. Show by geometrical considerations that D_n is a subgroup of D_{2n}.

'Dividing up' the group D_6

Let us now consider D_6 in a different aspect. Consulting the table on p. 192, we shall make the following changes of notation, the elements appearing in our new table in the order below:

notation in Table 13.01	e	r^2	r^4	a	c	f	r^3	r^5	r	d	g	b
replaced by Table 13.05	1	p	p^2	a	b	c	x	y	z	u	v	w

The table in its new version is shown in table 13.05.

The general dihedral group of order $2n$

The general dihedral group D_n, of order $2n$, is represented abstractly by the defining relations (see Ch. 15).

$$r^n = 1,\ a^2 = 1,\ ar = r^{n-1}a,$$

and these relations are enough to tell us all about the group.

Table 13.05

old notation →		e	r^2	r^4	a	c	f	r^3	r^5	r	d	g	b
↓	new notation	1	p	p^2	a	b	c	x	y	z	u	v	w
e	1	1	p	p^2	a	b	c	x	y	z	u	v	w
r^2	p	p	p^2	1	b	c	a	y	z	x	v	w	u
r^4	p^2	p^2	1	p	c	a	b	z	x	y	w	u	v
a	a	a	c	b	1	p^2	p	u	w	v	x	z	y
c	b	b	a	c	p	1	p^2	v	u	w	y	x	z
f	c	c	b	a	p^2	p	1	w	v	u	z	y	x
r^3	x	x	y	z	u	v	w	1	p	p^2	a	b	c
r^5	y	y	z	x	v	w	u	p	p^2	1	b	c	a
r	z	z	x	y	w	u	v	p^2	1	p	c	a	b
r													
d	u	u	w	v	x	z	y	a	c	b	1	p^2	p
g	v	v	u	w	y	x	z	b	a	c	p	1	p^2
b	w	w	v	u	z	y	x	c	b	a	p^2	p	1

(D_6).

Q. 13.19. Examine this form of the table, and the significance of the new arrangement, as far as possible. The full significance will be brought out in Ch. 21.
Can you discover another way of writing the table so that there is a different subgroup in the top left-hand corner, and which will also divide the table up into 'squares' in the way that this one will?

Q. 13.20. Prove from the above defining relations that $a = rar$, and interpret this geometrically in terms of the symmetries of a regular n-gon.

Q. 13.21. Find defining relations for D_n based on two generators of period 2.

The group D_n may be realised in a large variety of ways such as have been discussed in special cases above – groups of permutations, of symmetries of regular n-gons, of symmetries of certain solids, groups of matrices, of functions, and so on.

The group D_∞

In the previous chapter we introduced the group C_∞, and saw how it could be realised as a strip pattern by the repeated operation of a single translation.

Figure 13.141 shows the pattern produced by reflections in two mirrors inclined at 20°. Figure 13.142 shows part of the pattern when the angle between the mirrors is only 1°. In the first case we get the group D_9, in

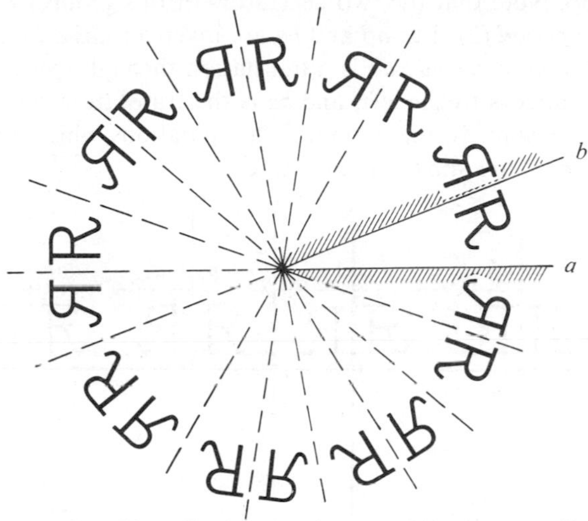

Fig. 13.141. Pattern produced by reflections in mirrors at 20°: the group D_9.

the second, D_{180}. In both cases we use the letter R as a typical operand, or 'motif', for the reflections.

If now the mirrors are made more and more parallel, the radius of the circle on which the R's lie becomes larger and larger. When the angle

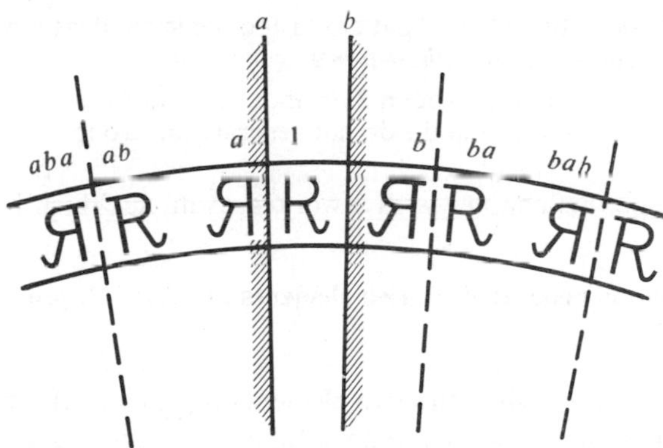

Fig. 13.142. Part of pattern produced by reflections in mirrors at 1°: the group D_{180}.

between the mirrors is 0.1°, we get the group D_{1800}, and finally, when the mirrors are actually parallel, we get the *strip pattern* shown in fig. 13.143, and the group is then known as D_∞.

Everyone is familiar, of course, with the infinite line of images that arise when two mirrors are placed in parallel positions, such as one sees in a barber's shop. Note that the two generators of this group,† a and b, are *both of finite period* (2). But ab and ba are inverse *translations*, and each of these has *infinite period: ab* is a translation through twice the distance between the mirrors to the left, and ba is the translation through this distance to the right. In the diagram, the operations which produce the respective images are shown in each case.

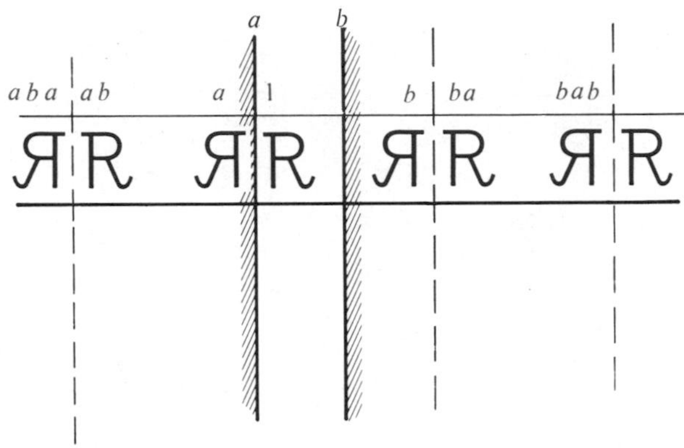

Fig. 13.143. The group D_∞.

Q. 13.22. Obtain the same strip pattern by using a translation and a reflection† in a single mirror perpendicular to it, as generators. Draw the pattern which would result if the mirror were not perpendicular to the translation.

Q. 13.23. Obtain another† strip pattern (using the letter R as a basic motif) by employing a translation and a *half-turn* as generators.

Q. 13.24. If r is a rotation through $2\pi/n$ about O, and i is an inversion in a circle centre O, show that r and i do not generate the group D_n.

We conclude this chapter with a worked example which has important consequences:

> **if a group is generated by two elements of period 2, prove that it is dihedral.**

Interpret in the case when the two elements represent (a) reflections in two axes, (b) half-turns about two points, (c) a reflection and a half-turn.
Let $a^2 = b^2 = 1$. Suppose first that the group is finite. Then ab has period (say) n. The only distinct 'words' that can be made with a and b are of the types $\ldots ababababababab$ and $\ldots bababababababa$.
Now $(ab)^n = 1 \Rightarrow a(ab)^n = a \Rightarrow a^2b(ab)^{n-1} = a \Rightarrow b(ab)^{n-1} = a = (ab)b$, i.e. $br^{n-1} = rb$, where $r = ab$. So we have the group generated by b and r, with $b^2 = r^n = 1$ and $br^{n-1} = rb$, and this is D_n.

† Bear in mind that these three groups, though *geometrically* distinct are nevertheless *abstractly* the same (isomorphic with D_∞).

As an illustration, in the case when $n = 7$, note that the fourteen elements may be written in alternative forms as below:

$$(ab)^7 = 1$$

period			period
2	$b(ab)^6 = a = (ab)b$	$b = a(ba)^6$	2
7	$(ab)^6 = ba$	$ab = (ba)^6$	7
2	$b(ab)^5 = aba$	$b(ab) = a(ba)^5$	2
7	$(ab)^5 = (ba)^2$	$(ab)^2 = (ba)^5$	7
2	$b(ab)^4 = a(ba)^2$	$b(ab)^2 = a(ba)^4$	2
7	$(ab)^4 = (ba)^3$	$(ab)^3 = (ba)^4$	7
2	$b(ab)^3 = a(ba)^3 = bababab = abababa.$		

Note that, for example, $b(ab)^4$ may be written in the alternative forms:

$$b(ab)^4 = a(ba)^2 = (ab)^2a = (ba)^4b = ababa = babababab.$$

Q. 13.25. If a and b are permutations given by the cycles $a = (AB)(CD)(EF)$; $b = (BC)(DE)$, find the period of ab and of ba, and identify Gp (a, b).† Find two other permutations each of period 2 of six letters which generate the same group. Repeat the above when a is $(BE)(CD)$ and b is $(AD)(BC)$; also when a is $(AB)(CD)(EF)(GH)$ and b is $(BC)(DE)(FG)$. Experiment with other pairs of permutations of period 2, and compare with the chapter on bell-ringing (Ch. 24).

Q. 13.26. Show that an infinite group cannot contain exactly two elements of period 2.

Q. 13.27. If $f(x) = 1/x$, $g(x) = (2 - x)/(1 + x)$, show that fg is of period 6, and find the other functions of Gp (f, g), and identify the structure of the group.

When ab is not of finite period, we obtain the group D_∞ as in the previous paragraph. The defining relations for D_n are $a^2 = b^2 = 1 = (ab)^n$, but for D_∞ they are simply $a^2 = b^2 = 1$.

If a and b are reflections in mirrors inclined at πn, we have the group D_n; when the mirrors are parallel, we have D_∞. If a and b are half-turns about two points, then ab is a translation, and we get the pattern generated as in fig. 13.15. Note that if $A^* = ab(A)$, then a half-turn about the point A^* (call this operation a^*) is given by:

$$a^* = (ab)\,a(ab)^{-1} \quad \text{(see pp. 200, 365)}$$
$$= abab^{-1}a^{-1} = ababa = (ab)^2a,$$

and this agrees with the geometrical interpretation in the above figure.

If on the other hand a is a reflection and b is a half-turn about B, then we shall get an infinite group so long as B is not on the axis of reflection.

† Gp(a, b) means the group generated by a and b – see pp. 216, 223.

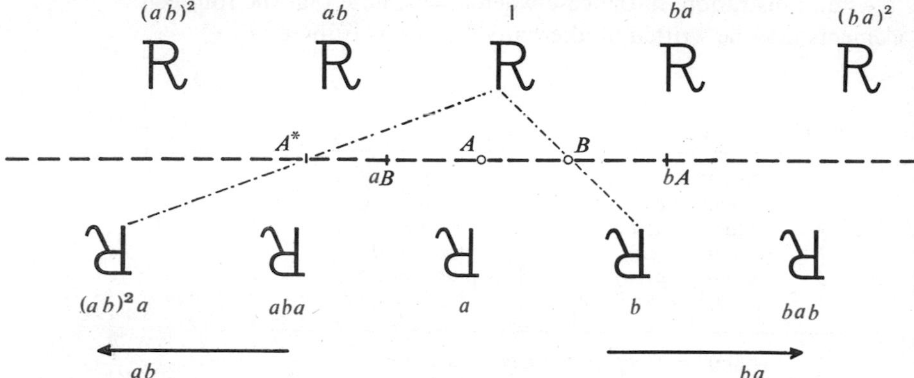

Fig. 13.15. Infinite strip pattern generated by half-turns about two points A, B.

It will be seen from fig. 13.16 that ab is a† *glide reflection* g in the line through B perpendicular to the mirror, and that its inverse g^{-1} is ba. Naturally this glide reflection must itself generate an infinite group, and in fact Gp (g) is the infinite cyclic group, which of course is a subgroup of D_∞. $(ab)^2$ is a translation t ($= g^2$), which also generates an infinite cyclic

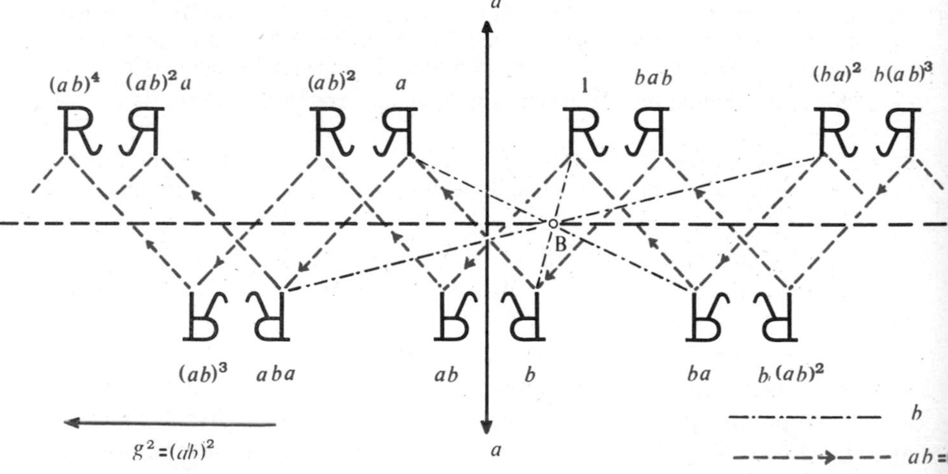

Fig. 13.16. Infinite strip pattern generated by half-turn about B and reflection in mirror a.

group; the images which belong to this subgroup Gp (t) consist of all those R's which are the right way up: R. Those R's in the pattern which are *not turned over* (i.e. either R or ꓤ) form the subgroup Gp (b, t), and consist of those images which result from transformations containing an *even number of reflections*. All those images on the top line of fig. 13.16

† See fig. 26.25; also Ex. 12.29.

contain an even number of half-turns, and these form the subgroup D_∞ (see fig. 13.143). If it should seem strange that D_∞ should be a subgroup of Gp (a, b) which itself has the structure D_∞, remember that C_∞ is a subgroup of C_∞ (see pp. 125, 181–2, 227). Moreover, D_n is a subgroup of D_{2n} (consider a regular $2n$-gon); (indeed, when n is odd, $D_{2n} \cong D_n \times D_1$ (see Ch. 16)). So we have D_{100} is a subgroup of D_{200}, and this remains true however large the order of the dihedral group...

Q. 13.28. Identify the cosets of these various subgroups (see Ch. 19).

When B lies on the axis a, we now have $ab = ba$, so that ab has period 2, and we find the finite group D_2 (see fig. 13.17).

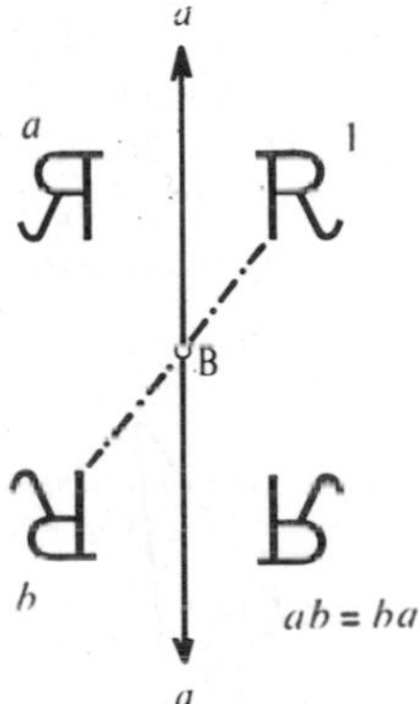

Fig. 13.17. The case when B lies on the mirror axis – the group D_2.

Dihedral groups are important aesthetically – they may be discerned in all the visual arts, particularly in pottery, in furniture and in buildings. For example, there is a ceiling in Culzean Castle, Ayrshire, in Scotland, with the symmetry group D_6; while the rose window in Durham Cathedral, in the North of England, has the group D_{24}.

EXERCISES 13

1 Show that those permutations of $ABCDE$ which *preserve cyclic order*, such as $\begin{pmatrix} A & B & C & D & E \\ D & C & B & A & E \end{pmatrix}$, form the group D_5. How may these be interpreted in terms of the symmetries of: (a) a regular pentagon; (b) a regular pentagonal pyramid; (c) a regular pentagonal prism; (d) a regular pentagonal bi-pyramid?

2 D_n is Abelian if and only if ...

3 The group D_n expresses the combination of the symmetries of the regular n-sided polygon. How do you explain D_2 in this sense? Can you give any meaning to D_1 by analogy?

4 Is it possible for a dihedral group to be obtained from a finite arithmetic under either addition or multiplication to some modulus?

5 Describe a polygon which has the symmetry group D_5, yet is not a regular pentagon.

6 In a certain group, $a^2 = p^3 = 1$, $(ap)^3 = 1$. Show that the group is not D_3.

7 The functions $1 - z$ and $1/z$ generate the group D_3. Find the remaining functions of the group, and give an interpretation in terms of Argand diagram transformations. (E.g. $z' = 1 - z$ is a half-turn about the point $\frac{1}{2}$.)

8 Show that $f(z) \equiv 2/(2 - z)$ is of period 4. Find a function $g(z)$ of period 2 such that i, f, f^2, f^3, g, gf, gf^2, gf^3 form the group D_4.

9 $f(x) \equiv (2x + 1)/(3x - 2)$, $g(x) \equiv (3 - x)/(1 + 2x)$, each of period 2. Find the period of the function $fg(x)$, and the structure of Gp (f, g). Find also the remaining functions of the group.

10 Find the group generated by the functions $2 - x$ and $2/x$.

11 Show that the function $f(z) \equiv (z - i)/(1 + zi)$ $(i^2 = -1)$ is of period 4, and find f^2 and f^3. How many functions can be formed by combining these functions with $g(z) \equiv -z$? Which group do they form? (See also Ex. 18.43.)

12 Find twelve functions forming the group D_6, one of which is $(x + 1)/(2 - x)$.

13 What is the group of the direct symmetries of the bi-tetrahedron? (Fig. 13.18.)

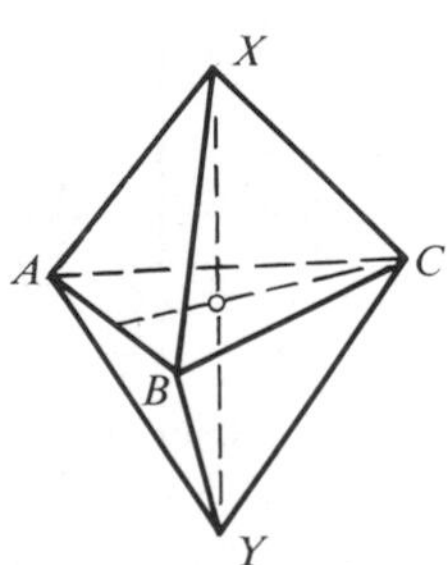

Fig. 13.18. Bi-tetrahedron.

Represent these by permutations of (a) the vertices, e.g. $\begin{pmatrix} A & B & C & X & Y \\ C & B & A & Y & X \end{pmatrix}$ where X and Y are the opposite vertices, (b) the faces. What is the full group of the bi-tetrahedron, that is, including the enantiomorphs, such as

$$\begin{pmatrix} A & B & C & X & Y \\ C & A & B & Y & X \end{pmatrix} \text{ and } \begin{pmatrix} A & B & C & X & Y \\ A & B & C & Y & X \end{pmatrix}?$$

14 Repeat the previous question for a square bi-pyramid which is not a regular octahedron. First give the structure of the rotation group, including permutations of vertices such as

$$\begin{pmatrix} A & B & C & D & X & Y \\ D & A & B & C & X & Y \end{pmatrix} \text{ and } \begin{pmatrix} A & B & C & D & X & Y \\ D & C & B & A & Y & X \end{pmatrix},$$

and then the group which includes enantiomorphs, such as

$$\begin{pmatrix} A & B & C & D & X & Y \\ A & B & C & D & Y & X \end{pmatrix}.$$

15 Show that the cycles (123)(456), (132)(465), (15)(24)(36), (14)(26)(35), and (16)(25)(34) together with the identity combine to give the group D_3, and that this group is a subgroup of S_6.

16 If $f(x) \equiv ax + b$, where $a \in \{1, 2\}$ and $b \in \{0, 1, 2\}$ show that the six functions under successive substitution form the group D_3, $+$ and $\times$ are mod. 3.

17 A penny and a halfpenny are placed on a table, and the following operations may be performed: either coin may be turned over; and the two coins may be interchanged. Thus there are eight possible states, represented Hh, Ht, Th, Tt, hH, tH, hT, tT. Show that the eight operations constitute the group D_4.

18 Consider three lines lying in a plane inclined at 60° to each other and all passing through a point O. Let reflections in these lines be denoted by a, b and c. Denote by $1, p, p^2$ rotations in the plane through 0, 120°, 240°, the positive direction being from a to b to c. Represent these six transformations by matrices, taking the a-line as the axis of x. Confirm that they give the group D_3 under matrix multiplication.

19 Find (a) two matrices, (b) two functions, of periods 2 which generate a dihedral group (see pp. 206 ff.). Try and obtain each of $D_2, D_3, D_4, \ldots$ in turn, and also D_∞.

20 Show that defining relations for D_3 may be written $a^2 = b^2 = 1$, $aba = bab$.

21 Give several sets of defining relations for D_4.

22 Can you find a chain of subgroups of D_6, each containing in the next such that $1 \subset \ldots \subset D_6$?

23 Consider automorphisms of the group D_4.

24 Can $D_3, D_4, D_5, \ldots$ be generated by *any* two elements? How can we decide whether a given pair of elements will serve as a pair of generators?

25 On pp. 145 ff. we saw how a slide rule could be constructed for the group C_n, either for its realisation as a group of residues under $+$ mod. n, or as a group of residues under multiplication to some modulus. Evidently a circular slide rule may be made to give products in any abstract cyclic group. Show how a slide rule may be constructed to give products in the group D_n by arranging for it to be possible for the 'slide', or movable scale, to be turned over. Try this, first with D_3, then D_4, and so on. Note that since these groups are not Abelian, you must take the precaution of specifying the order in which the product of the elements is read off.

26 Consider two mirrors at an angle π/n ($n \in \mathbf{Z}^+$). Are we bound to get the group D_n?

27 Show that the matrices

$$\begin{pmatrix} 1 & 0 \\ 0 & 1 \end{pmatrix}, \begin{pmatrix} -1 & 0 \\ 1 & 1 \end{pmatrix}, \begin{pmatrix} -1 & -1 \\ 1 & 0 \end{pmatrix}, \begin{pmatrix} 0 & -1 \\ -1 & 0 \end{pmatrix}, \begin{pmatrix} 0 & 1 \\ -1 & -1 \end{pmatrix} \text{ and } \begin{pmatrix} 1 & -1 \\ 0 & -1 \end{pmatrix}$$

form the group D_3.

Show also that $\begin{pmatrix} -1 & 0 \\ 1 & 1 \end{pmatrix}$ and $\begin{pmatrix} 0 & 1 \\ 1 & 0 \end{pmatrix}$ generate D_6, and find the other ten matrices.

28 By using Cayley's theorem, find eight permutations of 1, 2, 3, 4, 5, 6, 7, 8 which give the group D_4.

29 Show that the matrices

$$M = \tfrac{1}{2}\begin{bmatrix} -1 & 1 & 1 & 1 \\ 1 & -1 & 1 & 1 \\ 1 & 1 & -1 & 1 \\ 1 & 1 & 1 & -1 \end{bmatrix}, \quad J = \begin{bmatrix} -1 & 0 & 0 & 0 \\ 0 & -1 & 0 & 0 \\ 0 & 0 & -1 & 0 \\ 0 & 0 & 0 & -1 \end{bmatrix},$$

$$N = \tfrac{1}{2}\begin{bmatrix} -1 & 1 & 1 & -1 \\ 1 & -1 & 1 & -1 \\ 1 & 1 & -1 & -1 \\ -1 & -1 & -1 & -1 \end{bmatrix}$$

satisfy $M = JMJ$, and that the group D_3 is generated under multiplication.

30 An impoverished writer types four copies of a book using his only remaining three carbons. Each time he puts in a new sheet he switches the carbons by the following system: first the carbon paper that was on top is put into the middle, then that which was on the bottom is put in the middle, and so on alternately. Does this system ensure even wear of the three carbons? Describe another system. Which groups are associated with them?

31 If a and b are elements of a group each of period 2, show that the group must contain at least one more element of period 2. Show that D_n has $n + \tfrac{1}{2} + \tfrac{1}{2}(-1)^n$ elements of period 2, and that D_∞ has an infinite number.

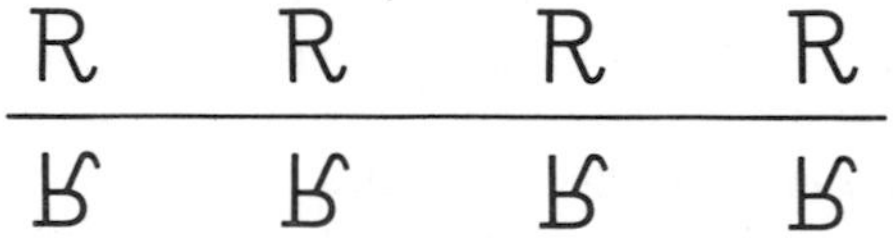

Fig. 13.19.

32 (a) If $f(x) \equiv 3 + (1 - x)\, f(1/x)$, find $f(x)$. (Replace x by $1/x$ in the given equation.)
(b) If $f(x) \equiv 2 + x\, f[1/(1 - x)]$, find $f(x)$. (Use the fact that, if $g(x) \equiv 1/(1 - x)$, then g is of period 3: $f(x) \equiv 2 + x\, fg(x)$; $fg(x) = 2 + g(x)\, f[g^2(x)]$, etc.)

33 The function $f(x) \equiv 1 + x$ (for example) generates the group C_∞. Find a function $g(x)$, of period 2, such that f and g generate D_∞. Find also two functions each of period 2 which generate D_∞.

34 In fig. 13.14 the cyclic group C_n is extended to D_n by adjoining the reflections in radial lines. Would we get D_n if, instead of reflecting in radial lines we (a) reflected in lines perpendicular to the radii or (b) inverted with respect to the circle?
Does the pattern shown in fig. 13.19 correspond to the group D_∞?

35 We have seen that reflections in mirrors at right-angles generate the group D_2. Show by inversion, that inversions in two orthogonal circles also commute and generate D_2. What must be true of two intersecting circles if the composition of successive inversions in them is to be of period 3?

36 If a and b are inversions in two given circles, Γ_1, Γ_2, show that they generate D_n if the circles intersect at an angle π/n. Consider special cases of this result. What happens when the circles have no point in common?

37 Show that the operation $*$ on the integers defined $p * q = p + (-1)^p q$ is associative but not commutative. Show that $(\mathbf{Z}, *)$ is a group isomorphic with D_∞. Show that the set $\{0, 1, 2, 3\}$ under the same operation except that the addition is to be modulo 4, form the group D_2. Can you generalise?

How may the connection between the group $(\mathbf{Z}, *)$ above and the group of the strip pattern of fig. 13.14 be established?

38 If $A = \begin{pmatrix} 1 & 0 \\ 0 & -1 \end{pmatrix}$ and $B = \begin{pmatrix} 1 & 1 \\ 0 & -1 \end{pmatrix}$, show that $A^2 = B^2 = I$, and that Gp. (A, B) has the structure D_∞. Interpret geometrically.

39 Discuss the pattern shown in fig. 13.20.

$\circ\, O$

Fig. 13.20. Part of pattern produced by rotation about O through $2\pi/n$ and reflections in tangents to circle centre O – group is only C_n, and the reflections do not belong to it.

14 *Groups within groups: subgroups*

Definition:

> **A subgroup of a group is a subset which is itself a group under the group operation.**

When we said that a group of prime order had no subgroups, we should have been more accurate by saying no *proper* subgroups, because every group contains the identity which, on its own, is a group of order 1; and besides this, the whole group can be regarded as a subgroup of itself. These are, however, trivial cases, and when we use the word 'subgroup' we shall always mean '. . . other than the group itself, or the identity', i.e. we shall always be referring to a 'proper' subgroup.

Examples

Already, we have met many cases of subgroups. In Ch. 12, we have seen that C_{12} has the subgroups C_2, C_3, C_4 and C_6; C_{10} has the subgroups C_5 and C_2; C_8 has subgroups C_4 and C_2 (three of the latter). While in Ch. 13, we observed that D_4 has a subgroup C_4, two subgroups D_2, and several subgroups D_1; D_5 has a subgroup C_5 and five subgroups C_2 (or D_1). Finally, D_6 contains subgroups C_6, C_4, C_3, C_2 together with three separate subgroups D_3 as well as one with structure D_2. In turn, all these groups mentioned are themselves subgroups of higher groups, and Cayley's theorem tells us that a group of order n is a subgroup of S_n, the group of all the permutations of n objects, which is a sort of 'master-group'. We saw, for example, that D_3 could be obtained by permutations of six letters, so that it is a subgroup of S_6 (of order 720). But later, on pp. 225–6, we shall derive D_3 by means of six permutations of only five symbols, thereby finding that D_3 was a subgroup of the much smaller group S_5, of order 120. Finally, we recognised (pp. 198 ff.) D_3 to be isomorphic to S_3 itself. C_{12} was first introduced as the group obtained by the twelve cyclic permutations of twelve symbols, and this revealed C_{12} as a subgroup of S_{12}, an enormous group of order 12! Later (see Q. 10.12), we saw that C_{12} could be obtained by permuting only seven symbols with two disjoint cycles of four and three. This shows that C_{12} is a subgroup of S_7 (or order 5040, still large!). One may well wonder: is it possible to find twelve permutations of *less than seven* objects which give C_{12}? Or, to put it another way, what is the least value of n such that C_{12} is a subgroup of

S_n? Is the answer seven, or could C_{12} be a subgroup of S_6, or even S_5? The answer to this question depends entirely on whether one can find a permutation x of six (or five) objects, which is of *period 12*. For if such a permutation exists, then it will generate C_{12}.

A larger question suggests itself: for any given finite group G of order n, what is the smallest value of m such that G is a subgroup of S_m? We know by Cayley's theorem that $m \leqslant n$. Do groups exist for which G is not a subgroup of S_{n-1}, for instance? Unfortunately, the answers to such apparently simple questions are far too complicated to be investigated here.

Lagrange's theorem

It is scarcely possible that you would have got as far as this in the reading of this book without having been struck by the fact that a group of order 12 may have subgroups of order 2, 3, 4 and 6, but none of order 5, 8, etc. A group of order 8 never has subgroups of order 3, and so on. If you have conjectured that the order of a subgroup must divide the order of the group, you will have anticipated a most important result – *Lagrange's theorem.* We shall use the notation $|G| = n$ to mean that the order of a finite group G is equal to n.

> ***Lagrange's theorem.* If H is a subgroup of a group G, and if $|G| = n$, $|H| = m$, then m is a factor of n.**

From this it follows immediately that all prime-order groups have no (proper) subgroups. The proof of Lagrange's theorem will be deferred till we have dealt with cosets (Ch. 19), but we will take the result of the theorem for granted from now onwards. It is immediately possible to dispose of the results on groups of prime order p (see pp. 170, 176): they can have no proper subgroups, and from this it follows that every element is of period p. For if x were of period q ($< p$), then we should have $x^q = 1$, and x would generate a subgroup $\{1, x, x^2, x^3, \ldots, x^{q-1}\}$ of order q. This contradiction compels us to accept that $q = p$, and this is true of all elements except the identity. Furthermore, there can be no group of prime order other than C_p. For if x ($\neq 1$) is an element of a group of order p (prime), then x is of period p. Therefore the elements $1, x, x^2, x^3, \ldots, x^{p-1}$ are all distinct, and so constitute the whole group, which is therefore cyclic.

Generation of subgroups

Much of the interest in groups lies in their subgroups, so one of our first jobs must be to be able to spot subgroups. Now it is very easy to pick out cyclic subgroups from any group, because *every element of a group may be used to generate a cyclic subgroup*. If the period of the element x is m, then x will evidently generate C_m $\{1, x, x^2, x^3, \ldots, x^{m-1}\}$. We shall use the notation: Gp (x) to denote the group generated by the element x. Suppose we have a group of order 18, and there is an element p of period 6. Then p generates the group $\text{Gp}(p) = \{1, p, p^2, p^3, p^4, p^5\} \cong C_6$.

Q. 14.01. What is $\text{Gp}(p^2)$; $\text{Gp}(p^3)$?
Q. 14.02. The identity is, of course, powerless to generate a proper subgroup. Under what circumstances is one bound to get a *proper* subgroup generated by an element x?

As an example of seeking cyclic subgroups, consider the group table 14.01.

Table 14.01

	1	*a*	*b*	*c*	*d*	*f*	*g*	*h*	*j*	*k*	*l*	*m*	period	
1	1	*a*	*b*	*c*	*d*	*f*	*g*	*h*	*j*	*k*	*l*	*m*	1	
a	*a*	*l*	1	***d***	*f*	***g***	***h***	***j***	***c***	*m*	*k*	*b*	6	
b	*b*	1	*m*	***j***	***c***	***d***	***f***	***g***	***h***	*l*	*a*	*k*	6	
c	*c*	***j***	***d***	*k*	***l***	***a***	1	***b***	***m***	*g*	***h***	***f***	4	
d	*d*	***c***	*f*	***m***	*k*	***l***	***a***	1	***b***	*h*	***j***	***g***	4	
f	*f*	***d***	***g***	***b***	***m***	*k*	***l***	***a***	1	*j*	***c***	***h***	4	
g	*g*	***f***	***h***	1	***b***	***m***	*k*	***l***	***a***	*c*	***d***	***j***	4	
h	*h*	***g***	***j***	***a***	1	***b***	***m***	*k*	***l***	*d*	***f***	***c***	4	
j	*j*	***h***	***c***	***l***	***a***	1	***b***	***m***	*k*	*f*	***g***	***d***	4	
k	*k*	*m*	*l*	*g*	*h*	*j*	*c*	*d*	*f*	1	*b*	*a*	2	
l	*l*	*k*	*a*	***f***	***g***	***h***	***j***	***c***	***d***	*b*	*m*	1	3	
m	*m*	*b*	*k*	***h***	***j***	***c***	***d***	***f***	***g***	*a*	1	*l*	3	(Q_6)

$$\text{Gp}(a) = \{1, a, a^2, a^3, a^4, a^5\} = \{1, a, l, k, m, b\} = \text{Gp}(b) \cong C_6$$
$$\text{Gp}(c) = \{1, c, c^2, c^3\} = \{1, c, k, g\} = \text{Gp}(g) \cong C_4$$
$$\text{Gp}(d) = \{1, d, d^2, d^3\} = \{1, d, k, h\} = \text{Gp}(h) \cong C_4$$
$$\text{Gp}(f) = \{1, f, f^2, f^3\} = \{1, f, k, j\} = \text{Gp}(j) \cong C_4$$
$$\text{Gp}(k) = \{1, k\} = C_2 \qquad \text{Gp}(l) = \{1, l, m\} = \text{Gp}(m) \cong C_3.$$

This exhausts all the cyclic subgroups. But of course, there may be other subgroups, particularly non-Abelian subgroups, and these can only be of order 2, 3, 4, or 6. The only remaining possibilities are D_2 and D_3.† But both these groups require three elements of period 2, yet our group has only the element k of period 2. Hence Q_6 certainly contains no other subgroups.

Q. 14.03. Consult various group tables in this book, and identify the groups generated by each element for each group.

† We are assuming here that the only groups of order 6 are C_6 and D_3 (see pp. 141ff).

Centre of a non-Abelian group

In the above group Q_6, there is something special about the element k, apart from the fact that it is the only element of period 2. It occurs in C_6 and in every one of the C_4 subgroups. Moreover, you will notice that the *column* headed k: $k, m, l, g, h, j, c, d, f, 1, b, a$ is exactly the same as *row* k, and this is true for no other element. This is also seen immediately from the fact that row k and column k contain no bold letters. This means that $ak = h = ka$; $bk = j = kb$; and so on, so that k *commutes with every element of the group*. Elements of a group which have this property form a special subset known as the *Centre* of the group. Thus the centre of the group Q_6 is $\{1, k\}$.

Q. 14.04. What is the centre of an Abelian group?

We shall now show that this set of elements which commute with every element of the group do in fact form a subgroup. But before doing this, we shall first establish a simple condition for a subset of a group to be a subgroup:

Theorem

A subset H of a finite group G is a subgroup of G if and only if it is *closed* under the operation of the group.

To prove this, let h be a typical element of H, of period m. Then in view of the closure property which is postulated, $hh = h^2$ must be in H; therefore $h^2h = h^3$ must be in H, and by continuing the process, we see that *every* power of h belongs to H. Finally we reach $h^m = 1$, showing that H contains the identity – an essential requirement. Moreover, $h^{m-1} = h^{-1} \in \text{H}$, so H contains the inverse of h. Thus, in view of the fact that h was *any* element of H, we have the requirements C, N and I of p. 73 satisfied. But A (associativity) requires no comment, since associativity in G implies associativity in H. Hence H is a subgroup.

Q. 14.05. Note that the proof is valid only for *finite* groups. At what point in the proof do we make this assumption?

Q. 14.06. Prove that a necessary and sufficient condition for H to be a subgroup is that $\forall h_1, h_2 \in \text{H}, h_1 h_2^{-1} \in \text{H}$.

Now suppose that H is the centre of G, i.e. it is the set of all those elements of G which commute with every element of G. We note first that the centre of every group is non-empty, since the identity element commutes with every element of a group. (A group whose centre consists of the identity *only* is described as having a *trivial* centre.) Suppose that h_1

and h_2 are any two elements of H. In view of the above theorem, it will only be necessary to prove that h_1h_2 also belongs to H, i.e. that the closure requirement is met. Let x be any element of G.

Then, $\quad h_1 \subset \mathrm{H} \Rightarrow h_1x = xh_1,$

and $\quad h_2 \in \mathrm{H} \Rightarrow h_2x = xh_2.$

$$\text{Then} \quad (h_1h_2)x = h_1(h_2x) = h_1(xh_2) = (h_1x)h_2 = (xh_1)h_2 = x(h_1h_2).$$

Thus h_1h_2 commutes with x (any element of G), and so $h_1h_2 \in \mathrm{H}$, by definition. Therefore the centre H is a subgroup of G, and is, of course, an Abelian subgroup.

Although Q_6 (see p. 216) was a group of order 12, it had a very small centre, of order only 2. There are some groups whose centre contains only the identity. It can be proved for example, that the centre of S_n is merely the identity, for all $n > 2$, and this may be surprising. Another way of saying this is that there is no permutation of n objects, other than the identity, which commutes with *every other* permutation; or, that for any given permutation, one can always find a permutation which does not commute with it.

Q. 14.07. Check that the centre of S_3 is the identity. Find the centre of D_4, D_5 and D_6. Check from the table on p. 219 that the centre of S_4 is the identity only.

Q. 14.08. Prove a general theorem about subgroups: if H_1 and H_2 are subgroups of G, then the intersection $H_1 \cap H_2$ is also a subgroup.

Subgroups of S_4

As a further exercise on subgroups, we might investigate the subgroups of S_4, in table 14.02. These subgroups must, by Lagrange's theorem, be of order 2, 3, 4, 6, 8 or 12. We leave it to you to begin the process by picking out the cyclic subgroups C_2, C_3 and C_4 generated by each element, there being nine of period 2 ($a, b, f, g, h, p, r, w, y$), eight of period 3 ($c, d, i, l, m, q, u, v$) and six of period 4 ($j, k, n, s, t, x$). There can be no other cyclic subgroups. (Why?)

★ We now search for subgroups D_2, D_3, D_4, and possibly other subgroups of order 8 and 12. D_2 requires three elements of period 2. The products of elements of period 2 are shown in sub-table, table 14.03. Having underlined the elements of period 2, it is a simple matter to pick out the D_2 subgroups. For example, if we want D_2 to contain 1 and g, row g shows that it must also contain a and h, and you may verify that 1, a, g, h gives D_2 structure.

Q. 14.09. Find three other D_2 subgroups.

Table 14.02 (The group S_4)

	1	a	b	c	d	f	g	h	i	j	k	l	m	n	p	q	r	s	t	u	v	w	x	y	period
1	1	a	b	c	d	f	g	h	i	j	k	l	m	n	p	q	r	s	t	u	v	w	x	y	1
a	a	1	c	b	f	d	h	g	j	i	l	k	n	m	q	p	s	r	u	t	w	v	y	x	2
b	b	d	1	f	a	c	i	k	g	l	h	j	p	r	m	s	n	q	v	x	t	y	u	w	2
c	c	f	a	d	1	b	j	l	h	k	g	i	q	s	n	r	m	p	w	y	u	x	t	v	3
d	d	b	f	1	c	a	k	i	l	g	j	h	r	p	s	m	q	n	x	v	y	t	w	u	3
f	f	c	d	a	b	1	l	j	k	h	i	g	s	q	r	n	p	m	y	w	x	u	v	t	2
g	g	h	m	n	t	u	1	a	p	q	v	w	b	c	i	j	x	y	d	f	k	l	r	s	2
h	h	g	n	m	u	t	a	1	q	p	w	v	c	b	j	i	y	x	f	d	l	k	s	r	2
i	i	k	p	r	v	x	b	d	m	s	t	y	1	f	g	l	u	w	a	c	h	j	n	q	3
j	j	l	q	s	w	y	c	f	n	r	u	x	a	d	h	k	t	v	1	b	g	i	m	p	4
k	k	i	r	p	x	v	d	b	s	m	y	t	f	1	l	g	w	u	c	a	j	h	q	n	4
l	l	j	s	q	y	w	f	c	r	n	x	u	d	a	k	h	v	t	b	1	i	g	p	m	3
m	m	t	g	u	h	n	p	v	1	w	a	q	i	x	b	y	c	j	k	r	d	s	f	l	3
n	n	u	h	t	g	m	q	w	a	v	1	p	j	y	c	x	b	i	l	s	f	r	d	k	4
p	p	v	i	x	k	r	m	t	b	y	d	s	g	u	1	w	f	l	h	n	a	q	c	j	2
q	q	w	j	y	l	s	n	u	c	x	f	r	h	t	a	v	d	k	g	m	1	p	b	i	3
r	r	x	k	v	i	p	s	y	d	t	b	m	l	w	f	u	1	g	j	q	c	n	u	h	2
s	s	y	l	w	j	q	r	x	f	u	c	n	k	v	d	t	a	h	i	p	b	m	1	g	4
t	t	m	u	g	n	h	v	p	w	1	q	a	x	i	y	b	j	c	r	k	s	d	l	f	4
u	u	n	t	h	m	g	w	q	v	a	p	1	y	j	x	c	i	b	s	l	r	f	k	d	3
v	v	p	x	i	r	k	t	m	y	b	s	d	u	g	w	1	l	f	n	h	q	a	j	c	3
w	w	q	y	j	s	l	u	n	x	c	r	f	t	h	v	a	k	d	m	g	p	1	i	b	2
x	x	r	v	k	p	i	y	s	t	d	m	b	w	l	u	f	g	1	q	j	n	c	h	a	4
y	y	s	w	l	q	j	x	r	u	f	n	c	v	k	t	d	h	a	p	i	m	b	g	1	2

Note: You will find that each of *h*, *r* and *y* occur in *two* of the four D_2 subgroups, and you notice that each of rows *h*, *r* and *y* in the above table have got six elements of period 2 in them, whereas the others have four. Therefore there is something different about those three elements – they are 'in a different class' from *a*, *b*, *f*, *g*, *p* and *w* (see pp. 298 ff, 384, 409).

Table 14.03 (not a group)

	a	b	f	g	h	p	r	w	y
a	1	c	d	h	g	q	s	v	x
b	d	1	c	i	k	m	n	y	w
f	c	d	1	l	j	r	p	u	t
g	h	m	u	1	a	i	x	l	s
h	g	n	t	a	1	j	y	k	r
p	v	i	r	m	t	1	f	q	j
r	x	k	p	s	y	f	1	n	h
w	q	y	l	u	n	v	k	1	b
y	s	w	j	x	r	t	h	b	1

Next, let us see if there are any D_3 subgroups. Such a subgroup requires an element of period 3 and its inverse, and three of period 2. Select q and v $(= q^2)$ to constitute the subgroup C_3, $\{1, q, v\}$. Which three elements of period 2 will go with these to form D_3? Considering the products of q with the elements of period 2, these are:

	a	*b*	*f*	*g*	*h*	*p*	*r*	*w*	*y*
q	$\underline{w}$	*j*	*s*	*n*	*u*	$\underline{a}$	*d*	$\underline{p}$	*i*.

and only those products of period 2 underlined are acceptable. So we consider testing whether $\{1, q, v, a, p, w\}$ form a group. It turns out that this set is closed and so we have succeeded in forming a subgroup D_3.

Q. 14.10. Find all the other D_3 subgroups (one of these is particularly easy in view of the way the table is printed).

Finally, there may possibly be a D_4 subgroup. Now D_4 requires for a start two elements of period 4, either of which will generate the C_4 subgroup. So let us take Gp $(j) = \{1, j, r, t\}$, and we shall try to annex four elements of period 2 to this to give the group D_4. Again, combining all the elements of period 2 with those selected (j, r and t), we get:

Table 14.04

	a	*b*	*f*	*g*	*h*	*p*	*r*	*w*	*y*
j	*l*	*q*	*y*	*c*	*f*	*h*	*t*	*i*	*p*
r	*x*	*k*	*p*	*s*	*y*	*f*	1	*n*	*h*
t	*m*	*u*	*h*	*v*	*p*	*y*	*j* ↑ already accepted.	*d*	*f*

Clearly, we may only accept f, h, p and y, and it will be found that $\{1, j, r, t, f, h, p, y\}$ do indeed form the subgroup D_4.

Q. 14.11. Find other D_4 subgroups.

Finally there is a subgroup of order 12. This is the 'alternating group' (A_4), and we shall study it later. For the moment, you may care to try and discover it, given that it contains all the elements of period 3.

Invariant properties associated with subgroups

The above empirical approach to finding subgroups is not recommended as a general method. More orthodox methods of identifying subgroups of abstract groups will be considered later. However, it is far easier to pick out subgroups when the group is realised in some concrete form. If, for example, one had known that the group S_4 may be realised as the group of symmetries of the *cube* (see Ch. 17), then the identification of the subgroups would have been a simpler matter. The reason for this is that *a subgroup may be associated with some invariant property within the parent group.* To show what this means, consider once again the group D_6 (see pp. 189–95). The subgroup C_6 may be thought of as *that subset of the symmetries of the regular hexagon* which *do not turn it over* (or which preserve the sense of the cyclic lettering). Again, which subgroup is associated with the diagonal AD being unchanged? The only symmetry, besides the identity, which will leave both A and D undisturbed is the half-turn about AD (fig. 13.04) (i.e. g). So the requirement that A and D are separately unchanged gives rise to the subgroup $\{1, g\}$. But if we are allowed to turn the diagonal AD round, with A and D changing places, then this interchange occurs as a result of r^3, and also as a result of c. Therefore the subgroup $\{1, g, r^3, c\}$ (D_2) is associated with the *invariance of the diagonal AD.* Furthermore, the subgroup D_3 of D_6 arises from concentrating upon the symmetries of one of the equilateral triangles (ACE or BDF) belonging to the regular hexagon.

Alternative approach to the subgroups of S_4

Resuming once again our study of the group S_4, its most obvious realisation is as the group of the twenty-four permutations of four symbols. Finding subgroups of S_4 is equivalent to finding *closed subsets* of permutations of four symbols. One such subset, for example, is the identity together with the double transpositions $(AB)(CD)$; $(AC)(BD)$; $(AD)(BC)$. [This is the group D_2 which leaves cross-ratio invariant, see p. 327.] Another closed subset would be those permutations which do not change one particular letter, e.g. D. These would consist of the six permutations on the remaining three letters, i.e. the cycles (BC); (CA); (AB); (ABC); (ACB) together with the identity permutation, and in this way, we may readily see that S_3 is a subgroup of S_4. Moreover, there will be four such subgroups: those which leave A where it was, those which leave B alone, those which leave C alone, and the one mentioned which permutes A, B and C without disturbing D.

Q. 14.12. Can you discover, from a consideration of permutations, how many D_2 subgroups there are of S_4 besides the one mentioned above?

Q. 14.13. How would you demonstrate that S_n is a subgroup of S_N when $n < N$? Find out how many subgroups S_5 has with structure (a) S_3, (b) S_4.

To find the D_4 subgroups of S_4, it is helpful to think of D_4 as the symmetry group of the square, and to interpret the eight symmetries of the square in terms of permutations of the vertices. The rotational symmetries of the square *ABCD* correspond to the cyclic permutations

$$\begin{pmatrix} A & B & C & D \\ A & B & C & D \end{pmatrix}, \quad \begin{pmatrix} A & B & C & D \\ D & A & B & C \end{pmatrix}, \quad \begin{pmatrix} A & B & C & D \\ C & D & A & B \end{pmatrix} \quad \text{and}$$

$$\begin{pmatrix} A & B & C & D \\ B & C & D & A \end{pmatrix}.$$

The reflections correspond to the permutations

$$\begin{pmatrix} A & B & C & D \\ A & D & C & B \end{pmatrix}, \quad \begin{pmatrix} A & B & C & D \\ C & B & A & D \end{pmatrix} \text{ (about diagonals), and}$$

$$\begin{pmatrix} A & B & C & D \\ B & A & D & C \end{pmatrix}, \quad \begin{pmatrix} A & B & C & D \\ D & C & B & A \end{pmatrix}.$$

Other D_4 subgroups may be found by imagining that the square, instead of being named *ABCD* (in cyclic order, with diagonals *AC*, *BD*), were renamed so that *AB*, *CD* were diagonals, or else that *AD*, *BC* were diagonals. Further subgroups may be formed by disregarding some particular property of the square and by thinking of the square as some more general type of quadrilateral, such as a rectangle or a rhombus.

Normalisers and centralisers

There is another theorem on subgroups which is of great assistance in identifying subgroups of non-Abelian groups. If we select one particular element a of a group, and write down the *set of all elements which commute with the given element*, i.e. $\{x : ax = xa, x \in G\}$, this set is called the *Centraliser* (or *Normaliser*) of a, and we can prove that it is bound to be a subgroup. For, if x_1 and x_2 are in the centraliser of a, then $ax_1 = x_1a$ and also $ax_2 = x_2a$.
Hence, $(x_1x_2)a = x_1(x_2a) = x_1(ax_2) = (x_1a)x_2 = (ax_1)x_2 = a(x_1x_2)$. Thus x_1x_2 also commutes with the given element a, and so is in the normaliser of a.† The theorem on p. 217 now establishes that the centraliser is a subgroup. Similarly, of course, x_2x_1 is also in the centraliser, but note that x_1x_2 is not necessarily equal to x_2x_1, so that one should not expect the centraliser to be Abelian.

† This proof should be compared with the proof on p. 218 that the centre of a group is a subgroup.

Use of centralisers in finding subgroups of S_4

Armed with this theorem, we may now seek subgroups of S_4. Take, for example, the centraliser of a. The only elements which do commute with a are 1, a, h, g and these form one of the D_2 subgroups already found. Again, the centraliser of c is $\{1, c, d\}$, and this is a C_3 subgroup. The centraliser of y is $\{b, h, k, n, r, w, y, 1\}$, one of the D_4 subgroups hitherto unidentified.

Q. 14.14. Use centralisers to find as many other subgroups as possible.

Q. 14.15. If an element belongs to the centre of any group, what is its centraliser?

Q. 14.16. Prove that the centre of any group is a subgroup of the centraliser of each of its elements.

In looking for centralisers, it is helpful in the case of small groups to write out the group table indicating which pairs of elements commute. This explains the elements in bold type in the table Q_6 on p. 216 as well as in the table for S_4. The centraliser of each element may then be read off at a glance, and the centre too is plainly seen. For example, the centraliser of m in the group Q_6 (p. 216) is the set $\{1, m, a, b, l, k\}$ whose periods are 1, 3, 6, 6, 3, 2, so this is the subgroup whose structure is C_6. The device of marking in prominently the products of pairs of elements which do not commute in non-Abelian groups will be followed in future work.

Non-cyclic subgroups generated by two or more elements

Another method of finding subgroups other than cyclic ones, which is an improvement on the method given above is as follows: To seek cyclic subgroups, we selected an element, say p, and found the group generated by p: $\{1, p, p^2, \ldots, p^{m-1}\}$ which we denoted Gp (p), the cyclic group C_m. To obtain non-cyclic groups, we simply select a *pair* of elements, say p and q, and find the group generated by these two elements, which we call Gp (p, q). This means every possible product formed by these two elements, such as p^3q^2; qp^4; $pqpq^2$; $p^2qpq^3p^4qp$, and so on. This is not so laborious a process as might be imagined, and we can simplify many such expressions by using the defining relations of the group. We illustrate this with some examples.

First, let us refer to the group Q_6 whose table is given on p. 216. Now it so happens that we found already that the only subgroups were cyclic, being generated by single elements, so we shall not in fact make any further discoveries by taking the subgroups generated by pairs of elements. For example, Gp (l, k) turns out to contain $\{1, a, k, l, b, m\}$ but this is C_6

and is more simply derived as Gp (a) or Gp (b). However, let us see what happens when we form, say, Gp (c, d). We start by drawing up a subgroup table, and as soon as a new element appears as a result of a product of elements already listed, we write it in its marginal position thus:

	1	c	d	k	l
1	1	c	d	...	
c	c	k	l	...	

and so on.

Often it is only necessary to complete a single row. Continuing we get:

Table 14.05

	1	c	d	k	l	g	h	b
1	1	c	d	k	l	g	h	b
c	c	k	l	g	h	1	b	d
d	d	m	......	......	......	......	......	......

After listing eight elements, no further elements appear to be generated in the c-row. Anyone lacking the knowledge of Lagrange's theorem might be forgiven for hoping that a subgroup of order 8 was about to manifest itself. But upon starting row d, we find $dc = m$, upon which the remaining four elements of the group appear. Of course, knowing Lagrange's theorem, our hopes of a subgroup evaporate as soon as we have found seven elements in row c.

Evidently the *whole group* Q_6 is generated by just two of its elements c and d. These generators are by no means unique.

Q. 14.17. Express all the other elements in terms of c and d.

Q. 14.18. Find all possible pairs of generators of Q_6.

We now turn our attention to S_4, and consider Gp (m, p), again with the notation of the table on p. 219. Forming the possible products, we get:

Table 14.06

	1	m	p	i	b	g
1	1	m	p	i	b	g
m	m	i	b	1	g	p
p	p	g	1	b	i	m.

The first two rows necessitate the inclusion of the three new elements i ($= m^2$), b ($= mp$) and g ($= m^2p$) and no more. The signs are hopeful that we are on the way to discovering a group of order 6, and in fact we did in Q. 14.10, show that $\{1, m, p, i, b, g\}$ is the subgroup D_3.

Again, let us find Gp (x, y):

Table 14.07

	1	x	y	h	a	s	r	g
1	1	x	y	h	a	s	r	g
x	x	h	a	s	r	1	g	y
y	....							

Having gone so far, the only possibilities are that these eight elements form a subgroup, or else that we may be heading for a subgroup of order 12, or possibly that x and y generate the whole group S_4. In point of fact, the eight elements do form a subgroup, and since there are but two of period 4 (x and s), the group is certainly D_4.

Q. 14.19. For S_4 find Gp (a, d); Gp (k, i); Gp (w, y); Gp (l, m); Gp (c, v); Gp (d, p); Gp (a, b, g); Gp (s, p); Gp (j, y); Gp (j, r); Gp (j, k).

Plainly the idea of the group generated by one or two elements may be extended for any number of generators; indeed, *any subset* of a group may be used to generate a subgroup. We shall, however, be mostly concerned with groups generated by up to three elements.

Q. 14.20. A single generator for the group $(\mathbf{Z}, +)$ is either $+1$ or -1. Find a set of two generators for this group, neither of which is ± 1. Describe Gp (10), Gp (-4).

Q. 14.21. In $(\mathbf{R}, \times)$ find Gp (3); Gp $(-\frac{1}{4})$; Gp (10); Gp (-1).

Q. 14.22. In any group, prove that

$$\text{Gp}\,(x_1, x_2, x_3, \ldots) = \text{Gp}\,(x_1^{-1}, x_2^{-1}, x_3^{-1}, \ldots).$$

Miscellaneous methods of discovering subgroups

The symmetries of a geometrical figure are of two kinds – direct and opposite. The former include the identity, and these direct symmetries will generally form a subgroup, since the product of two direct symmetries will also be a direct symmetry. We have observed this in the case of the subgroup C_6 of the full group D_6 of the regular hexagon (pp. 221, 189 ff.). For a three-dimensional figure, the direct symmetries (i.e. those which exclude the reflections or enantiomorphs) will also form a subgroup. The bitetrahedron (see fig. 13.18) has six direct symmetries – rotations 1, r, r^2 ($r^3 = 1$) about XY, and half-turns a, b, c about the medians of triangle ABC, that about the median through A permutes the vertices thus: $\begin{pmatrix} A & B & C & X & Y \\ A & C & B & Y & X \end{pmatrix}$. These six direct symmetries form a subgroup of the full group of twelve symmetries which include six enantiomorphs such as

$$\begin{pmatrix} A & B & C & X & Y \\ A & B & C & Y & X \end{pmatrix} \text{ and } \begin{pmatrix} A & B & C & X & Y \\ B & C & A & Y & X \end{pmatrix}.$$

The group of order 6 is D_3, and we note that it is represented here as a group of perms of the five vertices (cf. p. 214).

Consider another example of subgroups of permutations, this time subgroups of S_6. Those permutations which make transpositions (AB), or (CD), or (EF), or any combination of these three will evidently form a closed set of eight permutations and therefore a subgroup. For example, if

$$x = \begin{pmatrix} A & B & C & D & E & F \\ A & B & D & C & F & E \end{pmatrix} \text{ and } y = \begin{pmatrix} A & B & C & D & E & F \\ B & A & D & C & E & F \end{pmatrix},$$

then $$xy = \begin{pmatrix} A & B & C & D & E & F \\ B & A & C & D & F & E \end{pmatrix} = yx,$$

and this belongs to the set described. The property associated with this particular subgroup of S_6 is that the three transpositions are disjoint, i.e. that the three pairs (AB), (CD) and (EF) are being kept in watertight compartments.

Q. 14.23. Show that all eight of these permutations commute with each other, and construct their complete group table. (Compare Ch. 16, pp. 271, 279.) Find the subgroup which leaves A and B separately invariant. Invent further examples of subsets of the permutations of 4, 5, 6, . . . objects which are closed and therefore which form subgroups of S_4, S_5, S_6, . . .

We may also pick out subgroups of groups of matrices under multiplication. For example, the eight matrices $\begin{bmatrix} \pm 1 & 0 \\ 0 & \pm 1 \end{bmatrix}$ and $\begin{bmatrix} 0 & \pm 1 \\ \pm 1 & 0 \end{bmatrix}$ form the group D_4, since they perform the reflections and rotations which we noted in Ch. 13, p. 196 and illustrated in fig. 13.10. Now each matrix has a corresponding determinant, whose values may be either $+1$ or -1, thus: $\begin{bmatrix} 0 & -1 \\ -1 & 0 \end{bmatrix} \to \begin{vmatrix} 0 & -1 \\ -1 & 0 \end{vmatrix} = -1$. Consider the subset of those matrices whose determinant is $+1$. They will form a subgroup of D_4.

Q. 14.24. Obtain these matrices and state the structure of the subgroup they form. Why do the matrices whose determinant is -1 not form a group?

Q. 14.25. Consider the forty-eight matrices such as $\begin{bmatrix} 0 & -1 & 0 \\ 0 & 0 & 1 \\ -1 & 0 & 0 \end{bmatrix}$ each of which contains six zeros, the remaining elements being $+1$ or -1, no two in the same row or column. Show that the system form a group, and that those whose determinant has the value $+1$ form a subgroup of order 24. Show also that a subgroup of this subgroup may be found by taking all those matrices such as $\begin{bmatrix} 0 & 1 & 0 \\ 0 & 0 & 1 \\ 1 & 0 & 0 \end{bmatrix}$ in which no -1's occur. Identify this subgroup.

Of course, when we considered the group S_4 given by the table on p. 219, while we were aware that each small letter represented one of the twenty-four permutations of four symbols, we did not know which small letter represented which permutation. For example, a is of period 2, so it might possibly have represented a single transposition such as (AC), or a double transposition like $(AD)(BC)$. In the chapter on permutations, (Ch. 18), we shall consider the question of how a correspondence may be set up between the letters of the table S_4 and the permutations which they may represent. Such a correspondence is by no means unique, for the group will have many automorphisms (see pp. 147 ff. and Ch. 22).

Subgroups of infinite groups

Consider the set of integers which are multiples of 10, with the operation +. Clearly they form a group, and it is evidently a subgroup of the group of all integers under addition which we designate $(\mathbf{Z}, +)$. A rather surprising fact is that, while it is a subgroup of $(\mathbf{Z}, +)$ it is also isomorphic to it, though it would appear to contain only 10% of the membership of $\mathbf{Z}$. In point of fact, the last statement is nonsense, and we can not only set up a 1, 1 correspondence between the multiples of 10 and the integers, but we can also establish the isomorphism under addition. This is easy, for we have:

$$3 + (-7) \;=\; -4 \text{ in } (\mathbf{Z}, +)$$

and $30 + (-70) = -40$ in $(10\mathbf{Z}, +)$, the group of multiples of 10 under +.

Thus a correspondence which preserves sums (the group operation being addition) is achieved by associating with any integer $n \in \mathbf{Z}$ the number $10n$ in the group of multiples.

Q. 14.26. Establish the isomorphism in another way by associating with n a number different from $10n$ in the set $10\mathbf{Z}$.

Q. 14.27. Both these groups are C_∞. What are possible generators of the group of multiples of 10? Do the set $\{\ldots -17, -7, +3, +13, \ldots\}$ form a subgroup of $(\mathbf{Z}, +)$? Find other subgroups of $(\mathbf{Z}, +)$.

Q. 14.28. $(\mathbf{Z}, +)$ is a subgroup of $(\mathbf{Q}, +)$. Find other infinite groups of which $(\mathbf{Z}, +)$ is a subgroup.

The reals form a group under multiplication provided we exclude zero. Let us find some subgroups of $(\mathbf{R}, \times)$. First, the positive reals form a subgroup $(\mathbf{R}^+, \times)$. Test the truth of this statement. Next, the rationals, which are a subset of the reals, give us the group $(\mathbf{Q}, \times)$.

Q. 14.29. Do the following infinite subsets of **R** form a group under $\times$: (a) all the integers and their reciprocals; (b) all the irrational numbers together with 1; (c) the set $\{\ldots \frac{1}{16}, \frac{1}{8}, \frac{1}{4}, \frac{1}{2}, 1, 2, 4, 8, 16, \ldots\}$; (d) the set given in (c), but with the negative of each element included; (e) find other infinite subgroups of (**R**, $\times$).

Can we find *finite* subgroups of (**R**, +), of (**Q**, +), (**Z**, +), (**R**, $\times$), etc.? Of course, the identity *by itself* is always a subgroup, but this trivial case we have agreed not to bother about. Can we find a finite subgroup of, say, (**R**, +)? The answer is No, because suppose the group were to contain a non-zero element x, as well as 0, then, because of the closure property, the group would have to contain $x + x = 2x$, also $3x$, $4x$, ... and any multiple of x, and these are clearly unlimited in number. In fact x generates the group C_∞. A similar argument might appear to apply to the multiplicative groups (**R**, $\times$) and (**Q**, $\times$), but in fact these *do* both have a finite subgroup.

Q. 14.30. Can you find it? If not, it will become clearer after you have read the next paragraphs.

Subgroups of the additive group of complex numbers

So far, we have not considered the groups (**C**, +) and (**C**, $\times$) of complex numbers. These have a wealth of interesting subgroups. Not only do we have (**R**, +) (and its subgroups) as subgroups of (**C**, +), but we have the set known as the 'Gaussian Integers', i.e. the set $\{x + iy : x, y \in \mathbf{Z}\}$.

Q. 14.31. Prove that the Gaussian integers form a group under + but not under $\times$. Another subgroup of (**C**, +) might be all the complex numbers of the form $x - 2ix$ ($x \in \mathbf{R}$).

Q. 14.32. Prove that they form a group which is isomorphic to (**R**, +). Where do these numbers lie on the Argand diagram? Do all straight lines on the Argand diagram represent subgroups of (**C**, +)? Find other subgroups.

Subgroups of the multiplicative group of complex numbers

The group (**C**, $\times$) has infinite subgroups other than (**R**, $\times$) and (**Q**, $\times$). For example, all complex numbers with unit modulus, $\cos\theta + i\sin\theta$, are closed under multiplication, include unity ($\theta = 0$), and each possesses an inverse, $\cos\theta - i\sin\theta$. Here, then we have a subgroup of (**C**, $\times$), and these are all the points of the circle $|z| = 1$ in the Argand diagram.

Q. 14.33. Can you find any other infinite subgroups of (**C**, $\times$)? Show that the group of complex numbers with unit modulus, under $\times$, is isomorphic to the group of rotations of a circle about its centre, and also to the group (**S**, + mod. 1). Mention some subgroups of this group, and show that they each contain at least one element of any given finite period.

Just as in (**Z**, +), it is also impossible to have finite subgroups of (**C**, +), since, if z is a non-zero element, then $z + z$, $z + z + z$, and so on must all belong to this set, and there is no value of n such that $nz = 0$. Indeed, finding finite subgroups of infinite groups depends entirely on finding *elements of finite period.* Once one has found an element x of finite period, then one has succeeded in finding a finite subgroup, namely Gp (x). But the argument disproving the existence of finite (proper) subgroups breaks down in the case of (**C**, $\times$); for if z is a complex number belonging to a finite group $\{1, z, \ldots\}$, $\times$, then the set must contain all the powers of z. But this time, z *may* have finite period since there may exist a value of n such that $z^n = 1$, and so we should have the cyclic group C_n in this case. These groups, consisting of the n nth roots of 1 we have already discussed (Ch. 12, pp. 167ff.).

Subgroups of groups of matrices under multiplication

Let $\mathcal{M}_{m,n}$ be the set of $m \times n$ matrices with real elements, and + denotes matrix addition. Then, for any given m, n $(\mathcal{M}_{m,n}, +)$ is an infinite group with no particularly interesting properties. For matrix multiplication, we confine our attention to *non-singular square* matrices (Why?), and if $\mathcal{M}_n$ denotes the set of all such $n \times n$ matrices for some particular n, these form a group under matrix multiplication, which we shall call $(\mathcal{M}_n, \times)$.

Q. 14.34. Show that the subset of $\mathcal{M}_3$ consisting of matrices of the form $\begin{bmatrix} k & 0 & 0 \\ 0 & k & 0 \\ 0 & 0 & k \end{bmatrix}$ forms a subgroup of $(\mathcal{M}_3, \times)$, which is isomorphic with (**R**, $\times$).

As an example of one infinite subgroup of the set $\mathcal{M}_2$ (say), we shall consider all those 2×2 matrices whose determinant is $+1$, e.g. $\begin{bmatrix} -3 & 5 \\ 1 & -2 \end{bmatrix}$; $\begin{bmatrix} 1 & 0 \\ 6 & 1 \end{bmatrix}$. It is well known that, if A and B are square matrices of the same order, and $C = AB$, then $|C| = |A||B|$, where $|A|$ denotes the value of the determinant of the matrix A. It is clear from this, that if $|A| = 1$ and $|B| = 1$, then $|AB| = 1$, so that the set is closed, and it includes the unit matrix $I = \begin{bmatrix} 1 & 0 \\ 0 & 1 \end{bmatrix}$. Moreover, we are concerned only with non-singular matrices so that A^{-1} exists, where $AA^{-1} = I$. But $|A| = 1$, and $|I| = 1$, so $|A^{-1}| = 1$. All the requirements of a group are satisfied. Any square 2×2 matrix corresponds to some linear transformation in the plane. Those matrices which have unit determinant correspond to transformations which *preserve area*. For example, $\begin{bmatrix} 1 & k \\ 0 & 1 \end{bmatrix}$

represents a shear parallel to the x-axis (see fig. 10.02); $\begin{bmatrix} \cos\theta & \sin\theta \\ -\sin\theta & \cos\theta \end{bmatrix}$ represents a rotation about the origin.

Q. 14.35. Show that the matrices of the forms indicated in the last sentence form subgroups of the group of matrices with unit determinant. Find other subgroups.

Q. 14.36. Show that the matrices of the form $\begin{bmatrix} a & b \\ -b & a \end{bmatrix}$ form a subgroup of $(\mathcal{M}_2, \times)$, and that this is isomorphic with (C, $\times$).

Q. 14.37. Consider those 2×2 matrices whose determinant is either $+1$ or -1. What is the geometrical interpretation in the case when the value is -1? Extend your investigation to 3×3 matrices whose determinant is ± 1, and show that these form a subgroup of $(\mathcal{M}_3, \times)$.

Finite subgroups of $(\mathcal{M}_n, \times)$ abound. For example, in seeking finite subgroups of $(\mathcal{M}_2, \times)$, we are concerned to find matrices with *finite period*. Usually these will have a simple geometrical analogue: $a = \begin{bmatrix} 1 & 0 \\ 0 & -1 \end{bmatrix}$ represents a reflection in the x-axis, and so is of period 2; $r = \begin{bmatrix} \cos 40^\circ & \sin 40^\circ \\ -\sin 40^\circ & \cos 40^\circ \end{bmatrix}$ represents a rotation through 40°, and so has period 9, and so on. One may obtain a group of order 18 by combining all possible products of powers of a and r.

Q. 14.38. Identify this group.

Again, a possible finite subset of $\mathcal{M}_4$ which form a group is the set of permutation matrices, such as $\begin{bmatrix} 0 & 1 & 0 & 0 \\ 0 & 0 & 1 & 0 \\ 1 & 0 & 0 & 0 \\ 0 & 0 & 0 & 1 \end{bmatrix}$. We have already considered a subset of forty-eight matrices belonging to $\mathcal{M}_3$ which form a group under $\times$. The possibilities are endless.

Q. 14.39. If $A, B \in \mathcal{M}_2$, then in general $AB \neq BA$. But $AI = IA$, where I is the unit matrix, and if A is given, other matrices may be found which do commute with A. If $A = \begin{bmatrix} a & b \\ c & d \end{bmatrix}$, find out all you can about matrices which commute with A, first in the special case when $A = \begin{bmatrix} 1 & 1 \\ -2 & 3 \end{bmatrix}$, and then in the general case. Do these classes of matrices form a group? Generalise your result in some direction.

Geometrical transformations

Finally in this chapter, we consider sets of geometrical transformations which form infinite groups under successive application, and we think of their subgroups. It is obvious, for example, that all the rigid motions of a solid in three-dimensional space form a group – the group of direct isometries in 3-space. For if a body suffers a movement x and this is followed by a movement y, then evidently $y * x$ is itself a rigid movement of the body. Moreover, such movements are associative, they include an identity transformation, and every movement has its inverse by which the original position of the body is restored. A possible subgroup of this infinite group of movements might be those which leave a particular point in the body invariant – i.e. the body is tethered, as it were, to one point. Another subgroup would be associated with the invariance of a particular *line* in the body, and such movements of the body would consist of rotations about that line as an axis, or hinge.

Two-dimensional transformations

We shall, however, be chiefly concerned with motions in two dimensions, and it is only possible here to give a brief account of this large subject. (See Coxeter, *Introduction to Geometry*, Wiley, Chs. 3, 4, 5, 6; M. Jeger, *Transformation Geometry*, Allen and Unwin; I. M. Yaglom, *Geometric Transformations*, Random House.)

The various types of plane transformation may be summarised pictorially, the pictures showing the result of applying the transformations to a square (see fig. 14.01). Each of these types of transformation taken separately is an infinite group (e.g. the Affine Group, the Similarity Group, and so on). You will readily be able to distinguish some subgroups; for example, the group of Isometries is a subgroup of the group of Similarities; the group of Direct Similarities is a subgroup of the group of Similarities. The latter is a subgroup of the Affine Group, which itself is a subgroup of the Linear Group, and so on. Moreover, within the Isometric Group (broadly speaking, the group whose characteristic property is to preserve distance) we have subgroups such as the group of translations; the group of Rotations about a given point. In some cases, we may seem to be in trouble. For example, all the rotations in the plane do not form a group since the Closure requirement is lacking – two rotations through equal and opposite angles produce a *translation* (see fig. 14.03, p. 233). However, if R is the set of rotations and T is the set of translations, then $R \cup T$ *is* a group; in fact it is the group of *direct* isometries.

Q. 14.40. Show that the half-turns about two points do not form a group. Construct a group to contain these two half-turns.

Q. 14.41. Find a subgroup of the group of translations in the plane. Find a larger group, other than the plane isometries, of which the group of plane translations is a subgroup.

Q. 14.42. Is there a place in the above list for the transformation of inversion with respect to a given circle?

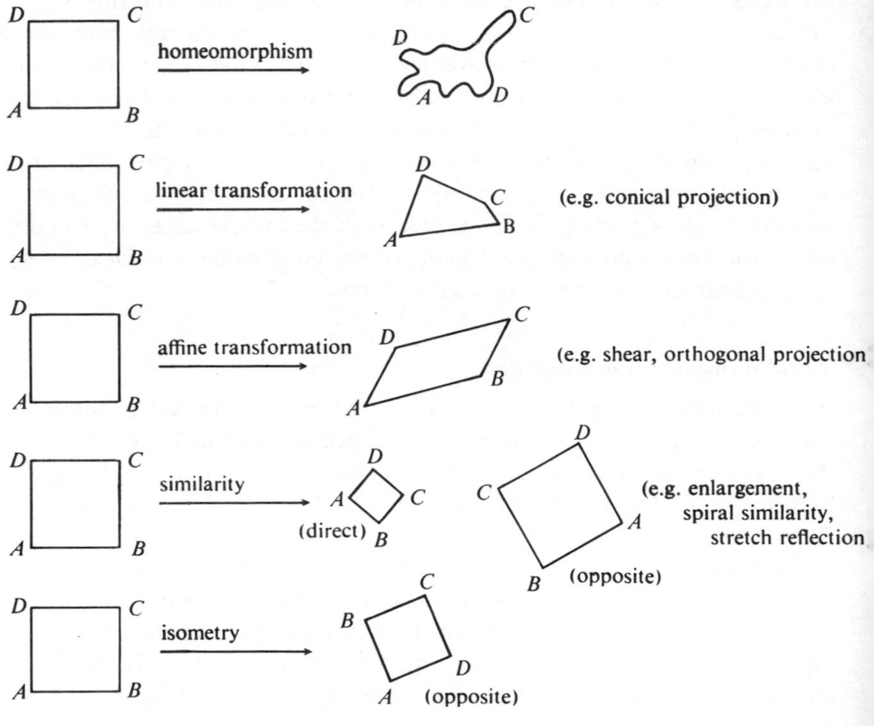

Fig. 14.01.

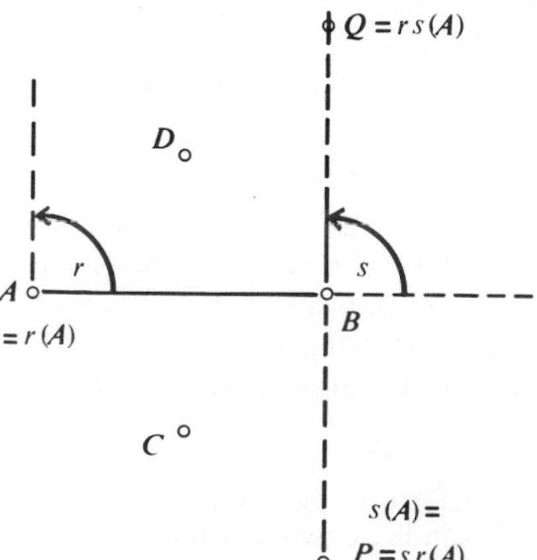

Fig. 14.02. Quarter-turns about A and B (r and s) do not commute. If $ACBD$ is a square, then $sr(C) = C$ and $rs(D) = D$. Find $rs(C)$ and $sr(D)$.

Some of these groups of transformations are Abelian, some are not. For example, the rotations about a fixed point do commute, but the rotations about *two different* points do not (see fig. 14.02 in which r and s are anticlockwise quarter-turns about A and B respectively).

Q. 14.43. In fig. 14.02, check that $P = sr(A)$, and $Q = rs(A)$. Also, if $ABCD$ is a square, then $sr(C) = C$; $rs(D) = D$; find $rs(C)$ and $sr(D)$.

Q. 14.44. Show that in general a rotation about a point and a translation do not commute.

Q. 14.45. Show that any set of transformations which includes a translation must be infinite. Such a set of transformations will contain C_∞ as a subgroup and will generate a *pattern* (see Ch. 26).

In the last question we observed that the presence of a translation in a set guarantees that it cannot be a finite group. Thus when we start looking for finite subgroups of transformations, we must be careful to exclude translations, or transformations which produce translations, such as reflections in two parallel mirrors (see p. 206), or rotations about two points through equal and opposite angles (see fig. 14.03).

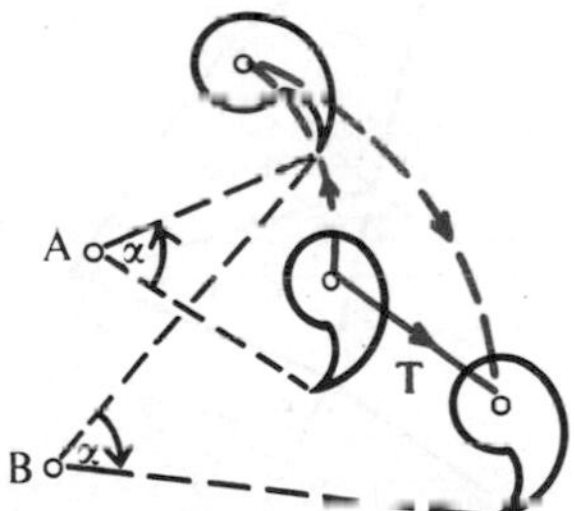

Fig. 14.03. $R_{B,-\alpha} * R_{A,\alpha} = T$.

Finite groups of plane transformations

Nevertheless, finite groups may be created by placing certain classes of transformation into equivalence classes. Let us explain what this means.† Suppose we had a triangular plate lying on a horizontal table. Then we may turn the triangle over about either of its three sides: Fig. 14.04 shows

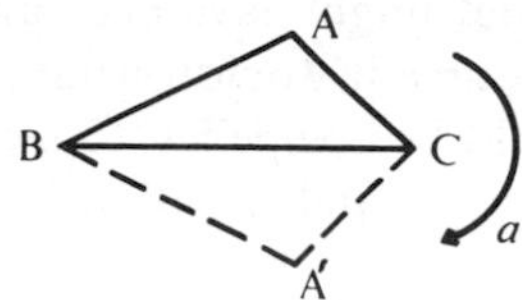

Fig. 14.04. Triangle reflected in one side.

† See also Ch. 19, pp. 351 ff.

the result of a half-turn about *BC* (or a reflection in *BC*, see p. 187), and this movement we shall call *a*. Similarly, let *b* and *c* be reflections in *CA* and *AB*. In this example, we shall understand *a*, *b* and *c* to refer to reflections in *BC*, *CA* and *AB in their moved positions*,† not about the original position of these axes. Evidently we can apply any number of successive transformations of the triangle corresponding to a series of instructions (or 'word') such as *abaccaabbbaca*. It is immediately apparent that, since each reflection is of period 2, this particular 'word' may be reduced to *abababca*. Evidently the three possible movements *a*, *b* and *c* generate an infinite set of movements of the triangle. Now in fig. 14.05, we show the effect on the triangle of the successive movements: *c* then *b*, then *a*, then *c*, then *b*, then *a*, i.e. of *abcabc*, or $(abc)^2$.

Q. 14.46. What is the effect of ...*bcbcbcbc*; of *abcabcbcabcacbacbcbacb*?

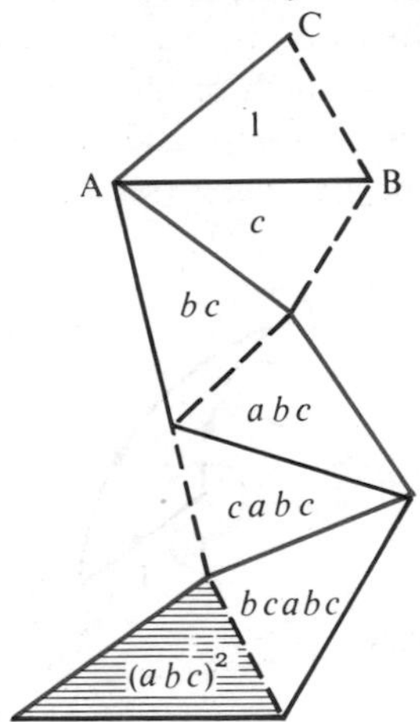

Fig. 14.05. Triangle reflected successively in its sides: $(abc)^2$ is a translation.

Now you may verify that $(abc)^2$ brings the triangle into the *same orientation* as in its initial position, and is thus equivalent to a parallel *translation*. But if we were not concerned with *where* the triangle is, but only with its orientation, we might agree to regard these positions as *equivalent*, in which case, the result of the operation $(abc)^2$ would be the *identity*. By this means, what might have been an infinite group may now be a finite group, but of course it would contain far more than six elements, because we should be compelled to include every possible 'word' that could be formed from *a*, *b* and *c*, with $(abc)^2$ replaced by '1'.

Q. 14.47. Which other 'words' are certain to be equivalent to the identity?

† See pp. 365–71; in Ch. 20 we prefer to use the notation a^*, b^*, c^* to denote reflections in the moved axes.

The order of the group, if it is finite at all, will depend on the shape of the triangle. For example, the motion *ba* rotates the triangle through an angle $2 \times \widehat{ACB}$, and clearly this rotation by itself could generate an infinite group, unless angle ACB were a rational multiple of π. For example, if ACB were $2\pi/5$, then *ba* would be of period 5. Consider the case when $\triangle ABC$ has $\hat{C} = \text{rt.} \angle$, $\hat{A} = \hat{B}$, so that we may perform the experiment with a set-square. We note from fig. 14.06 that *abc* appears to restore the orientation of the triangle, till we realise that the equal sides

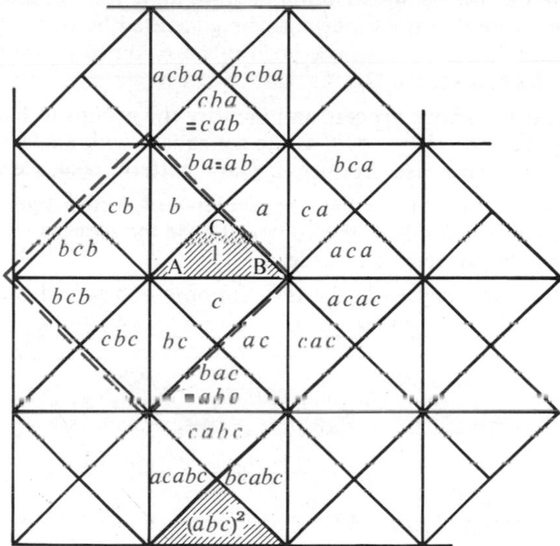

Fig. 14.06. Successive reflections of triangle with angles 90°, 45°, 45°.

have been interchanged (see fig. 14.07). To distinguish the equal sides, we have marked the '*a*' sides in each position of the triangle. (One of the equal sides of the set-square should be marked in some way to make the equal sides distinguishable.)

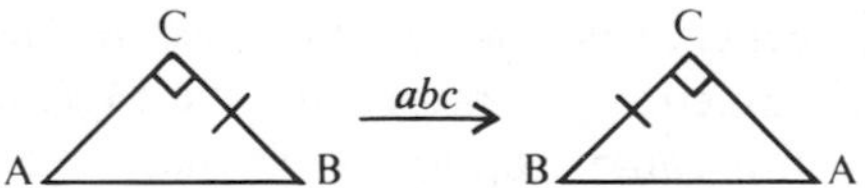

Fig. 14.07.

The diagram (fig. 14.06) shows that there are *eight* essentially different orientations of the triangle, and these may be represented by 1, *a*, *b*, *ab* (= *ba*), *c*, *bc*, *ac*, *abc*. From the diagram, we may *read off* such relations as $aca = bcb$; $caca = acac$, or $(ac)^4 = 1$; $ca = bc$, etc.

Q. 14.48. Express *a* in terms of *b* and *c* only.

To identify the group, it would be enough to find the period of each element: a, b and c have period 2, and so has ab ($= ba$), being a half-turn about C; the fact that abc also has period 2 has already been observed. Finally, ac has period 4, and so also has bc ($= ca$), each of these being quarter-turns. The group therefore has two elements of period 4, five of period 2 and the identity, and so it has D_4 structure – the dihedral group of the square.

Q. 14.49. Write out the complete multiplication table for this group, using the notation above. Note that the group may be generated by b and c only, and that the eight triangles of fig. 14.06 whose positions are named in terms of b and c only do in fact form a square.

Q. 14.50. Repeat the above process and identify the groups in the cases when: (a) $\triangle ABC$ equilateral; (b) $\triangle ABC$ is the set-square with angles $A = 90°$, $B = 60°$, $C = 30°$. (This case produces a plane pattern, **p6m**, see Ch. 25.)

Q. 14.51. In the example in the text, the elements of period 4 generate subgroups C_4. Interpret these subgroups in the context of the movements of the set-square. Do the same for the subgroups of order 2.

Q. 14.52. In the case analysed in the text, suppose one agreed to regard

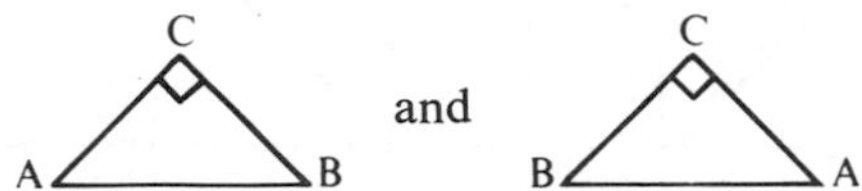

as equivalent, so that $abc = 1$.
Does this give a group of order 4?

The above process, of placing the positions of the triangle into equivalence classes by virtue of orientation, bears a marked resemblance to the process of placing the integers into equivalence classes to some modulus. Indeed, there is a closer correspondence than may at first sight appear to exist, for we see from fig. 14.06 that, in ignoring vertical or horizontal displacements of two whole units, we are, in effect, working with a modulo 2 arithmetic. It is rather like the Mercator map of the world in which Australia may appear twice: these two positions are equivalent, and here we are reckoning the scale of longitude modulo 360° (see also pp. 256 ff).

Q. 14.53. Show that *any* triangle subjected to the twenty-two successive reflections *bcabcbcabcacbacbcbacba* will be restored to its original positions. Show that the first six of these are a translation, and find other translations within the sequence. Show that we may start at any point in the sequence and return after completing the cycle of twenty-two operations. Find other sequences which produce the identity. (*Hint*: first try the above sequence with an equilateral triangle.)

Equivalence classes of matrices

The device of placing the objects of a set into equivalence classes and so contriving to get a finite instead of an infinite group, may be applied to matrices. For example, we might agree to regard the two matrices $\begin{bmatrix} a_1 & b_1 \\ c_1 & d_1 \end{bmatrix}$ and $\begin{bmatrix} a_2 & b_2 \\ c_2 & d_2 \end{bmatrix}$ as 'equivalent' if $a_2/a_1 = b_2/b_1 = c_2/c_1 = d_2/d_1 = k$, so that the identity matrix in this system would be of the form $\begin{bmatrix} k & 0 \\ 0 & k \end{bmatrix}$.

In such an event, the matrix (e.g.) $x = \begin{bmatrix} 2 & -1 \\ 1 & 1 \end{bmatrix}$ would no longer have infinite period. For

$$x^2 = \begin{bmatrix} 2 & -1 \\ 1 & 1 \end{bmatrix}\begin{bmatrix} 2 & -1 \\ 1 & 1 \end{bmatrix} = \begin{bmatrix} 3 & -3 \\ 3 & 0 \end{bmatrix}, \text{ i.e. equivalent to } \begin{bmatrix} 1 & -1 \\ 1 & 0 \end{bmatrix}$$

$$x^3 = \begin{bmatrix} 2 & -1 \\ 1 & 1 \end{bmatrix}\begin{bmatrix} 1 & -1 \\ 1 & 0 \end{bmatrix} = \begin{bmatrix} 1 & -2 \\ 2 & -1 \end{bmatrix}$$

$$x^6 = \begin{bmatrix} 1 & -2 \\ 2 & -1 \end{bmatrix}\begin{bmatrix} 1 & -2 \\ 2 & -1 \end{bmatrix} = \begin{bmatrix} -3 & 0 \\ 0 & -3 \end{bmatrix}, \text{ i.e. equivalent to } \begin{bmatrix} 1 & 0 \\ 0 & 1 \end{bmatrix}.$$

Compare with the work on bilinear transformations, Ex. 10.28.

Q. 14.54. Consider the matrix $\begin{bmatrix} 0 & 1 \\ -1 & 1 \end{bmatrix}$ whose period is 6 (see p. 120). What would be its period if the above equivalence relation were imposed on the set $\mathscr{M}_2$?

Q. 14.55. Is it possible to place 2 × 2 matrices into equivalence classes and to obtain intelligible results by the following devices:
(a) $\begin{bmatrix} a_1 & b_1 \\ c_1 & d_1 \end{bmatrix} \equiv \begin{bmatrix} a_2 & b_2 \\ c_2 & d_2 \end{bmatrix}$ if $a_1d_1 - b_1c_1 = a_2d_2 - b_2c_2$; (b) $a_1 + d_1 = a_2 + d_2$?
Try to invent other methods.

Finite subgroups of matrices which have arisen from time to time in the previous chapters may now be thought of as subgroups of $(\mathscr{M}_n, \times)$, and we have had examples of these in the cases $n = 2$, 3 and 4. An obvious case is the permutation matrices, and it is evident from this that S_n is a subgroup of $(\mathscr{M}_n, \times)$.

Of course, the link between the final two sections of this chapter – on transformations and matrices – is that it may be possible to represent a transformation by means of a matrix, and successive transformations by the multiplication of the appropriate matrices. For example, in two dimensional space, the matrix $\begin{bmatrix} 1 & k \\ 0 & 1 \end{bmatrix}$ represents a shear parallel to the

x-axis, while $\begin{bmatrix} -1 & 0 \\ 0 & 1 \end{bmatrix}$ represents a reflection in the y-axis. The matrix $\begin{bmatrix} 1 & k \\ 0 & 1 \end{bmatrix}\begin{bmatrix} -1 & 0 \\ 0 & 1 \end{bmatrix} = \begin{bmatrix} -1 & k \\ 0 & 1 \end{bmatrix}$ is the transformation of first reflecting in the y-axis and then shearing.

Q. 14.56. Do these transformations commute? If the two matrices are respectively called A and B, show that $(AB)^{-1} = B^{-1}A^{-1}$, and interpret this geometrically.

EXERCISES 14

1 Give examples of Abelian subgroups of non-Abelian groups.

2 Pick out from S_4 (table 14.02) those element which commute with a; which group do they form? Do the same for the elements h, i, n and r. How many elements give the identity only, how many give C_2, and how many C_3?

3 Show that D_n is a subgroup of D_{2n}. (Consider the symmetries of a regular $2n$-sided polygon.)

4 What is the subgroup of the group of permutations of A, B, C, D and E which leaves A and E invariant? Do the permutations of $A\ B\ C\ D\ E$ which make two transpositions, e.g. $(AD)(BC)$ form a subgroup of S_5?

5 **Prove that S_m is a subgroup of S_n for $m < n$.**

6 **Prove that if A and B are subgroups of G and if their orders are relatively prime, then $A \cap B = 1$.** (Any previous results may be used.)

7 Denote by $(V_3, +)$ the group of vectors, or displacements, in three dimensional space under vector addition. Discover various infinite subgroups. Can there be any finite subgroups?

8 Two tables for Q_6 have been given in the text (see pp. 108, 216). Exhibit the isomorphism between them by showing one possible pairing of their elements.

9 On p. 237 we remarked that S_n is a subgroup of $(\mathcal{M}_n, \times)$. Illustrate this by further examples.

10 Write down six 3×3 matrices which form a group under matrix multiplication, and show its subgroups.

11 **In any group, show that two commuting elements generate an Abelian subgroup.**

12 **Show that all subgroups of cyclic groups are cyclic, and that all subgroups of dihedral groups are either cyclic or dihedral.**

13 Find the subgroups of the group of polynomials under addition generated by:
(a) x; (b) $1, x, x^2, x^3$; (c) x, x^3, x^5, x^7; (d) $1, x^2, x^4, x^6$;
(e) $3x + 1, x^2$.

14 Show that the rationals of the form 2^x3^y $(x, y \in \mathbf{Z})$ under multiplication are a subgroup of $(\mathbf{Q}, \times)$, and that this subgroup is isomorphic to the group of Gaussian integers (i.e. complex numbers $x + iy$ where $x, y \in \mathbf{Z}$) under addition, which is itself a subgroup of $(\mathbf{C}, +)$.

15 What is the centre of S_4; of S_3; of S_2?

16 Prove that the set of all terminating decimals is a subgroup of (Q, +).

17 If H is a subgroup of K and K is a subgroup of G, prove that H is a subgroup of G.

18 Prove that a group of order p^m where p is a prime and $m = 2, 3, 4, \ldots$ must contain at least one subgroup of order p.

19 If H is a subgroup of G, prove that the centre of G is a subgroup of the centre of H. Illustrate from some of the groups whose tables are displayed in the text.

20 We shall see (p. 319) that S_4 is Gp (a, b, g) and also Gp (a, j) with the notation as on p. 219. Express b and g in terms of the generators a and j.

21 Consider $w = (az + b)/(cz + d)$ where $a, b, c, d \in \mathbf{C}$. This is the general bilinear transformation. Show that transformations of this form are a group under successive composition. Interpret in terms of the transformations of the Argand diagram. What are the various invariant properties associated with this when regarded as a subgroup of the group of all Argand diagram transformations? Discuss some subgroups of the group of bilinear transformations, such as (a) those given by $c = 0$, $d = 1$; (b) $a, b, c, d \in \mathbf{R}$, and find some finite subgroups.

22 In Ex. 13.07, we obtained six functions forming the group D_3 by generating them from two functions of period 2, namely $1 - z$ and $1/z$. Now S_4 may be generated by a, b and g, each of period 2, and we see from the table on p. 219 that $(ag)^2 = e$. Let $\mathrm{a}(z) \equiv 1 - z$, $\mathrm{b}(z) \equiv 1/z$, and $\mathrm{g}(z) \equiv (z + P)/Qz - 1)$, so that g is of period 2 (see the condition on p. 117). By obtaining the condition for ag to be of period 2, find P and Q (they are complex), and so obtain twenty-four functions to represent the group S_4. (*Note*: it can be proved that the only finite groups of bilinear transformations are either cyclic, or dihedral, or A_4, S_4 and A_5.)

23 Show that S_n has at least ${}_nC_r$ subgroups of order r $(< n)$. How many subgroups has S_5 of order 5?

24 If a proper subgroup is isomorphic to the whole group, what can you say about the group?

25 Is the group of quadratic polynomials under + a subgroup of the group of cubic polynomials under +? (Real coefficients.)
Is the group of cubic polynomials with rational coefficients a subgroup of the additive group of cubic polynomials with real coefficients?
Is the group of 2×2 matrices under addition a subgroup of the group of 3×3 matrices under addition?
Is the group of 3×3 matrices with integral elements under addition a subgroup of the group of 3×3 matrices with real elements?

26 Find some subgroups of:
(a) the group of polynomials with rational coefficients, under addition;
(b) the group of rational functions (see Ex. 8.31) with real coefficients under multiplication;
(c) the group of ordered pairs of reals under $\oplus$, where
$(a, b) \oplus (c, d) = (ac, bc + d)$ (see Exs. 4.14, 7.28 and 8.30).
Find also a group in each case above, of which the example given is a subgroup.

27 Prove that every infinite group has proper subgroups. (Consider Gp (x) in the cases when x has finite period, and when x generates C_∞.)

28 In the text, we considered subgroups of C_∞ by means of the realisation $(\mathbf{Z}, +)$. Do the same by considering the realisation of C_∞ as the group generated by a single translation (see p. 125).

29 Illustrate the subgroups, both finite and infinite, in the case of the group $D_\infty \times D_1$ generated by reflections in two parallel mirrors and a third mirror perpendicular to them (see Ch. 16, pp. 277–8 and fig. 26.167).

30 **Show that the following are subgroups of $(\mathscr{M}_2, \times)$:**

(a) $\begin{bmatrix} a & b \\ 0 & 1 \end{bmatrix}$ $(a, b \in \mathbf{R}, a \neq 0)$; (b) $\begin{bmatrix} 1 & b \\ 0 & 1 \end{bmatrix}$; (c) $\begin{bmatrix} a & 0 \\ 0 & 1 \end{bmatrix}$.

Find a subgroup of the first (a), which contains the second (b) as a subgroup.

31 Show that matrices $\begin{bmatrix} a & b \\ c & d \end{bmatrix}$ with $ad - bc = \pm 1$ $(a, b, c, d \in \mathbf{Z})$ are a subgroup of $(\mathscr{M}_2, \times)$. Do those matrices with $ad - bc = +1$ form a subgroup of those whose determinant is ± 1?

32 Show that the centre of $(\mathscr{M}_2, \times)$ contains only matrices of the form $\begin{pmatrix} k & 0 \\ 0 & k \end{pmatrix}$ $(k \in \mathbf{R})$.

33 What is the centre of the group of direct isometries of the plane?

34 What invariant property is associated with the subgroup of the plane isometries generated by the translations and a reflection in a single line?

35 Show that the group of translations is isomorphic with the group of vectors under vector addition. Find an infinite subgroup.

36 Show that, if a group of order p^2 (p prime) is not cyclic, then it is Abelian, contains $p^2 - 1$ elements of order p, and has $p + 1$ subgroups of order p.

15 *Group specifications: generators and defining relations*

We have seen that, in the cyclic groups, every element may be expressed in terms of a single element, which may be said to generate the whole group. Thus C_6 may be written $\{1, x, x^2, x^3, x^4, x^5\}$ with $x^6 = 1$. There is, of course, no reason why C_6 should not be expressed $\{1, x, y, xy, y^2, xy^2\}$ with $y^3 = 1$ and $y = x^2$, but it is more usual to express all the elements of a group in terms of as few generators as possible, or what is known as a *minimum set of generators*. The only groups which can be expressed in terms of a single generator are, indeed, the cyclic groups, and if the group is finite, and of order n, we have a generator satisfying the relation $x^n = 1$, which has the effect of specifying, or defining the group beyond all doubt.

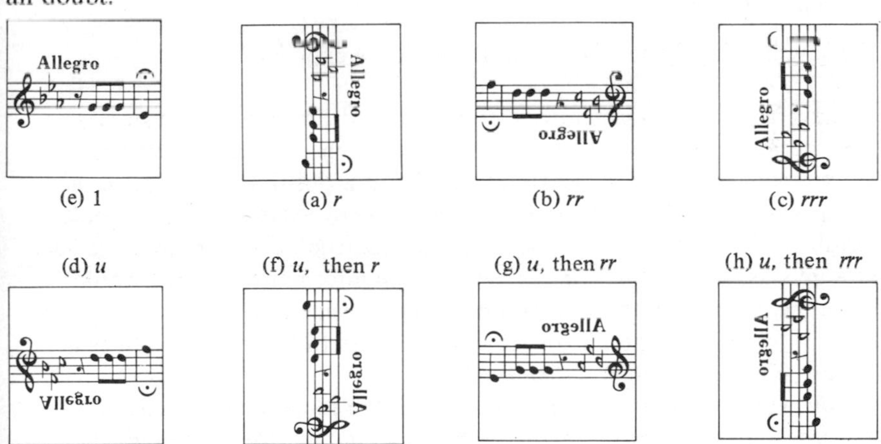

(e) 1 (a) r (b) rr (c) rrr

(d) u (f) u, then r (g) u, then rr (h) u, then rrr

Fig. 15.01. Eight ways of putting square film into projector.

Groups requiring two generators

So much for cyclic groups. Groups other than cyclic groups require a minimum of two or more generators. Take for example the group D_4. Suppose I am showing some square slides from an overhead projector which is being worked by an assistant. Each slide may be put into the projector in eight ways, only one of which is right (see fig. 15.01). Suppose the projectionist does not know whether the picture is right or wrong. I could say to him: 'No, it's back to front and upside-down, Turn it round ... No the other way ...', etc., and after much experimentation, he might get the picture to come out right. A better way would be to say to him: 'If the picture is wrong, carry out one of the following seven procedures labelled a, b, c, d, f, g, h, details of which I have set down on this card.' We go ahead, and the first slide comes out wrong, so I say 'd' to him. He then consults the card, carries out the instruction, and all is well. A still better way would be this: Say to the operator: 'If the picture is wrong, there are two things you may do to it to correct it. One is to turn it over about a horizontal axis, which I shall call "u"; the other is to give it a quarter-turn to the right, which I call "r".' Then if I wish to instruct the operator to change the picture from (e) to (h) in fig. 15.01, I would simply say 'u, then r then r then r'. Of course, in practice it would be the inverse, $(r^3u)^{-1} = ur$ which would be needed to get from (h) back to (e), but the point of this illustration is that *all the eight changes needed* may be expressed in terms of u and r alone. These two operations *generate the whole group.*

Defining relations

We saw in Ch. 13 that the group D_6 could be expressed as: $\{1, r, r^2, r^3, r^4, r^5, a, ar, ar^2, ar^3, ar^4, ar^5\}$ where $r^6 = 1$ and $a^2 = 1$. Here we have a set of two generators, and the two defining relations specify their periods. But this information is not enough to tell us all we need to know about this group – we should not be in a position to complete the group structure table. Armed only with the knowledge $r^6 = 1$, we could certainly fill in the whole of the C_6 subgroup generated by r, and also such products as $ar^2.r^3 = ar^5$; $ar^5.r^3 = ar^2$. The information $a^2 = 1$ would enable us to make further entries in the table, such as $a.a = 1$; $a.ar^3 = r^3$, but we should still be quite helpless to find such products as $ar^2.ar$. We need another *defining relation.*

You will recall that, in the group D_n, all the remaining elements not in the subgroup C_n† were of period 2, because, in fact, they represented reflections in the group of symmetries of the regular n-gon.

Suppose we were told $r^6 = a^2 = 1$, and in addition that *ar is of period 2.* Then $(ar)^2 = 1 \Rightarrow arar = 1 \Rightarrow aarar = a \Rightarrow a^2(rar) = a \Rightarrow rar = a$, since $a^2 = 1$. We may now proceed exactly as on p. 193, deducing that $ar^5 = ra$, $ar^4 = r^2a$ and $ar^3 = r^3a$. This enables us to go on and complete the table: for example, $(ar^2)(ar) = r^4a.ar = r^4a^2r = r^41r = r^5$; $(ar^4)^2 = ar^4.ar^4 = r^2a.ar^4 = 1$, and similarly each of the elements a, ar, ar^2, ar^3, ar^4, ar^5 is of period 2.

† The coset of C_n.

Q. 15.01. Do the work on similar lines for the group D_5 defined by $r^5 = 1 = a^2$, $ar^4 = ra$.

Q. 15.02. Find another set of defining relations for D_6 using two elements of period 2 (compare the example on reflections in two mirrors, pp. 196, 204 ff.).

Q. 15.03. Consider the general dihedral group D_n, showing that it may be defined by $r^n = 1$, $a^2 = 1$, $a = rar$, and find alternative defining relations with a and r, and also with a different pair of generators.

How many generators?

In previous chapters, we saw how subgroups may be found by picking two (or more) elements from the group, and finding the group which they generate. Of course, the fact that a group requires very few generators does not mean that it is necessarily a very small group – a single generator might generate a cyclic group of order 100, or 1000, ... Indeed, a single generator may generate the infinite cyclic group, C_∞.† Nor is it true that a group of small order will necessarily require very few generators, for there is a group of order 8 which requires three generators, and we shall meet this in the next chapter. The group S_4 requires two generators at least.

Q. 15.04. Can you find a set of two generators for S_4 (table 14.02). If you cannot find a set of two, then find a set of three generators. How many generators are needed for the group S_5; S_n?

Given a set of generators of a group, it is often quite a difficult task to discover whether they may be reduced in number, especially with large groups. Often one may discover that a given set are not a minimum set of generators by accident, perhaps during the process of expressing each element of a group in terms of the set which one supposed to be a minimum set. Sometimes one is fortunate, and it becomes easily apparent that a set of generators may be reduced in number. For example, in the group obtained by turning over the triangle with angles 90°, 45°, 45° (see p. 235), we gave each element in terms of three basic movements a, b, c, but we saw that, since a could be expressed in terms of b and c, it was actually necessary to use only *two* generators.

How many defining relations? redundant relations

A much more difficult task than finding a minimum set of generators is that of finding a minimum set of defining relations, that is, a set which do not contain any redundant information. For D_6, it would be superfluous

† C_∞ is a 'free group' which does not require any defining relations.
‡ See also Preface, pp. xii–xiii.

to state $(ar)^2 = 1$ in addition to $a^2 = r^6 = 1$, $ar^5 = ra$, because it is possible to *deduce* $(ar)^2 = 1$ from the other three.

Q. 15.05. Is it possible to deduce each one of the four relations $a^2 = 1$, $r^6 = 1$, $ar^5 = ra$, $(ar)^2 = 1$ from the remaining three?

As a further example, consider the group Q_6 discussed in Ch. 14. During the course of our work in discovering its subgroups, we found that there were several possible sets of generators, such as c and d (see table 14.01). As a start, in formulating a set of defining relations, it is usually essential to state the *period of each generator*, and in this case the relations are $c^4 = 1$, $d^4 = 1$. Sometimes one, sometimes more, further defining relations must then be drawn up to supplement this information, and this may be difficult. In fact, a possible set of defining relations could be $c^4 = 1$, $d^4 = 1$, $c^2 = d^2$ and $(cd)^3 = 1$, though $d^4 = 1$ is redundant here, since $c^4 = 1$ and $c^2 = d^2$ imply $d^4 = 1$. Note, however, that $c^4 = 1$, $d^4 = 1$ do *not* imply $c^2 = d^2$. Another pair of generators might be a and c, and we could then have defining relations $a^6 = 1$, $a^3 = c^2 = (ac)^2$.

Q. 15.06. Reconstruct the structure table on the basis of these defining relations, expressing every element in terms of a and c.

Q. 15.07. Find other sets of defining relations for Q_6, using other pairs of elements as generators. (You may deduce these from the defining relations given above connecting c and d, or a and c.)

Q. 15.08. Find a group in which $x^4 = y^4$, yet $x^2 \neq y^2$.

Form of defining relations; independence of generators

Note that when a defining relation is of the form $x^m = 1$, this usually implies that x is of period m, that is to say, if we state $x^6 = 1$, we intend it to be known that $x^2 \neq 1$, and $x^3 \neq 1$. For an Abelian group with generators x and y, one of the defining relations will generally be $xy = yx$, which states precisely that the group is Abelian.

Q. 15.09. What defining relations would be possible in an Abelian group with three generators, x, y and z?

Note also that it is desirable that generators shall be *independent*. This means that it must not be possible to express one in terms of another, so a relation like $y = x^3$ is inacceptable, though $x^2 = y^2$ is perfectly good – it is impossible in this case to express x in terms of y or vice versa. Similarly, if $xy = yx$, or if $x^2y = y^2x$, we cannot 'solve' these relations to obtain y in terms of x, and so in this case also, x and y are a pair of *independent* generators. Again, if defining relations specify, say,

$p^3 = q^3 = 1$, we should automatically exclude such a possibility as $q = p^2$. $xyx = y$ and $xyz = zyx$ are examples of valid defining relations, whereas $xyz = y$ is not, for the latter enables x to be expressed $x = yz^{-1}y^{-1}$, i.e. it serves to *reduce* the number of generators by expressing x in terms of y and z; whereas in the first two cases the generators are truly independent.

It is important, however, to realise that *any set of relations* define a group of uniquely determined structure. For example, $a^2 = 1$, $ara = r$ define an infinite group; $p^2 = q^2 = (pq)^3$ define a group of order 24; while the relations $r^5 = 1 = a^2$, $ar^3 = ra$ define the trivial group of order 1 (see p. 103). Consider the relations $p^3 = 1$, $pyp^{-1} = y^3$. The smallest finite group possessing elements which satisfy these relations is (surprisingly) of order 78, for the given relations may be shown to imply $y^{26} = 1$. But this is not to say that these two relations *define* the group of order 78; the group defined by them may be finite (of order a multiple of 78), or it may be infinite for all we know, as in the first example. If one wrote down a set of relations at random, the chances are that the group which these relations define uniquely might be of order 1, or equally well be infinite. The subject can really only be discussed satisfactorily in terms of 'free groups', outside the scope of this book.†

Groups of quaternions: dicyclic groups

The group Q_6 was selected for the discussion above, not because it has any special feature or merits, but because it suited our purpose, and contained elements of periods 2, 3, 4 and 6 and a good number of subgroups. Q_6 is, in fact, one of a special class of groups known as Dicyclic groups, of order a multiple of 4 (that of order $2n$ being designated Q_n with n even and > 3), which are given by the defining relations $r^n = 1$, $r^{n/2} = (ra)^2 = a^2$, the elements being $\{1, r, r^2, r^3, \ldots r^{n-1}, a, ra, r^2a, r^3a, \ldots r^{n-1}a\}$.

Q. 15.10. What happens when $n = 2$?

The simplest group in this class is Q_4, of order 8 ($n = 4$), and this is usually known as the **quaternion group,** though in fact all the dicyclic groups may be realised as groups of quaternions. These groups arise from the theory of quaternions, which are ordered quadruples of reals, (x_1, y_1, x_2, y_2). These are combined together by certain multiplication rules which may best be expressed in terms of matrices with complex elements.

† The whole question of generators and relations is a very difficult one. It is given a more expansive treatment in I. Grossman and W. Magnus, *Groups and their Graphs* (Random House). A more advanced and thorough account may be found in H. Coxeter and W. Moser, *Generators and Relations for Discrete Groups* (Springer-Verlag, Berlin, 1965). They show (Ch. 2) how an abstract definition for a group may be devised by a method known as the systematic enumeration of cosets, a process which can be done by computer.

If $x_1 + iy_1 = z_1, \quad x_1 - iy_1 = \bar{z}_1$ (the conjugate of z_1)

$x_2 + iy_2 = z_2, \quad x_2 - iy_2 = \bar{z}_2!$,

then we write the quaternion (x_1, y_1, x_2, y_2) as the matrix $\begin{pmatrix} z_1 & z_2 \\ -\bar{z}_2 & \bar{z}_1 \end{pmatrix}$.

Multiplication of quaternions then follows the usual matrix multiplication.

The group Q_4 is obtained from the general quaternion by the use of the fourth roots of 1, i.e. 1, i, -1, $-i$ to form the set of eight quaternions as follows:

$$e = \begin{pmatrix} 1 & 0 \\ 0 & 1 \end{pmatrix}; \quad a = \begin{pmatrix} -1 & 0 \\ 0 & -1 \end{pmatrix}; \quad b = \begin{pmatrix} i & 0 \\ 0 & -i \end{pmatrix};$$

$$c = \begin{pmatrix} -i & 0 \\ 0 & i \end{pmatrix}; \quad d = \begin{pmatrix} 0 & 1 \\ -1 & 0 \end{pmatrix}; \quad f = \begin{pmatrix} 0 & -1 \\ 1 & 0 \end{pmatrix};$$

$$g = \begin{pmatrix} 0 & i \\ i & 0 \end{pmatrix}; \text{ and } h = \begin{pmatrix} 0 & -i \\ -i & 0 \end{pmatrix}.$$

Q. 15.11. Find the period of each of these quaternions. Find a pair of generators.

Q. 15.12. Taking $\omega = \text{cis } 120°$, find a set of twelve quaternions (each expressed in the form of a matrix with complex elements) which form the group Q_6 under matrix multiplication. (Take two of the matrices to be $\begin{bmatrix} \omega & 0 \\ 0 & \omega^2 \end{bmatrix}$ and $\begin{bmatrix} 0 & \omega \\ -\omega^2 & 0 \end{bmatrix}$. Assign to these matrices the appropriate letter from the table (p. 216) so that the structure is identical. Work from the *period* of each).

Q. 15.13. Find other finite sets of quaternions which form groups.

Of course, the only possible subgroups of Q_4 are of order 2 and 4. Since there is only one element of period 2 ($= a$), this rules out the possibility of D_2 being a subgroup, and in fact we have only C_2 and C_4 as subgroups.

Q. 15.14. Determine the elements in each of the C_4 subgroups.

Q. 15.15. How many elements of period 4 are there in $Q_6, Q_8, \ldots, Q_n$? How many C_4 subgroups?

Q. 15.16. Q_4 and Q_6 are known to have only one element of period 2. Can you prove that this is true in all dicyclic groups?

Is $\begin{bmatrix} -1 & 0 \\ 0 & -1 \end{bmatrix}$ the only quaternion of period 2?

One of the chief characteristics of Q_4, and of the finite groups of quaternions is that they abound in elements of period 4. In Q_4 there are no fewer than six of these. (Remember that the number of elements of any given period, except 2, is *even* – see p. 107.)

While some of the other groups we have studied are capable of a large variety of realisations in all sorts of situations, Q_4 does not seem to 'crop up' with any great frequency. There is no figure in either two or three dimensions whose symmetry group is Q_4. One may seek a set of eight permutations which have Q_4 as their group, but it would appear to be necessary to take permutations on eight objects and to use Cayley's theorem.

★**Q. 15.17.** Can you find the smallest n for which Q_4 is a subgroup of S_n? It is certainly not a subgroup of S_4, and it certainly is a subgroup of S_8, so you are faced with answering whether Q_4 is a subgroup of S_5, S_6 or S_7. Is Q_4 a subgroup of $S_4 \times C_2$ (see p. 302).

Q. 15.18. Check that the matrices

$$x = \begin{bmatrix} 0 & 1 & 0 & 0 \\ -1 & 0 & 0 & 0 \\ 0 & 0 & 0 & -1 \\ 0 & 0 & 1 & 0 \end{bmatrix} \quad \text{and} \quad y = \begin{bmatrix} 0 & 0 & 1 & 0 \\ 0 & 0 & 0 & 1 \\ -1 & 0 & 0 & 0 \\ 0 & -1 & 0 & 0 \end{bmatrix}$$

generate the group Q_4, and establish the isomorphism with the eight 2×2 matrices $\{e, a, b, c, d, f, g, h\}$ above.

Cayley diagrams

All groups may be represented by a network (or 'graph') known as a Cayley diagram. Each element of the group is represented by a point, and the directed links between points represent generators of the group. Consider, for example, the group C_6. The six elements are arranged for convenience at the vertices of a convex hexagon – see fig. 15.02. Each of

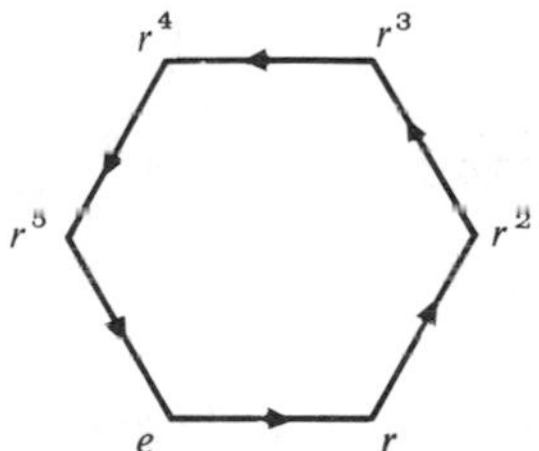

Fig. 15.02. Cayley diagram for C_6.

the directed links connecting the vertices represents the operation 'multiply by r', so that, for example, to get from r^4 to r^5, one moves along one link; to get from r^5 to r^3, one moves along four links, for $r^5 \times r^4 = r^3$, in each case in the direction of the arrows. To move along a link against the arrow represents the inverse operation, multiplication by r^{-1} or r^5.

For D_6, we shall have twelve points, and since the group requires two generators, r ($r^6 = 1$) and a ($a^2 = 1$), we shall need *two kinds of links*. The r-link is shown in fig. 15.03 with a single arrow-head, and the a-link is shown in red (since $a^2 = 1 \Rightarrow a = a^{-1}$, there is no need for an arrow on the a-link). Starting from e, one may reach the point ra either by performing first an a-link, then an r-link, or else by performing first r^5 and following this by a. This illustrates the relation $ra = ar^5$. If a and b ($= ar$) were used as generators, we should get (using a black line for b): fig. 15.032, or its equivalent version 15.033.

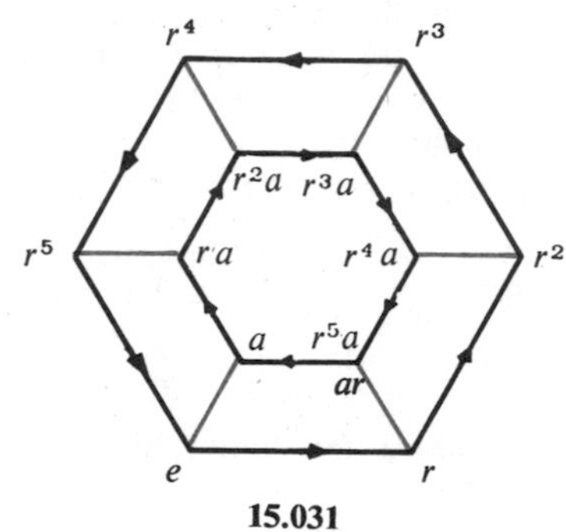

15.031

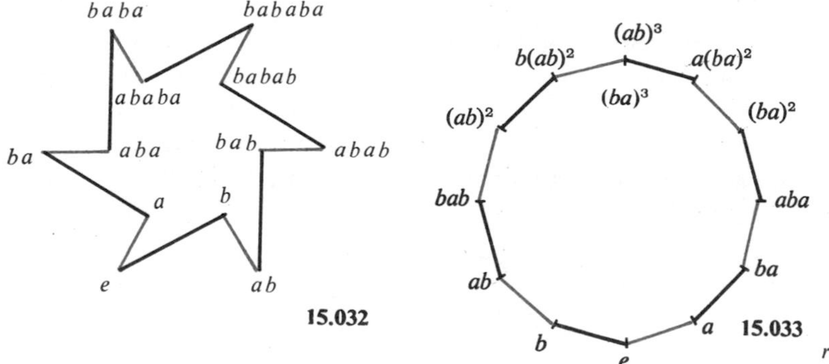

15.032 **15.033**

Fig. 15.03. Cayley diagram for D_6.

It is fascinating to pursue a study of these Cayley diagrams because they throw fresh light on the structure of groups. A very good treatment, in more detail of Cayley diagrams in a wide variety of cases is to be found in I. Grossman and W. Magnus, *Groups and their Graphs* (Random House, New Mathematical Library). The book also deals much more thoroughly with the whole question of generators and relations, and so enlarges the material of the present chapter. Moreover, Cayley diagrams are used to illuminate features such as subgroups and cosets. (*Warning*: in that book, the convention is used that ab means 'first a, then b'.)

Q. 15.19. Draw a Cayley diagram for D_2. Compare with the realisation of D_2 on p. 137. Draw Cayley diagrams for D_3, D_4, D_5, using relations $a^2 = b^2 = 1$, $(ab)^n = 1$.

Q. 15.20. Draw Cayley diagrams for C_∞ and D_∞. Investigate how Cayley diagrams may be used to reveal subgroups.

Cayley graph of the quaternion group

In view of the scant appearances made by Q_4 in other contexts, we conclude the chapter by giving it the special privilege of interpretation by a Cayley diagram. Taking the defining relations as $p^2 = q^2 = (pq)^2$, we may deduce further relations: $p^2 = pqpq \Rightarrow p = qpq = q(qpq)q = q^2pq^2 = p^2pp^2 = p^5 \Rightarrow p^4 = 1$.

Q. 15.21. Prove also that q and pq are of period 4, and that the group has elements $1, p, p^2, p^3, q, pq, qp, q^3$. Demonstrate the isomorphism between the group formed by these eight elements and the eight quaternion matrices (p. 246).

Q. 15.22. Show that an equivalent set of defining relations is $p^4 = 1$, $p^2 = q^2$, $qp = p^3q$; and from these prove that $qp^3 = pq$, $qp^2 = p^2q$.

Q. 15.23. In the automorphisms of Q_4, are all the elements of period 4 'interchangeable', i.e. could we call *any* pair of those elements p and q?

The subgroup $\{1, p, p^2, p^3\}$ may be represented by the vertices of a quadrilateral, and the generator p by a directed segment as shown by the full lines. Since $p^2 = q^2$, it is desirable to place the point q equidistant from the opposite corners 1 and q^2. The points q, pq, q^3 and qp may then be arranged in a smaller square whose sides are also p-segments, since these four elements are q, pq, p^2q and p^3q. (See fig. 15.04.)

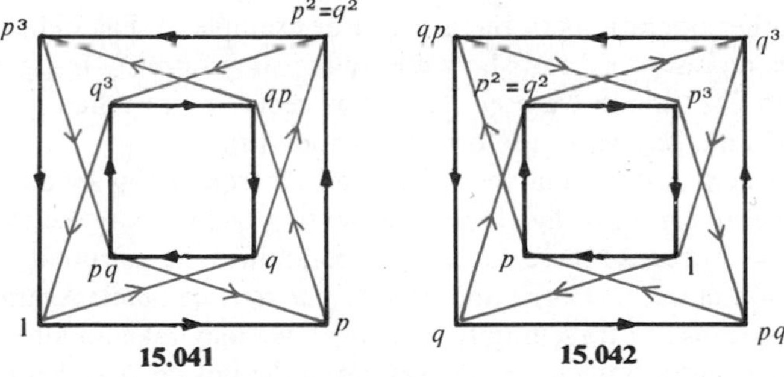

Fig. 15.04. The Quaternion group: $p = qpq$; $q = pqp$, ——→ p, ——→ q.

Q. 15.24. Invent defining relations for Q_4 other than those above. Another graphical representation of the quaternion group is given in Z. P. Dienes and E. W. Golding, *Groups and Coordinates* p. 30 (E.S.A. Hutchinson Educational). (*Note*: the table for the group Q_4 is given on pp. 97, 101.)

A further note on Cayley diagrams

(1) Any vertex of a Cayley diagram may be taken as the 'origin'. For example, fig. 15.042 is the same as fig. 15.041, only the points have been 'shuffled round'. In this sense, all the vertices of a Cayley diagram are of equal status.

(2) The Cayley diagram does not have to be two-dimensional. For example, the diagram for D_5 could consist of the vertices and edges of a pentagonal prism; the diagram for D_2 could be a tetrahedron, one pair of whose opposite edges represent a, another pair b and the third pair (if they were put in), ab or ba. Moreover, Cayley diagrams may be conveniently drawn on a two-dimensional surface which is not a plane, such as a cylinder, or even a torus. See Ex. 16.38.

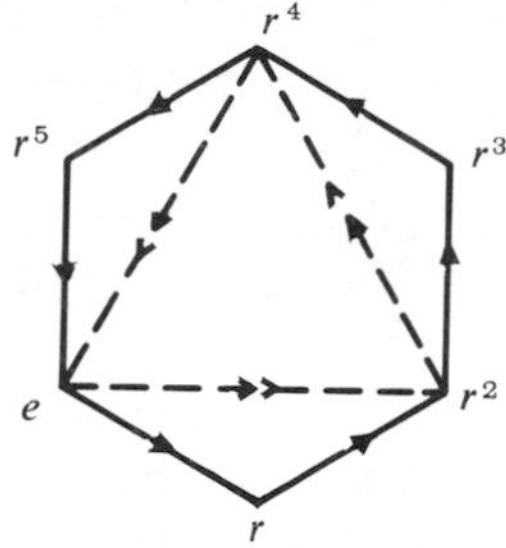

Fig. 15.05. The group C_6 showing Gp(r^2), the subgroup C_3.

(3) The edges, or line segments represent operations which are usually generators, but there is nothing to stop us putting in more edges to represent other operations in the group. For example, in fig. 15.05, the dotted lines represent r^2, and show clearly the subgroup C_3. In fig. 15.032, representing D_6, though we used a and b as generators, we had no need to erase the 'r' links shown in the original version (a).

(4) Equivalent 'words' may easily be read off from a Cayley diagram. For example, referring to fig. 15.04 representing Q_4, we may see that $p^3q^2pq^3p^2q = 1$ because these successive operations take us round the perimeter of a closed network and *back to our starting point.* Again, $p^3 = qpq^3$, because in travelling from 1 to p^3, we may take an alternative roundabout route following the arrows along the links q, q, q, p and q.

EXERCISES 15

1 Show that Q_4 may be generated by p and q where $p = qpq$ and $q = pqp$. Deduce from these two relations that $p^2 = q^2$ is the only element of period 2, and that all the other elements have period 4.

2 If x, y are elements of a group, and $x^2 = y^2$, is it true that each of x and y is of even period?

3 In looking for a set of (two) generators of a group, it is no good selecting two which lie in the same subgroup. Is this true?

4 If $x^2 = y^2 \neq 1$, $x \neq y$ and $x^3 = 1$, show that x and y are not independent, and that the group is C_6.

5 If $x^2 = y^2 \neq 1$, $x^5 = 1$, show that the group is C_{10} $(x \neq y)$. Consider other groups in which $x^2 = y^2 \neq 1$.

6 Describe groups with independent generators x, y satisfying:
(a) $x^2 = y^3$, $x^3 = 1$; (b) $x^2 = y^3$, $x^5 = 1$.

7 Show that the set of relations: $pq = qp^3$, $p^2 = q^2$; $p^4 = 1$ contains a redundancy.

8 Which group is specified by $a^2 = p^3 = 1$, $ap = pa$?

9 Which group is defined by the generating functions $z' = z \cos \frac{2}{3}\pi$, $z' = \bar{z}$, where z is a complex number and $\bar{z}$ is its conjugate? (Consider the geometrical meaning on the Argand diagram.)

10 The Hand Calculating Machine Group. (See Ex. 7.29, p. 71.)
If the multiplier stores, say, eight digits, then 10^8 rotations of the handle would again produce a row of 0's, i.e. $r^{100,000,000} = 1$. However, m does not have a finite period; we can perform m^8 but not m^9. Is there any sense in which the operations r and m may be said to generate a group? (Remember we established that $mr^{10} = rm$.)

11 What is the order of the group defined by $a^2 = p^4 = 1$, $ap = pa$? Attempt to construct the group table. (This is $C_4 \times C_2$, see p. 280.)

12 Add another defining relation to $a^2 = p^3 = 1$ so that: (a) Gp $(a, p) \cong C_6$; (b) Gp $(a, p) \cong D_3$; (c) Gp (a, p) is of order 12.

13 **Prove that (Z, +) is generated by any pair of co-prime integers (e.g. 8 and 19).**

14 Show that the relations $p^4 = 1$, $ap^3 = pa$ do not *define* a group of order 8 by finding two different groups of order 8 which contain elements satisfying them.

15 **Prove that all quaternions are closed under multiplication and that they form an infinite group. Consider subgroups of this group, such as those quaternions $\begin{bmatrix} z_1 & z_2 \\ -\bar{z}_2 & \bar{z}_1 \end{bmatrix}$ for which z_1 and z_2 are real (cf. Q. 20.33, p. 387); those quaternions for which $z_1 = 1$, $z_2 = x + iy$, etc.**

16 The quaternion expressed in the previous question as a 2×2 matrix with complex numbers may be considered to be an ordered pair of complex numbers, (z_1, z_2), or as an ordered quadruple of reals: (x_1, y_1, x_2, y_2). Show that addition and multiplication of these order quadruples may be stated:

$$(a_1, b_1, c_1, d_1) + (a_2, b_2, c_2, d_2) = (a_1 + a_2, b_1 + b_2, c_1 + c_2, d_1 + d_2)$$
$$\begin{aligned}(a_1, b_1, c_1, d_1) \times (a_2, b_2, c_2, d_2) = (&a_1a_2 - b_1b_2 - c_1c_2 - d_1d_2,\\ &a_1b_2 + a_2b_1 + c_1d_2 - c_2d_1, a_1c_2 + a_2c_1 - b_1d_2 + b_2d_1,\\ &a_1d_2 + a_2d_1 + b_1c_2 - b_2c_1).\end{aligned}$$

If (1, 0, 0, 0), (0, 1, 0, 0), (0, 0, 1, 0) and (0, 0, 0, 1) are denoted 1, i, j, k, find all the elements of Gp (i, j, k) under multiplication, draw up the group table, and demonstrate its isomorphism with Q_4. Show also how the ordered quadruples (a, b, c, d) may be represented by 4×4 matrices with real terms.

17 Is it possible to define Q_4 by supplementing the relations $a^2 = p^4 = 1$ with a third relation connecting a and p?

18 Using the relations $p^2 = q^2 = (pq)^2$, re-draw the Cayley diagram for Q_4, showing the elements at the vertices of a regular octagon.

19 How many automorphisms are there of Q_4? What is the group of these automorphisms?

20 Just as Q_4 may be defined $p = qpq$, $q = pqp$, can you find a similar pair of defining relations for Q_6?

21 How many elements of period 4 are there in Q_8, defined: $p^4 = q^2 = (pq)^2$? Discover whether Q_4 is a subgroup of Q_8.

22 Show that Q_6 may be defined $r^6 = 1$, $a^2 = (ra)^2 = r^3$ using a different notation from that on pp. 216, 244.

23 Show that the group Q_8 has four elements of period 8, ten of period 4, and that Q_4 is a subgroup of it.

24 In the dicyclic group of order 24 defined: $r^{12} = 1$, $r^6 = (rs)^2 = s^2$, prove that all those elements of the form $r^k s$ are of period 4 ($k \in \mathbf{Z}$).

25 Consider a group defined $p^2 = q^m = r^2 = pqr$. Redefine the group with only two generators. Can you identify the structure? (Take, for example, $m = 2$, $m = 3$, etc.)

26 Find the square of the matrix $\begin{bmatrix} 0 & z \\ -1/z & 0 \end{bmatrix}$. What do you conclude about the group Q_n, one generator of which has $z = \text{cis } 2\pi/n$?

27 Replace the set of relations: $a^2 = r^3 = 1$; $rar^{-1} = b$; $rbr^{-1} = ab$ by a set based upon two generators only.

28 A finite group has two independent generators p, q, each of period 3. Can you say anything about its order? Suppose in addition we know that the group is Abelian, i.e. $p^3 = q^3 = 1$, $pq = qp$. Does this define a finite group? Draw up the structure table.

Suppose we are given $p^3 = q^3 = 1 = (pq)^2$. Find out all you can about this group, and show that it can equally well be written in terms of generators p and a, where $p^3 = a^2 = 1 = (pa)^3$.

29 The group A_4 (see table 17.02) has elements a, b, c, of period 2, and p, q, r, s of period 3. By using the method of pp. 223 ff., find a minimum set of generators for the whole group, and show that these may either be both of period 3, or else one of period 2 and one of period 3. Give defining relations for A_4 in terms of (a) a and p; (b) p and q, and draw Cayley diagrams.

30 x, y and z are elements of a group. Let $p = xyz$, $q = xzy$, $r = yzx$, $s = yxz$, $t = zxy$, $u = zyx$. Then if the group is Abelian, $p = q = r = s = t = u$. If not, then p may be equal to some of the remaining products and not to others. Examine the case of the group D_3, and larger groups, for various choices of x, y and z. Can you find a group containing three elements for which all six products are different?

31 Draw a Cayley graph for a group of order 16 generated by the permutations $a = (24)(36)$, $b = (12)(35)(47)(68)$, $c = (13)(28)(45)(67)$.

32 A group has elements a, x satisfying $a^2 = 1$, $axa = x$, $x^8 = 1$. Given that there are sixteen elements, draw the Cayley diagram, and identify the structure of the group.

33 If $p^3 = r^4 = 1$, $pr^3 = rp$, prove that $r^2p = pr^2$, $r^3p = pr$ and $p = rpr$. Find the periods of pr, of rp, and of pr^2, and identify Gp (p, r).

34 (*a*) If a and r satisfy $a^2 = r^5 = 1$, $ar = r^2a$, prove that $r = 1$.
(*b*) Eliminate p from $q = rqp$, $r = prq$.

35 In Ex. 9.19, we showed that $p^3 = 1$, $pxp^{-1} = x^2 \Rightarrow x^7 = 1$. Show that Gp (p, x) is of order 21, and draw its Cayley graph.

36 Identity the groups defined: (a) $a^2 = 1$, $b^2 = 1$, $(ab)^4 = 1$; (b) $p^3 = r^4 = 1$, $(rp)^2 = 1$.

37 Show that the group $(\mathcal{M}_2, \times)$ is generated by the matrices

$$\begin{bmatrix} 0 & 1 \\ 1 & 0 \end{bmatrix}, \begin{bmatrix} 1 & 1 \\ 0 & 1 \end{bmatrix} \text{ and } \begin{bmatrix} a & 0 \\ 0 & 1 \end{bmatrix},$$ and interpret geometrically.

38 Draw Cayley graphs for C_6 and for D_3 based on different sets of generators from those given in the text.

39 In the Cayley diagrams so far noted (figs. 15.02, 15.03, 15.04, 15.05), any one of the vertices might have been taken as the point e, and the others could then be labelled. Does there exist a two- or three-dimensional graph whose vertices are indistinguishable, yet it is not a Cayley graph for any group?

40 If p, q and r in fig. 15.06 are quarter-turns about three mutually perpendicular axes, Ox, Oy, Oz, the senses being as indicated in the diagram, show that $rqp = q$, and write down other relations connecting p, q and r. Show that pqr is of period 2. What is its geometrical interpretation? Show that two

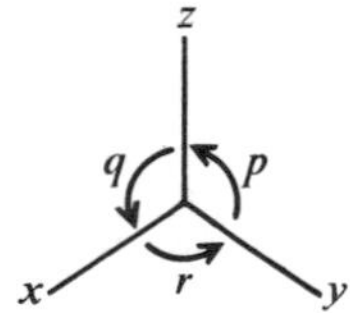

Fig. 15.06.

of the generators p, q and r are independent, and will generate the group S_4. Give a set of defining relations for S_4 beginning $p^4 = q^4 = 1$, ...

41 Show that S_4 may be constructed from generators R and S satisfying $R^3 = S^2 = (RS)^4 = 1$, and find pairs of elements from the table on p. 219 which would serve as R and S.

42 Given that $b^2 = c^2 = 1$, $ca = abc$, prove that $bc = cb$. If in addition $a^3 = 1$, $ba = ac$, which group is defined?

43 S_4 may be defined $a^2 = j^4 = 1$; $(aj)^3 = 1$ with the same notation as on p. 219. Let $a(z) \equiv (z + B)/(Cz - 1)$ (of period 2), and $j(z) \equiv (z - 1)/(z + 1)$ (of period 4). Find the condition for aj to be of period 3 (see Ex. 10.28, p. 129), and hence obtain $B = i$ and $C = -i$ as a possible pair of values for B and C. Go on to represent S_4 by twenty-four bilinear functions. (This question should be compared to Exs. 14.21, 14.22.)

44 Give defining relations for D_∞, and draw a Cayley diagram for D_∞ using two generators each of period 2.

45 Prove that a subset E of a group G is a system of generators for G if and only if no proper subgroup of G exists which contains all the elements of E.

46 Note that, in an infinite group which has x as a generator, we include x^{-1} as the same generator. Thus in the group $(\mathbf{Z}, +)$, we may regard ± 1 as the single generator. We also know that $(\mathbf{Z}, +)$ may be generated by any two coprime integers. How many generators does $(\mathbf{Q}, +)$ require? (Consider, for example, the subgroup of $(\mathbf{Q}, +)$ generated by a pair of fractions such as $\frac{2}{3}$ and $\frac{1}{4}$; or by $\frac{7}{30}$ and $\frac{15}{28}$; or by three fractions such as $\frac{1}{2}$, $\frac{1}{3}$ and $\frac{1}{5}$, and their additive inverses.) Describe the subgroup of $(\mathbf{R}, +)$ generated by $\sqrt{2}$ and $\sqrt{3}$; by 1 and $\sqrt{2}$; by 1, $\sqrt{2}$ and $\sqrt{3}$. Describe the subgroup of $(\mathbf{R}, \times)$ generated by $\sqrt{2}$; by $\sqrt{2}$ and $\sqrt{3}$; by $5^{\frac{1}{3}}$; by $\sqrt{2} + \sqrt{3}$, etc. How many generators does $(\mathbf{R}, +)$ require?

16 *Bigger and better groups: direct product groups*

Consider the two groups $\{1, i, -1, -i\}$ $(i^4 = 1)$ and $\{1, \omega, \omega^2\}$ $(\omega^3 = 1)$, both under multiplication. The first is of course C_4, and the second C_3. Now let us form *ordered pairs* of elements, the first member in each pair coming from the group C_4, the second from the group C_3. Then we have twelve possible ordered pairs:

$e\,(1, 1)$;	$a\,(i, 1)$;	$b\,(-1, 1)$;	$c\,(-i, 1)$;
$d\,(1, \omega)$;	$f\,(i, \omega)$;	$g\,(-1, \omega)$;	$h\,(-i, \omega)$;
$j\,(1, \omega^2)$;	$k\,(i, \omega^2)$;	$l\,(-1, \omega^2)$;	$m\,(-i, \omega^2)$.

We now define a binary operation $*$ on this set of ordered pairs, thus:

$$(x_1, y_1) * (x_2, y_2) = (x_1 x_2, y_1 y_2).$$

For example (omitting $*$ and using juxtaposition), we have:

$$jh = (1, \omega^2)(-i, \omega) = (-i, 1) = c$$
$$k^2 = (i, \omega^2)^2 = (i, \omega^2)(i, \omega^2) = (-1, \omega) = g.$$

It is evident that the set of twelve ordered pairs is closed under this operation. Associativity is immediately apparent, and the identity element is $(1, 1)$. Finally each element has an inverse, for $(x, y)(1/x, 1/y) = (1, 1)$, so that the inverse of (i, ω^2), for example, is $(1/i, 1/\omega^2) = (-i, \omega)$, i.e. $k^{-1} = h$. Thus the twelve elements form a group. Note that in the array of elements as listed above, the top row is the subgroup C_4, and the first column is the subgroup C_3.

Now $f = (i, \omega) \Rightarrow f^2 = (-1, \omega^2)$, $\quad f^3 = (-i, 1)$, $\quad f^4 = (1, \omega)$,
$f^5 = (i, \omega^2)$, $\quad f^6 = (-1, 1) \ldots$, $\quad f^{12} = (1, 1)$.

Hence f is of period 12, and the group must be C_{12}. We are now ready to give a definition:

Direct product groups

If G $\{1, g_1, g_2, g_3, \ldots, g_{n-1}\}$ and G′ $\{1', g_1', g_2', g_3', \ldots, g_{n'-1}'\}$ are two groups under the operations $*$ and $*'$, then the set of ordered pairs (g_p, g_q') with the binary operation defined:

$$(g_p, g_q')(g_r, g_s') = (g_p * g_r, g_q' *' g_s')$$

is a group of order nn', and is called the Direct Product† Groups of G and G′, and is abbreviated G × G′.

In the illustrative example, what we have done is to form the direct product of C_4 and C_3, and we have shown that the 'bigger and better group' so obtained is in fact the familiar C_{12}.

If we were to rewrite the elements of C_4 and C_3 in terms of generators, thus: $C_4\,\{1^0, i^1, i^2, i^3\}$; $C_3\,\{\omega^0, \omega^1, \omega^2\}$, our ordered pairs would then appear as

$$\begin{array}{llll} e\,(i^0, \omega^0); & a\,(i^1, \omega^0); & b\,(i^2, \omega^0); & c\,(i^3, \omega^0); \\ d\,(i^0, \omega^1); & f\,(i^1, \omega^1); & g\,(i^2, \omega^1); & h\,(i^3, \omega^1); \\ j\,(i^0, \omega^2); & k\,(i^1, \omega^2); & l\,(i^2, \omega^2); & m\,(i^3, \omega^2); \end{array}$$

and in applying the product rule, for example,

$$jh = (i^0, \omega^2)\,(i^3, \omega^1) = (i^3, \omega^3) = (i^3, \omega^0) = c$$
$$k^2 = (i^1, \omega^2)\,(i^1, \omega^2) = (i^2, \omega^4) = (i^2, \omega^1) = g,$$

it is clear that we are adding the indices modulo 4 and modulo 3 respectively; so that, if we represent our ordered pairs by depicting only the indices, thus:

$e\,(0, 0)$; $a\,(1, 0)$; $b\,(2, 0)$; $c\,(3, 0)$; $d\,(0, 1)$; $f\,(1, 1)$; $g\,(2, 1)$; $h\,(3, 1)$;

$j\,(0, 2)$; $k\,(1, 2)$; $l\,(2, 2)$; $m\,(3, 2)$,

we may now write $jh = (0, 2)(3, 1) = (3, 0) = c$
$k^2 = (1, 2)(1, 2) = (2, 1) = g$, and so on.

Thus the group $C_4 \times C_3$ is seen in a new guise, as the set of ordered pairs (a, b) with $a \in \{0, 1, 2, 3\}$, $b \in \{0, 1, 2\}$ combined according to the composition rule

$$(a_1, b_1)(a_2, b_2) = (a_1 + a_2 \text{ mod. } 4,\ b_1 + b_2 \text{ mod. } 3).$$

Now for ordinary two-dimensional vectors, we have:

$$\left.\begin{array}{l}(0, 2) + (3, 1) = (3, 3), \\ (3, 1) + (2, 2) = (5, 3),\end{array}\right\} \text{ vector addition.}$$

In the case being considered above, the x-coordinates have been reduced modulo 4, and the y-coordinates reduced modulo 3. It is as if one were constrained to work within the unshaded rectangle in fig. 16.01 where the points marked ● are all regarded as being *equivalent* to the point (0, 1).

† Sometimes the term used is Direct Sum – see F. M. Hall, *Abstract Algebra* I (C.U.P.), p. 249.

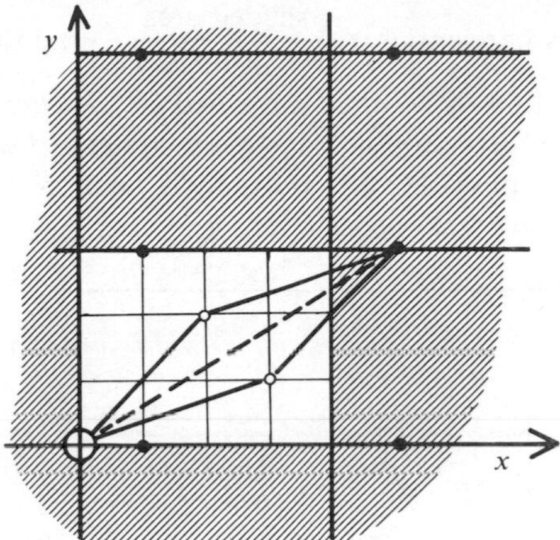

Fig. 16.01. Vector addition with modular arithmetic; (3, 1) + (2, 2) = (5, 3); 5 = 1 (mod. 4); 3 = 0 (mod. 3), etc.

Episodical illustration: Buffon's needle experiment

★An example will make this clearer. There is a famous experiment known as Buffon's needle experiment in which a needle of length l is thrown at random on to a table on which a set of parallel lines are ruled at a distance d apart. It can be shown that, provided $l < d$, the probability of the needle crossing one of the parallel lines is $\pi l/d$. Armed with this theoretical result, it is possible to obtain a value for π based upon the observed occurrence of this event in a large number of trials – an example of what is known as a Monte Carlo method.

The proof depends upon setting up what is known as a 'sample space'. Any particular throw of the needle is determined by two measurements: the distance y of the mid-point of the needle from the line 'just below' it (see fig. 16.021), so that $0 \leqslant y \leqslant d$; and the orientation of the needle, determined by a parameter θ which we may take to have any value between 0 and π. A random throw of the needle corresponds to a random choice of the numbers y and θ, and this in turn corresponds to the selection of a point at random in a rectangle whose length and breadth are d and π. For example, the point P represents the event that the centre of the needle falls at a distance $\frac{2}{3}d$ from the nearest line below it, and the needle is inclined at $\frac{3}{4}\pi$ to the direction $\theta = 0$ (see fig. 16.022).

Now for the needle to cut a line, we must have either $y < \frac{1}{2}l \sin \theta$, or $y > d - \frac{1}{2}l \sin \theta$. The point selected within the sample space (the rectangle of fig. 16.022) must lie within the shaded areas. The probability of this happening is $\frac{\text{shaded area}}{\text{area of rectangle}}$, and this may be verified by integration to be $\pi l/2d$ (provided $l < d$ – why?).

The point of this diversion is that we are content to work within a single rectangle, all possible vertical distances being reckoned modulo d, and the angle θ being reckoned modulo π.

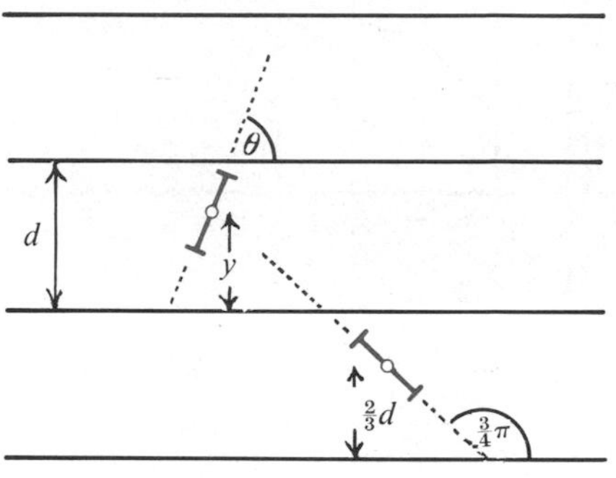

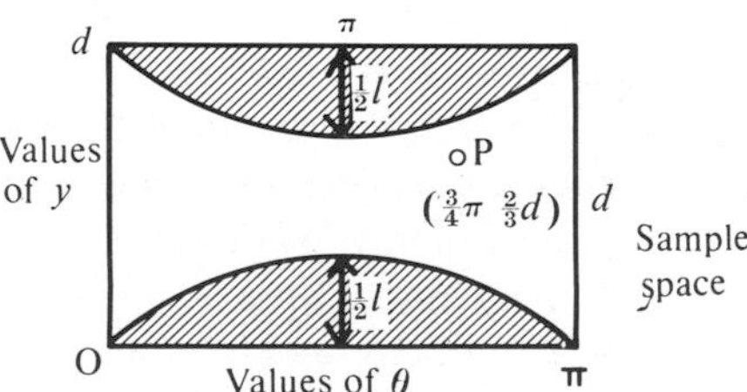

Fig. 16.02. Buffon's needle experiment.

Further illustration that $C_4 \times C_3 \cong C_{12}$

Resuming the original discussion of the realisation of the group $C_4 \times C_3$, the ordered pairs may also be written as matrices; the law of composition being matrix multiplication, and we illustrate by the same examples as above:

$$jh = \begin{bmatrix} 1 & 0 \\ 0 & \omega^2 \end{bmatrix} \begin{bmatrix} -i & 0 \\ 0 & \omega \end{bmatrix} = \begin{bmatrix} -i & 0 \\ 0 & 1 \end{bmatrix} = c;$$

$$k^2 = \begin{bmatrix} i & 0 \\ 0 & \omega^2 \end{bmatrix}^2 = \begin{bmatrix} i & 0 \\ 0 & \omega^2 \end{bmatrix} \begin{bmatrix} i & 0 \\ 0 & \omega^2 \end{bmatrix} = \begin{bmatrix} -1 & 0 \\ 0 & \omega \end{bmatrix} = g$$

and so on. However, we shall prefer the more compact notation of ordered pairs.

Q. 16.01. Prove that $C_2 \times C_2$ is isomorphic with D_2, or the Klein four-group.

Q. 16.02. Consider the direct product of $C_2 \times C_3$ in the same ways as $C_4 \times C_3$ was dealt with in the text. Experiment with other direct products of cyclic groups, such as $C_2 \times C_4$; $C_3 \times C_3$; $C_6 \times C_2$; $C_3 \times C_5$.

Q. 16.03. Prove from the definition that G and G′ are both subgroups of $G \times G'$.

Q. 16.04. Prove from the definition that the direct product of two Abelian groups is necessarily an Abelian group.

Q. 16.05. Prove from the definition that $G \times G' \simeq G' \times G$ even when G or G′ or both may not be Abelian.

The direct product of $C_4 \times C_3$ may also be illustrated as the product of the cyclic permutations of two independent sets, $\{A\ B\ C\ D\}$ and $\{P\ Q\ R\}$. The permutations which arise are as shown in table 16.01. The order in

Table 16.01

e	A	B	C	D	P	Q	R	
a	D	A	B	C	P	Q	R	$= a$
b	C	D	A	B	P	Q	R	$= a^2$
c	B	C	D	A	P	Q	R	$= a^3$
d	A	B	C	D	R	P	Q	$= d$
f	D	A	B	C	R	P	Q	$= da = ad$
g	C	D	A	B	R	P	Q	$= da^2 = a^2d$
h	B	C	D	A	R	P	Q	$= da^3 = a^3d$
j	A	B	C	D	Q	R	P	$= j$
k	D	A	B	C	Q	R	P	$= ja = aj$
l	C	D	A	B	Q	R	P	$= ja^2 = a^2j$
m	B	C	D	A	Q	R	P	$= ja^3 = a^3j.$

which these permutations have been written down matches exactly the scheme of the ordered pairs shown earlier. In fact, one might denote any particular permutation by an ordered pair specifying the position of the letters A and P: calling the positions in the cycles respectively 0, 1, 2, 3 and 0, 1, 2, h would be denoted (3, 1) since A occurs in the final position and P in the middle position. We are thus thrown back to the ordered pairs of 'indices'!

Q. 16.06. Consider $C_2 \times C_3$; $C_2 \times C_4$ and $C_3 \times C_3$ as the products of cycles of permutations. Compare with previous work in the text (e.g. pp. 114–15).

To show that $C_4 \times C_3$ is isomorphic with C_{12}, it is only necessary to detect an element of period 12. Using the cyclic permutations of the two independent sets $\{A\ B\ C\ D\}$ and $\{P\ Q\ R\}$, it is immediately obvious that the permutation $f\begin{pmatrix} A & B & C & D & P & Q & R \\ D & A & B & C & R & P & Q \end{pmatrix}$ is of period 12; and once closure for the set has been established, this leads at once to the identification of the direct product group as C_{12}.

A new group

Let us now consider $C_6 \times C_2$. We could do this by any of the above representations, but we shall prefer to think of the product of the disjoint cycles of permutations of $\{A\ B\ C\ D\ E\ F\}$ and $\{P\ Q\}$. There are, of course, twelve permutations in the set; table 16.02. Do we again get C_{12}? The

Table 16.02

e	A	B	C	D	E	F	P	Q
a	F	A	B	C	D	E	P	Q
a^2	E	F	A	B	C	D	P	Q
a^3	D	E	F	A	B	C	P	Q
a^4	C	D	E	F	A	B	P	Q
a^5	B	C	D	E	F	A	P	Q
b	A	B	C	D	E	F	Q	P
$ab = ba$	F	A	B	C	D	E	Q	P
$a^2b = ba^2$	E	F	A	B	C	D	Q	P
$a^3b = ba^3$	D	E	F	A	B	C	Q	P
$a^4b = ba^4$	C	D	E	F	A	B	Q	P
$a^5b = ba^5$	B	C	D	E	F	A	Q	P

answer is 'No', because this time *there is no permutation of period 12*. The underlying reason for this is, of course, that 2 is a factor of 6, so that when either of the twelve permutations has been performed six times, the letters P and Q are the right way round again, and the order of the whole set is restored. Thus no element can be of period greater than 6, and so we certainly do not have the group C_{12}. In fact, $C_6 \times C_2$ (or $C_2 \times C_6$) is a new group of order 12 which is not isomorphic with either C_{12}, or D_6, or Q_6, the groups of order 12 we have so far met. (There is another group, A_4, of order 12, which emerged as a subgroup of S_4 (see p. 220).) It is worth noting that, of the five groups of order 12, D_6, Q_6 and A_4 are all non-Abelian, whereas C_{12} and $C_6 \times C_2$ are Abelian.

Q. 16.07. Construct the group table for $C_2 \times C_6$, and show that there are three elements of period 2, two of period 3, none of period 4 and six of period 6.

Q. 16.08. Prove from the definition that $(A \times B) \times C = A \times (B \times C)$, where A, B and C are groups. Show that the elements of $A \times B \times C$ may be taken to be ordered triples of the form (a, b, c) where $a \in A$, $b \in B$, $c \in C$, and state the rule for composition of the triples.

Q. 16.09. Show that twelve of the residue classes modulo 21 under multiplication form the group $C_6 \times C_2$. Repeat for mod. 28. In each case, find the set of twelve numbers. Show that $\{1, 4, 16, 9, 29, 11\}$ form the group C_6 under multiplication mod. 35, and that $\{1, 6\}$ form C_2 under multiplication mod. 35. Form the direct product.

Q. 16.10. Represent the group $C_2 \times C_2 \times C_3$ as a set of twelve ordered triples; also as sets of ordered pairs of complex numbers under multiplication.

Q. 16.11. Describe a solid whose symmetry group is $C_6 \times C_2$.

Associativity and commutativity of direct product

Since $A \times B \simeq B \times A$ (see Q. 16.05), and also the direct product operation on groups is associative (see Q. 16.08), we may write $A \times B \times C$ without ambiguity, and, moreover, the letters may be permuted in any manner: $A \times B \times C \simeq B \times C \times A$, etc. For example,

$$C_2 \times C_6 \simeq C_6 \times C_2, \text{ and since } C_6 \simeq C_3 \times C_2, \text{ we may also write}$$
$$C_2 \times C_6 \simeq C_2 \times (C_2 \times C_3) \simeq C_2 \times C_2 \times C_3 \simeq C_3 \times C_2 \times C_2, \text{ etc.}$$

Also, in Q. 16.01, p. 259 we showed that $C_2 \times C_2 \simeq D_2$, therefore it follows that

$$C_2 \times C_6 \simeq D_2 \times C_3.$$

This means that this same group may be realised by the product of two disjoint sets of permutations one being those permutations of four letters $A\ B\ C\ D$ which give the group D_2, the other being the cyclic permutations of $P\ Q\ R$.

D_2 can be represented by the permutations

1	A	B	C	D
a	B	A	D	C
b	C	D	A	B
$ab = ba$	D	C	B	A

and C_3 by the permutations

1	P	Q	R
p	R	P	Q
p^2	Q	R	P.

Hence the group $D_2 \times C_3$ is the group of the set of twelve permutations obtained by combining these, such as:

$$p^2b = bp^2 = \begin{pmatrix} A & B & C & D & & P & Q & R \\ C & D & A & B & & Q & R & P \end{pmatrix}$$

$$pab \quad = \begin{pmatrix} A & B & C & D & & P & Q & R \\ D & C & B & A & & R & P & Q \end{pmatrix}.$$

Non-Abelian cases

All the examples considered so far have given Abelian direct products, since the constituent groups were themselves Abelian (see Q. 16.04). Consider now the direct product $D_3 \times C_2$. Since D_3 is not Abelian, we do not expect the direct product group to be Abelian. Using permutations, we need to combine the full set of six permutations of three letters A, B, C with the transpositions of two letters, P, Q. (The group could be called $S_3 \times S_2$, of course.) The permutations are as in table 16.03. (For the

Table 16.03

1	A	B	C	P	Q		d	A	B	C	Q	P
p	C	A	B	P	Q		dp	C	A	B	Q	P
p^2	B	C	A	P	Q		dp^2	B	C	A	Q	P
a	A	C	B	P	Q		da	A	C	B	Q	P
b	C	B	A	P	Q		db	C	B	A	Q	P
c	B	A	C	P	Q		dc	B	A	C	Q	P

subgroup D_3, the notation used in Ch. 13, p. 202 has been followed.) Instead of working out a complete group table, let us identify the direct product group $D_3 \times C_2$ by considering the period of each permutation. We find, without much trouble, that a, b, c, da, db, dc all have period 2; p, p^2 have period 3, and dp, dp^2 have period 6. Which group of order 12 is thus characterised? It is the group D_6, so we may write $D_3 \times C_2 \cong D_6$.

Q. 16.12. Which elements form the subgroup C_6? Check that they are generated by dp (or dp^2).

Q. 16.13. We have seen that $D_6 = D_3 \times C_2$. Is it true that $C_2 \times D_2 = D_4$? Find under what conditions D_{2n} is isomorphic to $D_n \times C_2$.

Q. 16.14. Show that the group of the twelve permutations of the above paragraph is isomorphic to the *full* group of symmetries of the equilateral triangular prism, i.e. including *enantiomorphs* – see fig. 13.1 and Ch. 17, pp. 292–3. (On pp. 201 ff. we considered the subgroup containing the direct symmetries, whose structure is D_3.)

Q. 16.15. Of which other solid is $D_3 \times C_2$ the full symmetry group?

Period of elements in a direct product group

Now fortunately there is a very quick way of finding the period of an element in a direct product group. Take an example first: suppose x is an element of period 4 in a group G ($x^4 = 1$), and y is an element of period 6 in a group G′ ($y^6 = 1'$).† We wish to find the period of the element (x, y) from the group G × G′.

Now, $\quad (x, y)^4 = (x^4, y^4) = (1, y^4); \qquad (x, y)^6 = (x^6, y^6) = (x^2, 1').$

† It is strictly necessary to use 1′ (not 1) for the identity element of G′, as there is no reason why it should be the same as 1 in the first group. Indeed, the groups do not have to have any elements in common. For example, if G is {0, 2, 4}, + mod. 6, and G′ is {1, −1}, ×, then we have the ordered pairs: (0, 1); (0, −1); (2, 1); (2, −1); (4, 1) and (4, −1) of which the first named is the neutral element. The rule for combination of these pairs is:

$(x_1, y_1)(x_2, y_2) = (x_1 + x_2 \text{ (mod. 6)}, y_1 y_2).$

But $(x, y)^{12} = (x^{12}, y^{12}) = (1, 1')$, the identity element. Hence (x, y) is of period 12. This suggests the following general result:

If an element x has period p in a group G, and an element y has period q in a group G′, then the element (x, y) has period k in G × G′ where k is the L.C.M. of p and q.

Q. 16.16. Prove this theorem.

Let us now apply the theorem to the identification of the group $D_3 \times C_2$. Along the top row of table 16.04 we show the periods of the

Table 16.04

L.C.M.		period of elements of D_3 (x)						
		1	3	3	2	2	2	
period of elements of C_2 (y)	1	1	3	3	2	2	2	period of (x, y) in $D_3 \times C_2$.
	2	2	6	6	2	2	2	

elements x of D_3, and down the side the periods of the elements y of C_2. The subsequent entries in the table show the periods of the corresponding elements in the direct product group, this being the L.C.M. of the appropriate two numbers in the border. We find there are two elements of period 6, two of period 3, seven of period 2, so that this confirms that $D_3 \times C_2 \cong D_6$.

Again, consider $D_3 \times C_4$. Table 16.05 shows the periods of the twenty-

Table 16.05

L.C.M.		period of elements x of D_3						
		1	3	3	2	2	2	
period of elements y of C_4	1	1	3	3	2	2	2	period of (x, y) in $D_3 \times C_4$.
	2	2	6	6	2	2	2	
	4	4	12	12	4	4	4	
	4	4	12	12	4	4	4	

four elements. The direct product group therefore has four elements of period 12, two of period 6, eight of period 4, two of period 3, and seven of period 2. Thus it is certainly not S_4, nor is it C_{24}. With four elements of period 12, one may easily suspect it to be D_{12}, the symmetry group of the regular dodecagon. This group certainly has four elements of period 12 (rotations through 30°, 150°, 210° and 330°). But you will realise that D_{12} must contain thirteen elements of period 2, these being the half-turns about the twelve axes of symmetry in the plane of the dodecagon, and the half-turn about the axis through its centre perpendicular to its plane. Therefore $D_3 \times C_4$ is certainly a different group from D_{12}.

Q. 16.17. Show by the same method that D_8 is different from $D_4 \times C_2$ but that $D_5 \times C_2$ is isomorphic to D_{10}. Also, study $C_3 \times C_5$; $C_4 \times C_4$ and show that $D_3 \times D_2$ is a different group from S_4.

Q. 16.18. We already know the groups C_{18} and D_9. Find two new direct product groups of order 18, one Abelian, the other non-Abelian.

Some groups of order 24

With this technique of forming direct product groups, we find ourselves with greatly increased powers of discovering new groups. It is already possible to find the following groups of order 24:

C_{24}; $C_{12} \times C_2$; $D_2 \times C_6$ (Abelian).
D_{12}; $D_2 \times D_3$; $D_4 \times C_3$; $D_3 \times C_4$; $Q_4 \times C_3$;
$Q_6 \times C_2$; Q_{12}; S_4; and $A_4 \times C_2$ (non Abelian) (see pp. 294, 316 ff. for A_4).

Q. 16.19. Obtain the number of elements of periods 2, 3, 4, 6, 8, 12, 24 in each of these groups using the tabulation method of the previous paragraph, and show thereby that they are all essentially different groups.

Q. 16.20. Show that $C_4 \times C_6 \cong C_2 \times C_{12} \cong C_2 \times C_3 \times C_4$; and that $D_2 \times D_3 \cong C_2 \times C_2 \times D_3 \cong C_2 \times D_6$.

In point of fact, there are fifteen essentially different groups of order 24, and already we have succeeded in uncovering twelve of them – eight by this powerful technique. Note, however, that it would be quite wrong to assume that most groups are direct products of easier groups; for example, the group discussed on pp. 100 ff., is not a direct product group. The formation of direct products is not the only way of achieving bigger groups – there are 'extensions' and 'split extensions' of groups; while H. Coxeter and W. Moser (*Generators and Relations for Discrete Groups*, Springer Verlag, Berlin) describe the method of 'adjoining' a new element to a given group to obtain a larger group; for example, the group D_n can be built from C_n (defined $r^n = 1$), by adjoining an element a of period 2 such that a transforms each element of C_n into its inverse: $ara^{-1} = r^{-1}$ (see Ch. 20). The three remaining groups of order 24 may not be classified using only our present resources.

The packing-case group

Suppose we have a heavy packing-case in the form of a cube, and we move it about by rolling it through a quarter-turn about one of the edges in contact with the ground. Let n be the operation of rolling it about its northernmost edge, so that it afterwards rests upon the square that was facing north; e is the operation of rolling it about the eastern edge of the face in contact with the ground, and s and w are similarly defined, though superfluous, since $s = n^{-1}$ and $w = e^{-1}$. When the packing-case is again in its original orientation, even though not in its original position, we shall regard this as being equivalent to the identity, i.e. we may say that $e^4 = 1$ and $n^4 = 1$. The diagram (fig. 16.03) shows the positions of the packing-

		ne				
e^3	1	e	e^2	e^3	1	e
n^3e^3	n^3	n^3e	n^3e^2	n^3e^3	n^3	n^3e
n^2e^3	n^2	n^2e	n^2e^2	n^2e^3	n^2	n^2e
ne^3	n	$ne = en$	ne^2	ne^3	n	ne
e^3	1	e	e^2	e^3	1	e

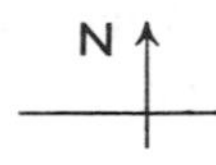

Fig. 16.03

case after carrying out all possible sequences of operations; e.g. n^2e $(= en^2)$ takes the packing-case to the square marked thus, making a sort of knight's move.

It is immediately evident that we have group structure here, a group of order 16. The group is manifestly Abelian, and has the generators n and e, with defining relations $n^4 = e^4 = 1$, $en = ne$. It should be apparent to you that the group is in fact the direct product of the group $\{1, e, e^2, e^3\}$ with the group $\{1, n, n^2, n^3\}$, and so has the structure $C_4 \times C_4$. We may check the truth of this by the periods of its elements (table 16.06). There are, in

Table 16.06

	L.C.M.	C_4: 1	2	4	4
C_4	1	1	2	4	4
	2	2	2	4	4
	4	4	4	4	4
	4	4	4	4	4

$C_4 \times C_4$, twelve elements of period 4, three of period 2. In the packing-case group, we have e^2, n^2 and e^2n^2 with period 2, and the remaining twelve with period 4. Note that n^2e^2 ($= e^2n^2$) is the only operation other than the identity by which the packing-case is brought again into an *upright* position.

Warning

Since the group D_3 has the subgroup C_3 $\{1, p, p^2\}$ (following the notation of the table on p. 199), while the other elements may be expressed $\{a, ap, ap^2\}$, where a satisfies $a^2 = 1$, and generates the group $\{1, a\}$ (C_2), it is tempting to think that D_3 must be a direct product of C_3 with C_2. Of course this is not so, since we know that $C_3 \times C_2 \cong C_6$. But what is the fallacy? The answer is that the direct product group of $\{1, p, p^2\}$ with $\{1, a\}$ is NOT the set of products of elements of the first group with those of the second:

×	1	p	p^2
1	1	p	p^2
a	a	ap	ap^2

What we do is to form *ordered pairs*. Refer back to the definition so that you have this distinction quite clear. In fact, in constructing the direct product group, we are really forming the Cartesian product of the two sets, hence the term direct 'product'. To explain briefly the general meaning of the term Cartesian product, suppose there is a set of n boys and a set of m girls; then the Cartesian product of these two sets is the set of all possible boy-and-girl couples, and there are mn of them. The term 'direct *sum*' is often used in the case of Abelian groups when the operation is described as '+'. See for example, W. Lederman, *Theory of Finite Groups*, p. 139 (O. and B.); Papys, *Groupes*, p. 26 (O.U.P.). 'Direct sum' is also used with particular reference to vector spaces and to rings and fields: see *A Survey of Modern Algebra*, G. Birkhoff and S. Maclane, pp. 172, 304, 346 (Collier Macmillan).

Groups of order eight

In preceding chapters, we have discovered C_8, D_4 and Q_4. In the present chapter, the construction of two new groups of order 8 has been suggested in the Questions. These are $C_4 \times C_2$, and $D_2 \times C_2$, or $C_2 \times C_2 \times C_2$. These are both Abelian groups, so that there are three Abelian groups of order 8; these with the two non-Abelian groups D_4 and Q_4 exhaust all the possible groups of order 8.† The truth of the above statement may be easily verified in the following way. If a group of order 8 has an element

† For a proof that there are only five groups of order 8, see Lederman, *Theory of Finite Groups* (Oliver and Boyd), pp. 48–55; for a proof that there are fifteen groups of order 24, see Burnside, *Theory of Groups of Finite Order* (Dover), pp. 157–161.

of period 8, then it is C_8. If not, then, by Lagrange's theorem, it may have elements of period 2 and 4 only (these elements generate subgroups). But the number of elements of period 4 must be *even* (they occur in inverse pairs – see Ch. 10, p. 107). So the only possibilities are:

Table 16.07

period	No. of elements			
2	1	3	5	7
4	6	4	2	0
	Q_4	$C_4 \times C_2$	D_4	$C_2 \times D_2$

This shows briefly that there are only five possible groups of order 8, but it is based on the (unproved) theorem stated on p. 143.

Q. 16.21. Find generators and defining relations for $C_4 \times C_2$; $C_3 \times C_3$; $C_6 \times C_2$; $D_3 \times C_3$. Draw Cayley diagrams based on your selection of generators.

The group $C_2 \times C_2 \times C_2$

Now $C_2 \times C_2$ contains elements of period 1, 2, 2, 2 $\{1, a, b, ab\}$ and C_2 contains elements of period 1, 2 $\{1, c\}$. Table 16.08 shows that

Table 16.08

L.C.M.	1	2	2	2
1	1	2	2	2
2	2	2	2	2

$C_2 \times C_2 \times C_2$ contains the identity and *seven elements all of period 2*. We have the ordered pairs $(1, 1)$, $(a, 1)$, $(b, 1)$, $(ab, 1)$, $(1, c)$, (a, c), (b, c), (ab, c) and we name these as follows:

$$\left.\begin{aligned} e &= (1, 1) \\ a &= (a, 1) \\ b &= (b, 1) \\ c &= (1, c) \\ d &= (b, c) \\ f &= (a, c) \\ g &= (ab, 1) \\ h &= (ab, c) \end{aligned}\right\}$$

and we note that $g = ab = ba$; $f = ac = ca$; $d = bc = cb$; $h = abc$.

In seeking defining relations, we bear in mind that a and b are required to generate the subgroup $C_2 \times C_2$, the elements of which are e, a, b and g.

Therefore c cannot be expressed in terms of a and b, and so the group *requires 3 generators*, say a, b and c. Suitable defining relations would be:

$$a^2 = b^2 = c^2 = e; \qquad ab = ba,\ bc = cb,\ ac = ca.$$

Q. 16.22. Show that $ac = ca$ may not be deduced from the first five of these.

We *need* all three defining relations $ab = ba$, $bc = cb$ and $ac = ca$ in addition to the statements that a, b and c is each of period 2. For if we merely had $ab = ba$ and $bc = cb$ (which in fact implies that b belongs to the *centre* of the group), then the group could be dihedral D_4, and this also has two subgroups D_2.

Q. 16.23. In this case, give an interpretation in terms of the symmetries of the square —b must represent . . .; a and c may represent . . .

So this rather innocent little group has the distinction of requiring three generators, and no fewer than six defining relations! Here is its table:

Table 16.09

	e	a	b	c	bc	ca	ab	abc	period
e	e	a	b	c	d	f	g	h	1
a	a	e	g	f	h	c	b	d	2
b	b	g	e	d	c	h	a	f	2
c	c	f	d	e	b	a	h	g	2
bc	d	h	c	b	e	g	f	a	2
ca	f	c	h	a	g	e	d	b	2
ab	g	b	a	h	f	d	e	c	2
abc	h	d	f	g	a	b	c	e	2

Now since the group is the direct product of three groups, $C_2 \times C_2 \times C_2$, we could just as well obtain it as a set of *ordered triples*, each component taken from a group C_2, e.g. $\{+1, -1\}, \times$. In this case, the elements would be:

$$(1, 1, 1); \quad (-1, 1, 1); \quad (1, -1, 1); \quad (1, 1, -1); \quad (1, -1, -1);$$
$$(-1, 1, -1); \quad (-1, -1, 1) \text{ and } (-1, -1, -1).$$

The elements might also be represented alternatively by 3×3 matrices, such as

$$(1, -1, -1) \rightarrow \begin{bmatrix} 1 & 0 & 0 \\ 0 & -1 & 0 \\ 0 & 0 & -1 \end{bmatrix}.$$

The importance of the matrix representation is that each of these matrices may be interpreted as a transformation matrix, and these transformations (of three-dimensional space) link up with our realisation of the group in terms of the symmetries of the cuboid (see p. 270). For example, the matrix

$$\begin{bmatrix} 1 & 0 & 0 \\ 0 & -1 & 0 \\ 0 & 0 & -1 \end{bmatrix}$$

represents a reflection in the plane $x = 0$, for

$$\begin{bmatrix} x' \\ y' \\ z' \end{bmatrix} = \begin{bmatrix} 1 & 0 & 0 \\ 0 & -1 & 0 \\ 0 & 0 & -1 \end{bmatrix} \begin{bmatrix} x \\ y \\ z \end{bmatrix} \Rightarrow \left.\begin{aligned} x' &= x \\ y' &= -y \\ z' &= -z \end{aligned}\right\},$$

(see also pp. 277–8, the illustration of the three perpendicular mirrors).

However, it will be simpler to use the elements from the set $\{0, 1\}$ under addition modulo 2, and our ordered triples are now:

$e = (0, 0, 0)$; $a = (1, 0, 0)$; $b = (0, 1, 0)$; $c = (0, 0, 1)$;

$d = (0, 1, 1)$; $f = (1, 0, 1)$; $g = (1, 1, 0)$; $h = (1, 1, 1)$.

(In adding these triples modulo 2, it is exactly as if we were adding the numbers 0, 1, 2, 3, 4, 5, 6, 7 expressed as three-digit *binary* numerals, but ignoring 'carrying'.)

This representation of the group suggests the connection with the three coins game mentioned on pp. 24, 137. Three coins are placed on a table, and we have eight operations, symbolised by the ordered triples above: a '0' means 'leave the coin in that place alone'; a '1' means 'turn it over'. Thus $f = (1, 0, 1)$ means 'turn the coins in the two outside positions over and leave the middle one alone'. You will see that this game gives rise to the group $C_2 \times C_2 \times C_2$ by virtue of the way in which the possible moves combine.

Q. 16.24. Extend the game to four pennies. What group appears now?

You may also be reminded of the operations on the trousers! (see pp. 89, 131). Each digit of the ordered triple might be interpreted:

First digit: 1 means 'turn the trousers inside out'; 0 means 'leave alone'.

Second digit: 1 means 'turn the trousers back to front'; 0 means 'leave alone'.

Third digit: 1 means 'select the wrong pair of trousers'; 0 means 'select the right pair'.

Thus, (1, 0, 1) means 'put Peter's trousers on John inside-out',
(1, 1, 0) means 'put John's own trousers on back to front and inside-out'.

If these two operations were carried out successively, John would still have the wrong trousers, and they would be back-to-front, but the effect of the two turning-inside-out operations would be that they would now be right-side-out.

Q. 16.25. If both twins are being dressed we shall now get five ways of going wrong: either or both of John's and Peter's trousers may be inside-out or back-to-front. Moreover, the trousers may be interchanged. How many elements are there in the group, and what structure has it?

Q. 16.26. Show that the quadratic polynomials of the form $ax^2 + bx + c$ with $a, b, c \in \{0, 1\}$ and x an intermediate, when added modulo 2 give the group $C_2 \times C_2 \times C_2$.

Next we look at the group $C_2 \times C_2 \times C_2$ as the symmetry group of a rectangular box (see fig. 16.04). We have already noted that the group

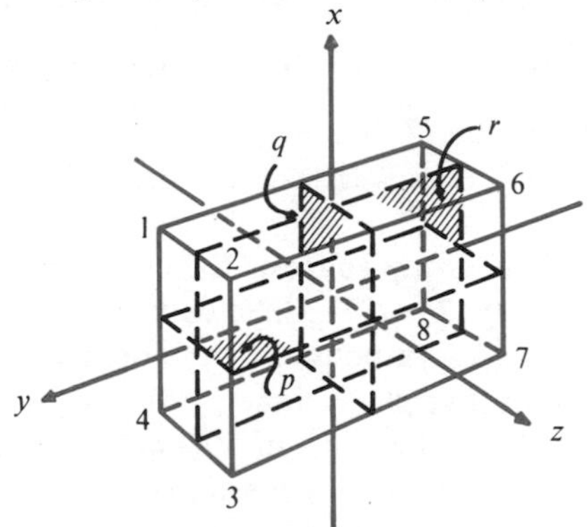

Fig. 16.04. Full group of symmetries of Cuboid – three half-turns, three reflections and central inversion; the group $C_2 \times C_2 \times C_2$.

$C_2 \times C_2$ ($\cong D_2$) describes the *direct* symmetries of the cuboid, but we now include the enantiomorphs as well.

There are reflections in three planes through the centre of the box parallel to the faces. These are shown in fig. 16.04, the operations of reflecting in these three planes are called p, q and r; for example, p is the reflection in the plane which interchanges the faces 1265 and 4378. The half-turns about the three axes are denoted x, y and z, the axis for the x rotation being perpendicular to the plane of the p reflection, and so on.

As we saw on p. 131, we get the group $\{e, x, y, z\} \cong D_2$ as the group of the direct symmetries, and this will, of course, be a subgroup of the full group of symmetries. Altogether, we appear to have only seven symmetries, and we have not yet checked on the *closure* requirement. It will be best to do this by considering *permutations* of the vertices. For the half-turn y performs the transpositions (1 3)(2 4)(5 7)(6 8), and corresponds to the permutation $\begin{pmatrix}1 & 2 & 3 & 4 & 5 & 6 & 7 & 8\\ 3 & 4 & 1 & 2 & 7 & 8 & 5 & 6\end{pmatrix}$. The reflection r interchanges the pairs (1 2)(3 4)(5 6)(7 8), and corresponds to the permutation $\begin{pmatrix}1 & 2 & 3 & 4 & 5 & 6 & 7 & 8\\ 2 & 1 & 4 & 3 & 6 & 5 & 8 & 7\end{pmatrix}$. Consider yq. We see that yq is the permutation $\begin{pmatrix}1 & 2 & 3 & 4 & 5 & 6 & 7 & 8\\ 7 & 8 & 5 & 6 & 3 & 4 & 1 & 2\end{pmatrix}$. This is a new transformation, not in our set of seven

e	1	2	3	4	5	6	7	8
q	5	6	7	8	1	2	3	4
y	3	4	1	2	7	8	5	6

listed so far. You will note that it *interchanges the opposite vertices* (1 7)(2 8) (3 5)(4 6) of the box; in other words, each vertex is reflected through the centre O of the box by an operation known as a 'central inversion'. We shall call this operation i, and we have $i = yq$ (check also that $i = qy$, and in fact that all the operations commute). Our group contains the eight permutations shown in table 16.10, which should be compared with that on

Table 16.10

e	1 2 3 4	5 6 7 8
r	2 1 4 3	6 5 8 7
y	3 4 1 2	7 8 5 6
p	4 3 2 1	8 7 6 5
q	5 6 7 8	1 2 3 4
x	6 5 8 7	2 1 4 3
i	7 8 5 6	3 4 1 2
z	8 7 6 5	4 3 2 1

and the group table is

	e	r	y	p	q	x	i	z	period
e	e	r	y	p	q	x	i	z	1
r	r	e	p	y	x	q	z	i	2
y	y	p	e	r	i	z	q	x	2
p	p	y	r	e	z	i	x	q	2
q	q	x	i	z	e	r	y	p	2
x	x	q	z	i	r	e	p	y	2
i	i	z	q	x	y	p	e	r	2
z	z	i	x	q	p	y	r	e	2

p. 268 with which it is isomorphic. The isomorphism is not immediately obvious because, in the first version the elements were arranged in the order 1, a, b, c, d, f, g, h where a, b, c were generators, whereas in the second version, we had the subgroup D_2 $\{e, x, y, z\}$ in the top left-hand corner, with $z = xy$. In this latter table, p, q and r would serve as a set of generators. The reason that the permutations have been written in this peculiar order is that they correspond exactly to the permutations of the rows in the accompanying group table, for you will observe that the two squares have exactly the same pattern. Cayley's theorem thus stands out plainly to be seen. Note that in the version above, each of the four 4×4 'blocks' (cosets) has the same internal pattern as D_2.

Q. 16.27. Establish an isomorphism between the two tables, starting with $a \leftrightarrow x$, and $b \leftrightarrow y$.

Automorphisms of $C_2 \times C_2 \times C_2$

With this group having seven elements all of period 2, it is evident th there is going to be a large group of automorphisms. But, though we may reshuffle those seven elements in 7! ways, there will certainly not be as many automorphisms as all that, and you will have found, in attempting to solve the above problem, that one's 'freedom of movement' is surprisingly restricted.

In seeking automorphisms, we are certainly free to choose to replace any *pair* of elements by any two others, as generators. This may be done in $7 \times 6 = 42$ ways. A third generator may then be chosen in four ways, since it must be independent of the first two, and this rules out one element, namely their product. Thus, there are $7 \times 6 \times 4 = 168$ automorphisms of $C_2 \times C_2 \times C_2$, and they form a group of order 168. This is, in fact, a particularly interesting group, having a peculiar rare

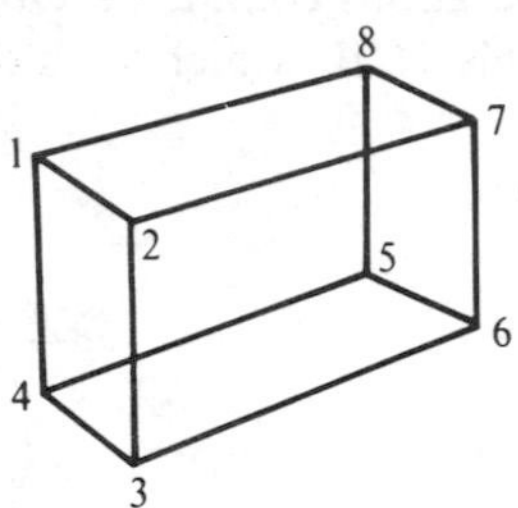

Fig. 16.05. The cuboid with renumbered vertices.

property of being what is called 'simple'. This property is shared by the group A_5 of order 60, and by no other groups of order less than 168. It means that the group has no normal subgroups (see Chs. 19, 20, 21).

In spite of the fact that all seven of the elements of period 2 are of 'democratically equal status' in the group – and we shall find a striking illustration of this presently – it would appear from the realisation as the group of symmetries of the cuboid that this is not so. For p, q and r appear to be in a class of their own (reflections), x, y and z are half-turns, whereas i is an odd man out being, apparently, a different sort of transformation altogether. To see that this is an illusion, consider the group as that of the permutations of the vertices which are listed on p. 271. Now renumber the cuboid as in fig. 16.05. You will see that the permutations now represent altogether different kinds of symmetries. For instance,

$i = \begin{pmatrix} 1 & 2 & 3 & 4 & 5 & 6 & 7 & 8 \\ 7 & 8 & 5 & 6 & 3 & 4 & 1 & 2 \end{pmatrix}$ effects the four interchanges (17)(28)(35)(46), and, for the renumbered cuboid, this is a half-turn about the vertical axis; while the permutation which we call $x = \begin{pmatrix} 1 & 2 & 3 & 4 & 5 & 6 & 7 & 8 \\ 6 & 5 & 8 & 7 & 2 & 1 & 4 & 3 \end{pmatrix}$, and which previously represented a half-turn, now represents the central inversion which interchanges opposite vertices. Again, you may check that z, which previously was a half-turn, now represents one of the three reflections. Therefore, it is an illusion, dependent on the particular method of numbering the vertices, that the elements of the group seem to differ from each other in character. Indeed, the group of automorphisms, mentioned above describes the ways of renumbering the vertices of the cuboid.

Illustration of the subgroups of $C_2 \times C_2 \times C_2$

Consider now the subgroups of $C_2 \times C_2 \times C_2$ other than the seven subgroups of order 2. We find the group $C_2 \times C_2$ ($\cong D_2$) also occurs seven times as a subgroup. Taking the group as $\{e, x, y, x, i, p, q, r\}$ we

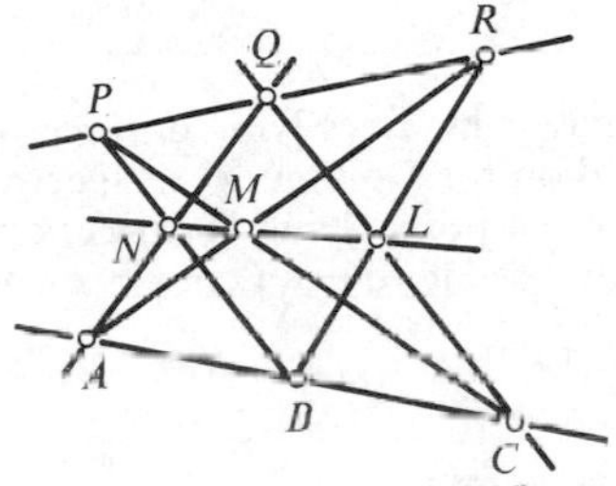

Fig. 16.06. Pappus' theorem.

have the following subgroups: $\{e, x, q, r\}$, $\{e, y, r, p\}$, $\{e, z, p, q\}$, $\{e, x, y, z\}$, $\{e, i, x, p\}$, $\{e, i, y, q\}$, $\{e, i, z, r\}$. Let us now omit 'e' and write down the triads of elements which occur together in these subgroups:

$$\left\{\begin{array}{ccc} x & q & r \\ y & r & p \\ z & p & q \\ x & y & z \\ i & x & p \\ i & y & q \\ i & z & r. \end{array}\right.$$

These have the remarkable property that each contains three of the seven letters, and each letter occurs in exactly three of the seven 'lines'. Now if you have done a little projective geometry you may be familiar with the configuration of fig. 16.06, in which ABC, PQR are two sets of collinear points, and L, M, N are the intersections $L(BR \cap CQ)$, $M(CP \cap AR)$, $N(AQ \cap BP)$. Then LMN is a straight line (Pappus' theorem). The configuration has the interesting property that there are nine points lying by threes on nine lines. It *appears* as if there were something special about the two given sets $\{ABC\}$ and $\{PQR\}$ but in fact it is possible to choose any pair of lines (such as *PMC* and *RLB*) and to derive the remaining three collinear points from them. (This question of the relettering of geometrical configurations is dealt with at more length in Ch. 25.) We have here, in fact, a completely egalitarian system – nine points on nine lines, all equal in status; each point has three lines through it, each line contains three points. Think, if you like, of nine people who serve on nine committees, each of three persons, arranged in such a way that each person serves on exactly three committees, while any pair of committees has exactly one member in common. Or, reverting to the geometrical aspect, think of the Sunday newspaper type puzzle about a farmer who has to plant nine trees in such a way that there are nine rows of three trees each, each tree being in three rows. The configuration of Pappus' theorem provides the solution to these problems.

A	*B*	*C*
A	*M*	*R*
A	*N*	*Q*
B	*N*	*P*
B	*L*	*R*
C	*M*	*P*
C	*L*	*M*
P	*Q*	*R*
L	*M*	*N*

Q. 16.28. Is it possible to have nine trees lying in threes on ten rows? Next consider the configuration for Desargues' perspective triangle theorem (see fig. 25.20), which consists of ten points lying by threes on ten lines, with each point lying on three lines, and see if you can convince yourself that all the points are of equal status.

An excursion into finite geometry

Now it is surprisingly difficult to succeed in arranging n points in a plane in such a way that they lie m on each line with m lines passing through each point, and with n lines altogether. We have seen how it can be done for $n = 9$, $m = 3$ (Pappus), and also for $n = 10$, $m = 3$ (Desargues) fig. 25.20. Reverting to the consideration of the group $C_2 \times C_2 \times C_2$, we appear to have succeeded also in the case $n = 7$, $m = 3$, for our seven members (points) x, y, z, i, p, q, r, are arranged in seven committees (lines), each committee containing three members, † and each member belonging to three committees. However, if we try to *draw* this using straight lines with three points on to represent committees with three members, we run into trouble; for however you may try to draw the figure, you will find that three points will refuse to lie on a line (see fig. 16.07 where x, y and z provide the exception). What we do is to pretend that a line may be so drawn, and this is shown dotted through x, y and z in fig. 16.07. The *pretence* is only necessary when we are working in the framework of ordinary Euclidean geometry, but the device of *inventing* a

† Note that collinearity of points corresponds to the relationship in the group that one of the three elements is the product of the other two; e.g. *pyr* is a straight line $\Leftrightarrow p = yr$, $y = pr$, $r = py$, or $pyr = e$!

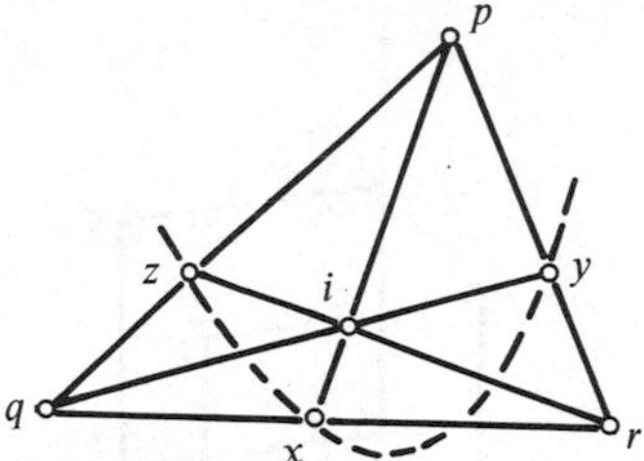

Fig. 16.07. Seven-point finite geometry.

line xyz is perfectly acceptable so long as we regard ourselves as being freed from the restrictions of Euclidean space.

We have here what is known as a finite geometry of seven points. There is another finite geometry which contains thirteen points and thirteen lines each point lying on four lines and each line containing four points. There is another with $n = 31$, $m = 6$. But such systems are by no means common.

The group $C_2 \times C_2 \times C_2$ has revealed one such system by means of its subgroups $C_2 \times C_2$. Is it feasible that the group $C_2 \times C_2 \times C_2 \times C_2$ of order 16, which has fifteen elements of period 2, might lead to another finite geometry? The question, and others, we shall consider later.

You will recall that the elements of our $C_2 \times C_2 \times C_2$ group could be represented by ordered triples, each component of which was either 0 or 1, and the operation of combining these triples was performed by adding corresponding elements modulo 2. Now ordered triples suggest a coordinate system. Consider the cube whose vertices referred to three mutually perpendicular axes are

$e\,(0, 0, 0)$; $a\,(1, 0, 0)$; $b\,(0, 1, 0)$; $c\,(0, 0, 1)$
$d\,(0, 1, 1)$; $f\,(1, 0, 1)$; $g\,(1, 1, 0)$; $h\,(1, 1, 1)$.

Then we may regard the edges and vertices of this cube as a Cayley diagram for $C_2 \times C_2 \times C_2$, with those edges parallel to the x-axis representing the generator a, and those parallel to the y and z axes representing b and c respectively (see fig. 16.08).

★**Q. 16.29.** Draw a Cayley diagram for $C_2 \times C_2 \times C_2$ in a plane, with no edges overlapping.

★ But this is not all. Suppose we take abc as a triangle of reference and use *areal coordinates*. (Those of you unfamiliar with areal coordinates may prefer to omit this paragraph, but we give a very short account referring to fig. 16.09 below.) Briefly, a point p in the plane of triangle

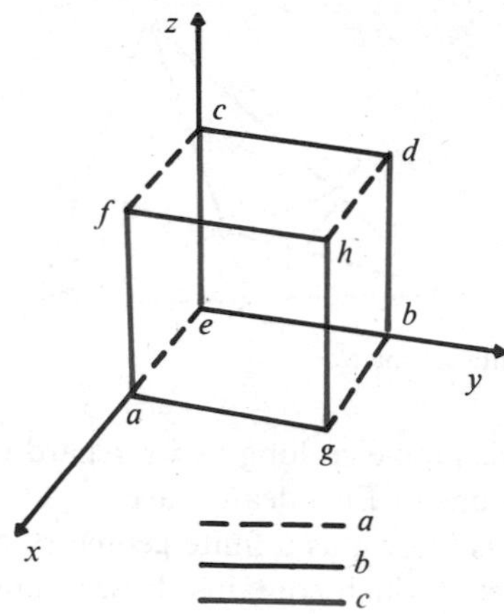

Fig. 16.08. The cube as a Cayley diagram,

abc may be specified by a triple of coordinates (x, y, z) such that these coordinates are proportional to the *areas* of triangles, *bcp*, *cap*, *abp*. When *p* is outside the triangle, one or more of these areas carries a negative sign. If *p* were inside the triangle, and *bcp* were $\frac{4}{9}$ of area *abc*, and *cap* were $\frac{1}{3}$ area *abc*, then we should say that *p* had areal coordinates (4, 3, 2). There is no such point as (0, 0, 0), while the points *a*, *b* and *c*

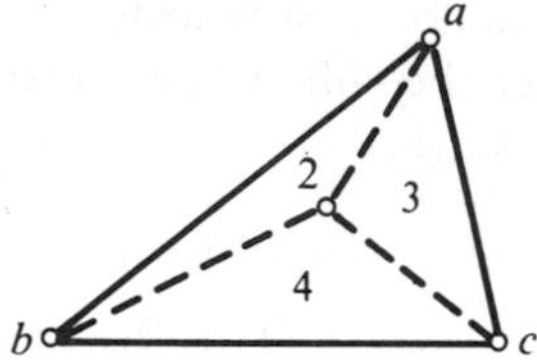

Fig. 16.09. Areal coordinates.

have coordinates exactly as specified above. Marking in fig. 16.10 the remaining points, we find that *d*, with coordinates (0, 1, 1) is the mid-point of *bc*, while *f* and *g* are the midpoints of *ca* and *ab*. The point *h* (1, 1, 1) is the intersection of the medians *ad*, *bf*, *cg*. What we have here is nothing more or less than our seven-point geometry, but with the letters replaced as in the answer to Q. 16.27.

But again, the three points *d f g* do not appear to lie on a line. However, you should bear in mind that we are working with coordinates in the *finite arithmetic modulo 2*, so that we regard all even numbers as being equivalent to 0, and all odd numbers as equivalent to 1. Hence the point *d* (0, 1, 1) is equivalent to the point (2, 1, 1), and the point whose areal coordinates are (2, 1, 1) is in fact half way along the median *ad* – the shaded areas in the figure are equal. Hence we may regard the point *d* as being either at the mid-point of *bc*, *or* equally well, at the mid-point of *ad*, and this equivalent position does indeed lie on the line *gf*.

This explanation of a finite geometry – in terms of a finite arithmetic – may help to convince you that it is not just idle chatter to talk about putting a 'straight line' through the points *d*, *f* and *g* in fig. 16.10, or through *x*, *y* and *z* in fig. 16.07.

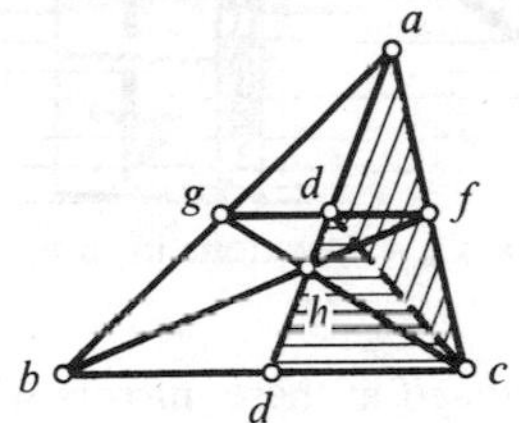

Fig. 16.10. Seven-point geometry. The points *d* are equivalent with areal coordinates modulo 2.

In the latter version of the figure, it would appear that *i* is a 'special' point, because it is 'on its own' inside the triangle. But in fact this is just as much an illusion as that *i* (see p. 271) was a special kind of transformation in a class of its own, and we dispelled this effectively on pp. 272–3.

Further realisations of $C_2 \times C_2 \times C_2$

The group $C_2 \times C_2 \times C_2$ is so interesting that it really deserves a chapter on its own! The symmetry group of the cuboid suggests a further context in which the group may appear. We saw on p. 135 that the group $C_2 \times C_2$ ($\simeq D_2$) arises by the reflections in two perpendicular plane mirrors, forming four images. (When the mirrors were inclined at 60°, we should get the group D_3.) Suppose now we add another mirror, perpendicular to each of the others. Imagine, for example, that you are in a room in which two adjacent perpendicular walls are covered by mirrors, and *also the floor* were a mirror. Then you would not only see the images *a*, *b* and *ab* in the wall mirrors (as in fig. 11.06), but also there would be images *c*, *ac*, *bc* and *abc* due to reflections from the floor. Figure 16.11 is an attempt to show the images of a cubical block with numbers on its faces produced by these reflections.

Fig. 16.11. The group $C_2 \times C_2 \times C_2$ – reflections in three mutually perpendicular mirrors.

Q. 16.30. If you looked at *yourself* in these mirrors where would you see the images *ab*, *ac*, *bc* and *abc*?

Q. 16.31. Suppose the two wall mirrors were inclined at 60°, both being at right-angles to the floor mirror, show that we now get the group $C_2 \times D_3$. What connection is there between the set of images produced in these mirrors and the symmetries of certain prisms? What group would be obtained if the wall mirrors were inclined at 30°, 72°, 5°, parallel?

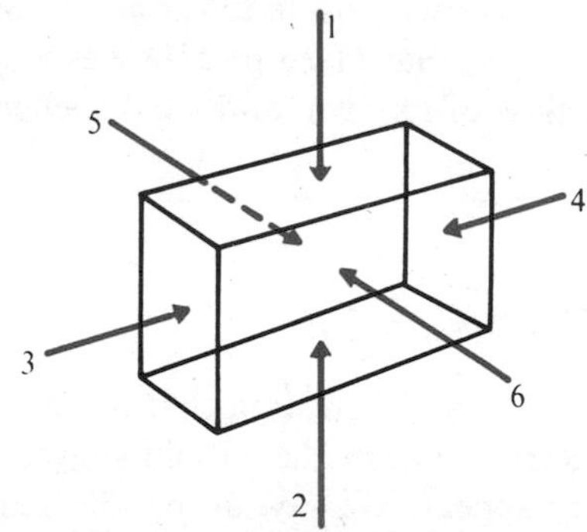

Fig. 16.12. The faces of the cuboid numbered.

For our three mutually perpendicular mirrors, you will observe that the eight images of the cubical block form a three-dimensional figure which has the same symmetry group as the cuboid.

When we dealt with the symmetries of the cuboid, we thought in terms of the permutations of the vertices. A simpler way is to name (or number) the *faces*, and to consider the permutations of the six faces. Suppose the two faces parallel to the plane of reflection *p* are numbered 1 and 2, those parallel to the plane of reflection *q* are numbered 3 and 4, and the other pair of parallel faces are 5 and 6 (see fig. 16.12). What is the effect of the reflection p? It is to interchange 1 and 2 while the other faces remain unchanged. Similarly, q and r correspond respectively to the transpositions (34) and (56). Again, the half-turn x leaves 1 and 2 unchanged, but interchanges the pairs of faces (34) and (56). Similarly with y and z. Finally, i interchanges every pair of opposite faces. Thus the eight symmetries of the cuboid correspond to the permutations:

e	1 2 3 4 5 6
x	1 2 4 3 6 5
y	2 1 3 4 6 5
z	2 1 4 3 5 6
i	2 1 4 3 6 5
p	2 1 3 4 5 6
q	1 2 4 3 5 6
r	1 2 3 4 6 5

What we have here is the product of the disjoint cycles (12), (34) and (56). The fact that the transpositions are disjoint guarantees commutativity, and it is immediately evident that $x = qr$, $p = yz$, $i = pqr$, etc.

This representation of $C_2 \times C_2 \times C_2$ as a group of permutations of six objects is evidently much simpler than the method using eight symbols, but the value to us of using the numbered vertices was chiefly as an illustration of Cayley's theorem.

The discussion of $C_2 \times C_2 \times C_2$ as a group of permutations of six figures serves to show that it is a subgroup of S_6. We know that $C_2 \times C_2 \times C_2$ is not a subgroup of S_4 (even though S_4 has seven elements of period 2). Is it possible that $C_2 \times C_2 \times C_2$ could be a subgroup of S_5? See if you can find three permutations of *A B C D E* each of period 2, which generate $C_2 \times C_2 \times C_2$.

The full group of the cube (including enantiomorphs) is of order 48. Evidently $C_2 \times C_2 \times C_2$ must be a subgroup of this group, since it is the symmetry group of any cuboid.

Q. 16.32. Is the full group of the cube $S_4 \times C_2$?

Q. 16.33. Write down eight 6×6 permutation matrices which correspond to the permutations generated by the cycles (12), (34) and (56). Check some of the products of these to see that they agree with the results obtained from the composition of the permutations. Repeat for the 8×8 matrices corresponding to the permutations on p. 271. Note that we now have sets of eight matrices forming the group $C_2 \times C_2 \times C_2$ in three sizes: (a) 3×3; (b) 6×6; and (c) 8×8. (For (a), see pp. 267–8.)

The group $C_2 \times C_2 \times C_2$ as a group of sets under symmetric difference

If we have three objects in our universal set, a, b, c, then there are the following possible subsets: $\emptyset$; $\{a\}$, $\{b\}$, $\{c\}$, $\{b, c\}$, $\{c, a\}$, $\{a, b\}$, $\{a, b, c\}$ $(= \mathscr{E})$. Remember the meaning of the operation Δ on sets (p. 9), and apply it to the pairs of these eight subsets, e.g.

$$\{c, a\} \,\Delta\, \{b, c\} = \{a, b\}; \qquad \{a\} \,\Delta\, \{b\} = \{a, b\}; \qquad \{b, c\} \,\Delta\, \{b\} = \{c\};$$
$$\{c, a\} \,\Delta\, \{c, a\} = \phi.$$

Q. 16.34. Verify closure, and that each element is its own inverse, and that we do get the group $C_2 \times C_2 \times C_2$. Show the isomorphism with the abstract version of the group given on p. 268. (Note the curious resemblance:
$\{a, c\} \Delta \{c, b\} = \{a, b\} \quad ac \,.\, cb = ac^2b = aeb = ab$.)

The subsets of the sets of three objects may be likened to the game with the three coins (p. 269): any particular combination of heads and tails may be reckoned to correspond to one of the above sets. For example, if we agreed that 'heads' means 'accept', and 'tails' means 'reject', then the state *t h h* of the coins would correspond to the exclusion of *a* and the selection of *b* and *c*.

The true nature of the group $C_2 \times C_2 \times C_2$ is revealed in several of the realisations we have studied; – it is the *group that is concerned with three independent interchanges* (the three pairs (AB), (CD), (EF); or the three coins; or the three reflections, etc.). Again, the essence of $C_2 \times C_2 \times C_2 \times C_2$ would be that it is concerned with four interchanges; and so on.

Q. 16.35. Show that a group in which every element is of period 2 must be: (a) **Abelian**; (b) **of order** 2^n ($n \in \mathbf{Z}^+$); (c) **isomorphic with** $C_2 \times C_2 \times C_2 \times \ldots \times C_2$ (**n factors**).

Our final illustration of $C_2 \times C_2 \times C_2$ is – perhaps not unexpectedly, since it is an Abelian group – from finite arithmetics. Under addition modulo n, we always get the group C_n. But under multiplication modulo n, the set of residues which are prime to n form a group which is necessarily Abelian. It can, in fact, be proved that any Abelian group may be realised as a set of residues under multiplication to some modulus. In the case of $C_2 \times C_2 \times C_2$, the smallest modulus for which this happens is 24, and the set $\{1, 5, 7, 11, 13, 17, 19, 23\}$ has this structure: note that every element is of period 2, e.g. $19^2 = 361 = 1$ (mod. 24).

The group $C_4 \times C_2$

This group is not quite so interesting as $C_2 \times C_2 \times C_2$, and we shall not go into it in so much detail. The most obvious realisation would be as the direct product of $C_4 \{1, i, -1, -i\}, \times$ and $C_2 \{1, -1\}, \times$

or of $\qquad C_4 \{0, 1, 2, 3\}, +$ mod. 4, and $C_2 \{0, 1\}, +$ mod. 2.

Taking the latter, we have the ordered pairs: (0, 0), (1, 0), (2, 0), (3, 0), (0, 1), (1, 1), (2, 1), (3, 1). These may be labelled $e, r, r^2, r^3, a, ar, ar^2, ar^3$ respectively, and we have the group table 16.11. Note that each of the four 4×4 'blocks' (cosets) has the same internal pattern as C_4.

Table 16.11

	e	r	r^2	r^3	a	ar	ar^2	ar^3	period
e	e	r	s	t	a	b	c	d	1
r	r	s	t	e	b	c	d	a	4
r^2	s	t	e	r	c	d	a	b	2
r^3	t	e	r	s	d	a	b	c	4
a	a	b	c	d	e	r	s	t	2
ar	b	c	d	a	r	s	t	e	4
ar^2	c	d	a	b	s	t	e	r	2
ar^3	d	a	b	c	t	e	r	s	4

($C_4 \times C_2$)

Q. 16.36. How many C_4 subgroups are there? State the elements in each of these subgroups. Find a further subgroup of order 4. How many automorphisms can you find? Identify their group.

$C_4 \times C_2$ may also be realised as the product of the cyclic permutations of the disjoint cycles of $A\ B\ C\ D$ and $P\ Q$. You should establish the isomorphism between the eight permutations and the eight symbols e, r, s, t, a, b, c, d above.

For a geometrical figure which has $C_4 \times C_2$ as its symmetry group, we may combine rotational symmetry through quarter-turns with a reflection, as in the case of the right prism whose section is shown in fig. 16.13.

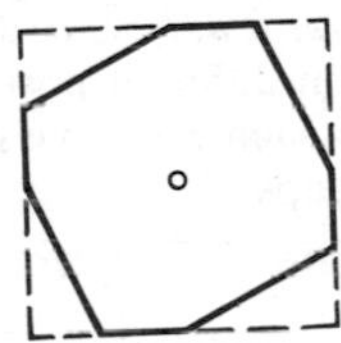

Fig. 16.13. Cross section of prism whose symmetry group is $C_4 \times C_2$.

(Compare with p. 160 and fig. 12.04 where we invented a prism having the symmetry group $C_3 \times C_2 \cong C_6$.)

With the notation of the above abstract group table, we might let r represent a quarter-turn through $\frac{1}{2}\pi$ about the central axis, and a represent a reflection in the plane of symmetry midway between the ends. These would serve as generators of the group.

Q. 16.37. Give a set of defining relations, and draw the Cayley diagram.

Q. 16.38. Why is $C_4 \times C_2$ not a subgroup of S_4, yet it is a subgroup of the full group of the cube, of order 48?

Q. 16.39. In the group $C_4 \times C_2$, one of the three elements of period 2 is 'different'. In what way?

In view of what was said in the previous section, we may expect to find $C_4 \times C_2$ as a group of residue classes under multiplication to some modulus. For example, you may verify that {1, 2, 4, 7, 8, 11, 13, 14}, $\times$ mod. 15 is one possibility.

Q. 16.40. A certain set of eight residues form the group $C_4 \times C_2$ under multiplication modulo 16. They contain 3 and 13. Find them, and verify the structure.

Q. 16.41. Find a set of eight numbers which form $C_4 \times C_2$ under $\times$ mod. 20. Find another set, under a higher modulus.

Q. 16.42. Give a set of eight 2×2 matrices which form the group $C_4 \times C_2$ under matrix addition modulo 4. Give also a set of eight 2×2 matrices which give this same group under matrix multiplication.

Finally, the alternate-corresponding-vertically opposite angles game (see p. 25) which led to the group $C_2 \times C_2$ may be extended to give the group $C_4 \times C_2$. First of all, we may get the group C_4 by using the operation 'D', which means 'move round anticlockwise to the adjacent angle'. Then evidently, $D^2 = V$ ('move to the vertically opposite angle'), and $D^4 = I$, and we have C_4 structure. If in fig. 4.02 we now draw in the parallel line, we obtain eight angles.

Q. 16.43. Verify such relations as $C = AV$, $D^2C = A$, etc. Obtain the period of each of the eight elements, and verify $C_4 \times C_2$ structure by showing the isomorphism with the group e, r, s, t, a, b, c, d, above. Draw the diagram where the 'starting angle' (I) is selected in different positions, and indicate the interpretation of 'allied' angles in each case. We started with the subgroup $\{I, D, D^2, D^3\}$. Find other subgroups.

Exercise on the group $C_3 \times C_3$

Q. 16.44. Consider the group $C_3 \times C_3$.

(a) As the product of disjoint cycles of (ABC) and (PQR).
(b) As the addition of ordered pairs (or of 2×2 matrices) mod. 3.
(c) As the multiplication of ordered pairs (or 2×2 matrices) (let $\omega = \cos \frac{2}{3}\pi$).
(d) As the addition mod. 3 of linear polynomials $ax + b$, where a and b are drawn from {0, 1, 2}, and x is an indeterminate.
(e) Find other instances of $C_3 \times C_3$ to be found in this book.
(f) Using the (unproved) theorem on p. 143, and Lagrange's (as yet unproved) theorem, show that there can be only two distinct groups of order 9.
(g) Can you find a set of residues under $\times$ mod. n which gives the group $C_3 \times C_3$? Also a set which gives C_9.
(h) Draw a Cayley diagram for $C_3 \times C_3$.

Q. 16.45. Consider the group $C_3 \times C_3 \times C_3$ and give a set of defining relations. Prove that this group has twenty-six elements of period 3. Describe another Abelian group of order 27 other than C_{27}.

Q. 16.46. Show that the set of ordered triples (x, y, z) where $x, y, z \in \{0, 1, 2\}$ form the group $C_3 \times C_3 \times C_3$ under addition mod. 3. If (0, 0, 0) is excluded, and the remaining twenty-six elements are considered to be equivalent when 1's and 2's are interchanged (e.g. (0, 2, 1) is the 'same' as (0, 1, 2); (2, 2, 1) is the same as (1, 1, 2), show that there are thirteen objects, and see if you can from these construct a finite geometry containing thirteen points.

Q. 16.47. Referring to p. 102, show that $\text{Gp}(a, b) \cong C_3 \times C_3$ is a subgroup of the non-Abelian group of order 27 discussed there.

Direct products of infinite groups

The general idea of a direct product embodied in the definition applies just as well to infinite groups. A simple example is the direct product of the group $(\mathbf{Z}, +)$ with itself. This gives us ordered pairs of integers which are added by the ordinary 'vector' rule: $(a, b) + (c, d) = (a + c, b + d)$ $(a, b, c, d \in \mathbf{Z})$. On the Argand diagram, these correspond to the Gaussian integers, $a + ib$ $(a, b \in \mathbf{Z}, i = \sqrt{-1})$ under addition (see Q. 14.31). Note that the direct product has *added another dimension* by providing us with two-dimensional entities, which we may regard as displacements, or vectors. Thus we have constructed a new infinite group which may be designated $C_\infty \times C_\infty$. It may also be regarded as a two-dimensional vector space.

Q. 16.48. What are generators of this group? Draw a Cayley diagram (the 'network of one-way streets').

Q. 16.49. Give a realisation of $C_\infty \times C_\infty \times C_\infty$.

Q. 16.50. Show that the set $\{a + b\sqrt{2}, a, b \in \mathbf{Z}\}$ under addition, is isomorphic to $C_\infty \times C_\infty$.

Q. 16.51. Show that $C_\infty \times C_\infty$ may be used to generate a plane pattern by taking the generators to be translations in two directions.

Consider now the group obtained as the direct product of $(\mathbf{R}, +)$ and $(\mathbf{R}, +)$. A typical element will be (x, y) $(x, y \in \mathbf{R})$, and we have the law of combination $(x_1, y_1) + (x_2, y_2) = (x_1 + x_2, y_1 + y_2)$, where we denote the law of combination here as '+'. These ordered pairs of reals may be taken to represent *vectors*, or displacements (or complex numbers) in two-dimensional space, the two original sets of reals being taken as measurements of length in two (perpendicular) directions. The reason we use the + sign is thus seen to be that we are really doing ordinary vector 'addition'. The group of two-dimensional vectors under addition is usually denoted $(\mathbf{V}_2, +)$. Structurally, it is isomorphic with the direct product of $(\mathbf{R}, +)$ with itself.

Q. 16.52. Does this group have generators? Interpret the group $(\mathbf{V}_3, +)$ as a direct product.

Q. 16.53. Show that ordered pairs of complex numbers under addition are isomorphic with the quaternions (see p. 246).

Note: infinite groups containing no elements of finite period, such as those discussed in the above paragraph are described as 'torsion-free'.

Direct product of infinite with finite group

It is also possible to have the direct product of an infinite group with a finite group. For example, the positive reals form a group under $\times$, and so do the set $\{-1, +1\}$. The direct product of these two groups would consist of ordered pairs such as $(x, -1)$; $(y, +1)$, etc. The laws of combination:

$$(x_1, y_1) \times (x_2, y_2) = (x_1x_2, y_1y_2) \text{ would give}$$

$$\left.\begin{array}{l}(x, 1) \times (y, 1) = (xy, 1)\\ (x, 1) \times (y, -1) = (xy, -1)\\ (x, -1) \times (y, -1) = (xy, +1)\\ (x, -1) \times (y, 1) = (xy, -1)\end{array}\right\}$$

and the direct product group would in fact be isomorphic with the whole set of reals positive and negative, under multiplication, the correspondence being: $(x, 1) \leftrightarrow +x$; $(x, -1) \leftrightarrow -x$ (the elements $(0, 1)$ and $(0, -1)$ being excluded).

Again, if we take the groups $(\mathbf{R}, +)$ and $\{0, 1, 2\}$, + mod. 3, we shall have ordered pairs of the form (x, y) where $x \in \mathbf{R}$ and $y = 0$, 1 or 2. The elements $(x, 0)$ $(x \in \mathbf{R})$ form the subgroup $(\mathbf{R}, +)$. The elements $(0, 0)$, $(0, 1)$ and $(0, 2)$ form the subgroup C_3. Referred to two perpendicular axes, all the elements of the form $(x, 0)$ lie on the x axis. The elements of the form $(x, 1)$ give us the whole of the horizontal line $y = 1$, and the elements $(x, 2)$ give the line $y = 2$. Any two elements on the latter line would combine to give an element on the line $y = 1$, e.g. $(-3{\cdot}2, 2)$ $(1{\cdot}4, 2)$ $= (-1{\cdot}8, 1)$ and so on (compare with Ch. 19, Cosets, pp. 340–1).

The group $C_\infty \times C_2$ is an important special case. This may be realised by taking C_∞ as the integers under addition, and C_2 as the group $\{0, 1\}$ + mod. 2. Or we may think of two generators, r (of infinite period), and a (of period 2), and the group will contain elements

$$\begin{array}{lllll}\ldots (r^{-2}, 1), & (r^{-1}, 1), & (1, 1), & (r, 1), & (r^2, 1) \ldots\\ \ldots (r^{-2}, a), & (r^{-1}, a), & (1, a), & (r, a), & (r^2, a) \ldots\end{array}$$

This can be seen more vividly on a Cayley diagram:

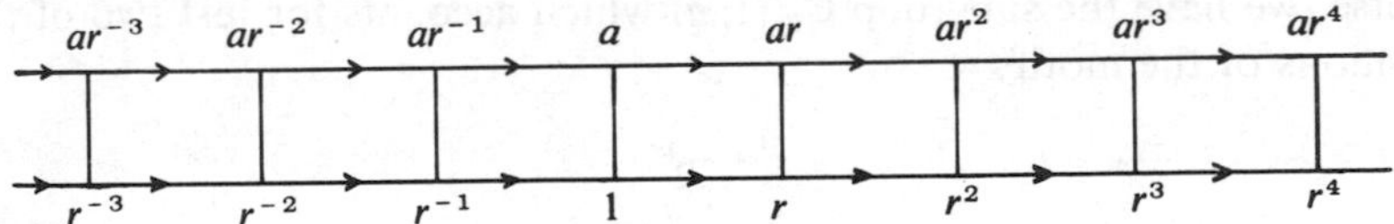

Fig. 16.14.

Here the arrowed link represents the operation r, the unarrowed links represent a.

Q. 16.54. What group does the following Cayley diagram represent?

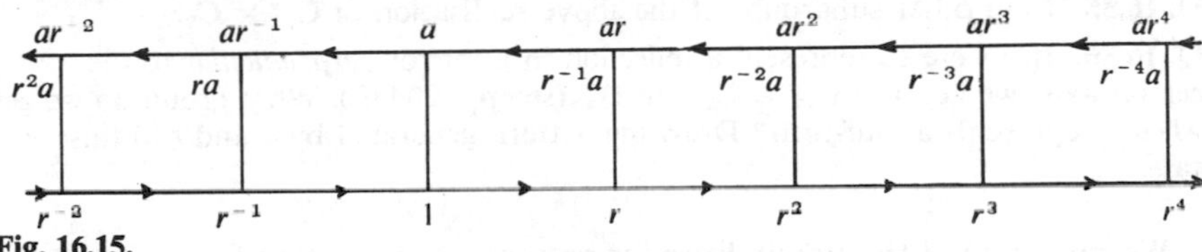

Fig. 16.15.

But a much more realistic representation of $C_\infty \times C_2$ may be seen by letting r denote a translation, and a a geometric transformation of period 2, such as a mirror image in a line, or a half-turn about a point:

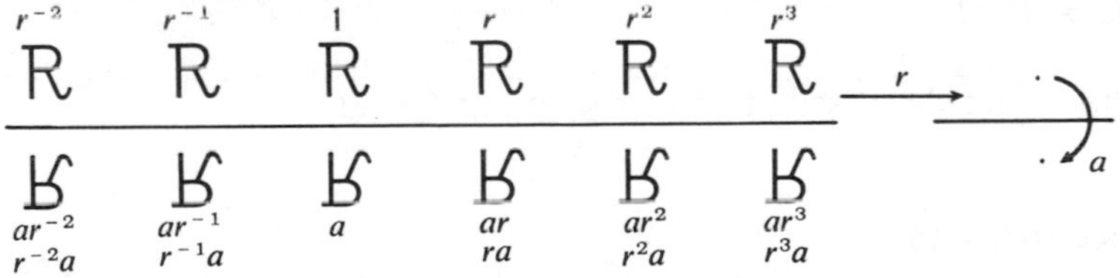

r^{-2} r^{-1} 1 r r^2 r^3

R R R R R R $\xrightarrow{r}$

a

ar^{-2} ar^{-1} a ar ar^2 ar^3

$r^{-2}a$ $r^{-1}a$ ra r^2a r^3a

Fig. 16.16.

Here a represents a reflection in the central axis of the strip pattern. Note that ar is a 'glide reflection', and if $ar = g$ $(= ra)$, then this element itself generates the group C_∞ $\{\ldots g^{-1}, 1, g, g^2, \ldots\}$, which is a subgroup of Gp (a, r). So we observe that both Gp (r) and Gp (g) are subgroups, each being isomorphic to C_∞. The latter accounts for the part of the pattern below:

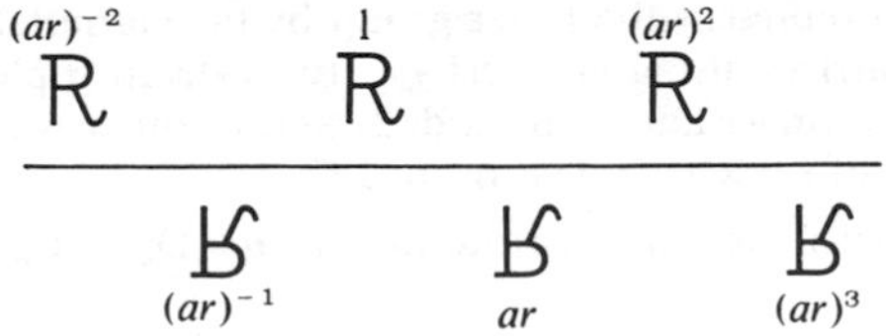

$(ar)^{-2}$ 1 $(ar)^2$

R R R

$(ar)^{-1}$ ar $(ar)^3$

Fig. 16.17.

The former corresponds to the upper half of the original pattern. And of course, we have the subgroup C_2 $\{1, a\}$ which accounts for just two of the positions of the motif:

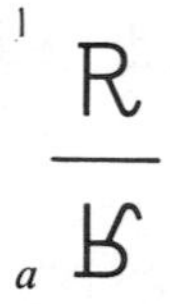

1 R

a

Fig. 16.18.

Q. 16.55. Find other subgroups of the above realisation of $C_\infty \times C_2$.

Q. 16.56. If a were to represent a reflection in a mirror *perpendicular* to the central axis, we get, not $C_\infty \times C_2$, but D_∞ (see pp. 204 ff.). What group do we get when a represents a half-turn? Draw the pattern generated by a and r in this case.

We may extend the possibilities for patterns by including further generators in the group. We have had examples of infinite patterns described by the groups C_∞, $C_\infty \times C_2$, $C_\infty \times C_\infty$, and D_∞.

Q. 16.57. Can you invent an infinite strip pattern whose group is $D_\infty \times C_2$ (or $D_\infty \times D_1$)?

Patterns are discussed at greater length in Ch. 26.

EXERCISES 16

1 Show that $C_2 \times C_3 \times C_4 \cong C_6 \times C_4 \cong C_2 \times C_{12}$.

2 Draw a Cayley diagram to illustrate that $C_3 \times C_2 \cong C_6$.

3 Give defining relations for $C_6 \times C_2$; $C_n \times C_2$.

4 What is the structure of the group suggested by Ex. 3.15.

5 Show that $C_3 \times C_3 \times C_6$ may be generated by the disjoint cycles (1 2 3), (4 5 6), (7 8 9) and (10 11).

6 Give defining relations for $C_3 \times C_3$, and draw a Cayley graph?

7 How many subgroups of structure $C_3 \times C_3$ does the group $C_3 \times C_3 \times C_3$ have? You should be able to represent the latter group by (a) permutations of nine letters (b) 3×3 matrices, using $\omega = \text{cis } \frac{2}{3}\pi$ (c) ordered triples (x, y, z), with $x, y, z \in \{0, 1, 2\}$ under addition mod. 3, and in other ways suggested by the realisations of $C_3 \times C_3$ given on p. 282.

8 How would you use three mirrors to demonstrate the groups $D_3 \times C_2$; $D_4 \times C_2$; $D_6 \times C_2$?

9 Show that a ray of light reflected from three mutually perpendicular plane mirrors emerges parallel to its original direction.

10 Show that the 'quadratic equation' $x^2 = 1$ has eight solutions in the group $C_2 \times C_2 \times C_2$.

11 Show that the set {1, 9, 16, 22, 29, 53, 74, 79, 81} under multiplication modulo 91 is the group $C_3 \times C_3$. (*Note*: it can be proved that any Abelian group may be realised as a set of residues under multiplication to some modulus. 91 is the smallest possible modulus which enables the group $C_3 \times C_3$ to be constructed in this way.)

12 In Ch. 9 (p. 102), we met a group defined $a^3 = b^3 = c^3 = 1$, $ab = ba$, $ac = ca$, $bc = cba$ with twenty-seven elements, including twenty-six of period 3. The group $C_3 \times C_3 \times C_3$ also has twenty-six elements of period 3. Find two different groups each of which has the identity, an element of period 2, twenty-six of period 3, and twenty-six of period 6.

13 Describe another group of order 27 besides C_{27}, $C_3 \times C_3 \times C_3$ and the non-Abelian group of p. 102 quoted in the previous question.

14 Give examples of realisations of the group $C_n \times C_2$, and draw a Cayley diagram.

15 Show that addition of four-digit binary numerals *without carrying* (e.g. $1011 + 0110 = 1101$) gives the group $C_2 \times C_2 \times C_2 \times C_2$. Generalise this result. Show also that the numbers 1, 2, 3, . . ., 8 expressed in ternary (base 3), and added without carrying (e.g. $7 + 8 = 21 + 22 = 10 = 3$) give the group $C_3 \times C_3$. Generalise this result, and consider the numbers 0, 1, 2, 3, . . ., 99 (base 10), and added without carrying (e.g. $86 + 29 = 05$).

16 Find out how many subgroups of $C_2 \times C_2 \times C_2 \times C_2$ there are: (i) isomorphic to $C_2 \times C_2$; (ii) isomorphic to $C_2 \times C_2 \times C_2$. Can we form a finite geometry as we did in the case of the subgroups of $C_2 \times C_2 \times C_2$?

17 What are the orders of non-cyclic Abelian groups of odd order less than 30?

18 Describe all the Abelian groups of order 18, 20, 36 and 40. (You may assume the theorem that any Abelian group is either cyclic or the direct product of cyclic groups.)

19 Find all the Abelian groups of order 16 and of order 18, expressing each as a direct product. (It can be proved that every Abelian group is isomorphic to the direct product of cyclic groups. Attempt to prove this, and then search for a proof in one of the references.)

20 Show that the number of distinct Abelian groups of order p^r (p prime, r a positive integer), is equal to the number of *partitions* of r (e.g. if $r = 4$, the partitions are $1 + 1 + 1 + 1$, $1 + 1 + 2$, $2 + 2$, $1 + 3$ and 4). *Hint*: consider each group as a direct product of cyclic groups.

21 **Show that any Abelian group is the direct product of cyclic groups.**

22 Show that it is impossible to construct $C_2 \times C_2 \times C_2$ as a group of eight bilinear functions of the form $(az + b)/(cz + d)$ with a, b, c, d real.

23 If $n = pq$ where p and q are primes, is it true that there exists only one group of order n, as it is in the case when $p = 5$, $q = 3$? Is it true for Abelian groups only? Is it true if p and q are both odd primes?

24 A group of order 48 has thirteen elements of period 2, eight of period 3, six of period 4, eight of period 6 and twelve of period 8. Show that it is not the direct product of C_8 with any group of order 6. By considering the periods of its elements, show that it is not a direct product group at all.

25 Consider a group with generators x and y of periods p and q respectively, and such that xy is of period r. For example, we saw in Ch. 13, p. 206, that if $p = 2$, $q = 2$, then the group is D_r. Fill in the table as far as possible for various selections of values for p, q and r, showing that the same group arises for given p, q and r when these indices are permuted in any manner.

p q r	2 2 r	2 3 3	2 3 4	. . . . . .
possible structure	D_r	.	.	. .

26 What is the structure of the group of residues which are prime to 100 under multiplication modulo 100 (see Ex. 12.18 and Ex. 12.26).

27 Show that the group $S_4 \times C_2$ has nineteen elements of period 2, eight of period 3, twelve of period 4, and eight of period 6, and that these are the same as for the full group of the cube (direct and opposite symmetries) (see Ch. 17).

28 A certain group has generators a, b and p with defining relations $p^4 = a^2 = b^2 = 1$, $pb = bp$, $ab = ba$, $(ap)^2 = 1$. Prove $ap = p^3a$; $ap^2 = p^2a$. Show that the group has order 16, and identify it as one of the groups described on p. 291. Construct the Cayley diagram.

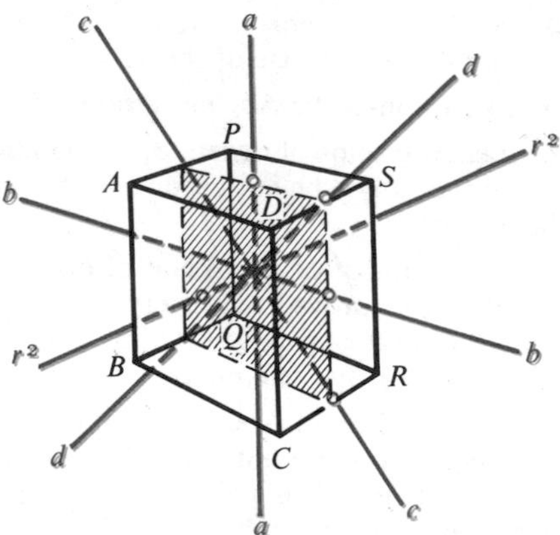

Fig. 16.19. The half-turn symmetries of a square prism.

29 Consider the full group of the symmetries of the square prism. Fig. 16.19 shows the direct symmetries, with the subgroup $\{1, r, r^2, r^3\}$ and the half-turns a, b, c, d together forming the group D_4. Let reflection in the shaded plane $\begin{pmatrix} A & B & C & D & P & Q & R & S \\ P & Q & R & S & A & B & C & D \end{pmatrix}$ be denoted m. Express the seven enantiomorphs in terms of m, r, a, b, c, d, and represent all sixteen elements of the group as (a) permutations of the vertices (b) permutations of the faces. Show that the full group has the structure $D_4 \times C_2$. Name a solid whose group is $D_6 \times C_2$, and another solid whose group is $D_4 \times C_2$.

Identify some subgroups of $D_4 \times C_2$. Show also that the full group of symmetries of the square prism may be represented as a subgroup of S_6 by labelling the *faces* a, b, c, d, p, q (p and q being the two squares), and considering permutations of these faces.

30 Reconcile the alternative definition of direct product group below with the one given in this chapter: G_1, G_2 are two groups such that $G_1 \cap G_2 = 1$, $g_1g_2 = g_2g_1 \ \forall \ g_1 \in G_1, \forall \ g_2 \in G_2$, then Gp (G_1, G_2) is the direct product group.

31 Show that the transpositions (12)(34)(56) ... *ad. inf.* on the set of natural numbers generate an infinite group, every element of which has period 2. What is its structure?

32 Show that the group of rotations about a fixed point through angles which are rational multiples of π (see Ex. 12.31) contains C_n as a subgroup for all $n \in \mathbf{N}$, but is different from the group $C_2 \times C_3 \times C_4 \times C_5 \times$ ... *ad. inf.*

33 What is the structure of the groups described in Ex. 3.15, 12.21?

34 Investigate Gp. (X, Y) where $X = \begin{pmatrix} 1 & 1 \\ 1 & 0 \end{pmatrix}$ and $Y = \begin{pmatrix} 0 & -1 \\ -1 & 0 \end{pmatrix}$, the operation being matrix multiplication (compare p. 183).

35 If A is the group of rotations about a fixed point O, and B is the group $(\mathbf{R}, \times)$ show that $(\mathbf{C}, \times) \simeq A \times B$.

36 What is the direct product of the group of plane translations with the group containing: (a) the identity and a reflection in a single line; (b) the identity and the half-turn about a single point; (c) the rotations about a fixed point?

37 Give defining relations for $C_\infty \times C_\infty$.

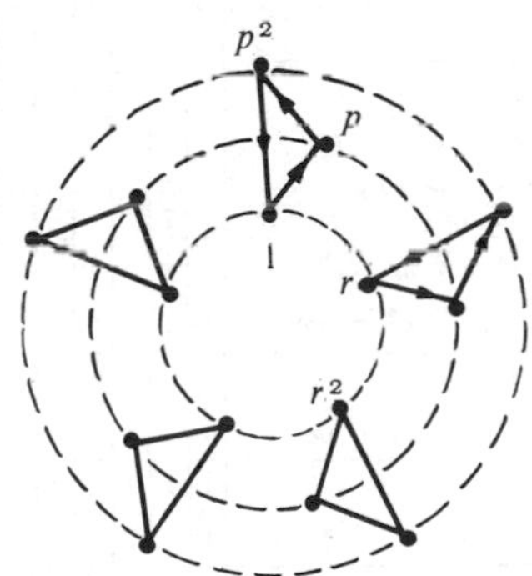

Fig. 16.20.

38 Which group has a Cayley diagram as in fig. 16.20? Label the remaining vertices. Show how the Cayley graph for $C_3 \times C_n$ may be drawn on a torus, and how the Cayley graph for $C_3 \times C_\infty$ may be drawn on a cylinder.

39 Show that $C_2 \times C_2 \times C_\infty$ has exactly three elements of period 2. Draw a Cayley graph. Repeat for $C_4 \times C_\infty$.

40 Describe a Cayley diagram for $C_\infty \times C_\infty \times C_2$; $D_2 \times C_\infty$; $C_\infty \times C_\infty \times C_\infty$.

41 Find an infinite group with five elements of period 2.

17 *Catalogue of groups: symmetry groups*

We are now in a position to take stock of finite groups with which we have familiarised ourselves so far. Table 17.01 shows characteristics of groups up to order 12.

Table 17.01

Order	Symbol for group	Defining relations†	No. of elements of period 1	2	3	4	5	6	7	8	9	10	11	12	Abelian ?	Automorphisms
1	C_1		1												√	C_1
2	$C_2 \cong D_1$	$a^2 = 1$	1	1											√	C_1
3	C_3	$p^3 = 1$	1		2										√	C_2
4	C_4	$r^4 = 1$	1	1		2									√	C_2
	$D_2 \cong C_2 \times C_2$	$a^2 = b^2 = 1,\ ab = ba$	1	3											√	D_3
5	C_5	$r^5 = 1$	1				4								√	C_4
6	C_6	$r^6 = 1$	1	1	2			2							√	C_2
	$D_3 \cong S_3 \cong A_3$	$p^3 = a^2 = 1,\ ap = p^2a$	1	3	2										×	$D_3 \lhd D_3$‡
7	C_7	$r^7 = 1$	1						6						√	C_6
8	C_8	$r^8 = 1$	1	1		2				4					√	D_2
	$C_4 \times C_2$	$r^4 = a^2 = 1,\ ar = ra$	1	3		4									√	D_4
	$C_2 \times C_2 \times C_2$ $(\cong D_2 \times C_2)$	$a^2 = b^2 = c^2 = 1,$ $ab = ba,\ ac = ca,\ bc = cb$	1	7											√	168
	D_4	$a^2 = r^4 = 1,\ ar = r^3a$	1	5		2									×	$D_2 \lhd D_4$
	Q_4	$p^4 = 1,\ p^2 = a^2 = (ap)^2$	1	1		6									×	$D_2 \lhd S_4$
9	C_9	$r^9 = 1$	1		2						6				√	C_6
	$C_3 \times C_3$	$p^3 = q^3 = 1,\ pq = qp$	1		8										√	48
10	C_{10}	$r^{10} = 1$	1	1			4					4			√	C_4
	D_5	$r^5 = a^2 = 1,\ ar = r^4a$	1	5			4								×	20
11	C_{11}	$r^{11} = 1$	1										10		√	C_{10}
12	C_{12}	$r^{12} = 1$	1	1	2	2		2						4	√	D_2
	$C_6 \times C_2 \cong C_3 \times D_2$	$r^6 = a^2 = 1,\ ar = ra$	1	3	2			6							√	D_6
	$D_6 \cong D_3 \times C_2$	$r^6 = a^2 = 1,\ ar = r^5a$	1	7	2			2							×	$D_3 \lhd D_6$
	Q_6	$r^6 = 1,\ r^3 = (ra)^2 = a^2$	1	1	2	6		2							×	$D_3 \lhd D_6$
	A_4	$p^3 = a^2 = 1,\ (ap)^3 = 1$	1	3	8										×	$A_4 \lhd S_4$

† That is, a '*possible* set of defining relations for the given *choice* of generators'.
‡ Indicates the group of inner automorphisms as a normal subgroup ($\lhd$) of the full group of automorphisms.

The problem of identifying all the groups of a particular order is a difficult and laborious one, and there is no simple way of organising a method. It is, indeed, quite a tedious process to prove that there are only two groups of such a low order as 6, even when we use Lagrange's theorem. (See pp. 141 ff.) Imagine how heavy a task it would be to show that there are just five groups of order 12. Note that there is only one group of order 15 (C_{15}, of course) so that just because a number is composite, it does not mean that there will be more than one group. The orders which produce the highest numbers of groups tend to be those which have many factors, particularly high powers of prime factors. For example, there are fourteen groups of order 16, and no fewer than 267 of order 64, and 238 of order 160. We have already mentioned that there exist 15 distinct groups of order 24, of which only three are Abelian. This is proved in Burnside, *Theory of Finite Groups* (Dover), pp. 157–61.

The groups of order 16 with which we are already familiar comprise:

$$C_{16},\quad C_8 \times C_2,\quad C_4 \times C_4,\quad C_4 \times C_2 \times C_2,\quad C_2 \times C_2 \times C_2 \times C_2$$
$$\text{(Abelian)};\qquad D_8,\quad Q_8,\quad D_4 \times C_2,\quad Q_4 \times C_2,$$

and there are also five other non-Abelian groups.

Q. 17.01. Find the number of elements of period 2, 4 and 8 in the groups mentioned above. (Q^8 may be defined by $r^8 = 1$, $r^4 = a^2 = (ra)^2$.)

There is one important group that we have said very little about, and that is A_4, the 'alternating group' of order 12. This will come into its own in Ch. 18, when we consider permutations in more detail. But we shall consider this group in the present chapter, the remainder of which is devoted to a further study of symmetry groups.

Symmetry groups

We have already considered several examples of the symmetry groups of plane and three-dimensional figures. Note that *every* configuration has the trivial symmetry group, consisting of the identity alone. The group of order 2 (which was denoted C_2 or D_1 in the table above) would refer to the possession of a single symmetry of period 2, but it is usual to use C_2 to denote the half-turn symmetry for a plane figure, and D_1 to denote a bilateral symmetry. Thus the letter 'S' possesses C_2 symmetry; 'T' possesses D_1 symmetry.

Q. 17.02. Classify the capital letters of the alphabet according to their symmetry groups (regard O as being an ellipse).

Q. 17.03. Give the symmetry groups of the plane figures in fig. 17.01.

Q. 17.04. Crossword puzzles. Make a collection of crossword puzzles, and classify the patterns according to their symmetry groups: D_1, C_2, D_2, C_4, D_4 (are any others possible?) and find which occur most frequently. If the group were D_4, what fraction of the pattern would one need to know in order to fill up all the rest? Which symmetry groups may a crossword puzzle have if the numbers of clues across and down are equal?

Fig. 17.01. What are the symmetry groups of these plane figures?

Q. 17.05. A 15 × 15 crossword has the symmetry group C_4. Reading from one corner, the following squares are black: (1, 1), (3, 1), (5, 1), (7, 1), (9, 1), (9, 2), (1, 3), (3, 3), (5, 3), (7, 3), (9, 3), (9, 4), (1, 5), (3, 5), (5, 5), (7, 5), (9, 6), (6, 7), (7, 7). Fill up the pattern over the whole crossword square. Repeat in the case when the pattern has the symmetry group D_2.

Q. 17.06. Which quadrilaterals are described by the symmetry groups D_1, C_2, D_2, C_4 and D_4? Draw a hexagon which has the group C_6 only; C_3 only; D_3 only; D_2, D_1, C_2.

Q. 17.07. Prove that if a plane figure is symmetrical about two perpendicular axes, then it is also symmetrical by a half-turn about a perpendicular to its plane through their intersection.

Q. 17.08. Give examples of how the symmetry group of a plane figure may be increased or decreased by altering it; e.g. if an isosceles right-angled triangle is cut off the corner of a square, the symmetry group reduces from D_4 to D_1; while if the right-angled triangles *BCX*, etc. of fig. 12.4 were shaved off to become isosceles triangles, we might improve the symmetry group from C_3 to D_6.

Symmetries of three-dimensional figures

We have already noted that the plane symmetries may be divided into those which 'turn the lamina over' (opposite), and those which do not (direct). The former are equivalent to reflections,† while the latter will be rotations. The direct symmetries are always a subgroup of the full group. For three-dimensional bodies, the distinction between direct and opposite

† See discussion on p. 187.

symmetries is more important than with plane figures, since the opposite symmetries in three dimensions – the enantiomorphs – cannot be effected by a physical movement of the body. When we speak of the 'symmetry group' of a solid, e.g. the regular tetrahedron, we shall normally refer to the direct symmetries only.† If we wish to include the enantiomorphs, we shall describe the group as the 'full group of symmetries' of the tetrahedron, or whatever.

We have already examined the symmetries of a number of solid objects. For example, the symmetry group of a prism whose cross-section is a regular n-gon is D_n, the same as for a lamina in the shape of a regular n-gon, which may be regarded as a very thin prism. But the full group, which includes the reflection in the central plane of symmetry perpendicular to the longitudinal axis, will be $D_n \times C_2$, which is D_{2n} *only* when n is odd.

Q. 17.09. Which other solids have the symmetry group D_n, and full group $D_n \times C_2$?

Q. 17.10. Right pyramids on a regular polygon of n sides have C_n symmetry, but the full group is D_n. Illustrate, and explain.

Q. 17.11. Which classes of solids will have the full group $C_n \times C_2$?

Q. 17.12. By considering the symmetries of the bi-tetrahedron, show that $D_3 \times C_2$ is a subgroup of S_5.

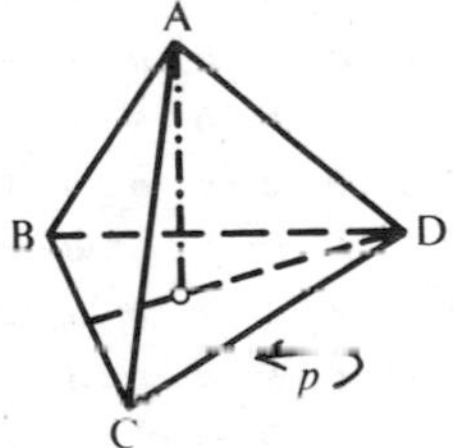

Fig. 17.02. Regular tetrahedron. Rotation through 120° about altitude.

The group of the regular tetrahedron

Consider first the direct symmetries of the regular tetrahedron (see fig. 17.02). Evidently there are rotations through $\pm\frac{2}{3}\pi$ about each of the four altitudes, (eight rotational symmetries of period 3) as well as the identity. It might be thought that these will combine to give the group $C_3 \times C_3$, but this is not so as we shall see. Suppose we denote by p the rotation which produces the cycle (BDC) of the vertices,‡ and p^2 is represented by

† This is also termed the 'Rotation group' of the solid, since all the direct symmetries do in fact consist of rotations.

‡ For simplicity, so that we can consider permutations of vertices, we shall work here with *axes which move with the tetrahedron* (see Ch. 13, p. 200).

(BCD). Similarly q is (ACD), r (ADB) and s (ABC). Now consider the composition of the rotations first p, then r^2.

You may verify that r^2p is the permutation $\begin{pmatrix} A & B & C & D \\ B & A & D & C \end{pmatrix}$ so that the set of nine permutations is not, as it stands, closed. This particular permutation interchanges the pairs (AB) and (CD), and geometrically represents a *half-turn* about an axis of rotation joining the mid-points of the edges AB and CD. Evidently there are three such half-turns, and we denote these: a (AB)(CD); b (AC)(BD) and c (AD)(BC). Check that $pr^2 = c$.

Thus we obtain a group which is different from the other groups of order 12: C_{12}, D_6, $C_6 \times C_2$ and Q_6 (see table 14.01). The structure table for the tetrahedral group A_4 is shown in table 17.02. Those twelve

Table 17.02

	first: 1	a	b	c	p	p^2	q	q^2	r	r^2	s	s^2	period	perm.
1	1	a	b	c	p	p^2	q	q^2	r	r^2	s	s^2	1	$ABCD$
a	a	1	c	b	s	r^2	r	s^2	q	p^2	p	q^2	2	$BADC$
b	b	c	1	a	q	s^2	p	r^2	s	q^2	r	p^2	2	$CDAB$
c	c	b	a	1	r	q^2	s	p^2	p	s^2	q	r^2	2	$DCBA$
p	p	r	s	q	p^2	1	r^2	b	s^2	c	q^2	a	3	$ADBC$
second p^2	p^2	s^2	q^2	r^2	1	p	c	s	a	q	b	r	3	$ACDB$
q	q	s	r	p	s^2	b	q^2	1	p^2	a	r^2	c	3	$CBDA$
q^2	q^2	r^2	p^2	s^2	c	r	1	q	b	s	a	p	3	$DBAC$
r	r	p	q	s	q^2	c	s^2	a	r^2	1	p^2	b	3	$DACB$
r^2	r^2	q^2	s^2	p^2	a	s	b	p	1	r	c	q	3	$BDCA$
s	s	q	p	r	r^2	a	p^2	c	q^2	b	s^2	1	3	$BCAD$
s^2	s^2	p^2	r^2	q^2	b	q	a	r	c	p	1	s	3	$CABD$ (A_4).

permutations of the vertices which also provide a realisation of the tetrahedral group are known as the 'even' permutations of the four letters. The meaning of this will become clear later (pp. 312 ff.).

Note that the 'centre' of this group consists of the identity alone.

Q. 17.13. Draw a Cayley diagram for A_4. Show that this group may be defined $p^3 = 1$, $a^2 = 1$, $(ap)^3 = 1$, and obtain the other elements in terms of the generators a and p.

Q. 17.14. Does A_4 have D_3 as a subgroup? Use centralisers to find subgroups. Find Gp. (a, p), Gp. (p, q), Gp. (p, s). Using p and q as a pair of generators, set up defining relations, and express every element in terms of p and q. Repeat with c and p^2.

Q. 17.15. Is it possible to express A_4 as the direct product of two groups?

Q. 17.16. Since the permutations of four objects may be represented by 4×4 permutation matrices, we may discover twelve matrices which form this group A_4, e.g.

$$\begin{bmatrix} 0 & 0 & 1 & 0 \\ 0 & 0 & 0 & 1 \\ 1 & 0 & 0 & 0 \\ 0 & 1 & 0 & 0 \end{bmatrix} \quad \text{and} \quad \begin{bmatrix} 0 & 1 & 0 & 0 \\ 0 & 0 & 0 & 1 \\ 0 & 0 & 1 & 0 \\ 1 & 0 & 0 & 0 \end{bmatrix}$$

Find the other ten matrices, showing that the whole group is generated by the two given. Label these twelve matrices with the letters used for the symmetries of the regular tetrahedron so as to show the isomorphism.

Matrices to represent the group of symmetries of the regular tetrahedron

It may be shown that, of the twenty-four 4×4 permutation matrices, those twelve which concern us here, and which were found in the question immediately preceding, are those whose determinant is $+1$ and not -1.

The group may also be represented by the 3×3 matrices:

$$\begin{bmatrix} 1 & 0 & 0 \\ 0 & 1 & 0 \\ 0 & 0 & 1 \end{bmatrix}, \begin{bmatrix} 1 & 0 & 0 \\ 0 & -1 & 0 \\ 0 & 0 & -1 \end{bmatrix}, \begin{bmatrix} -1 & 0 & 0 \\ 0 & 1 & 0 \\ 0 & 0 & -1 \end{bmatrix}, \begin{bmatrix} -1 & 0 & 0 \\ 0 & -1 & 0 \\ 0 & 0 & 1 \end{bmatrix},$$

$$\begin{bmatrix} 0 & 1 & 0 \\ 0 & 0 & 1 \\ 1 & 0 & 0 \end{bmatrix}, \begin{bmatrix} 0 & 0 & 1 \\ 1 & 0 & 0 \\ 0 & 1 & 0 \end{bmatrix}, \begin{bmatrix} 0 & 0 & 1 \\ -1 & 0 & 0 \\ 0 & -1 & 0 \end{bmatrix}, \begin{bmatrix} 0 & -1 & 0 \\ 0 & 0 & -1 \\ 1 & 0 & 0 \end{bmatrix},$$

$$\begin{bmatrix} 0 & 1 & 0 \\ 0 & 0 & -1 \\ -1 & 0 & 0 \end{bmatrix}, \begin{bmatrix} 0 & 0 & -1 \\ 1 & 0 & 0 \\ 0 & -1 & 0 \end{bmatrix}, \begin{bmatrix} 0 & 0 & -1 \\ -1 & 0 & 0 \\ 0 & 1 & 0 \end{bmatrix}, \begin{bmatrix} 0 & -1 & 0 \\ 0 & 0 & 1 \\ -1 & 0 & 0 \end{bmatrix}.$$

Since these matrices may be taken as representing a transformation in 3-space, we should attempt to give an explanation.

In fig. 17.03, we show the tetrahedron so that its vertices are at the points $A\ (1, 1, -1)$, $B\ (1, -1, 1)$, $C\ (-1, -1, -1)$, $D\ (-1, 1, 1)$. It will be found that each of the twelve matrices represents transformations which take the tetrahedron into itself. For example, the transform of the point A by the matrix

$$\begin{bmatrix} 0 & 0 & 1 \\ -1 & 0 & 0 \\ 0 & -1 & 0 \end{bmatrix} \quad \text{is} \quad \begin{bmatrix} 0 & 0 & 1 \\ -1 & 0 & 0 \\ 0 & -1 & 0 \end{bmatrix} \begin{bmatrix} 1 \\ 1 \\ -1 \end{bmatrix} = \begin{bmatrix} -1 \\ -1 \\ -1 \end{bmatrix},$$

i.e. A moves into position C. But we may go further and find the effect of

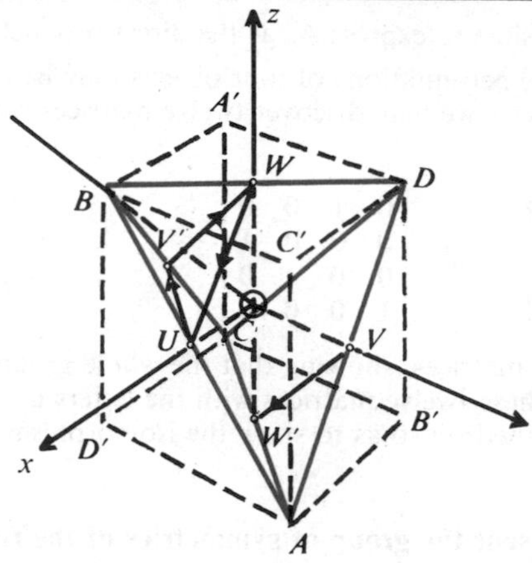

Fig. 17.03. The effect of the matrix $\begin{bmatrix} 0 & 0 & 1 \\ -1 & 0 & 0 \\ 0 & -1 & 0 \end{bmatrix}$ on the tetrahedron $ABCD$.

this matrix on every vertex more compactly by arranging the coordinates of the four vertices as four-column vectors to form a 3×4 matrix:

$$\begin{bmatrix} 0 & 0 & 1 \\ -1 & 0 & 0 \\ 0 & -1 & 0 \end{bmatrix} \overset{\begin{matrix} A & B & C & D \end{matrix}}{\begin{bmatrix} 1 & 1 & -1 & -1 \\ 1 & -1 & -1 & 1 \\ -1 & 1 & -1 & 1 \end{bmatrix}} = \overset{\begin{matrix} C & B & D & A \end{matrix}}{\begin{bmatrix} -1 & 1 & -1 & 1 \\ -1 & -1 & 1 & 1 \\ -1 & 1 & 1 & -1 \end{bmatrix}}$$

and these columns are the coordinates of C, B, D, A. Hence the matrix $\begin{bmatrix} 0 & 0 & 1 \\ -1 & 0 & 0 \\ 0 & -1 & 0 \end{bmatrix}$ sends A, B, C, D to the positions previously occupied by C, B, D, A, and this is the permutation of the vertices $\begin{pmatrix} A & B & C & D \\ D & B & A & C \end{pmatrix}$. '*A* becomes *D*', or '*A is replaced by D*', and, since B is unchanged, represents a rotation about OB. This rotation takes the points U (1, 0, 0), (mid-point of AB), V (0, 1, 0) (mid-point of AD) and W (0, 0, 1) (mid-point of BD) to the points V' (0, −1, 0), W' (0, 0, −1) and U (1, 0, 0), as shown by the arrows in the figure. These three sets of coordinates will, of course, form the columns of the transformation matrix:

$$\begin{bmatrix} 0 & 0 & 1 \\ -1 & 0 & 0 \\ 0 & -1 & 0 \end{bmatrix}$$
$$(V') \quad (W') \quad (U)$$

The second, third and fourth of our list of twelve (each of period 2) will interchange two pairs of vertices; for example, $\begin{bmatrix} -1 & 0 & 0 \\ 0 & 1 & 0 \\ 0 & 0 & -1 \end{bmatrix}$ changes the point (x, y, z) into the point $(-x, y, -z)$, and so is a half-turn about the y-axis, which is one of our three symmetries a, b, c of the tetrahedron.

Q. 17.17. Examine the remaining 3×3 matrices from the same viewpoint.

We may also consider the symmetries of the tetrahedron to be the permutations of the four *faces*, or as permutations of the six *edges*. The latter will reveal A_4 as a subgroup of S_6, and could be represented by 6×6 permutation matrices.

Q. 17.18. Consider the subgroups of A_4 in relation to the direct symmetries of the tetrahedron, and state which invariant property is associated with each subgroup (e.g. $(AD)(BC)$ leaves the line joining the mid-points of AD and BC invariant).

The *full* group of the regular tetrahedron will include not only six reflections in planes of symmetry (e.g. the reflection in the plane COD which will interchange A and B), but also transformations of period 4, such as the one represented by the permutation $\begin{pmatrix} A & B & C & D \\ D & A & B & C \end{pmatrix}$ of the vertices. Clearly this transformation cannot be effected by a physical movement. There will be six of these symmetries of period 4, making the total of twelve enantiomorphs.

Indeed, the full group of the tetrahedron contains all those twenty-four transformations which permute the four vertices (or faces) in any manner. Thus it is isomorphic with the group S_4. Consult the S_4 table to verify the periods of the elements.

The cube and octahedron

The regular tetrahedron, being the simplest of the five regular solids, has the simplest symmetry group. We now consider the symmetries of the cube, and its dual† solid, the regular octahedron.

A clumsy method of doing this would be to consider the permutations of the eight vertices (compare the corresponding work on the cuboid in Ch. 16, pp. 271, 278). A less clumsy way would be to consider the permutations of the six faces of the cube, and you will recall that this greatly simplified the representation of the group $C_2 \times C_2 \times C_2$ as a

† The centres of the six faces of a cube form a regular octahedron, and the centres of the eight faces of a regular octahedron form a cube.

group of permutations in Ch. 16. However, far the simplest method of thinking of the direct symmetries of the cube is by means of *permutations of its four diagonals*. Since the diagonals of the cube (see fig. 17.03) were in fact the *altitudes* of the tetrahedron, we may anticipate a close connection between their symmetry groups.

Let the diagonals of the cube be called a, b, c, d (AA', BB', CC', DD' in fig. 17.04). We shall show that every direct transformation of the cube into

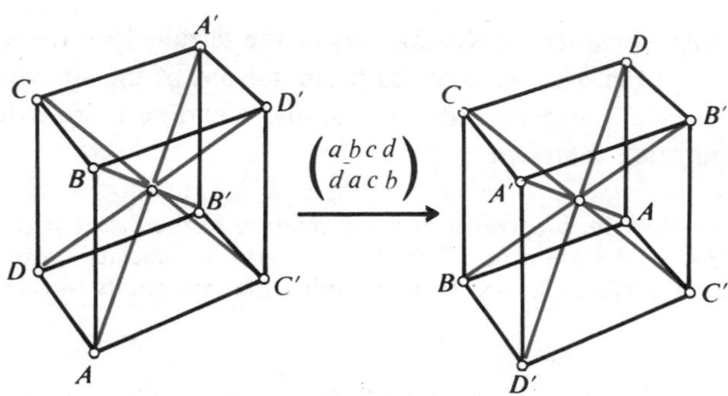

Fig. 17.04. Each symmetry of the cube corresponds to a permutation of its diagonals.

itself corresponds to just one of the permutations of the diagonals, and vice versa. In the first place, there are altogether twenty-four permutations of four objects, and it is easy to see that there are also twenty-four symmetries of the cube. For consider first the symmetries of the cube which carry a particular vertex (say A) into itself. Evidently if A is invariant under a transformation, so must the other end A' of the diagonal through A, and in fact the entire diagonal a is not moved. The three points B, D, C' at the ends of the edges through A form an equilateral triangle, and the only possible symmetries of the cube with A fixed are those which rotate about the fixed diagonal a through angles O, $\frac{2}{3}\pi$ and $-\frac{2}{3}\pi$, and so cause the cyclic permutations of the vertices of triangle BCD' and hence a cyclic permutation of the diagonals (bdc) (the subgroup C_3). But the point A may be mapped into any of the eight vertices of the cube, and since there are three symmetries for each of these possibilities, it follows that the total number is $8 \times 3 = 24$, it being noted that we have not counted any of them twice.

Q. 17.19. Obtain the result by considering the number of ways that the cube may be placed on a table (like a dice being thrown) and then orientated.

Let us now consider the types of movement which are included in these twenty-four symmetries. There are rotations through $\pm\frac{2}{3}\pi$ about each of the four diagonals, so we have eight elements of period 3 (see figs. 17.051, 17.052). There are six half-turns about axes joining mid-points of opposite

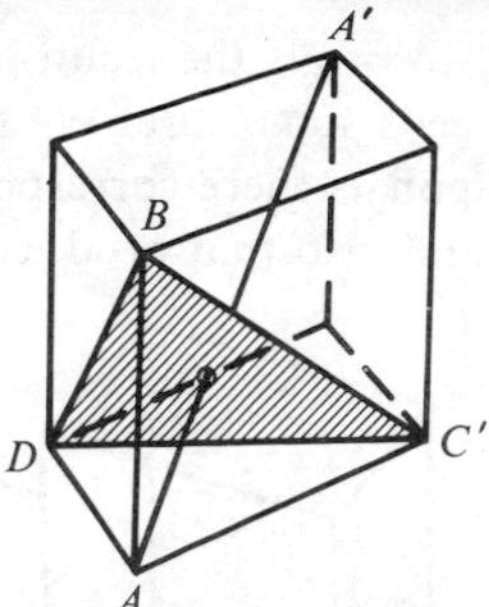

Fig. 17.051. Symmetries of the cube which leave A invariant cause rotation of equilateral triangle BDC'.

edges, such as that shown in fig. 17.053. Finally there are quarter-turns and half-turns about the joins of the centres of opposite faces – altogether, since there are three such axes – six of period 4 and three of period 2 (see

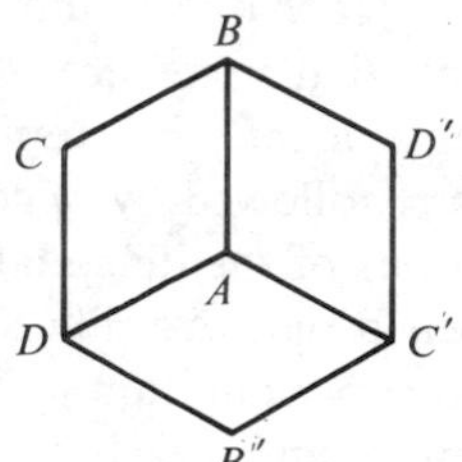

Fig. 17.052. View of cube as seen from a point on AA' produced.

fig. 17.054). Thus the group contains nine elements of period 2, eight of period 3, six of period 4, and the identity. These are the same as the numbers of elements of periods 2, 3 and 4 in the group S_4, and establishes the isomorphism, so long as we are prepared to accept the unproved theorem on p. 143.

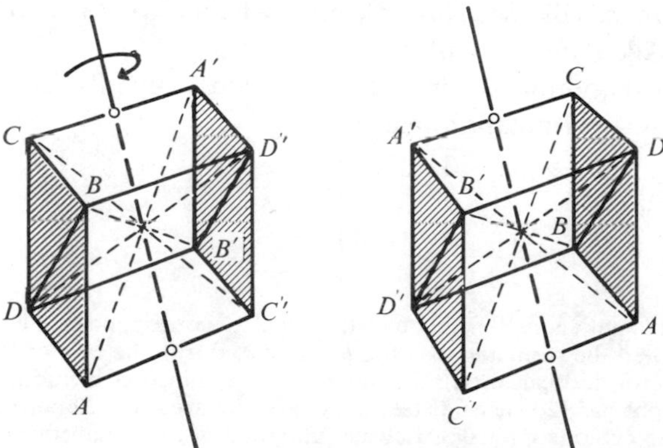

Fig. 17.053. The half-turn about an axis joining the mid-points of $A'C$, AC' (perpendicular to plane $BDB'D'$) interchange the faces $ABCD$ and $C'B'A'D'$.

However, it is not difficult to prove the result independently of that theorem, though the proof needs some care. We need to show that to every permutation of the diagonals there corresponds exactly one symmetry of the cube, and vice versa, and also that products are preserved.†

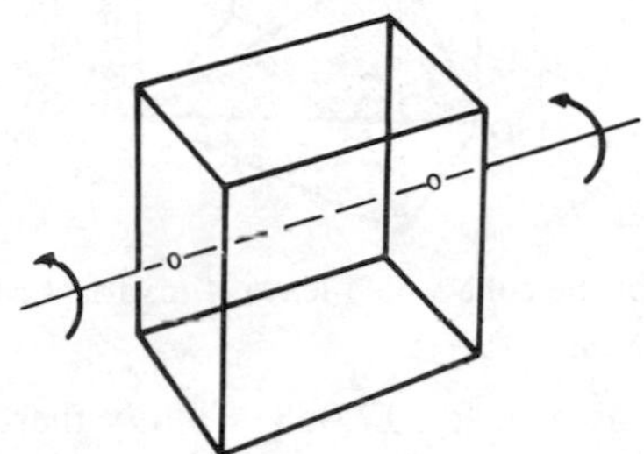

Fig. 17.054. Rotations about joins of centres of opposite faces.

The latter follows from the fact that if r_1 and r_2 are rotations of the cube, and suppose for example that r_1 replaces diagonal a by diagonal b, while r_2 replaces b by c. Then evidently r_2r_1 replaces a by c, so that the composition of the rotations r_1 followed by r_2 corresponds exactly to the composition of the permutations of the diagonals.

Suppose now that it happened that two different rotations r_1 and r_2 were both to correspond to the same permutation p. Then $r_1r_2^{-1}$ corresponds to pp^{-1} = the identity permutation, and clearly the only rotation of the cube which retains the positions of all four diagonals is the identity.‡ Hence $r_1r_2^{-1} = e \Rightarrow r_1 = r_2$. This establishes that two different rotations cannot bring about the same permutation. Moreover, it is self-evident that two different permutations cannot both correspond to the same rotation, and the argument is complete.

For an alternative approach, see P. S. Alexandroff, *Introduction to the Theory of Groups* (Blackie Hafner), pp. 55–62 and A. Bell, *Algebraic Structures* (Allen and Unwin), Ch. 8.

Now the twenty-four symmetries of the cube may also be represented by the twenty-four permutation matrices, e.g.

$$\begin{pmatrix} a & b & c & d \\ d & a & c & b \end{pmatrix} \leftrightarrow \begin{bmatrix} 0 & 0 & 0 & 1 \\ 1 & 0 & 0 & 0 \\ 0 & 0 & 1 & 0 \\ 0 & 1 & 0 & 0 \end{bmatrix}.$$

† Note that it is not enough merely to establish a 1,1 correspondence between the symmetries and the permutations of a, b, c and d. For in the case of the regular icosahedron (or dodecahedron), the full group of symmetries including enantiomorphs has 120 elements (see p. 306). There are also 120 permutations of the five inscribed cubes (see p. 306). But the full group of the icosahedron is not S_5, but $A_5 \times C_2$.

‡ This is not true in the *full* group, for the *central inversion* leaves all four diagonals invariant.

But, more interestingly, they may be represented by 3 × 3 matrices, each of which carries out the appropriate transformation. These will include, as the subgroup A_4, the twelve matrices listed above on p. 295. Those matrices had their (±) 1's in the following three positions:

$$\begin{bmatrix} 1 & \cdot & \cdot \\ \cdot & 1 & \cdot \\ \cdot & \cdot & 1 \end{bmatrix} \text{ or } \begin{bmatrix} \cdot & 1 & \cdot \\ \cdot & \cdot & 1 \\ 1 & \cdot & \cdot \end{bmatrix} \text{ or } \begin{bmatrix} \cdot & \cdot & 1 \\ 1 & \cdot & \cdot \\ \cdot & 1 & \cdot \end{bmatrix}$$

and either three or one of these 1's carried a '+' sign. It will be found that the remaining twelve matrices which complete the full group of the tetrahedron (or the group of direct symmetries of the cube or octahedron), will have their (±) 1's in the three remaining possible positions.

Q. 17.20. Find these twelve remaining matrices, and check the whole set for closure.

We should find that every one of these twenty-four matrices will have a corresponding determinant whose value is +1 or −1. For example,

$$\begin{bmatrix} 0 & 1 & 0 \\ -1 & 0 & 0 \\ 0 & 0 & 1 \end{bmatrix} \to \begin{vmatrix} 0 & 1 & 0 \\ -1 & 0 & 0 \\ 0 & 0 & 1 \end{vmatrix} = +1.$$

In fact, of the forty-eight† possible matrices with six zeros and three (±) 1's, those with determinant +1 correspond to the direct symmetries of the cube, those whose determinant is −1 to the opposite symmetries. We may ask: to which enantiomorph does the matrix (say) $\begin{bmatrix} 0 & -1 & 0 \\ 1 & 0 & 0 \\ 0 & 0 & -1 \end{bmatrix}$ correspond? This may be decided by remembering that the points $U(1, 0, 0)$, $V(0, 1, 0)$ and $W(0, 0, 1)$ are mapped into the points whose coordinates are the *columns* of the matrix, namely $V(0, 1, 0)$, $U'(-1, 0, 0)$ and $W'(0, 0, -1)$, and by examining the geometrical interpretation of this. Or, we may write:

$$\begin{bmatrix} 0 & -1 & 0 \\ 1 & 0 & 0 \\ 0 & 0 & -1 \end{bmatrix} = \begin{bmatrix} -1 & 0 & 0 \\ 0 & -1 & 0 \\ 0 & 0 & -1 \end{bmatrix} \begin{bmatrix} 0 & 1 & 0 \\ -1 & 0 & 0 \\ 0 & 0 & 1 \end{bmatrix}.$$

† We can easily show that there are forty-eight possible matrices containing six zeros and three (±) 1's like that just mentioned. For column 1 will contain either a 1 or a −1 in one of the three places so that there are six possible ways of filling up column 1. Having done this, only two places in column 2 may contain ±1, so this column may be completed in four ways. Once columns 1 and 2 are completed the only possible place in column 3 may be filled with either +1 or −1. Thus the number of such matrices is 6 × 4 × 2 = 48.

Now the matrix $\begin{bmatrix} 0 & 1 & 0 \\ -1 & 0 & 0 \\ 0 & 0 & 1 \end{bmatrix}$ represents one of the direct symmetries of the cube, sending the points U, V, W to the points V', U, W. The point W being unchanged reveals that the line WOW' must be an axis of rotation.

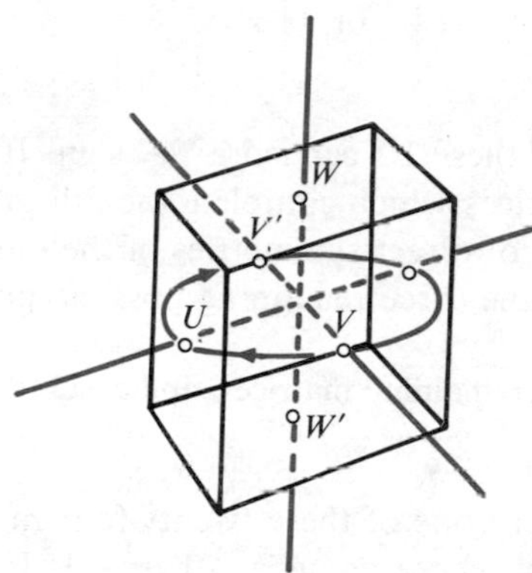

Fig. 17.06. The matrix $\begin{bmatrix} 0 & 1 & 0 \\ -1 & 0 & 0 \\ 0 & 0 & 1 \end{bmatrix}$ represents a quarter-turn about WOW'.

Figure 17.06 shows that we have a quarter-turn through $\frac{1}{4}\pi$ about the line WOW' (the z-axis).

Furthermore, the matrix $\begin{bmatrix} -1 & 0 & 0 \\ 0 & -1 & 0 \\ 0 & 0 & -1 \end{bmatrix}$ moves the point (x, y, z) to the point $(-x, -y, -z)$, and so is the 'central inversion' in O (see also p. 271).

Hence the matrix $\begin{bmatrix} 0 & -1 & 0 \\ 1 & 0 & 0 \\ 0 & 0 & -1 \end{bmatrix}$ represents the composition of a quarter-turn about the z-axis together with the central inversion. You can easily verify that these two elements may be performed in either order. In fact, the central inversion commutes with every one of the remaining forty-seven symmetries, and so forms part of the 'centre' of the full group, $S_4 \times C_2$.

Any of the twenty-four enantiomorphs of the cube (or octahedron) may be considered to be the product of one of the direct symmetries with the central inversion.

Q. 17.21. Represent the following by matrices: two rotations through $\pm\frac{2}{3}\pi$ about COC' where C is $(1, -1, 1)$; a half-turn as shown in fig. 17.053; the four rotations through 0, $\frac{1}{2}\pi$, π and $1\frac{1}{2}\pi$ about the y-axis. Combine each with the central inversion, and interpret the result. Which matrix represents a reflection in

the plane $z = 0$? Which direct symmetry must be combined with the central inversion to give the reflection in the plane $z = 0$? Interpret

$$\begin{bmatrix} 0 & 0 & 1 \\ -1 & 0 & 0 \\ 0 & -1 & 0 \end{bmatrix} \begin{bmatrix} -1 & 0 & 0 \\ 0 & -1 & 0 \\ 0 & 0 & -1 \end{bmatrix} = \begin{bmatrix} 0 & 0 & -1 \\ 1 & 0 & 0 \\ 0 & 1 & 0 \end{bmatrix}$$

in terms of the symmetries of (a) the cube (b) the regular octahedron.

Q. 17.22. Show that $C_2 \times C_2 \times C_2$ is a subgroup of the full group of the cube. Find other subgroups which include opposite symmetries. Are the groups $C_4 \times C_2$ and Q_4 among the subgroups of $S_4 \times C_2$?

Q. 17.23. Show that the 'scalar' matrix $\begin{bmatrix} k & 0 & 0 \\ 0 & k & 0 \\ 0 & 0 & k \end{bmatrix}$ commutes with all 3×3 matrices (in particular when $k = -1$).

The icosahedral group, A_5

Given a regular icosahedron (fig. 17.07), there is a 'dual' regular dodecahedron whose vertices are the centres of the faces of the original solid; and vice versa. This is what we mean by saying that the regular icosahedron and the regular dodecahedron are dual solids. Now for every symmetry of the icosahedron, that is, for every transformation which carries it into itself, the inscribed dodecahedron will also be carried into itself.

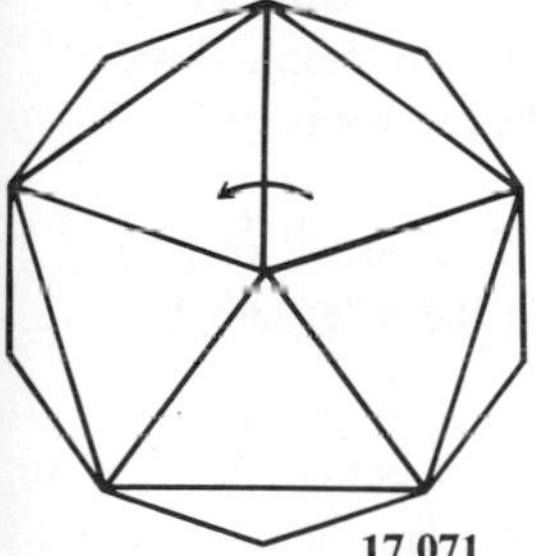
17.071

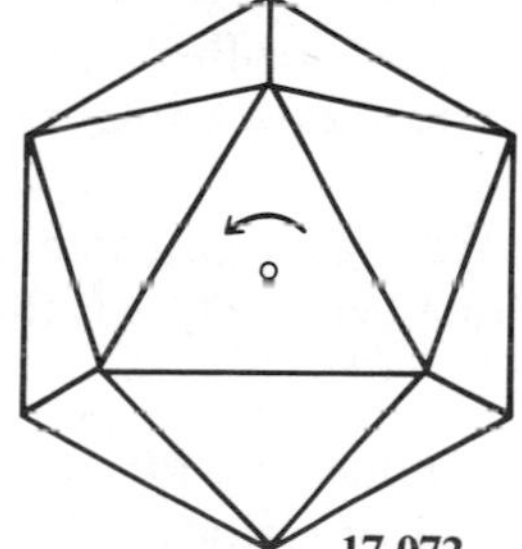
17.072

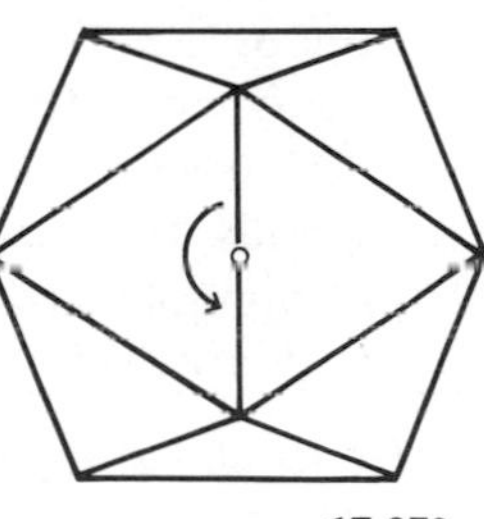
17.073

Fig. 17.07. Rotations of regular icosahedron.

17.071. Four rotations through $72°n$ about joins of opposite vertices: total 24.

17.072. Two rotations through $\pm 120°$ about joins of centres of opposite faces, total 20.

17.073. Half-turns about joins of mid-points of opposite edges: total 15.

Hence, in the first place, the *numbers* of symmetries of the dual solids are equal. But this is not sufficient for isomorphism, which must also preserve products. If a, b, c are three symmetries of the icosahedron, such that a followed by b produces the symmetry c, i.e. $c = ba$, and if a', b', c' are the corresponding symmetries induced in the dodecahedron by the symmetries *a*, *b*, *c* of the icosahedron, then *ba* induces the symmetry $b'a'$ in the dodecahedron. But *c* induces c' in the dodecahedron, and therefore, since $c = ba$, we must also have $c' = b'a'$, and isomorphism is proved.

How many elements has this group of symmetries? Imagine the *dodecahedron* placed on a horizontal table so that a pentagonal face is fitted on to an outline on the table congruent to the face. Then for any particular face which is fitted to the outline, there are five possible positions of the solid corresponding to rotations through 0, $\pm\frac{2}{5}\pi$, $\pm\frac{4}{5}\pi$. But any one of the twelve faces might be fitted to the pentagonal outline, so the number of (direct) symmetries is $5 \times 12 = 60$.

Q. 17.24. Derive the number 60 in another way by considering the symmetries of the dodecahedron which leave a particular vertex (instead of a face) invariant. Repeat both arguments for the regular icosahedron. How many elements are there in the *full* group of the dodecahedron (icosahedron)? Prove that C_5 is a subgroup of the icosahedral group A_5. Which other cyclic groups are subgroups of A_5?

We may analyse these sixty symmetries as follows. We consider the case of the icosahedron, and note that all the direct symmetries are *rotations*:

(a) Rotations through multiples of $\frac{2}{5}\pi$ about lines joining opposite vertices. Since there are four such rotations, of period 5, about each of the six 'diagonals', we have twenty-four such elements (see fig. 17.071).

(b) Rotations through multiples of $\frac{2}{3}\pi$ about the joins of the centres of opposite faces. For each such axis there are two rotations, and, there being ten of these axes, this makes a total of twenty rotations of period 3 (fig. 17.072).

(c) Finally, there are half-turns about axes joining the mid-points of pairs of opposite edges. The number of edges is thirty, so the number of such axes is fifteen (see fig. 17.073).

Hence the group contains an identity, fifteen elements of period 2, twenty of period 3, twenty-four of period 5. These figures are the same for the group of even permutations (see p. 312) of five letters, known as the alternating group of order 60 (A_5) (the group S_5 contains, of course, $6! = 120$ permutations). This observation does not prove that the icosahedral group is isomorphic with A_5, though this is in fact the case.

Q. 17.25. Analyse the rotation group of the regular dodecahedron in the same way as we have done with the icosahedron.

Q. 17.26. Prove, by a consideration of the symmetries of these regular solids that C_6 is not a subgroup of A_5.

The group A_5 is of enormous interest to mathematicians, because it is one of the few 'simple' groups, having no invariant subgroups (see pp. 411 ff.)

One consequence of this property is the remarkable fact that the solution of the general quintic equation cannot be expressed in terms of its coefficients by means of algebraic formulae involving radicals (square and cube roots, etc.).

★ It is possible to inscribe five cubes in a dodecahedron. One such is shown in fig. 17.08. One edge of one of these cubes may be taken to be a diagonal (say AC) of the pentagonal face $ABCDE$ which is being viewed directly in the figure. By selecting appropriate diagonals of successive adjacent faces, a cube is formed. Evidently there must be five

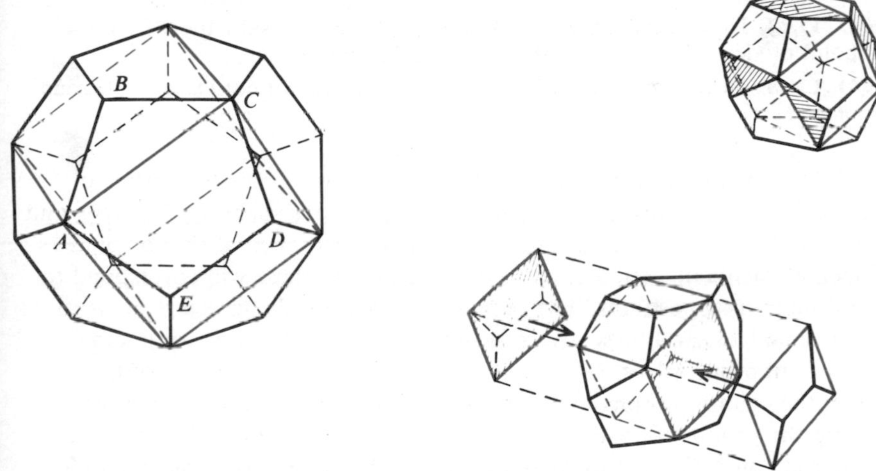

Fig. 17.08. One of the five cubes which may be inscribed in the regular dodecahedron.

of these cubes, since the choice of diagonal of $ABCDE$ might have been made in five ways. Of the sixty symmetries of the dodecahedron, some will cause one particular cube to be mapped into itself. Those which do so will form a subgroup of A_5. Obviously, all twenty-four direct symmetries of the cube cannot be in this subgroup, by Lagrange's theorem, since 24 is not a factor of 60. But the rotations through $\pm\frac{2}{3}\pi$ about those diagonals of the dodecahedron which are also diagonals of the cube, will certainly be included, and there are eight of these. Also included will be the half-turns about the joins of mid-points of opposite edges (such as ED) of the dodecahedron. Of these, there are three which take the cube into itself, being the three mutually perpendicular axes through the centre and perpendicular to the faces of the cube. These eleven symmetries, together with the identity, form the group A_4. Thus we have shown *geometrically* that A_4 is a subgroup of A_5. The particular property associated with this subgroup is that it takes one particular cube into itself. Since there are five cubes, it follows that A_5 has five subgroups A_4.

It is further possible to show that every symmetry of the full group of the regular dodecahedron is associated with one of the permutations of these five cubes. To study this, it is desirable to have a model of the five cubes, each cube being coloured differently. Each face of the dodecahedron will then have five coloured diagonals, being edges of the five cubes. It will be found that the *order* in which these colours occur as one goes round the pentagram is different for every face, being an *even*† permutation of the five colours.

★**Q. 17.27.** How many subgroups A_3 does A_4 have? Consider these from the point of view of the symmetries of the regular tetrahedron.

★**Q. 17.28.** How would you convince yourself whether or not D_5 is a subgroup of A_5?

Now we know that there are thirty permutations of the five letters of period 2 – fifteen consisting of double transpositions such as $(BE)(AC)$, and fifteen of a single transposition, e.g. (AD). The icosahedral group contains fifteen elements of period 2, as we have seen, and these will correspond to the double transpositions. The remainder represent the fifteen enantiomorphs, and these are reflections in planes of symmetry joining pairs of opposite edges. Notice also that, as with the full group of the cube, the enantiomorphs may be paired off with the direct symmetries by taking the product of each direct symmetry with the central inversion. In a later chapter (Ch. 21) we shall deal more thoroughly with the group S_5 and its subgroups and 'classes' (see pp. 412 ff.). We shall also prove the important result that S_5 has no normal subgroups other than A_5, and that the latter is a simple group, i.e. does not have any normal subgroups (see pp. 411 ff.).

To round off the chapter, it would be appropriate to give, without proof, an important result.

The only possible finite symmetry groups in three dimensional space are:

$$\mathbf{C}_n;\quad \mathbf{D}_n;\quad \mathbf{A}_4;\quad \mathbf{S}_4;\quad \mathbf{A}_5;$$

for rotation groups.

To these may be added:

$$\mathbf{C}_n \times \mathbf{C}_2;\quad \mathbf{D}_n \times \mathbf{C}_2;\quad \mathbf{A}_4 \times \mathbf{C}_2;\quad \mathbf{S}_4 \times \mathbf{C}_2;\quad \mathbf{A}_5 \times \mathbf{C}_2;$$

when opposite symmetries are included.

† See p. 312.

Q. 17.29. Describe solids whose rotation groups are each of the five possible abstract groups, supposing *n* to be given.

Describe solids whose full groups are each of the ten possible groups listed above.

For further discussion of the above, see H. Weyl, *Symmetry* (Princeton U.P.), especially pp. 77–80, 119, and the Appendices, where proofs are given of the theorem enunciated above, also 'Symmetry of Solids', Peter Wells, in *Mathematics Teaching*, No. 55, p. 48.

This concludes our discussion of symmetry groups. Further work which you may wish to pursue might be to study, from the point of view of group theory, the symmetries of the Archimedean semi-regular solids and their duals (see Cundy and Rollett, *Mathematical Models* (O.U.P.), Ch. 3), as well as of the antiprisms and their duals, the rhombohedra.

Q. 17.30. Consider which of the five regular solids and also which of the thirteen semi-regular solids will serve as Cayley-graphs.

Further discussion of the symmetry groups of the regular solids will be found in A. Bell, *Algebraic Structures* (Allen and Unwin); H. Coxeter, *Regular Polytopes* (Ch. 3); L. Fejes Toth, *Regular Figures* (Pergamon); and P. Alexandroff, *Introduction to the Theory of Groups* (Blackie/Hafner). Bell (pp. 124–6) shows the rotation group of the regular icosahedron to be A_5 by considering the five tetrahedra which may be circumscribed to it, and also be considering the colouring of its faces. Coxeter obtains the same result (p. 50).

As a postscript to this chapter, it is worth pointing out that, just as it is easier to destroy than to create, so it is easy to *reduce* the symmetry group of a geometrical figure (see Exs. 8, 27, 32 below). Symmetry is easily destructible – even fig. 26.03 may lose *all* its symmetry by the insertion of a single blot!

EXERCISES 17

1 The rectangle and rhombus are dual plane figures with the group D_2. Mention other plane figures with the symmetry group D_2.

2 What are the symmetry groups (rotation groups and full groups) of a forty-eight-toothed gear-wheel; a fifteen-toothed gear-wheel; a three-start worm-wheel. (Note that in the latter case, a rotation of 120° will cause the worm-wheel to take a position indistinguishable from its previous one.)

3 What is the symmetry group of a figure consisting of two coplanar circles: (a) when they are unequal; (b) when they are equal? Does it make any difference whether they intersect?

4 Add one line to fig. 17.09 to give it as much symmetry as possible.

5 What is the symmetry group of a chessboard with alternate black and white squares: (a) for an 8 × 8 board; (b) for a 7 × 7 board?

Fig. 17.09.

6 Draw a diagram with D_2 symmetry containing two circles.
Draw a diagram with D_3 symmetry containing three equilateral triangles.
Draw a diagram with D_4 symmetry containing two rectangles.
Draw a diagram with C_4 symmetry containing two squares; three squares.
Draw a diagram with C_3 symmetry containing three congruent irregular triangles.

7 What are the symmetry groups of the following loci:
(a) $y = 8x^4 - 8x^2 + 1$; (b) $y^4 = x^6$; (c) $|x| + |y| = 1$; (d) $3(\sqrt{x} + \sqrt{y}) = 2\sqrt{(x + y)}$; (e) $x^{\frac{2}{3}} + y^{\frac{2}{3}} = a^{\frac{2}{3}}$; (f) $x^4 + y^4 + ax^2y^2 + b(x^2 + y^2) + c = 0$; (g) $r = a \cos n\theta$ $(n = 1, 2, 3, 4, \ldots)$; (h) $r^2 = a^2 \cos 2\theta$; (i) the deltoid; (j) the five-cusped hypocycloid; (k) the curve $r = a \cos \frac{1}{2}\theta$. Discuss the symmetry of various curves which may be drawn with *Spirograph*.

8 Discuss the effect of placing arrows on the edges of figures in two and three dimensions; e.g. if arrows are placed along the edges AB, BC, CD, AF, FE, ED of a regular hexagon, the group is reduced from D_6 to D_1. Also, consider the effect of placing arrows on some edges and not on others.

9 What are the rotation groups and the full groups of the 'house' whose plan, front and side elevations are shown in fig. 17.10?

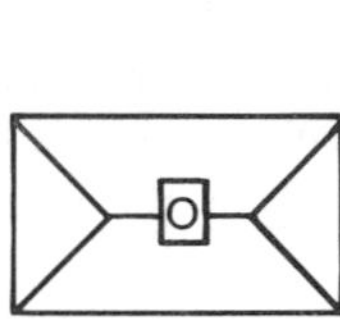

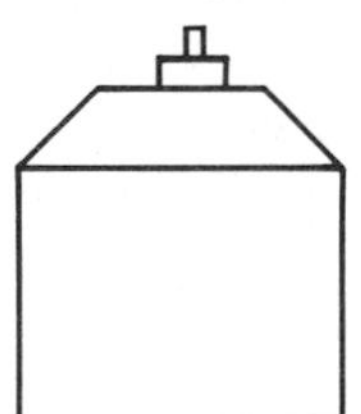

Fig. 17.10.

10 What are the rotation groups and the full groups of:
(a) the general parallelepiped; (b) a parallelepiped four of whose faces are rectangles; (c) a parallelepiped four of whose faces are congruent rhombi; (d) a parallelepiped all six of whose faces are rhombi; (e) a parallelepiped all the faces of which are congruent rhombi.

11 A tetrahedron has its opposite edges equal so that all its faces are congruent triangles. Show that its symmetry group is D_2, the same as that of the cuboid (Consider the parallelepiped formed by three pairs of parallel planes through its opposite edges.) Answer the same question by considering the octahedron formed by the mid-points of the six edges.

12 What is the rotation group of a tetrahedron which has each pair of opposite edges perpendicular (the 'orthocentric' tetrahedron)?

13 A tetrahedron $ABCD$ has $AB = BC = CD$, $AB \perp BC \perp CD$. What is its symmetry group?

14 A tetrahedron $ABCD$ has $AB = AC = AD = a$; $BC = CD = DB = b$. What is its (a) rotation group; (b) full group? Would there be any further symmetry if $BAC = CAD = DAB = \text{rt} \angle$?

15 A tetrahedron $ABCD$ has $CA = AB = BD = a$; $AD = DC = CB = b$. What is its rotation group, full group?

16 What is the symmetry group of the stella octagula? (This solid consists of two congruent regular tetrahedra whose eight vertices lie at the vertices of a cube.)

17 The tetrahedral group (p. 249) has exactly three elements of period 2, and these, together with the identity, form the group D_2. If a finite group has exactly three elements of period 2, these do not always generate a subgroup with D_2 structure. Find a counter-example to show this. Can you find two distinct counter-examples?

18 A tetrahedron has vertices (0, 0, 0), (1, 0, 0), (0, 1, 0), (1, 0, 1). Find its rotational symmetry(s). The group of rotations of any tetrahedron must be a subgroup of A_4. (Why?) Describe the classes of tetrahedron whose rotation groups are: (a) C_2, (b) C_3, (c) D_2.

19 The rotation group of any parallelepiped must be a subgroup of the group of rotations of the cube. Now S_4 has subgroups with structure C_2, C_3, C_4, D_2, D_3, D_4 and A_4 (see Ch. 14). Can you find which types of parallelepiped has each of these subgroups as its symmetry group? If you cannot find a parallelepiped find some type of hexahedron.

20 By considering the symmetry groups of plane figures, find the groups of automorphisms of C_8 and of D_4.

21 Interpret the cycles (*abcd*), (*abc*), (*ab*) and (*ab*)(*cd*) as symmetries of the regular octahedron by taking *a*, *b*, *c* and *d* to be four pairs of opposite faces.

22 In fig. 17.04 the diagonals of the cube are denoted *a*, *b*, *c* and *d*. Which symmetries of the cube are represented by the permutations:

$$\begin{pmatrix} a & b & c & d \\ d & a & c & b \end{pmatrix}, \quad \begin{pmatrix} a & b & c & d \\ c & d & a & b \end{pmatrix}, \quad \begin{pmatrix} a & b & c & d \\ c & a & d & b \end{pmatrix}, \quad \begin{pmatrix} a & b & c & d \\ a & c & b & d \end{pmatrix}.$$

Which permutations of the diagonals correspond to a half-turn about the join of the mid-points of DC, $D'C'$? a quarter-turn about the axis joining the centres of the faces $AC'B'D$, $BD'A'C$?

23 Consider one of the subgroups D_4 of S_4. What invariant property is associated with such a subgroup in the realisation of S_4: (a) as a group of permutations of four objects; (b) as the group of rotations of the cube?

24 Consider the square prism of fig. 16.04. Let r be the quarter-turn (1234)(5678), and a the reflection (15)(26)(37)(48). Consider Gp (a, r) both from the point of view of the permutations of the vertices, of the symmetries of the figure and also from the abstract point of view. What is the central inversion in terms of a and r? Show that this group cannot be represented by permutations of the four diagonals.

Repeat the above treatment for the group $C_2 \times C_6$.

25 Discuss the symmetry groups of the Archimedean semi-regular solids and their duals.

26 Show that the symmetry group of a regular prism is isomorphic with that of the bi-pyramid (both based on a regular polygon of n sides). Is this true for the full group of symmetries? Repeat for the antiprisms and their duals.

27 Show that, if a regular tetrahedron is truncated by removing from each vertex a regular tetrahedron, all four of these being identical, then the symmetry group is unchanged. Would it be possible to remove tetrahedra from the vertices which were not regular without altering the symmetry group of the solid? Suggest how various subgroups of A_4 could be obtained by cutting corners from a regular tetrahedron (for example, if two identical tetrahedra are removed, the group is reduced to C_2). Can you generalise about the truncation of other regular solids?

28 A solid is bounded by four pairs of parallel planes $x = \pm 1$, $y = \pm 1$, $z = \pm 1$, $x + y + z = \pm 1$. Find the coordinates of its vertices, describe it, and find its volume. What is its rotation group, and also its full group of symmetries?

29 Consider the effect on the symmetry group of various solids by different ways of colouring their faces; for example, if opposite faces of a cube are coloured alike, three different colours meeting at each vertex, then the cube retains only those symmetries which retain or interchange opposite pairs of faces, and the group reduces from S_4 to D_2.

30 Draw Cayley diagrams for S_4 using various sets of generators given in Chs. 14, 15.

31 Using defining relations $p^3 = a^2 = 1 = (pa)^3$, draw a Cayley diagram for A_4.

32 Consider the effect on the symmetry groups of various plane and solid figures of cutting off some or all of the corners, when the portions removed may be isosceles triangles or tetrahedron, or where the triangles or tetrahedra may be irregular in shape. For example, if congruent isosceles triangles are removed from alternate vertices of a regular hexagon, the group is reduced from D_6 to D_3; if the triangles are not isosceles, but congruent to each other, the group may be C_3; if only one isosceles triangle is removed, the group is now merely D_1. How many isosceles tetrahedra should be removed from a cube so as to reduce the rotation group from S_4 to A_4; from S_4 to C_4; and so on.

33 Prove geometrically that A_4 has no subgroups of order 6.

34 Show that A_4 is a subgroup of A_5 by considering the regular tetrahedron inscribed in one of the five cubes which may be inscribed in the regular dodecahedron.

35 The rotation group of the cube is S_4, its full group is $S_4 \times C_2$. (1)
The rotation group of reg. tetrahedron is A_4, full group is S_4, NOT $A_4 \times C_2$. (2)
The rotation group of reg. icosahedron is A_5, full group is $A_5 \times C_2$, NOT S_5. (3)
How do you account for the apparent discrepancy between these? Verify (3) by showing that the full group contains elements of period 10. Can you describe a solid whose full symmetry group is $A_4 \times C_2$?

36 If R_1 and R_2 are coplanar rotations about A and B through angles α and β, prove that $R_1 R_2 R^{-1} R_2{}^{-1}$ is a translation. Under what circumstances could this be the identity? Deduce that a finite plane figure cannot have more than one centre of rotational symmetry. (See also Chs. 20, 25.)

18 *Permutations*

Resumé of how permutations have arisen in previous chapters

The symmetric group S_n (or Permutation group) is the group of all the permutations of n letters, and has order $n!$ It is the subgroups of the symmetric group which are of great interest, and we have analysed all the subgroups of S_3 and S_4. We have also noted – and proved – Cayley's theorem, that every group of order n is a subgroup of S_n. Every subgroup of S_n corresponds to a group formed by a subset of the permutations of n letters; for example, the subgroup $C_4 \times C_2$ of S_6 corresponds to those permutations of *A B C D E F* in which (*ABCD*) and (*EF*) are permuted as disjoint cycles. Since any of the fifteen ($= {}_6C_2$) pairs of the six letters might have been selected to form the cycle of two, it follows that we must have (at least) fifteen such subgroups of S_6 with the structure $C_4 \times C_2$. In the previous chapters, we have seen how symmetry groups may correspond to sets of permutations. For example, the group of direct symmetries of the regular tetrahedron is isomorphic to the group of even permutations of four letters; the full group of symmetries of the equilateral triangular prism is a subgroup of S_6 containing those permutations of *A B C P Q R* such that the first three places are occupied by *A*, *B*, *C* and the last three by *P*, *Q*, *R*, or vice versa; and moreover that each of the pairs *AP*, *BQ* and *CR* are separated by two letters, as for example in

$$\begin{pmatrix} A & B & C & P & Q & R \\ Q & P & R & B & A & C \end{pmatrix}$$

(compare pp. 201 ff.). Again, the group of the regular pentagon corresponds to the permutations of *A B C D E* which preserve 'adjacency', such as $\begin{pmatrix} A & B & C & D & E \\ C & B & A & E & D \end{pmatrix}$. Moreover, we saw that symmetry group of a three-dimensional figure might be considered as permutations of its faces, its vertices, or its edges, or indeed, of its diagonals. The icosahedral group A_5 was seen to correspond to the even permutations of the five inscribed cubes of the dodecahedron. It could, of course, be seen as a subgroup of S_{12}, being the group of the permutations of the faces of the dodecahedron; or of S_{15} when the symmetries are thought of as a 'reshuffling' of the fifteen lines joining the mid-points of opposite edges. And so on.

Thus the importance of groups of permutations cannot be overemphasised, and we have tried to stress its importance throughout. In this chapter we shall look a little more closely into the question of permutations.

Odd and even permutations: inversions of order†

Consider the permutation of the numbers

$$\begin{pmatrix} 1 & 2 & 3 & 4 & 5 & 6 & 7 & 8 & 9 & 10 \\ 5 & 6 & 2 & 10 & 1 & 8 & 3 & 7 & 4 & 9 \end{pmatrix}.$$

We count up how many pairs are *out of their natural order*. For example, in the natural order, 6 comes after 3, but in this permutation, the order of 3 and 6 is reversed, so we have here what is known as an 'inversion of order', or a 'reversal'. We wish to enumerate all such reversals. The best way is to take each digit in turn, and count up the number of reversals as indicated:

5 is followed by 2, 1, 3 and 4, so contributes	4 reversals
6 is followed by 2, 1, 3, 4, so contributes	4 reversals
2 is followed by 1, so contributes	1 reversal
10 is followed by 1, 8, 3, 7, 4, 9, so contributes	6 reversals
1 contributes no reversals	
8 is followed by 3, 4, 7, so contributes	3 reversals
3 contributes no reversals	
7 is followed by 4, so contributes	1 reversal
9 contributes no reversals	—
Total number of inversions of order	19

This number is odd, and so this particular permutation is described as being an *odd permutation*; or we may say that the *parity* of the permutation is odd. Alternatively, if the number of inversions of order is r, we say that the 'sign' of the permutation is $(-1)^r$, so that in this case, with $r = 19$, the sign of the permutation is minus.

Q. 18.01. Find, by counting the inversions of order, the parity of the permutations

$$\begin{pmatrix} \text{EVIL} \\ \text{VILE} \end{pmatrix}, \quad \begin{pmatrix} 1 & 2 & 3 & 4 & 5 \\ 5 & 2 & 4 & 1 & 3 \end{pmatrix}, \quad \begin{pmatrix} 1 & 2 & 3 & 4 & 5 \\ 5 & 4 & 3 & 2 & 1 \end{pmatrix}, \quad \begin{pmatrix} \text{BRUSH} \\ \text{SHRUB} \end{pmatrix},$$

$$\begin{pmatrix} 1 & 2 & 3 & 4 & 5 & 6 \\ 6 & 4 & 3 & 5 & 1 & 2 \end{pmatrix}, \quad \begin{pmatrix} 1 & 2 & 3 & 4 & 5 & 6 \\ 3 & 2 & 1 & 6 & 5 & 4 \end{pmatrix}, \quad \begin{pmatrix} 1 & 2 & 3 & 4 & 5 & 6 & 7 & 8 \\ 5 & 6 & 8 & 7 & 1 & 2 & 4 & 3 \end{pmatrix}.$$

Q. 18.02. Consider the permutation $\begin{pmatrix} 1 & 2 & 3 & 4 & 5 & 6 & 7 & 8 & 9 & 10 \\ 5 & 6 & 2 & 10 & 1 & 8 & 3 & 7 & 4 & 9 \end{pmatrix}$ which has nineteen inversions. What is the effect on the parity of: (a) interchanging the following pairs: (5 9), (2 10), (6 8); (b) applying the cycle (4 10 9)?

Q. 18.03. In the table of the group $C_3 \times C_3$ (p. 123), show that each row is an *even* permutation of the top row.

† Some readers may prefer the quite different approach to odd and even permutations to be found for example in A. Bell, *Algebraic Structures* (Allen and Unwin), pp. 102–5, and F. M. Hall, *Introduction to Abstract Algebra*, I (C.U.P.), p. 235.

Change of parity due to a single transposition

Let us next consider the effect on the parity of the permutation quoted as a consequence of a single interchange, say of the numbers 2 and 7. In the first place, the interchange of these two numbers themselves causes a change in the parity, for there is now a new reversal, (7 2). A little thought will convince you that the only other difference that can be made is on account of those numbers lying in position between 2 and 7, and of these, only those which are greater than 2 and less than 7 need to be taken into account. So in the present example, the only number which could cause any alteration in parity is the number 3; for in changing from

$$\ldots 2 \quad 10 \quad 1 \quad 8 \quad 3 \quad 7 \ldots$$

to $$\ldots 7 \quad 10 \quad 1 \quad 8 \quad 3 \quad 2 \ldots,$$

we get two new reversals, (7, 3) and (3, 2). The final effect of the interchange, therefore, is to cause three extra inversions of order to appear. In changing back again, from 7 ... 2 to 2 ... 7, these three reversals would be lost.

Next, suppose that there are several numbers lying between the pair which are being switched whose numerical value lies between them. To be definite, suppose we interchange the numbers 12 and 58, and that there are r numbers lying between them which are greater than 12 and less than 58. For example, one of them might be 49:

$$\ldots 12 \ldots 49 \ldots 58 \ldots$$

becomes $$\ldots 58 \ldots 49 \ldots 12 \ldots .$$

The effect of the switch – so far as the number 49 is concerned – is to cause two extra inversions to appear, and the same will happen for every one of the r numbers mentioned. Thus we get $2r$ extra reversals – an even number – on account of these r intermediate numbers. There is a further extra reversal due to the interchange of 12 and 58 themselves. Thus, altogether an *odd number of inversions* have appeared.

Q. 18.04. When two numbers are interchanged of which the larger one is on the left, show that an odd number of inversions is lost.

We see therefore that in all cases, *an interchange of two symbols changes the parity of the permutation.*

Permutations as the product of transpositions

Consider again the permutation $\begin{pmatrix} 1 & 2 & 3 & 4 & 5 & 6 & 7 & 8 & 9 & 10 \\ 5 & 6 & 2 & 10 & 1 & 8 & 3 & 7 & 4 & 9 \end{pmatrix}$.
Starting from the numbers in their natural order, we may obtain this

permutation by *successive interchanges of pairs* of numbers, or 'transpositions'. We may start with (1 5), which brings the number 5 into the desired position: 5 2 3 4 1 6 7 8 9 10. Next, since we require 6 in second position, we make the transposition (2 6), obtaining the order:

		5	6	3	4	1	2	7	8	9	10.
Then,	(2 3)	5	6	2	4	1	3	7	8	9	10,
next,	(10 4)	5	6	2	10	1	3	7	8	9	4,
next,	(3 8)	5	6	2	10	1	8	7	3	9	4,
next,	(3 7)	5	6	2	10	1	8	3	7	9	4,
and finally,	(4 9)	5	6	2	10	1	8	3	7	4	9.

Thus we see that the permutation may be 'resolved' into a finite number of transpositions. It is clear that the method may be extended to cover any permutation of any number of objects. The procedure is far from being unique; and the 'number of moves' will not always be the same. But though the process is laborious, we can always reduce the amount of work by using a 'system'. In the above example, we built up the required permutation symbol by symbol starting from the extreme left, and clearly such a method is applicable in the general case.

Q. 18.05. Obtain the same permutation, but by building up from the right, i.e. starting with (9 10) to get 9 in position, and so on.

Q. 18.06. Obtain the following permutations as the product of transpositions in a variety of ways, and find whether they are even or odd:

$$\begin{pmatrix} 1 & 2 & 3 & 4 & 5 & 6 \\ 4 & 6 & 3 & 5 & 1 & 2 \end{pmatrix}; \quad \begin{pmatrix} a & b & c & d & e & f & g & h \\ e & f & a & c & b & h & g & d \end{pmatrix}; \quad \begin{pmatrix} 1 & 2 & 3 & 4 & 5 & 6 & 7 \\ 6 & 5 & 7 & 1 & 3 & 4 & 2 \end{pmatrix};$$

the permutation obtained from the letters *ABCDEFGHI* by the cycles (*AFDHI*)(*CGB*)(*E*).

It should also be evident that we may always achieve the desired permutation by not more than $n - 1$ moves (less than this if we are lucky). However, it is not the *number* of moves that is important, but the *parity* of that number. In the above case, the number of transpositions was seven, and this was an odd permutation with nineteen inversions of order. We shall presently show that any *odd* permutation (i.e. having an odd number of inversions of order) is the product of transpositions the number of which will always be *odd*. Similarly, an even permutation will decompose (in an unlimited variety of ways) into transpositions, the number of which is bound to be even.

The theorem follows immediately from the 'Interchange theorem'. Starting from the natural order, suppose that a given permutation is built up from k transpositions. Then, since each transposition changes the parity of the permutation, and since the parity of the original order is even, it follows that the parity of the resulting permutation is equal to the parity of k – an even number of transpositions will restore the parity to even, an odd number will cause the final permutation to be odd.

You may like to follow through the gradual process in the case of the permutation considered above as the product of seven transpositions (table 18.01), the final number, 19, agreeing with the count on p. 312. The

Table 18.01

transposition	permutation	number of inversions	parity	sign
	1 2 3 4 5 6 7 8 9 10	0	even 0	+
(1 5)				
	5 2 3 4 1 6 7 8 9 10	0 + 7 = 7	odd 1	–
(2 6)				
	5 6 3 4 1 2 7 8 9 10	7 + 5 = 12	even 0	+
(2 3)				
	5 6 2 4 1 3 7 8 9 10	12 – 1 = 11	odd 1	–
(4 10)				
	5 6 2 10 1 3 7 8 9 4	11 + 7 = 18	even 0	+
(3 8)				
	5 6 2 10 1 8 7 3 9 4	18 + 3 = 21	odd 1	–
(3 7)				
	5 6 2 10 1 8 3 7 9 4	21 – 1 = 20	even 0	+
(4 9)				
	5 6 2 10 1 8 3 7 4 9	20 – 1 = 19	odd 1	–

permutation may also be expressed as the product of transpositions by first expressing it as the product of disjoint cycles (see pp. 111, 114 ff.): (1 5)(2 6 8 7 3)(4 10 9), and we see that this is clearly the product of the seven transpositions (reading from right to left): (4 9)(4 10)(2 3)(2 7)(2 8) (2 6)(1 5). The method may be generalised.

Q. 18.07. Repeat the above analysis for the permutations in Q.'s 10.13–17, 18.01

Q. 18.08. Consider the product of the cycles $(AIPGDM)(FRLN)(EHO)(BK)(C)$ $(J)(Q)$. Find the parity of each separate cycle. Write the composition of the cycles as a permutation of the letters $ABCD...QR$. Determine the parity (a) by a count of inversions of order, and (b) by obtaining the permutation as the product of transpositions in several ways. What is the period of this permutation?

Q. 18.09. Consider successive permutations of 1 2 3 4 5 6 produced by applying the cycle (1 4 6 3 2 5), and show that they are successively even and odd. Repeat for a cycle of period 7. Can you generalise the result?

Q. 18.10. If n symbols are written down backwards, is the result an even or an odd permutation?

Q. 18.11. Prove that the product of two permutations of like parity is even, of unlike parity is odd.

Q. 18.12. Classify the even and odd permutations of 1 2 3; 1 2 3 4; 1 2 3 4 5.

The pairing of odd with even permutations: the alternating groups

Now we have already noted that of the twenty-four permutations of four symbols, there are twelve even and twelve odd. If you pursued the above Q. 18.12 to the bitter end, you will have discovered that the number of even and of odd permutations of five symbols are also equally balanced. We shall now show that, **of the *n*! permutations of *n* objects, there are an equal number of odd and even permutations.**

Let us do this in the case of five letters *ABCDE*. Suppose that a list of all the *even* permutations is written out in any order, e.g.

1	*A*	*B*	*C*	*D*	*E*
2	*A*	*B*	*D*	*E*	*C*
3	*A*	*B*	*E*	*C*	*D*
4	*B*	*A*	*D*	*C*	*E*
5	*C*	*D*	*A*	*B*	*E*
6	*A*	*C*	*B*	*E*	*D*, etc.

Then an odd permutation may be obtained from each even permutation by making a single transposition. Thus, if the transposition (*DE*) is performed on No. 1, we get *ABCED*; and if (*BC*) is performed on No. 6, we get *ABCED*, a repeat of the odd permutation produced from line 1. What we are going to do is to manage these transpositions in such a way that there are *no repetitions* of odd permutations. This is easy, for all we have to do is to be consistent by always interchanging the same pair. This may mean that we always decide to switch (say) *A* and *B* wherever they may be, which would give table 18.02. Alternatively, we may agree always

Table 18.02

	even perms.						odd perms.				
1	*A*	*B*	*C*	*D*	*E*	→	*B*	*A*	*C*	*D*	*E*
2	*A*	*B*	*D*	*E*	*C*	→	*B*	*A*	*D*	*E*	*C*
3	*A*	*B*	*E*	*C*	*D*	→	*B*	*A*	*E*	*C*	*D*
4	*B*	*A*	*D*	*C*	*E*	→	*A*	*B*	*D*	*C*	*E*
5	*C*	*D*	*A*	*B*	*E*	→	*C*	*D*	*B*	*A*	*E*
6	*A*	*C*	*B*	*E*	*D*	→	*B*	*C*	*A*	*E*	*D*, etc.

to interchange the pair of letters occurring at the beginning (table 18.03). Now for every even permutation, we must get a 'matching' odd permutation; and since all the even permutations are different, it is easy to see that all the odd permutations must also be different. Therefore the number of even permutations must be *60 or less*. (If, for example, there had been fifty-five even permutations, there would be fifty-five matching odd permutations, and it is feasible that there could be ten more unmatched odd permutations.) But if we now repeat the above argument, interchanging the words 'even' and 'odd' (so that the arrows in the lists point the other way), it is clear that we can equally well say that the number of odd permutations must be sixty or less. The only possible conclusion therefore is that there are sixty of each. The argument may evidently be imitated to prove that for any value of n, the number of even and the number of odd permutations is $\frac{1}{2}n!$

Table 18.03

	even perms.		odd perms.
1	$A\ B\ C\ D\ E$	→	$B\ A\ C\ D\ E$
2	$A\ B\ D\ C\ E$	→	$B\ A\ D\ E\ C$
3	$A\ B\ E\ C\ D$	→	$B\ A\ E\ C\ D$
4	$B\ A\ D\ C\ E$	→	$A\ B\ D\ C\ E$
5	$C\ D\ A\ B\ E$	→	$D\ C\ A\ B\ E$
6	$A\ C\ B\ E\ D$	→	$C\ A\ B\ E\ D$, etc.

Now we have seen (Q. 18.11) that the even permutations are *closed* under successive application. Therefore these even permutations form a group (see theorem on subgroups, p. 217), and this is the *alternating group*, A_n.

Q. 18.13. One could *define* the parity of a permutation to be the parity of the number of transpositions of which it may be composed. But this would involve proving that if a given permutation can be built up in one way from p transpositions, and in another way from q transpositions, then p and q have the same parity (i.e. that $|p - q|$ is even). Prove this, by first proving that the identity permutation will always be even in the sense defined.

Q. 18.14. By using the fact that two even permutations combine to give an even permutation, and two odd permutations combine to give an even one, prove that the numbers of even and of odd permutations is each $\frac{1}{2}n!$

A set of permutations to represent the group S_4

The table on p. 219 is an abstract table showing the group S_4, and a rewarding exercise will be to realise it as the group of permutations of A, B, C and D. There will be a large number of ways of doing this (the group having a large number of automorphisms), but we shall be content with producing one solution.

In the first place, $\{1, a, b, c, d, f\}$ is a subgroup D_3 (or S_3), so it would appear to be a good start to let these represent the six permutations which leave D invariant, i.e. the six permutations of A, B and C. Since c and d are of period 3, we shall use these to denote the cycles (ABC) and (ACB), though not necessarily respectively; while a, b, f (of period 2) must represent the cycles (BC), (CA) and (AB), also not necessarily respectively. In fact the group S_3 requires two generators, which means that we are free to make two arbitrary selections. Suppose we decide that a shall be the transposition (AB) and b (BC). Then since ab is the cycle (ABC) while from the table $c = ab$, we must have c and d representing respectively the cycles (ABC) and (ACB). This leaves f to stand for (AC), and to check this we may note that $ca = f$ from the table, and the product of the cycles first (AB), then (ABC) is the transposition (AC), which is indeed represented by f.

Next we step outside the subgroup $\{1, a, b, c, d, f\}$ and come to the element g, which is of period 2. We must now move the letter D, and let us try making g the transposition (CD). Since $h = ga$, $i = dh = dga$, $j = fh = fga$, $k = bh = bga$, $l = ch = cga$, this set of elements $\{g, h, i, j, k, l\}$ are all determined. For example, $j = fga$ is the product of transpositions $(AC)(CD)(AB)$, and this is the permutation $\begin{pmatrix} A & B & C & D \\ B & C & D & A \end{pmatrix}$, i.e. the cycle ($ABCD$) of period 4. All the rest of the set are now determined, and you may verify that the permutations represented are as shown in table 18.04.

Table 18.04

	A	B	C	D		A	B	C	D		A	B	C	D		A	B	C	D
e	A	B	C	D	*g*	A	B	D	C	*m*	A	D	B	C	*t*	D	A	B	C
a	B	A	C	D	*h*	B	A	D	C	*n*	B	D	A	C	*u*	D	B	A	C
b	A	C	B	D	*i*	A	C	D	B	*p*	A	D	C	B	*v*	D	A	C	B
c	B	C	A	D	*j*	B	C	D	A	*q*	B	D	C	A	*w*	D	B	C	A
d	C	A	B	D	*k*	C	A	D	B	*r*	C	D	A	B	*x*	D	C	A	B
f	C	B	A	D	*l*	C	B	D	A	*s*	C	D	B	A	*y*	D	C	B	A

Q. 18.15. Write each of these as cycles or product of cycles. Obtain a different realisation of S_4 by starting with the permutations which leave A invariant to be the subgroup $\{1, u, l, g, f, w\}$.

Q. 18.16. Show how each of the elements in the structure table for S_4 may be made to represent one of the direct symmetries of the cube by considering permutations of the diagonals. (Compare Ch. 17, p. 298). For example, the permutations prepresented by h, r and y above, each of which causes *two transpositions*, must represent half-turns about the three axes through the centre parallel to the edges; whereas the other elements of period 2 will represent half-turns about joins of mid-points of opposite edges. Note that half-turns about three mutually perpendicular axes together with the identity will give subgroups D_2, of which one is $\{1, h, r, y\}$. Identify the other Klein subgroups. What is it that is different about $\{1, h, r, y\}$? To which motions of the cube do the subgroups C_4: Gp (j), Gp (k) and Gp (s) correspond?

We shall see in the next chapter that the four columns of the table above are *cosets* of the subgroup $\{1, a, b, c, d, f\}$ – you will observe that the coset $\{g, h, i, j, k, l\}$ moves D to third position, $\{m, n, p, q, r, s\}$ moves D to second position, and $\{t, u, v, w, x, y\}$ moves D to the front. Observe also, that in any *row*, the order of the letters A, B and C is preserved; e.g.

$$d\,(\mathbf{C\ A\ B}\ D), \qquad k\,(\mathbf{C\ A}\ D\ \mathbf{B}), \qquad r\,(\mathbf{C}\ D\ \mathbf{A\ B}), \qquad x\,(D\ \mathbf{C\ A\ B}).$$

But these rows are not (disappointingly!) cosets, for the first row $\{e, g, m, t\}$ is not a subgroup. (Check this!)

The process demonstrated above has revealed that the whole group S_4 may be generated by the three elements a, b and g.† Any element of the group may be expressed in terms of these three generators. For instance, we may obtain x in terms of a, b and g by many different methods. From the table, $x = kd = (dg)d = bagba$. This may be rearranged in various ways, using such relations as $ag = ga$,‡ $(ab)^2 = ba$, etc. Using permutations, we could obtain x as a combination of transpositions (AB), (BC), (CD) thus:

	A	B	C	D	
a	B	A	C	D	$\}a$
ba	C	A	B	D	$\}b$
gba	D	A	B	C	$\}g$ ← thus bringing D to the front
$agba$	D	B	A	C	$\}a$
$x = bagba$	D	C	A	B	$\}b$ ← leave D alone!

One could do the same sort of thing in terms of another realisation of S_4, e.g. by using the set of 3×3 matrices (see p. 301), or by considering the symmetries of the cube.

Q. 18.18. Give a set of defining relations on a, b and g. Find other sets of three generators. Express all the other elements in terms of a, b and g: do this both by using the abstract table, and also by using the transpositions (AB), (BC) and (CD) on the letters $ABCD$.

Q. 18.19. Find which of the twenty-four letters in the table for S_4 represent even permutations of A, B, C, D, and check that they do form the group A_4.

Q. 18.20. Our example gave generators a, b and g each of period 2. Give other sets of generators and defining relations of S_4, of different periods (3, 4). S_4 may be generated by two elements only. Can you find such a minimum set? (*Hint:* one must be of period 4, and the other may be of period 2, 3 or 4.)

† This should be expected! Any permutation can be effected as a product of transpositions, and the transpositions (AB), (BC) and (CD) enable us to move the four letters about to wherever we please.

‡ **Q. 18.17.** Give a quick reason why a and g commute . . .

Q. 18.21. Show that a cycle of length n is an odd permutation when n is even, but an even permutation when n is odd. Hence show that A_4 contains no elements of period 4, but that A_5 does contain elements of period 5, i.e. that C_5 is a subgroup of A_5. Illustrate these results by the groups of symmetries of the tetrahedron and the icosahedron. Can you generalise the result for A_n?

Q. 18.22. The permutations of the three letters A, B, C give the group S_3. They contain three even and three odd permutations. The six permutations (of six letters) on p. 76 also give S_3, and also contain three odd and three even. Find a group of six *even* permutations which give the group S_3. (*Hint*: the elements of period 2 must be double transpositions; use the work on p. 206 ff. on the construction of D_n by means of two generators of period 2.)

Q. 18.23. Is it possible for a group of permutations some of which are odd to contain unequal numbers of even and odd permutations?

Q. 18.24. Prove that if a group of permutations is of odd order, then every permutation of the group *is even.*

Two generators only needed for S_4

The next question we consider is: how can the group S_4 be constructed with only two generators? Or are a, b and g a *minimum* set of generators? The answer to the latter question is 'No', and indeed, the group S_4 may be constructed entirely from a cycle of period 4 as well as one of period 2; for example, $j = (ABCD)$ and $a = (AB)$ (see Q. 14.19, 15.04, 18.29, 18.39, 20.02 and Exs. 15.40, 15.41, 18.14, 18.27). For we may proceed in the above example to find Gp (j, a).

	1	j	r	t	a	l	x	m	i	s	u	k	y	...
j	j	r	t	e	l	x	m	a	...					
a	a	i	s	u	e	k	y	...						

Having found thirteen elements generated by j and a, we may be sure they generate the whole group. (Why?)

Another way of seeing this is to consider the following products:

$$\left.\begin{aligned} jaj^{-1} &= jaj^3 = (ABCD)(AB)(DCBA) = (BC) = b \\ j^2aj^{-2} &= j^2aj^2 = (AC)(BD)(AB)(AC)(BD) = (CD) = g \\ j^{-1}aj &= j^3aj = (DCBA)(AB)(ABCD) = (DA) = w \end{aligned}\right\} \text{compare pp. 363–5.}$$

and, since any three of the single transpositions a, b, g or w will generate S_4, then so will the two elements a and j.

Q. 18.25. Express the elements of S_4 in terms of a and j. Can you find defining relations to supplement $a^2 = j^4 = 1$? Is S_4 generated by *any* element of period 4 and any element of period 2 other than its square? Can you generate S_4 by an element of period 4 and one of period 3? Show that S_4 may also be generated by two independent cycles of period 4, and interpret in terms of the rotations of the cube.

Cycles

We have seen (Ch. 10, pp. 114 ff.) that any permutation whatever may be resolved into *disjoint* cycles. For example, the permutation

$$\begin{pmatrix} A & B & C & D & E & F & G & H & I & J & K & L & M & N & O & P & Q & R \\ I & K & C & M & H & R & D & O & P & J & B & N & A & F & E & G & Q & L \end{pmatrix}$$

is the product of the cycles ($AIPGDM$), ($FRLN$), (EHO) and (BK) (see Q. 18.08). These cycles may be described as being of 'length' 6, 4, 3 and 2 respectively, and these are the periods of the individual subgroups which each generates. What is the period of the permutation? Evidently it is 12 in this case, for when the permutation is applied twelve times, the first cycle will have gone round twice, the second will have made three complete 'revolutions', the cycle (EHO) will have been completed four times, while the letters B and K will have changed places twelve times. This is an illustration of a general theorem:

The period of a permutation is the L.C.M. of the lengths (or periods) of its *disjoint* cycles. (See also p. 115.)

Note that the word *disjoint* is essential: the product of the non-disjoint (overlapping) cycles (AB) and (BC) is the cycle (ACB) or (ABC) according to the order in which the two cycles are applied, and either of these is of period 3.† (*Note:* disjoint cycles, being completely independent of each other, always commute.)

Q. 18.26. Apply this result to show that $C_5 \times C_3 \cong C_{15}$, but $C_4 \times C_6 \not\cong C_{24}$.

Example

If $a = (1\ 6\ 3)$, $b = (1\ 3\ 5\ 7)$, $c = (6\ 7)$, $d = (1\ 2\ 3\ 4\ 5)$, represent the permutation $abcd$ as the product of disjoint cycles. We have

$$1\,2\,3\,4\,5\,6\,7 \xrightarrow{d} 2\,3\,4\,5\,1\,6\,7 \xrightarrow{c} 2\,3\,4\,5\,1\,7\,6 \xrightarrow{b} 2\,5\,4\,7\,3\,1\,6$$
$$\xrightarrow{a} 2\,5\,4\,7\,1\,6\,3.$$

Hence $abcd$ is composed of the cycles (125)(347), and is of period 3.

† Compare p. 117.

Q. 18.27. Repeat for the product (12)(25)(34)(45)(37)(62), and also for the same product in the reverse order.

Q. 18.28. Find the permutations of highest possible period in S_5, S_6, S_7, S_8, S_9, ... (For example, a permutation of seven objects could be the product of cycles of length 3 and 4, or 2 and 5, or 2, 2 and 3, or 7 ... the L.C.M.'s are 12, 10, 6, 7, ... Hence S_7 contains cyclic subgroups C_{12}, C_{10}, C_6, C_7, ... The highest is C_{12}, and a permutation which generates C_{12} could be the product of cycles (1 2 3 4) and (5 6 7).

Q. 18.29. Show that S_5 is generated by the cycles (1 2 3 4) and (4 5). Can you generalise?

Q. 18.30. Find how many permutations of *A B C D E* there are of period 2, 3, 4, 5, etc. Compare with the discussion in Ex. 10.12. Repeat for permutations of six letters.

Q. 18.31. Prove that every permutation of period 6 of five letters is odd, and hence that C_6 is a subgroup of S_5 but not of A_5.

Q. 18.32. Show that A_4 is generated by (*A B C*) and (*B C D*) or by any two independent cycles of period 3. Find a similar result for A_5, and attempt to generalise.

Q. 18.33. Find the least value of n so that C_m is a subgroup of S_n in the cases when $m = 2, 3, 4$, etc. For example, when $m = 5$, $n = 5$; when $m = 6$, $n = 5$, since $C_2 \times C_3 \cong C_6$.

Q. 18.34. We know that S_3 may be generated by two transpositions (12) and (23), and S_4 may be generated by three transpositions (12), (23) and (34). Find a set of generators for the general symmetric group S_n.

Q. 18.35. By considering a certain set of permutations of five letters *ABCPQ*, show that D_6 is a subgroup of S_5.

Manipulation of cycles

You are reminded that the notation for cycles is as follows: (*ABCDEFG*) means that *A* becomes (is replaced by) *B*, *B* becomes *C*, ..., *F* becomes *G*, *G* becomes *A*, i.e. the permutation is $\begin{pmatrix} A & B & C & D & E & F & G \\ B & C & D & E & F & G & A \end{pmatrix}$.† The cycle notation is not entirely satisfactory since it does not readily draw attention to the fact that the last letter in the list (*G*) is replaced by the first (*A*). The possible improvement, (*ABCDEFGA*) suffers from the defect that its 'length' is eight letters, whereas the period is in fact 7, and this is

† *Note*: the cycle (*GFEDCBA*) means the permutation $\begin{pmatrix} A & B & C & D & E & F & G \\ G & A & B & C & D & E & F \end{pmatrix}$. A common error for the beginner is to imagine that the cycle (321) means the permutation $\begin{pmatrix} 1 & 2 & 3 \\ 3 & 2 & 1 \end{pmatrix}$; it is nothing of the sort, the cycle (321) means the permutation $\begin{pmatrix} 1 & 2 & 3 \\ 3 & 1 & 2 \end{pmatrix}$.

apt to be misleading. Perhaps the best device would be to arrange the letters of the cycle in a circle as in fig. 18.01. With some such convention as that the 'arrows' indicate the direction of movement of the letters: $B \rightarrow A$ meaning 'B replaces A', or 'A is replaced by B'. Even this has its disadvantages, for some books have the arrows in the reverse direction. However, the traditional 'cycle notation' has an advantage over the permutation notation, that the latter may be written down in $n!$ ways (for a cycle of length n), whereas the cycle notation admits of only n

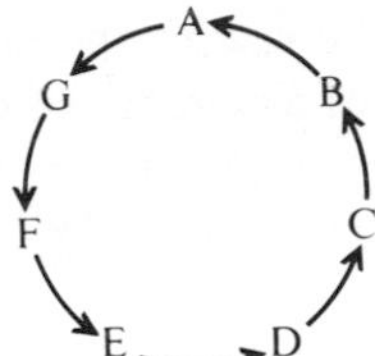

Fig. 18.01.

alternatives; e.g. when $n = 7$, the cycle $(ABCDEFG)$ may be written in seven ways, such as $(DEFGABC)$; whereas when written as a permutation

$$\begin{pmatrix} A & B & C & D & E & F & G \\ B & C & D & E & F & G & A \end{pmatrix} \quad \text{or} \quad \begin{pmatrix} A & D & G & F & C & B & E \\ B & E & A & G & D & C & F \end{pmatrix}$$

$$\text{or} \quad \begin{pmatrix} F & C & A & D & G & E & B \\ G & D & B & E & A & F & C \end{pmatrix}, \text{ etc.}$$

there are 5,040 ways! Moreover, when a permutation contains several disjoint cycles, the cycle notation calls instant attention to the independence of these cycles.

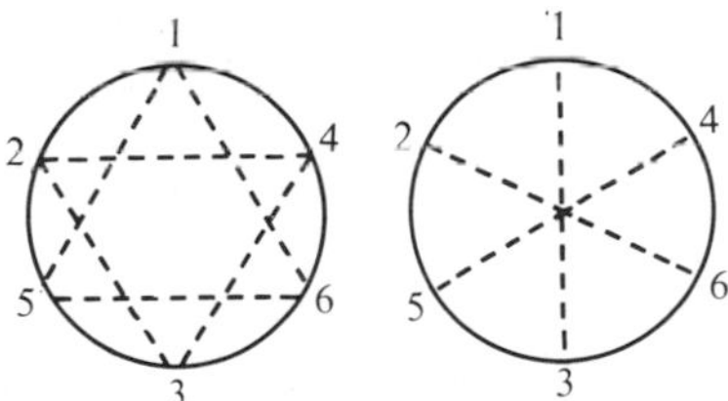

Fig. 18.021. The cycles x^2.
18.022. The cycles x^3.

Consider the permutation of period 6, $x = \begin{pmatrix} 1 & 2 & 3 & 4 & 5 & 6 \\ 4 & 1 & 5 & 6 & 2 & 3 \end{pmatrix}$. This is formed by the cycle (1 4 6 3 5 2), and if we require x^2, we might argue thus: x replaces 1 by 4, and then 4 by 6, so x^2 replaces 1 by 6; 6 is replaced by 3, then 3 by 5; 5 is replaced by 2, then 2 by 1 – so we have the cycle (1 6 5) of period 3. Similar reasoning shows the cycle (4 3 2), so x^2 is the product of the cycles (1 6 5)(4 3 2) – see fig. 18.021. Again, x^3 replaces 1 by 4, 4 by 6, 6 by 3; and 3 by 5, 5 by 2, 2 by 1, so x^3 contains the cycle (1 3), and similarly the cycles (4 5) and (6 2) – see fig. 18.022. The work should be compared with the discussion of cyclic groups of composite order, pp. 170 ff. With practice, one can easily read off the powers of a given cycle:

$$y = (1\ 2\ 3\ 4\ 5\ 6\ 7\ 8) \Rightarrow y^2 = (1\ 3\ 5\ 7)(2\ 4\ 6\ 8);$$
$$y^3 = (1\ 4\ 7\ 2\ 5\ 8\ 3\ 6); \qquad y^4 = (1\ 5)(2\ 6)(3\ 7)(4\ 8),$$

and so on. Inverses can also be written down readily: $x^{-1} = (2\ 5\ 3\ 6\ 4\ 1)$; $y^{-1} = (8\ 7\ 6\ 5\ 4\ 3\ 2\ 1)$, and the rule is obvious.

The inverse of the product of a number of cycles should give no difficulty when the cycles are disjoint: if $w = (1\ 2\ 3\ 4)(5\ 6\ 7)(8\ 9)(10)$, then $w^{-1} = (10)\ (9\ 8)(7\ 6\ 5)(4\ 3\ 2\ 1) = (4\ 3\ 2\ 1)(7\ 6\ 5)(8\ 9)$.

Q. 18.36. Show that in general, if x is a permutation, then x^{-1} contains the same independent cycles as x.

Q. 18.37. Write the permutation $\begin{pmatrix} 1 & 2 & 3 & 4 & 5 & 6 \\ 5 & 6 & 4 & 1 & 2 & 3 \end{pmatrix}$ as a cycle, and hence write down its various 'powers'.

Overlapping cycles

When the cycles overlap, say $s = (1\ 2\ 3\ 4\ 5)(2\ 4\ 7)(6\ 1)$, we may proceed as follows. Note first of all, that the cycles are applied in the reverse order from the order in which they are written down (see pp. 28 ff.), so that the permutations produced are as follows:

$$1\,2\,3\,4\,5\,6\,7 \xrightarrow{(6\ 1)} 6\,2\,3\,4\,5\,1\,7 \xrightarrow{(2\ 4\ 7)} 6\,4\,3\,7\,5\,1\,2$$
$$\xrightarrow{(1\ 2\ 3\ 4\ 5)} 6\,5\,4\,7\,1\,2\,3$$

so we see that s resolves itself into the *independent* cycles (1 6 2 5)(3 4 7) and it follows that $s^{-1} = (5\ 2\ 6\ 1)(7\ 4\ 3)$.

The same result might equally well have been obtained by considering the effect of s on each digit in turn, e.g. $6 \xrightarrow{(6\ 1)} 1 \xrightarrow{(2\ 4\ 7)} 1 \xrightarrow{(1\ 2\ 3\ 4\ 5)} 2$. Again, 2 is replaced first by 4 and then by 5, and so on.

Q. 18.38. Find the cube of $(AB)(CDE)$; show that $(ABCDE)(ABCED) = (AC)(BD)$. (You should be able to do this without writing anything down.)

Q. 18.39. Show that the cycles $p = (1\ 2\ 3\ 4\ 5)$ and $q = (2\ 3\ 4\ 5)$ generate the whole of S_5.

Finally, consider the permutation $y \begin{pmatrix} 1 & 2 & 3 & 4 & 5 & 6 & 7 & 8 \\ 7 & 6 & 8 & 1 & 4 & 3 & 5 & 2 \end{pmatrix}$ which is the product of the cycles (1 7 5 4) (2 6 3 8), and is evidently of period 4.

Applying the work of the previous paragraph, we have

$$y^2 = (1\ 5)(7\ 4)(2\ 3)(6\ 8); \qquad y^3 = (1\ 4\ 5\ 7)(2\ 8\ 3\ 6).$$

Again, if z is the permutation $\begin{pmatrix} 1 & 2 & 3 & 4 & 5 & 6 & 7 & 8 \\ 4 & 7 & 5 & 8 & 1 & 3 & 2 & 6 \end{pmatrix}$, containing the cycles (1 4 8 6 3 5)(2 7), then

$$z^2 = (1\ 8\ 3)(4\ 6\ 5)(2)(7); \qquad z^3 = (1\ 6)(4\ 3)(8\ 5)(2\ 7), \text{ and so on.}$$

Taking x as (1 4 6 3 5 2)(7)(8), and remembering $y = (1\ 7\ 5\ 4)(2\ 6\ 3\ 8)$ we now form the product xy (first y, then x, with x and y containing cycles which are no longer disjoint), and set out the work as follows:

$$\left.\begin{matrix} 1 \xrightarrow{y} 7 \xrightarrow{x} 7 \\ 7 \rightarrow 5 \rightarrow 2 \\ 2 \rightarrow 6 \rightarrow 3 \\ 3 \rightarrow 8 \rightarrow 8 \\ 8 \rightarrow 2 \rightarrow 1 \end{matrix}\right\} \text{ so we have the cycle } (1\ 7\ 2\ 3\ 8) \text{ of length 5}$$

$$4 \xrightarrow{y} 1 \xrightarrow{x} 4 \quad \text{so 4 is unchanged}$$

$$\left.\begin{matrix} 5 \rightarrow 4 \rightarrow 6 \\ 6 \rightarrow 3 \rightarrow 5 \end{matrix}\right\} \text{ so we have the transposition } (5\ 6).$$

Hence xy is the product of cycles (1 7 2 3 8)(4)(5 6), and this has period 10.

Q. 18.40. Repeat for the product yx in the reverse order.

The group of a polynomial

Consider the expression $f(x_1, x_2, x_3, x_4) \equiv (x_1 - x_2)^2 + (x_2 - x_3)^2 + (x_3 - x_4)^2 + (x_4 - x_1)^2$. The value of this function is unaltered when the subscripts undergo any of the following permutations:

$$\begin{pmatrix} 1 & 2 & 3 & 4 \\ 1 & 2 & 3 & 4 \end{pmatrix}, \quad \begin{pmatrix} 1 & 2 & 3 & 4 \\ 4 & 1 & 2 & 3 \end{pmatrix}, \quad \begin{pmatrix} 1 & 2 & 3 & 4 \\ 3 & 4 & 1 & 2 \end{pmatrix},$$

$$\begin{pmatrix} 1 & 2 & 3 & 4 \\ 2 & 3 & 4 & 1 \end{pmatrix}, \quad \begin{pmatrix} 1 & 2 & 3 & 4 \\ 4 & 3 & 2 & 1 \end{pmatrix}, \quad \begin{pmatrix} 1 & 2 & 3 & 4 \\ 1 & 4 & 3 & 2 \end{pmatrix},$$

$$\begin{pmatrix} 1 & 2 & 3 & 4 \\ 2 & 1 & 4 & 3 \end{pmatrix}, \text{ or } \begin{pmatrix} 1 & 2 & 3 & 4 \\ 3 & 2 & 1 & 4 \end{pmatrix}.$$

For example, the fourth of these permutations replaces the expression by:

$$f(x_2, x_3, x_4, x_1) \equiv (x_2 - x_3)^2 + (x_3 - x_4)^2 + (x_4 - x_1)^2 + (x_1 - x_2)^2,$$

and the last one gives

$$f(x_3, x_2, x_1, x_4) \equiv (x_3 - x_2)^2 + (x_2 - x_1)^2 + (x_1 - x_4)^2 + (x_4 - x_3)^2,$$

and in both cases the value of the expression is unaltered. These permutations of suffices are precisely the same as the permutations of the vertices of a square under its eight symmetries (see fig. 18.03), and they correspond to the group D_4.

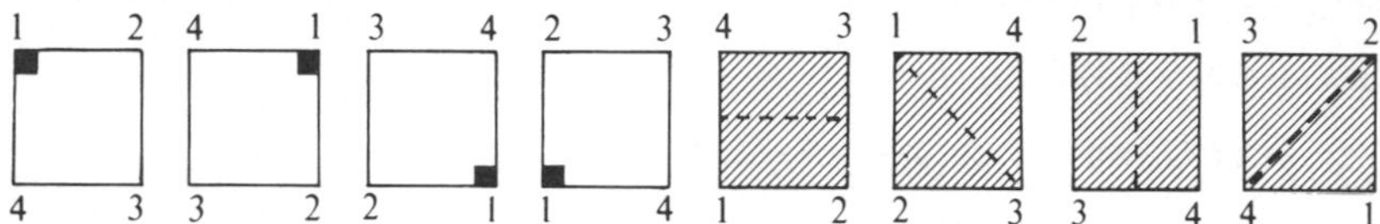

Fig. 18.03.

Just as we say that the group D_4 *describes the symmetries of the square, so we may also say that this same group describes the symmetry of this polynomial* $f(x_1, x_2, x_3, x_4)$. Note that

$$\begin{aligned} f(x_1, x_2, x_3, x_4) &= \Sigma(x_1 - x_2)^2 = \Sigma x_1^2 - 2\,\Sigma x_1x_2 + x_2^2 \\ &= 2\,\Sigma x_1^2 - 2\,\Sigma x_1x_2 \\ &= 2\,\Sigma x_1^2 - (x_1x_2 + x_2x_3 + x_3x_4 + x_4x_1) \\ &= 2\,\Sigma x_1^2 - (x_1 + x_3)(x_2 + x_4) \qquad \ldots \quad (1). \end{aligned}$$

The first part of this expression, Σx_1^2, is invariant under *any* permutation of the four subscripts – it is 'completely symmetrical' in x_1, x_2, x_3, x_4, and its symmetry group is S_4. In fact the term 'symmetric group' is used because this group describes the symmetries of a polynomial which is symmetric under all the possible permutations of the variables x_1, x_2, x_3, x_4. Thus, S_n describes the symmetries of such expressions as $x_1 + x_2 + x_3 + \ldots + x_n$; $x_1x_2x_3 \ldots x_n$; $x_1x_2 + x_1x_3 + x_1x_4 + \ldots + x_1x_n + x_2x_3 + x_2x_4 + \ldots + x_2x_n + x_3x_4 + \ldots + x_3x_n + \ldots + x_{n-1}x_n$; $\sum_{r=1}^{n} x_r^3$, and so on. The second part of the expression (1), $(x_1 + x_3)(x_2 + x_4)$, is not completely symmetric in x_1, x_2, x_3, x_4 but naturally enough has the same symmetries as the original expression, forming the group D_4.

Since this particular polynomial is invariant under eight of the twenty-four possible permutations of the four variables, the polynomial may take *three* distinct values, and so we say it is 'three-valued'. A polynomial such as $x_1 + 3x_2 - x_3 - 2x_4$ which possesses no symmetry at all would, of course, be twenty-four-valued; that is, by permuting the four suffices the resulting expression would take twenty-four distinct values. A completely symmetrical polynomial, such as $x_1^2 + x_2^2 + x_3^2 + x_4^2$ is one-valued, for however the suffices are reshuffled, the value of the expression is unaltered.

Again, the expression $a^2b + b^2c + c^2a$ is invariant under the group C_3 but not under the group D_3 – the cycles (abc) or (acb) leave the expression unchanged, but the transpositions (bc), (ca) or (ab) do not. The same applies to the expression $\sum_{a,b,c} ab(a - b)$. Both expressions are evidently two-valued.

Q. 18.41. If $x_1 < x_2 < x_3 < x_4$, which set of permutations on the subscripts give the expression $f(x_1, x_2, x_3, x_4)$ above its least value? Interpret the results in terms of four points lying on a line *ABCD*: let $OA = x_1$, etc. where O is an origin.

Cross-ratio

The above remarks are not restricted to polynomials only.

$$\text{Let}\quad F(x_1, x_2, x_3, x_4) \equiv \frac{(x_1 - x_2)(x_3 - x_4)}{(x_1 - x_4)(x_3 - x_2)}$$

The permutations of the subscripts under which the value of this function is invariant are easily seen to be

$$\begin{pmatrix}1 & 2 & 3 & 4\\ 1 & 2 & 3 & 4\end{pmatrix}, \begin{pmatrix}1 & 2 & 3 & 4\\ 2 & 1 & 4 & 3\end{pmatrix}, \begin{pmatrix}1 & 2 & 3 & 4\\ 3 & 4 & 1 & 2\end{pmatrix} \text{ and } \begin{pmatrix}1 & 2 & 3 & 4\\ 4 & 3 & 2 & 1\end{pmatrix},$$

and these permutations form the group D_2 so that we may expect the function to take not twenty-four different values, but six, each of the six ratios being duplicated under *two interchanges* of suffices. This function is the important *Cross-ratio* of the four numbers x_1, x_2, x_3, x_4. Thus we see that four numbers have *six* distinct† cross-ratios according to the order in which they are taken. It is possible for each of the cross-ratios to be expressed in terms of the others. For example, if $F(x_1, x_2, x_3, x_4) = \lambda_1$, suppose we require:

$$F(x_2, x_4, x_3, x_1) = \frac{(x_2 - x_4)(x_3 - x_1)}{(x_2 - x_1)(x_3 - x_4)} = \lambda_2.$$

We may start with the identity:

$$(x_1 - x_2)(x_3 - x_4) + (x_2 - x_3)(x_1 - x_4) + (x_3 - x_1)(x_2 - x_4) \equiv 0‡$$

† Distinct when the four numbers are also distinct (no two equal).

‡ Note that the group of the function on the left-hand side is C_3, since it is invariant under the cycles (1 2 3) and (1 3 2) of the suffices. If x_1, x_2, x_3, x_4 were distances from an origin along a fixed line to points A, B, C, D, then the expression would be $BA.DC + CB.DA + AC.DB$. The fact that it is equal to zero for all positions of A, B, C and D may be deduced as a limiting case of Ptolemy's theorem for cyclic quadrilaterals (see p. 490).

$$\Rightarrow 1 + \frac{(x_2 - x_3)(x_1 - x_4)}{(x_1 - x_2)(x_3 - x_4)} + \frac{(x_3 - x_1)(x_2 - x_4)}{(x_1 - x_2)(x_3 - x_4)} = 0$$

$$\Rightarrow 1 + \frac{1}{-\lambda_1} + (-\lambda_2) = 0 \Rightarrow \lambda_2 = \frac{\lambda_1 - 1}{\lambda_1}.$$

If you deal in a similar way with the other possible values of the cross-ratio, you will find:

$$\mathrm{F}(x_1, x_2, x_3, x_4) = \lambda_1 \qquad \mathrm{F}(x_1, x_3, x_2, x_4) = \lambda_4 = 1 - \lambda_1$$

$$\mathrm{F}(x_3, x_1, x_2, x_4) = \lambda_2 = \frac{\lambda_1 - 1}{\lambda_1} \qquad \mathrm{F}(x_3, x_2, x_1, x_4) = \lambda_5 = \frac{1}{\lambda_1}$$

$$\mathrm{F}(x_2, x_3, x_1, x_4) = \lambda_3 = \frac{1}{1 - \lambda_1} \qquad \mathrm{F}(x_2, x_1, x_3, x_4) = \lambda_6 = \frac{\lambda_1}{\lambda_1 - 1}.$$

Now you should recognise these six functions as themselves forming the group D_3 under successive substitution: λ_2 and λ_3 are of period 3 and the last three of period 2. Note that we have arranged the suffices in each case in such a way that the '4' comes last. (We may do this since we are free to make two interchanges; thus when we found that:

$$\lambda_2 = \mathrm{F}(x_2, x_4, x_3, x_1) = \frac{\lambda_1 - 1}{\lambda_1},$$

and when we completed the above list, we replaced the order of the suffices by (3 1 2 4) so that '4' comes last.) The arrangement of the suffices will be seen to be as follows:

1 2 3 4 (λ_1)   1 3 2 4 (λ_4)
3 1 2 4 (λ_2)   3 2 1 4 (λ_5)
2 3 1 4 (λ_3)   2 1 3 4 (λ_6),

and this shows that each value of λ is associated with one of the six permutations of the three suffices 1, 2 and 3. This makes plain the connection between the six cross-ratios and the group S_3 ($\simeq D_3$) of the permutations of the suffices 1, 2, 3. (*Note:* Since cross-ratio is invariant under the group D_2, and the six cross-ratios themselves form the group D_3 under successive substitution, and since there are altogether $4! = 24$ possible cross-ratios, it might be thought that $S_4 \simeq D_3 \times D_2$ but this is *not so*, though, of course, both D_3 and D_2 are subgroups of S_4 (see p. 264).)

Q. 18.42. What is the group of the following expressions:

$x_1x_2 + x_3x_4$; $x_1x_2 - x_3x_4$; $(x_1 + x_2)(x_2 + x_3)(x_3 + x_1)$;
$(x_1 - x)(_2x_2 - x_3)(x_3 - x_1)$; $(x_1x_2 - x_3x_4)^2 + (x_1x_3 - x_2x_4)^2$;
$(ab - cd)^2 + (ac + bd)^2$; $(s - a)(s - b)(s - c)$, where $s = \frac{1}{2}(a + b + c)$;

$\sqrt{\left\{\dfrac{(s-b)(s-c)}{s(s-a)}\right\}}$; $\quad x_1x_3 + x_2x_4x_5$; $\quad \sum\limits_{a,b,c} b^2(a-c)$; $\quad \Sigma b^2(a+c)$;

$(x_4 - x_1)(x_4 - x_2)(x_4 - x_3)$; $\quad (p + x_1)(p + x_2)(p + x_3)$;

$x_1^2x_2 + x_3x_4^2 + x_1^2x_3 + x_2x_4^2$; $\quad a^2 + b^2 - 2ab\cos C$;

$\sqrt{\{(ab+cd)(ac+bd)(ad+bc)\}}$; $\quad \dfrac{a^2+b^2-c^2-d^2}{2(ab+cd)}$; $\quad \dfrac{x_1x_2 - x_3x_4}{x_1x_2 + x_3x_4}$;

$|(m_1 - m_2)/(1 + m_1m_2)|$; $\quad ac + ad + bc + bd$; $\quad ac + ad - bc - bd$;
$ab + c + d$.

EXERCISES 18

1 Prove that the parity of the permutation x is equal to that of x^{-1}.

2 Discover whether the following products of overlapping cycles are even or odd:

(a) $(ABC)(BDF)(EDCB)$; (b) $(EDCB)(BDF)(ABC)$;
(c) $(1\ 8\ 9)(6\ 7)(3\ 4\ 5)(1\ 2)$.

3 Prove that a permutation of period 10 on seven symbols is odd.

4 If $p = \begin{pmatrix} A & B & C & D & E \\ A & E & D & C & B \end{pmatrix}$, $\quad q = \begin{pmatrix} A & B & C & D & E \\ E & C & A & D & B \end{pmatrix}$,

$r = \begin{pmatrix} A & B & C & D & E \\ B & A & D & E & C \end{pmatrix}$,

find pqr and r^2p, and check the parity of each.

5 Analyse into cycles the following permutations of the top row in each case:

PETALS	EMIGRANTS	SAINTED	GONURSE	SPARELY	SALTER	GRANDIOSE
PASTEL	REMASTING	DETAINS	SURGEON	PARSLEY	SLATER	ORGANISED
PLATES	STREAMING	STAINED		PARLEYS	STALER	SIDE-GROAN
STAPLE	MASTERING	INSTEAD		PLAYERS	ALERTS	O-DEAR-SING
PALEST					ALTERS	E-ROAD-SIGN.

6 If $y = (1\ 2\ 3)(4\ 5)$ and $z = (2\ 4\ 3)(1\ 5)$, find x to satisfy $xyx^{-1} = z$.

7 Consider the permutation: $\begin{pmatrix} 1 & 2 & 3 & 4 & 5 \\ 3 & 5 & 1 & 4 & 2 \end{pmatrix}$.

Five lines are drawn, one joining the original to the final position of each digit. These lines intersect in 6 points, and there are 6 inversions of order. Investigate whether this method of counting inversions of order and thereby determining the parity of a permutation may be applied generally.

8 Show that the number of cycles of 6 digits of length 6 is 120. Find the number of cycles of n digits of length n.

9 If $x = (1\ 2)(3\ 4\ 5\ 6)$, $y = (1\ 2)(3\ 6\ 4\ 5)$, $z = (1\ 2\ 3)(4\ 5\ 6)$, show that $yxy^{-1}x^{-1} = (3\ 5\ 6)$ and $yzy^{-1}z^{-1} = (1\ 4\ 2\ 6\ 3)$.

10 Express the permutation $\begin{pmatrix} p & q & r & s & t & u & v & w \\ r & s & p & q & t & u & w & v \end{pmatrix}$ in terms of

$x = (p\ r\ t\ w\ q\ s\ u\ v)$ and $y = (p\ q)(r\ u\ v\ s\ t\ w)$.

11 Find permutations x and y such that $xa = b = ay$, where

$$a = \begin{pmatrix} 1 & 2 & 3 & 4 & 5 & 6 & 7 \\ 2 & 5 & 3 & 4 & 7 & 6 & 1 \end{pmatrix}, \quad b = \begin{pmatrix} 1 & 2 & 3 & 4 & 5 & 6 & 7 \\ 1 & 2 & 4 & 7 & 6 & 3 & 5 \end{pmatrix},$$

12 **Show that in a group of permutations, either all are even, or else one half are even and one half odd. What do you conclude about a group of permutations of odd order?**

13 Show that any two different permutations of four letters of period 4 combine to give one of period 3 provided they are not inverses. Check from the table for S_4 (p. 219).

14 If $p = (1\ 2\ 3\ 4)$ and $q = (1\ 3\ 2\ 4)$, find r such that $rqp = q$, and show that the group S_4 is generated by two of p, q and r (cf. Exs. 15.40, 20.15).

15 Find a cycle of length 3 of four symbols which combines with a single transposition to give a cycle of length 4. (These will generate S_4 – see Exs. 15.36, 18.41 (*b*).)

16 Prove that the cycle (1 2 3 4) will not commute with any of the permutations of 1, 2, 3 and 4 of period 2 except its square.

17 By taking $C_4 \times C_2$ to be Gp (r, a) where $r = (1\ 2\ 3\ 4)$ and a is some permutation of period 2, deduce from the previous question that a must be either of the form $(5\ X)$, or else of the form $(5\ X)(Y\ Z)$, where X, Y and Z are taken from 1, 2, 3 and 4. Show, however, that none of these commutes with r, so that $C_4 \times C_2$ cannot be a subgroup of S_5.

18 Prove that every even permutation is the product of ternary cycles (i.e. cycles of length 3) and that these cycles in general overlap. For example the permutation $\begin{pmatrix} 1 & 2 & 3 & 4 & 5 & 6 \\ 5 & 2 & 6 & 1 & 3 & 4 \end{pmatrix}$ resolves into the cycles (6 4 2)(5 3 1)(4 3 2). (Show first that a single cycle of three may be resolved into two transpositions.)

19 Suppose $x = (p\ q\ r)$ is a ternary cycle, and y is a perm. which replaces p, q and r with s, t and u. Prove that yxy^{-1} is the ternary cycle (stu).

20 Show that the even permutations of six symbols which displace all of the symbols consist either of two disjoint 3-cycles, or of two disjoint cycles of lengths 2 and 4.

21 Show that A_n may be generated by the ternary cycles (1 2 3), (1 2 4), . . . $(1\ 2\ n)$.

22 D_3 is a subgroup of S_4 though not of A_4. D_4 is a subgroup of S_4. Is D_4 a subgroup of A_5?

23 How many distinct 3-cycles are there in S_5, in S_6?
How many distinct 4-cycles are there in S_5, in S_6?

24 Show that S_6 may be generated by the two cycles (1 2) and (1 2 3 4 5 6).

25 Show that the cycle $(1\ 2\ 3\ 4\ 5\ \ldots\ n)$ is the product of the transpositions $(1\ 2)\ (1\ 3)\ (1\ 4)\ \ldots\ (1\ n{-}1)\ (1\ n)$ (applied in the order of writing down).

26 Simplify (1 2)(1 3)(2 3), and deduce that any permutation of 1, 2, 3, . . ., n may be expressed as a product of (1 2), (1 3), (1 4), . . . (1 n), which may thus be taken as generators of S_n.
Show that S_n may also be generated by the two cycles (1 2 3 . . . n) and (1 2).

27 **Show that S_n is generated by the following sets of cycles:**
(a) (1 2), (2 3), . . . (n −1 n); (b) (1 2), (1 2 3 . . . n −1 n);
(c) (1 2 3 . . . n − 1), (n −1 n).

What is the minimum number of generators of S_n?

28 Show that $A_5 \times C_2$ contains elements of period 10. (Combine a cycle) (1 2 3 4 5) with a transposition (6 7.)

29 Why does $a \to a, b \to c, c \to b, p \to q$ (table 17.02) not give an automorphism of A_4?

30 If $p = (1\ 2\ 3\ 4\ 5)$, $q = (2\ 3\ 4\ 5)$, show that pq is of period 6. What do you conclude about Gp (p, q)? Why can this group not possibly be A_5?

31 Show that $u = (1\ 2\ 3\ 4\ 5)$ and $s = (2\ 5\ 4)$ satisfy $(us)^2 = 1$, and that they generate A_5. If $a = (1\ 2)(4\ 5)$ and $p = (1\ 3\ 4)$, show $(ap)^5 = 1$, and that a and p also generate A_5.

32 Find Gp (p, q), where $q = (1\ 2\ 3\ 4\ 5)$ and p is now a cycle of period 3, such as (1 2 4), and in various other cases. Repeat when p is a transposition.

33 If $a = (1\ 2)$, $b = (2\ 3)$ and $c = (4\ 5)$, find Gp (a, b, c).

34 If $p = (1\ 2\ 3\ 4)(5\ 6\ 7\ 8)$ and $q = (1\ 5\ 3\ 7)(2\ 8\ 4\ 6)$, show that Gp (p, q) is Q_4.

35 The cycles $p = (1\ 2\ 3\ 4\ 5)$, $p^2 = (1\ 3\ 5\ 2\ 4)$, $p^3 = (1\ 4\ 2\ 5\ 3)$, $p^4 = (1\ 5\ 4\ 3\ 2)$ with the identity are the group C_5. They may be obtained by starting with 1 and following with permutations of the digits 2, 3, 4 and 5. Thus we may associate the four cycles p, p^2, p^3, p^4 with the permutations by the isomorphism:

$$p \leftrightarrow e,\ p^2 \leftrightarrow a,\ p^3 \leftrightarrow b,\ p^4 \leftrightarrow c,$$

where $e = \begin{pmatrix} 2 & 3 & 4 & 5 \\ 2 & 3 & 4 & 5 \end{pmatrix}$, $a = \begin{pmatrix} 2 & 3 & 4 & 5 \\ 3 & 5 & 2 & 4 \end{pmatrix}$,

$b = \begin{pmatrix} 2 & 3 & 4 & 5 \\ 4 & 2 & 5 & 3 \end{pmatrix}$, $c = \begin{pmatrix} 2 & 3 & 4 & 5 \\ 5 & 4 & 3 & 2 \end{pmatrix}$

and this establishes that the automorphism group of C_5 is C_4. Imitate the above procedure to show that the automorphism group of C_7 is C_6. Show how the procedure breaks down when the original cycle p is of non-prime order.

36 Show that the group of the expression $x_1x_2 + x_3 + x_4$ is a subgroup of that of $x_1x_2 + x_3x_4$.

37 Consider the symmetry of some of the formulae in the trigonometry of the triangle and quadrilateral,

e.g. $\tan \tfrac{1}{2}A = \sqrt{\left\{\dfrac{(s-b)(s-c)}{s(s-a)}\right\}}$; $r_1 = 4R \sin \tfrac{1}{2}A \cos \tfrac{1}{2}B \cos \tfrac{1}{2}C$, etc.

38 We define the 'elementary symmetric polynomials' as follows, in the case where there are four variables; x_1, x_2, x_3 and x_4:

$$p_1 = x_1 + x_2 + x_3 + x_4;$$

$$p_2 = x_1x_2 + x_1x_3 + x_1x_4 + x_2x_3 + x_2x_4 + x_3x_4;$$

$$p_3 = x_2x_3x_4 + x_1x_3x_4 + x_1x_2x_4 + x_1x_2x_3; \qquad p_4 = x_1x_2x_3x_4,$$

and each has the symmetry group S_4.
Express Σx_i^2, $\Sigma x_i^2x_j$, Σx_i^3 in terms of the p_i. (An important theorem states that any polynomial in x_1, x_2, x_3, x_4 whose group is S_4 may be expressed rationally in terms of the p_i's. Indeed, any two functions which belong to the same subgroup (such as $x_3/x_1 + x_3/x_2$ and $x_1x_2/(x_3 + 2x_4)$) may be expressed rationally each in terms of the other. The proof is difficult. The result is capable of generalisation for any number of variables.

39 Show that the even permutations of 1, 2, 3 and 4 are those which do not change the sign of the expression

$$(x_1 - x_2)(x_1 - x_3)(x_1 - x_4)(x_2 - x_3)(x_2 - x_4)(x_3 - x_4) = \prod_{i>j} (x_i - x_j).$$

For example, (1 2)(3 4) leaves the expression invariant, whereas (1 2 3 4) changes its sign. Show that any permutation may be regarded as the product of a number of transpositions, which may be done in an unlimited number of ways, all of like parity. Work out the theory of odd and even permutations on the basis of a generalisation of the above, showing that A_n is a subgroup of S_n of index 2.

40 For the group defined $p^3 = a^2 = (pa)^3 = 1$, represent a and p by permutations of 1, 2, 3 and 4. What is the structure of the group?

41 Repeat the above question in the case of groups defined:

(a) $a^2 = p^3 = (ap)^6 = 1$; (b) $a^2 = p^3 = (ap)^4 = 1$,

in each case representing the group as a set of permutations of as few digits as possible.

42 Some of the numbers on my phone dial have been rearranged, but occasionally when I dial I get the number correctly. If I don't, I dial the number I have just obtained and continue the process until I have the number originally required.

Sometimes I have to dial two, three or four times in this way, sometimes even more. I have noticed that I can get '05' by dialling twice in this manner, but to get '10' or '62' I have to dial six times. I have also established that to get '0138' directly, I have to dial '6984'. What number do I have to dial in order to obtain '27654' directly?

43 Show that the twenty-four functions:

$$i^n z; \qquad \frac{i^n}{z}; \qquad i^n \frac{z - i^m}{z + i^m};$$

where n and m take each of the values 0, 1, 2, 3 independently form the group S_4. Find which of them make the subgroup A_4.

19 *Cosets in finite and infinite groups: equivalence classes*

Cosets in the group D_6

Consider the set of permutations in table 19.01. The period of each permutation has been noted, and they have been expressed in terms of

Table 19.01

	perm.					period	cycle
1	A	B	C	P	Q	1	——
p	C	A	B	P	Q	3	(CBA)
p^2	B	C	A	P	Q	3	(BCA)
x	A	B	C	Q	P	2	(PQ)
px	C	A	B	Q	P	6	$(ACB)(PQ)$
p^2x	B	C	A	Q	P	6	$(ABC)(PQ)$
a	A	C	B	P	Q	2	(BC)
b	C	B	A	P	Q	2	(CA)
c	B	A	C	P	Q	2	(AB)
ax	A	C	B	Q	P	2	$(BC)(PQ)$
bx	C	B	A	Q	P	2	$(CA)(PQ)$
cx	B	A	C	Q	P	2	$(AB)(PQ)$

generators p, x, a, b and c (by no means a minimum set of generators, since in fact the group D_6 requires only two). The arrangement is that the first six are the product of the disjoint cycles of the two sets $\{A, B, C\}$ and $\{P, Q\}$ forming the subgroup C_6 (see p. 160). The last six are single or double transpositions, and all of them keep the sets $\{A, B, C\}$ and $\{P, Q\}$ distinct.

Consider those permutations *which do not move A*. These clearly form the subgroup D_2, (D_6 does, in fact, have three subgroups D_2, as we saw on p. 221) and we may associate with this subgroup the *property that A is not moved.* Now consider those permutations which *move A to second place.* These are certainly not a subgroup, because the identity – by the very fact that A is moved – is absent. But they do form what is called a *Coset* of the subgroup $\{1, x, a, ax\}$. The elements of this coset $\{p, px, c, cx\}$ may be written $\{p, px, ap, axp\}$ (since $c = ap$),

1	A	B	C	P	Q
x	A	B	C	Q	P
a	A	C	B	P	Q
ax	A	C	B	Q	P

p	C	A	B	P	Q
p	C	A	B	Q	P
c	B	A	C	P	Q
cx	B		C	Q	P

or $\{p, xp, ap, axp\}$ since $px = xp$ or $\{1, x, a, ax\}p$ where we accept the latter notation to mean that post-multiplication by p *distributes* over the whole set. In fact, if we had denoted the subgroup by the single capital letter H, the coset which corresponds to those permutations which send A to second place would be denoted Hp.

This is called the *Right Coset* of the subgroup H by the element p. Consider next the Left Coset of the subgroup H by the element p:

$$pH = p\{1, x, a, ax\} = \{p, px, pa, pax\} = \{p, px, b, bx\},$$

p	C	A	B	P	Q
px	C	A	B	Q	P
b	C	B	A	P	Q
bx	C	B	A	Q	P

a different set from Hp, and consisting of the permutations which evidently all do the job of *sending C to first place*.

We may form the coset (left or right) of a given subgroup by any element, e.g. the left coset of H by b: $b\text{H} = \{b, ba, ba, bax\} = \{b, bx, p, px\}$, and this turns out to be the *same coset* as pH. But if we look at $c\text{H} = \{c, cx, ca, cax\} = \{c, cx, p^2, p^2x\}$, we obtain a new coset, the remaining elements of the group that were neither in H nor in pH.

Q. 19.01. Find all the remaining left cosets and right cosets of H. Describe the special feature associated with each.

Q. 19.02. Consider the subgroup which leaves P and Q undisturbed. Describe the cosets of this subgroup, and show that in this case both left and right cosets are the same.

Q. 19.03. Find the cosets of the subgroup $\{1, c\}$. Are the left and right cosets the same?

Q. 19.04. Write down the subset of these permutations which includes two transpositions. Is it a subgroup? Find as small as possible a subgroup which includes them, and identify the right and left cosets.

Q. 19.05. Investigate right and left cosets of various subgroups by each element in various groups which have so far been discussed in the text. This is a huge task, so we suggest you concentrate on the groups D_3 (p. 202), C_6 (p. 158), Q_4 (p. 101), $C_4 \times C_2$ (p. 281), A_4 (p. 294); in each case taking a selection of subgroups. Or the task may be organised as a class activity, the labour being divided up between small subclasses of pupils.

Resumé of some examples of cosets from previous chapters

We have already used the concept of cosets in this book. For example, in Ch. 18 when we paired off the odd and even permutations. Let us consider this classification again. Denote by 'a' the permutation which interchanges A and B; and by $p_1, p_2, p_3, \ldots$ the various even permutations:

Table 19.02

even perms. $\xrightarrow{a}$ odd perms.

	even perms.		odd perms.
1	$A\ B\ C\ D\ E$	a	$B\ A\ C\ D\ E$
p_1	$C\ E\ A\ B\ D$	ap_1	$C\ E\ B\ A\ D$
p_2	$E\ B\ D\ C\ A$	ap_2	$E\ A\ D\ C\ B$
	.		.
	.		.
	.		.

and so on (cf. pp. 316–7).
It will be seen that what we are doing is, in fact, finding the left coset of the subgroup† $\{1, p_1, p_2, p_3, \ldots\}$ by the element a. Call this subgroup of even permutations E. We were trying to show that the number of even permutations was 60 ($= \frac{1}{2}.5!$). Since they form a subgroup, the number could be 60, 40, 30, 24, ..., any aliquot part of 120, by Lagrange's theorem, and our problem was to show why we must reject 40, 30, 24, ... and accept 60. Suppose E had contained 40 even permutations. Then aE would contain 40 odd permutations – all different, and so there would have to be 40 extra odd permutations (which would be a coset of E by another element, b, say bE). Now take any odd permutation, say c. This would combine with each of the 80 supposed odd permutations to give 80 distinct even permutations, and so we arrive at a contradiction.

Again, when we considered the full group of the cube, we said that any of the enantiomorphs could be obtained from one of the direct symmetries by the application of the central inversion (see p. 302). If therefore, the twenty-four direct symmetries of the cube are denoted $x_1, x_2, x_3, \ldots$, and the central inversion is denoted i, then the opposite symmetries are $ix_1, ix_2, ix_3, \ldots$, i.e. are the left‡ coset of the subgroup $\{x_1, x_2, x_3, \ldots\}$ by the element i.

The same could be interpreted in terms of the set of forty-eight 3×3 matrices which contain six zeros and three ± 1's (no two in the same row or column). Those with determinant $+1$ form the subgroup S_4, and multiplication of each by $\begin{bmatrix} -1 & 0 & 0 \\ 0 & -1 & 0 \\ 0 & 0 & -1 \end{bmatrix}$ produces the coset consisting of the remaining 24 matrices. In terms of the permutations of the diagonals of the cube (see pp. 298 ff.), the central inversion is a transformation which leaves all four diagonals in the same place but 'turns them all round'.

† We now know that the *even* permutations do form a subgroup.
‡ Or right, since i commutes with all the symmetries.

Reading off cosets from the group table

It is easy to find cosets of a given subgroup when a group table is available. For example, consider the group A_4, the direct symmetries of the regular tetrahedron, or the even permutations of four letters. The table is printed on p. 294 so that the subgroup $\{1, a, b, c\}$ appears in the top left-hand corner, and it is a simple matter to read off the cosets. For example, $q\{1, a, b, c\} = \{q, s, r, p\}$, these being the first four elements in row q; $s\{1, a, b, c\} = \{s, q, p, r\}$, so that the left cosets of $\{1, a, b, c\}$ by q and by s are *equal* (the order in which the elements appear in the set is immaterial). Indeed, we find that $p\text{H} = q\text{H} = r\text{H} = s\text{H} = \{p, q, r, s\}$, where H is the subgroup $\{1, a, b, c\}$; while $p^2\text{H} = q^2\text{H} = r^2\text{H} = s^2\text{H} = \{p^2, q^2, r^2, s^2\}$, and $1\text{H} = a\text{H} = b\text{H} = c\text{H} = \text{H}$. Thus the whole group is 'partitioned', or divided up, into the three cosets: $\{1, a, b, c\}$, $\{p, q, r, s\}$ and $\{p^2, q^2, r^2, s^2\}$.

Q. 19.06. Check that these are also right cosets. What may you say because of this?

For the other subgroups, it requires more care to pick out the cosets. For example, if $\text{H} = \{1, p, p^2\}$ we may see the left cosets by reading off under columns $1, p, p^2$. Thus, $s\{1, p, p^2\} = \{s, r^2, a\}$. The four left cosets are: $\{1, p, p^2\}$; $\{a, s, r^2\}$; $\{b, q, s^2\}$; and $\{c, r, q^2\}$. Similarly the right cosets may be read off from *rows* $1, p, p^2$ (see table 19.03).

Q. 19.07. Make a list of the right cosets.

Interpretations of cosets

Do the cosets have a physical meaning in the case of A_4? We may make use of the permutations of the vertices to describe the symmetries of the regular tetrahedron which concern us. The subgroup $\{1, a, b, c\}$ corresponds to the permutations $\begin{Bmatrix} A & B & C & D \\ B & A & D & C \\ C & D & A & B \\ D & C & B & A \end{Bmatrix}$ representing the half-turns of the tetrahedron about the joins of mid-points of opposite edges. The coset $\{p, q, r, s\}$ corresponds to the permutations $\begin{Bmatrix} \mathbf{A} & D & B & C \\ C & \mathbf{B} & D & A \\ D & A & \mathbf{C} & B \\ B & C & A & \mathbf{D} \end{Bmatrix}$

	1 ...	p	p^2 ...
1	1 ...	p	p^2 ...
a	a ...	s	r^2 ...
b	b ...	q	s^2 ...
c	c ...	r	q^2 ...
p	p ...	p^2	1 ...
p^2	p^2 ...	1	p ...
q	q ...	s^2	b ...
q^2	q^2 ...	c	r ...
r	r ...	q^2	c ...
r^2	r^2 ...	a	s ...
s	s ...	r^2	a ...
s^2	s^2 ...	b	q ...

Table 19.03

If the solid were resting on a horizontal table, these would be *anticlockwise* turns through $\frac{2}{3}\pi$ about each altitude in turn.

The subgroup $\{1, p, p^2\}$ corresponds to the permutations $\left\{\begin{matrix} \mathbf{A} & B & C & D \\ \mathbf{A} & D & B & C \\ \mathbf{A} & C & D & B \end{matrix}\right.$ which leave A invariant, and these rotate the tetrahedron about the altitude through A. The left coset $\{a, s, r^2\}$ corresponds to the permutations $\left\{\begin{matrix} \mathbf{B} & A & D & C \\ \mathbf{B} & C & A & D \\ \mathbf{B} & D & C & A \end{matrix}\right.$, i.e. to those symmetries of the figure which cause the vertex B to replace the vertex A.

Q. 19.08. Interpret all the cosets (left and right) of other subgroups, including subgroups of order 2.

Now look at the structure table for the group S_4, for which we have already described all the subgroups (see Ch. 14). The subgroup D_3 $\{1, a, b, c, d, f\}$ is in the top left-hand corner, so it is easy to read off the cosets. Indeed, the elements of the group have been arranged in a convenient order so that the right cosets of this subgroup are $\{g, h, i, j, k, l\}$, $\{m, n, p, q, r, s\}$, $\{t, u, v, w, x, y\}$.

Q. 19.09. Find the left cosets; also make a catalogue of both left and right cosets of the other subgroups. Interpret these cosets in terms of the permutations of $A\ B\ C\ D$ listed on p. 318. Observe, for example that when the elements of S_4 are interpreted in that sense, then each permutation in the coset $\{g, h, i, j, k, l\}$ *sends D into third position.*

As an example of a coset which is not so straightforward to spot, let us find the right cosets of the subgroup $\{1, j, r, t\}$ (see table 19.04). Thus the

Table 19.04

	1	*a*	*b*	*c*	*d*	*f*	*g*	*h*	*i*	*j*	*k*	*l*	*m*	*n*	*p*	*q*	*r*	*s*	*t*	*u*	*v*	*w*	*x*	*y*
1	1	*a*	*b*	*c*	*d*	*f*	*g*	*h*	*i*	*j*	*k*	*l*	*m*	*n*	*p*	*q*	*r*	*s*	*t*	*u*	*v*	*w*	*x*	*y*
j	*j*	*l*	*q*	*s*	*w*	*y*	*c*	*f*	*n*	*r*	*u*	*x*	*a*	*d*	*h*	*k*	*t*	*v*	1	*b*	*g*	*i*	*m*	*p*
r	*r*	*x*	*k*	*v*	*i*	*p*	*s*	*y*	*d*	*t*	*b*	*m*	*l*	*w*	*f*	*u*	1	*g*	*j*	*q*	*c*	*n*	*a*	*h*
t	*t*	*m*	*u*	*g*	*n*	*h*	*v*	*p*	*w*	1	*q*	*a*	*x*	*i*	*y*	*b*	*j*	*c*	*r*	*k*	*s*	*d*	*l*	*f*

right cosets are $\{e, j, r, t\}$; $\{a, l, x, m\}$; $\{b, q, k, u\}$; $\{c, s, v, g\}$; $\{d, w, i, n\}$; and $\{f, y, p, h\}$. Notice that the coset $\{f, y, p, h\}$ contains elements all of period 2, whereas the other four, apart from the subgroup itself, each contain one element of period 2, one of period 4 and two of period 3. Thus there is 'something different' about the coset $\{f, y, p, h\}$.

Q. 19.10. What are the left cosets of $\{1, j, r, t\}$? What are the left and the right cosets of the centraliser (see p. 222) of r; of w in S_4?

Preview of normal subgroups

When the left cosets and the right cosets of a subgroup are identical, that subgroup is known as an Invariant, or Self-Conjugate, or *Normal* subgroup. We shall use the latter term, and normal subgroups will be discussed more fully in Chs. 20 and 21. At this stage we will say no more than that the existence of normal subgroups, and their properties and treatment is perhaps the most important single part of Group Theory. We introduce the idea at this premature stage merely so that you will have a chance to get used to the concept before being launched into detailed treatment.

Q. 19.11. Show that for a group of order $2n$ with a subgroup of order n, the latter must be a normal subgroup. Give instances.

Q. 19.12. Show that every subgroup of an Abelian group is normal.

Q. 19.13. List the normal subgroups of S_4 (p. 219), A_4 (p. 294), Q_4 (p. 101), D_6 (p. 192), Q_6 (p. 216).

Q. 19.14. Show that every subgroup of Q_4 is normal (use the result of Q. 19.11 above). Is this true of Q_6? Can you find a larger non-Abelian group all of whose subgroups are normal? Can you find any groups which have no normal subgroups?

Another illustration

Consider the group $\{0, 1, 2, \ldots, 11\}$, $+$ mod. 12, i.e. $\mathbf{Z}_{12}, +$. This may be illustrated by the rotational symmetries of the regular dodecagon (fig. 19.01), or by the multiplication of the 12th roots of unity on the Argand diagram. Now one subgroup is C_6, $\{0, 2, 4, 6, 8, 10\}$ corresponding to the regular hexagon in the diagram. The other regular hexagon, $\{1, 3, 5, 7, 9, 11\}$ is the coset of this subgroup. (In many ways, these two hexagons resemble the two bishops of one player in the game of chess – one always moves on white squares, and the other always moves on black squares, so that they occupy disjoint sets of positions on the board, though between them they may cover all the squares on the board.)

Q. 19.15. List the other subgroups of C_{12} above, and give geometrical illustrations of their cosets. Repeat for C_{10}, C_{16}, C_{18}, C_{20}, etc.

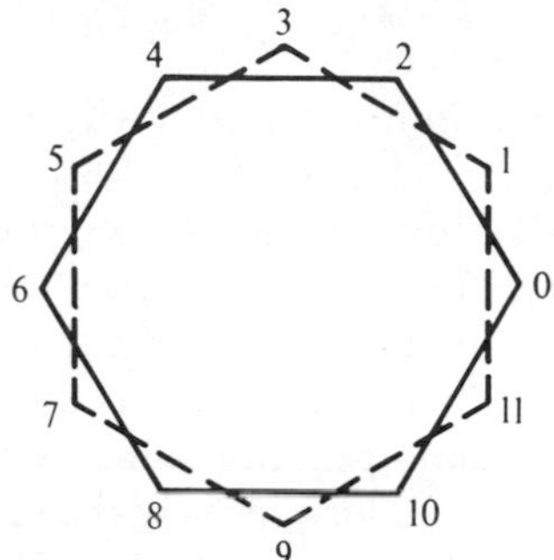

Fig. 19.01. The group C_{12}: subgroup C_6 and coset.

Properties of cosets

In all the examples of cosets discussed so far in this chapter, the following properties have come to light:

(a) If H is a subgroup of order m of a group G of order mr, then H has r left cosets each containing m elements (r is called the 'index' of H in G).

(b) These r left cosets are disjoint, i.e. no pair of them have an element in common.

(c) One of the cosets is H itself.

(d) Every element of the group is in one and only one of the left cosets.

These properties remain true, of course, under the transposition (left, right)! Note that the properties, as stated, contain some redundancies – for example, the phrase 'one and only one' in (d) really makes (b) unnecessary. These properties we shall prove later in this chapter, and from there we shall be in a position to prove Lagrange's theorem, as promised earlier in this book.

Cosets in infinite groups

Before proceeding to establish these important results, however, we shall look for some further examples of cosets, this time in infinite groups, and this may help you to gain more insight into the concept.

We have seen how infinite groups may have infinite subgroups, or, in cases when they contain elements of finite period, finite subgroups. Whenever there is a subgroup, this subgroup partitions the group into cosets, whether the group is finite or infinite. Let us now consider examples.

Residue classes

Consider first the group C_∞ in its realisation (**Z**, +). One subgroup, also with structure C_∞, is H{... −10, −5, 0, 5, 10 ...}, the set of multiples of 5.† What is the coset (left or right, since the group is Abelian) of H by the element +1? By our definition of cosets, we see that it consists of those numbers obtained by adding 1 to each number in H, i.e. the set {... −9, −4, 1, 6, 11 ...}, which is known as the 'residue class' congruent to 1 (mod. 5). Since our group operation is +, we might refer to this coset as 1 + H. Evidently 6 + H is precisely the same coset, and so is H − 4, H − 9, etc. The other cosets are H + 2, H + 3 and H + 4, these being the residue classes congruent to 2, 3 and 4 (mod. 5).

Q. 19.16. Specify the cosets of the subgroups: (a) even integers; (b) multiples of 3; (c) (6**Z**, +).

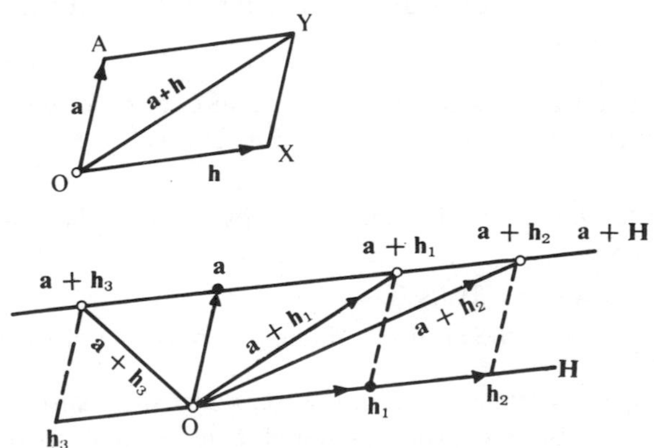

Fig. 19.02. Coset of ($\mathbf{V}_1$, +) in ($\mathbf{V}_2$, +).

Groups of vectors

Next consider the group ($\mathbf{V}_2$, +) of two-dimensional vectors under vector addition. Let us think of these vectors as displacements from some fixed origin *O* in the plane. Now one subgroup **H** of this group ($\mathbf{V}_2$, +) consists of all the vectors along one particular line, e.g. all the displacements in an east–west direction. These vectors form a one-dimensional system, the group ($\mathbf{V}_1$, +) ≅ (**R**, +). Now let **a** be a fixed vector of ($\mathbf{V}_2$, +), not in **H**, and let **h** be a typical vector of **H**. Then the (left or right) coset of **H** by **a** consists of the set of vectors **a** + **h**, where **h** runs through the whole set **H**. It will be clear from fig. 19.02 that the displacements of the coset

† This subgroup will be indicated (5**Z**, +), a notation which is capable of generalisation.

a + **H** take us on to the line through the extremity of vector **a** and parallel to the direction of **h**. Other choices of the fixed vector **a** may lead to different cosets, all of which will be represented by lines parallel to **h**. Of course, if one found the coset of **H** by any one of those vectors lying in a particular coset (e.g. if $\mathbf{a} + \mathbf{h}_1$ were selected in place of **a**), then the result would be the same coset: $(\mathbf{a} + \mathbf{h}_1) + \mathbf{H} = \mathbf{a} + \mathbf{H}$. You may like to re-interpret the above in terms of the group of complex numbers under addition, which is isomorphic to $(\mathbf{V}_2, +)$ – take **H** to be the subgroup (**R**, +).

Q. 19.17. Consider the group $(\mathbf{V}_3, +)$ with a subgroup $(\mathbf{V}_2, +)$. Identify the cosets.

Cosets in groups of real and of complex numbers

Now one subgroup of (**R**, +) is (**Z**, +). Consider the coset of (**Z**, +) by the real number 0·68. This consists of the numbers 0·68 + {... −3, −2, −1, 0, 1, 2, 3, ...}, i.e. the numbers {... $\bar{3}$·68, $\bar{2}$·68, $\bar{1}$·68, 0·68, 1·68, 2·68, 3·68, ...} and this is illustrated on the number line below:

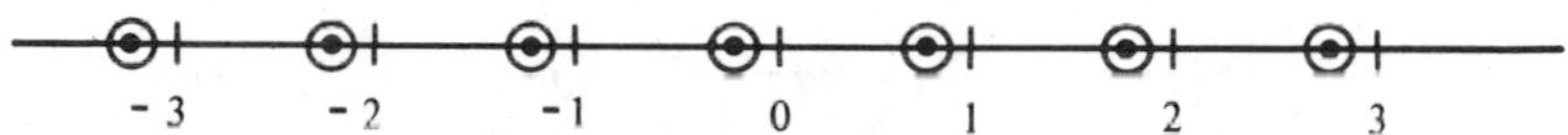

Fig. 19.03.

Q. 19.18. What is the coset of the real number −9·32 relative to this same subgroup?

Q. 19.19. What are the cosets of the Gaussian integers (see p. 228), with structure $C_\infty \times C_\infty$, regarded as a subgroup of (**C**, +).

Q. 19.20. Another subgroup of (**C**, +) consists of all those complex numbers of the form $(2 - 3i)k$, where k is real. These lie on the line $y = -1\frac{1}{2}x$. What are the cosets? Is it possible to show *any* straight line in the Argand diagram to be a coset of some subgroup of (**C**, +) by a suitable choice of subgroup?

Next think of the group $(\mathbf{R}^+, \times)$, i.e. of positive reals under multiplication. One subgroup of this group might be the integral powers of 10: $\{\ldots 10^{-2}, 10^{-1}, 1, 10, 10^2, 10^3, \ldots\}$ which you may easily show to be isomorphic to (**Z**, +). What is the coset of this subgroup by the element 4·79? By definition, it is the set 4·79 $\{\ldots, 10^{-2}, 10^{-1}, 1, 10, 10^2, 10^3, \ldots\}$,

i.e. {... 0·0479, 0·479, 4·79, 47·9, 479, 4790, ...},

or $\{\ldots 10^{\bar{2}.68}, 10^{\bar{1}.68}, 10^{0.68}, 10^{1.68}, 10^{2.68}, 10^{3.68}, \ldots\}$.

(Compare with the previous example!)

Q. 19.21. Find another infinite subgroup of ($\mathbf{R}^+$, $\times$), and discuss the cosets.

Q. 19.22. Find the cosets in the group $a + b\sqrt{2}$ ($a, b \in \mathbf{Q}$) under addition relative to the subgroup for which $b = 0$. Repeat when the operation is $\times$.

A *finite* subgroup of ($\mathbf{R}$, $\times$) is $\{+1, -1\}$, $\times$. We consider next the coset by the real number x. This is the set $\{x, -x\}$, and so all the cosets are thus seen to be the sets consisting of pairs of negatives, and there are an infinity of them.

Q. 19.23. Consider the group of all polynomials with rational coefficients under addition. A subgroup of this is the set of quadratic polynomials, including the null element $0x^2 + 0x + 0$. What is the coset of this subgroup relative to the cubic polynomial x^3? Discuss other cosets of the subgroup of polynomials of the second degree.

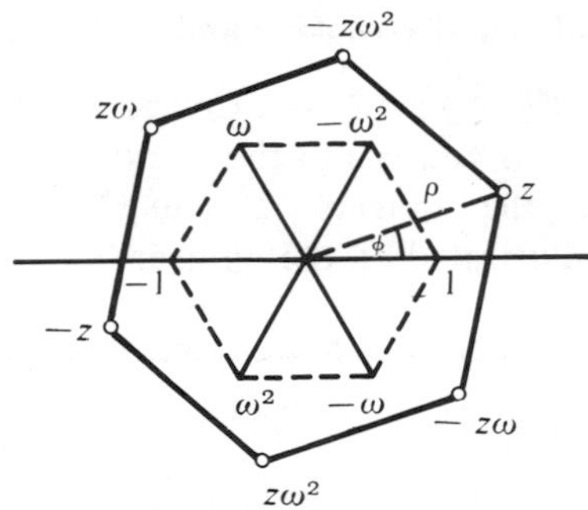

Fig. 19.04. Coset of $\{\pm 1, \pm\omega, \pm\omega^2\}$ in ($\mathbf{C}$, $\times$).

An interesting infinite group from the point of view of cosets is the group ($\mathbf{C}$, $\times$). One possible subgroup consists of all those complex numbers whose modulus is 1, of the form $\cos\theta + i\sin\theta = \operatorname{cis}\theta$. To form the coset by a complex number $\rho(\cos\phi + i\sin\phi)$, we consider all complex numbers $\rho \operatorname{cis}\phi \times \operatorname{cis}\theta$, where θ takes all possible values. This gives all the complex numbers whose modulus is ρ, and these all lie on a circle centre O, radius ρ. Thus the coset of the subgroup ($\operatorname{cis}\theta$, $\times$) by any number of modulus ρ will be this same circle $|z| = \rho$, while for different values of ρ we shall get the other cosets which therefore consist of a system of concentric circles.

A finite subgroup of ($\mathbf{C}$, $\times$) might be $\{\pm 1, \pm\omega, \pm\omega^2\}$ where $\omega = \operatorname{cis} 120°$. The coset by the number $\rho \operatorname{cis}\phi$ consists of the six numbers $\rho \operatorname{cis}(\phi + k\pi/3)$ where $k = 0, 1, 2, 3, 4, 5$, and these lie at the vertices of a regular hexagon (fig. 19.04).

Q. 19.24. Find other subgroups of ($\mathbf{C}$, $\times$) and investigate their cosets.

Q. 19.25. Consider the set of functions of the form $f(x) = (ax + b)/(cx + d)$ ($a, b, c, d \in \mathbf{R}$). These form a group under successive substitution. One subgroup contains those functions for which $c = 0$, $d = 1$. Find the coset of this subgroup by the function $1/x$; $1/(1 - x)$, etc.

Q. 19.26. Consider the group of all non-singular 2×2 matrices with real terms under matrix multiplication. A subgroup is the set of matrices of the form $\begin{bmatrix} a & b \\ -b & a \end{bmatrix}$. Discuss the cosets of this subgroup by the matrices

$\begin{bmatrix} 0 & 1 \\ 1 & 0 \end{bmatrix}$; $\begin{bmatrix} -1 & 0 \\ 0 & -1 \end{bmatrix}$; $\begin{bmatrix} 1 & 2 \\ 0 & 1 \end{bmatrix}$; etc.

Q. 19.27. (**R**, $\times$) is a subgroup of (**C**, $\times$). What is the coset of this subgroup by the number i?

Q. 19.28. The set of all rational functions $f(x) = P_1(x)/P_2(x)$ where P_1 and P_2 are polynomials form a group under multiplication (see Ex. 8.31, 14.26). Find a subgroup, and identify the cosets.

Q. 19.29. Consider the strip patterns discussed in Ch. 26, pp. 515 ff. where some subgroups, both finite and infinite, were found in the cases where the full group had structure D_∞, $C_\infty \times C_2$ and $D_\infty \times C_2$. Identify cosets of subgroups in some of these cases, and indicate the cosets on the Cayley diagram.

Subgroups and cosets in groups of transformations

All the plane rotations about a fixed point O form a group, of which rotations through $2\pi/n$ ($n \in \mathbf{Z}^+$) are a subgroup. The cosets of this by a rotation through α will resemble the cosets of the subgroup (cis $2\pi/n$, $\times$) of (**C**, $\times$) by a complex number cis α of unit modulus – they will represent rotations through angles α, $\alpha + (2\pi/n)$, $\alpha + (4\pi/n)$, ..., $\alpha + 2\pi(n - 1)/n$. This group of rotations about O is, in its turn, a subgroup of the group of all the plane isometries (see pp. 231 ff., and remember that the rotations do not themselves form a group). We may ask: what is the coset of this subgroup by a translation t? To answer this question, we need to combine the translation with all the rotations about O, and must remember that these transformations do not commute. Consider, then, the left coset, so that we first perform a rotation (through any angle) about O, and then the fixed translation. Clearly, a rotation r_θ about O followed by a translation t will, intuitively, combine to give a rotation about some point – which point? Figure 19.05 shows that, if the translation is indicated by the vector $\overline{OA}$, and if, for reference we choose coordinate axes along and perpendicular to OA, as shown, then the point C_1 ($\frac{1}{2}a$, $\frac{1}{2}a \cot \frac{1}{2}\theta$) will be mapped by the rotation r_θ into the point $r_\theta(C_1)$ ($-\frac{1}{2}a$, $\frac{1}{2}a \cot \frac{1}{2}\theta$); while the translation $\overline{OA}$ will carry this point back to its original position C_1. Thus the point C_1 is invariant under the combined transformations, and so must be the centre of rotation for the single rotation. The diagram shows that in fact tr_θ(F) is equivalent to a single rotation about the point C_1.

In a similar way, the rotation $r_{-\phi}$ followed by the translation t is equivalent to a single rotation about the point C_2, for the figure shows that this point is unchanged under the composite movements $r_{-\phi}$, then t. Thus the elements of the set of transformations obtained by applying the translation t to the rotations about O, i.e. the left coset of this subgroup of rotations, is a set of rotations about points *on the perpendicular bisector of OA*.

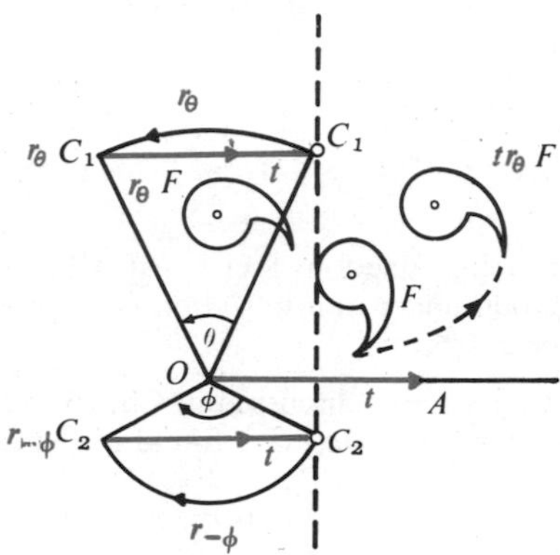

Fig. 19.05. Coset of group of rotations about O by a translation in the group of plane isometries.

Q. 19.30. What is the right coset?

Q. 19.31. Discuss the left and right cosets of the subgroup of rotations about O relative to a rotation through angle α about *another* point A. (*Note*: fig. 19.06 gives a hint as to how the combined centre of rotation may be found.)

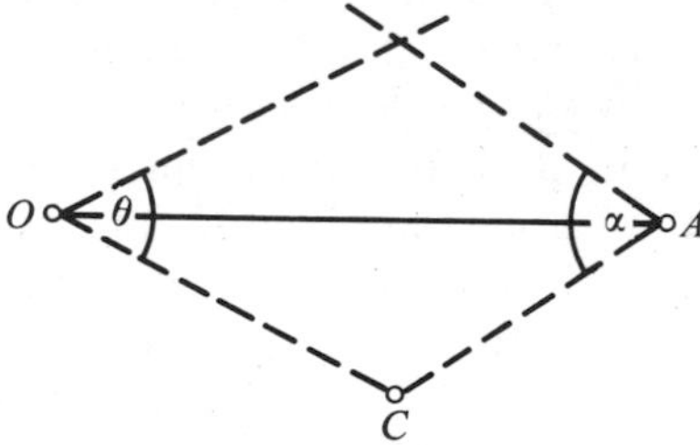

Fig. 19.06. Successive rotations.

Q. 19.32. If a and b are reflections in parallel mirrors, which generate the group D_∞ (see p. 206), then $\{1, a\}$ is a subgroup of order 2. One of its cosets is $b\{1, a\} = \{b, ba\}$. Indicate the other cosets, and show them in a diagram.

★ Again, let G be the group of all isometries in the plane, i.e. transformations which preserve the metrical properties of a figure – distance, angle, etc. – and which include rotations, translations and reflections. Let H be the subgroup of all *direct* isometries, and let m be a reflection in a single fixed line. Then the left coset mH will be the entire set of opposite isometries (reflections and glide reflections), in much the same way as we saw, in considering the symmetries of the cube (p. 302) that the coset of the direct symmetries by the central inversion was the set of all the enantiomorphs. In the present case, the set H can be represented by the set of functions of the form $z' = az + b$, where a and b are complex constants with $|a| = 1$, and z and z' are complex numbers corresponding to the point and its image. These functions correspond to the direct symmetries, and are combined by successive composition of functions. The opposite isometries may be obtained by taking the (left) coset by the reflection in the real axis, i.e. the transformation given by $z' = \bar{z}$, and if we let $\mathrm{f}(z) \equiv az + b$, and $\mathrm{g}(z) \equiv \bar{z}$, then:

$$\mathrm{gf}(z) \equiv \bar{a}\,\bar{z} + \bar{b} \quad \text{(to give the left coset)}$$

or $$\mathrm{fg}(z) \equiv a\,\bar{z} + b \quad \text{(to give the right coset).}$$

In either case, the resulting function provides the entire set of opposite isometries.

★**Q. 19.33.** In the above case, may we say that $\mathrm{G} = \mathrm{H} \times C_2$?

★**Q. 19.34.** Considering the set of isometries in the plane as a subgroup of the group of similarities show that the left coset of this subgroup by a single similarity which enlarges by double the linear dimensions (represented by $z' = 2z$ or by the matrix $\begin{bmatrix} 2 & 0 \\ 0 & 2 \end{bmatrix}$) consists of all those similarities in the plane which result in a doubling of linear dimensions. Show that the left and right cosets are identical.

★**Q. 19.35.** Considering the group of translations in the plane as a subgroup of the group of direct isometries, describe the coset of this subgroup by a rotation about a single point O through an angle α. Is the group of translations a *normal* subgroup?

★**Q. 19.36.** Consider the group of matrices $\begin{bmatrix} a & 0 & 0 \\ 0 & b & 0 \\ 0 & 0 & c \end{bmatrix}$ $(a \neq 0, b \neq 0, c \neq 0)$ of $(\mathscr{M}_3, \times)$, what is its left coset by the 3×3 matrix $\begin{bmatrix} 0 & 0 & -1 \\ 0 & -1 & 0 \\ -1 & 0 & 0 \end{bmatrix}$?

★**Q. 19.37.** A rigid body has two points A and B fixed, and is capable of rotation about the axis AB. These rotations form a subgroup of the group of all those movements of the rigid body which leave only the point A fixed and where the restraint on the point B is removed. Investigate the coset of this subgroup with respect to a half-turn about an axis through A and perpendicular to AB.

★**Q. 19.38.** Consider the subgroup of the group $(\mathscr{M}_2, \times)$ consisting of those non-singular 2×2 matrices which commute with a *given matrix* A. This is the centraliser of A, the set $\{M : MA = AM,\ M \in \mathscr{M}_2,\ |M| \neq 0\}$. Given a matrix B such that $AB \neq BA$, identify the left and right cosets by taking particular values of A and B, e.g.

$$A = \begin{bmatrix} -1 & 0 \\ 1 & 1 \end{bmatrix}; \qquad B = \begin{bmatrix} 1 & 2 \\ 1 & 0 \end{bmatrix}, \text{ etc.}$$

★**Q. 19.39.** Consider the set of 2 × 2 matrices $\begin{bmatrix} a & b \\ c & d \end{bmatrix}$ such that $ad - bc = 1$. These form a group under matrix multiplication (see p. 229) which is a subgroup of $(\mathscr{M}_2, \times)$. What are the left and right cosets of this subgroup by the matrices $\begin{bmatrix} 2 & 0 \\ 0 & 1 \end{bmatrix}, \begin{bmatrix} 0 & 1 \\ 1 & 0 \end{bmatrix}, \begin{bmatrix} 2 & 0 \\ 0 & 2 \end{bmatrix}$?

Having 'broadened your outlook' on cosets by taking the above excursion into infinite groups, we now return to consider how the properties of cosets may be proved, and also to provide the long-awaited proof of Lagrange's theorem.

Lagrange's theorem: proof

Suppose H is a subgroup of G, and let the elements of H be $\{1, h_1, h_2, h_3, \ldots, h_{m-1}\}$ where H is of order m. Take an element x of G which is not in H, and form the left coset xH:

$$x, xh_1, xh_2, xh_3, \ldots, xh_{m-1}.$$

Now the elements of this coset must be all distinct, for if $xh_r = xh_s$, then the cancellation law would lead to $h_r = h_s$ which is untrue since the elements of the subgroup must be distinct. Moreover, H and xH cannot possibly have any element in common. For suppose that $xh_r = h_s$. Then $xh_rh_r^{-1} = h_sh_r^{-1}$, or $x = h_sh_r^{-1}$. But h_r^{-1} is in H, and so is the product $h_sh_r^{-1}$ since H is a subgroup. But we deliberately chose x not to be in H, so we have reached a contradiction. Therefore $\text{H} \cap x\text{H} = \phi$.

Now it is possible that $\text{H} \cup x\text{H} = \text{G}$, i.e. that all the elements of G have appeared either in H or in xH. If this is the case, then the order of G would be $2m$, since both H and xH each contain m elements, and Lagrange's theorem would then be proved. If, however, there are still elements in G which are not in H or in xH, choose any one of these, y, and form the left coset yH:

$$y, yh_1, yh_2, yh_3, \ldots, yh_{m-1}.$$

The cancellation law again convinces us that these are all distinct, but we must still guard against appearances of elements already listed in the first two cosets H and xH. An argument similar to the one which we used to show that there was no overlapping between H and xH will be used:

suppose if possible that $xh_r = yh_s$. Then $y = xh_rh_s^{-1}$, and since $h_rh_s^{-1} \in$ H, we have $y \in x$H, which is contrary to our certain knowledge that y is not in xH. Thus $x\mathrm{H} \cap y\mathrm{H} = \phi$, and similarly $\mathrm{H} \cap y\mathrm{H} = \phi$. Thus there is no overlapping between these three cosets H, xH and yH – they are disjoint, and contain $3m$ distinct elements of the group. If by now the group is exhausted, its order is $3m$, and Lagrange's theorem is proved. If not, choose an element z not in H, xH nor yH, and form the left coset zH ... The argument proceeds as above, at each stage it being necessary to prove (a) no repetitions in the newly formed coset and (b) no duplications in the new coset of any elements in the previously formed cosets.

Finally, *if the group is finite*, after say r cosets, H, xH, yH, zH, ... have been formed, there remain no unused elements of G from which to form new cosets, and we conclude that G contains exactly mr elements, so that m must be a factor of the order n of the group, and Lagrange's theorem is proved. *The number of cosets r* $(= n/m)$ is called the *index* of the subgroup in the group G.

Of course, when we selected $x, y, z, \ldots$ to form the new cosets, we had a free choice on each occasion from all the remaining elements of the group, and had we made different choices, we should have arrived at the same cosets, but in a different order. We might, for example, have selected y first, then x; in this case the coset yH would have been second instead of third on the list. Or we might have selected the element zh_7 with which to form our second coset. In that case, we should have had $\{zh_7, zh_7h_1, zh_7h_2, zh_7h_3, \ldots, zh_7h_{m-1}\}$,

i.e. $\quad z\{h_7, h_7h_1, h_7h_2, h_7h_3, \ldots, h_7h_{m-1}\}$.

But the latter set { } is simply a rearrangement of the elements of the subgroup H, so that the left coset of H by zh_7 is in fact zH.

As a final assurance – not only that the cosets are distinct – but also that every element in the group appears in *some* coset, this is obvious, since if g is any element whatever of G, then it most certainly appears in the coset gH, being the product of itself with the identity, which is undoubtedly in H. The above consideration points to the general result that if y is any element in xH, then xH and yH are the same set.

The main lines of the above argument are still sound when applied to infinite groups so far as the proof of the disjoint property of the cosets is concerned; but of course either r or m or both must now be infinite, and we do not have any result for infinite groups corresponding to Lagrange's theorem. Note that the non-overlapping (disjoint) property of cosets (i.e. $x\mathrm{H} \cap y\mathrm{H} = \phi$) is entirely dependent upon H being a subgroup. For taking the table for D_3 on p. 202, suppose $\mathrm{H} = \{e, p\}$ (not a subgroup). Then $\mathrm{H}a = \{a, c\}$, $\mathrm{H}c = \{c, b\}$ and these sets are *not disjoint*.

The construction of group structure tables

In the previous chapters we have on many occasions constructed tables to show all possible products of elements of a group. Armed now with a knowledge of the properties of cosets, the task is greatly facilitated.

Consider first the construction of the table for $C_2 \times C_2 \times C_2$ (see also pp. 267 ff.). We have the identity 1 and seven elements of period 2: a, b, c, d, f, g, h. Let us construct the abstract table row by row, and agree that the product ab ($= ba$) shall be the element called c, so that the portion of the table shown can be completed. Now $\{1, a\}$ is a subgroup, and its right coset $\{1, a\}b = \{b, c\}$, as may be seen from the two elements at the head of column b. It is then easy to see that ac must be b, because the coset of $\{1, a\}$ which contains c must also contain b. In fact this method of constructing the table will give us the subgroup $\{1, a, b, c\}$ in the top left-hand corner.

	1	*a*	*b*	*c*
1	1	*a*	*b*	*c*
a	*a*	1	*c*	?

Taking the next element in the border to be d, and selecting ad to be called f, we are in a position to fill the portion:

	1	*a*	*b*	*c*	*d*	*f*	...
1	1	*a*	*b*	*c*	*d*	*f*	...
a	*a*	1	*c*	*b*	*f*	?	...

of the table. Consideration of the coset $\{d, f\}$ shows that af is bound to be d, and we can now complete the first two rows:

	1	*a*	*b*	*c*	*d*	*f*	*g*	*h*
1	1	*a*	*b*	*c*	*d*	*f*	*g*	*h*
a	*a*	1	*c*	*b*	*f*	*d*	*h*	*g*
b	*b*	*c*	1	.	.	.	.	.

Continuing with the third row, the subgroup $\{1, b\}$ has the right coset $\{a, c\}$ and this enables us to complete $bc = a$. We are then faced with bd, which must be g or h, and we may select either to be the product bd.

Having decided that $bd = g$, we have cosets $\{d, g\}$ and $\{f, h\}$ which enable the rest of row b to be completed:

	1	*a*	*b*	*c*	*d*	*f*	*g*	*h*
1	1	*a*	*b*	*c*	*d*	*f*	*g*	*h*
a	*a*	1	*c*	*b*	*f*	*d*	*h*	*g*
b	*b*	*c*	1	*a*	*g*	*h*	*d*	*f*
c	*c*	*b*	*a*	1	.	.	.	.

Our next problem is to decide upon the last four elements in row c. The Latin square requirement leaves us no alternative but h, g, f, d respectively. Half the table is now complete. Commutativity now enables us to get as far as:

	1	a	b	c	d	f	g	h
1	1	a	b	c	d	f	g	h
a	a	1	c	b	f	d	h	g
b	b	c	1	a	g	h	d	f
c	$\underline{c}$	b	a	1	h	$\underline{g}$	f	d
d	d	f	g	h	1	.	.	.
f	f	d	h	g	.	1	.	.
g	$\underline{g}$	h	d	f	.	?	1	.
h	h	g	f	d	.	.	.	1.

The remaining twelve products may be completed by using cosets: for example $\{c, g\}$ is a left coset of the subgroup $\{1, f\}$ (the appropriate elements are underlined), and so we are immediately able to supply c in the position gf, and the rest of the missing products may be decided by similar considerations. In the completed table 19.05, you will see that it has been possible to divide into sixteen squares, each containing a

Table 19.05

	1	a	b	c	d	f	g	h
1	1	a	b	c	d	f	g	h
a	a	1	c	b	f	d	h	g
b	b	c	1	a	g	h	d	f
c	c	b	a	1	h	g	f	d
d	d	f	g	h	1	a	b	c
f	f	d	h	g	a	1	c	b
g	g	h	d	f	b	c	1	a
h	h	g	f	d	c	b	a	1

coset. Compare this arrangement of the table with the versions of C_{12} given on pp. 171–3, where each square also contains a coset of the subgroup in the top left-hand corner. The reason for these methods of drawing up the group table will emerge in Ch. 21.

Q. 19.40. Rewrite the table A_4 (see p. 294) so that the cosets of the subgroup D_2 appear in nine 'blocks' in a similar way. Try to do this for one of the subgroups C_3.

Again, if we were constructing the table for D_4 from defining relations $r^4 = 1 = a^2$, $rar = a$, we could begin by filling up the subgroup C_4 in the top left-hand corner. The elements ra, r^2a and r^3a may be named b, c, d, and so we are able to fill up the first four rows straight off as in table 19.06. (*Note:* the right cosets $\{1, r^2\}$, $\{r, r^3\}$, $\{a, c\}$, $\{b, d\}$.)

We cannot begin to fill in the last four rows till we have used the defining relations $rar = a$ and $a^2 = 1$. The fact that we have not done so means that these first four rows would do just as well for any group of order 8 which contains elements r and a with $r^4 = 1$, and these include both Q_4, $C_2 \times C_4$, and C_8 besides D_4.

Table 19.06

	1	r	r^2	r^3	a	ra	r^2a	r^3a
1	1	r	r^2	r^3	a	b	c	d
r	r	r^2	r^3	1	b	c	d	a
r^2	r^2	r^3	1	r	c	d	a	b
r^3	r^3	1	r	r^2	d	a	b	c
a	a							
b	b							
c	c							
d	d							

Now $rar = a \Rightarrow ar = r^3a = d$, and we are now ready to proceed. $a^2 = 1$ gives us the subgroup $\{1, a\}$ with a right coset $\{r, d\}$ which provides the information $ad = r$. The left cosets $\{1, a\}$, $\{r, b\}$, $\{r^2, c\}$, and $\{r^3, d\}$ are already determined, and enable us to fill up the whole of the column headed a. Furthermore, since $d = ar = r^3a$, we have $d^2 = 1$, and a further subgroup $\{1, d\}$ whose left cosets are already determined as $\{1, d\}$, $\{r, a\}$, $\{r^2, b\}$, $\{r^3, c\}$ permitting us to complete the entire column headed d, as well as the product $dr^3 = a$ from the left coset $\{a, r^3\}$.

In completing the table (see table 19.07), we do not want to be slaves

Table 19.07

	1	r	r^2	r^3	a	ra	r^2a	r^3a
1	1	r	r^2	r^3	a	b	c	d
r	r	r^2	r^3	1	b	c	d	a
r^2	r^2	r^3	1	r	c	d	a	b
r^3	r^3	1	r	r^2	d	a	b	c
a	a	d	.	.	1	.	.	r
b	b	.	.	.	r	.	.	r^2
c	c	.	.	.	r^2	.	.	r^3
d	d	.	.	a	r^3	.	.	1

to the coset method, so we go back to the defining relations again, and note that $c = r^2a$ (by definition), so that $c^2 = r^2ar^2a = r(rar)ra = rara = (rar)a = aa = 1$. Similarly we may prove $b^2 = 1$. Thus we have subgroups $\{1, c\}$ and $\{1, b\}$, and every one of the remaining blanks may be filled in from a consideration of cosets if desired.

Q. 19.41. Complete the table.

Q. 19.42. Leaving the first four rows exactly as they are, try and complete the table so as to get each of the groups $C_4 \times C_2$, Q_4 and C_8, each of which contains an element of period 4 which we call r. (In the case of C_8, you may take $a^2 = r$; remember that $C_4 \times C_2$ and C_8 are Abelian; and that Q_4 has all its remaining elements of period 4.)

Q. 19.43. If D_4 is realised as the group of the symmetries of the square with r representing a quarter-turn, what is the interpretation of the coset a, b, c, d above?

Q. 19.44. What is the geometrical interpretation of the subgroups and their cosets in the group C_8?

Subgroups of order 2 are particularly useful to harness to this coset technique. We now give a final illustration using a subgroup of order 3, in an imaginary case. Suppose the subgroup is H: $\{1, p, p^2\}$, and that we are filling in rows p and p^2 and have already worked out table 19.08. Then

Table 19.08

	1	a	b	c	d	f	g	...	p	p^2	...	r	s	...
1	1	a	b	c	d	f	g	...	p	p^2	...	r	s	...
.									.	.				
.									.	.				
.									.	.				
p	p	r	d	f	.	.	b	...	p^2	1	...	.	.	...
p^2	p^2	s	.	.	.	.	.	...	1	p	...	.	.	...

Ha = $\{a, r, s\}$ so Hr and Hs must also contain a, r and s. This enables us to fill in pr, ps, p^2r and p^2s in the only possible way as shown in the sub-table, table 19.09. Also we have Hb = $\{b, d, ?\}$ and Hg = $\{g, b, ?\}$.

Table 19.09

	a	r	s
1	a	r	s
p	r	s	a
p^2	s	a	r

Evidently therefore b, d and g must all be in the same coset, so we can fill in

Table 19.10

	b	d	g
1	b	d	g
p	d	g	b
p^2	g	b	d.

Further properties of cosets

Given a subgroup H of a group G, two elements of G (not in H) may or may not lie in the same coset of H. For example, taking H to be the subgroup $\{1, a, b, c, d, f\}$ of the group S_4 (table, p. 219), the elements p

and s are in the same right coset, whereas p and w are not. Again, in the same group, we have the subgroup $\{1, i, m\}$ with the right coset $\{a, k, t\}$ and the left coset $\{a, j, n\}$. So for example, a and t *do* occur in the same right coset, but are *not* in the same left coset.

The relationship between two elements of belonging to the same coset of a particular subgroup, is an important one. We shall presently discuss some theorems about this relationship, but first will say something about *equivalence relations.*

Equivalence relations: equivalence classes

Consider the set of all the boys in a certain school. There are various ways in which this set of boys can be 'partitioned', or decomposed, into disjoint subsets. We might divide them up by houses, by forms, by age, by marks in an exam, by the colours of their eyes, by their Christian names, and in a large number of other ways. Suppose we have occasion to partition them into subsets according to the letter that their surname begins with. Then the set A contains {Allen, Atkins, Alexander, etc.}; B contains {Butcher, Blake, Bertram, etc.}, ... X is the empty set; Y contains {Young, Yeo, etc.} and Z {Zeeman}. We say that the set of boys has been partitioned into 'equivalence classes', membership of these classes being described by the *equivalence relation* defined thus.

Two boys belong to the same equivalence class if they have surnames beginning with the same letter. The relationship between the boys is that of *having names beginning with the same letter*, and we shall denote this relationship by the symbol $\sim$ (pronounced 'twiddles'), so that in this case, Butcher $\sim$ Blake, but Blake $\nsim$ Atkins (these two do *not* stand in the relationship of mutual twiddling).

Consider now a more mathematical example. This time the set will be the set **Z** of integers, and the equivalence relation will be, let us say, that of being separated by a multiple of 100, e.g. $953 \sim 253 \sim -347$ (because these pairs of numbers are separated by 700 and by 600 respectively, and these differences are divisible by 100). On the other hand, $953 \nsim 952$, etc., though $952 \sim 752$, etc. It will be seen that the set of integers has, by this equivalence relation been partitioned into classes:

$$\{\ldots, -347, -247, -147, 53, 153, 253, \ldots\}$$
$$\{\ldots, -348, -248, -148, 52, 152, 252, \ldots\}, \text{ etc.,}$$

and there are evidently exactly one hundred classes. These are the *residue classes modulo 100*, and each class may be characterised, or identified, by a typical member and it is natural to use the numbers **0**, **1**, **2**, ..., **99** as *representatives* of each class. Thus the above two classes may be called the **53** class and the **52** class. The class of boys whose names begin with S could be labelled with Smith class, or simply the S-class. If the plane polygons are classified by the number of sides, we refer to the members of a particular class by the collective name for that class: triangle, quadrilateral, pentagon, etc., the equivalence relation here being that $a \sim b$ means 'a has the same number of sides as b'.

Properties of equivalence relation

An Equivalence Relation $\sim$ on a set† must satisfy the following properties:

(1) $a \sim a$	**(Reflexive)**	**R**	**easily remembered:**
(2) $a \sim b \Rightarrow b \sim a$	**(Symmetric)**	**S**	***R, S, T!***
(3) $a \sim b$ and $b \sim c \Rightarrow a \sim c$	**(Transitive)**	**T**	

where *a*, *b* and *c* are any members of the set.

These three requirements are also sufficient, and constitute a definition for an equivalence relation. The relation $>$ between real numbers is not an equivalence relation since, although the transitive requirement is satisfied ($a > b$, $b > c \Rightarrow a > c$), the other two break down. The relation $||$ (= 'is parallel to') between lines *is* an equivalence relation, but the relation $\perp$ (= 'is perpendicular to') is not (Where does it fail?). The first requirement may appear somewhat trivial, but it is essential, and it is sometimes necessary to 'stretch' definitions to make it work; e.g. the relation 'is the brother of' between males is an equivalence relation so long as one accepts the convention that a man is his own brother.‡

Q. 19.45. Invent examples which fail to be equivalence relations for various combinations of the requirements *R*, *S* and *T* above.

The relation between words 'rhymes with' is an equivalence relation, for we may accept that a word rhymes with itself (and indeed, some rhymesters use the same word for a rhyme when short of ideas!).

† In some books a different notation may be used to denote an equivalence relation (e.g. '*R*' in place of '$\sim$').

‡ It might be thought that *S* and *T* imply *R*. For if $x \sim y$, then $y \sim x$ (by *S*), and hence by *T* it must follow that $x \sim x$. However, the above reasoning breaks down in the case *when there does not exist any other element* (such as y above) *which is related to* x, i.e. when x is in a class all on its own. Thus the reflexive requirement *R* is *not* redundant, being required for the sake of logical completeness. Suppose for example we take the group D_3 $\{1, p, q, a, b, c\}$ as on p. 202, and set up the relation $x \sim y$ when x and y *have the same period*. This is clearly an equivalence relation which places $\{a, b, c\}$ in one equivalence class, these being the elements of period 2, satisfying *S* and *T*, and $\{p, q\}$ in another class. But unless we postulated the reflexivity axiom *R*, the identity element would be an outcast – one could *not* prove it is equivalent to itself by an argument such as the one at the beginning of this footnote, since there is no *other* element equivalent to it.

Symmetry and transitivity are evident. The relation 'is near to' between places is an equivalence relation, but 'is far from' is not (Why?); the relation 'is synonymous with' between words is, but 'is antonymous with' is not an equivalence relation. Thus equivalence relations express a *quality of sameness* between objects of the set, and those members which share this quality fall in the same equivalence class.† A very good example, described in Ch. 23, is the relation between musical notes separated by an exact number of octaves. Such notes are all called by the *same name*, and have *a very similar sound.* Another mathematical example is the set of rationals, which are placed into equivalence classes such as $\{\frac{2}{3} = \frac{4}{6} = \frac{6}{9} = \ldots = \frac{200}{300} = \ldots\}$, the equivalence relation being $(a/b) \sim (c/d) \Leftrightarrow ad = bc$.

Q. 19.46. Which of the following are equivalence relations. In the case of those which are not, point out why they fail:

(a) equality between a set of weights, lengths, etc.;
(b) congruence between geometrical figures;
(c) similarity between geometrical figures;
(d) $AB \sim CD \Leftrightarrow AB$ equal and parallel to CD;
(e) 'is contemporary with' between boys of a school;
(f) 'has the same name as' between days of the week;
(g) 'are separated by an integral multiple of 360° between angles;
(h) are separated by an integral multiple of 90° between angles;
(i) 'have the same colour' between cars;
(j) 'are brothers' between human beings; between men. 'is friendly with', ditto;
(k) $(x, y, z) \sim (kx, ky, kz)$ (e.g. $(1, -3, -2) \sim (-3, 9, 6)$) between points in 3-space;
(l) the same as (k), only using homogeneous coordinates for points in 2-space;
(m) $\begin{bmatrix} a & b \\ c & d \end{bmatrix} \sim \begin{bmatrix} a' & b' \\ c' & d' \end{bmatrix} \Leftrightarrow ad - bc = a'd' - b'c'$ between 2×2 matrices;
(n) $\begin{bmatrix} a & b \\ c & d \end{bmatrix} \sim \begin{bmatrix} ka & kb \\ kc & kd \end{bmatrix}$ between 2×2 matrices;
(o) 'is skew to' between lines in 3-space.

We may now enunciate a general theorem:

An Equivalence Relation partitions a set into disjoint subsets.

What this amounts to is that every element of the set belongs to one, and exactly one, equivalence class; this class consisting (by definition) of all those elements which are related to it by the equivalence relation, or, that no member of the set can belong to two different equivalence classes (*no*

† However, just because a relation expresses a 'quality of sameness', it does not mean it must be an equivalence relation. For example, the relation between towns of being 'less than 50 miles apart' (assuming it were well-defined) is not an equivalence relation (Why?). However, the relation 'is near to' may be made into an equivalence relation.

overlapping of classes). Indeed, if an element were to belong to two classes, then those classes would necessarily be identical.

To prove this result, suppose b and c are distinct members of two equivalence classes B and C respectively, and if possible, let a belong to both B and C.

Then $a, b \in B \Rightarrow a \sim b \Rightarrow b \sim a$ (by S).

$a, c \in C \Rightarrow a \sim c$.

But $b \sim a$ and $a \sim c \Rightarrow b \sim c$ (by T), and this would mean that b and c belong to the same equivalence class, which is contrary to supposition. Hence B and C can have no element in common, and the theorem is proved.

Cosets as equivalence classes

We now go on to show that, in a group, membership of the same coset (left or right) relative to a given subgroup is an equivalence relation. But first we express membership of the same coset in a slightly different manner.

Suppose a and b belong to the same left coset xH. Then it is possible to find elements h_r, h_s in H such that $a = xh_r$; $b = xh_s$. Therefore $a^{-1}b = (xh_r)^{-1}(xh_s) = h_r^{-1}x^{-1}xh_s = h_r^{-1}h_s$, so that, since h_r and h_s belong to the *subgroup* H, so does h_r^{-1}, and so does the product $h_r^{-1}h_s$. Thus $a^{-1}b \in$ H. This condition is also sufficient to guarantee membership of the same left coset. For $a^{-1}b \in \text{H} \Rightarrow a^{-1}b = h$, (where $h \in$ H) $\Rightarrow b = ah$, so that b is in aH.†

Q. 19.47. Show that a necessary and sufficient condition for a and b to belong to the same right coset is $ab^{-1} \in$ H.

Next, we show that $a \sim b \Leftrightarrow a^{-1}b \in$ H is an equivalence relation. First, $a \sim a$, since $a^{-1}a = 1 \in$ H (since H is a subgroup, it must contain 1), so R is satisfied. Secondly, $a \sim b \Leftrightarrow a^{-1}b \in$ H. But H is a subgroup, and so contains the inverse of $a^{-1}b$, namely $b^{-1}a$, i.e. $b^{-1}a \in \text{H} \Rightarrow b \sim a$; thus S is fulfilled. Finally, $a \sim b$ and $b \sim c \Rightarrow a^{-1}b \in$ H and $b^{-1}c \in$ H. But H, being a subgroup, is closed, so that $(a^{-1}b)(b^{-1}c) \in$ H, i.e. $a^{-1}(bb^{-1})c \in$ H, or $a^{-1}c \in \text{H} \Rightarrow a \sim c$ and the third requirement, T, is in order.

Having established that membership of the same coset is an equivalence relation, the disjoint property of cosets (p. 339) now follows immediately using the disjoint property of equivalence classes proved above, and this leads to the establishment of Lagrange's theorem for finite groups.

† Some books *define* the left coset of a relative to a subgroup H as the set $\{b : a^{-1}b \in \text{H}\}$, or equally well, the set $\{b : b^{-1}a \in \text{H}\}$. See, for example, F. M. Hall, *Abstract Algebra I* (C.U.P.), p. 276 ff., II, p. 22 ff.

Q. 19.48. Give examples of equivalence classes which, unlike the cosets of a group, do not contain equal numbers of elements.

Multiplication of subsets

If A $\{a_1, a_2, a_3, \ldots, a_r\}$ and B $\{b_1, b_2, b_3, \ldots, b_s\}$ are two subsets of a group G, we shall use the notation AB to denote the set consisting of all possible products $\{a_1b_1, a_2b_1, \ldots, a_rb_1, a_1b_2, a_2b_2, \ldots, a_rb_2, \ldots, a_rb_s\}$ the number of which is clearly rs, though *some values may be repeated.* These products could be clearly set out as an array in a multiplication table as in table 19.11. The object of the notation AB is for the purpose of

Table 19.11

	b_1	b_2	b_3	$\ldots$	b_s
a_1	a_1b_1	a_1b_2	a_1b_3	$\ldots$	a_1b_s
a_2	a_2b_1	a_2b_2	a_2b_3	$\ldots$	a_2b_s
$\cdot$					$\cdot$
$\cdot$					$\cdot$
$\cdot$					$\cdot$
a_r	a_rb_1	a_rb_2	a_rb_3	$\ldots$	a_rb_s

abbreviation. We shall call AB the '*product set*' of the subsets A and B. Note that in general AB and BA are different unless the group is Abelian.†

Now consider a subgroup H $\{1, h_1, h_2, \ldots, h_{m-1}\}$. If we form the product set of H with itself, HH (which we may denote H^2), it is clear that, since H is a subgroup, every element of the form h_rh_s is in H, and so there are not m^2 different products, but only m distinct products, being the elements of H itself. We may say, in fact, that $H^2 = H$. This condition does in fact express the *closure* of H under internal products, and in view of the theorem on p. 217 this means that $H^2 = H$ is also a *sufficient* condition for a subset H to be a subgroup of a finite group.

Products of subsets from an infinite group

In the case of an infinite group, we may observe the same property for subgroups. For example, H $\{\ldots -8, -4, 0, 4, 8, \ldots\}, +$ is a subgroup of $(\mathbf{Z}, +)$, and it is clear that the product‡ of any two elements of H is also an element of H; for example, $-64 + 24$ is an element of H. On the other hand, consider the product of the two subsets of Z:

† Note that we have a binary operation on subsets which is non-commutative (see Ch. 2). We shall see in Ch. 21 how cosets may themselves form a group under this operation.

‡ We continue to use the term 'product' even though the group operation is +.

A $\{\ldots -5, -2, 1, 4, 7, \ldots\}$ and B $\{\ldots -6, -2, 2, 6, 10, \ldots\}$ (table 19.12). It will be seen that this time, the product set is the *whole set* **Z**,

Table 19.12

		Set B						
	...	−10	−6	−2	2	6	10	...
⋮		⋮	⋮	⋮	⋮	⋮	⋮	
−5	...	−15	−11	−7	−3	1	5	...
−2	...	−12	−8	−4	0	4	8	...
Set A 1	...	−9	−5	−1	3	7	11	...
4	...	−6	−2	2	6	10	14	...
7	...	−3	1	5	9	13	17	...
⋮		⋮	⋮	⋮	⋮	⋮	⋮	

and in fact that each integer occurs an infinite number of times (e.g. $5 = 7 + (-2) = 10 + (-5) = 13 + (-8)$, etc.), and this infinity of 5's is strung out along a line of which only two may be seen in the above table).

If one takes the finite subset A $\{3, 4, 5, 6\}$ from the set **Z**, with the operation multiplication, one obtains the set A^2 $\{9, 12, 15, 16, 18, 20, 24, 25, 30, 36\}$ containing exactly ten elements, there being six repetitions among the sixteen possible products. Again, if B is the set $\{2, 3, 4, 5, 6, \ldots$ *ad inf.*$\}$, then B^2 (again under the operation $\times$) consists of the set $\{4, 6, 8, 9, 10, 12, 14, \ldots\}$ of all composite positive integers.

Q. 19.49. Experiment with products of subsets taken from (a) sets of residue classes under + mod. n and also under $\times$ mod. n, for selected values of n; (b) subsets of the set **Q** of rationals under $\times$.

As a further example, this time from an infinite group $(V_2, +)$, suppose A refers to the set of all vectors $\overline{OP}$ along the unit line segment OU, and B refers to the set of vectors $\overline{OQ}$ such that Q is between O and V on the unit segment OV. Then AB is the set of all vectors such as $\overline{OR}$ whose extremities (R) lie within the rhombus $OUWV$.

Products of subsets from a finite group

We now take some examples from the group Q_6, whose table is shown on p. 216. Let A be the subset $\{a, d, m, h\}$, then A^2 (= AA) is given by the elements in sub-table 19.13, i.e. A^2 is the set $\{l, f, b, j, c, k, g, 1\}$ and we may note that, even if B were limited to the elements $\{a, d\}$, AB would still

Table 19.13

	a	d	m	h
a	l	f	b	j
d	c	k	g	1
m	b	j	d	f
h	g	1	c	k

be the same set, the first two rows of the table containing all eight elements of A^2. Again, if we take C = $\{a, b, k\}$ and D = $\{d, g, j\}$, then CD contains the elements in table 19.14. In this case, it seems that we get the *minimum*

Table 19.14

(*a*) CD

	D: d	g	j
C: a	f	h	c
b	c	f	h
k	h	c	f

(*b*) DC

	C: a	b	k
D: d	c	f	h
g	f	h	c
j	h	c	f

possible number of elements – only three, and that although Q_6 is non-Abelian (so that, in fact, none of the above pairs of elements commute), yet here we have the same set in both cases with CD equal to DC. Is this a fluke, or is it inevitable, we may ask?

You may quickly convince yourself that it is not inevitable: take E = $\{a, b, c\}$, F = $\{d, f, g\}$. The sub-tables are:

Table 19.15

	F: d	f	g
E: a	f	g	h
b	c	d	f
c	l	a	1

and

	E: a	b	c
F: d	c	f	m
f	d	g	b
g	f	h	1

so that EF = $\{1, a, c, d, f, g, h, l\}$ whereas FE = $\{1, b, c, d, f, g, h, m\}$. On the other hand, if G = $\{a, b, k\}$, H = $\{a, d, m, b\}$ we have sub-table 19.16, and we note that GH = HG = $\{1, a, b, c, d, f, h, k, l, m\}$. This

Table 19.16

	H: a	b	d	m
G: a	l	1	f	b
b	1	m	c	k
k	m	l	h	a

and

	G: a	b	k
H: a	l	1	m
b	1	m	d
d	c	f	h
m	d	k	a

time, the 'product of the subsets' is commutative, but here we get ten different elements of the possible twelve products, two of the latter having been repeated.

Products of cosets

Some of the subsets selected above were cosets of some of the subgroups analysed on p. 216 (e.g. A, C, D, G, H). You will notice that there was one case, CD, when only three – the minimum possible number – of elements were present in the product set. We saw that the product of a subgroup with itself has this property, yet neither C nor D is a subgroup. The answer to the question 'when will the product of two sets each of m elements contain only m distinct elements?' is, 'When the two sets are *cosets of a normal subgroup*.' In the above case, C and D were cosets of the normal subgroup $\{1, l, m\}$. You will appreciate also that the product set, $\{c, f, h\}$ is another of the cosets of this same subgroup. The theory behind this will be discussed in Ch. 21.

The notation A^2 suggests that we may invent a meaning for A^{-1}. If A is the set $\{a, b, c, d, \ldots\}$, then we may define A^{-1} to be the set $\{a^{-1}, b^{-1}, c^{-1}, d^{-1}, \ldots\}$. In Q_6, with the notation above,

$A^{-1} = \{a^{-1}, d^{-1}, m^{-1}, h^{-1}\} = \{b, h, l, d\}$; $E^{-1} = \{a^{-1}, b^{-1}, c^{-1}\} = \{b, a, g\}$; $D^{-1} = \{d^{-1}, g^{-1}, j^{-1}\} = \{h, c, f\}$ (= CD or DC, remember);

finally, if $J = \{1, d, h, k\}$ (a subgroup), then $J^{-1} = \{1, h, d, k\} = J$. In the case of a subgroup, it is plain that, since the inverse of every element of the subgroup must also lie in the subgroup, that no new elements can arise.

Q. 19.50. If H and K are subgroups, prove that HK is a subgroup if and only if $HK = KH$.

Q. 19.51. With the notation above, for Q_6, find AJ, JA, AC, CA, AC^{-1}, CA^{-1}, FC, GF, $F^{-1}G$, $G^{-1}F$. Find other subsets such that the product set has as few members as possible.

Q. 19.52. Take the cosets $A\{\ldots -5, -2, 1, 4, 7, \ldots\}$ and $B\{\ldots -4, -1, 2, 5, \ldots\}$ of the subgroup $H\{\ldots -6, -3, 0, 3, 6, \ldots\}$ of $(\mathbf{Z}, +)$ and find their product set. What do you notice? Do the same with the cosets of $\mathbf{0}\{\ldots -10, -5, 0, 5, 10, \ldots\}$ letting the cosets be **1**, **2**, **3**, **4** (the residue classes modulo 5).

Q. 19.53. If H and K are subgroups, prove that the number of elements in HK is equal to the number in H multiplied by the number in K and divided by the number common to H and K.

Q. 19.54. If H and K are subgroups of G, and one or both of them is a normal subgroup, prove that HK (= KH) is a subgroup of G.

Q. 19.55. If A, B, C are subsets of a group G, prove $(A \cup B)C = AC \cup BC$, but that $(A \cap B)C \neq AC \cap BC$. What may we say in the latter case?

Q. 19.56. Find a counter-example to show that $AB = AC$ does not imply $B = C$; show, however, that the conclusion is true when A contains a single element.

EXERCISES 19

1 Given a subgroup H, the only coset of H which contains the identity is H itself.

2 What is the index of the subgroup D_3 in the group S_4?

3 Go through the examples of subgroups in Ch. 14 including those in the text as well as those in the questions and exercises, and elsewhere in this book, and describe the cosets in as many cases as possible, stating whether or not the subgroups are normal.

4 We have already proved that the centre of a group is a subgroup (p. 217). Prove that it is a *normal* subgroup.

5 Give the cosets of the subgroup of S_4 generated by the cycle $(ABCD)$ with the notation on p. 318.

6 In S_4 (p. 219), let $A = \{j, k, n, s, t, x\}$, i.e. the set of elements of period 4. Show that A^2 contains all the elements of the alternating group.

7 Examine the cosets of the subgroups of the symmetries of (a) the regular tetrahedron, (b) the cube, which leave a vertex fixed.

8 If H_1 and H_2 are subgroups of G, and g is any element of G, show that $gH_1 \cap gH_2 = g(H_1 \cap H_2)$.

9 Invent as simple a counter-example as possible to show that, if H_1, H_2 are subgroups of G, then H_1H_2 is not necessarily a subgroup.

10 If $x, y \in$ G, and H is a subgroup of G, so that xH and yH are cosets, prove (a) $y \in x\text{H} \Rightarrow y\text{H} = x\text{H}$; (b) $y\text{H} = x\text{H} = x^{-1}y \in \text{H}$.

11 Suppose H is a subgroup of a group G, of index r, and that G decomposes into left cosets H, xH, yH, zH. Show that a possible decomposition into right cosets is H, $\text{H}x^{-1}$, $\text{H}y^{-1}$, $\text{H}z^{-1}$, ...

12 We have seen (p. 355) that $a^{-1}b \in \text{H}$ is a necessary and sufficient condition for a and b to belong to the same left coset. When the group is Abelian and the operation is addition, this may be written: $b - a \in \text{H}$. Interpret this condition in the case of the subgroup (10**Z**, +) of the group (**Z**, +).

13 Suppose G is a group of permutations of m objects (a subgroup of S_m), and x is a permutation not in G, so that xG is a left coset. Is $x\text{G} \cup \text{G}$ necessarily a group? Consider special cases, e.g. when G is Gp (g) where $g = (1\ 2\ 3\ 4)$ and $x = (1\ 2)$. If G is generated by $g_1, g_2, g_3, \ldots$, how can the order of Gp $(g_1, g_2, \ldots, x)$ be determined?

14 In the previous question we considered the effect of 'adjoining' an extra generator to a given group. Consider a group of functions under successive substitution, such as Gp (f), where $f(z) \equiv 2/(2 - z)$, or where $f(z) \equiv 1/(1 - z)$, and consider the effect of adjoining the function $1/z$, or th function $-z$. Give a geometrical interpretation from the Argand diagram.

15 Show that the set of rationals which are perfect squares (e.g. 4/49), form a group under multiplication which is a subgroup both of (**Q**, $\times$) and also of ($\mathbf{Q}^+$, $\times$). Describe the cosets in each case.

16 The functions of the form $az + b$ (where a and b are constants belonging to some field) form a group under successive substitution. Consider the subgroups for which (a) $a = 1$, (b) $b = 0$. Identify the cosets. Are these subgroups normal? Interpret them as groups of transformations on the Argand diagram when a and b are complex constants and $z' = az + b$.

17 We have shown that, if H is a subgroup of G, and g is any element of G not in H, then $g\mathrm{H} \cap \mathrm{H} = \phi$. Invent further counter-examples to show that if H is a sub*set* of G (but not a subgroup), then gH and H may have elements in common.

18 Describe the cosets of $\begin{bmatrix} a & b \\ c & d \end{bmatrix}$ with $ad - bc = 1$ under matrix multiplication regarded as a subgroup of ($\mathcal{M}_2$, $\times$).

19 Show that the matrices $\begin{bmatrix} a & b \\ c & d \end{bmatrix}$ with $a, b, c, d \in 2\mathbf{Z}$ form a group under addition. Describe the cosets: (a) in the group of matrices where $a, b, c, d \in \mathbf{Z}$; (b) in the group of matrices where $a, b, c, d \in \mathbf{R}$.

20 Show that the matrices $\begin{bmatrix} a & b \\ c & d \end{bmatrix}$ where $a, d \in 2\mathbf{Z} + 1$, and $c, d \in 2\mathbf{Z}$ are a group under multiplication, provided $ad - bc = 1$. Identify the cosets when regarded as a subgroup of ($\mathcal{M}_2$, $\times$).

21 Which of the following are equivalence relations:

(a) between points P, Q in a plane, the property that $PA + PB = QA + QB$, where A, B are two given fixed points;
(b) between men, the property of having the same mother-in-law;
(c) between integers, the property of having a common factor;
(d) between the line in a plane, the property of being separated by an angle less than 5°.

22 A and B are subsets of a group G, each containing m elements. Then AB contains k elements, where $m \leqslant k \leqslant m^2$. Consider circumstances under which $k = m$, and under which $k = m^2$.

23 **If H is a subgroup of G and x is an element of G not in H, prove that $x\mathrm{H}x^{-1}$ is a subgroup of G which is isomorphic to H.** What can you say when $x\mathrm{H}x^{-1}$ and H contain the same elements?

24 A group G of order 16 has elements a, b, p, where $p^4 = a^2 = b^2 = (pa)^2 = 1$, $ab = ba$, $pb = bp$. Show that H$\{1, a, b, ab\}$ and $K\{1, p^2, b, p^2b\}$ are subgroups. Decompose G into right cosets relative to H and to K. Verify that H and K commute, and that their product set is a group of order 8. What is its structure, and what is the structure of the group G?

25 If $p = (1\ 2\ 4)$ and $q = (1\ 3\ 4)$, we know that Gp $(p, q) = \mathrm{A}_4$. Find a permutation r of period 3 which does not belong to Gp (p, q), and so must include a fifth digit, such that $(qr)^2 = 1 = (rp)^2$. What is the index of Gp (p, q) in Gp (p, q, r)?

26 For a given subgroup H of a group G, where $|\mathrm{H}| = m$, $|\mathrm{G}| = n = mr$, the number of left cosets is r, the index of H in G. A set of r elements taken one from each left coset is called a *left transversal*, or a set of left coset

representatives for H. A right transversal is similarly defined. For example, in S_4 (p. 219), we have the subgroup H = $\{1, a, b, c, d, f\}$. An example of a right transversal would be the set $S = \{d, j, s, t\}$, one from each of the four right cosets. How many possible right (or left) transversals are there? Note that j and s are in the same *left* coset of H, and so are *alternatives* as representatives for the left cosets. A possible left transversal might be $\{a, h, v, j\}$, and another example $\{c, t, x, j\}$.

For any group, show that if x and y are in different left cosets, then x^{-1} and y^{-1} are in different right cosets. Deduce that, if $S = \{x_1, x_2, \ldots, x_r\}$ is a left transversal, then $S^{-1} = \{x_1^{-1}, x_2^{-1}, \ldots, x_r^{-1}\}$ is a right transversal. (Note that in the above example, $S = \{d, j, s, t\}$ and $S^{-1} = \{c, t, x, j\}$.) The above provides a method of setting up a 1,1 correspondence between the left and right cosets of a given subgroup.

27 In the group defined $r^4 = p^3 = (rp)^2 = 1$, show that $\{1, r^2, p^2r^2p, pr^2p^2\}$ is a normal subgroup.

28 Is S_4 a subgroup of $A_5 \times C_2$?

20 *Conjugate elements: normal subgroups (1)*

The reader who does not require further development of the theory, but is anxious to study the applications of group theory to music, bell-ringing and geometry, may omit Chs. 20, 21 and 22 and proceed directly to chapters 23 ff., for which a knowledge of the two preceding chapters is not essential. Homomorphism is, however, mentioned briefly in Ch. 23.

Examples of the combined operation of the form xyx^{-1} have been met before in several contexts (see Ex. 10.10, pp. 152, 200, 320, 461, 464). This phenomenon occurs so frequently in group theory that we devote a whole chapter to it:

> **the element xyx^{-1} is called the transform of the element y by the element x.**

Also, if $z = xyx^{-1}$, then y and z are called *conjugate elements* of the group. The term 'conjugate' implies that the relationship between conjugate elements is 'reciprocal', 'mutual', 'two-way', or 'symmetric' – if z is conjugate to y, then one has a right to expect that y will be conjugate to z. In fact,

$$z = xyx^{-1} \to zx = xyx^{-1}x = xy \to x^{-1}zx = x^{-1}xy = y.$$

Hence $y = x^{-1}zx$, so that y is the transform of z by the element x^{-1}, and the symmetric nature of the relationship is proved. It would be possible to distinguish xyx^{-1} and $x^{-1}yx$ by describing them as 'right' and 'left' conjugates of y by x. However, in the absence of any specification, we shall normally mean the former.

Note that in an Abelian group the whole concept of the transform of an element becomes trivial; for in this case, $xyx^{-1} = xx^{-1}y = y$, and so each element is its own transform – the process of first doing x^{-1} before doing y, and afterwards doing x (thus undoing the original x^{-1}) makes no difference to the operation y. This is extremely uninteresting, so we shall not be concerned with this process in Abelian groups.

But in a non-Abelian group, xyx^{-1} will generally be different from the transformed element y, though it will have some features in common with it. If, however, by chance xyx^{-1} does equal y, then we have $xy = yx$, so that *the two elements commute*, though the group as a whole may be non-Abelian.

Conjugate elements have the same period

One of the features which conjugate elements have in common is that they have the *same period.* If $y^k = 1$, then $(xyx^{-1})^k = (xyx^{-1})(xyx^{-1}) \ldots (xyx^{-1})$ (k factors) $= xy(x^{-1}x)y(x^{-1}x)y(x^{-1}x)y \ldots y(x^{-1}x)yx^{-1} = xyyyy \ldots yx^{-1} = xy^kx^{-1} = x1x^{-1} = 1$ and k is the least integer for which this is true. *Hence xyx^{-1} is also of period k.*

Similar permutations

We first illustrate xyx^{-1} from a group of permutations.

Let $\quad y \quad$ represent the product of cycles (1 5 3 2 8)(4 7 6)(9 10),
$\quad x \quad$ represent the product of cycles (5 6 8 9)(2 7)(3 10),
so that $\quad x^{-1}$ represents the product of cycles (10 3)(7 2)(9 8 6 5).

Consider the effect of xyx^{-1}:

$$\left.\begin{array}{l} 1 \xrightarrow{x^{-1}} 1 \xrightarrow{y} 5 \xrightarrow{x} 6 \\ 6 \longrightarrow 5 \longrightarrow 3 \longrightarrow 10 \\ 10 \longrightarrow 3 \longrightarrow 2 \longrightarrow 7 \\ 7 \longrightarrow 2 \longrightarrow 8 \longrightarrow 9 \\ 9 \longrightarrow 8 \longrightarrow 1 \longrightarrow 1 \end{array}\right\} \quad \text{the cycle } (1\ 6\ 10\ 7\ 9)$$

$$\left.\begin{array}{l} 2 \longrightarrow 7 \longrightarrow 6 \longrightarrow 8 \\ 8 \longrightarrow 6 \longrightarrow 4 \longrightarrow 4 \\ 4 \longrightarrow 4 \longrightarrow 7 \longrightarrow 2 \end{array}\right\} \quad \text{the cycle } (2\ 8\ 4)$$

$$\left.\begin{array}{l} 3 \longrightarrow 10 \longrightarrow 9 \longrightarrow 5 \\ 5 \longrightarrow 9 \longrightarrow 10 \longrightarrow 3 \end{array}\right\} \quad \text{the cycle } (3\ 5).$$

Therefore xyx^{-1} contains the cycles (1 6 10 7 9)(2 8 4)(3 5). We have here an example of a general theorem.

The transform of a given permutation by another permutation will contain cycles of the same lengths as the given permutation.

(In the above case, the cycles were of lengths 5, 3 and 2 both for y and for xyx^{-1}.) Here is another way in which an element may resemble its transform. The permutations y and xyx^{-1} are often described as '*similar permutations*'.

Furthermore, we observe that

$$y = (1\ 5\quad 3\ 2\ 8)(4\ 7\ 6)(9\ 10),$$
$$xyx^{-1} = (1\ 6\ 10\ 7\ 9)(4\ 2\ 8)(5\quad 3),$$

and the elements in each cycle of y are ***replaced according to the***

permutation x which replaces 1 by 1, 5 by 6, 3 by 10, etc. This is illustrated by the diagram in fig. 20.01.

Q. 20.01. Using x and y to mean the permutations of the previous paragraph, and letting $z = (1\ 2\ 3\ 4\ 5\ 6)(8\ 9\ 10)$, and $w = (1\ 2)(3\ 4)(5\ 6)(8\ 9)$, and $s = (1\ 3\ 5\ 7)(2\ 4\ 6\ 8)$, find $z^{-1}yz$; zyz^{-1}; wxw^{-1}; $w^{-1}xw$; xwx^{-1}; $x^{-1}wx$; $s^{-1}ys$; and sys^{-1}, and check the properties described.

Q. 20.02. If r is the cycle $(1\ 2\ 3\ 4)$, and a is the transposition $(1\ 2)$, find the meaning of rar^{-1}, r^2ar^{-2}, $r^{-1}ar$. Deduce that r and a generate the group S_4. Also interpret ara^{-1}, ar^2a^{-1} and ar^3a^{-1}. Generalise the first result to show that S_n may be built up from two generators, and exhibit this in the case of S_5.

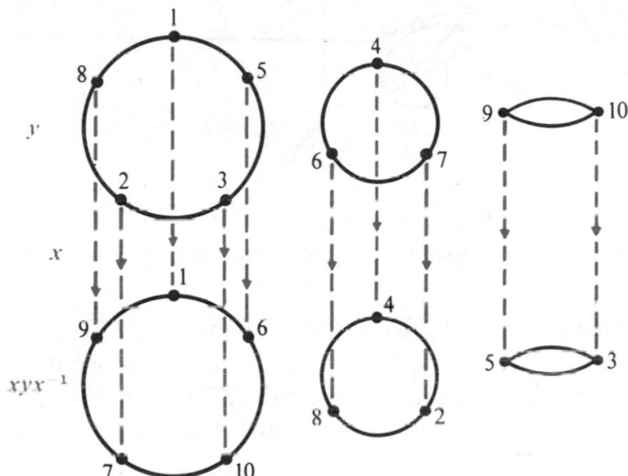

Fig. 20.01. $y = (1\ 5\ 3\ 2\ 8)(4\ 7\ 6)(9\ 10)$
$x = (5\ 6\ 8\ 9)(2\ 7)(3\ 10)$
$xyx^{-1} = (1\ 6\ 10\ 7\ 9)(4\ 2\ 8)(5\ 3)$ and takes (e.g.) 6 to 5, then 5 to 3, then 3 to 10, i.e. 6 moves round to 10.

Reflection in a moved axis

In fig. 20.02 let m and m^* denote the operations of reflecting in the axes labelled m and m^*, where the second axis is derived from the first by a rotation r about a certain point O. Now consider the effect of reflecting a figure (F) in the moved (*) axis. The result of this operation, $m^*(F)$ is marked on the diagram. Now $m^*(F)$ can be obtained in another way: apply the inverse rotation, r^{-1}, to the figure, taking it back to the position $r^{-1}(F)$, then reflect in the *original* position of the mirror, so obtaining $mr^{-1}(F)$, then apply the rotation r to carry this image, with the mirror, into its final position, $rmr^{-1}(F)$.

Hence we obtain $\mathbf{m^* = rmr^{-1}}$,

i.e. the operation of reflecting in the rotated mirror is the transform by the rotation, of the operation of reflection in its original position.

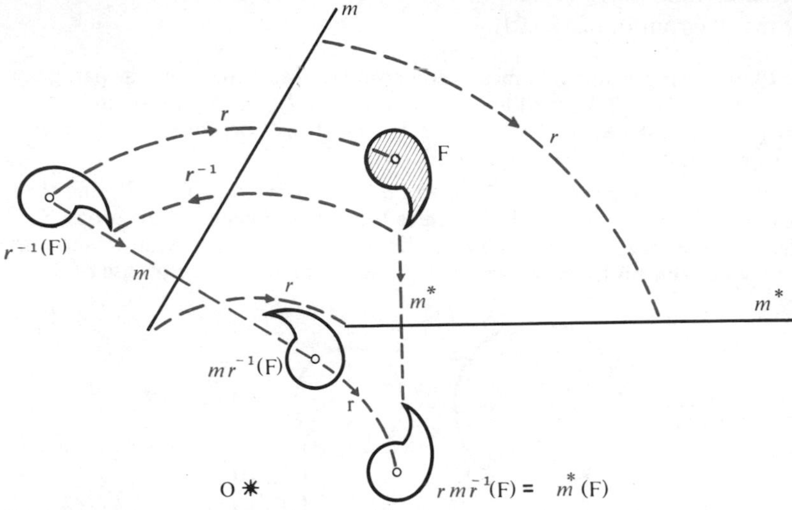

Fig. 20.02. Reflection in the rotated mirror is the conjugate of m by r, i.e. $m^* = rmr^{-1}$.

To illustrate this in a specific case, consider the group D_4 as the group of symmetries of the square. With the notation of p. 94, we may let f denote an anticlockwise rotation through $90°$, $f^2 = g$, $f^{-1} = h$; the reflections are shown in fig. 20.03, these axes being fixed in space.

$$\text{Now} \quad \left. \begin{aligned} b &= fdf^{-1} \\ d &= fbf^{-1} \end{aligned} \right\}$$

or, reflection in one diagonal is conjugate, under a quarter-turn, to reflection in the other diagonal, and this is precisely because the quarter-turn transforms one diagonal into the other.

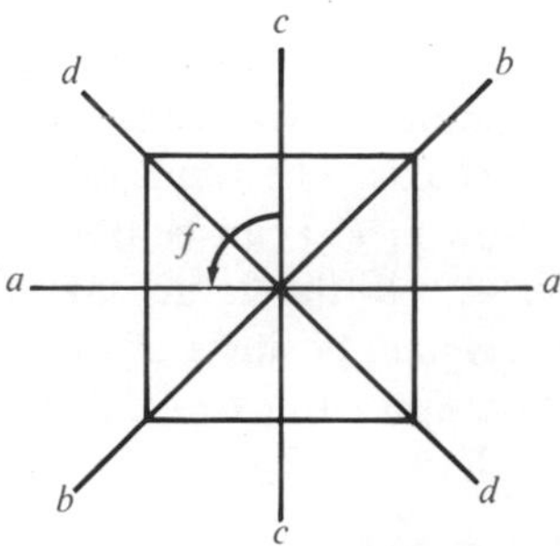

Fig. 20.03. Symmetries of square $b = fdf^{-1}$; $d = fbf^{-1}$.

In the particular example above, we used r to denote a rotation of the mirror, *but it might have represented any transformation.* The above statement may thus be generalised thus:

The operation of reflecting in the transformed mirror is equivalent to applying the inverse transformation (i.e. first taking the mirror back to its original position), reflecting in the original position, and finally applying the transformation (to restore the final position).

Symbolically: $\boldsymbol{m}^* = \boldsymbol{tmt}^{-1}$, where $\boldsymbol{t}$ is any transformation of the mirror axis.

Successive reflections in moved axes

In Ch. 14, pp. 233 ff. we considered the question of the successive reflections of a triangle in one of its sides, and we used a, b and c to denote reflections in the *moved* positions of the sides. According to our present convention, we should have denoted these by a^*, b^* and c^*, and this we do now, reserving the letters a, b, c to denote reflections in the fixed lines which were the positions originally occupied by the sides of the triangle. a^*, b^*, c^* will mean, then, reflections in the axes *BC, CA, AB carried by the triangle.*

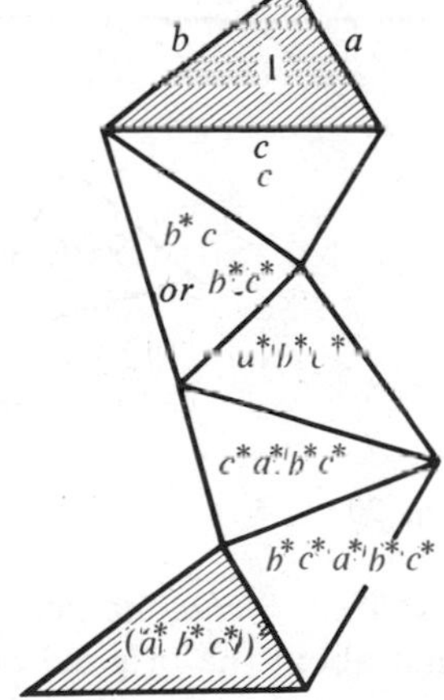

Fig. 20.04. Successive reflections of triangle in moved positions of its sides.

Now suppose the operations c^*, b^*, a^*, c^*, b^*, a^*, are applied successively to the triangle (see fig. 20.04). Now the second operation in this sequence is b^*, a reflection about the position of b after it has suffered the reflection in c, so that, by the above theory, we have

$$b^* = c\,b\,c^{-1}, \qquad \text{and hence } b^*c = c\,b\,c^{-1}c = cb.$$

The third reflection, a^* is in the position of the line a after it has suffered the transformation b^*c ($= cb$). Therefore, by the above, we have

$$a^* = (cb)\,a\,(cb)^{-1} = c\,b\,a\,b^{-1}c^{-1},$$

so that $\quad a^*b^*c = (c\,b\,a\,b^{-1}c^{-1})(c\,b\,c^{-1})c = c\,b\,a\,b^{-1}c^{-1}c\,b\,c^{-1}c = c\,b\,a.$

Proceeding in this way, we find that the successive transformations undergone by the triangle are:

$$\begin{aligned} c^* &= c, & c^*a^*b^*c^* &= c\,b\,a\,c,\\ b^*c^* &= c\,b, & b^*c^*a^*b^*c^* &= c\,b\,a\,c\,b,\\ a^*b^*c^* &= c\,b\,a, & (a^*b^*c^*)^2 &= (c\,b\,a)^2, \end{aligned}$$

and the latter, a *translation*, we may regard as being equivalent to the identity if we are interested only in the orientation of the triangle (compare p. 234).

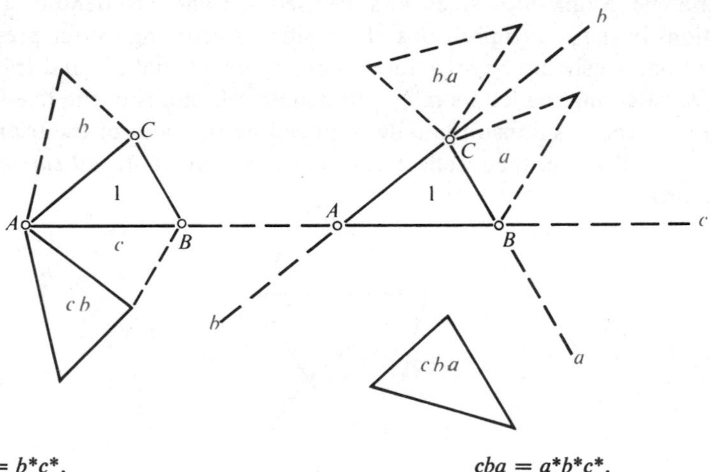

Fig. 20.05.
This should be compared with Fig. 20.04.

The fact that $b^*c^* = cb$, $a^*b^*c^* = cba$, etc., can be seen geometrically. For cb means a reflection first about the original position of b, followed by one about the original position of c – reflections about the *fixed* lines b and c. cba means respective reflections in the original (fixed) lines a, b, c in turn. The final image here (see fig. 20.05) should be compared with $a^*b^*c^*$ in fig. 14.05.

Q. 20.03. Show that $a^*b^*c^*$ is a glide reflection. What is its inverse? Compare with the glide reflection $b^*c^*a^*$. How can you justify the relation $(a^*b^*c^*)^2(b^*c^*a^*)^2 = (b^*c^*a^*)^2(a^*b^*c^*)^2$? Deduce that $a^*b^*c^*a^*b^*c^*b^*c^*a^*b^*c^*a^*c^*b^*a^*c^*b^*c^*b^*a^*c^*b^* = 1$. Compare Q. 14.46, 14.53 Draw the diagram for a scalene triangle, and also for an equilateral triangle.

Successive rotations about moved vertices

A further example is provided by considering the movement of a triangle in a plane as a result of performing successive half-turns about its vertices.

In fig. 20.06 the triangle starts from position ABC (marked 1), and is

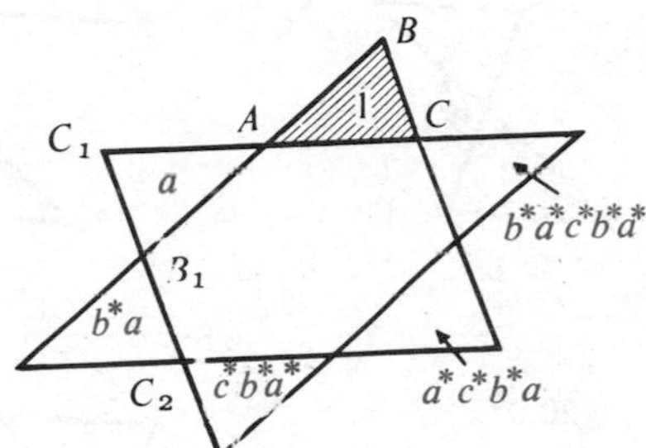

Fig. 20.06. Successive half-turns about vertices of triangle in their moved positions.

first given a half-turn a about A to bring it into position AB_1C_1. This is followed by the half-turn b^* about the moved position B_1, so that the third position of the triangle is the result of the composite operation b^*a. Next a half-turn c^* about the second position C_2 of C, giving c^*b^*a, and so on. Evidently six half-turns about successive positions of the vertices in sequence restore the triangle to its former position, and we may say $c^*b^*a^*c^*b^*a^* = 1$ ($a = a^*$ for the first operation).

But b^* is the operation b transformed by a, i.e. $b^* = a\,b\,a^{-1}$, so that $b^*a^* = ab$.

Again, $\quad c^*$ is the operation c transformed by b^*a^* (or ab),

i.e. $\quad c^* = (ab)c(ab)^{-1} = a\,b\,c\,b^{-1}a^{-1},$

so that $\quad c^*b^*a^* = a\,b\,c\,b^{-1}a^{-1}.a\,b = a\,b\,c\,b^{-1}1\,b = a\,b\,c$

and we have $\quad c^*b^*a^* = a\,b\,c.$

Proceeding in this way, we obtain:

$$
\begin{aligned}
b^*a^* &= a\,b \\
c^*b^*a^* &= a\,b\,c \\
a^*c^*b^*a^* &= a\,b\,c\,a \\
b^*a^*c^*b^*a^* &= a\,b\,c\,a\,b \\
c^*b^*a^*c^*b^*a^* &= a\,b\,c\,a\,b\,c = 1.\dagger
\end{aligned}
$$

Thus all the motions of the triangle have been expressed in terms of half-turns a, b, c about the fixed points A, B, C. We illustrate geometrically

† The groups generated by $\{a, b, c, \ldots\}$ and $\{a^*, b^*, c^*, \ldots\}$ are sometimes referred to as 'opposite groups'.

that $b^*a^* = a\,b$ and that $c^*b^*a^* = a\,b\,c$, also that $(a\,b\,c)^2 = 1$ in fig. 20.07. Fig. 20.072 illustrates that *abc* is a half-turn about *O*, and this is enough to convince us that $(abc)^2 = 1$.

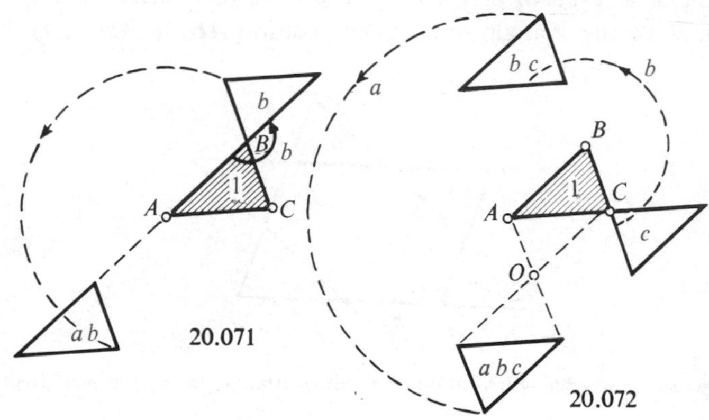

Fig. 20.071. $b^*a^* = ab$.
20.072. $c^*b^*a^* = abc$. This should be compared with Fig. 20.06.

Transformed operations in general

Consider next a rotation *r* through angle θ about *P* and a translation *t* represented by the vector $\overline{PQ}$ (fig. 20.08). Let us find out what is the significance of the *operation r transformed by t*, i.e. by trt^{-1}. The diagram

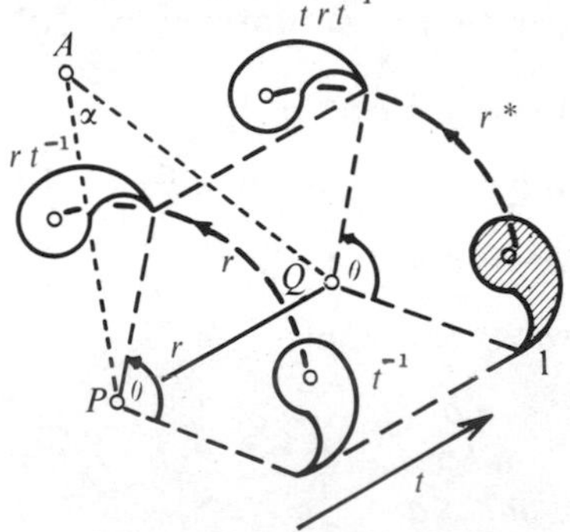

Fig. 20.08. Rotation transformed by a translation.

makes it clear that the combined operation trt^{-1} is equivalent to a single rotation about *Q* (i.e. the new position of *P* transformed by the operation *t*, which carries *P* to *Q*). This is also apparent from other considerations: for the effect of trt^{-1} on the point *Q* itself is to leave it where it is:

$$\begin{aligned} trt^{-1}(Q) &= tr(P) \quad (\text{since } t^{-1}(Q) = P) \\ &= t(P) \quad (\text{since } r(P) = P) \\ &= Q. \end{aligned}$$

And since the combined operation which consists of a parallel movement, a rotation through θ, and then another parallel movement is intuitively equivalent to a single rotation through θ, it is evident that trt^{-1} is a rotation through θ about Q.

Q. 20.04. Show that rtr^{-1} is a translation through the same distance as t, but in a direction inclined at angle θ to the direction of t. What is $t^{-1}rt$?

Q. 20.05. Suppose s is a rotation through α which maps P into Q (about a point on the perpendicular bisector of PQ). Show that srs^{-1} is also a rotation through θ about Q.

Q. 20.06. If a and b are reflections in two intersecting lines, and $c = bab^{-1}$, what does c represent?

Q. 20.07. If r is a rotation and a is a mirror reflection, what is ara^{-1}?

Q. 20.08. If r_1 and r_2 are rotations about A and B through angles α and β, interpret the operations $r_2r_1r_2^{-1}$; $r_1^{-1}r_2r_1r_2^{-1}$.

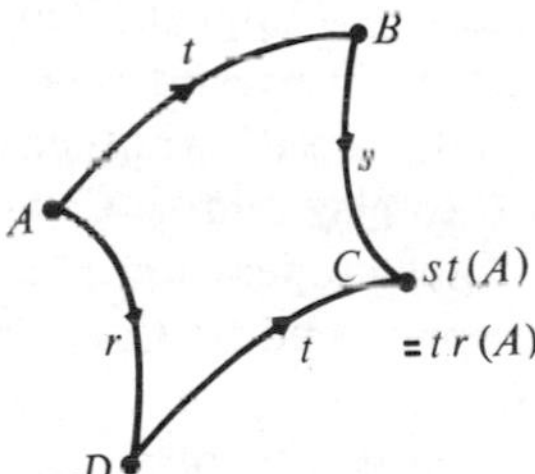

Fig. 20.09. $s = trt^{-1}$.

The above discussion may be summarised by reference to fig. 20.09 illustrating $s = trt^{-1}$. A point A is transformed by t into the point B, i.e. $B = t(A)$. B is then transformed by s into C, so that $C = s(B) = st(A)$. On the other hand, A is transformed by r into D, so $D = r(A)$, while $C = t(D) = tr(A)$. Hence we have $st(A) = tr(A)$, or $st = tr$, that is, $s = trt^{-1}$. The point of the diagram is to show that to go from B to C (operation s), one may go via A and D, performing the operations t^{-1}, r and t consecutively. Or, looking at it another way, the transformation s (the conjugate of r under t) is obtained from r by replacing each point and its image under r (e.g. A and D) by their images under t (i.e. B and C).

Further informal illustrations

The above examples illustrate in several contexts how the operation of the form xyx^{-1} may arise. Further examples will appear, e.g. in the chapter on bell-ringing (Ch. 24). Why is this special form so common, one may ask? Why is one not so much concerned with, say, $xyxy$, or with yxy, or xyx^2, etc.? The answer may be seen in many homely contexts: it is common to perform an operation, then a different operation, and then to undo the first operation (assuming, that is, that the first operation is reversible!). One opens a door, walks through it, and then closes it; one gets up, dresses, does a day's work, undresses and then goes to bed; one turns the steering-wheel of a car clockwise, drives round a right-hand bend, and then turns the wheel anticlockwise to its centre position; a message is coded, then transmitted, then decoded; a machine is switched on, a job is done, and the current then switched off; a navigator will read off a course from a chart, manoeuvre the ship, and then plot the ship's position on to the chart, and in reading information off the chart he will have to allow for magnetic variation one way, but when putting information back on to the chart, allowance for magnetic variation will work the opposite way. Then there is the type of problem where you 'think of a number, add six, ... do a lot of other computations, ...' take away the number you first thought of' ... An English text is translated into German; a composer sets the German words to music, the German words are then translated back into English! This is roughly what happened in the case of Haydn's *Creation*. The result was not quite the same as if the original text had been undisturbed! Another example from music: a composer wants a horn in F to play middle C. So he transposes up a fifth (operation x), and writes it down (operation y) as the G above middle C. The horn player plays it to sound middle C, so transposing *down* a fifth (operation x^{-1}). And so on ...

Further mathematical examples will follow to show that xyx^{-1} may crop up in mathematical situations not immediately concerned with groups.

Suppose we have a transformation $(w - a)/(w - b) = c(z - a)/(z - b)$ connecting two variables z and w, where a, b and c are constants. Then if we let

$$z_1 = \frac{z - a}{z - b} = \mathrm{f}(z), \quad \text{and} \quad z_2 = cz_1 = \mathrm{g}(z_1) = \mathrm{gf}(z),$$

we shall also have

$$\frac{w - a}{w - b} = z_2 = \mathrm{f}(w),$$

which may be rewritten $\quad w = \mathrm{f}^{-1}(z_2).$

But, $\quad z_2 = \mathrm{gf}(z);\quad$ so finally $\quad w = \mathrm{f}^{-1}[\mathrm{gf}(z)] = \mathrm{f}^{-1}\mathrm{gf}(z),$

so we have an example of the *function g transformed by the function* f^{-1}.

Suppose from a height h above sea level the distance of the horizon is given by the formula $d = \mathrm{f}(h)$. Figure 20.10 shows an observer at height x ($= BA$) viewing the horizon in the direction AH, and in a direct line with a second observer at C, who is at height y above the earth, and at a horizontal distance a from A.

Then we have $\quad BH = \mathrm{f}(x); \quad DH = \mathrm{f}(y).$

So $\quad a = BH - DH = \mathrm{f}(x) - \mathrm{f}(y) \Rightarrow \mathrm{f}(y) = \mathrm{f}(x) - a$

$$\Rightarrow y = \mathrm{f}^{-1}[\mathrm{f}(x) - a].$$

Letting $\mathrm{g}(x) \equiv x - a$, so that $\mathrm{g}[\mathrm{f}(x)] \equiv \mathrm{f}(x) - a$, we arrive at the formula $y = \mathrm{f}^{-1}\mathrm{gf}(x)$.

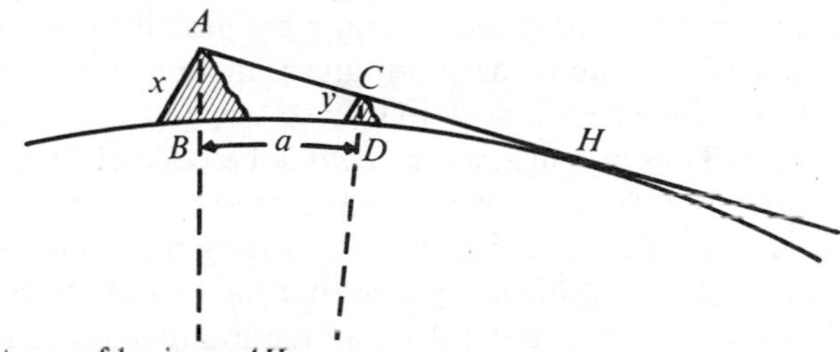

Fig. 20.10. Distance of horizon, AH.

For example, taking measurements of height in feet, and distances of horizon in miles, $\mathrm{f}(x)$ takes the form $\sqrt{(3x/2)}$ (approximately), so that $\mathrm{f}^{-1}(x) \equiv \frac{2}{3}x^2$. If $x = 3210$ (summit of Scafell Pike), and $a = 30$, we have $\mathrm{f}(x) = \sqrt{(1\tfrac{1}{2} \times 3210)} = 69.4$

$$\mathrm{gf}(x) = 69.4 - 30 = 39.4,$$
$$y = \mathrm{f}^{-1}\mathrm{gf}(x) = \tfrac{2}{3}\,39.4^2 = 1035$$

So if there is a peak 2,500 feet high, 30 miles away from an observer on Scafell Pike, it will appear to that observer as if 2500 − 1035 = 1465 ft of that peak were 'sticking up above the horizon'. [In fact, when standing on Scafell Pike, the summit of Scafell (3,162 ft) at a horizontal distance of about 0.7 miles does appear a little higher.]

Q. 20.09. If $\mathrm{f}(x) = ax + b$; $\mathrm{g}(x) = 1/x$; $\mathrm{h}(x) = (ax + b)/(cx + d)$; $\mathrm{j}(x) = -x$, find the functions gfg^{-1}; ghg^{-1}; jfj^{-1}; jhj^{-1}; hgh^{-1}; $\mathrm{h}^{-1}\mathrm{gh}$.

Q. 20.10. If $\mathrm{g}(x) = 1/(1 - x)$ and $\mathrm{f}(x) = x^2$, find $\mathrm{g}^{-1}\mathrm{fg}(x)$ and $\mathrm{gfg}^{-1}(x)$.

Q. 20.11. If $R = \begin{bmatrix} \cos\theta & \sin\theta \\ -\sin\theta & \cos\theta \end{bmatrix}$, $M = \begin{bmatrix} \cos 2\theta & \sin 2\theta \\ \sin 2\theta & -\cos 2\theta \end{bmatrix}$, find RMR^{-1} and $R^{-1}MR$, and interpret geometrically.

Q. 20.12. If $A = \begin{bmatrix} a & b \\ c & d \end{bmatrix}$, $M = \begin{bmatrix} 0 & 1 \\ 1 & 0 \end{bmatrix}$ find MAM^{-1}, and interpret geometrically for various matrices A. Repeat when A is the general 3×3 matrix, and M is a permutation matrix.

Q. 20.13. A matrix A is said to be 'similar to' a matrix B if there exists a matrix P such that $B = PAP^{-1}$, and we write $A \sim B$. Prove that $\sim$ is an equivalence relation. (This is important in the theory of eigenvectors and characteristic polynomials – similar matrices have the same characteristic polynomial, the same eigenvalues, and the values of their determinants are equal. See G. Matthews, *Matrices*, II (Edward Arnold), Ch. 3; T. Brand and A. Sherlock, *Matrices Pure and Applied* (Contemporary Mathematics, Edward Arnold), Ch. 5.)

Conjugacy classes

If, for two elements y and z of a group, it is possible to find an x in the group such that $z = xyx^{-1}$ (or that $y = x^{-1}zx$, or that $zx = xy$, or that $x^{-1}z = yx^{-1}$), then, as we have seen y and z are described as a pair of 'conjugate elements'. If y and z are both given, there may not exist in the whole group a single element x such that $zx = xy$, and in this case, y and z are not conjugate. Thus conjugacy is a *relation* between elements of a group. It is evidently reflexive (since $y = 1y1^{-1}$) and we have already shown that it is a symmetric relation which justifies the name 'conjugate'. To show that it is also transitive, suppose that the relation 'is conjugate to' is denoted 'R'. Then if $y \text{ R } z$ and $z \text{ R } w$, we require to show that $y \text{ R } w$.

Now $\quad y \text{ R } z \Rightarrow yx_1 = x_1z \quad$ for some x_1 in G.

And $\quad z \text{ R } w = zx_2 = x_2w \quad$ for some x_2 in G.

Hence, $\quad yx_1 = x_1x_2wx_2^{-1}, \qquad$ so that $yx_1x_2 = x_1x_2w$.
Thus if $x_1x_2 = x_3$ (the group G being closed), we have $yx_3 = x_3w$ and therefore $y \text{ R } w$. We have thus proved that conjugacy is an equivalence relation.

We may now proceed to set up equivalence classes. For a fixed element y, all those elements $z_1, z_2, z_3, \ldots$ which are conjugate to y will be in the same equivalence class with y, which we shall denote $\{y\}$. This is called the 'conjugacy class' of y. Note that in an Abelian group, as we have already seen, the question of conjugacy is trivial: the only element which is conjugate to y is y itself, so each element is the sole member of its particular conjugacy class.

In all groups, including non-Abelian groups, it is obvious that the identity is going to be in a class on its own – it cannot be 'transformed into' another element of the group because it has a different character from all the others. Indeed, it is the only element of period 1, and since we have seen that the period of an element is unchanged by conjugation by any other element, it is clear that no other element of the group can possibly be conjugate to the identity.

An element is said to be 'self-conjugate' when it commutes with every element of the group, i.e. belongs to the centre. For if $xy = yx$ for all y

in the group, then $x = yxy^{-1}$ for all y, i.e. x is transformed into *itself* by all the elements of the group – no other element of the group is conjugate to it – its conjugacy class contains it alone.

Finding conjugacy classes – the 'Snap' method

Let us find the conjugacy classes in a particular group, say the tetrahedral group A_4 (table 17.02). Now it is evidently going to be a very long and

Table 20.01

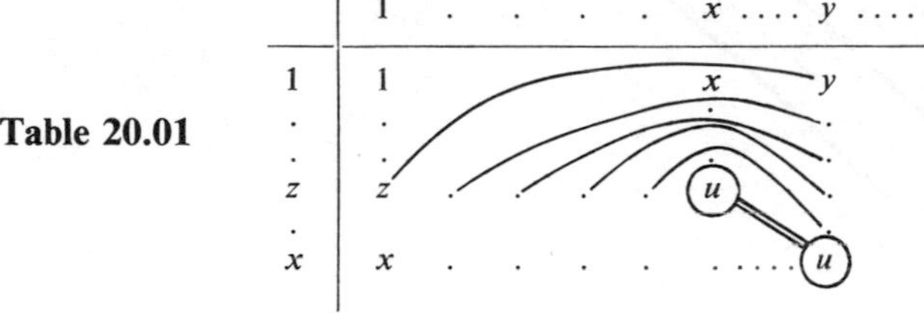

tedious process to work out xyx^{-1} for every possible pair of values of x and y (up to 144 such computations would be needed for a group of order 12). We shall see that the work may be greatly reduced. We may easily find pairs of conjugate elements as follows: find any element p in row x (say $p = xy$). Next find p in column x, so that $p = zx$. Then $p = xy = zx$, so that y R z. This is a suitable procedure only when the transforming element x is given.

Suppose we are trying to discover whether or not two given elements y and z are conjugate. This means we are seeking an x in the group such that $zx = xy = u$ (say). The element u will occur in row z, column x, and also in column y, row x. To find it, we look *along row* z and *down column* y at each element in turn till we find a pair the same – 'Snap!' In the diagram it is supposed that the first five elements in row z and column y disagree till we reach the sixth, which are both u's (corresponding elements of row z and column y are 'linked'). This common element u has the property that it lies in row z *column* x (say), and in column y *row* x, so that $zx = xy$, or $z = xyx^{-1}$ and z R y. (It would also be possible to have used column z in conjunction with row y, since $z \text{ R } y \Leftrightarrow y \text{ R } z$.) Having found a single example of $zx = xy$, we need go no further, though we shall do so in the example which follows for the sake of completeness.

Now one way in which the work can be reduced is that we know that y R z implies that y and z have the same period (see p. 364). Therefore it is a waste of time even to make the above check for elements of different periods. In the group A_4, we begin by checking whether or not a R b. (*Note:* both a and b are of period 2.) (See table 20.02.)

Table 20.02

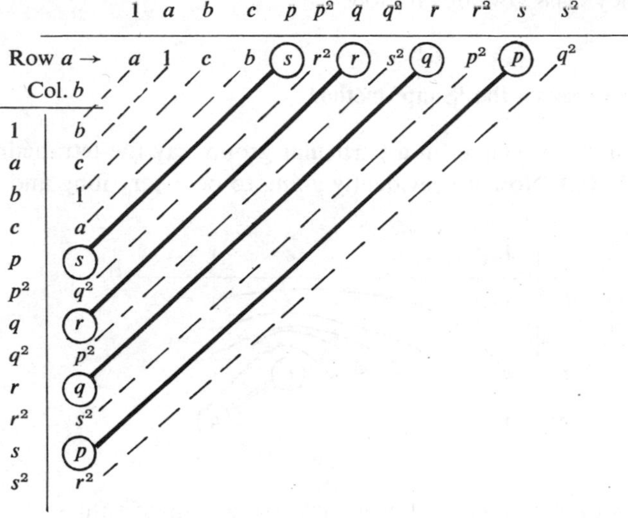

From this we see that:

$$\left.\begin{aligned} ap &= pb = s \\ aq &= qb = r \\ ar &= rb = q \\ as &= sb = p \end{aligned}\right\} \text{ and this means that } a \mathrm{R} b.$$

Moreover, we may write:

$$\left.\begin{aligned} bp^2 &= p^2a = s^2 \\ bq^2 &= q^2a = r^2 \\ br^2 &= r^2a = q^2 \\ bs^2 &= s^2a = p^2 \end{aligned}\right\} \text{ which confirms that } b \mathrm{R} a.$$

Similarly the following may be verified:

$$\left.\begin{aligned} ap^2 &= p^2c = r^2 \\ aq^2 &= q^2c = s^2 \\ ar^2 &= r^2c = p^2 \\ as^2 &= s^2c = q^2 \\ cp &= pa = r \\ cq &= qa = s \\ cr &= ra = p \\ cs &= sa = q \end{aligned}\right\} a \mathrm{R} c \qquad \left.\begin{aligned} bp &= pc = q \\ bq &= qc = p \\ br &= rc = s \\ bs &= sc = r \\ cp^2 &= p^2b = q^2 \\ cq^2 &= q^2b = p^2 \\ cr^2 &= r^2b = s^2 \\ cs^2 &= s^2b = r^2 \end{aligned}\right\} b \mathrm{R} c.$$

Therefore we have the conjugacy class $\{a, b, c\}$. Now it so happens that this class contains all the elements of period 2. But do not imagine that just because two elements have the same period they are bound to be conjugate.† For instance, consider p and q^2 (both of period 3).

Row $p \to p$	r	s	q	p^2	1	r^2	b	s^2	c	q^2	a .
Col. $q^2 \to q^2$	s^2	r^2	p^2	b	s	1	q	a	p	c	r .

(Column q^2 has been given an anticlockwise quarter-turn for convenience.) As we go along, we are unable to say 'snap', so we conclude that p and q^2 are *not* conjugate elements, even though they have the same period. On the other hand, p and q are conjugate, there being three occasions for saying 'snap' as we go along row p and down column q.

Row $p \to p$	$\boxed{r}$	s	q	p^2	1	r^2	b	$\boxed{s^2}$	c	q^2	$\boxed{a}$
Col. $q \to q$	$\boxed{r}$	p	s	r^2	c	q^2	1	$\boxed{s^2}$	b	p^2	$\boxed{a}$.

Hence $\left.\begin{array}{l} pa = aq = r \\ pr = rq = s^2 \\ ps^2 = s^2q = a \end{array}\right\}$ whence $p \mathrm{R} q$ and the companion relations: $\left\{\begin{array}{l} qa = ap = s \\ qr^2 = r^2p = a \\ qs = sp = r^2 \end{array}\right.$.

It turns out that $p \mathrm{R} q \mathrm{R} r \mathrm{R} s$, giving a class $\{p, q, r, s\}$, and also that $p^2 \mathrm{R} q^2 \mathrm{R} r^2 \mathrm{R} s^2$, giving the class $\{p^2, q^2, r^2, s^2\}$.

The group A_4 is therefore partitioned into the following conjugacy classes:

$$\{1\}; \quad \{a, b, c\}; \quad \{p, q, r, s\} \text{ and } \{p^2, q^2, r^2, s^2\}.$$

For the sake of completeness, and of interest, we give the full results in the investigations which establish these last two classes:

$$\left.\begin{array}{l} pb = br = s \\ pq^2 = q^2r = b \\ ps = sr = q^2 \\ rb = bp = q \\ rq = qp = s^2 \\ rs^2 = s^2p = b \end{array}\right\} p \mathrm{R} r \qquad \left.\begin{array}{l} pc = cs = q \\ pq = qs = r^2 \\ pr^2 = r^2s = c \\ sc = cp = r \\ sq^2 = q^2p = c \\ sr = rp = q^2 \end{array}\right\} p \mathrm{R} s$$

$$\left.\begin{array}{l} ra = as = p \\ rp = ps = q^2 \\ rq^2 = q^2s = a \\ sa = ar = q \\ sp^2 = p^2r = a \\ sq = qr = p^2 \end{array}\right\} r \mathrm{R} s \qquad \left.\begin{array}{l} qc = cr = p \\ qp = pr = s^2 \\ qs^2 = s^2r = c \\ rc = cq = s \\ rp^2 = p^2q = c \\ rs = sq = p^2 \end{array}\right\} q \mathrm{R} r$$

† In an Abelian group, for example, each element belongs to a class on its own. Even in such democratic systems as C_p (p prime), and $\mathrm{C}_2 \times \mathrm{C}_2 \times \mathrm{C}_2$ (see p. 272), no element may be transformed into any other!

$$\left.\begin{array}{lll} qb &= bs &= r \\ qp^2 &= p^2s &= b \\ qr &= rs &= p^2 \\ sb &= bq &= p \\ sp &= pq &= r^2 \\ sr^2 &= r^2q &= b \end{array}\right\} q \,\mathrm{R}\, s$$

$$\left.\begin{array}{lll} p^2a &= aq^2 &= s^2 \\ p^2r &= rq^2 &= a \\ p^2s^2 &= s^2q^2 &= r \\ q^2a &= ap^2 &= r^2 \\ q^2r^2 &= r^2p^2 &= s \\ q^2s &= sp^2 &= q \end{array}\right\} p^2 \,\mathrm{R}\, q^2 \qquad \left.\begin{array}{lll} p^2b &= br^2 &= q^2 \\ p^2q^2 &= q^2r^2 &= s \\ p^2s &= sr^2 &= b \\ r^2b &= bp^2 &= s^2 \\ r^2q &= qp^2 &= b \\ r^2s^2 &= s^2p^2 &= q \end{array}\right\} p^2 \,\mathrm{R}\, r^2$$

$$\left.\begin{array}{lll} p^2c &= cs^2 &= r^2 \\ p^2q &= qs^2 &= c \\ p^2r^2 &= r^2s^2 &= q \\ s^2c &= cp^2 &= q^2 \\ s^2q^2 &= q^2p^2 &= r \\ s^2r &= rp^2 &= c \end{array}\right\} p^2 \,\mathrm{R}\, s^2 \qquad \left.\begin{array}{lll} q^2c &= cr^2 &= s^2 \\ q^2p &= pr^2 &= c \\ q^2s^2 &= s^2r^2 &= p \\ r^2c &= cq^2 &= p^2 \\ r^2p^2 &= p^2q^2 &= s \\ r^2s &= sq^2 &= c \end{array}\right\} q^2 \,\mathrm{R}\, r^2$$

$$\left.\begin{array}{lll} q^2b &= bs^2 &= p^2 \\ q^2p^2 &= p^2s^2 &= r \\ q^2r &= rs^2 &= b \\ s^2b &= bq^2 &= r^2 \\ s^2p &= pq^2 &= b \\ s^2r^2 &= r^2q^2 &= p \end{array}\right\} q^2 \,\mathrm{R}\, s^2 \qquad \left.\begin{array}{lll} r^2a &= as^2 &= q^2 \\ r^2p &= ps^2 &= a \\ r^2q^2 &= q^2s^2 &= p \\ s^2a &= ar^2 &= p^2 \\ s^2p^2 &= p^2r^2 &= q \\ s^2q &= qr^2 &= a \end{array}\right\} r^2 \,\mathrm{R}\, s^2.$$

These results should be interpreted in the context of the symmetries of the regular tetrahedron. We have seen that p, q, r and s are in one class, and p^2, q^2, r^2 and s^2 in another. Interpreted as symmetries of the tetrahedron, these conjugate elements represent rotations through $\frac{2}{3}\pi$, one class being clockwise and the other class anticlockwise. Those which belong to the same conjugacy class (e.g. p, q, r, s) *may be transformed into each other*, each being a rotation about an altitude of the tetrahedron *in the same sense*. (Note however that, while in A_4 the elements of period 3 fall into two classes, in S_4 these eight elements are all in the same class.)

More generally, it is true that rotations of the same magnitude and sense about different axes are conjugate if an operation exists *in the group* which would map one axis into the other. Thus, in A_4 for example, p and q are conjugate with $xpx^{-1} = q$, and the transforming element x is one which maps the directed altitude p into the directed altitude q of the tetrahedron. But there is no element *in the group* which transforms p into

q^2, in other words, there is no operation *in the group* which transforms the p-altitude into the q-altitude *and reverses its sense.*

Q. 20.14. Find other examples and counterexamples.

Q. 20.15. Show that the conjugacy classes of Q_6 (table 14.01) are $\{1\}$; $\{b\}$; $\{k, l\}$; $\{a, m\}$; $\{c, f, h\}$ and $\{d, g, j\}$. Which elements transform a into m; c into f; c into h; d into j; f into h; k into l; d into g?

Q. 20.16. Find the conjugacy classes and interpret in terms of the symmetries of a geometric figure for (a) D_3 (table 13.04), (b) D_4 (p. 94), (c) D_6 (p. 192), (d) Q_4 (p. 101) (this may not be realised as the symmetry group of any figure). Interpret these classes also in terms of groups of permutations.

The transforming elements

Now when we were looking for conjugacy classes in A_4, we did not bother to check, say, row p against column p, because we know that $p \mathrm{R} p$ (i.e. conjugacy is reflexive). However, in a non-Abelian group, row x and column x are in general different.

Q. 20.17. Under what circumstances would they be identical?

It would be interesting to note which elements of a group *can* transform an element into itself. In the case of p in the group A_4:

row p →	p	r	s	q	p^2	1	r^2	b	s^2	c	q^2	a
col. p →	p	s	q	r	p^2	1	s^2	c	q^2	a	r^2	b .

Hence p is transformed into itself by the elements 1, p and p^2; or $xpx^{-1} = p$ when $x = 1$, $x = p$ or $x = p^2$.
Again, consider the values of x for which $c = xcx^{-1}$:

row c →	c	b	a	1	r	q^2	s	p^2	p	s^2	q	r^2
col. c →	c	b	a	1	q	r^2	p	s^2	s	p^2	r	q^2

so c is self-conjugate under transformation by the elements 1, a, b and c.

You will realise that, for a given element y of any group, if y is self-conjugate under transformation by an element x so that $y = xyx^{-1}$, then this means that $yx = xy$, in other words that x will commute with y. All those elements which do commute with the given element y constitute what we have described as the 'centraliser' or 'normaliser' of y (see p. 222), and we proved there that the centraliser is a subgroup. Confirmation of this is seen in the case of A_4 by noting that the centraliser of p is the subgroup $\{1, p, p^2\}$, and that of c is the subgroup $\{1, a, b, c\}$.

Q. 20.18. In Ch. 14, you found the centralisers of various elements of certain groups. Check that each element of one of these centralisers is transformed by another such element into an element of the same centraliser subgroup.

Q. 20.19. Show that the conjugacy classes of S_4 (table 14.02) are:

$$\{1\};\ \{a, b, f, g, p, w\};\ \{h, r, y\};\ \{c, d, i, l, m, q, u, v\};\ \{j, k, n, s, t, x\}.$$

If a is the cycle (AB), b is (BC) and g is (CD), write down the corresponding permutations for each class.

Conjugacy classes and cosets

Now *cosets of a group only exist in relation to some pre-selected subgroup.* In the above work, we have not made any reference to any subgroup except the brief mention of centralisers in the last paragraph. Nevertheless, cosets seem somehow to be emerging even though no subgroups have been identified. Look at the detailed work on the group A_4 on p. 377 and you will see that those elements which transform r into p by the relations

$$\begin{aligned} pb &= br \quad (= s) \\ pq^2 &= q^2r \quad (= b) \\ ps &= sr \quad (= q^2) \end{aligned}$$

are b, s and q^2, and that these are also a permutation of the three products (in brackets on the right-hand side). Now $\{b, s, q^2\}$ is in fact one of the cosets in A_4, being a left coset of $\{1, r, r^2\}$:

$$\{b, s\ q^2\} = b\{1\ r, r^2\} = s\{1, r, r^2\} = q^2\{1, r, r^2\},$$

or a right coset of $\{1, p, p^2\}$:

$$\{b, s, q^2\} = \{1, p, p^2\}b = \{1, p, p^2\}s = \{1, p, p^2\}q^2.$$

Now just as subgroups arise from a consideration of centralisers, so their cosets arise in an analogous way from a consideration of conjugacy classes. We now proceed to investigate this more fully.

★General results connected with centralisers

Let a be a fixed element of a group G, and let H be the centraliser of a, i.e. the set H $\{1, h_1, h_2, h_3, \ldots\}$ such that $ah = ha$ for all $h \in$ H. Now let b be a second fixed element of G which is not in H, and let $bab^{-1} = c$. Consider the transform of a by any element bh of the left coset bH:

$$\begin{aligned} (bh)\,a\,(bh)^{-1} &= bhah^{-1}b^{-1} \\ &= b(hah^{-1})b^{-1} \\ &= bab^{-1} \quad (\text{since } h \in \mathrm{H} \Rightarrow hah^{-1} = a) \\ &= c. \end{aligned}$$

Hence all elements of the left coset bH transform a into c. For example, in A_4, the centraliser of r is the subgroup $\{1, r, r^2\}$ and its left coset by b is $b\{1, r, r^2\} = \{b, s, q^2\}$, and each of these three elements transforms r into p:

$$\begin{aligned} brb^{-1} &= brb = p, \\ srs^{-1} &= srs^2 = p, \\ q^2r(q^2)^{-1} &= q^2rq = p. \end{aligned}$$

Q. 20.20. Which are the elements which transform p into r? Interpret this in terms of cosets. Consider $\{b, s, q^2\}$ as the right coset $\{1, p, p^2\}b$, and apply the above treatment.

We have shown that in the general case, all the elements of the left coset bH transform a into bab^{-1} $(= c)$. We shall also prove the converse, that if $xax^{-1} = c$, then $x \in b$H where H is the centraliser of a. For $c = bab^{-1} = xax^{-1}$ (given) so that $ab^{-1} = b^{-1}xax^{-1} \Rightarrow a = b^{-1}xax^{-1}b = (b^{-1}x)a(b^{-1}x)^{-1}$ (see p. 66). This means, by definition, that $b^{-1}x$ lies in the centraliser of a, i.e. $b^{-1}x \in$ H so finally $x \in b$H, which was what we wished to prove.

We may add that, since all the cosets of a particular subgroup contain the same number of elements, it follows that the *number* of elements which will transform a given element a alike is equal to the *order of the centraliser of a*.

★**Q. 20.21.** Produce counterexamples from D_6 to prove that the centraliser of a given element is not necessarily a normal subgroup (table for D_6, p. 192).

★**Q. 20.22.** Give the centraliser of a in D_6 (p. 192), and its left and right cosets. Test the truth of the statements of the previous paragraph by finding the transform of a by each element of each coset in turn. Repeat for the groups D_4, Q_4 and Q_6.

★**Q. 20.23.** If elements lie in different cosets of N, then they transform a differently, and the index of N is equal to the number of elements in the class of a. (N is the centraliser of a.)

The transform of a given subgroup: conjugate subgroups

When H is a *given* subgroup of a group G, we now consider the transform of H by a *fixed* element a of G (not in H). This will be denoted $a\text{H}a^{-1}$, and refers to the set $\{a1a^{-1}, ah_1a^{-1}, ah_2a^{-1}, \ldots\}$, where $1, h_1, h_2, \ldots$ are the elements of H. We shall show that this new set is also a subgroup of G, and to do this (in the case of finite groups) it will only be necessary to establish *closure* (see p. 217).

If ah_ra^{-1} and ah_sa^{-1} are any two elements of $a\text{H}a^{-1}$, then

$$(ah_ra^{-1})(ah_sa^{-1}) = ah_r(a^{-1}a)h_sa^{-1} = ah_rh_sa^{-1} \quad (\text{since } a^{-1}a = 1).$$

But h_r and h_s are in the subgroup H, so that $h_r h_s = h_t \in$ H. Hence the product is $ah_t a^{-1}$, and this is undoubtedly one of the elements of $a\text{H}a^{-1}$. *Thus* $a\text{H}a^{-1}$ *is a subgroup of* G. It will be of the same order as H provided there are no repetitions of elements. This possibility is ruled out by the fact that $ah_r a^{-1} = ah_s a^{-1}$ would imply that $h_r = h_s$ by two applications of the cancellation law. Therefore *H and* $a\text{H}a^{-1}$ *are subgroups of the same order*, and the group $a\text{H}a^{-1}$ is called the *Conjugate Subgroup* of the group H by the element a.

The remaining question is: 'Are they the same subgroup?' The answer to this question is, in general, 'No.' For example, in the group A_4, we have the subgroup $\{1, p, p^2\}$. Calling this H, then $p\text{H}p^{-1} = \{1, p, p^2\} = \text{H}$; however, $a\text{H}a^{-1} = \{1, q, q^2\}$ and this is *not* the subgroup H.

Q. 20.24. Prove that conjugate subgroups are isomorphic.

Q. 20.25. Consider which elements of D_6 (table p. 204) transform H $\{1, a, u, x\}$ into the same subgroup, and which transform it into a different subgroup.

Clearly H is transformed into itself by any of its own elements, i.e. $h\text{H}h^{-1} = \text{H}$. However, one may ask whether there are any elements of G *not in* H which transform H into itself? The answer is that this may well be possible, and we may illustrate the truth of this statement by taking H to be the subgroup $\{1, a, b, c\}$ of A_4. Consider $p\text{H}p^{-1} = p\text{H}p^2$ $= \{1, pap^2, pbp^2, pcp^2\} = \{1, c, a, b\} = \text{H}$. Again, $r^2\text{H}(r^2)^{-1} = r^2\text{H}r$ $= \{1, r^2ar, r^2br, r^2cr\} = \{1, b, c, a\} = \text{H}$.†

Why does this happen in some cases and not in others?

We may write the relation $a\text{H}a^{-1} = \text{H}$ in another way: $a\text{H} = \text{H}a$, and this may be interpreted in the language of cosets: every element of the left coset aH is an element of the right coset Ha. Note that this does *not* require that $ah = ha$ for every element h in H. It *does* require that the elements of aH, i.e. $a, ah_1, ah_2, ah_3, \ldots$, are a rearrangement of the elements of Ha, i.e. $a, h_1a, h_2a, h_3a, \ldots$ For example, taking H to be $\{1, a, b, c\}$ in the group A_4, we have $p\text{H} = \{p, r, s, q\}$ and $\text{H}p = \{p, s, q, r\}$ – the elements turn up in a different order. But the order in which they turn up does not concern us – we are only interested in the *set*, and in which particular elements it contains.

Normal subgroups

Now the requirement $a\text{H} = \text{H}a$, that the left and right cosets of H by the element a are the same, endows the subgroup H with the vastly important

† Note that the transform of H by p and then by r^2 produces in each case H, but with the elements rearranged, and this is an automorphism of H. When H is the *whole* group G, we get $x\text{G}x^{-1}$, and as we saw in Ch. 11, p. 152, we get the various inner automorphisms of G by the various elements x.

property of being a *Normal Subgroup* (see p. 338), provided it is true for *all* elements a in G. It may not be generally realised by students that if $a\text{H} = \text{H}a$ for *some particular* a in G, there may yet exist another element b for which the left and right cosets are different, i.e. $a\text{H} = \text{H}a$ does not imply $b\text{H} = \text{H}b$. For example, in D_6 (p. 204), we have the subgroup $\text{H} = \{1, u\}$.

Now $\quad x\text{H} = x\{1, u\} = \{x, a\}$
while $\quad \text{H}x = \{1, u\}x = \{x, a\}$ $\Big\}$ so $x\text{H} = \text{H}x$.

But $\quad v\text{H} = v\{1, u\} = \{v, p\}$
and $\quad \text{H}v = \{1, u\}v = \{v, p^2\}$ $\Big\}$ so $v\text{H} \neq \text{H}v$,

and so $\{1, u\}$ is not a normal subgroup. It was an easy matter to find this counterexample, for with a subgroup of order 2, $x\{1, u\} = \{1, u\}x$ requires that x commutes with u, while $y\{1, u\} \neq \{1, u\}y$ requires that y does not commute with u. It was only necessary to select the element x from the *centre* of the group (see p. 217), and the element y to be one of those (many) elements which do not commute with u, which include p, p^2, b, c, y, z and w, as well as v.

Further observations on normal, invariant, or self-conjugate subgroups

Hitherto we have concentrated chiefly on one aspect of normal subgroups, namely that their left and right cosets are the same, i.e. $x\text{H} = \text{H}x \;\forall\; x \in \text{G}$. When H is a normal subgroup of G, this is abbreviated $\text{H} \lhd \text{G}$. You will remember that the *centraliser* or *normaliser* of a particular element h is the set of all elements of G which commute with h. The name 'normal' is connected with 'normaliser' by the fact that the relation $x\text{H} = \text{H}x \;\forall\; x \in \text{G}$ states that the *subgroup* H commutes with all elements of G. This does not mean of course that H is necessarily the *centre* of G, i.e. that each individual element of H commutes with every x in G, but it does mean that the set $x\text{H}$ contains the same elements as the set $\text{H}x$, that is, that the elements $x1x^{-1}$, xh_1x^{-1}, $xh_2^2x^{-1}$, ... form a rearrangement of the elements $1, h_1, h_2, h_3, \ldots$ The term 'self-conjugate', as an alternative to 'normal' may also be explained at this stage. If H is any sub*set* of G, then the set $x\text{H}x^{-1}$ is another subset, the conjugate of H by x (x being a selected element of G). Now if H is a sub*group*, we have seen that $x\text{H}x^{-1}$ is also a subgroup – the *conjugate subgroup* of H by the element x. In general, $x\text{H}x^{-1}$ will be a *different* subgroup from H, though isomorphic to it.† When, however, $x\text{H}x^{-1}$ coincides with the subgroup H itself for *all* elements x of the group, then H may be described as *self-conjugate*. However, the term 'normal' is far more usual. The term 'invariant' is also

† See the preliminary work above, and Q. 20.24.

meaningful in the context $xHx^{-1} = H$, for it suggests that the subgroup H may not be 'changed' into any other subgroup under conjugation by any element of the group. Indeed, we may say that H is normal in G if and only if it is invariant under all inner automorphisms of G, i.e. contains with any element all its conjugates.

Another way of stating the result in the previous paragraph is that if h is any element in a *normal* subgroup H, then H will contain *all the members of the class of h*, i.e. all elements of the form xhx^{-1} (x any element of G, see p. 374). For example, if G is the group S_4, then using the notation of the table on p. 219, the classes are: $\{1\}$; $\{a, b, f, g, p, w\}$; $\{h, r, y\}$; $\{c, d, i, l, m, q, u, v\}$; $\{j, k, n, s, t, x\}$ (see Q. 20.19, p. 380). Now S_4 has two normal subgroups: A_4, which contains the whole of the classes $\{1\}$, $\{h\}$ and $\{c\}$ above; and D_2 which contains the whole of the classes $\{1\}$ and $\{h\}$. Now both these contain the class of h, that is, the double transpositions (1 2)(3 4), (1 3)(2 4) and (1 4)(2 3); while the class of c, which is contained in A_4 includes all those cycles of period 3. The above theorem has a variety of applications. For example we may *immediately* deduce that S_3 is not a normal subgroup of S_4. For S_3 contains two elements of period 3, whereas a normal subgroup would have to contain all the other six permutations of period 3, the eight cycles being (1 2 3), (1 3 2), (1 2 4), (1 4 2), (1 3 4), (1 4 3), (2 3 4) and (2 4 3).

Q. 20.26. Prove that D_4 and C_4 are not normal subgroups of S_4. Prove that C_3 is not a normal subgroup of A_4.

Q. 20.27. In Q. 20.15 we found the conjugacy classes of Q_6 (table 14.01). Use these to discover whether C_4 is a normal subgroup of Q_6.

When H is a normal subgroup of *order 2*, say $\{1, a\}$, it is easy to show that in this case a does have to belong to the centre of the group, for $\{1, a\}x = x\{1, a\} \ \forall \ x \in G$, i.e. $\{x, ax\} = \{x, xa\}$, hence $ax = xa \ \forall \ x \in G$.

Q. 20.28. When H is a normal subgroup $\{1, p, p^2\}$ of order 3, show that either p belongs to the centre, or else that $x = pxp \ \forall \ x \in G$.

To discover whether a given subgroup is normal or not†

Examples are given of a variety of methods. Note first, however, that whenever one can show that the *index of a subgroup is 2*, then there is no problem – that subgroup is normal. This is, however, merely a special case of a more general theorem of Frobenius: if $|H| = m$ and $|G| = mr = n$,

† Coxeter and Moser (*Generators and Relations for Discrete Groups*, p. 12, Ch. 2) describe a technique for the systematic enumeration of cosets, and the authors go on to show a method of determining whether a given subgroup is normal (p. 17).

and if m has no prime factor less than r, then H is normal in G. For example, in a group of order 20, a subgroup of order 5 must be normal.

Method 1

By a consideration of conjugacy classes, as in the example above on p. 384.

Method 2

Using the Structure Table.

In a finite group whose table has been constructed, it is a simple matter to list the right and left cosets of the given subgroup, as in Ch. 19. If they are all the same, then the subgroup is normal. If there is any disagreement, the subgroup is not normal.

Q. 20.29. Find which of the subgroups of D_6 (table 13.01), A_4 (p. 294), and S_4 (p. 219) are normal.

Method 3

From the Defining Relations.

It may be a laborious task to construct the group table, and indeed, this is usually quite unnecessary. Once defining relations in terms of a given set of generators have been laid down, then any further information about the group may theoretically be deduced. We shall illustrate by examples, and in each case, the investigation into the question whether a given subgroup H is or is not normal in G, will depend on *whether* xHx^{-1} *is the same or a different subgroup for all* x *in* G.

Consider the group defined $p^3 = r^4 = 1$; $rpr^{-1} = p^{-1}$. Show without constructing the table that Gp (p) is normal and Gp (r) is not.

First, let $H = \{1, p, p^2\}$, so that, clearly

$$pHp^{-1} = H, \text{ and } p^2Hp^{-2} = H.$$

Now $rHr^{-1} = \{1, rpr^{-1}, rp^2r^{-1}\} = \{1, p^{-1}, p\}$ since $rpr^{-1} = p^{-1}$, while rpr^{-1} and rp^2r^{-1} are inverses.

Hence $rHr^{-1} = H$.

Moreover, $r^2Hr^{-2} = r(rHr^{-1})r^{-1} = rHr^{-1} = H$, and similarly $r^3Hr^{-3} = H$.

Finally, since $pHp^{-1} = H$ and $p^2Hp^{-2} = H$, then if x is any element in G, such as $p^2r^3pr^2$, then

$$\begin{aligned} xHx^{-1} &= p^2r^3pr^2\, H\, (p^2r^3pr^2)^{-1} \\ &= p^2r^3pr^2\, H\, r^{-2}p^{-1}r^{-3}p^{-2} \quad \text{(see p. 66)} \\ &= p^2r^3p(r^2Hr^{-2})p^{-1}r^{-3}p^{-2} = p^2r^3pHp^{-1}r^{-3}p^{-2} \\ &= p^2r^3(pHp^{-1})r^{-3}p^{-2} = p^2r^3Hr^{-3}p^{-2} \\ &= p^2(r^3Hr^{-3})p^{-2} = p^2Hp^{-2} = H. \end{aligned}$$

Hence H is invariant under transformation by *any* element of G, and this is what we mean by saying the H is a normal subgroup.

Next, if $H = \{1, r, r^2, r^3\}$, then $pHp^{-1} = \{1, prp^{-1}, pr^2p^{-1}, pr^3p^{-1}\}$.
Now $rpr^{-1} = p^{-1}$ (given) $\Leftrightarrow r = prp$.
Hence $prp^{-1} = p(prp)p^{-1} = p^2r$, and this cannot be either of the elements of H.

$$\left.\begin{array}{ll} \text{For} & p^2r = 1 \Rightarrow r = p \\ & p^2r = r \Rightarrow p^2 = 1 \\ & p^2r = r^2 \Rightarrow p^2 = r \\ \text{and} & p^2r = r^3 \Rightarrow p^2 = r^2 \end{array}\right\} \text{ in every case contradicting the given relations } p^3 = 1 = r^4.$$

Hence pHp^{-1}, which contains prp^{-1}, cannot be the subgroup H which contains 1, r, r^2 and r^3. Thus H may be transformed (by p) into a *different* subgroup, and so Gp (r) is *not* normal in Gp (r, p).

In the case of Gp (p), the relation $rpr^{-1} = p^{-1}$ gives the game away, of course, because it really guarantees that the element r transforms Gp (p) into itself. However, the same relation might well have been given in a variety of disguised forms:

$$rpr^{-1} = p^{-1} \Leftrightarrow r = prp \Leftrightarrow rp\, p^2r \Leftrightarrow pr^3 = r^3p^2$$
$$\Leftrightarrow (pr)^2 = r^2 \Leftrightarrow (pr)^4 = 1,$$

these deductions making use of the other two relations specifying the periods of p and r.

Q. 20.30. Discover the number of elements in this group, and reveal its structure.
Q. 20.31. In the group G defined $p^3 = a^2 = (pa)^3 = 1$, prove without constructing the table that $\{1, a\}$ is not normal, but that $\{1, a, pap^{-1}, p^{-1}ap\}$ is normal in G, with structure D_2. Prove also that Gp (p) is not normal. What is the structure of G? (*Note:* in the proof it will be necessary to make various deductions from $(pa)^3 = 1$, such as $(pa)^3 = 1 \Rightarrow papapa = 1 \Rightarrow papap = a \Rightarrow pap^{-2}ap = a \Rightarrow (pap^{-1})(p^{-1}ap) = a$, etc.)

The above apply to abstract groups. When we have a realisation of a group, the task may be easier, and we illustrate with the following examples, in which some of the groups are infinite.

Method 4

Matrices of the form $\begin{bmatrix} a & b \\ c & d \end{bmatrix}$ $(a, b, c, d \in \mathbf{R}, ad - bc = 1)$ are a subgroup H of $(\mathcal{M}_2, \times)$. We investigate whether this subgroup is normal.

Let $A = \begin{bmatrix} a & b \\ c & d \end{bmatrix}$ be any matrix which belongs to the subgroup H, so that $ad - bc = 1$, and let $M = \begin{bmatrix} p & q \\ r & s \end{bmatrix}$ be any other 2×2 (non-singular) matrix with real terms. Consider the matrix MAM^{-1}. By a well-known theorem, the determinant of this matrix is the product of the determinants of the three matrices:

$$\text{i.e.} \quad \det(MAM^{-1}) = \det M \times \det A \times \det M^{-1}$$
$$= \det M \times \det M^{-1} \times \det A$$
$$= \det(MM^{-1}) \times \det A$$
$$= \det \begin{bmatrix} 1 & 0 \\ 0 & 1 \end{bmatrix} \times (ad - bc) = 1 \times 1 = 1.$$

Hence MAM^{-1} also belongs to the set of matrices with unit determinant, which we have called H. But A was *any* element of H, and M was *any* element of G. Hence the whole of the subgroup H is transformed into itself by any element of G. Therefore H is a normal subgroup.

Q. 20.32. Show that those 3 × 3 matrices whose determinant is ± 1 are a normal subgroup of ($\mathscr{M}_3$, ×).

Q. 20.33. Find whether, in the group of all quaternions $\begin{bmatrix} x_1 + iy_1 & x_2 + iy_2 \\ -x_2 + iy_2 & x_1 - iy_1 \end{bmatrix}$ under multiplication, those with real terms, i.e. of the form $\begin{bmatrix} x_1 & x_2 \\ -x_2 & x_1 \end{bmatrix}$ form a normal subgroup. (*Note:* the latter is isomorphic with the group (C, ×).)

A further example, this time on permutations, illustrates the same method. Those permutations of A, B, C, P and Q which keep P and Q in fourth and fifth place (see p. 333) form the group D_6, and this is a subgroup of S_5. Is it a normal subgroup? To prove that it is not, it is only necessary to find a single counter-example in the shape of a permutation which will transform an element of the group into an element outside it.

Let $\quad y = (B\ C\ A)(P\ Q), \qquad a = (A\ P) \quad (y \in H, a \notin H).$

Then $\quad a\,ya^{-1} = (B\ C\ P)(A\ Q)$ (see pp. 364–5), and this is not in the subgroup, whose conjugate subgroup is therefore different from it.

Q. 20.34. Use a similar method to find whether C_3 and C_4 are normal subgroups in S_4; and whether S_4 is a normal subgroup in S_5.

Method 5

Next, we consider examples from transformation geometry in two dimensions. First, we ask whether the group of plane translations is a normal subgroup of the group of direct isometries (rotations and translations). If R is any rotation and T is any translation, then it is easy to see on geometrical grounds that RTR^{-1} is also a translation, since it restores the orientation of the figure, and this is evidence to convince us that the group of translations is indeed normal in the group of direct isometries.

But the question is not so easily answered in the following cases, and we use complex number methods (see F. J. Budden, 'Transformation Geometry in the Plane by Complex Number methods', *Mathematical Gazette*, pp. 19–31, Feb. 1969) for solving them. In each case we use G and H to denote the parent group and the subgroup respectively.

Is the group of direct isometries a normal subgroup of the group of direct similarities? Any direct isometry may be denoted

$$r(z) = z \operatorname{cis} \phi + a \quad (a \in \mathbf{C}),$$

and any direct similarity may be denoted

$$s(z) = pz + q \quad (p, q \in \mathbf{C}),$$

so that $\quad s^{-1}(z) = (z - q)/p.$

$$\text{Then} \quad srs^{-1}(z) = sr\left(\frac{z-q}{p}\right) = s\left(\frac{z-q}{p}\operatorname{cis}\phi + a\right)$$

$$= (z - q)\operatorname{cis}\phi + ap + q$$

$$= z \operatorname{cis}\phi + b,$$

and so srs^{-1} is also in the subgroup of direct similarities. But s was any direct similarity, and the invariant property follows.

Again, we ask if the group of rotations about a fixed point is a normal subgroup of the group of all isometries. Taking the fixed point as the origin of our Argand diagram, any element of the subgroup may be represented: $r(z) = z \operatorname{cis} \theta$. Suppose first that $s(z) = z \operatorname{cis} \phi + a$ so that $s^{-1}(z) = (z - a) \operatorname{cis} (-\phi)$.

$$\text{Then} \quad srs^{-1}(z) = sr[(z - a)\operatorname{cis}(-\phi)]$$

$$= s[(z - a)\operatorname{cis}(\theta - \phi)]$$

$$= (z - a)\operatorname{cis}\theta + a,$$

and this is a rotation about A, not about O. Thus H is transformed by s into a different subgroup of G. There is no need to consider opposite symmetries of the form $s(z) = \bar{z} \operatorname{cis} \phi + a$, because we have already enough to convince ourselves that H is *not* normal in G.

Q. 20.35. Is the group of plane translations a normal subgroup of the full group of plane isometries? (Consider sts^{-1}, where $s(z) = z \operatorname{cis} \phi + p$ and also when $(sz) = \bar{z} \operatorname{cis} \phi + p$.)

Q. 20.36. Are the functions $z \to z + b$ a normal subgroup of $z \to az + b$ under successive composition?

Q. 20.37. Is the group of direct similarities a normal subgroup of the general bilinear transformations? (Consider $srs^{-1}(z)$, where $r(z) \equiv pz + q$ and $s(z) \equiv (az + b)/(cz + d)$.)

Method 6

Finally, we consider a group of symmetry operations, and illustrate first with the symmetries of the regular hexagon (see Ch. 13, pp. 187 ff.), and then of the cube (see Ch. 17, pp. 297 ff.).

With the notation of Ch. 13, p. 192, we have the subgroup H $\{1, a\}$ of D_6. Is it a normal subgroup? Let x be any of the symmetries of the group. Then $xHx^{-1} = \{1, xax^{-1}\}$, and this will only be a normal subgroup if $a = xax^{-1}$. But, as we saw on p. 365 in this chapter, xax^{-1} is a *reflection in the new position of the reflection axis a* to which that axis is moved by the operation x. If xax^{-1} is to be equal to a, therefore, x must preserve the axis of a in its original position. Since all the operations of G (other than 1, a and r^3) do in fact *move* the axis of a, it follows immediately that $\{1, a\}$ is not a normal subgroup. On the other hand, $\{1, r^3\}$ is a normal subgroup, for this time the rotation axis of the half-turn r^3 is invariant under all the operations of the group.

Q. 20.38. Find another normal subgroup of D_6 by a similar argument.

Now we turn our attention to the *rotation* group of the cube. Suppose that H $\{1, r_1, r_2, \ldots\}$ is a normal subgroup, and s is an operation of the group not in H. Then $sHs^{-1} = \{1, sr_1s^{-1}, sr_2s^{-1}, \ldots\}$, and contains rotations about the axes of the rotations $r_1, r_2, \ldots$ in the new positions to which they have been moved by s. Therefore, if H is to be a normal subgroup, *the system of rotation axes must be invariant under any operation of the group*. We consider three examples of subgroups.

First C_3, containing rotations through $\pm\frac{2}{3}\pi$ about one of the diagonals. This sole axis is moved to a new position by most of the symmetry operations of the rotation group, so clearly the subgroup is not normal, and neither is any of the four subgroups of order 3.

Q. 20.39. Consider which operations do not move a particular diagonal.

Q. 20.40. Show that the C_4 subgroups are also not normal. Find a set of rotations which comprise the subgroup D_3, and show that this too is not a normal subgroup.

Next, we consider the half-turns about the three axes through the centre and parallel to the edges which form the subgroup D_2 (see also Ch. 3, p. 17). This time we find that any one of the remaining symmetry operations of the cube leaves this *set of axes* invariant. For example, a quarter-turn about one of them will interchange the other two; a third-turn about a diagonal will permute them cyclically and so on. Here, then, we have *geometrical evidence* of the fact that the subgroup is normal.

Q. 20.41. In the table for S_4 (see p. 219), which letters represent the operations of this normal subgroup?

Q. 20.42. Consider the subgroup which leaves the rectangle $CD'C'D$ in fig. 17.04 in the same position. This also has D_2 structure, and contains three half-turns about mutually perpendicular axes. Show that this (and the other five isomorphic subgroups) are *not* normal in the group S_4.

The above method may evidently be extended to cover the investigation of the invariance of any subgroup of symmetries in a group of symmetry operations which may include reflections and central inversion, and may be extended still further to cover the case of infinite groups connected with patterns (see Ch. 26), which will include translations, and possibly glide reflections. Briefly, the method is to consider the system of reflection axes, rotation centres and glide axes of the subgroup (the 'Symmetry Chart' of the subgroup, as it is called) and decide what happens to this system under the operations of the rest of the group. If *all* the symmetry operations of the group map the framework on to itself, then the subgroup is normal in the symmetry group. Much light can be thrown on the subgroups of plane patterns by adopting a system of *colouring*. Those elements of the subgroup under consideration are coloured alike, and each coset is given its own colour (so that the number of colours is equal to the index of the subgroup). If the subgroup is not normal, the colouring of the left cosets will produce a different aspect from the colouring of the right cosets. If it is normal, the two coloured patterns will be indistinguishable.

Method 7

We have seen that a given subgroup H is transformed by any element x into a subgroup xHx^{-1} which is isomorphic to H. It follows that, if H is the *only* subgroup of a particular type, then it can only be transformed into itself, and so must be self-conjugate. Thus, for example, in Q_6 there are only two elements of period 3, and these with the identity constitute the *only* subgroup of order 3. This is $\{1, k, l\}$ with the notation of Ch. 14, p. 216, and the above reasoning shows conclusively that this subgroup is normal in Q_6. Similarly, the subgroup $\{1, a, b, c\}$ of A_4 (p. 294) is the only one with structure D_2, and must, for the same reason be a normal subgroup.

Q. 20.43. Find other examples of groups with unique subgroups of a given type.

Sylov subgroups

Some knowledge of the number of subgroups of a given order is available from an important result known as Sylov's theorem. If the order of a group G is $n = 2^\alpha 3^\beta 5^\gamma 7^\delta \ldots p^\mu, \ldots$ when expressed in prime factors, then a subgroup of order 2^α, or of 3^β, etc. ... is called a Sylov subgroup of G, that of order p^μ being called the Sylov subgroup corresponding to (or belonging to) p. The theorem of Sylov states the following properties:

(a) Every group possesses at least one Sylov subgroup for each prime factor of its order n.
(b) All Sylov subgroups corresponding to a given prime factor p are conjugate to each other.
(c) The number of Sylov subgroups belonging to p is congruent to 1 modulo p, and is a factor of n.
(d) Every subgroup of order p^κ $(1 \leqslant \kappa < \mu)$ is normal in some subgroup of order $p^{\kappa+1}$.

A deduction from part (c) is that if the number of Sylov subgroups, $kp + 1$ is a factor of the order of G for $k = 0$ only, then that Sylov subgroup must be normal in G. For example, when $n = 20$ $(= 2^2.5)$, we have Sylov subgroups of order 4 $(p = 2)$ and 5 $(p = 5)$. For $p = 2$, the only possible values of k to make $kp + 1$ a factor of 20 are $k = 0$ or 2, showing that there are either 1 or 5 subgroups of order 4. While when $p = 5$, the only possible value of k is 0, and so the *unique* subgroup of order 5 must be normal. This is true of any one of the five types of group of order 20.

Sylov's theorem is a very powerful tool in the enumeration and identification of all the groups of a given order. (See for example, W. Burnside, *Theory of Groups of Finite Order* (Dover), pp. 157–61, where all fifteen groups of order 24 are identified in this way.)

Normaliser of a subset

If S is a subset (not necessarily a subgroup)† of a group G, the *normaliser of the set* S consists of all those elements of G which commute with S, i.e. it is the set of elements such as g satisfying $gS = Sg$. This does not mean that g commutes with each element of S individually. For example, in A_4 (table 17.02), if $S = \{p, q\}$, then $aS = \{ap, aq\} = \{s, r\}$ while $Sa = \{pa, qa\} = \{r, s\}$ so that a is in the normaliser of the set $\{p, q\}$, though a does not commute with either p or q individually. In Q_6, on the other hand (see

† In some books a subset of a group is called a 'complex'. We have studiously avoided this grotesque and misleading description.

p. 216), the normaliser of $\{c, k\}$ contains only two elements, e and c. It so happens in this case that these form the intersection of the normaliser of c and that of k; indeed, they are both in the centre of Q_6. To avoid confusion, the normaliser of a single element is preferably known as the 'centraliser' of that element.

Postscript

The above discussion leads us to the next chapter in which we make a more detailed study of normal subgroups. As a short postscript to the present chapter, however, note finally that $a\text{H}a^{-1}$ may be thought of as a post-multiplication by a^{-1} of the set $a\text{H}$, which is the left coset of H by a. One could generalise this concept by forming a set $x\text{H}y$ from any subgroup H by any pair of selected elements x and y not in H. For example, in A_4:

$$a\text{H}b = a\{1, p, p^2\}b = \{a, s, r^2\}b = \{c, p, s^2\},$$

$$a\text{H}r^2 = a\{1, p, p^2\}r^2 = \{a, s, r^2\}r^2 = \{p^2, b, r\}.$$

Each of the sets obtained is a coset of a *different subgroup*, the first being $c\{1, r, r^2\}$, or $\{1, q, q^2\}c$, and the second being $b\,\{1, s, s^2\}$ or $\{1, q, q^2\}b$.

It is interesting to pursue these discoveries in the cases of other subgroups, and to find what emerges. When the subgroup selected is a *normal* subgroup, we find that $x\text{H}y$ will always turn out to be a coset of H itself, and not of a different subgroup as happened in the examples of the previous paragraph. For example, if $\text{H} = \{1, a, b, c\}$ in A_4, then $p\text{H}r^2 = (p\text{H})r^2 = \{p, q, r, s\}r^2 = \{c, a, 1, b\} = \text{H}$ while $p\text{H}r = (p\text{H})r = \{p, q, r, s\}r = \{s^2, p^2, r^2, q^2\} = p^2\text{H} = \text{H}p^2$.

That this is bound to happen is an immediate consequence of the property of normal subgroups, that all their left and right cosets contain the same elements. For then we see that, since $x\text{H} = \text{H}x$ $(\forall x \in \text{G})$, therefore $x\text{H}y = \text{H}xy = \text{H}(xy) = (xy)\text{H} = \text{H}z = z\text{H}$, the coset of H by the element $z = xy$.

The above discussion was a useful appendage to the chapter, since it leads on to the discussion of quotient groups, or groups of cosets, which we pursue in the following chapter.

Q. 20.44. Prove that the normaliser of a subset S is a subgroup; also that it contains the intersection of the centralisers of its individual elements as a subgroup.

Q. 20.45. Prove that the centre of a group is the intersection of the centralisers of the individual elements.

Q. 20.46. Show that the normaliser of a subset S may be described as the set of all elements of G which transform S into itself.

Q. 20.47. Find other examples in various groups of elements which commute with a subset though not with the individual elements in it.

Q. 20.48. If a subset S is such that its normaliser contains every element of G, what can you say about S?

Q. 20.49. Form the sets xHx^{-1} and xHy, H being a subgroup of G in the following cases: (a) G is Q_4 (Table p. 101). Take H to be various subgroups, and take various values of x and y. (We showed in Q. 19.14 that all the subgroups of Q_4 are normal.) (b) Repeat for Q_6 (p. 216), D_4 (p. 94), and D_6 (p. 204).

Q. 20.50. Show that if an element h is in a normal subgroup H, then every element of the class of h: $\{xhx^{-1} : x \in G\}$ is in H.

Q. 20.51. Using the result of the previous question, prove that every normal subgroup of A_5 which contains a ternary cycle contains all the ternary cycles (e.g. if a normal subgroup of the group A_5 of even permutations of 1 2 3 4 5 contains the cycle (2 5 3), then it will contain (1 2 3), etc. – all the cycles of length 3. Is this true of the general alternating group, A_n?

Q. 20.52. Elements of the form $x^{-1}y^{-1}xy$ are called commutators. Form commutators with pairs of elements selected from certain groups, e.g. D_4, Q_4, etc. After this preliminary experimental work, prove that the set of all commutators (that is, as x and y run through the whole group) generate a normal subgroup of G.

Q. 20.53. Use the previous result to show that, if H_1 and H_2 are both normal subgroups of G which have no element in common other than the identity, then $h_1h_2 = h_2h_1$ for all $h_1 \in H_1$ and $h_2 \in H_2$. (First prove that $h_1h_2h_1^{-1}h_2^{-1}$ is in both H_1 and in H_2.)

Q. 20.54. In Ex. 10.40, it was proved that, if every element of a group other than 1 is of period 3, then for every pair of elements x, y we have $(xyx^{-1})y = y(xyx^{-1})$. Interpret this result in terms of the work of the present chapter. Is the converse true, that if $xyx^{-1} = yxyx^{-1}$ for every pair of elements x and y, then every element of the group is of period 3?

EXERCISES 20

1 If a is the only element of period 2 in a finite group G, prove that a belongs to the centre of G. (Compare Ex. 10.17.)

2 If $y = (1\ 2\ 3)(4\ 5)$, $z = (2\ 4\ 3)(1\ 5)$, find x so that $z = xyx^{-1}$.

3 **Prove that xy is conjugate to yx in any group.**
If x is any self-conjugate element of G (i.e. belongs to the centre of G), prove that Gp (x) is normal in G.

4 If r is a quarter-turn about the x-axis and s is a quarter-turn about the y-axis, what is srs^{-1}: (a) when r and s are both clockwise when looking away from the origin; and (b) when r is clockwise and s is anticlockwise?

5 In the group Q_6, with the notation as on p. 216, show that the classes of the elements of period 4 are C $\{c, f, h\}$ and D $\{g, j, d\}$. Verify that $C^2 = D^2 = \{a, b, k\}$ whereas $CD = DC = \{1, l, m\}$.

6 Show that, if an element of a group is transformed alike by a second element and by the inverse of the latter, then the square of the latter commutes with the former.

7 Produce a counterexample to show that, if two elements are in the same class, then neither need to be in the centraliser of the other.

8 If x and y are overlapping cycles, show that they can commute only if one is a power of the other. (Consider xyx^{-1}.)

9 Show that in D_n, with $a^2 = r^n = 1$, each element of the subgroup Gp $(r) \cong C_n$ is transformed by each element of its coset into its own inverse.

10 Since an element and its inverse have the same period, it is feasible that an element might be transformed into its inverse. Discover conditions under which $xyx^{-1} = y^{-1}$ in various cases: (a) when y is of period 2; (b) when y is of period 3 or more. In the group A_4 (Table 17.02), $xpx^{-1} = p^2$ has no solution for x. Find other examples of cases when an element of a non-Abelian group is not conjugate to its own inverse. Can you find an example of an element of period 4 which is not conjugate to its inverse? (See a previous question in this exercise.)

11 **Prove that the index of the centraliser of x is equal to the number of elements in the class of x, and that the cosets of the centraliser can be placed into 1,1 correspondence with the elements in the class of x. Deduce that the number of elements in the class of x is a factor of the order of the group.**

12 If A is a sub*set* of a group G, make a study of the set xAx^{-1} in various cases.

13 Discuss the examples of subgroups to be found elsewhere in the text, especially in chapter 14, and say whether they are normal subgroups.

14 Let T be a translation and R an isometry (including rotation, reflections and glide reflections) in a plane. By showing that RTR^{-1} is a translation, prove that the translations are a normal subgroup of the group of all plane isometries. (You may use pure geometry, or matrices, or complex numbers.)

15 p, q and r are generators of a group which satisfy

$$p^4 = 1 \quad (1) \qquad q^4 = 1 \quad (2) \qquad r^4 = 1 \quad (3)$$
$$qpr = p \quad (4) \qquad rqp = q \quad (5) \qquad prq = r \quad (6).$$

Show that (2) and (4) imply (3); also that (1), (4) and (5) imply (2) and (3). However, it is not possible to deduce (6) from (1), (2), (3), (4) and (5), for in the group Q_6, there are elements satisfying the first five relations, yet where $prq = r^{-1}$. Find elements from Q_6 to produce this counterexample, using the notation of the table on p. 216.

16 If p, q and r are elements of S_4 satisfying the relations of the previous question, then necessarily each of pq, qp, qr, rq, rp, pr is of period 3. Do the relations $(pq)^3 = 1$, etc. *necessarily* follow from relations (1) to (6) above?

17 Is the group of plane rotations about a fixed point a normal subgroup of the group of direct isometries or the group of all isometries?

18 Is Q_4 a normal subgroup of the infinite group of all quaternions under multiplication?

19 For any given subgroup H of a group G, show that those elements which transform H into itself are a subgroup of G. Does H have to be a normal subgroup of G?

20 Given an element a of a group, consider the relation between x and y defined $x\mathrm{R}y \Leftrightarrow x^{-1}yay^{-1}x = a$. Show that R is an equivalence relation. (Note that $x\mathrm{R}y$ may be written $yay^{-1} = xax^{-1}$, i.e. *x and y transform a*

alike.) What is the set of all elements which transform a particular element alike? In other words, for given a and b, describe the set $\{x_r: x_r a x_r^{-1} = b\}$.

21 What is the meaning of $rsr^{-1}s^{-1}$ when r and s are geometric transformations? Consider special cases.

22 Given that $s^3 = 1$, $st = t^2 s$, prove that $t^7 = 1$. What is the order of Gp (s, t)? (*Hint*: $t^2 = sts^{-1} \Rightarrow t^4 = sts^{-1}sts^{-1} = st^2s^{-1} = s(sts^{-1})s^{-1}$ etc.) (Compare Exs. 9.19 and 15.35.)

23 **The number $a_{11} + a_{22} + a_{33}$ is called the trace† of the matrix**

$$\mathbf{A} = \begin{bmatrix} a_{11} & a_{12} & a_{13} \\ a_{21} & a_{22} & a_{23} \\ a_{31} & a_{32} & a_{33} \end{bmatrix}.$$

If A and B are two 3 × 3 matrices, show that the traces of the matrices AB and BA are equal.

If the matrix **AB** represents a rotation through an angle ϕ about the directed axis U, and **A** represents a rotation interchanging the axes U and V, explain why **BA** represents a rotation through the angle ϕ about V.

Given that the matrix $\mathbf{M} = \begin{bmatrix} \cos\phi & \sin\phi & 0 \\ -\sin\phi & \cos\phi & 0 \\ 0 & 0 & 1 \end{bmatrix}$ represents a rotation through the angle ϕ about the z-axis, and that the matrix **C** represents a rotation about *some* axis, find a formula for the angle of rotation in terms of the trace of **C**. (Cambridge Schol., 1968.)

24 Show that the result in the first part of the previous question, which may be written: tr (**AB**) = tr (**BA**) may be used to deduce the result: tr $(\mathbf{ABA}^{-1})$ = tr (**B**), i.e. that the trace of any matrix is invariant under conjugation by another matrix; or that similar matrices have equal traces. Take a set of matrices which form a finite group under multiplication, and separate them into conjugacy classes, and evaluate the trace of each class. For example, the group of twelve 3 × 3 matrices representing A_4 on p. 295; or the group of 4 × 4 permutation matrices representing S_4 (p. 318); or the representation of D_4 by 2 × 2 matrices (Q. 13.09).

† Or 'spur' or 'character' of the matrix.

21 *Homomorphism: quotient groups: normal subgroups (2)*

We have met many instances of isomorphism between groups, and now add to these a further example – the isomorphism between the group $(\mathbf{Z}, +)$ and the group $\{10^n, n \in \mathbf{Z}\}, \times$. For example, in the group $(\mathbf{Z}, +)$, we have $-1 + 6 = 5$; the corresponding result in the second group is $10^{-1} \times 10^6 = 10^5$, and we remind you that the elements of the two groups may be put into 1, 1 correspondence in such a way that *products are preserved*, the result being that, whether the groups are finite or infinite, their structure is identical. In the above example, the structure of both groups was C_∞.

What we were really doing in the above case was mapping from the integers on to their *antilogarithms* (base 10).

Q. 21.01. Which multiplicative group is isomorphic to $(\mathbf{R}, +)$?

Homomorphic mappings

Suppose that, instead of base 10, we had used base i $(= \sqrt{-1})$; that is, suppose we had performed the mapping $n \rightarrow i^n$. Then all those integers which divide exactly by 4 will map into the number 1, for $i^{-8} = i^{-4} = i^0 = i^4 = i^8 = \ldots = 1$; the set $\{\ldots -3, 1, 5, 9, \ldots\}$ will map into i; the set $\{\ldots -6, -2, 2, 6, \ldots\}$ into the number -1, and the remaining integers, $\{\ldots -5, -1, 3, 7, \ldots\}$ map into the number $-i$. The *image* of the set $\mathbf{Z}$ under this mapping consists therefore of four complex numbers only, $\{1, i, -1, -i\}$. The correspondence is no longer 1 to 1, but *many-to-one* – in fact, infinitely many-to-one in this case. Nevertheless, *products are still preserved*, for the result:

$-1 + 6 = 5$ in $(\mathbf{Z}, +)$ induces the result:

$i^{-1} \times i^6 = i^5, \quad \text{or } i^3 \times i^2 = i^1,$

that is $(-i) \times (-1) = +i$ in the 'image group' $\{1, i, -1, -i\}, \times$.

We have here an example of a *Homomorphic Mapping* from the group $(\mathbf{Z}, +)$ on to the *smaller* group $\{1, i, -1, -i\}, \times$. The properties of a homomorphism are similar to those of an isomorphism, namely that products are preserved, but the difference is that in a homomorphism, *many* elements of the original set may be mapped into a particular element of the image set; that is to say, while structural properties are preserved, the *individuality* of the elements is destroyed.

If the mapping, or function, is denoted by ϕ, we have in the case above,

$$\phi(-4) = 1; \quad \phi(0) = 1; \quad \phi(4) = 1, \text{ etc.,}$$

$$\phi(-1) = -i; \quad \phi(6) = -1; \quad \phi(1) = i, \text{ and so on.}$$

In fact we may say: $\phi\{\ldots -8, -4, 0, 4, 8, \ldots\} = 1$, that is to say, the homomorphism carries the whole set of integers divisible by 4 into the number 1, which is the identity of the group $\{1, i, -1, -i\}, \times$. That subset which is mapped into the null element of the image group is called the *Kernel* of the homomorphism.

Now the group $\{1, i, -1, -i\}, \times$ is isomorphic to the group $\{0, 1, 2, 3\}$, + mod. 4, and we can also map the group $(\mathbf{Z}, +)$ homomorphically on to the latter realisation of the group C_4 by placing the set $\mathbf{Z}$ into four *equivalence classes*, namely the residue classes modulo 4. If the homomorphism function is again denoted ϕ, we have:

$$\phi\{\ldots -8, -4, 0, 4, 8, \ldots\} = 0$$
$$\phi\{\ldots -7, -3, 1, 5, 9, \ldots\} = 1$$
$$\phi\{\ldots -6, -2, 2, 6, 10, \ldots\} = 2$$
$$\phi\{\ldots -5, -1, 3, 7, 11, \ldots\} = 3.$$

Many-to-one mappings in general

Indeed, the partitioning of a set into equivalence classes by an equivalence relation induces a homomorphism of the whole set on to the set of equivalence classes. For example, when we classify cars by their *make*, we are mapping the whole set of cars on to the set of manufacturers {Ford, Jaguar, Volkswagen, etc. ...}, and this set of 'makes' is a homomorphic image of the set of cars. It is true that in this rather commonplace example we are not mapping a group on to a group, but a set with no structure on to a smaller amorphous set.† The principle is the

† Some mathematicians would object to the term 'homomorphism' being stretched to cover cases when no operation is involved. In this sense, the days may be mapped on to the set {Sun, Mon, Tues, Weds, Thurs, Fri, Sat}; while phone numbers containing letters may be mapped homomorphically on to all-figure phone numbers by the correspondence: {A, B, C} → 2, {D, E, F} → 3, {G, H, I} → 4, {J, K, L} → 5, {M, N} → 6, {P, R, S} → 7, {T, U, V} → 8, {W, X, Y} → 9, {O} → 0. The correct term is 'many-to-one mapping'. However, homomorphisms are not necessarily many-to-one mappings, since they include isomorphisms as a special case.

Q. 21.02. Can a homomorphism be one-to-one and yet not be an isomorphism?

same, however, that we have a homomorphic mapping ϕ which maps my car on to the element 'Jaguar' of the set of makes: ϕ(my car) = Jaguar, if you like.

In Ch. 16, p. 256 we considered the group $C_4 \times C_3$ to be realised by the addition of ordered pairs of integers by the rule $(x_1, y_1) \oplus (x_2, y_2) = (x_1 + x_2, \text{mod. } 4, y_1 + y_2, \text{mod. } 3)$, and we noted that the points of the infinite lattice consisting of points with integral coordinates have been placed into equivalence classes such as {(6, 1), (10, 1), (2, −2), (6, −2), etc. . . .}, all of these belonging to the classes represented by (2, 1). The ordered pairs of integers under addition form the group $C_\infty \times C_\infty$, a subgroup of $(\mathbf{V}_2, +)$ (see p. 340). So here we have a homomorphic mapping from $C_\infty \times C_\infty$ on to $C_4 \times C_3$. The property of preservation of products is illustrated:

$$\text{in } C_\infty \times C_\infty,\ (-13, -7) + (-14, +13) = (-27, +6),$$

$$\text{in } C_4 \times C_3,\quad (3, 2) \oplus (2, 1) = (1, 0),$$

and (−27, 6) is indeed in the class (1, 0).

As a further example of a homomorphic map, consider the group $(\mathscr{M}_2, \times)$ of non-singular 2×2 matrices under matrix multiplication, and let the homomorphism be that which maps each matrix $\begin{bmatrix} a & b \\ c & d \end{bmatrix}$ into its determinant $\begin{vmatrix} a & b \\ c & d \end{vmatrix} = ad - bc$, so that if $\mathrm{M} = \begin{bmatrix} a & b \\ c & d \end{bmatrix}$, then $\phi(\mathrm{M}) = ad - bc = |\mathrm{M}|$. Here we have a mapping from the set of non-singular 2×2 matrices on to the non-zero reals (it being assumed that a, b, c and d are drawn from $\mathbf{R}$). Now it is well known that if $A, B \in \mathscr{M}_2$, then $|AB| = |A|\,|B|$, so that *products are preserved* by this mapping, for this relationship may be written $\phi(\mathbf{AB}) = \phi(\mathbf{A})\,\phi(\mathbf{B})$,† which is the very essence of a homomorphism.

Thus the group $(\mathscr{M}_2, \times)$ has been mapped on to the group $(\mathbf{R}, \times)$, the latter being the homomorphic image of the group $(\mathscr{M}_2, \times)$ under this homomorphic mapping. The kernel of this homomorphism is the set of all matrices which map into the null element of $(\mathbf{R}, \times)$, i.e. unity. Thus the kernel is the set of all matrices $\begin{bmatrix} a & b \\ c & d \end{bmatrix}$ such that $ad - bc = 1$. These matrices represent those linear mapping in the plane which *preserve area.* (Compare also pp. 229 ff., Ex. 14.31, pp. 386 ff.)

† Note that $\phi(AB) = \phi(A)\,\phi(B)$ is really an abbreviation for $\phi(A * B) = \phi(A) *' \phi(B)$, where $*$ and $*'$ are the operations of the given group and the image group. In the present example, $*$ is matrix multiplication, while $*'$ is ordinary multiplication.

Homomorphisms of finite groups

All the examples of homomorphic mappings so far considered have been illustrated by infinite groups. Many examples of homomorphic maps of finite groups will appear in this and later chapters, but we give one at this stage. Consider the set of permutations of three letters: $e = \begin{pmatrix} \text{ABC} \\ \text{ABC} \end{pmatrix}$, $p = \begin{pmatrix} \text{ABC} \\ \text{CAB} \end{pmatrix}$, $q = \begin{pmatrix} \text{ABC} \\ \text{BCA} \end{pmatrix}$, $a = \begin{pmatrix} \text{ABC} \\ \text{ACB} \end{pmatrix}$, $b = \begin{pmatrix} \text{ABC} \\ \text{CBA} \end{pmatrix}$, $c = \begin{pmatrix} \text{ABC} \\ \text{BAC} \end{pmatrix}$ which form the group S_3 ($\simeq D_3$), the table being as on p. 202. Now the first three of these are even permutations and the last three are odd permutations. Consider the function ϕ such that:

$$\phi(e) = I; \quad \phi(p) = I; \quad \phi(q) = I$$
$$\phi(a) = X; \quad \phi(b) = X; \quad \phi(c) = X.$$

This means that we are making no distinction between the permutations ABC, CAB and BCA – they lose their individuality – but are placing them all in a class I, and likewise the odd permutations are placed into the class X. We may now say that $I^2 = I = X^2$; $XI = IX = X$, meaning that the composition of two even permutations or of two odd permutations is an even permutation, while the product of an odd with an even permutation is an odd permutation. This may be summarised in the table:

	I	X
I	I	X
X	X	I

which has the structure of the 2-group. Here, then, we have a homomorphic mapping of the group S_3 onto the smaller group C_2 ($\simeq S_2$). If one thinks of an equilateral triangle, and interprets the above permutations as permutations of its vertices, then the even permutations correspond to the movements of the triangle which do not turn it over ('heads' – see p. 198); while the odd permutations correspond to those symmetries in which the opposite face ('tails') of the triangle is exposed. The homomorphism may be summarised:

Table 21.01

	e	p	q	a	b	c
e	e	p	q	a	b	c
p	p	q	e	b	c	a
q	q	e	p	c	a	b
a	a	c	b	e	q	p
b	b	a	c	p	e	q
c	c	b	a	q	p	e

$\xrightarrow{\phi}$

	I	X
I	I	X
X	X	I

.

See also fig. 2.101, and p. 80.

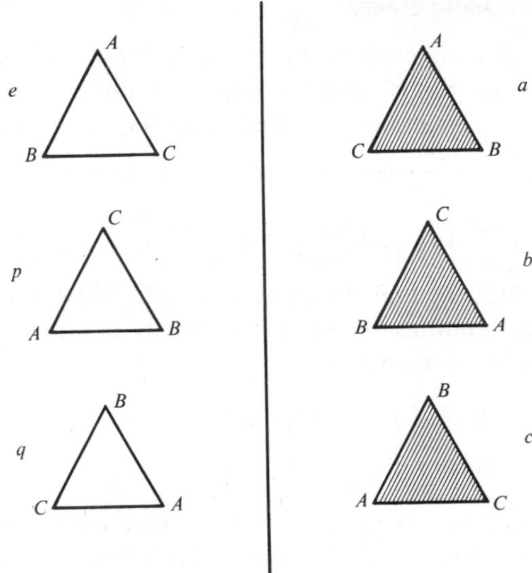

Fig. 21.01.

The kernel of a homomorphism

An important theorem is that, if a group G is mapped homomorphically onto a group G′, then the kernel K of G is a *normal subgroup* of G. For example, the kernel of the group S_3 above is the set of even permutations $\{e, p, q\}$ for these map into the identity I of the image group. These three elements do in fact form a normal subgroup (see p. 338) of S_3.

To prove this in general, let the kernel K contain the elements $\{k_1, k_2, k_3, \ldots\}$ so that, if the homomorphic function is ϕ and the identity of the image group is I, we have

$$\phi(k_1) = I, \phi(k_2) = I, \text{ etc., and } \phi(k_r) = I \text{ for all } r.$$

Now $\phi(k_r k_s) = \phi(k_r)\phi(k_s)$ (since products are preserved by a homomorphic mapping)

$$= I.I = I,$$

and this means that $k_r k_s$ also maps into I. Hence $k_r \in$ K and $k_s \in$ K $\Rightarrow k_r k_s \in$ K, and *closure* is established. We cannot yet be sure that K contains the null element e of G, so suppose $\phi(e) = X$. Then if k is any element of K, $ke = k$ so that $\phi(ke) = \phi(k) = I$. But $\phi(ke) = \phi(k)\,\phi(e) = IX = X$. It follows that $X = I$ and so the kernel contains e. Finally, to show that $k \in$ K $\Rightarrow k^{-1} \in$ K we have $\phi(kk^{-1}) = \phi(e) = I$, as proved above. But $\phi(kk^{-1}) = \phi(k)\,\phi(k^{-1}) = I(k^{-1})$. Hence $I\phi(k^{-1}) = I$, and so $\phi(k^{-1}) = I$ and $k^{-1} \in$ K, and the inverse of each element k of K is contained in K. *Therefore* K *is a subgroup.*

To show that, in addition it is a *normal* subgroup, we must show that the left and right cosets of K coincide; i.e. that if x is any element whatsoever in G, then $x\mathrm{K} = \mathrm{K}x$, or that $x\mathrm{K}x^{-1} = \mathrm{K}$. This means that for any element k_r in K, the transform of k_r by x, that is xk_rx^{-1} must also be in K.

Now: $\phi(xk_rx^{-1}) = \phi(x)\phi(k_r)\phi(x^{-1})$ (properties of homomorphism).

But $\phi(k_r) = I$ since $k_r \in \mathrm{K}$.

Hence, $\phi(xk_rx^{-1}) = \phi(x)\phi(x^{-1}) = \phi(xx^{-1}) = \phi(e) = I$, as proved above.

Thus the kernel K is a normal subgroup of the parent group.

Q. 21.03. Show that differentiation is a homomorphic mapping from the additive group of polynomials on to itself. What is the kernel of the mapping?

Q. 21.04. A homomorphism is very often a *many-to-one* mapping – each element of the image is the transform of several elements of the original (domain) set. Suppose the kernel K contains m elements, then each of these elements maps into the neutral element of the image group. Is *every* element of the image group the transform of exactly m elements of the domain?

Normal subgroups: groups of cosets: quotient groups

Next, consider a group of even order, $2n$, which has a subgroup of order n, for example D_n with the subgroup C_n. We shall take the case when $n = 4$ and rearrange the table on p. 94 using the new notation $e \to 1; f \to r; g \to r^2; h \to r^3$. Table 21.02 is printed with the subgroup $\{1, r, r^2, r^3\}$ (C_4)

Table 21.02

	1	r	r^2	r^3	a	b	c	d
1	1	r	r^2	r^3	a	b	c	d
r	r	r^2	r^3	1	b	c	d	a
r^2	r^2	r^3	1	r	c	d	a	b
r^3	r^3	1	r	r^2	d	a	b	c
a	a	d	c	b	1	r^3	r^2	r
b	b	a	d	c	r	1	r^3	r^2
c	c	b	a	d	r^2	r	1	r^3
d	d	c	b	a	r^3	r^2	r	1

in the top left-hand corner, and the elements of the coset $\{a, b, c, d\}$ follow this. It is divided into four blocks, and it will be seen that the bottom right-hand block also consists entirely of the elements of the subgroup $\{1, r, r^2, r^3\}$. The remaining two squares are entirely filled with the elements from the coset $\{a, b, c, d\}$ so that the table possesses the broad overall structure of the C_2 group: $\begin{array}{|cc} \mathrm{K} & \mathrm{A}, \\ \mathrm{A} & \mathrm{K} \end{array}$ where K† denotes the subset

† The reason for the choice of the letter K rather than H as the subgroup is that it will be seen to be the kernel of a homomorphic mapping.

$\{1, r, r^2, r^3\}$ and A denotes the set $\{a, b, c, d\}$. The reason for this is not difficult to find. Since the top left-hand square contains the subgroup $\{1, r, r^2, r^3\}$, each of its four rows must be a permutation of these four elements. Therefore, for the whole group, since each of the first four rows must be a permutation of all eight elements, it follows that the remaining four elements of each row must be drawn from $\{a, b, c, d\}$, and thus the whole of the top right-hand block must consist of a's, b's, c's and d's. A similar consideration of the Latin-square property applied to the first four columns convinces us that the bottom left-hand block must also consist entirely of a's, b's, c's and d's. Finally, the remaining four elements of the last four rows (or columns) are bound to be $1, r, r^2$ or r^3 – again by the Latin-square property.

You will notice that other group tables in the text have been printed so as to reveal the same overall C_2 structure,† for example:

D_3 ($K \cong C_3$) (see p. 151)
D_6 ($K \cong C_6$) (see p. 192)
Q_4 ($K \cong C_4$) (see p. 101)
Q_6 ($K \cong C_6$) (see p. 216).

It is obvious that this arrangement can always succeed when the order of the subgroup K is *one half the order of the group G*, i.e. when the *index* of the subgroup is 2 (see p. 339). As a *tour de force*, you might write out the table S_4 (p. 219) of order 24 with the subgroup A_4 in the top left-hand corner, having first identified the elements of A_4. [The simplest method of identifying these is to note that A_4 is generated by two elements of period 3 (e.g. Gp (c, i)).]

Normal subgroups of index 3

We next ask the question: if the index of the subgroup is 3, will it always be possible to 'map' the group onto the group C_3 in a similar way, as is done in table 21.03 where the group is A_4 and the subgroup is D_2. Here A_4 is mapped homomorphically onto the group C_3 by the correspondence:

$$\{1, a, b, c\} \to K; \quad \{p, q, r, s\} \to P; \quad \{p^2, q^2, r^2, s^2\} \to Q.$$

(For the *internal* structure of normal subgroups to be reproduced in all the cosets, the order of the elements in the borders has to be different across from the order down. This must be so, otherwise two different groups, such as C_6 and Q_6, each of which has a normal subgroup of

† We may say that the group has been mapped on to the group C_2 by a homomorphism which takes the elements of the subgroup K into the identity element of C_2, while the elements of its coset are carried into the other element of C_2.

Table 21.03

	1	a	b	c	p	q	r	s	p^2	q^2	r^2	s^2
1	1	a	b	c	p	q	r	s	p^2	q^2	r^2	s^2
a	a	1	c	b	s	r	q	p	r^2	s^2	p^2	q^2
b	b	c	1	a	q	p	s	r	s^2	r^2	q^2	p^2
c	c	b	a	1	r	s	p	q	q^2	p^2	s^2	r^2
p	p	r	s	q	p^2	r^2	s^2	q^2	1	b	c	a
q	q	s	r	p	s^2	q^2	p^2	r^2	b	1	a	c
r	r	p	q	s	q^2	s^2	r^2	p^2	c	a	1	b
s	s	q	p	r	r^2	p^2	q^2	s^2	a	c	b	1
p^2	p^2	s^2	q^2	r^2	1	c	a	b	p	s	q	r
q^2	q^2	r^2	p^2	s^2	c	1	b	a	r	q	s	p
r^2	r^2	q^2	s^2	p^2	a	b	1	c	s	p	r	q
s^2	s^2	p^2	r^2	q^2	b	a	c	1	q	r	p	s

(A_4)

$\longrightarrow$

	K	P	Q
K	K	P	Q
P	P	Q	K
Q	Q	K	P

(C_3)

index 2, could be shown to be isomorphic by writing each table with identical pattern! For a full discussion, see F. H. Francis, 'Patterns in Group Tables', *Mathematical Gazette*, p. 354, Dec. 1968.)

We now attempt to do the same sort of thing in the case of the subgroup $\{e, b, c, g\}$ (C_4) of the group Q_6 (table 10.02) Now $Ha = \{e, b, c, g\}a = \{a, l, j, f\}$, so we place these elements next in order across the top margin of table 21.04, and complete the row with d, h, k and m. Everything

Table 21.04

	e	b	c	g	a	l	j	f	d	h	k	m
e	e	b	c	g	a	l	j	f	d	h	k	m
b	b	e	g	c	l	a	f	j	h	d	m	k
c	c	g	b	e	j	f	l	a	k	m	h	d
g	g	c	e	b	f	j	a	l	m	k	d	h
a	a	l	d	h								
l	l	a	h	d								
j	j	f	k	m								
f	f	j	m	k								
d	d											
h	h											
k	k											
m	m											

(boxed block $d\ h\ /\ h\ d\ /\ k\ m\ /\ m\ k$) ← contains eight elements.

(Q_6)

goes smoothly for the first four rows, but when we come to rows a, l, j, f we run up against trouble, for the left-hand block of the second row (arrowed) contains not only the elements a, l, j and f, but also four alien elements d, h, k and m. The title of the present chapter, and the preliminary work of the last one, will have suggested to you that the reason the above case 'does not work' is precisely because $\{e, b, c, g\}$ is not a *normal* subgroup of Q_6: the left coset $a\{e, b, c, g\} = \{a, l, d, h\}$ is different from the right coset $\{a, l, f, j\}$, and it is from this left coset that two of these new 'alien' elements d and h have come; the other two, k and m appearing in the left cosets jH and fH.

Homomorphic images of Abelian groups

Now *every subgroup of an Abelian group is obviously normal*, and so it is always possible to produce homomorphic images of an Abelian group of composite order, whereby *any* subgroup may be mapped into the identity. We have shown this in the case of C_{12} by the subdivisions, on pp. 171–3. There, in version (c), the group $\{0, 1, 2, \ldots, 11\}$ + mod. 12 is mapped onto the group C_6:

Table 21.05

	K	*A*	*B*	*C*	*D*	*F* †
K	*K*	*A*	*B*	*C*	*D*	*F*
A	*A*	*B*	*C*	*D*	*F*	*K*
B	*B*	*C*	*D*	*F*	*K*	*A*
C	*C*	*D*	*F*	*K*	*A*	*B*
D	*D*	*F*	*K*	*A*	*B*	*C*
F	*F*	*K*	*A*	*B*	*C*	*D*

by the correspondence:

$\{0, 6\} \to K$; $\{1, 7\} \to A$; $\{2, 8\} \to B$;

$\{3, 9\} \to C$; $\{4, 10\} \to D$; $\{5, 11\} \to F$.

In version (d), the index of the subgroup C_3 $\{0, 4, 8\}$ is 4, and we get a homomorphic mapping onto C_4 by the correspondence:

$\{0, 4, 8\} \to K$

$\{1, 5, 9\} \to A$

$\{2, 6, 10\} \to B$

$\{3, 7, 11\} \to C.$

Table 21.06

	K	*A*	*B*	*C*
K	*K*	*A*	*B*	*C*
A	*A*	*B*	*C*	*K*
B	*B*	*C*	*K*	*A*
C	*C*	*K*	*A*	*B*

The homomorphic images in the last two versions (e) and (f) are easily seen to be C_3 and C_2 respectively.

† We have so far been adhering fairly carefully to the custom of using *small* letters as the elements of groups, and capitals for complexes, or sets of elements. Here we are torn between using capitals and small letters, but it seems preferable to use capitals because this notation stresses the fact that the *elements* of our new groups are, in fact, *sets* (indeed, *cosets*) of elements of the 'parent' group.

Q. 21.05. Write out tables for other cyclic groups of composite order such as C_6, C_8, etc. and also direct product Abelian groups such as $C_4 \times C_2$ (table 16.11), $C_2 \times C_2 \times C_2$ (table 16.10), $C_3 \times C_3$ (table 10.07), $C_6 \times C_2$ (p. 260), showing in each case a homomorphic image of the group. (If the table in the text is already so arranged as to reveal a homomorphic image, then try and show a different one.)

Q. 21.06. In the case of the groups $\{K, A, B, C, D, F\}$ and $\{K, A, B, C\}$ above, reconstruct their tables showing that they in their turn may yield homomorphic images by their own subgroups.

Another case of failure when subgroup is not normal

Before drawing together the threads of the foregoing paragraphs, let us give one more example to demonstrate the failure of a group to yield a homomorphic image by a *non-normal* subgroup, by considering $\{1, p, p^2\}$ in the group A_4. The left cosets are $\{1, p, p^2\}$; $\{a, s, r^2\}$; $\{b, q, s^2\}$ and $\{c, r, q^2\}$, and the right cosets are $\{1, p, p^2\}$; $\{a, r, s^2\}$; $\{b, s, q^2\}$; $\{c, q, r^2\}$.

That the system of left cosets is not closed when their product sets are formed is revealed by a single example:

Table 21.07

		a	s	r^2
		⋮	⋮	⋮
c	...	b	q	s^2
r	...	p	p^2	1
q^2	...	r^2	a	s

the product of the sets $\{c, r, q^2\}$ and $\{a, s, r^2\}$ is complete chaos!

By contrast, the product of the right (or left) cosets of $\{1, a, b, c\}$ is perfectly orderly, as the table 21.03 shows.

Products of cosets of normal subgroups – general theory

That this must always happen in the case of the cosets of normal subgroups is easily proved. For if K is a normal subgroup of G, and aK and bK are two of its (left) cosets, then we have $b\mathrm{K} = \mathrm{K}b$. Hence the product set

$a\mathrm{K}\, b\mathrm{K} = a\mathrm{K}\, \mathrm{K}b$. But $\mathrm{K}\,\mathrm{K} = \mathrm{K}$ (as K is a subgroup, see pp. 217, 356).

Hence $a\mathrm{K}\, b\mathrm{K} = a\mathrm{K}b = ab\mathrm{K} = (ab)\mathrm{K}$, and this is one of the cosets of K. Hence the system of cosets is closed under the operation of the formation of product sets, and this, together with the fact that K is an identity element, and that a^{-1}K is the inverse coset of aK (since aK a^{-1}K = K$a\,a^{-1}$K = KK = K), shows that these cosets form a group.

Q. 21.07. Prove that the product rule: aK . bK = abK is associative.

Now in case you are still not fluent with the above condensed notation, you may like to read the argument in expanded form. The product set aK bK means the set of all elements such as $ak_r\, bk_s$. But, since bK = Kb, we know that bk_s is in Kb, so $bk_s = k_tb$ for some element k_t of K. Hence a typical element of aK bK is $ak_rbk_s = ak_rk_tb = ak_ub$ (since K is a subgroup, $k_rk_t = k_u$). But again, we require the 'normal' property of K to be able to say that $k_ub = bk_v$ for some k_v in K. Hence a typical element of aK bK is $ab\; k_v$, and this is an element of the left coset (ab) K.

You observe how much longer the argument becomes when written down in detail in this way. **It is for this reason that you should make the effort to accustom yourself to the notation of the argument in its greatly condensed form: *a*K*b*K = *a*KK*b* = *a*K*b* = *ab*K,** so as to be able to bypass the laborious details.

At this stage you should revise the work of the chapter on cosets, and go back over the work on pp. 356 ff., where we experimented with the formation of product sets and noted, without proof that only when two cosets of a *normal* subgroup of order m are used do we get a product set containing m elements only. The illustrations in that chapter were all drawn from the group Q_6. We have now generalised the result, and examined the theory behind it, and provided the proof promised on p. 359.

Quotient groups, or factor groups

When H is a normal subgroup of a group G, we shall write H $\lhd$ G.

Q. 21.08. Is the relation $\lhd$ between groups transitive?

We now return to the question of the mapping of a group G which contains a normal subgroup K of index m onto a smaller group of order m, as in the examples on pp. 396 ff. The *group of cosets* is called the *Quotient Group*, or the *Factor Group*, and is indicated G/K. Thus, for example, when G is A_4 and K is D_2, the quotient group G/K is C_3. Again, when G is $C_4 \times C_2$, and K is C_2, the quotient group is C_4, and it is this latter sort of example which probably accounts for the name 'quotient'. But the name, and the notation, are most unfortunate, for if the quotient group of G by the normal subgroup K is a group H (i.e. G/K $\simeq$ H), the notation strongly suggests that G = K $\times$ H, but this is *not always so*. Thus, $A_4/D_2 \simeq C_3$ (see p. 403), but A_4 is *not* the direct product of the groups

D_2 and C_3, since in any case, the latter is Abelian, whereas A_4 is not. Tables 12.093 and 12.096 show that $C_{12}/C_2 \cong C_6$ and $C_{12}/C_6 \cong C_2$, yet C_{12} is different from $C_6 \times C_2$. Again, $S_4/A_4 \cong C_2$, as we saw on p. 402, but S_4 is *not* the direct product of A_4 and C_2. More remarkable is that both C_6/C_3 *and* D_3/C_3 give the same quotient group, C_2. However, it may be proved that $A \lhd G$, $B \lhd G$, $A \cap B = 1$ and $|A| \,.\, |B| = |G| \Rightarrow G \cong A \times B$, i.e. that if A and B are two normal subgroups of a group G which have *only the identity in common*, then the group G is the direct product of A and B.

Q. 21.09. Show that the map $r \rightarrow 2r$ is a homomorphism of the group C_{2n}: $\{0, 1, 2, \ldots, r, \ldots, 2n\}$, $+$ mod. $2n$. What is the quotient group?

Q. 21.10. Verify the above in the case $D_6 \cong D_3 \times C_2$. Show that Q_6 cannot be expressed as a direct product. Note that in the group C_{12} $\{0, 1, 2, \ldots, 11\}$, $+$ mod. 12, we have subgroups C_3 $\{0, 4, 8\}$ and C_4 $\{0, 3, 6, 9\}$ such that $C_3 \cap C_4 = \{0\}$, so that according to the theorem $C_{12} = C_3 \times C_4$; whereas the subgroups C_2 $\{0, 6\}$ and C_6 $\{0, 2, 4, 6, 8, 10\}$ have two elements in common, so $C_2 \times C_6$ is *not* C_{12}, as we saw earlier (p. 260).

★Quotient groups of direct product groups

We have noted that $C/A = B$ does not imply that $C \cong A \times B$. One may wonder, however, whether the converse is true: if $C \cong A \times B$, is it true that $C/A = B$? Before we can answer this question, it is first necessary to discover whether A is a *normal* subgroup of C, *otherwise* C/A *is meaningless.*

Suppose A is $\{1, a_1, a_2, a_3, \ldots\}$ and B is $\{1, b_1, b_2, b_3, \ldots\}$. The elements of $A \times B$ are:

$$\begin{array}{lllll}
(1, 1), & (a_1, 1), & (a_2, 1), & (a_3, 1), & \leftarrow \text{subgroup A} \\
(1, b_1), & (a_1, b_1), & (a_2, b_1), & (a_3, b_1), \ldots & \\
(1, b_2), & (a_1, b_2), & (a_2, b_2), & (a_3, b_2), \ldots & \\
(1, b_3), & \ldots & & & \\
\vdots & & & & \\
\uparrow & & & & \\
\text{subgroup B} & & & &
\end{array}$$

Now the first row of this array constitutes an isomorphic image of the subgroup A, while the successive rows are the cosets by the elements $(1, b_1)$, $(1, b_2)$, etc. For example, the left coset of A by $(1, b_1)$ consists of the elements $(1, b_1)\{(1, 1), (a_1, 1), (a_2, 1), (a_3, 1) \ldots\}$, i.e. $\{(1, b_1), (a_1, b_1), (a_2, b_1), (a_3, b_1), \ldots\}$, and this is the second row of the array. It is evident that the left and right cosets are identical, since in any case we have $(a_r, 1)(1, b_s) = (1, b_s)(a_r, 1) = (a_r, b_s)$ for all r and s, so that such elements commute. We are therefore certain that A is a normal subgroup of C, and so C/A is meaningful, and we now need to know whether C/A has the structure of B.

Suppose the rows of the above array are denoted A, B_1, B_2, B_3, ..., so that B_r denotes the set of ordered pairs $\{(1, b_r), (a_1, b_r), (a_2, b_r), (a_3, b_r), \ldots\}$. Consider the product of the cosets B_rB_s: letting (a_i, b_r) be any element of B_r, and (a_j, b_s) be any element of B_s, we have:

$$(a_i, b_r)\,(a_j, b_s) = (a_ia_j, b_rb_s) = (a_k, b_t),$$

for a_ia_j must be an element of the subgroup A, which we choose to call a_k, and similarly $b_rb_s = b_t$. Therefore the product of any two elements of B_r and B_s is an element of B_t, and we may express this: $B_rB_s = B_t$. The behaviour of the cosets of A in the direct product group therefore exactly mimics the behaviour of the elements of the group B itself, i.e. the group of cosets C/A is isomorphic to the group B. Hence it is true that $(A \times B)/A \cong B$, and also that $(A \times B)/B \cong A$.

★**Q. 21.11.** Verify the truth of the above in the cases when A and B are: (a) D_3 and C_2; (b) D_4 and C_2; (c) D_3 and C_3; (d) Q_4 and C_2; (e) C_n and C_2; (f) D_n and C_2; (g) C_n and C_3. Write out the table for $D_2 \times C_3$, taking the elements to be $\{1, a, b, c, x, ax, bx, cx, x^2, ax^2, bx^2, cx^2\}$ where $a^2 = b^2 = c^2 = 1 = x^3$, and reconstruct the table in three ways to depict the quotient group by the subgroups C_2, C_3 and C_6, and identify the structure of the quotient group in each case.

★**Q. 21.12.** Show that, for a subgroup H of index 2 in a group G, the quotient group G/H is C_2. Show also that, for a normal subgroup of index 3, the quotient group is isomorphic to C_3.

★**Q. 21.13.** We noted on p. 407 that $G/C_3 \cong C_2$ is satisfied when G is either D_3 or C_6. What structure may G possess to satisfy: $G/C_4 \cong C_2$; $G/C_5 \cong C_2$; $G/C_6 \cong C_2$; $G/D_2 \cong C_2$; $G/C_2 \cong C_3$; $G/D_2 \cong C_3$; $G/C_2 \cong D_2$?

★**Q. 21.14.** In most of the examples considered so far, we have arrived at quotient groups which were themselves cyclic (usually C_2 or C_3). This is common, and of great importance. Give, however, examples of groups whose cosets form a non-cyclic quotient group.

★**Q. 21.15.** Show that, if m and n are integers, then C_m is a normal subgroup of C_{mn}, and $C_{mn}/C_m \cong C_n$.

✱**Q. 21.16. If G is a group of order mn, and H is a normal subgroup of order m, show that the mapping such that $g \to gH$ for all g is n to 1.**

Chains of normal subgroups

We have seen that $D_2 \lhd A_4$, the subgroup consisting of $\{1, a, b, c\}$ in the case of the table on p. 403. But D_2 is Abelian, so any of its subgroups is normal to it, e.g. $\{1, a\} \lhd \{1, a, b, c\}$. So we have $C_2 \lhd D_2 \lhd A_4$, a 'chain' of normal subgroups. We may extend the chain one stage further to the left by including the identity 1 as a trivial (normal) subgroup of C_2. An extension to the right is also possible since we know that $A_4 \lhd S_4$. Hence we have the chain:

$$1 \lhd C_2 \lhd D_2 \lhd A_4 \lhd S_4.$$

Note the curious fact that, though $C_2 \lhd D_2 \lhd S_4$, yet C_2 is *not* normal in S_4. Referring to the group table for S_4, p. 219, we have the normal subgroup $\{1, h, r, y\}$. Though each of the subgroups $\{1, h\}$, $\{1, r\}$, $\{1, y\}$ is normal in D_2, neither of them is normal in S_4. This must be so, since we have seen that if $\{1, h\}$ is to be normal in a group, then h must commute with every element of the group,† i.e. must belong to the centre. But the centre of S_4 consists of the identity alone (p. 218). This counter-example illustrates that $\lhd$ is not a transitive relation between groups.

In the chain $1 \lhd C_2 \lhd D_2 \lhd A_4 \lhd S_4$, the respective quotient groups are $C_2/C_1 \simeq C_2$; $D_2/C_2 \simeq C_2$; $A_4/D_2 \simeq C_3$; $S_4/A_4 \simeq C_2$. These quotient groups are called 'composition factors' of the chain, and a group which resolves into a succession of *cyclic* quotient groups in this way is called a 'soluble'‡ group.

Q. 21.17. Find another counter-example to illustrate that $\lhd$ is not transitive.

Q. 21.18. Is S_4 normal in S_5? (Consider permutations of five letters which leave one of the letters invariant.)

Q. 21.19. **Show that $H_1 \lhd G, H_2 \lhd G \Rightarrow (H_1 \cap H_2) \lhd G$. Show also that $H_1H_2 \lhd G$.**

Q. 21.20. Using the table of Ch. 13, p. 204 for D_6, specify elements in each of the normal subgroups of the chain $1 \lhd C_3 \lhd C_6 \lhd D_6$. Find another chain of normal subgroups in D_6. Generalise for D_{2n}.

Q. 21.21. In the group Q_6, we have two elements of period 6, either of which generates C_6, so we have $C_6 \lhd Q_6$. Q_6 also contains six elements of period 4. How many normal subgroups are there of Q_6 with structure C_4?

Q. 21.22. Consider the group $G = C_3 \times D_4$, where D_4 is generated by $a = (1\ 2)(3\ 4)$ and $b = (1\ 3)$, and C_3 is generated by $p = (5\ 6\ 7)$. Find the structure of G/H when H is (a) the centre, (b) a subgroup of structure C_4; and find chains of normal subgroups. Investigate in a similar way the group $D_2 \times D_3$ (or $C_2 \times C_2 \times D_3$; or $C_2 \times D_6$): this group may be realised by using the transpositions (1 2), (3 4), (5 6), (6 7).

Q. 21.23. Make a similar investigation for the group $D_2 \times D_3$ (or $C_2 \times C_2 \times D_3$, or $C_2 \times D_6$). Take D_2 to be $\{1, a, b, c\}$, D_3 to be $\{1, r, s, u, v, w\}$ with $rs = 1$. Or else consider the product of permutations, either as $S_2 \times S_2 \times S_3$ or as $S_2 \times D_6$, where the permutations of six letters which give the group D_6 are those which preserve 'adjacency', such as $\begin{pmatrix} A & B & C & D & E & F \\ D & C & B & A & F & E \end{pmatrix}$.

† *N.B.* only true for subgroups *of order* 2.
‡ See p, 411.

Simple groups

It is comparatively rare for a finite group to have no normal subgroups, apart, of course from the case of cyclic groups of prime order which have no proper subgroups anyway. Such rare groups of composite order with no normal subgroups are called 'simple'. Every group of composite order up to 59 has one or more normal subgroups. There are upwards of a dozen groups of order 60, all of which *save one* have normal subgroups. The exception is the group A_5 which we discussed in Ch. 17. A more general important result, known as Galois' theorem, is that **the groups A_n have no normal subgroups for $n \geqslant 5$.** We have already noted that A_4 does have a normal subgroup (D_2), and so Galois' theorem states that all the alternating groups of higher order are simple. This theorem has the important consequence that an equation of degree 5 and above cannot be solved to give the roots explicitly in terms of the coefficients by a formula involving radicals. Whereas for the equation $ax^4 + bx^3 + cx^2 + dx + e = 0$ it is possible to express x in terms of a, b, c, d and e by a formula involving square and cube roots, it would not be possible to find such a formula for any equation of degree beyond the fourth. Of course, the formula for the solution of the general quadratic equation:

$$ax^2 + bx + c = 0 \Rightarrow x = \frac{-b \pm (b^2 - 4ac)^{\frac{1}{2}}}{2a},$$

is very simple and well-known. The formula for the solution of the general quartic equation would be extremely complicated, and indeed it may be found only in the most advanced books on algebra. Even the formula for the cubic equation is not at all simple. Nobody ever bothers to learn formulae for the solution of 3rd and 4th-degree equations, and it is not the formula itself which is of interest. What concerns us is not the formula but *the fact that it would be possible to find one* if one persevered sufficiently.

A non-simple group (like A_4) which has one or more normal subgroups, and therefore admits of the corresponding homomorphic image onto a quotient group, resembles a composite natural number which may be resolved into factors. A 'simple' group, which admits no homomorphic images may be likened to a prime number which may not be factorised. The fact that A_2, A_3 and A_4 do allow themselves homomorphic images may be shown to account for the fact that quadratic, cubic and quartic equations may be solved to obtain the unknown in terms of the coefficients, and the background to this is that in this case it is possible for the process of the solution of the equation to be 'broken up' into simpler steps, each involving the solution of equations of lower degree. This breaking up into irreducible steps each of which in fact involves nothing more than the extraction of a square root or a higher root, depends upon being able to find just such a chain of normal subgroups as in the examples we gave in the last section. To solve the general equation of the nth degree it would be necessary to find a chain of subgroups $1 \lhd G_1 \lhd G_2 \lhd G_3 \lhd \ldots \lhd S_n$. Unfortunately when $n \geqslant 5$, the only intermediate subgroup that can be found is A_n: $1 \lhd A_n \lhd S_n$, and the quintic equation is not 'solvable' just because the symmetric group S_5 is not a 'solvable' group.

Thus when faced with the problem of solving the quintic equation, the underlying group theory involved leads us to the group A_5, and the fact that this is a simple group (perhaps 'prime' might have been a more apt term) accounts for the impossibility in this case of resolving the problem into simpler basic steps. The problem of the solution of the quintic equation occupied mathematicians for centuries, and they felt, with some justification, that it *ought* to be possible to solve the general quintic equation by extending the methods used for cubic and quartic equations. All their efforts were of no avail, and it was only in the eighteenth century when N. Abel (1802–29) proved the impossibility of the task that attempts were finally abandoned. Concurrently, Galois (1811–32) who was the chief pioneer of the theory of groups, further developed the applications of group theory to the theory of equations.

It was mentioned above that simple groups are rare. After A_5 (of order 60), the next one is of order 168, and this one has been mentioned as the group of automorphisms of $C_2 \times C_2 \times C_2$ (see p. 272). We shall meet it again in Ch. 25. The orders of other simple groups are 360, 504, 660, ... there being only five of order less than 1000.

Q. 21.24. Identify the structure of one of the last three mentioned.

As long as fifty years ago, some fifty-three simple groups of order less than one million were known, and it is a remarkable fact that these include *two of the same order*, namely 20,160 ($= 2^6 \times 3^2 \times 5 \times 7$). There are a number of interesting theorems concerning simple groups, including one which was not proved until 1963: every simple group must be of even order. To the list of known simple groups more are gradually being added as the years pass, but the theory is far from complete, and it is not even known whether the number of such groups is finite.

To prove that A_5 is a simple group

We shall show two results: (1) **that A_5 has no normal subgroups,**
(2) **A_5 is the only normal subgroup of S_5.**

Considering the even permutations of five letters which form the group A_5, these fall into the following classes; and in each case we interpret them in terms of the direct symmetries of the regular icosahedron in order to make the work more vivid:

(1) the identity;
(2) 24 cycles of period 5, such as $(ABCDE)$, each of these corresponding to a rotation through $2\pi k/5$ about the joins of opposite vertices (see also fig. 17.07);
(3) 20 cycles of period 3, such as (ABC), each of these corresponding to a rotation through $\pm\frac{2}{3}\pi$ about joins of mid-points of opposite faces;
(4) 15 double transpositions, such as $(AB)(CD)$, corresponding to half-turns about joins of mid-points of opposite edges,

making a total of 60 even permutations.

Now we know that, if a normal subgroup contains a particular element, then it contains every one of its conjugate elements (see p. 384). For example, if there does exist a normal subgroup of A_5, then if it contains the cycle (ABC) it must also contain all the other nineteen cycles of period 3. It follows that the order of such a normal subgroup of A_5 must be of the form $1 + 24n_1 + 20n_2 + 15n_3$, where n_1, n_2 and n_3 are each 0 or 1 (but not all 0, nor all 1). But this number must be a factor of 60, by Lagrange's theorem, and must therefore be 30, 20 or 15. (It can only be less than 15 when $n_1 = n_2 = n_3 = 0$, which would yield the identity only.) The fact that it is impossible to find values of the n's to satisfy these severe restrictions concludes the argument.

To prove that A_5 is the only normal subgroup of S_5, we may argue on similar lines, but naturally enough there are more possibilities to be considered. Permutations of the same class will contain cycles of the same length (see p. 364), and so we have the following classes:

(1) the identity
(2) 24 cycles of period 5, such as $(ABCDE)$
(3) 20 cycles of period 3, such as (ABC)
(4) 15 double transpositions, such as $(AB)(CD)$

} the even permutations;

(5) 30 cycles of period 4, such as $(ABCD)$
(6) 20 cycles of period 2 and 3, such as $(ABC)(DE)$
(7) 10 single transpositions, such as (AB)

} the odd permutations.

This time we must find values of $n_1, n_2, n_3, n_4, n_5, n_6$ (each either 0 or 1) so that $1 + 24n_1 + 20n_2 + 15n_3 + 30n_4 + 20n_5 + 10n_6$ is a factor of 120, i.e. either 60, 40, 30, 24, 20, 15, 12, or 10 (again, we are not interested in the cases when every n is zero, or every n is 1, so clearly there is no possibility of getting a total of 8, 6, 5, 4, 3 or 2). It will be found by trial and error that only the following selections are favourable:

(a) $n_1 = 1, n_2 = 1, n_3 = 1, n_4 = n_5 = n_6 = 0$, giving $1 + 24 + 20 + 15 = 60$, and this is the normal subgroup A_5.

(b) $n_1 = 1, n_2 = 0, n_3 = 1, n_4 = n_5 = n_6 = 0$, giving $1 + 24 + 15 = 40$.

(c) $n_1 = 1, n_2 = 0, n_3 = 1, n_4 = 0, n_5 = 1, n_6 = 0$, giving $1 + 24 + 15 + 20 = 60$.

It is a simple matter to dismiss (b) and (c) on the grounds of *closure*. For (b), when one of the twenty-four permutations of type $(ABCDE)$ is combined with one of the fifteen of type $(AB)(CD)$, we get one of the 20 cycles of period 3. The set of 40 do not therefore even form a group.

Q. 21.25. Check that $(ABCDE)(AB)(CD) = (ACE)$, and that $(AB)(CD)(ABCDE) = (BDE)$. You are left to check for yourself that (c) is inadmissible on similar grounds.

Q. 21.26. Prove that S_5 has no subgroup of order 60 other than A_5.

The insolubility of the quintic equation is a consequence of these two results, in the sense that in order to solve the quintic equation one would need to find a 'chain' of normal subgroups of S_5 of the form $1 \lhd G_1 \lhd G_2 \lhd \ldots \lhd S_5$. The first result of this paragraph admits that we can fill in one intermediate normal subgroup, A_5, but says that the gap between 1 and A_5 cannot be filled. The second result tells us that to search for an alternative means of descending to 1 via a chain of normal subgroups would be fruitless.

Infinite groups

Normal subgroups, and the resulting quotient groups of cosets, occur in infinite groups as well as in finite groups. We may find the subgroup and the induced factor group to be both infinite, or only one of these may be an infinite group. For example, consider the group $(\mathbf{Z}, +)$ which has the structure C_∞, and take the subgroup H $\{\ldots -15, -10, -5, 0, 5, 10, 15, \ldots\} \cong (5\mathbf{Z}, +)$. Then the cosets are

H + 1: $\{\ldots -14, -9, -4, 1, 6, 11, \ldots\}$
H + 2: $\{\ldots -13, -8, -3, 2, 7, 12, \ldots\}$
H + 3: $\{\ldots -12, -7, -2, 3, 8, 13, \ldots\}$
and H + 4: $\{\ldots -11, -6, -1, 4, 9, 14, \ldots\}$.

These are the residue classes modulo 5, and we may use the least positive residue as a representative of each class, and denote the classes **0**, **1**, **2**, **3**, **4** (also p. 340). Table 21.08 shows the addition of these cosets: when we say that **4** + **2** = **1**, we mean that any number selected from the coset **4** added to a number from the coset **2** will give a number in the coset **1**. (This is, indeed, what we were really doing when we considered the sets of odds and evens on the very first page of the book.)

The above casts a new light on the finite arithmetic tables which we have met many times in the previous chapters. We now see the group $\{0, 1, 2, 3, \ldots, n-1\}$, + mod. n as the quotient group in the group (**Z**, +) formed by the cosets of the subgroup consisting of multiples of n, these cosets being the residue classes which we might denote **0**, **1**, **2**, **3**, ... $\boldsymbol{n-1}$. Thus when H is the normal subgroup (n**Z**, +), consisting of

Table 21.08

+	0	1	2	3	4
0	0	1	2	3	4
1	1	2	3	4	0
2	2	3	4	0	1
3	3	4	0	1	2
4	4	0	1	2	3

multiples of n, we see that the quotient group (**Z**, +)/H is the group C_n. Another example is provided by (**Q**\0, ×)† a normal subgroup of which is ($\mathbf{Q}^+$, ×) (i.e. all the positive rationals under multiplication). There are just two cosets, being the partition of the rationals in positive and negative rationals. These cosets combine according to the table so the quotient group is C_2. In fact, in this case, the group (**Q**\0, ×) is the direct product of the group ($\mathbf{Q}^+$, ×) with the group C_2.

×	+	−
+	+	−
−	−	+

We may not interpret the group $\{1, 2, 3, \ldots, n-1\}$, × mod. n in quite a comparable way, since (**Z**, ×) itself is not a group. However, (taking $n = 5$ as an example), when we say that $\mathbf{4} \times \mathbf{3} = \mathbf{2}$ (mod. 5),‡ this may be interpreted in terms of product sets as follows.

Any number in the set $\{\ldots -1, 4, 9, 14, \ldots\}$ multiplied by
any number in the set $\{\ldots -2, 3, 8, 13, \ldots\}$ gives
a number in the set $\{\ldots -3, 2, 7, 12, \ldots\}$, and if these sets are

indicated by the least positive residue in bold type, this abbreviates to $\mathbf{4} \times \mathbf{3} = \mathbf{2}$.‡

Infinite groups with infinite quotient groups

Next, let us consider an infinite group with a *finite* normal subgroup, e.g. H $=\{1, \omega, \omega^2\}$, × is a subgroup of (**C**, ×). The coset zH is $\{z, z\omega, z\omega^2\}$ and this is an *equilateral triangle* in the Argand diagram whose centre is at the

† **Q**\0 means the set of rationals *excluding zero*, see p. xvii.
‡ Here the '=' sign is used, because we are saying that these *sets* (or classes) are equal, i.e. contain the same elements. When however, we say $12 \equiv 2$ (mod. 5), the '≡' sign expresses an *equivalence relation*, for it tells us that 12 and 2 belong to the same equivalence class.

origin (compare fig. 19.04). The product set of the cosets z_1H and z_2H is the set z_1z_2H, as you may easily verify; the inverse of zH is z^{-1}H, and the cosets form the factor group (C, ×)/H. This time the factor group itself is infinite.

Q. 21.27. Show that it is isomorphic with (C, ×) itself.

Q. 21.28. Consider the quotient group of (C, ×) by other finite subgroups.

Q. 21.29. Let t be a translation and m a reflection in a line perpendicular to t. Then t and m generate the group D_∞ (p. 206), the elements being $\{\ldots mt^{-1}, t^{-1}, m, 1, mt, t, \ldots\}$. Show C_∞ $\{\ldots t^{-1}, 1, t, t^2, \ldots\}$ is a normal subgroup, and find the structure of D_∞/C_∞ by reference to the repeating strip pattern shown in fig. 26.163.

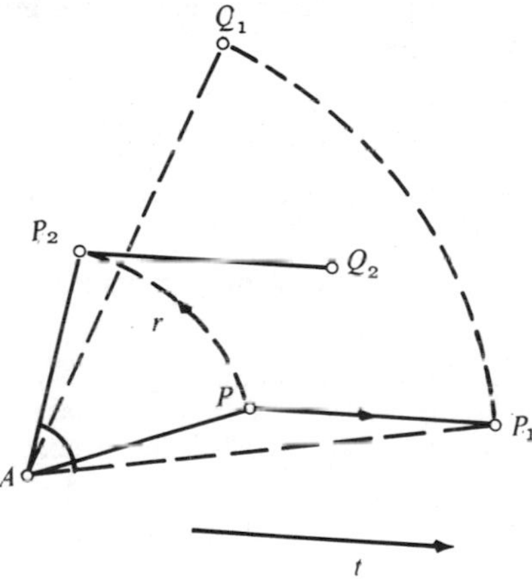

Fig. 21.02. $Q_1 = rt(\text{P})$, $Q_2 = tr(\text{P})$.

Q. 21.30. In fig. 21.02, the rotation r about A through α is combined with a translation t:

$$rt(\text{P}) = r(\text{P}_1) = \text{Q}_1;$$
$$tr(\text{P}) = t(\text{P}_2) = \text{Q}_2, \quad \text{so } tr \neq rt.$$

Show however that tr and rt both represent a rotation through α about different points, C_1, C_2 (say). Now consider the group T of all plane translations, and the set R of all rotations in the plane. Show that the left coset rT and the right coset Tr are the same, and identify the quotient group K/T, where K is the group R ∪ T of all direct isometries. Now let G be the group of *all* isometries in the plane (direct and opposite). Show that K (the group of direct isometries) is a normal subgroup of G, and that G/K ≃ C_2.

For an example of an infinite group with an infinite normal subgroup which leads to an infinite quotient group, consider the group G $(a + b\sqrt{2}, \times)$, where $a, b \in$ Q (a, b not both zero). Then we have the infinite subgroup (**Q**, $\times$) consisting of all those elements of G for which $b = 0$. A typical coset of this subgroup would be all rational multiples of (say) $3\frac{1}{3} - \frac{1}{4}\sqrt{2}$. There are an infinity of such cosets, so clearly the quotient group itself is infinite. Again, the complex numbers with unit modulus, of the form $\cos\theta + i\sin\theta$, under multiplication, are a subgroup H of (**C**, $\times$). The cosets are concentric circles on the Argand diagram (see p. 342). If r_1 and r_2 are real numbers, the product of the cosets r_1H and r_2H is r_1r_2H, and it is clear that the quotient group is isomorphic with the group (**R**$^+$, $\times$), since r_1 and r_2 are any two positive reals, and the product r_1r_2 in the group (**R**$^+$, $\times$) induces the product $r_1\text{H}.r_2\text{H} = r_1r_2\text{H}$ in the quotient group of (**C**, $\times$)/H.

Q. 21.31. Show that (C, $\times$) is the direct product of the group (R, + mod. 2π) with the group (R$^+$, $\times$).

Q. 21.32. Consider the quotient group of the group (**V**$_2$, +) by the subgroup (**V**$_1$, +); and of the group (**V**$_3$, +) by the subgroup (**V**$_2$, +).

Q. 21.33. Consider the quotient group of the group (**C**, +) by the subgroup of Gaussian integers (i.e. complex numbers of the form $a + ib$ where $a, b \in \mathbf{Z}$).

Q. 21.34. Describe quotient groups for the other strip patterns, taking various normal subgroups (see p. 515, fig. 26.16).

Q. 21.35. Consider the group ($\mathscr{M}_2$, $\times$) of non-singular 2×2 matrices under matrix multiplication, and in each of the following cases find whether the subgroup described is normal, and if so, find the structure of the quotient group:

(1) the subgroup of the matrices $\begin{bmatrix}1 & 0\\0 & 1\end{bmatrix}, \begin{bmatrix}-1 & 0\\0 & -1\end{bmatrix}$;

(2) the subgroup of the matrices $\begin{bmatrix}1 & 0\\0 & 1\end{bmatrix}, \begin{bmatrix}0 & 1\\1 & 0\end{bmatrix}$;

(3) the subgroup of the matrices $\begin{bmatrix}1 & 0\\0 & 1\end{bmatrix}, \begin{bmatrix}1 & 0\\0 & -1\end{bmatrix}, \begin{bmatrix}-1 & 0\\0 & 1\end{bmatrix}, \begin{bmatrix}-1 & 0\\0 & -1\end{bmatrix}$;

(4) the subgroup of the matrices $\begin{bmatrix}\cos\theta & -\sin\theta\\\sin\theta & \cos\theta\end{bmatrix}$ where $\theta = 2\pi/n$;

(5) the subgroup generated by $\begin{bmatrix}0 & -1\\1 & 0\end{bmatrix}$;

(6) the subgroup generated by matrices $\begin{bmatrix}0 & -1\\1 & 0\end{bmatrix}$ and $\begin{bmatrix}1 & 0\\0 & -1\end{bmatrix}$;

(7) the subgroup of matrices for which $ad - bc = 1$.

Q. 21.36. Show that the group (**Z**, +) is isomorphic to the group H{... 0·001, 0·01, 0·1, 1, 10, 100, ...} under multiplication. The latter is a normal subgroup of (**R**$^+$, $\times$). The coset of the real number 2·6 is the set;

{... 0·0026, 0·026, 0·26, 2·6, 26, 260, ...},

i.e. {... $10^{\bar{3}.415}$, $10^{\bar{2}.415}$, $10^{\bar{1}.415}$, $10^{0.415}$, $10^{1.415}$, $10^{2.415}$, ...}

Show that the quotient group formed by the cosets is isomorphic to the group of real numbers in the interval (0, 1) under addition modulo 1 (e.g. $0{\cdot}7 + 0{\cdot}6 = 0{\cdot}3$).

Looking ahead

The substance of the present chapter is of enormous importance if you are going more deeply into the theory of groups. Indeed, it is no exaggeration to say that, in studying a particular group, those properties which are of chief interest to the mathematician are first: which normal subgroups does the group possess, and second, what quotient groups do these normal subgroups produce, i.e. what homomorphic images does the group admit? For note that every single normal subgroup partitions the group into cosets, and these form the quotient group peculiar to that particular subgroup, so that there is a homomorphic image of a group relative to each one of its normal subgroups. This is known as a 'Natural Homomorphism' of the group.

Postscript: The first chapter of H. Coxeter and W. Moser, *Generators and Relations for Discrete Groups* (Springer-Verlag, Berlin, 1965), is about *adjoining* relations to obtain homomorphic images of a group. The method is clarified and illustrated in I. Grossman and W. Magnus, *Groups and their Graphs* (Random House), Chs. 11, 12.

The work of this chapter has been varied and difficult, and a brief summary is given in view of its importance to those who are going further into group theory.

We began by giving instances of groups which may be mapped into (smaller) groups by a homomorphism. The meaning of homomorphism is expressed most concisely as on pp. 396 ff., or, in words, as a mapping which preserves products. Those elements in the given group G which map into the identity of the image group constitute the kernel K, and this is a normal subgroup of G.

Next we saw how the cosets of a *normal* subgroup themselves form a group under the product set law of composition described on pp. 356, 401 ff., and we gave examples from finite groups with the quotient group displayed by 'division' of the group table as on pp. 204, 403. We also showed how an attempt to 'divide' the group will fail if the subgroup is not normal. We then proved the result $(A \times B)/B \simeq B$, and warned that there is no valid converse of this result.

A brief discussion of chains of normal subgroups, of simple groups, and of the insolubility of the general quintic equation followed, a proof being given of the theorem that A_5 is simple. Finally, quotient groups derived from infinite groups were considered in the two cases when the normal subgroup was finite and also when it was infinite.

EXERCISES 21

1. Show that D_4 (p. 94) has a normal subgroup H of order 2. What is the structure of D_4/H? Is H the centre of D_4?
2. Find G/H where G is Q_4 and H is its centre.
3. Find all the homomorphic images by their normal subgroups of D_4, D_6, Q_4, Q_6, A_4.
4. Which of the following mappings are homomorphisms of the group $(\mathbf{R}, \times)$? State whether the homomorphism is an isomorphism in each case: $x \to -x$; $x \to x^r$ ($r = 1, 2, 3, 4, -1$); $x \to -x^2$; $x \to 2x$; $x \to 3x$; $x \to -1/x$; $x \to 10^x$; $x \to +\sqrt{|x|}$; $x \to$ integral part of x?
5. The method of checking a multiplication by 'casting out the nines' depends on finding the least positive residue of each number modulo 9, and checking the product, e.g.

   $347 = 5$ (mod. 9); $1875 = 3$ (mod. 9)
   $347 \times 1875 = 15 = 6$ (mod. 9)
   $652{,}625 = 26 = 8$ (mod. 9)

   so that 'answer' is wrong – the test is certain when used negatively. When the result of the test is positive, we merely establish that the answer is in the right class modulo 9. Here we have a homomorphic mapping from the set of positive integers on to the set of residues 0, 1, 2, . . ., 8. The operation is multiplication in each set, but neither is a group, of course.

   Many other *checks* are really homomorphisms. Find other examples.
6. Show that the mapping $x \to x^2$ which associates each element of a group with its square is an automorphism of the group if and only if the group is Abelian and contains no element of even period. If the group does contain elements of even period, show that, provided it is Abelian, we get a homomorphic mapping. Illustrate in the case of D_3, Q_4, etc.
7. If x and y belong to the same left coset of H, show that xz and yz lie in the same left coset for all $z \in G$, if $H \lhd G$.
8. Invent a counter-example to show that $G/A \cong B \nRightarrow G/B \cong A$.
9. Find a homomorphism from $(\mathbf{Z}_8, +)$ to $(\mathbf{Z}_4, +)$, and also to $(\mathbf{Z}_2, +)$.
10. Find the quotient group of $(\mathbf{Z}_{20}, +)$ by the subgroup C_5.
11. Show, by arriving at a contradiction, that there cannot exist a homomorphism from the group D_3 whose kernel is one of the subgroups of order 2.
12. Show that the homomorphic mapping $S_4 \to S_3$ may be realised in terms of the symmetries of the cube as follows: S_4 is the group of permutations of the diagonals of the cube (see p. 298), and S_3 is the group of permutations of the three axes through the centre parallel to the edges. What is the structure of the kernel? Show how the cube can be used to demonstrate a homomorphism from S_6 on to S_4.
13. Show that, if G is $(\mathbf{R}, +)$ and H is $(\mathbf{Z}, +)$, then G/H is isomorphic with $(\mathbf{S}, +$ (mod. 1)).
14. Show how $(\mathbf{R}, +)$ may be mapped homomorphically on to the group of plane rotations about a fixed point. [Consider the mapping which takes each

real number into its positive decimal part, e.g. $-2{\cdot}7 \to 0{\cdot}3$, with the operation + (mod. 1), e.g. $-2{\cdot}7 + 8{\cdot}9 \equiv 0{\cdot}3 + 0{\cdot}9 \equiv 0{\cdot}2$ (mod. 1).] What is the kernel of this mapping?

15 $z \to z^6$ is a homomorphism from (**C**, $\times$) on to itself. What is the kernel?

16 Which of the mappings (**C**, $\times$) $\to$ (**C**, $\times$) by the functions $z \to z^8$; $z \to \bar{z}$; $z \to 2z + 1$, $z \to 1/z$ are homomorphisms, isomorphisms?

17 Describe ways of setting up homomorphisms from (**C**, $\times$) on to (**R**, $\times$).

18 Does the mapping $z \to$ arg. z effect a homomorphism of the group (**C**, $\times$)?

19 Let $R(x)$ be the remainder when a polynomial $P(x)$ with rational coefficients is divided by $x^2 + 1$. Show that $P(x) \to R(x)$ is a homomorphism from the additive group of polynomials with rational coefficients on to the linear polynomials. What is the kernel? Find suitable moduli other than $x^2 + 1$.

20 Find a homomorphism of ($\mathscr{M}_2$, $\times$) on to (**R**, $\times$) other than that described in the text (p. 398), and identify its kernel.

21 Prove, for a homomorphic mapping ϕ, that $\phi(x^{-1}) = [\phi(x)]^{-1}$ and that $\phi(x) = \phi(y) \Rightarrow (\phi xy^{-1}) = \phi(x^{-1}y)$.

22 **a** is a fixed vector in 3-space, and **x** is a general vector. Does the mapping ($\mathbf{V}_3$, +) $\to$ (**R**, +) defined by $\mathbf{x} \to \mathbf{x}.\mathbf{a}$ (the scalar product) represent a homomorphism?

23 G is $\{2^x 3^y 5^z\}$, $\times$, where $x, y, z \subset \mathbf{Z}$; H is $\{2^x\}$, $\times$, where $x \in \mathbf{Z}$. Construct G/H.

24 Describe the quotient groups: (**C**/**R**, $\times$); (**R**/**Q**, $\times$); (**C**/**Q**, $\times$).

25 Establish a homomorphism from (**Q**, $\times$) on to the group C_2.

26 If G is the group of isometries of the plane, and H is the group of translations, what is the structure of G/H, (a) when G contains the direct isometries; (b) when G contains direct and opposite isometries?

27 If G is the group (**Z**, +) and H is (2**Z**, +) (each having the structure C_∞), show that G/H = C_2, and interpret the table $\begin{array}{|cc} E & 0 \\ 0 & E \end{array}$. What other possible structures may C_∞/C_∞ have?

28 If G is $\{a + bi\ (a, b \in \mathbf{Z})\}$, + and H is $\{2a + 2bi\ (a, b \in \mathbf{Z})\}$, +, find G/H, and illustrate by a Cayley diagram. (For the Cayley diagram of $C_\infty \times C_\infty$, see Q. 16.48, Ex. 16.40.)

29 Establish a homomorphism from the group $\begin{bmatrix} a & b \\ 0 & d \end{bmatrix}$, $\times$ on to the group (**R**, +).

30 Describe how the group $\begin{bmatrix} a & b \\ c & d \end{bmatrix}$, + with $a, b, c, d \in \mathbf{Z}$ may be mapped homomorphically on to the group D_2.

31 Show how it is possible to establish a homomorphism from the group of linear transformations in the plane on to a group of plane transformations which includes similarities and inversions.

Hint: e.g. $\begin{bmatrix} x' \\ y' \end{bmatrix} = \begin{bmatrix} 2 & -1 \\ 3 & 1 \end{bmatrix} \begin{bmatrix} x \\ y \end{bmatrix} \to z' = \dfrac{2z - 1}{3z + 1}$.

What is the kernel of this homomorphism?.

32 A given matrix $\begin{bmatrix} a & b \\ c & d \end{bmatrix}$ $(a, b, c, d \in \mathbf{R})$ maps the set of two dimensional vectors on to itself:

$$\begin{bmatrix} x \\ y \end{bmatrix} \to \begin{bmatrix} a & b \\ c & d \end{bmatrix} \begin{bmatrix} x \\ y \end{bmatrix}.$$

Show that this is an isomorphism of the group $(\mathbf{V}_2, +)$ so long as $ad - bc \neq 0$, but that when $ad - bc = 0$, we have a homomorphism. Find the kernel in the cases when the transforming matrix is

$$\begin{bmatrix} 1 & 1 \\ 1 & 1 \end{bmatrix}; \begin{bmatrix} 1 & 0 \\ 0 & 0 \end{bmatrix}; \begin{bmatrix} 2 & 1 \\ 6 & 3 \end{bmatrix}; \begin{bmatrix} 0 & 0 \\ 0 & 1 \end{bmatrix}.$$

33 Show that the group $(\mathbf{V}_3, +)$ may be mapped homomorphically on to $(\mathbf{V}_2, +)$ by

$$\begin{bmatrix} x' \\ y' \end{bmatrix} = \begin{bmatrix} 1 & 1 & 1 \\ 2 & -1 & 3 \end{bmatrix} \begin{bmatrix} x \\ y \\ z \end{bmatrix}$$

Find the kernel, and interpret geometrically. Generalise.

34 Those quaternions $\begin{bmatrix} z_1 & z_2 \\ -\bar{z}_2 & \bar{z}_1 \end{bmatrix}$ which have $z_1, z_2 \in \mathbf{R}$ are a subgroup of the multiplicative group of quaternions which is isomorphic to $(\mathbf{C}, \times)$. Letting

$$q = \begin{bmatrix} z_1 & z_2 \\ -\bar{z}_2 & \bar{z}_1 \end{bmatrix} \quad \text{and} \quad z = \begin{bmatrix} x & y \\ -y & x \end{bmatrix},$$

show that $q\,z\,q^{-1}$ is not in the subgroup. What do you conclude?

35 Using the results:

(a) $\mathrm{H}_1 \lhd \mathrm{G}, \mathrm{H}_2 \lhd \mathrm{G} \Rightarrow \mathrm{H}_1 \cap \mathrm{H}_2 \lhd \mathrm{G}$;
and (b) A_5 has no normal subgroups;
and (c) $\mathrm{A}_5 \lhd \mathrm{S}_5$,

prove that S_5 has no subgroups of order 60 other than A_5. Generalise.

36 Prove that if H is a normal subgroup of order m of a group G of order mr, then the group table can be constructed with H in the top left-hand corner and the r cosets arranged in sequence so as to show the structure of the quotient group, and in such a way that the internal pattern of each $m \times m$ coset square is identical.

22 *Automorphisms*

In Ch. 11 we considered some simple cases of isomorphisms of groups with themselves, or automorphisms. We saw, for example, that the group $\{e, x, x^2, x^3\}$ (C_4) has the single automorphism which interchanges x with x^3. This, with the identity gives the group C_2. Again, $\{e, a, b, c\}$ ($a^2 = b^2 = e$, $ab = c$) with structure D_2, has the automorphism group D_3, these automorphisms corresponding to every possible permutation of the elements a, b and c between themselves.

Inner and outer automorphisms

When it came to non-Abelian groups, we noted that there are two types of automorphism, 'inner' and 'outer'. An inner automorphism of a group is obtained by selecting an element from the group (say p), and conjugating every element of the group by it; that is, the general element g of the group is transformed into pgp^{-1}. If the group is of order n, then there are n *possible* inner automorphisms, but some of these may well overlap. For example, in the group D_4 (table 9.06), conjugation by a gives:

$$e \to aea^{-1} = e; \quad a \to aaa^{-1} = a; \quad b \to aba^{-1} = d;$$
$$c \to aca^{-1} = c; \quad d \to ada^{-1} = b; \quad f \to afa^{-1} = h;$$
$$g \to aga^{-1} = g; \quad h \to aha^{-1} = f,$$

so that we have the permutation $\begin{pmatrix} e & a & b & c & d & f & g & h \\ e & a & d & c & b & h & g & f \end{pmatrix}$, i.e. the transpositions $(b\ d)(f\ h)$. On the other hand, conjugation by c produces the same automorphism.

Q. 22.01. Verify that there are four inner automorphisms: $(b\ d)(f\ h)$; $(b\ d)(a\ c)$; $(a\ c)(f\ h)$, and the identity. Which group do they form?

Thus the group D_4 has not eight but four inner automorphisms. However, there are four further automorphisms, and these are the so-called 'outer' automorphisms. The elements of period 2 (other than g, which is the very special element that belongs to the centre), that is, the set $\{a, b, c, d)$, may be subjected to permutations other than $(b\ d)$, $(a\ c)$ and $(a\ c)(b\ d)$ described above, namely $(a\ b\ c\ d)$, its inverse $(a\ d\ c\ b)$, as well as $(a\ b)(c\ d)$ and $(a\ d)(b\ c)$. These two latter have the effect of interchanging f and h. Of the eight automorphisms of D_4, there turn out to be two of period 4, five of period 2 and the identity. The full automorphism group is thus D_4, and contains the inner automorphisms as a normal subgroup with structure D_2.

Q. 22.02. Interpret the above automorphisms in terms of the symmetries of the square. For example $\begin{pmatrix} e & a & b & c & d & f & g & h \\ e & b & c & d & a & f & g & h \end{pmatrix}$ might be interpreted as a re-labelling of the four mirror symmetry axes, as in figure 22.01.

Fig. 22.01.

We also remarked that the only inner automorphism of an Abelian group is the identity, since every element is self-conjugate. All the automorphisms of *Abelian* groups are therefore *outer*. We also found that the group D_3 ($\cong S_3$) has six automorphisms whose group is D_3, and these are all inner ones. This group therefore has no outer automorphisms. These examples show the two extreme cases.

That all the automorphisms – regarded as permutations of the elements of the given group – themselves form a group, is readily appreciated. For an automorphism is a permutation of the elements which preserves products. When one such permutation is followed by another, the result will be another permutation, with products preserved throughout. Associativity of permutations under successive composition is assured, and the identity is present. The inverse of an automorphism also preserves structure, so all the requirements of a group are fulfilled.

Q. 22.03. Prove that the inner automorphisms of a group are a subgroup of the group of automorphisms.

The stronger result, that the inner automorphisms are a *normal* subgroup, will be proved in due course (see p. 428). At present, we shall accept the truth of this statement.

In what follows, we shall denote the group of automorphisms of a given group G by aut. G. Thus we have seen that aut. $C_4 \cong C_2$; aut. $D_2 \cong D_3$; aut. $D_3 \cong D_3$; aut. $D_4 \cong D_4$. The second example quoted is an illustration of an Abelian group whose automorphisms are a non-Abelian group, a phenomenon which is not uncommon. However, the automorphisms of a *cyclic* group do form an Abelian group which, however, need not be cyclic. For example, aut. $C_{12} \cong D_2$ (see below).

Q. 22.04. Show that aut. $C_8 \cong D_2$.

We have noted elsewhere that aut. $C_p \cong C_{p-1}$ when p is prime, and this result is of considerable importance (see pp. 167 ff., and Ex. 11.21, Ex. 12.35, 18.35).

Automorphisms of Abelian groups

We proceed now to identify the automorphism groups of two Abelian groups, first C_{12}, then $C_6 \times C_2$. The former we shall take to be represented by the set {0, 1, 2, 3, 4, 5, 6, 7, 8, 9, 10, 11} under addition modulo 12. Then either of the primitive elements 5, 7, 11 may be taken as generator in place of 1. These give rise to the following permutations:

0	1	2	3	4	5	6	7	8	9	10	11
0	5	10	3	8	1	6	11	4	9	2	7
0	7	2	9	4	11	6	1	8	3	10	5
0	11	10	9	8	7	6	5	4	3	2	1.

which have the cycles

$$\left.\begin{array}{l}(1\,5)(2\,10)(4\,8)(7\,11);\\ (1\,7)(3\,9)(5\,11);\\ (1\,11)(2\,10)(3\,9)(4\,8)(5\,7)\end{array}\right\}\text{ each of period 2,}$$

and with the identity give the group D_2. Hence aut. $C_{12} \cong D_2$.

Q. 22.05. Show that aut. $C_9 \cong C_6$, and aut. $C_{10} \cong C_4$.

Next we take $C_6 \times C_2$, defined by $r^6 = 1 = a^2$, $ar = ra$, where r and a are independent generators of periods 6 and 2 respectively. In terms of these generators, the group contains:

$$\left.\begin{array}{l}r,\ r^5\\ ar,\ ar^5\\ ar^2,\ ar^4\end{array}\right\}\text{ (six elements of period 6, arranged here in inverse pairs),}$$

r^2, r^4 (two elements of period 3),
r^3, a and ar^3 (three elements of period 2).

To seek the automorphisms, we replace the generators r and a by other pairs of independent generators of period 6 and 2. For example, in the sixth row of table 22.01, r and a have been replaced respectively by ar (of period 6) and r^3 (of period 2). The other elements in that row may then be filled in. For example, corresponding to the product $r^5.ar^2 = ar$ in row 1, we have the product $ar^5.r^5 = ar^4$ in row 6. The table exhausts the possibilities, so the automorphisms are a group of order 12. Its structure may be deduced from the periods of the permutations which are given in the right-hand column. We see that aut. $C_6 \times C_2 \cong D_6$.

Table 22.01 (showing automorphisms of $C_6 \times C_2$)

	periods of element												period of permutation
row	1	6	3	2	3	6	2	6	6	2	6	6	
1	1	r	r^2	r^3	r^4	r^5	a	ar	ar^2	ar^3	ar^4	ar^5	1
2	1	r	r^2	r^3	r^4	r^5	ar^3	ar^4	ar^5	a	ar	ar^2	2
3	1	r^5	r^4	r^3	r^2	r	a	ar^5	ar^4	ar^3	ar^2	ar	2
4	1	r^5	r^4	r^3	r^2	r	ar^3	ar^2	ar	a	ar^5	ar^4	2
5	1	ar	r^2	ar^3	r^4	ar^5	a	r	ar^2	r^3	ar^4	r^5	2
6	1	ar	r^2	ar^3	r^4	ar^5	r^3	ar^4	r^5	a	r	ar	3
7	1	ar^5	r^4	ar^3	r^2	ar	a	r^5	ar^4	r^3	ar^2	r	2
8	1	ar^5	r^4	ar^3	r^2	ar	r^3	ar^2	r	a	r^5	ar^4	6
9	1	ar^2	r^4	a	r^2	ar^4	ar^3	r^5	ar	r^3	ar^5	r	6
10	1	ar^2	r^4	a	r^2	ar^4	r^3	ar^5	r	ar^3	r^5	ar	2
11	1	ar^4	r^2	a	r^4	ar^2	ar^3	r	ar^5	r^3	ar	r^5	3
12	1	ar^4	r^2	a	r^4	ar^2	r^3	ar	r^5	ar^3	r	ar^5	2

Q. 22.06. Find the automorphism group of $C_4 \times C_2$.

Q. 22.07. Show that the correspondence $x \to x^k$ is an automorphism of an *Abelian* group of order n provided that k is prime to n. Which of the automorphisms of $C_6 \times C_2$ listed in the above table are of this type?

In Ch. 16, p. 272 we noted that the group $C_2 \times C_2 \times C_2$ has 168 automorphisms. The reason for this large number is that its seven elements of period 2 are, so to speak, indistinguishable. Now $C_3 \times C_3$ has eight elements of period 3, these being four pairs of inverses: $p, p^{-1}; q, q^{-1}; r, r^{-1}; s, s^{-1}$. Evidently this group will also have a large number of automorphisms, since any pair of the eight elements may be taken as generators provided they are not inverses.

Q. 22.08. Show that aut. $(C_3 \times C_3)$ is of order 48. Write out the forty-eight permutations of $p, p^{-1}, q, q^{-1}, r, r^{-1}, s, s^{-1}$, taking $pq = r$ and $pr = s$. Identify this group with the one described in Ex. 16.24, p. 288.

Automorphisms of non-Abelian groups

In the case of non-Abelian groups, we may obtain all the *inner* automorphisms systematically by transforming all the elements of the group by each element in turn. An interesting example is provided by Q_4, and we change the notation of the table on p. 101 by the following scheme: $\begin{pmatrix} 1 & r & r^2 & r^3 & a & b & c & d \\ 1 & r & a & s & t & u & v & w \end{pmatrix}$, the letters in the top row being replaced by those in the bottom row. The idea is that a shall now be the sole element of period 2, while the six elements of period 4 are r and s, t and w, u and v, three inverse pairs. The transformation is shown in table 22.02. The results

Table 22.02

	1	r	a	s	t	u	v	w
1	1	r	a	s	t	u	v	w
r	r	a	s	1	v	t	w	u
a	a	s	1	r	w	v	u	t
s	s	1	r	a	u	w	t	v
t	t	u	w	v	a	s	r	1
u	u	w	v	t	r	a	1	s
v	v	t	u	w	s	1	a	r
w	w	v	t	u	1	r	s	a

(Q_4)

of the transformation (conjugation) of the elements may be tabulated as in table 22.03. The elements 1 and a are unchanged throughout, each being

Table 22.03

x	1	a	r	s	t	w	u	v
axa^{-1}	1	a	r	s	t	w	u	v
rxr^{-1}	1	a	r	s	w	t	v	u
sxs^{-1}	1	a	r	s	w	t	v	u
txt^{-1}	1	a	s	r	t	w	v	u
wxw^{-1}	1	a	s	r	t	w	v	u
uxu^{-1}	1	a	s	r	w	t	u	v
vxv^{-1}	1	a	s	r	w	t	u	v

unique in the group. There are three distinct automorphisms: $(t\ w)(u\ v)$; $(r\ s)(u\ v)$ and $(r\ s)(t\ w)$, and with the identity they form the group D_2.

However, with six elements all of period 4, one expects a large number of automorphisms, and we now proceed to find the outer ones.

We may regard the group as generated by a pair of elements of period 4, such as r and t, and in the above notation the other elements are: $s = r^{-1}$, $w = t^{-1}$, $u = tr$, $v = rt$. To find automorphisms, we replace r and t by any pair selected from the set $\{r, s, t, w, u, v\}$ provided they are not a pair of inverses. For example, when r and t are replaced by w and u respectively, then their inverses, s and w, are automatically replaced by t and v; while u $(= tr)$ is replaced by uw $(= s)$, and v $(= rt)$ is replaced by wu $(= r)$. Thus we have the permutation $\begin{pmatrix} r & s & t & w & u & v \\ w & t & u & v & s & r \end{pmatrix}$; i.e. the cycle $(r\ w\ v)(s\ t\ u)$, of period 3. Evidently the selection of a pair of generators may be made in 6×4 ways (having selected the first, only *four* choices remain for the second). Thus there are twenty-four automorphisms of Q_4, including the four inner ones already found. You may write out these twenty-four permutations by following the procedure above.

It remains to identify the structure of this group. It will be found that there are nine of period 2, eight of period 3, six of period 4, and this leaves no doubt that the full group of automorphisms of Q_4 is S_4, the inner automorphisms being the normal subgroup with structure D_2.

The way in which the group S_4 arises may be seen in another way. For the three pairs of elements of period 4 may be likened to the three pairs of opposite faces of a cube. If we label these faces r, s; t, w; u, v in opposite pairs, each of the twenty-four rotations of the cube will correspond to exactly one of the permutations in the list of automorphisms of Q_4.

Q. 22.09. To which rotations of the cube do the *inner* automorphisms correspond?

The interest in this lies in the fact that, though the group Q_4 is not the symmetry group of any geometrical figure, yet it has this curious association with the cube (or octahedron).

Q. 22.10. Show that there are twenty automorphisms of D_5, and that the inner ones themselves have the group D_5.
Q. 22.11. Investigate the automorphisms of the non-Abelian groups of order 12: D_6 (table 13.05); Q_6 (p. 216); A_4 (p. 294).

Now Q_6 is also a group with six elements of period 4, and one might expect that its automorphisms would also number twenty-four. This is not so, because one's freedom of choice is now more restricted in the following way. The group contains two elements of period 6 (a and a^5),† and two of period 3 (a^2 and a^4). It is essential that an automorphism shall either leave each of these pairs alone, or else interchange them. Thus, for example, $dg = a$, while $fc = a^5$, so an automorphism which replaces d by f and g by c is feasible, and indeed one of the automorphisms of Q_6 is $(c\ g)(d\ f)(h\ j)$. But since, for example, $jh = a^4$, we do *not* have an automorphism which replaces d by j and g by h.

Automorphisms of S_4

Finally, we consider the automorphisms of S_4, and, using the notation of p. 219, we may take two of the elements of period 4, $\{j, t, k, n, s, x\}$ to be generators of the whole group so long as they are not inverses. We are

† Notation as on p. 109.

therefore in a similar position to the one we were in when obtaining the automorphisms of Q_4, and the group of these is of order 24. Not unexpectedly, its structure is S_4, but what is surprising is that all the automorphisms are inner. This may be seen by considering the effect of conjugation upon the subset $\{j, t, k, n, s, x\}$, as table 22.04 shows. (Note

Table 22.04

inner automorphisms of S_4 elements of period 4	j	t	k	n	s	x	period of permutation
X	j	t	k	n	s	x	1
aXa^{-1}	k	n	j	t	x	s	2
bXb^{-1}	s	x	n	k	j	t	2
cXc^{-1}	x	s	t	j	k	n	3
dXd^{-1}	n	k	s	x	t	j	3
fXf^{-1}	t	j	x	s	n	k	2
gXg^{-1}	n	k	t	j	x	s	2
hXh^{-1}	t	j	n	k	s	x	2
iXi^{-1}	k	n	x	s	t	j	3
jXj^{-1}	j	t	s	x	n	k	4
kXk^{-1}	x	s	k	n	j	t	4
lXl^{-1}	s	x	j	t	k	n	3
mXm^{-1}	x	s	j	t	n	k	3
nXn^{-1}	s	x	k	n	t	j	4
pXp^{-1}	t	j	s	x	k	n	2
qXq^{-1}	n	k	x	s	j	t	3
rXr^{-1}	j	t	n	k	x	s	2
sXs^{-1}	k	n	t	j	s	x	4
tXt^{-1}	j	t	x	s	k	n	3
uXu^{-1}	k	n	s	x	j	t	3
vXv^{-1}	s	x	t	j	n	k	3
wXw^{-1}	x	s	n	k	t	j	2
xXx^{-1}	n	k	j	t	s	x	4
yXy^{-1}	t	j	k	n	x	s	2

that in every case, the period of the transforming element is equal to the period of the induced permutation.)

It will readily be seen that the six elements of period 4 may be associated with the six faces of a cube, and that the automorphism may be thought of as a re-labelling of the six quarter-turns.

Q. 22.12. Investigate aut. S_4 from the point of view of the eight elements of period 3.

Q. 22.13. Trace the effect of the automorphisms of S_4 upon its subgroup A_4, verifying that the full automorphism group of both S_4 as well as A_4, is S_4. How is it that the twelve *outer* automorphisms of A_4 correspond to *inner* automorphisms of the whole group S_4?

Proof that the inner automorphisms are a normal subgroup of the full group of automorphisms

Finally we give the proof that the inner automorphisms of any group are a normal subgroup of all the automorphisms.

Suppose that X is any outer automorphism, and Y is any inner automorphism. It will be sufficient to show that XYX^{-1} is an inner automorphism. Let Y be produced by conjugation by the element y, so that $x \xrightarrow{Y} yxy^{-1}$; and suppose that X has the effect shown below:

$$X \quad \begin{array}{l} 1, g_1, g_2, \ldots, g_r, \ldots, y, \ldots, y^{-1}, \ldots, w, \ldots, w^{-1}, \ldots \\ 1, x_1, x_2, \ldots, x_r, \ldots, z, \ldots, z^{-1}, \ldots, y, \ldots, y^{-1}, \ldots \end{array} \quad X^{-1}$$

that is, the general element g_r is changed by X into x_r, y is changed into z, and w into y. The inevitable consequence of the latter is that y^{-1} is changed into z^{-1}, and w^{-1} into y^{-1}.

Now,

$$XYX^{-1}(x_r) \quad \text{(where } x_r \text{ may be regarded as any selected element of G)}$$
$$= XY(g_r) = X(yg_ry^{-1}) = X(y)\,X(g_r)\,X(y^{-1}) \quad \text{(in view of the fact that } X \text{ is an automorphism, and so preserves products)}$$
$$= zx_rz^{-1}$$

Hence XYX^{-1} is that automorphism which is brought about by transforming the elements of the group, each in turn, by z. It is therefore an *inner* automorphism, so the subgroup of inner automorphisms is invariant under conjugation by X. Seeing that X was *any* automorphism, it follows that the inner automorphisms are an invariant subgroup.

Q. 22.14. Show that, **if H is the centre of a group G, then the group of inner automorphisms is isomorphic with G/H.** Verify in the cases D_4, Q_4, D_5, D_6, Q_6, A_4. What may be deduced about the automorphism group of S_n if it is assumed that the centre of S_n consists of the identity alone?

Q. 22.15. Show that every one of the inner automorphisms of Q_6 (table 10.04) changes the set of conjugate elements $\{c, f, h\}$ into itself, but that the outer automorphisms interchange the two sets of conjugate elements $\{c, f, h\}$ and $\{d, g, j\}$. Prove in general that a set of conjugate elements is unchanged by an inner automorphism of the group. What can be said of the effect of an outer automorphism on a set of conjugate elements?

Q. 22.16. The subgroup $\{1, a, a^2, a^3, a^4, a^5\}$ of Q_6 is unchanged by every one of its automorphisms, but the subgroup $\{1, c, a^3, g\}$ is changed into either Gp (d) or else Gp (f) by the outer automorphisms. Consider the effect on various subgroups of various groups by their various automorphisms.

23 *Groups and music*

Musical pitch

The pitch of a musical note is defined by its frequency, measured in cycles per second. The frequencies of pure musical tones form an infinite set of real numbers lying between the lower and upper limits of audibility. The notes of the pianoforte form a finite subset of this infinite 'spectrum' containing usually eighty-eight members. Other instruments provide different subsets of available notes, these subsets being infinite in the case of the strings and the trombone, whereas in the case of most other instruments, we have finite subsets of discrete musical tones.

Interval between two notes

Given two notes of frequencies a and b, if $b > a$, we say that b is the higher note and a the lower note. The relation 'higher than' between two notes is an order relation, antisymmetric and transitive. When $a = b$, we have the 'unison'.

For two given notes of frequencies a and b, with $b > a$, the *interval* between them is defined: ${}_aI_b = b/a$ (> 1). Thus we have a mapping from the pairs of musical notes onto the reals greater than 1. For example, the two notes whose frequencies are 200 and 350 Hz map into the real number 1·75. The interval does not provide a proper 'distance function', since the unison is represented by the ratio unity, whereas for a distance function it should be zero. It is therefore convenient to take logs. The value of $\log {}_aI_b$ ($= \log b/a = \log b - \log a$) will be known as the 'logarithmic interval'. The logarithmic interval *is* a distance function on the set. (We shall see that, when the base of the logs is 2, the interval will be reckoned in *octaves*; while if $2^{1/12}$ is taken as the base of the logs, the interval will be reckoned in tempered semitones.) Intervals may be *combined* by multiplication. Thus, in the case $a < b < c$, we have ${}_aI_c = {}_aI_b \times {}_bI_c$, or $c/a = b/a \times c/b$. Alternatively by addition of logarithmic intervals.

The octave

The simplest interval between two tones (other than the unison) is the *octave*, for which the ratio is 2:1, i.e.

$b/a = 2 \Leftrightarrow b$ is an octave higher than a.

$b/a = 2^n$ $(n \in \mathbf{Z}^+) \Leftrightarrow b$ is n octaves higher than a.

For example, middle C is 260 Hz. Three octaves above middle C has frequency $260 \times 2 \times 2 \times 2 = 260 \times 2^3$. Two octaves below middle C has frequency $260 \times 2^{-2} = 65$ Hz.

The notes may be partitioned into *equivalence classes* by the equivalence relation:

$$p \sim q \Leftrightarrow {}_pI_q = q/p = 2^n \quad (n \in Z),$$

i.e. two notes are equivalent in this sense if they are separated by an exact number of octaves. This equivalence relation is recognised by musicians by the fact that all the notes in a particular equivalence class are *called by the same name*. For example, the notes with frequencies $32\frac{1}{2}$, 65, 130, 260, 520, 1040, ... are all called C. Middle C (= 260) may be taken as the *representative* of this equivalence class. All notes in a particular equivalence class *sound rather alike*, so much so, in fact that, it is often extremely difficult to distinguish them.† In fact, it is not uncommon for the human voice to transpose an octave if the melody goes beyond the reach of the voice. (See for example, the version of the hymn in fig. 23.01 as sung by

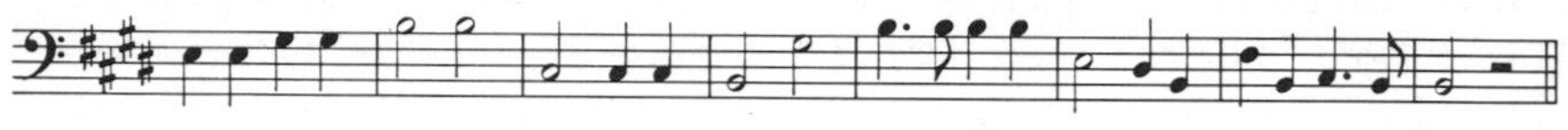

Fig. 23.01.

the old chap in the pew behind.) The similarity of sound is not surprising when one realises that, with a ratio of 2:1, the vibrations from the two notes coincide exactly, the higher note simply doing an extra vibration in between each vibration of the lower note, rather like a child walking beside an adult at the same speed, with the adult's paces exactly twice the length of those of the child.

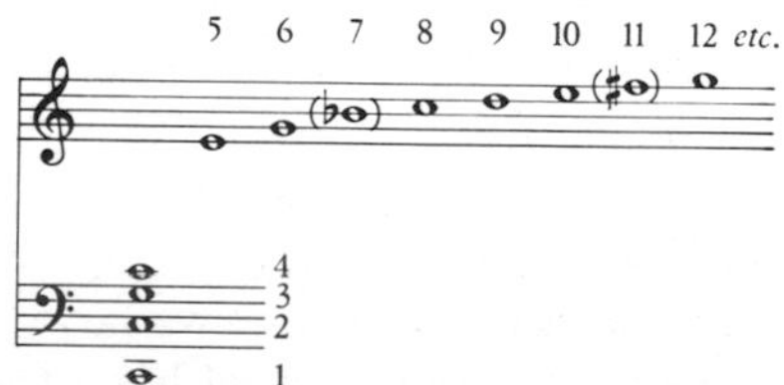

Fig. 23.02. The natural harmonic series.

The harmonic series

The notes of the natural 'harmonic series' (see fig. 23.02) have frequencies which are exact multiples of the 'fundamental'. These notes are the basis of the natural scale (with the exception of the 7th and 11th harmonics),

† Sir Malcolm Sargent was known to have found great difficulty in distinguishing the octave of a *whistle*.

and the intervals which are most pleasing to the ear are those given by the simplest rational numbers, such as 4/3, 6/5, 8/3, etc. Thus consonance is associated with the ratio of *small* integers. More precisely, natural ratios are of the form $2^p.3^q.5^r$ ($p \in \mathbf{Z}$, $q \in \{-3, -2, -1, 0, 1, 2, 3\}$, $r \in \{-1, 0, 1\}$), the 7th and 11th harmonics always being excluded from consideration, as stated.

Perfect fifth

After the octave, the next simplest interval is the 'perfect fifth' (ratio = 3:2). This interval is so important that it is used as a yardstick to generate the other notes of the musical scale, as we shall presently see. In the case of the perfect fifth, we have three vibrations from the higher note in the same time as two from the lower note, and the result is very smooth. The 'perfect fourth' (ratio 4:3) is the 'inverse' of the fifth, in the

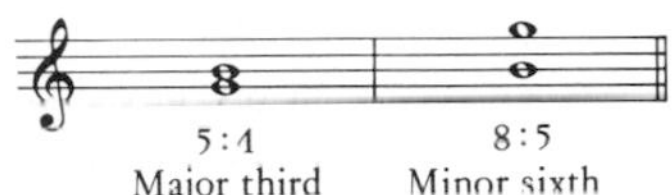

Fig. 23.03. Inversions.

sense that when the two intervals are combined ($3/2 \times 4/3 = 2$), the resultant interval is the octave, which is equivalent† to the unison, or identity. Musicians are mathematically accurate here, for they have their terminology almost correct: two intervals which combine to give the octave are known in musical terms as 'inversions' of each other. For example, the major third (5/4) and the minor sixth (8/5) are such a pair (see fig. 23.03). The full range of natural musical intervals is shown in table 23.01. The common major chord (doh-me-soh-doh) has its notes in the ratio 4:5:6:8. See fig. 23.04, where the ratios for the minor chord are also shown.

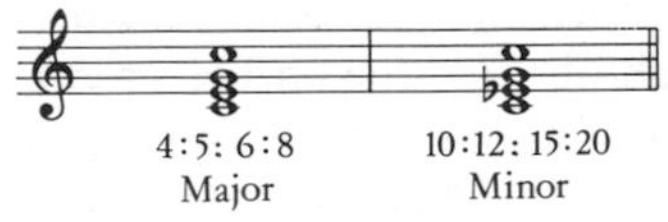

Fig. 23.04. The common chord.

† See p. 437 the group of musical intervals, and the homomorphic mapping onto intervals less than the octave.

Table 23.01

name	frequency ratio	tempered interval	no. of semitones
Unison	1:1 = 1·0000	1·0000	0
Octave	2:1 = 2·0000	2·0000	12
Perfect fifth	3:2 = 1·5000	1·4983	7
Perfect fourth	4:3 = 1·3333	1·3348	5
Major third	5:4 = 1·2500	1·2599	4
Minor sixth	8:5 = 1·6000	1·5874	8
Minor third	6:5 = 1·2000	1·1892	3
Major sixth	5:3 = 1·6667	1·6818	9
Tone (major second)	9:8 = 1·1250	1·1225	2
	or 10:9 = 1·1111		
Minor seventh	16:9 = 1·7778	1·7818	10
	or 9:5 = 1·8000		
Semitone (minor second)	16:15 = 1·0667	1·05946	1
Major seventh	15:8 = 1·8750	1·8877	11
Chromatic semitone	25:24 = 1·0417	1·05946	1
Augmented fourth	45:32 = 1·4063	1·4142	6
Diminished fifth	64:45 = 1·4222	1·4142	6
Diminished seventh	128:75 = 1·7067	1·6818	9

The Pentatonic scale

If we start on a particular note (say middle C), and pile up intervals of the perfect fifth, we arrive in turn at the notes C, G, D, A, E, B, F♯ (G♭),† D♭, A♭, E♭, B♭, F, C. The first five of these‡ constitute the pentatonic scale (see fig. 23.05), the basis of many folk-songs, particularly those of Scotland, and, perhaps more ancient still, those influenced by the Chinese culture. In fig. 23.05 each of the notes in the original series of fifths has been replaced (as shown by the arrows) by its equivalent within the octave above middle C. It will be seen that the note E in the original sequence has

† Here we cross, as Sir Walford Davies so aptly put it, the 'Musical Date-Line'. Strictly, we should continue ... C♯, G♯, D♯, A♯, E♯, B♯.

‡ Or equally well, those between F♯ and B♭ inclusive, i.e. the *black* notes of the keyboard.

frequency $(3/2)^4$† of the frequency of middle C, and so when brought down two octaves, the interval is $3^4/2^6 = 81/64$, which is not quite the simple ratio 5/4 (= 80/64) of the major third referred to above. It is not, indeed, a very simple ratio at all, and the musical effect is not pleasant. The total interval of twelve fifths is $(3/2)^{12} \simeq 129{\cdot}7$. This is just over seven

Fig. 23.05. The Pentatonic scale.

octaves ($= 2^7 = 128$). The discrepancy between the two accounts for many of the difficulties in obtaining an organised system of pitch. Indeed, it is most unfortunate that the system generated by successive multiplication by 3/2 is not closed under successive multiplication by 2. This is something that can never be altered. The interval 129·7/128 is known as the Comma of Pythagoras, after its famous discoverer. Musically, it represents the interval between B♯ and C.

The Pythagorean scale

When six fifths are piled on top of each other, the frequencies so obtained, i.e. $(3/2)^r$, $r = 0, 1, 2, 3, 4, 5, 6$, give the ratios for the Pythagorean scale, represented diagrammatically in fig. 23.06. The first step in the scale, from

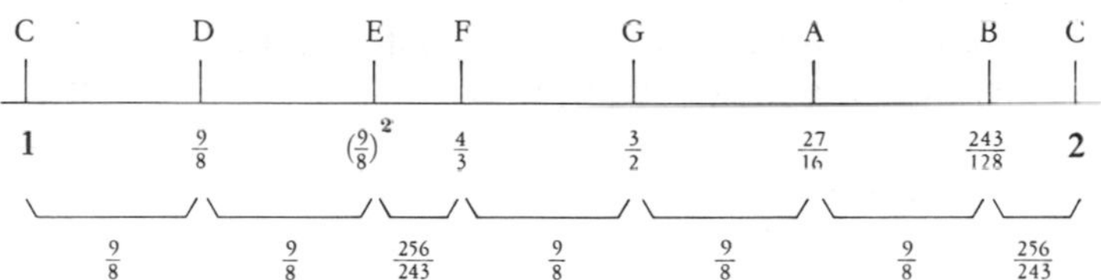

Fig. 23.06. The Pythagorean scale.

† If the four strings of a violin, viola or cello are tuned to *perfect* fifths, the interval between the highest and lowest string will be $(3/2)^3 = 27/8 = 3{\cdot}375$, whereas the interval of the major sixth should have a frequency ratio $5/3 \sim 10/3 = 3{\cdot}333$. This discrepancy is of little moment in the case of the violin, since the A string which is tuned first is not an outside string. In the case of the cello and viola, however, A is the top string, and tuning in perfect fifths will render the C string (the bottom string) slightly flat. How much flat is to be seen from the ratio $3{\cdot}375/3{\cdot}333 = 1{\cdot}0126$; since a semitone represents a rise of about 6%, we see that the C string would be about one fifth of a semitone flat. Sometimes, therefore a cello may prefer to tune its bottom string to the C of the keyboard, or possibly of the bassoon. The difficulty is reduced by the equal tempered system, whereby the same interval becomes $2 \times 2^{9/12} = 2^{7/4} = 3{\cdot}362$.

C to D, is the result of compounding two fifths, C to G and G to D, i.e. $(3/2)^2 = 9/4 \sim 9/8$,† and this is the familiar whole-tone. Unfortunately, some of the intrevals in this scale are anything but 'simple'. It turns out that all the wider intervals are 9/8, but the narrower ones (E to F and B to C) are the rather unpleasantly narrow interval 256/243, while the interval from C to E, as well as from F to A and from G to B are still not the pure major third (5/4) of the harmonic series. Thus, if the note E of the Pythagorean scale were sounded *simultaneously* with C, the effect would be most unpleasant because of the clash between E (81/64) and the fifth harmonic of C ($5/1 \sim 5/4$)† which would insist on sounding as one of the harmonic ingredients of the note produced by instrument or voice. The scale, though suitable for *melodic* writing, is therefore not satisfactory for *harmonic* writing.

However, this was the scale used by the Greeks and followed by the early mediaeval composers as the basis of the 'Ecclesiastic Modes'. The same seven notes would always be used, but the 'home' or 'key' note – the centre of perspective, the final resting note – might be any one of them. Of the seven Greek modes, only the Ionian (beginning and ending on C in fig. 23.06), and the Aeolian (key-note A in fig. 23.06) survive in most tone-centred music nowadays. These are what we usually know as the ordinary major and minor scales respectively.

Just intonation

In the fulness of time, as harmonic writing developed, it was found that a more satisfactory intonation was achieved if the intervals C to E, F to A and G to B were narrowed down and made into exact major thirds (5/4). In this way, each of the major triads, F–A–C, G–B–D, C–E–G becomes a perfect major triad with the ratio 4:5:6. The resulting scale is known as 'just intonation', and the intervals are shown in fig. 23.07. Some of the

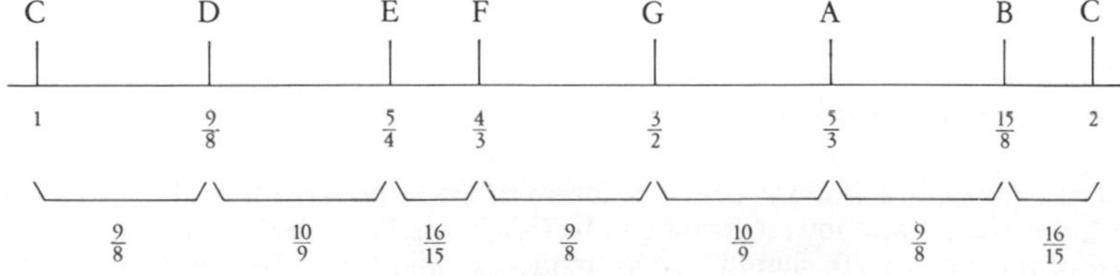

Fig. 23.07. The major scale; just intonation.

whole tones (9/8) have become narrowed to 10/9 (minor tones). Both the very narrow intervals, which were 256/243, have consequently become wider, with ratio 16/15. These are 'natural semitones', but they are not *exactly* half either type of tone, since $(16/15)^2$ is neither 9/8 nor 10/9.

† See p. 437, the homomorphism of the intervals onto the intervals less than an octave.

Why did the Ionian mode gain the distinction above the six other modes of fathering our major scale? There are probably two reasons. First, the note below the keynote (B) is at a *small* interval (256/243) below it, and this gives it the character of a 'leading-note', i.e. it has the tendency to *rise to the tonic* thus making possible a full close with more of a feeling of finality.

The second reason is connected with the position of the major triads within the natural scale (see fig. 23.08). It will be seen that the two major

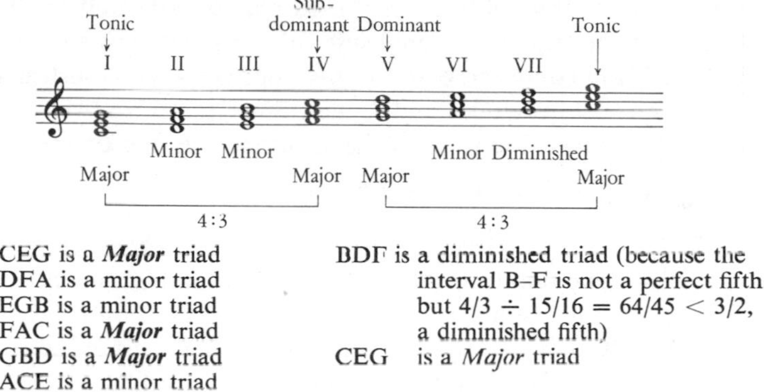

Fig. 23.08. The position of the major triads.

triads other than the tonic triad, C–E–G, that is F–A–C and G–B–D, are *symmetrically situated* with respect to the tonic note – their roots F and G are respectively a fifth below (subdominant) and a fifth above (dominant) the keynote. The relationship confers upon this system a tonal symmetry which partly accounts for the ear's ready acceptance of the scale.

Disadvantages of just intonation

When playing within one particular key, just intonation provides the ideal relation between the notes of the scale, in the sense that the intervals are natural intervals based on the natural harmonic series, the ratios being as simple as possible (see p. 431).

When, however, it is desired to *modulate*, i.e. to move the centre of tonality, to pass into a new key, serious difficulties arise. It will be seen from fig. 23.07 that the interval D–A in the scale of C is $\frac{5}{3}/\frac{9}{8} = 40/27 < 3/2$ so that the interval is a slightly *flat* fifth. If therefore one wanted to play in the key of D with just intonation, the note A would be out of tune. The more 'remote' the key, the greater the difficulties. The 'black notes' may be added to the scale; for example, F$^{\#}$ would be a major third above D, and so its frequency, should be $9/8 \times 5/4 = 45/32$ times the frequency of C. If, however, one were playing in the key of A, this would not be satisfactory, since the interval F♯ – A would be 32/27 ($< 6/5$) – a *flat* minor third. All these difficulties stem primarily from the fact that $(3/2)^{12} \neq 2^7$, or $3^{12} \neq 2^{19}$.

Equal temperament

The most acceptable solution of the problem of making a keyboard instrument on which it would be possible to play equally well in all keys, is that known as the 'equal tempered system'. This system made *all semitones equal*, and conferred equal status upon all keys – illustrated by Bach in his forty-eight preludes and fugues. Since there are twelve semitones to the octave, the ratio of the semitone was chosen to be:

$$\textbf{1 tempered semitone} = 2^{1/12} \simeq 1{\cdot}05946\ldots$$

Note that $25/24 < 2^{1/12} < 16/15$, i.e. the tempered semitone lies between the natural chromatic semitone (25/24) and the natural diatonic semitone (16/15) already met in the major scale. The interval C–G on the keyboard is now $2^{7/12} \simeq 1{\cdot}4983$, a *slightly flat* fifth. Indeed, *none* of the intervals (except for the octave) between the tempered notes is a true natural interval – obviously, because $2^{1/12}$ is irrational, and so are its integral powers. The scales in *all* keys are *slightly* out of tune, but mercifully to an extent which the average ear is unable to detect.

In one sense, the achievement of equal temperament was the placing of eleven geometrical means between 1 and 2. But we are more concerned with the achievement as no less than that of causing the musical intervals to form a *group* under combination. The simple natural intervals are *not even closed* under combination, for as we have seen, four fifths give a ratio $(3/2)^4 = 81/16 \sim 81/64$, and this is slightly sharp on the natural major third 5/4. Again, when other natural intervals are combined, more often than not we move out of the set of simple natural intervals.

Equal temperament surmounts this difficulty (while sacrificing the perfection of individual intervals in any particular key). The major third, for example, is 4 semitones, i.e. $(2^{1/12})^4 = 2^{1/3}$. The (perfect) fourth is 5 semitones, i.e. $(2^{1/12})^5 = 2^{5/12}$. When these two intervals are combined, we get a frequency ratio $2^{4/12} \times 2^{5/12} = 2^{9/12} = (2^{1/12})^9$, i.e. 9 semitones, which is a major (tempered) sixth. Again, $(2^{1/12})^7 \times (2^{1/12})^{10} = (2^{1/12})^{17} = 2 \times (2^{1/12})^5$, i.e. an interval of 7 semitones (fifth) together with an interval of 10 semitones (minor seventh), gives an interval of 17 semitones, i.e. an octave plus a 'fourth'.

It is more convenient to take the logarithm of the interval to base 2, in which case, we have in the latter example, the addition $7/12 + 10/12 = 17/12 = 1 + 5/12$, the interval of 7 semitones being thought of now as

7/12 of an octave. Indeed, the simplest plan is to use $2^{1/12}$ as the base of the logs, and the above will now read: $7 + 10 = 17 = 12 + 5$, the log of the interval to base $2^{1/12}$ being a measure of the interval in (tempered) semitones.

Thus the combination of intervals in the equal tempered system is isomorphic to the multiplication of integral powers of $2^{1/12}$, and this is isomorphic to the addition of the set of integers. The latter infinite group has neutral element 0, and is generated by $+1$ (or by its inverse -1). In the same way, the group of tempered intervals have neutral element the Unison, and are generated by the interval of the semitone. One might even say that ♮ is the neutral element of this group, ♯ (or its inverse ♭) is a generator of the whole group.

Tempered intervals: the group C_{12}

At this point, we think again of the equivalence classes into which the notes have been divided by the equivalence relation of the octave. In like manner, it is possible to partition the musical intervals into equivalence classes. For example, the interval from A ($f = 440$) to C♯ ($f = 550$) is a major third (5/4). The interval from A (440) to C♯ (1100) is really a 'tenth', but it may be regarded as a major third – and certainly sounds like one – if the extra octave is dispensed with. Furthermore, the intervals 5:1, 10:1, 20:1, etc. would all rank as major thirds, and we may say:

$$5/4 \sim 5/2 \sim 5/1 \sim 10/1 \sim \ldots.$$

The ratio 5/4 may be taken as the *representative* of the equivalence class of intervals, the equivalence relation being recognised by the fact that the intervals may all be *called* major thirds, and they all sound alike.

In the equal tempered system, the major third has frequency ratio $(2^{1/12})^4$ and we say:

$$(2^{1/12})^4 \sim (2^{1/12})^{16} \sim (2^{1/12})^{20} \sim \ldots.$$

Taking logarithmic intervals, and using the base $2^{1/12}$, the interval is reckoned by the number of semitones, and it is evident that we are now using addition in the set $\{0, 1, 2, 3, 4, 5, 6, 7, 8, 9, 10, 11\}$ of integers modulo 12. Again, 7 semitones + 8 semitones = 15 semitones = 3 semitones (mod. 12); while $7 + 5 = 0$ (mod. 12), so that the fourth and the fifth are inverse, as we saw on p. 431.

The result of this is that the whole set of tempered intervals $(2^{1/12})^k$ ($k \in \mathbf{Z}^+$) has been mapped onto the set of 12 intervals $(2^{1/12})^k$, $k = 0, 1, 2, \ldots, 11$. This is a *homomorphism* of the infinite group of tempered intervals onto the group of the equivalence classes of intervals represented by intervals less than the octave. The kernel of this homomorphism is the set of intervals of an exact number of octaves, since all these intervals map into the unison.

Now the arithmetic modulo 12 may conveniently be represented by a clockface (see fig. 23.09; compare also fig. 11.08). In this diagram, C is

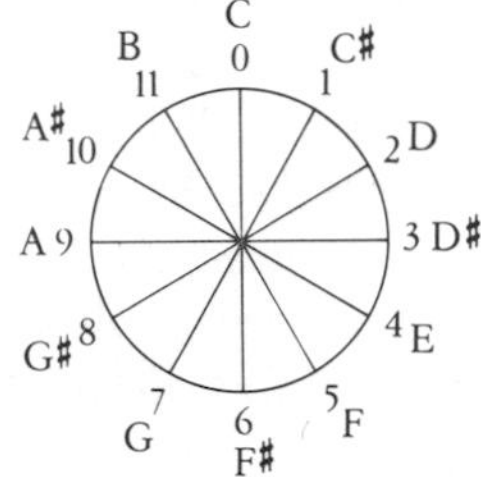

Fig. 23.09. Musical intervals on the clockface.

Note: A 'tour' of the regular dodecagon represents an ascent of the chromatic scale.
A tour of the regular hexagon represents the ascent of the whole-tone scale.
A tour of the square represents the chord of the diminished seventh.
A tour of the equilateral triangle represents the chord of the augmented fifth.

taken as the datum, and the figure illustrates that, for example, F is 5 semitones above C, this interval (the fourth) corresponding to a (clockwise) rotation of $5 \times 2\pi/12$. In ignoring any superfluous octaves ($7 + 8 = 3$ (mod. 12)), we are in effect ignoring superfluous rotations through 2π in the diagram. *All* the notes whose names are F♯ appear on the clockface at six o'clock. Geometrically, we are thinking of the isomorphism of the group of rotations of the regular dodecagon with addition in the finite arithmetic modulo 12, which as we have seen is isomorphic with that of the tempered musical intervals under combination.

One subgroup of the cyclic group $\{0, 1, 2, \ldots, 11\}$ (addition modulo 12) is $\{0, 3, 6, 9\}$, and this corresponds to the square in fig. 23.09 representing the chord of the diminished seventh, D♯, F♯, A, C (fig. 23.10). The *cosets* of this subgroup are shown up as the other squares $\{1, 4, 7, 10\}$ and $\{2, 5, 8, 11\}$ in fig. 23.09, representing the other chords of the diminished seventh shown in fig. 23.10.

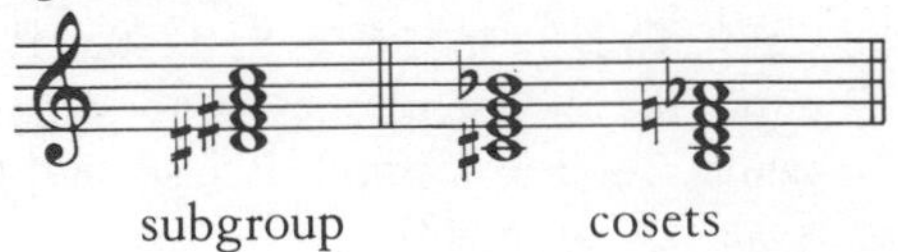

Fig. 23.10. Diminished seventh chords.

Had we been using natural instead of tempered intervals, the major third (5/4) would have been represented not by a rotation through $2\pi/3$ on the clockface, but through an angle $\frac{1}{6}\pi \log_{2^{1/12}}(5/4) = 2\pi \log_2 (5/4)$ $= 360° \{\log_{10} 5/\log_{10} 2 - 2\} = 0{\cdot}322$ rev. $=$ *116°*. Again, the perfect fifth

(3/2) would be represented by a rotation through $2\pi \log_2 3/2 = 360° (\log_{10} 3/\log_{10} 2 - 1) = 0{\cdot}585$ rev. $=$ *210·8°*, whereas the tempered fifth corresponds to 210°. While 12 *tempered* fifths make *exactly* 7 octaves, 12 *perfect* fifths are shown by a rotation through $210{\cdot}8 \times 12 = 2529{\cdot}6°$ $= 7$ revolutions $+ 9{\cdot}6°$ (the comma of Pythagoras).

An illustration

We may think of a three-dimensional picture of the ascent of the (chromatic) scale in terms of the ascent of a spiral staircase. Suppose that the ascent of each step causes a rotation through 30°, and we think of each step as a semitone in the tempered scale. Then after ascending twelve steps, we are again facing in the same direction – say we start on the step pointing due north, and this is the note C, then when we next stand on a step pointing due north we shall have risen an octave up the scale. The panorama that we see looking out of the window from any of these 'C steps' will 'look the same', just as all the notes C 'sound the same'. Moreover, the clockface of fig. 23.09 is simply the orthogonal projection of our spiral staircase on to any horizontal plane. This orthogonal projection 'maps' all the steps marked C on to the same point of the clockface, and this is a homomorphic map of the spiral staircase onto the regular dodecagon, representing the homomorphic map of the set of all notes of the chromatic scale onto the twelve equivalence classes of the notes represented by the names C, C♯, D, D♯, E, F, F♯, G, G♯, A, A♯, B.

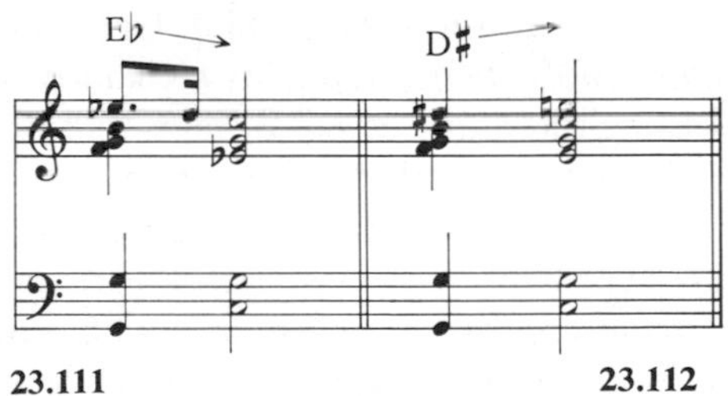

Fig. 23.11. Illustrating the different mental impression produced by E flat and D sharp.

The question remains: what is the difference between, for example, D sharp and E flat? This is really a matter of musical *grammar* rather than of the underlying frequency relationships. On the equal tempered scale, for example on a keyboard instrument, there is no difference at all, one note serving for both E♭ and D♯. Nevertheless, if you play the sequences of chords in fig. 23.11 on a keyboard instrument, you will get the feeling of the different *mental impression* of the notes D♯ and E♭ in the two different contexts, even though on the pianoforte, the first chord of each pair is identical.

The first example is in C minor, and the E♭ in the top part belongs to that scale, being the minor thirteenth above the dominant. In the second, in C major, the D♯ is a chromatically altered note, not belonging to the scale of C. It forms with the dominant the interval of the augmented fifth.

If one were playing a stringed instrument, or singing, then one would attempt to get just intonation within any particular key: in fig. 23.111 the E♭ should be $6/5 = 1{\cdot}2$ of the frequency of the C; while in fig. 23.112 the D♯ ought to be a diatonic semitone below the E, i.e. $15/16 \times 5/4$ of the frequency of C, $= 75/64 = 1{\cdot}172$, so the D♯ would be *flatter* than the E♭. But we have already seen that even a 'white note', like A, would have to change its frequency when playing just intonation in various keys. In

Fig. 23.12.

most music prior to the introduction of equal temperament, just intonation would have been used. When a keyboard instrument was included, modulations would only be possible into closely related keys. Since equal temperament became established, just intonation may continue to be used only in unaccompanied vocal music, and also in string music (assuming open strings are avoided – see footnote, p. 433). One reason why the piano has never been popular with string players as a chamber music instrument is that in the more romantic music of say Brahms and Franck, the players have to adjust their intonation to that of the piano when modulations are made into remote keys. This too is probably the reason why the piano, apart from its percussive effect, has never been a permanent member of the orchestra. In a piano concerto the clash is less objectionable, since the essence of this form is in the opposition and conflict between solo and orchestra.

Groups and musical form

Groups may also occur in the way a piece of music is constructed. One of the simplest cases is that of a 'round', in which the same melody is sung by several voices (or 'parts') coming in one after the other, usually at a fixed time interval of so many bars. For example, 'Great Tom is Cast' is a three-part round (see fig. 23.12).

With the second and third voices entering at the points marked, we get (omitting the words for brevity):

Fig. 23.13.

The round may continue *ad infinitum!* Now if a fourth voice were to enter at the point marked 4, it would simply be 'doubling' the first voice. Thus, if x denotes the operation 'come in two bars later' (here the round is at a 'distance' of two bars), and x^2 means 'come in four bars later', it is evident that x^3, which means 'come in six bars later' must be *the identity*, since anyone carrying out that instruction is merely joining in with the original (1st) voice. Thus we have a cyclic group C_3, generated by the operation x.

We might have abbreviated the above by the following notation:

Let '*A*' stand for the music of the first two bars: *A*

Let '*B*' stand for the music of the next two bars: *B*

Let '*C*' stand for the music of the last two bars: *C*

Fig. 23.14.

Then we could set out as follows:

Table 23.02

(finally)

1	*A*	*B*	*C*	*A*	*B*	*C*	*A*	*B*	*C* . . . etc.	*C*	—	—
2	—	*A*	*B*	*C*	*A*	*B*	*C*	*A*	*B* . . . etc.	*B*	*C*	—
3	—	—	*A*	*B*	*C*	*A*	*B*	*C*	*A* . . . etc.	*A*	*B*	*C*

Of course, after the 3rd voice has entered, every subsequent stretch of two bars would sound the same,† since *A*, *B* and *C* are all being sung simultaneously, and the notes that would be heard would be:

Fig. 23.15.

The only difference would be that a different permutation of the voices would be singing *A*, *B* and *C* in each case. Of course, anyone who only knew the first two bars, or who was idle, could get away with singing *AAAAAA*... 'Great Tom is cast' over and over again, but this would be contrary to the spirit of the round! He would be dodging from one voice to the other, following respectively voices 1, 2, 3, 1, 2, ..., and so on.

Figure 23.15 was an example of a three-part round. Many rounds are in four parts, and you may examine ones that you know. Here is one example, which we abbreviate:

Fig. 23.16.

The complete air is *AABBCCDD*. When sung as a round we get:

1	*A A B B C C D D A A B B C C D D A A B B C C D D* ...
2	– – *A A B B C C D D A A B B C C D D A A B B C C* ...
3	– – – – *A A B B C C D D A A B B C C D D A A B B* ...
4	– – – – – – *A A B B C C D D A A B B C C D D A A* ...

1	*A A B B C C D D* – – – – – –
2	*D D A A B B C C D D* – – – –
3	*C C D D A A B B C C D D* – –
4	*B B C C D D A A B B C C D D*

Here, if x denotes the operation 'come in two bars later', we have $x^4 = e$ (identity), and we get the cyclic group C_4.

† Provided the three parts were indistinguishable in quality and loudness.

Perhaps the most remarkable example of a round (too lengthy to quote) is 'Summer is i-cumen in', reputed to date from about 1200. In this, the air in the upper voices is sung as a four-part round over a 'ground' or 'ostinato' by the lower two voices forming a two-part round. The scheme works out thus:

$$
\left.
\begin{array}{ll}
\text{Upper voices} & \left\{
\begin{array}{cccccccc}
A & B & C & D & A & B & C & D\ldots \\
 & A & B & C & D & A & B & C\ldots \\
 & & A & B & C & D & A & B\ldots \\
 & & & A & B & C & D & A\ldots
\end{array}\right. \\
\text{Lower voices} & \left\{
\begin{array}{cccccccc}
P & Q & P & Q & P & Q & P & Q\ldots \\
 & P & Q & P & Q & P & Q & P\ldots
\end{array}\right.
\end{array}
\right\} \text{etc.}
$$

and the group is again C_4.

The Canon and Fugue: imitation and inversion

The round is one of the simplest 'contrapuntal' devices, i.e. a way of combining simultaneous tunes. Of course, tunes have to be composed specially so that the parts *A*, *B*, *C*, etc. will sound well when played simultaneously: most tunes would not sound satisfactory if one attempted to treat them as rounds. Another simple contrapuntal form is the 'Canon'. This differs from the round in the following respects. First, only two parts are in canon, the theme being 'shadowed' by itself some bars later (the 'consequent'). There may be other parts besides the two parts in canon, but these would be free, though it is possible to have a double canon with two themes. Second, whereas in a round the imitation is at the same pitch, in the case of a canon, the consequent may enter at any interval chosen by the composer. If the canon is not strict, a more intricate web of sound may be created by subtle changes in treatment of the theme and the counterpoints. In a sense the round, in spite of its simplicity, may be regarded as a development of the canon.

Frequently, a composer makes use of 'imitation', whereby one voice or instrument imitates, or repeats, an entry of another voice a few beats or a few bars later, but no attempt is made to continue the canon; the resemblance is only in the first few notes of the entries.

A common device used by composers to heighten excitement at a climax of a Fugue is that known as a 'stretto'. Here various voices close on each others' heels, so to speak, perhaps only one or two beats later, in tight sequence. Good examples are to be found throughout the 'Amen' chorus of Handel's *Messiah*. In the B minor Mass of Bach, for instance, a passage of very close imitation appears in the chorus *Cum Sancto Spirito* (see Novello Vocal Score, p. 88).

One of the most highly developed musical forms is the Fugue. Not only do the voices enter in sequence, but it is usual for the key of the entry to be changed. Thus, if y denotes the operation 'transpose one fifth higher', and y^{-1} denotes 'transpose a fifth lower, or a fourth higher', and x denotes 'come in four bars after the first entry', we might get successive voices following the operations:

First voice (altos)	e	(come in at bar 1 in the home key).
Second voice (soprano)	xy	(come in at bar 5 a fifth higher).†
Third voice (bass)	x^2y^{-1}	(come in at bar 9, and go back into the original key).
Fourth voice (tenor)	x^4y	(come in at bar 17, a fifth higher),

and so on.

The structure may contain other operations, such as:
z: 'turn the theme upside-down'.

Thus, the following motif

if inverted would become

Fig. 23.17.

Geometrically, this appears as a reflection in the bottom line of the stave.

It is not suggested that, in constructing either fugues, or lower musical structures, composers would wittingly conform to a mathematical pattern, or even that they would be consciously aware of the use of devices which could remotely be described as mathematical. The point of this description is merely to point out that the ingredients of group structure are to be found in musical art in much the same way that symmetry can be discerned in the visual arts.

Sequences

When a group of notes (a fragment) is repeated at a higher (or lower) pitch several times in succession, we have what is known as a *sequence*.

† The 'answer' may have to be modified in various ways in order to control the direction of tonality. If no modification other than transposition is needed, the answer is described as 'real'. Much has been written by theorists on the subject of the answer in a fugal exposition. See for example, C. H. Kitson, *Studies in Fugal Construction* (O.U.P.).

Figure 23.18 is a simple example in which the fragment containing four notes is repeated in a descending sequence. The operator x means 'repeat the fragment one step lower', and it is clear that x is of period 7, since after the seventh appearance of the fragment, the equivalent notes (CFGA) would reappear an octave lower, for we shall again consider notes of the same name to be equivalence classes.

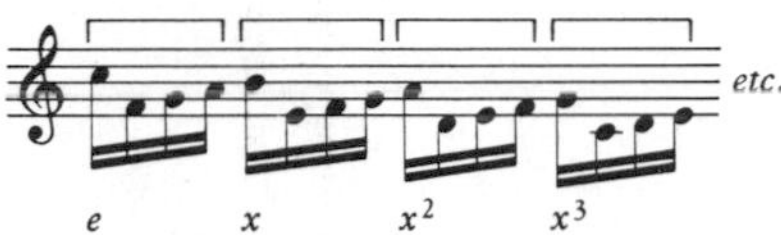

Fig. 23.18.

It is rare for a sequence to 'go the full circle', but again, we see music using devices which are basically mathematical, and familiar in our study of groups.

Fig. 23.19.

Figure 23.19 shows an example where the sequential figure consists of eight notes, but the second group of four semiquavers is an inversion of the first four at an interval one degree higher in the scale. Here the operators are

r: repeat the four-note motif one degree higher in the scale;

a: invert the fragment.

Evidently, a is of period 2, while r is of period 7, though the pattern of eight notes CDECFEDF would not occur until we had risen *two* octaves in this case.

Even if r is not considered to be periodic, and we were to disregard the equivalence of notes with the same name, we should still have a repeating pattern, with r now taking on the character of a translation. You should compare these types of musical sequence with the strip patterns described in Ch. 26.

Sequences may be *exact* (with identical intervals), or *diatonic* (remaining in the same key and using only the notes of that key). The basic motif may be as short as two or three notes; or a whole phrase may be treated sequentially, as in figs. 23.202, 23.203, 23.204. The harmony of a passage may be reproduced sequentially, every one of the parts moving in sequence, as in fig. 23.204. Do not imagine that the use of sequence is

'The Irish Washerwoman'

Fig. 23.201.

Chopin: Mazurka in C.

Fig. 23.202.

Chopin: Fantasie in F minor.

Exact sequence at an interval of one tone.

Fig. 23.203.

Hymn tune

Harmonic sequence at an interval of a third. *Note:* for a diatonic sequence, the D sharp would have to be a D natur

Fig. 23.204.

Mozart: Clarinet Quintette.

Fig. 23.205.

confined to the classical composers – it is common in both Romantic and in Modern music, and three examples of its use by Chopin are included. Note that in fig. 23.202 there is a subtle change in the fourth bar quoted.

In fig. 23.201 we have a two-bar phrase repeated in diatonic sequence. The first four of the examples shown are *falling* sequences. The last is a *rising* sequence chiefly remarkable for the fact that the sequence is not broken till the four-note fragment has been treated five times, after which it is inverted. Sequence may be thought of as a sort of musical 'rhyme'.

Another 'mathematical' device is that of *augmentation*, or playing the same portion of melody in notes of greater length (usually twice the length), the effect of this being to give prominence to the theme, sometimes to dignify it with majesty – the music has, so to speak, been 'stretched'. The inverse process is known as '*diminution*' (see fig. 23.21).

Fig. 23.21.

Here is a two-part passage where a short theme (in the top part) is combined with its own augmentation:

Fig. 23.22.

There are instances among the works of the great composers of ways in which they would amuse themselves by allowing their musical genius to be led away by their mathematical instincts – Mozart wrote 'permutation minuets'; Haydn wrote duets for violins consisting of a single line of music in which one of the two players read from a mirror; and in other cases one finds themes which are played upside-down, backwards and otherwise disguised. Some of the fourteenth and fifteenth-century Flemish composers, such as Ockeghem,† indulged in this sort of thing. The well-known chant by Dr W. Crotch, shown in fig. 22.23, is interesting in that in each of the four parts, the third quarter is the reverse of the first and the fourth quarter is the reverse of the second. You may also care to examine the

† Quote from P. Scholes, *Oxford Companion to Music*: 'It has been usual to say that the music of the Netherlands in the fourteenth and fifteenth centuries degenerated into a scholastic art, with a cultivation of crabbed canonic writing showing remarkable ingenuity but lacking musical expression, and the name of Ockeghem (d. 1513) has been linked with this charge. Recent research has proved that whilst there were many ingenious "puzzle canons" written by him and by his contemporaries, there was likewise much beautiful music.'

Fig. 23.23.

sentence shown in fig. 23.24 and discern its cryptic design.

Before leaving this subject, it should again be emphasised that, though, as we have shown, the elements of group structure are present in music, and can be discerned in a variety of contexts, composers did not generally write with any express mathematical intentions: the sort of instances we have noticed 'happened' spontaneously and almost inevitably as a result of the composer's architectural instinct for balance and unity. However, some

Fig. 23.24.

twentieth-century composers have deliberately set out to work to mathematical 'formulae', more often than not with less regard for the pleasantness of the resulting sound! Notable experiments of this kind were made in the twelve-tone system, pioneered by Schoenberg,† and further developed by the Serialists, chief of whom is Webern.

The student who is a musician and wishes to study in greater detail the incidence of mathematical pattern in music at its highest level cannot do better than to study the works of J. S. Bach, the supreme master. The fugues of the *Well-tempered Clavier*, a collection of forty-eight Preludes and Fugues for keyboard by J. S. Bach are fine examples of this art form, providing a breadth of musical expression unequalled anywhere. Every device of counterpoint is to be found in the course of this masterpiece. The reader who has access to a copy will find Fugue 2 of Book 2 (in C minor) particularly rewarding to study. The *Musical Offering*, the *Goldberg Variations* and the *Art of Fugue* represent the culmination and perfection of the marriage of the art of the musician and the mathematician, but again, the music always came first in Bach's mind, and the mathematical perfection arose inevitably and naturally, and the works mentioned offer a most rewarding field for study.

† It is said that the group $C_{12} \times D_2$ is to be discerned in Schoenberg's works.

EXERCISES 23

1 The works of Bach referred to in the last paragraph may not be easily accessible, so some examples are given below of portions of well-known works which may be consulted, and these are referred to as follows:

B Bach: *Mass in B minor* (Novello)
M Handel: *Messiah* (Novello)
C Haydn: *Creation* (Novello)
V R. Vaughan Williams: *Mass in G minor* (Curwen)

} references to pages of vocal scores in each case. Reference is also made to the numbers in the lists of contents.

Examples of sequence abound; indeed, it is scarcely possible to look at the works of any composer without finding sequential treatment of some sort. Good examples of sequences which repeat many times may be found especially in Handel (e.g. Chandos Anthem No. 6, *O Praise the Lord with one Consent* (Novello), a good instance being in the bass part on pp. 47–8); while in B, p. 46 the violin has an ascending sequence in which a three-note motif appears ten times before the sequence is broken.

Fugues B Nos. 1, 3, 4 (p. 37), 6, 12, 13, 19, 20 (p. 170), 24
M Nos. 1, 25, 28, 53 (p. 204)
C Nos. 11 (p. 34), 27 (p. 99), 33 (the latter two are double fugues).

In addition, there are many examples of fugal type entries in cases where a fugal development is not subsequently worked out, as in:

V pp. 9, 14;
B pp. 88, 155, 167, 173, 180, 196;
M pp. 15, 62, 99, 101, 107, 138, 148, 202, 206;
C pp. 8, 9, 37, 49, 105.

Canonical writing may be found in:

V pp. 10, 17, 33, 36;
B pp. 18, 29, 56, 63, 106, 123, 193 (violin imitating voice in this last example);
M pp. 20 (cello imitates voice), 65 (voice closely shadowed by 1st violin), 148;
C p. 45;

though the canons are by no means strict in these examples.

Examples of *augmentation* may be found in C p. 35, bar 23, and in V, where the altos in the last three bars (p. 46) augment a figure from the opening *Kyrie*. Another striking example from an earlier period may be found in the bass part of the madrigal *As Vesta was from Latmos hill descending* by Thomas Weelkes, to the words 'Long live fair Oriana' on the final page (Stainer and Bell, No. 2870). Finally, a beautiful example of a device not mentioned in the text – the ground bass – is to be found in No. 16 (*Crucifixus*) of the *B minor Mass*.

2 Examine the sequence of keys in (a) the Bach forty-eight Preludes and Fugues; and (b) the twenty-four Preludes of Chopin. Find other works which use the whole range of keys. How are the key-schemes related to the group C_{12}?

3 What is the effect on the pitch of a record being played at 45 r.p.m. which should be played at $33\frac{1}{3}$ r.p.m.? (Note that the standard speeds of *tapes* are usually in a ratio $2^n : 1$.)

4 If a brass instrument has a total effective length of tube of 4 feet, what extra length of tube is needed in the valves to lower the pitch by (a) a tone, (b) a semitone?

5 The length of the F-crook of a French horn is 50 inches, and the D-crook is 78 inches long. Find the effective length of the open tube, assuming that the interval D to F is a minor third in just intonation. What should be the length of the extra tube for the valve which depresses the pitch by one whole tone?

6 A man invests £260 at compound interest, and finds that, after twelve years it is worth £520 (i.e. doubled its value). Find the rate per cent, and compute the amount at the end of each of the twelve years. Your answer will give the frequencies of vibration of C, C#, D, D#, E, F, F#, G, G#, A, A#, B and C′ in the equal tempered scale.

7 The following round might deserve to be more widely known:

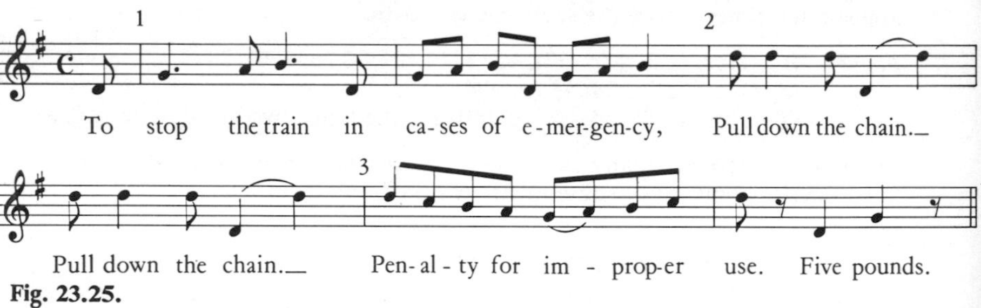

Fig. 23.25.

24 *Ringing the changes: groups and campanology*

The following gives the actual sound of the first five sequences of changes in a peal of eight bells:

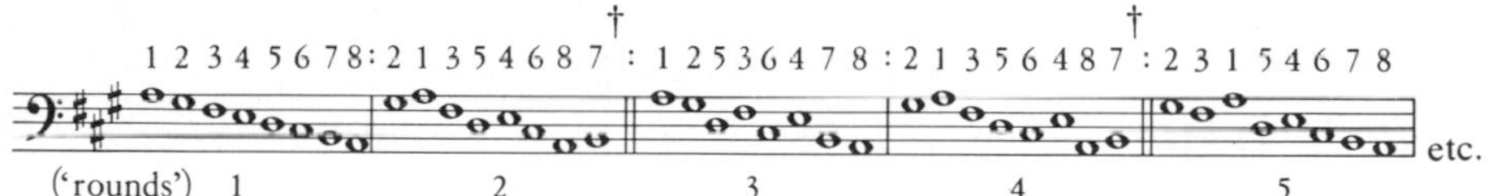

Fig. 24.01.

The first sequence is 'rounds' in which the bells sound in their natural descending order. The bells are numbered from No. 1 (the 'treble') to No. 8 (the 'tenor'). Obviously it would be tedious to write out the notes on the musical stave for a large sequence of some thousands of changes, so we dispense with the musical notation altogether, and simply write down the *numbers* of the bells in the order in which they sound, each permutation being written on a new line, or 'row' (in table 24.01 we continue as far as the 13th row). For example, in the fifth change (row 5), we have the bells sounding in the order 2 3 1 5 4 6 7 8, as indicated in the 5th 'bar' of the music. Here we say that bell No. 2 is 'in the lead', or in 'first place'; No. 3 is striking second, the treble is striking third, and so on. So as to avoid confusion between the number of the bell and the number of the position in the sequence, we shall prefer to use *letters*‡ for the positions or places, and these are shown across the top. Thus in row 5, bell No. 2 is associated with place *A*, bell No. 3 with place *B*, and so on.

Let us now consider the rules by which these various permutations are formed. The object of change-ringing is to produce the maximum possible *variety*, so one wants to go through all the possible permutations (in this case 8! of them) with no repetitions. But the permutations have to be constructed according to a system or 'method', and the above example is illustrative of the type of pattern which may be produced. The change from one permutation to the next must be as smooth as possible, and this means that *no bell may move more than one place up or down.* The

† After every two changes, a gap should occur in the ringing to give it a sort of punctuation. This gap lasts the time which one bell would take to strike.

‡ It might have been thought to be more logical to use letters for the sounds of the bells, namely the names of the musical notes, in this case A G♯ F♯ E D C♯ B A, and numbers for the places, but this is liable to lead to a different kind of confusion!

Table 24.01

row	A	B	C	D	E	F	G	H
	place							
1	1	2	3	4	5	6	7	8
2	2	1	3	5	4	6	8	7
3	1	2	5	3	6	4	7	8
4	2	1	3	5	6	4	8	7
5	2	3	1	5	4	6	7	8
6	3	2	5	1	4	7	6	8
7	2	3	1	5	7	4	8	6
8	3	2	5	1	7	8	4	6
9	2	3	5	7	1	8	6	4
10	3	2	7	5	8	1	4	6
11	3	7	2	5	1	8	6	4
12	7	3	5	2	8	1	4	6
13	3	7	5	8	2	4	1	6 etc.

following would therefore not be allowed: $\begin{matrix} 1 & 2 & 3 & 4 & 5 & 6 & 7 & 8 \\ 2 & 1 & 4 & 3 & 7 & 5 & 6 & 8 \end{matrix}$ because No. 7 has jumped two places from *G* to *E*. Nor would it be possible to have $\begin{matrix} 1 & 2 & 3 & 4 & 5 & 6 & 7 & 8 \\ 2 & 3 & 4 & 5 & 6 & 7 & 8 & 1 \end{matrix}$ since No. 1 would have to move 7 places. The consequence of this rule is that the only possible changes consist of *transpositions of pairs of adjacent bells.* While some pairs of bells are thus being interchanged, others will 'rest' in the same place, and these have been indicated in the above list. For instance in rows 8 and 9, bell No. 5 rests in position *C* (or rings third). There is another rule, that a bell shall not rest in the same position for more than two consecutive pulls. This is in the interest of variety, but there are occasions when this rule has to be sacrificed, for sometimes there is no way of avoiding a bell resting in the same place for three or even four consecutive pulls.

Campanologists use various notations for recording the sequences of changes. One is to write down the complete list of rows showing the order in which the bells sound, as we have just done. Another is to show which pairs of bells 'cross', so $\begin{matrix} 1 & 2 & 3 & 4 & 5 & 6 \\ 1 & 3 & 2 & 5 & 4 & 6 \end{matrix}$ would be shown: | × × |, to indicate that the outside bells are unchanged while the two inside pairs are interchanged.

An even briefer notation is simply to record the numbers of the bells which are *unchanged*, so that the above would be noted as 1, 6, to show that bells Nos. 1 and 6 rest in the same places. The sequence shown on p. 452 (the opening of a method known as *Glasgow Surprise Major*) would be recorded:

3 6
×
6 4
2 5
4 8
×
7 6
5 8
×
3 5
×
5 6
etc.,

the bells named being those which do not move to a new position; when all the bells are changed, a cross is recorded. Once one knows which bells rest in the same place, one knows everything about the change, since all the other pairs of adjacent bells change places. However, it is more usual in campanological circles to indicate not the bells which are unchanged, but the places or positions. We should then have:

our notation for places →	campanologists' notation →
C F	3 6
×	×
E F	5 6
A D	1 4
E H	5 8
×	×
E H	5 8
C F	3 6
×	×
A D	1 4
×	×
C H	3 8

$$\begin{pmatrix} A & B & C & D & E & F & G & H \\ 1 & 2 & 3 & 4 & 5 & 6 & 7 & 8 \end{pmatrix}.$$

This way of indicating the sequence records not the *numbers* of the bells, but the *positions* of those bells which are unchanged. Thus from row 7 to row 8, the bells in places *E* and *H* remain unchanged. From the mathematician's point of view this has advantages as a method of recording the changes, as we shall see. There might also be some mathematical advantages in recording transpositions of places rather than of bells. So again, from row 7 to row 8, we have the transpositions (2 3)(1 5)(8 4) of bells, or equivalently, the transposition (*A B*)(*C D*)(*F G*) of places.

A further requirement of a good method is that each bell should do the 'same amount of work', which means that we want to avoid one bell having a lot of interesting movement while another hovers between two positions for a long spell, or simply moves up and down with no interesting variety. From the point of view of the individual ringer, he is interested in the 'path' of his own bell as the sequence of changes proceeds. The path of two particular bells is traced in the table on p. 452 – Nos. 1 and 7. The 'kinks' or zig-zags in the straight lines are called 'dodges'. The movement of a bell forwards or backwards is called 'hunting', and when there is a steady movement up or down with no dodges, and no consecutive blows in the same position, i.e. when the path is a straight line, this is called 'plain hunting'; for example, bell No. 7 is plain hunting from row 8 as far as row 12. Note that No. 3 'comes into the lead' (position *A*) in row 6, and thereafter is occupied in a great deal of dodging.

Let us now consider how sequences of changes may be built up *so as to run through every possible permutation*, and suppose we start by taking a small number of bells. The name given to any particular method is partly derived from the number of bells in the peal, and these names are as follows:

4 Minimus (singles)
5 Doubles
6 Minor
7 Triples
8 Major
9 Caters
10 Royal
11 Cinques
12 Maximus.

The complete description of a set of changes requires the specification not only of the number of bells, as above, but also of an 'identifier' (e.g. St Clements, Kent, Trinity Sunday, etc.), and a 'Method Type' (e.g. Surprise, Bob, Delight, Court, etc.)

Three bells

We may quickly dismiss the possibilities with three bells. There being only six changes, we may easily show the complete sequence in staff notation:

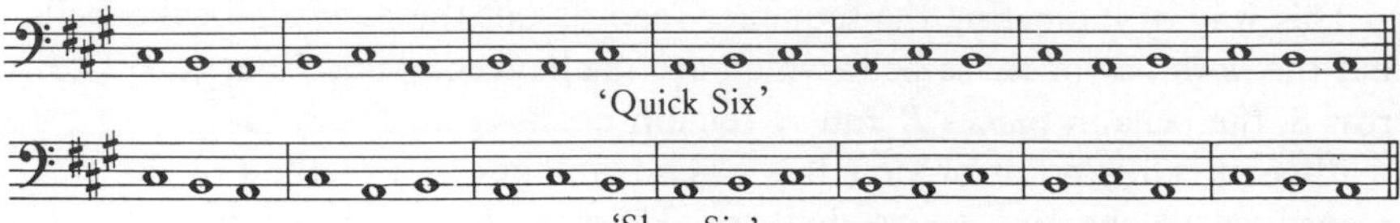

Fig. 24.02. Note that the changes occur in the reverse order in these two sets.

In table 24.02 we use ***a*** and ***b*** to denote the two possible interchanges, (*AB*) and (*BC*) respectively, and we see here all six permutations generated by these two transpositions. Here it is more convenient to use transpositions of *places*, rather than of *bells*, because in using the latter we should have required three generators, (1 2), (1 3) and (2 3).

Table 24.02

	row	A B C	change	perm.
	1	1 2 3		*1*
		×	*a*	
	2	2 1 3		*a*
		×	*b*	
	3	2 3 1		*ba*
'quick six'		×	*a*	
	4	3 2 1		*aba*
		×	*b*	
	5	3 1 2		*baba*
		×	*a*	
	6	1 3 2		*ababa*
		×	*b*	
	7	1 2 3		*1*

row	A B C	change	perm.	
1	1 2 3		*1*	
	×	*b*		
2	1 3 2		*b*	
	×	*a*		
3	3 1 2		*ab*	
	×	*b*		'slow six'
4	3 2 1		*bab*	
	×	*a*		
5	2 3 1		*abab*	
	×	*b*		
6	2 1 3		*babab*	
	×	*a*		
7	1 2 3		*1*	

Four bells

There is nothing more that can be said about three bells, so we pass to consider changes on four bells ('singles' or 'minimus'). In the sequence shown in table 24.03 we use a hard-and-fast system – alternately changing

Table 24.03

row	A B C D	cycles of bells	perm. of bells produced
1	1 2 3 4		1
	× ×	(12)(34) *h*	
2	2 1 4 3		*h*
	×	(14) *w*	
3	2 4 1 3		*wh* = *n*
	× ×	(13)(24) *r*	
4	4 2 3 1		*rwh* = *w*
	×	(23) *b*	
5	4 3 2 1		*brwh* = *y*
	× ×	(12)(34) *h*	
6	3 4 1 2		*hbrwh* = *r*
	×	(14) *w*	
7	3 1 4 2		*whbrwh* = *k*
	× ×	(13)(24) *r*	
8	1 3 2 4		*rwhbrwh* = *b*
	×	(23) *b*	
9	1 2 3 4		$(brwh)^2 = 1$

notation as on pp. 219, 318.

the two outside pairs of bells and then the inside pair, and we do not succeed in producing all twenty-four permutations but only eight of them, namely {1, *b*, *h*, *k*, *n*, *r*, *w*, *y*}. These form one of the D_4 subgroups of S_4. But it is easier to see *why* we do not obtain the complete set of twenty-four permutations if we reckon with permutations of *places* instead of permutations of bells, for the only transpositions of places that we use are (*AB*)(*CD*) and (*BC*). Calling these (as per notation on p. 318) ***h*** and ***b*** respectively, we obtain table 24.04. (Note that the permutations of places

Table 24.04

row	A B C D	cycles of places		perm. of *places* produced
1	1 2 3 4			***1***
		(*AB*)(*CD*)	***h***	
2	2 1 4 3			***h***
		(*BC*)	***b***	
3	2 4 1 3			***bh*** = ***k***
		(*AB*)(*CD*)	***h***	
4	4 2 3 1			***hbh*** = ***w***
		(*BC*)	***b***	
5	4 3 2 1			***bhbh*** = ***y***
		(*AB*)(*CD*)	***h***	
6	3 4 1 2			***hbhbh*** = ***r***
		(*BC*)	***b***	
7	3 1 4 2			***bhbhbh*** = ***n***
		(*AB*)(*CD*)	***h***	
8	1 3 2 4			***hbhbhbh*** = ***b***
		(*BC*)	***b***	
9	1 2 3 4			***1***

are *inverses* of permutations of bells, so that *k* and *n* (each of period 4) are interchanged between these two tables.) The scheme of eight changes is thus seen to be the group generated by ***b*** and ***h***, or Gp (***b***, ***h***). This cannot be the whole group S_4, for we know that S_4 requires three generators of period 2 (see p. 319), and also that the group generated by two elements of period 2 is dihedral (see p. 206). Indeed, the only possible changes which we may perform are (*AB*) (= ***a***), (*BC*) (= ***b***), (*CD*) (= ***g***) and (*AB*)(*CD*) (= ***h*** = ***ag*** = ***ga***), and since the group may be generated by using, for example, ***a***, ***b*** and ***g***, it should be possible to obtain a complete peal of twenty-four changes by using these transpositions alone (see p. 319).

For example, if we use the operations ***a***, ***g*** and ***b*** in sequence repeatedly, we get 1234

1234
) ***a***
2134
) ***g***
2143
) ***b***
2413
) ***a***
4213
) ***g***
4231
) ***b***
4321
) ***a***
3421
) ***g***
3412
) ***b***
3142
) ***a***
1342
) ***g***
1324
) ***b***
1234 and this brings us back to 'rounds' in twelve changes.

Q. 24.01. Try and produce a sequence of all twenty-four permutations, using (*AB*), (*BC*) and (*CD*) only, each permutation occurring once only, of course.

Even if successful, we still have the problem of lack of variety, due to bells remaining in the same position for three consecutive blows. For the sake of variety, it is more satisfactory to use ***h*** (= ***ag***) in conjunction with ***b*** (as we did in our preliminary attempt), but to *get out of the group* D_4 by the occasional use of ***g***, that is, occasionally to interchange the bells in third and fourth place.

Table 24.05 shows one method† of getting all twenty-four changes on four bells. At row 8, to have used ***b*** would have produced rounds again, as on p. 456. So we use ***g*** instead, and this enables us to get into a new set of eight permutations, {***m***, ***c***, ***f***, ***t***, ***v***, ***l***, ***j***, ***p***}. Not unexpectedly, this is a coset of the subgroup H {**1**, ***h***, ***k***, ***w***, ***y***, ***r***, ***n***, ***b***}, the reason emerging when we inspect the working in the final column. For example, row 13 (permutation ***v***) is obtained from row 12 (permutation ***t***) by applying the transposition ***b***, so that ***bt*** = ***v***. But ***t*** = ***wm*** (already obtained), so that ***v*** = ***bwm***. But row 5 (= 13 − 8) shows that ***bw*** = ***y***, hence ***v*** = ***ym***, and so is in the right coset H***m***. Similarly the final eight permutations belong to the coset H***i***, the introduction of the permutation ***g*** at row 16 taking us across into this coset. Note that ***g*** was introduced each time *the treble* (No. 1) *returned to the lead* (position *A*).

Note that in our method worked out in table 24.06, all the rules were obeyed – no bell jumped more than one place; no permutation was repeated, every bell struck once in each row; and finally, no bell struck more than two consecutive blows in the same position. The method falls

† There are exactly eleven methods on four bells with the treble plain hunting that give a true set of twenty-four changes. Plain Bob Minimus is one of these.

Table 24.05

row	cycle of places	place *A*	*B*	*C*	*D*	perm of places produced	
1		1	2	3	4	1	H
	h						
2		2	1	4	3	*h*	
	b						
3		2	4	1	3	*bh* = *k*	
	h						
4		4	2	3	1	*hk* = *w*	
	b						
5		4	3	2	1	*bw* = *y*	
	h						
6		3	4	1	2	*hy* = *r*	
	b						
7		3	1	4	2	*br* = *n*	
	h						
8		1	3	2	4	*hn* = *b*	
	g	~~1~~	~~2~~	~~3~~	~~4~~		
9		1	3	4	2	*gb* = *m* = 1*m*	H*m*
	h						
10		3	1	2	4	*hm* = *c* = *hm*	
	b						
11		3	2	1	4	*bc* = *f* = *bhm* = *km*	
	h						
12		2	3	4	1	*hf* = *t* = *hkm* = *wm*	
	b						
13		2	4	3	1	*bt* = *v* = *bwm* = *ym*	
	h						
14		4	2	1	3	*hv* = *l* = *hym* = *rm*	
	b						
15		4	1	2	3	*bl* = *j* = *brm* = *nm*	
	h						
16		1	4	3	2	*hj* = *p* = *hnm* = *bm*	
	g	~~1~~	~~3~~	~~4~~	~~2~~		
17		1	4	2	3	*gp* = *i* = 1*i*	H*i*
	h						
18		4	1	3	2	*hi* = *q* = *hi*	
	b						
19		4	3	1	2	*bq* = *s* = *bhi* = *ki*	
	h						
20		3	4	2	1	*hs* = *x* = *hki* = *wi*	
	b						
21		3	2	4	1	*bx* = *u* = *bwi* = *yi*	
	h						
22		2	3	1	4	*hu* = *d* = *hyi* = *ri*	
	b						
23		2	1	3	4	*bd* = *a* = *bri* = *ni*	
	h						
24		1	2	4	3	*ha* = *g* = *hni* = *bi*	
	g						
		1	2	3	4	1	

Q. 24.02. Construct a peal of four bells to give all the permutations by a different method, e.g. either using $\boldsymbol{a} = (AB)$ instead of $g = (CD)$, or by not starting with $\boldsymbol{h}$.

Q. 24.03. Reletter the above sequence of changes according to the permutations of bells instead of permutations of places, using the notation as in table 24.03, i.e. $h = (12)(34)$; $a = (12)$; $b = (23)$; $g = (34)$; $w = (14)$, etc., starting as in table 24.06. Show that the cosets are H $\{1, h, n, w, y, r, k, b\}$, iH and mH.

Table 24.06

row	*A*	*B*	*C*	*D*	change	perm. of bells	
1	1	2	3	4		1	
2	2	1	4	3	*h*	*h*	
3	2	4	1	3	*w*	*n*	
4	4	2	3	1	*r*	*w*	
5	4	3	2	1	*b*	*y*	etc.

short in one respect, however, that the bells do not all do the same work – No. 1, the treble, carried out a plain hunt between *A* and *D* and back again to the lead, whereas the other bells have more interesting paths: that of No. 3 is traced into the diagram. Notice that No. 4 carries out the identical path eight rows later, and part of this is shown for comparison, starting at row 9. In fact, the effect of passing from the subgroup H into the coset H$\boldsymbol{m}$ is that bell No. 3 becomes replaced by bell No. 4, No. 2 by No. 3 and No. 4 by No. 2, while No. 1 does exactly the same work throughout each coset (or 'lead' as each block of eight would be called by campanologists). The reason for this is that the permutation $\boldsymbol{m}$ replaces row 1 (1 2 3 4) by row 9 (1 3 4 2) to start each plain course.

Q. 24.04. Compare the paths of the working bells 2, 3 and 4 in the final coset of eight permutations.

Doubles (five bells)

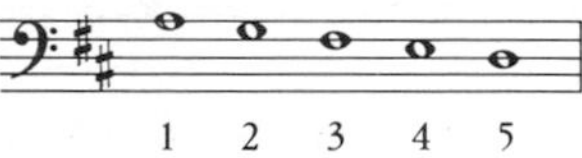

Fig. 24.03.

With five bells, we may perform the following available transpositions of places: $\boldsymbol{e}$ $(AB)(CD)$; $\boldsymbol{c}$ $(AB)(DE)$; $\boldsymbol{a}$ $(BC)(DE)$; s_1 (AB); s_2 (BC); s_3 (CD); s_4 (DE). The notation is selected because the letter $\boldsymbol{e}$, for example, suggests that the bell in fifth place (E) is unchanged. The last four types of changes are called 'singles' for obvious reasons. Evidently $\boldsymbol{e} = \boldsymbol{s}_1\boldsymbol{s}_3 = \boldsymbol{s}_3\boldsymbol{s}_1$; $\boldsymbol{c} = \boldsymbol{s}_1\boldsymbol{s}_4 = \boldsymbol{s}_4\boldsymbol{s}_1$; $\boldsymbol{a} = \boldsymbol{s}_2\boldsymbol{s}_4 = \boldsymbol{s}_4\boldsymbol{s}_2$. This time, however, we shall work out the changes in terms of permutations of *bells*. You may repeat the work using permutations of places, and compare with the previous section on peals of four bells.

The notation used for transpositions of bells will be as follows: a small letter will be used for each permutation as it turns up, and the appropriate small letter for the change from one row to the next, thus:

$$\left.\begin{matrix} 3 & 1 & 2 & 4 & 5 \\ 1 & 3 & 2 & 5 & 4 \end{matrix}\right] c \quad \left|\begin{matrix} d \\ f \end{matrix}\right.$$

The cycle c, containing transpositions (13)(45) converts the permutation $d \begin{pmatrix} 1 & 2 & 3 & 4 & 5 \\ 3 & 1 & 2 & 4 & 5 \end{pmatrix}$ into the permutation $f \begin{pmatrix} 1 & 2 & 3 & 4 & 5 \\ 1 & 3 & 2 & 5 & 4 \end{pmatrix}$, so that $f = cd$. As we proceed, the later permutations will be indicated as products rather than by using new letters of the alphabet.

We shall try first the effect of applying the permutations of places, $\boldsymbol{c}$, $\boldsymbol{a}$, $\boldsymbol{c}$, $\boldsymbol{a}$, $\boldsymbol{c}$, ... alternately, see table 24.07. It is obvious that this method will

Table 24.07

row	place change	places A B C D E	bell change	perm. of bells resulting
1		1 2 [3] 4 5		1
	$\boldsymbol{c}$		a	
2		[2] 1 [3] 5 4		$a = a1$
	$\boldsymbol{a}$		c	
3		[2] 3 [1] 4 5		$b = ca$
	$\boldsymbol{c}$		f	
4		[3] 2 [1] 5 4		$c = fb$
	$\boldsymbol{a}$		a	
5		[3] 1 [2] 4 5		$d = ac$
	$\boldsymbol{c}$		c	
6		[1] 3 [2] 5 4		$f = cd$
	$\boldsymbol{a}$		f	
7		[1] 2 3 4 5		$1 = (fca)^2$

notation:
a (12)(45)
b (123)
c (13)(45)
d (132)
f (23)(45).

never shift Nos 4 and 5 from places D and E, and in fact we find rounds are produced after only six changes (forming the subgroup D_3). We *must* get a dihedral group, because, as we showed on p. 206, the group generated by two elements of period 2 is bound to be dihedral.

So at row 6, in order to avoid rounds, we replace the change $\boldsymbol{a}$ by the change $\boldsymbol{e}$ which leaves the bell in place E unchanged. On this, and on subsequent occasions when a new type of change is introduced, we shall, before proceeding, write down the row which we sought to avoid, so that the last two rows will now appear:

$$\begin{array}{c|c|c|c} 6 & & \left[\begin{array}{ccccc} 1 & 3 & 2 & 5 & 4 \\ \not1 & \not2 & \not3 & \not4 & \not5 \\ 3 & 1 & 5 & 2 & 4 \end{array}\right] & f \\ & e & gf^{-1} = gf & \\ 7 & & & g \end{array}$$

We now show, in table 24.08, the sequence of leads produced by this system, known as 'Erin'. The effect of the use of ***e*** (bell in fifth place resting) is to throw us into another coset, these cosets being as marked on the right of the table. Moreover, we can be sure that *we do not make any repetitions*, since these cosets are all disjoint (see Ch. 19, p. 347).

The system still does not succeed in producing all 120 possible permutations and it is rather discomforting here to find that these thirty rows are *not* a subgroup. We may easily demonstrate that the set fails for closure, for gag^{-1} will be found to be the permutation $\begin{pmatrix} 1 & 2 & 3 & 4 & 5 \\ 3 & 4 & 1 & 2 & 5 \end{pmatrix}$, and this is missing from the set. Indeed, we can hardly expect to get a subgroup, for though we are using only three generators of period 2, (***a***, ***c*** and ***e***), we are not making *all possible words* with these three generators: only a subset of Gp (***a***, ***c***, ***e***) is present.

In changing from row 7 (permutation g) to row 8 (permutation ga), the transposition applied to the bells has been denoted p. So $pg = ga$, or $p = gag^{-1}$. Here we see another instance of the conjugate of an element (see Ch. 20). Again, q is the permutation which changes row 8 (ga) into row 9 (gb). Therefore $qga = gb$, so that $q = gba^{-1}g^{-1} = g(ba)g^{-1} = gcg^{-1}$.

Q. 24.05. Obtain r, s and t in a similar way.

We find that the first five changes are produced by applications of the permutations a, c, f, a, c. The sixth change took us into the coset, and the successive changes then were gag^{-1}, gcg^{-1}, gfg^{-1}, gag^{-1}, gcg^{-1}, i.e. the transforms of a, c, f, a, c by g.

Q. 24.06. Find in the same way the changes of bells which take us through the other cosets. Investigate the coursing of the several individual bells in their various paths, and check that, in this method all the bells do the same work. (Such a method is called a 'principle'.)

In our next attempt (see table 24.09) to produce all 120 changes, we start in the same way with place changes ***c***, ***a***, ***c***, ***a***, ***c***, ***e***, but then resume the next lead by applying ***a*** first instead of ***c***, and subsequently we commence each set of six with ***a*** and ***c*** alternately. The result of this is to

Table 24.08

row	place change	A	B	C	D	E	bell change	perm. of bells resulting	
1		1	2	3	4	5		1	
	c						*a*		
2		2	1	3	5	4		*a*	
	a						*c*		
3		2	3	1	4	5		*b*	subgroup H
	c						*f*		
4		3	2	1	5	4		*c*	
	a						*a*		
5		3	1	2	4	5		*d*	
	c						*c*		
6		1	3	2	5	4		*f*	
	e	~~1~~	~~2~~	~~3~~	~~4~~	~~5~~	*gf*		
7		3	1	5	1	4		*g*	
	c						*p*		
8		1	3	5	4	2		*ga*	
	a						*q*		
9		1	5	3	2	4		*gb*	coset *g*H
	c						*r*		
10		5	1	3	4	2		*gc*	
	a						*s*		
11		5	3	1	2	4		*gd*	
	c						*t*		
12		3	5	1	4	2		*gf*	
	e	~~3~~	~~1~~	~~5~~	~~2~~	~~4~~			
13		5	3	4	1	2		*h*	
	c								
14		3	5	4	2	1		*ha*	
	a								
15		3	4	5	1	2		*hb*	coset *h*H
	c								
16		4	3	5	2	1		*hc*	
	a								
17		4	5	3	1	2		*hd*	
	c								
18		5	4	3	2	1		*hf*	
	e	~~5~~	~~3~~	~~4~~	~~1~~	~~2~~			
19		4	5	2	3	1		h^{-1}	
	c								
20		5	4	2	1	3		$h^{-1}a$	
	a								
21		5	2	4	3	1		$h^{-1}b$	coset h^{-1}H
	c								
22		2	5	4	1	3		$h^{-1}c$	
	a								
23		2	4	5	3	1		$h^{-1}d$	
	c								
24		4	2	5	1	3		$h^{-1}f$	
	e	~~4~~	~~5~~	~~2~~	~~3~~	~~1~~			
25		2	4	1	5	3		g^{-1}	
	c								
26		4	2	1	3	5		$g^{-1}a$	
	a								
27		4	1	2	5	3		$g^{-1}b$	coset g^{-1}H
	c								
28		1	4	2	3	5		$g^{-1}c$	
	a								
29		1	2	4	5	3		$g^{-1}d$	
	c								
30		2	1	4	3	5		$g^{-1}f$	
	e								
		1	2	3	4	5		1	

	1	*a*	*b*	*c*	*d*	*f*
1	1	*a*	*b*	*c*	*d*	*f*
a	*a*	1	*f*	*d*	*c*	*b*
b	*b*	*c*	*d*	*f*	1	*a*
c	*c*	*b*	*a*	1	*f*	*d*
d	*d*	*f*	1	*a*	*b*	*c*
f	*f*	*d*	*c*	*b*	*a*	1

(subgroup D_3)

leads to rounds after sixty changes. However, g^{-1} does not turn up till row 55, so there is never any question of a subgroup till we have completed these sixty changes. (In this table, the column showing the permutations of bells is set down before the column showing the change of bells.)

A little reflection will show you that we have the group A_5 by these sixty changes, because *every one of the permutations is even*. (Why?) In order to obtain the odd permutations, it will be necessary to move into the coset by applying *an odd permutation*, and for this, the only available type of change is one containing a *single* transposition. These 'singles' were indicated on p. 459 as changes of places by s_1, s_2, s_3 and s_4. It would not matter which of these was selected to take us into the remaining 60 odd permutations, so that rows 60 and 61 might be

60		2	1	4	3	5			2	1	4	3	5
	s_1	×					or	s_2		×			
61		1	2	4	3	5			2	4	1	3	5

60		2	1	4	3	5			2	1	4	3	5
or	s_3			×			or	s_4				×	
61		2	1	3	4	5			2	1	4	5	3

Rows 61 to 120 would then be obtained by applying the identical sequence of place transpositions as were used to construct rows 1 to 60, namely **c**, **a**, **c**, **a**, **c**, **e**, etc. The complete method, known as 'Stedman Doubles' is now shown in table 24.10, but the operations are omitted. Note that row 61 is obtained from row 1 by the transposition (34) of bells. Calling this transposition *s*, we obtain in turn the permutations *s*, *sa*, *sb*, *sc*, *sd*, *sf*, *sg*, ..., i.e. the left coset by *s* of the subgroup A_5. In fact, note that row *n* and row $n \pm 60$ may be obtained from each other by the transposition (3 4).

Observe that throughout this sequence of changes, there is only one occasion when a bell rests in the same place even for three consecutive blows.† (Can you find it?)

The path of No. 1 has been traced in the first 60 rows, but has been omitted in rows 61–120 since it is identical. No. 2 has been traced through rows 61–120. Note that bell No. 2 copies exactly what No. 1 was doing only 48 rows later. For example, from rows 29 to 38, No. 1 carries out plain hunting from *E* to *A* and back to *E* again, resting in the lead

† This is avoided in the usually recognised order, where the single s_4 (*DE*) is made at row 57, leading, as you may check, to 1 2 5 4 3 at row 61.

Table 24.09

No.		Row		
1		1 2 3 4 5	1	
	c			*a*
2		2 1 3 5 4	*a*	
	a			*c*
3		2 3 1 4 5	*b*	
	c			*f*
4		3 2 1 5 4	*c*	
	a			*a*
5		3 1 2 4 5	*d*	
	c			*c*
6		1 3 2 5 4	*f*	
	e	~~1 2 3 4 5~~		gf^{-1}
7		3 1 5 2 4	*g*	
	a			gfg^{-1}
8		3 5 1 4 2	*gf*	
	c			gcg^{-1}
9		5 3 1 2 4	*gd*	
	a			gag^{-1}
10		5 1 3 4 2	*gc*	
	c			gfg^{-1}
11		1 5 3 2 4	*gb*	
	a			gcg^{-1}
12		1 3 5 4 2	*ga*	
	e	~~3 1 5 2 4~~		*c*
13		3 1 4 5 2	*i*	
	c			iai^{-1}
14		1 3 4 2 5	*ia*	
	a			ici^{-1}
15		1 4 3 5 2	*ib*	
	c			ifi^{-1}
16		4 1 3 2 5	*ic*	
	a			iai^{-1}
17		4 3 1 5 2	*id*	
	c			ici^{-1}
18		3 4 1 2 5	*if*	
	e	~~3 1 4 5 2~~		
19		4 3 2 1 5	*j*	
	a			jfj^{-1}
20		4 2 3 5 1	*jf*	
	c			jcj^{-1}
21		2 4 3 1 5	*jd*	
	a			jaj^{-1}
22		2 3 4 5 1	*jc*	
	c			jfj^{-1}
23		3 2 4 1 5	*jb*	
	a			jcj^{-1}
24		3 4 2 5 1	*ja*	
	e	~~4 3 2 1 5~~		
25		4 3 5 2 1	*k*	
	c			kak^{-1}
26		3 4 5 1 2	*ka*	
	a			kck^{-1}
27		3 5 4 2 1	*kb*	
	c			kfk^{-1}
28		5 3 4 1 2	*kc*	
	a			kak^{-1}
29		5 4 3 2 1	*kd*	
	c			kck^{-1}
30		4 5 3 1 2	*kf*	
	e	~~4 3 5 2 1~~		
		(5 4 1 3 2)	1	

No.		Row	
31		5 4 1 3 2	1
	a		
32		5 1 4 2 3	
	c		
33		1 5 4 3 2	
	a		
34		1 4 5 2 3	
	c		
35		4 1 5 3 2	
	a		
36		4 5 1 2 3	
	e	~~5 4 1 3 2~~	
37		5 4 2 1 3	k^{-1}
	c		
38		4 5 2 3 1	
	a		
39		4 2 5 1 3	
	c		
40		2 4 5 3 1	
	a		
41		2 5 4 1 3	
	c		
42		5 2 4 3 1	
	e	~~5 4 2 1 3~~	
43		2 5 3 4 1	j^{-1}
	a		
44		2 3 5 1 4	
	c		
45		3 2 5 4 1	
	a		
46		3 5 2 1 4	
	c		
47		5 3 2 4 1	
	a		
48		5 2 3 1 4	
	e	~~2 5 3 4 1~~	
49		2 5 1 3 4	i^{-1}
	c		
50		5 2 1 4 3	
	a		
51		5 1 2 3 4	
	c		
52		1 5 2 4 3	
	a		
53		1 2 5 3 4	
	c		
54		2 1 5 4 3	
	e	~~2 5 1 3 4~~	
55		1 2 4 5 3	g^{-1}
	a		
56		1 4 2 3 5	
	c		
57		4 1 2 5 3	
	a		
58		4 2 1 3 5	
	c		
59		2 4 1 5 3	
	a		
60		2 1 4 3 5	
	e	~~1 2 4 5 3~~	
Rounds		1 2 3 4 5	

Table 24.10 (Stedman Doubles)

1	1 2 3 4 5	31	5 4 1 3 2	61	1 2 4 3 5	91	5 3 1 4 2
2	2 1 3 5 4	32	5 1 4 2 3	62	2 1 4 5 3	92	5 1 3 2 4
3	2 3 1 4 5	33	1 5 4 3 2	63	2 4 1 3 5	93	1 5 3 4 2
4	3 2 1 5 4	34	1 4 5 2 3	64	4 2 1 5 3	94	1 3 5 2 4
5	3 1 2 4 5	35	4 1 5 3 2	65	4 1 2 3 5	95	3 1 5 4 2
6	1 3 2 5 4	36	4 5 1 2 3	66	1 4 2 5 3	96	3 5 1 2 4
7	3 1 5 2 4	37	5 4 2 1 3	67	4 1 5 2 3	97	5 3 2 1 4
8	3 5 1 4 2	38	4 5 2 3 1	68	4 5 1 3 2	98	3 5 2 4 1
9	5 3 1 2 4	39	4 2 5 1 3	69	5 4 1 2 3	99	3 2 5 1 4
10	5 1 3 4 2	40	2 4 5 3 1	70	5 1 4 3 2	100	2 3 5 4 1
11	1 5 3 2 4	41	2 5 4 1 3	71	1 5 4 2 3	101	2 5 3 1 4
12	1 3 5 4 2	42	5 2 4 3 1	72	1 4 5 3 2	102	5 2 3 4 1
13	3 1 4 5 2	43	2 5 3 4 1	73	4 1 3 5 2	103	2 5 4 3 1
14	1 3 4 2 5	44	2 3 5 1 4	74	1 4 3 2 5	104	2 4 5 1 3
15	1 4 3 5 2	45	3 2 5 4 1	75	1 3 4 5 2	105	4 2 5 3 1
16	4 1 3 2 5	46	3 5 2 1 4	76	3 1 4 2 5	106	4 5 2 1 3
17	4 3 1 5 2	47	5 3 2 4 1	77	3 4 1 5 2	107	5 4 2 3 1
18	3 4 1 2 5	48	5 2 3 1 4	78	4 3 1 2 5	108	5 2 4 1 3
19	4 3 2 1 5	49	2 5 1 3 4	79	3 4 2 1 5	109	2 5 1 4 3
20	4 2 3 5 1	50	5 2 1 4 3	80	3 2 4 5 1	110	5 2 1 3 4
21	2 4 3 1 5	51	5 1 2 3 4	81	2 3 4 1 5	111	5 1 2 4 3
22	2 3 4 5 1	52	1 5 2 4 3	82	2 4 3 5 1	112	1 5 2 3 4
23	3 2 4 1 5	53	1 2 5 3 4	83	4 2 3 1 5	113	1 2 5 4 3
24	3 4 2 5 1	54	2 1 5 4 3	84	4 3 2 5 1	114	2 1 5 3 4
25	4 3 5 2 1	55	1 2 4 5 3	85	3 4 5 2 1	115	1 2 3 5 4
26	3 4 5 1 2	56	1 4 2 3 5	86	4 3 5 1 2	116	1 3 2 4 5
27	3 5 4 2 1	57	4 1 2 5 3	87	4 5 3 2 1	117	3 1 2 5 4
28	5 3 4 1 2	58	4 2 1 3 5	88	5 4 3 1 2	118	3 2 1 4 5
29	5 4 3 2 1	59	2 4 1 5 3	89	5 3 4 2 1	119	2 3 1 5 4
30	4 5 3 1 2	60	2 1 4 3 5	90	3 5 4 1 2	120	2 1 3 4 5
e	5 4 1 3 2	s_1	1 2 4 3 5	e	5 3 1 4 2	s_1	1 2 3 4 5

for consecutive blows at rows 33 and 34. No. 2 performs the same path from rows 77 to 86. Thus bells Nos. 1 and 2 do *exactly the same work*.

Q. 24.07. Do the other three bells do exactly the same work?

Note also that in each lead of six changes in Stedman doubles, we have the three bells in positions *A*, *B* and *C* performing a 'quick six', while the bells in positions *D* and *E* remain in those positions to dodge until the next lead. For example, in rows 103–108, bells 1 and 3 dodge in positions *D* and *E* while Nos. 2, 5 and 4 follow the quick six sequence.

The above observations serve to show that this method is very satisfactory from the point of view of all the requirements mentioned at the outset. (In actual practice, Stedman doubles usually starts with ***c***, ***a***, ***e***, ***c***, ***a***, ***c***, ***a***, ***c***, ***e***, . . ., so that in effect one enters the sequence shown on p. 465 at a different point.)

Q. 24.08. With the notation for place changes already adopted (p. 459), construct a peal starting:

Table 24.11

Row		
1		1 2 3 4 5
	e	
2		2 1 4 3 5
	a	
3		2 4 1 5 3
	e	
4		4 2 5 1 3
	a	
5		4 5 2 3 1
	e	
6		5 4 3 2 1
	a	
7		5 3 4 1 2
	e	
8		3 5 1 4 2
	a	
9		3 1 5 2 4
	e	
10		1 3 2 5 4
	s_3	×
11		1 3 5 2 4

Continue the pattern till rounds are reached, the 'single' being introduced at rows 10, 20, 30, . . . Note that the paths of the bells will consist mostly of plain hunting, and also that the bells come 'into the lead' (place A) in the sequence 2, 4, 5, 3 apart from the intervention of the treble, which comes into the lead at rows 10, 20, 30, . . . Mention any drawbacks of this method.

Contrive a continuation after row 40 so as to complete the 120 changes. (Use s_2 instead of s_3, for example. This device is known as a 'bob'.)

Q. 24.09. Write a plain course for Grandsire Doubles, which uses ***a*** and ***c*** alternately and ***e*** at each treble lead, beginning thus:

	1 2 3 4 5
c	
	2 1 3 5 4
a	
	2 3 1 4 5, etc.

Devise a continuation beyond the plain course.

Six bells ('Minor')

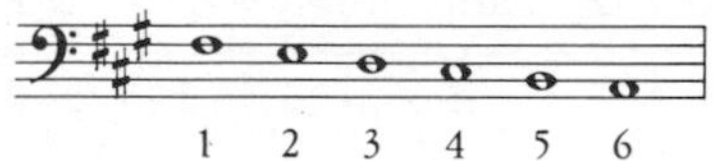

Fig. 24.04.

In table 24.12 we give a sequence of changes known as 'Plain Bob Minor' which lead back to rounds after 60 changes. The sequence of 60 is known as a 'plain course'. It is based on the alternation of the changes X X X, i.e. $(AB)(CD)(EF)$ which we call ***a*** and I X X I, i.e. $(BC)(DE)$, which we call ***b***. At row 12, in order to avoid rounds, ***b*** is replaced by I I X X, i.e. $(CD)(EF)$, which we call ***c***. This type of change is again used at rows 12, 24, 36, 48, 60, i.e. whenever the treble comes into the lead. The signs in the first column are to indicate the *parity* of the permutations. ***a*** will always change the parity (since it contains three transpositions), while ***b*** and ***c***, with two transpositions, will each leave the parity unaltered. This plain course of 60 changes, does in fact contain 30 even and 30 odd permutations.

Is this set of 60 permutations a subgroup of S_6, we may well† ask? The available changes are as follows, using the various notations:

a,	X X X	or $(AB)(CD)(EF)$	the basic moves.
b,	I X X I	or $(BC)(DE)$	
c,	I I X X	or $(CD)(EF)$	

The new devices which are possible are:

x,	X I X I	or $(AB)(DE)$
y,	I X I X	or $(BC)(EF)$
z,	X I I X	or $(AB)(EF)$

and in addition, we may have recourse to the 'singles': (AB), (BC), (CD), (DE) and (EF), by which only one pair of bells is interchanged.

Certainly the first twelve are a subgroup, for they constitute the group Gp (a, b), all possible 'words' in these two generators having been used,

† It may be wondered whether it makes sense to talk about a number of rows forming a subgroup, for there appears to be no law of composition – there is no sense in which one may combine say, row 23 with row 31 to obtain row 46. But if we use the ordinary law of composition for permutations, there *is* some point in knowing whether or not a sequence of changes form a subgroup. For if they *do* (as in the case of rows 1–12 here), then the application of certain devices at the end of each lead will produce cosets, and this means that, unless a previous lead is completely reproduced, we shall be sure of getting *no repetitions*, which is, after all, the object of the exercise. If however a sequence of changes do not form a subgroup, as in the case of rows 1–60, then this happy situation is not guaranteed. Thus if H is a subgroup, then $aH \neq bH \Rightarrow aH \cap bH = \phi$, but if H is not a subgroup, then aH and bH *may* contain *some* elements in common (see p. 347).

Table 24.12 (Plain Bob Minor)—Plain Course

row	place change	place *A B C D E F*	parity	row	*A B C D E F*	row	*A B C D E F*
1		1 2 3 4 5 6	+	13	1 3 5 2 6 4	25	1 5 6 3 4 2
	a	× × ×					
2		2 1 4 3 6 5	−	14	3 1 2 5 4 6	26	5 1 3 6 2 4
	b	| × × |					
3		2 4 1 6 3 5	−	15	3 2 1 4 5 6	27	5 3 1 2 6 4
	a	× × ×					
4		4 2 6 1 5 3	+	16	2 3 4 1 6 5	28	3 5 2 1 4 6
	b	| × × |					
5		4 6 2 5 1 3	+	17	2 4 3 6 1 5	29	3 2 5 4 1 6
	a	× × ×					
6		6 4 5 2 3 1	−	18	4 2 6 3 5 1	30	2 3 4 5 6 1
	b	| × × |					
7		6 5 4 3 2 1	−	19	4 6 2 5 3 1	31	2 4 3 6 5 1
	a	× × ×					
8		5 6 3 4 1 2	+	20	6 4 5 2 1 3	32	4 2 6 3 1 5
	b	| × × |					
9		5 3 6 1 4 2	+	21	6 5 4 1 2 3	33	4 6 2 1 3 5
	a	× × ×					
10		3 5 1 6 2 4	−	22	5 6 1 4 3 2	34	6 4 1 2 5 3
	b	| × × |					
11		3 1 5 2 6 4	−	23	5 1 6 3 4 2	35	6 1 4 5 2 3
	a	× × ×					
12		1 3 2 5 4 6	+	24	1 5 3 6 2 4	36	1 6 5 4 3 2
	c	~~1 2 3 4 5 6~~			~~1 3 5 2 6 4~~		~~1 5 6 3 4 2~~
13		(1 3 5 2 6 4)		25	(1 5 6 3 4 2)	37	(1 6 4 5 2 3)

row	*A B C D E F*	row	*A B C D E F*
37	1 6 4 5 2 3	49	1 4 2 6 3 5
38	6 1 5 4 3 2	50	4 1 6 2 5 3
39	6 5 1 3 4 2	51	4 6 1 5 2 3
40	5 6 3 1 2 4	52	6 4 5 1 3 2
41	5 3 6 2 1 4	53	6 5 4 3 1 2
42	3 5 2 6 4 1	54	5 6 3 4 2 1
43	3 2 5 4 6 1	55	5 3 6 2 4 1
44	2 3 4 5 1 6	56	3 5 2 6 1 4
45	2 4 3 1 5 6	57	3 2 5 1 6 4
46	4 2 1 3 6 5	58	2 3 1 5 4 6
47	4 1 2 6 3 5	59	2 1 3 4 5 6
48	1 4 6 2 5 3	60	1 2 4 3 6 5
	~~1 6 4 6 2 3~~		
49	1 4 2 6 3 5		1 2 3 4 5 6 (rounds)

and by p. 206, we know that this group must be dihedral. The period of the permutation $\begin{pmatrix} A & B & C & D & E & F \\ C & A & E & B & F & D \end{pmatrix}$ in row 3 is 6, so we know that the group must be D_6. The second, third, fourth and fifth leads, each of twelve changes, are cosets of this subgroup, so we can be sure, in view of the disjoint property of cosets (see p. 347) that we shall get no repetitions till we reach row 61.

Q. 24.10. Verify the D_6 structure of the first 12 rows.
Q. 24.11. Find ways of discovering that the first 60 rows are *not* a subgroup. (We shall verify this later (p. 471, footnote).)

We make the following observations on the structure of this plain course. The work of the treble is plain hunting from A to E and back again, coming into the lead every twelve changes. The sequence of twelve permutations between treble leads is called a 'treble lead', or simply a 'lead'. Here we have a plain course of sixty changes consisting of five leads. The same system applied to n bells would in fact result in a plain course containing $n - 1$ treble leads.

The work of the other bells is *almost* plain hunting. This must be so, for the inside bells (in places B, C D and E) are constantly being changed, two consecutive blows in the same position only occurring in positions A and F, except that occasionally position B is unchanged at the treble lead with the use of c. The result of this is that, at treble lead, there is a 'dodge' in position C. For example, in rows 35, 36, 37, bell No. 5 moves from position D to C and then back to D again where it dodges again, after which it continues plain hunting till reaching the lead at row 40. Bells 2, 3, 4, 5 and 6 all perform identical work during the plain course, but differ only in the points at which they begin that work. For example the path of No. 5 is shown during the second and third leads (rows 13–37) and this is seen to be identical with the path of No. 4 during the fourth and

Table 24.13

row	A	B	C	D	E	F	call this permutation	
1	1	2	3	4	5	6	1	
13	1	3	5	2	6	4	p	$(p^5 = 1)$
25	1	5	6	3	4	2	$q = p^2$	
37	1	6	4	5	2	3	$r = p^3$	
49	1	4	2	6	3	5	$s = p^4$	

fifth leads (rows 37–61). Again, the work of bell No. 3 in the first lead is identical with that of No. 5 during the second lead, and also with the work of No. 6 during the third lead, and so on. The reason for this is not difficult to see. The permutations occurring at the head of each lead are as shown in table 24.13. Thus p replaces 3 by 5, and so the work of No. 3

in the first lead is now exactly imitated by No. 5. Again, in passing from the fourth to the fifth lead, No. 4 is replaced by No. 2, and so is all its subsequent work.

Between treble leads, the bells come into the lead in the order 2, 4, 6, 5, 3, and this 'natural coursing order' is maintained throughout the plain course. But the bell which follows the treble into the lead is No. 2 in the the first lead (at row 2), 3 in the second lead (at row 14), then 5, then 6, and then 4, as we may confirm by consulting the permutations p, q, r, s above (2 is replaced by 3, then 3 by 5, then 5 by 6 and finally 6 by 4). These properties may be summarised by the following extract from the table:

Table 24.14

	first lead								*second lead*						
row	*A*	*B*	*C*	*D*	*E*	*F*	perm. of bells	row	*A*	*B*	*C*	*D*	*E*	*F*	perm. of bells
1	1	2	3	4	5	6	1	13	1	3	5	2	6	4	p
·								·							
·								·							
·								·							
5	4	6	2	5	1	3	t	17	2	4	3	6	1	5	pt

The notation used here is that p is the permutation $\begin{pmatrix} 1 & 2 & 3 & 4 & 5 & 6 \\ 1 & 3 & 5 & 2 & 6 & 4 \end{pmatrix}$ which takes us from any row in the first lead across to the corresponding row in the second lead. t is the permutation $\begin{pmatrix} 1 & 2 & 3 & 4 & 5 & 6 \\ 4 & 6 & 2 & 5 & 1 & 3 \end{pmatrix}$ by which row 1 is transformed into row 5. Then we see that the permutation which transforms row 13 into row 17 is ptp^{-1}, i.e. the conjugate of t by p

★**Q. 24.12.** Referring to the twenty-four changes on four bells (p. 458), we have in the three 'lead-heads' (rows 1, 9 and 17) the permutations of places $\mathbf{1}$, $\boldsymbol{m}$, $\boldsymbol{m}^2$, making the group C_3. Also rows 7, 15, 23 have the permutations $\boldsymbol{n}$, $\boldsymbol{nm}$, $\boldsymbol{nm}^2$ and so form a left coset of this subgroup. We have just seen that in Plain Bob Minor, the lead-heads (rows 1, 13, 25, 37, 49) of the plain course are the subgroup C_5 generated by p. Do rows r, $r + 12$, $r + 24$, $r + 36$, $r + 48$ ($r = 2, 3, 4, \ldots, 12$) make cosets of this subgroup? Investigate reasons.

★ Now a full peal of six bells should contain 720 changes, and the question now is, what new move shall we make at row 60 to avoid rounds, so as to take us into a coset† of the plain course; and shall we employ this same device, whatever it is, whenever we reach rows 120, 180, ...? The device $(BC)(EF)$ is known as a 'bob', and is commonly used as a linkage between the plain courses of 60 changes, which are the 'cosets'† referred to. Other devices containing two transpositions are known as reverse bobs, shunts, double bobs, etc. If we now use $(BC)(EF)$ as a link between successive sets of 60 changes, we find that rounds are reached after 180 changes (table 24.15). Here the bob $(BC)(EF)$ (denoted

Table 24.15

row		A	B	C	D	E	F
1		1	2	3	4	5	6
	c						
60		1	2	4	3	6	5
	y						
61		1	4	2	3	5	6
	c						
120		1	4	3	2	6	5
	y						
121		1	3	4	2	5	6
	c						
180		1	3	2	4	6	5
	y						
181		1	2	3	4	5	6
			(rounds)				

y) is used at rows 60, 120 and 180. Note that row 60 is the permutation $(CD)(EF)$ of places, for, as we saw, if we use ***c*** at row 60, we would get rounds (row 1) again. Also, subsequently, ***c*** takes us from row 61 to row 120, and from row 121 to row 180. Note also that, while No. 1 is in the lead at rows 1, 60, 61, 120, 121, 180, 181, the remaining five bells carry out a pattern very reminiscent of Stedman doubles, for the jump from row 1 to row 60, from row 61 to row 120, and from 121 to 180 in which the last two pairs are changed (***c***), alternates with the bob at rows 60, 120 and 180 by which the middle bell of those five is unchanged. And of course, we know from p. 460 that rounds are reached after only

† We are stretching the meaning of the word coset, since in fact the plain course is not a subgroup of order 60. For in row 17 we have the permutation $\begin{pmatrix}1&2&3&4&5&6\\2&4&3&6&1&5\end{pmatrix}$, i.e. the cycle u (1 2 4 6 5) of period 5. Now u^2 is the cycle (1 4 5 2 6), i.e. the permutation $\begin{pmatrix}1&2&3&4&5&6\\4&6&3&5&2&1\end{pmatrix}$ and this is *not* in the plain course, which therefore fails to be a group. This is a pity, because it means that when we employ a new type of change at row 60 we cannot be sure of not repeating some permutations that have already occurred (compare footnote p. 467).

six rows forming the group D_3, so we may not proceed beyond row 180 by this method. Compare with table 23.07.

Now so far we have had the permutations that are shown in table 24.16, with 1 in the lead, and these are 30 of the 120 permutations with

Table 24.16

row		*A B C D E F*	row		*A B C D E F*	row		*A B C D E F*
1		1 2 3 4 5 6	61		1 4 2 3 5 6	121		1 3 4 2 5 6
	b			***b***			***b***	
12		1 3 2 5 4 6	72		1 2 4 5 3 6	132		1 4 3 5 2 6
	c			***c***			***c***	
13		1 3 5 2 6 4	73		1 2 5 4 6 3	133		1 4 5 3 6 2
	b			***b***			***b***	
24		1 5 3 6 2 4	84		1 5 2 6 4 3	144		1 5 4 6 3 2
	c			***c***			***c***	
25		1 5 6 3 4 2	85		1 5 6 2 3 4	145		1 5 6 4 2 3
	b			***b***			***b***	
36		1 6 5 4 3 2	96		1 6 5 3 2 4	156		1 6 5 2 4 3
	c			***c***			***c***	
37		1 6 4 5 2 3	97		1 6 3 5 4 2	157		1 6 2 5 3 4
	b			***b***			***b***	
48		1 4 6 2 5 3	108		1 3 6 4 5 2	168		1 2 6 3 5 4
	c			***c***			***c***	
49		1 4 2 6 3 5	109		1 3 4 6 2 5	169		1 2 3 6 4 5
	b			***b***			***b***	
60		1 2 4 3 6 5	120		1 4 3 2 6 5	180		1 3 2 4 6 5
	y			***y***			***y***	
61		1 4 2 3 5 6	121		1 3 4 2 5 6			1 2 3 4 5 6

1 in position *A*. This is as it should be, for we have so far produced 180, or one quarter, of the 720 changes. Moreover, the 30 permutations are all *even* permutations, since they result from the changes ***b***, ***c*** and ***y***, each of which contains two transpositions.

If now at row 180 we were to use an alternative form of bob, ***x*** $(AB)(DE)$, this would give:

row		*A B C D E F*
180		1 3 2 4 6 5
	x	
181		3 1 2 6 4 5
	a	
182		1 3 6 2 5 4
	b	
183		1 6 3 5 2 4,

etc.

with two *new* permutations with 1 in the lead at rows 182 and 183. It is hopeful now that we may achieve all the 720 changes. We show the beginning and end of each set of 60 (i.e. each plain course) in table 24.17.

Table 24.17

row		*A*	*B*	*C*	*D*	*E*	*F*
1		1	2	3	4	5	6
	c						
60		1	2	4	3	6	5
	y						
61		1	4	2	3	5	6
	c						
120		1	4	3	2	6	5
	y						
121		1	3	4	2	5	6
	c						
180		1	3	2	4	6	5
	x						
181		3	1	2	6	4	5
	c						
240		3	1	6	2	5	4
	y						
241		3	6	1	2	4	5
	c						
300		3	6	2	1	5	4
	y						
301		3	2	6	1	4	5
	c						
360		3	2	1	6	5	4

(no repetitions so far!)

We achieve 360 different permutations without mishap, but they do not form a subgroup, since the only subgroup of S_6 of index 2 is A_6 which contains all the even permutations (cf. pp. 411 ff.).

★**Q. 24.13.** Verify that, of these 360 rows, there are 60 with 1 in the lead, and that of these, 42 are even and 18 odd permutations – find a quick way of making this count! Why is this fact sufficient to ensure that the 360 do not form a subgroup?

★**Q. 24.14.** In table 24.17, fill in the intermediate permutations for rows $60 + 12\,k$ and $61 + 12k$ ($k \in \mathbf{Z}$).

How do we proceed from row 360? (It is useless to use y, for we should get 3 1 2 6 4 5 – row 181 again.) It would appear to be a promising move to use x, as we did at row 180, but we are again to be disappointed!

Row		Permutation
360		3 2 1 6 5 4
	x	
361		2 3 1 5 6 4
	b	
372		2 1 3 6 5 4
	c	
373		2 1 6 3 4 5,

and this permutation has already occurred, though none of the previous twelve has.

★**Q. 24.15.** Find at which row 2 1 6 3 4 5 occurred previously.

★**Q. 24.16.** Experiment to show that the use of z at row 360 would not succeed in leading to all the permutations either.

The problem may be solved, as you might have expected, by using *singles*, and it is indeed found that the use of the three types of device – normal changes (like ***a***, ***b*** and ***c***), bobs (like x, y and z), and singles can be used to produce all the changes not only with Minor (six bells) but even with twelve bells (Maximus). Unless a single is used, it will only be possible to construct one half of the available permutations.

Note that the insertion of a single at row 360 will have the effect of interchanging odd and even permutations in the next 360 changes: $+ - - + + - -$ ..., etc., will be replaced by $- + + - - + +$ ..., etc.

The above discussion is given for the sake of its mathematical interest and is not orthodox campanology. For example, it would be considered preferable to use only one type of bob if possible. In fact the 720 changes as described above are unlikely to be rung in practice. It was at one time thought that it would be essential to use a single to achieve a complete peal of 720 changes, but this is not in fact the case, and a complete extent of 720 changes has in fact been rung by using bobs alone.

Other minor methods

The above is only one of a host of 'methods' for producing all the changes on six bells. We do not pursue this in detail any further, except to show an example of an entirely different method for Minor in table 24.18. You will see that the normal changes ***a*** and ***b*** are used in conjunction with two bobs. The effect is that No. 1 carries out plain hunting and returns to the lead at row 12. From row 3, bells Nos. 2 and 4 have been left 'in front' (places A and B) to dodge there until the treble returns to the lead. In the meantime, the other three bells (3, 5 and 6) hunt between third and sixth place (*C* and *F*). In the first continuation (the 'College' method), the system is continued as in the first lead. In the second alternative ('St Clements'), ***c*** is introduced at row 12 instead of ***a***.

Q. 24.17. Construct these continuations.

Q. 24.18. Experiment with the construction of a minor method using a quick six in the first three places and a slow six in the last three places, linking these leads with **a**; and other ways.

Table 24.18

```
row  |      A B C D E F
-----|------------------
 1   |      1 2 3 4 5 6
     |  a    X   X   X
 2   |      2 1 4 3 6 5
     |  b      X   X
 3   |      2 4 1 6 3 5
     |  a    X   X   X
 4   |      4 2 6 1 5 3
     |  x      X   X
 5   |      2 4 6 5 1 3
     |  a    X   X   X
 6   |      4 2 5 6 3 1
     |  x      X   X
 7   |      2 4 5 3 6 1
     |  a    X   X   X
 8   |      4 2 3 5 1 6
     |  x      X   X
 9   |      2 4 3 1 5 6
     |  a    X   X   X
10   |      4 2 1 3 6 5
     |  b      X   X
11   |      4 1 2 6 3 5
     |  a    X   X   X
12   |      1 4 6 2 5 3                     1 4 6 2 5 3
     |  b      X   X               or   c          X X
13   |      1 6 4 5 2 3, etc.               1 4 2 6 3 5, etc.
```

'Symmetry'

Before leaving this chapter, we consider one further aspect of change-ringing. Look again at the sequence 'Glasgow Major', the first 13 rows of which were given at the beginning of the chapter. These are reproduced in table 24.19, continuing to the point where the treble is again in the lead at row 30 and just beyond. A feature of this system is that, after reaching the middle of the section (row 16), the same changes are produced *in the reverse order*. They do not, of course, reproduce the same permutations, for that would be the last thing we should want to happen. It is the list of 'instructions' which has been turned upside-down, thereby producing the feature of symmetry. With such a system, only half the lead needs to be shown in order to be able to complete it.

Table 24.19

row		*A B C D E F G H*	row		*A B C D E F G H*
1		1 2 3 4 5 6 7 8	17		2 5 3 7 4 6 8 1
	a	X | X | X		*h*	
2		2 1 3 5 4 6 8 7	18		2 3 5 4 7 6 1 8
	b	X X X X		*b*	
3		1 2 5 3 6 4 7 8	19		3 2 4 5 6 7 8 1
	c	X X | | X		*h*	
4		2 1 3 5 6 4 8 7	20		3 4 2 6 5 7 1 8
	d	| X | X X		*g*	
5		2 3 1 5 4 6 7 8	21		4 3 2 5 6 1 7 8
	f	X X | X |		*b*	
6		3 2 5 1 4 7 6 8	22		3 4 5 2 1 6 8 7
	b			*d*	
7		2 3 1 5 7 4 8 6	23		3 5 4 2 6 1 7 8
	f			*b*	
8		3 2 5 1 7 8 4 6	24		5 3 2 4 1 6 8 7
	a			*a*	
9		2 3 5 7 1 8 6 4	25		3 5 2 1 4 6 7 8
	b			*f*	
10		3 2 7 5 8 1 4 6	26		5 3 1 2 4 7 6 8
	d			*b*	
11		3 7 2 5 1 8 6 4	27		3 5 2 1 7 4 8 6
	b			*f*	
12		7 3 5 2 8 1 4 6	28		5 3 1 2 7 8 4 6
	g			*d*	
13		3 7 5 8 2 4 1 6	29		5 1 3 2 8 7 6 4
	h			*c*	
14		3 5 7 2 8 4 6 1	30		1 5 2 3 8 7 4 6
	b			*b*	
15		5 3 2 7 4 8 1 6	31		5 1 3 2 7 8 6 4
	h			*a*	
16		5 2 3 4 7 8 6 1	32		1 5 3 7 2 8 4 6
	g				
17		(2 5 3 7 4 6 8 1)	33		1 3 5 2 7 4 8 6 (= Glasgow Surprise Major)
			or 33		1 5 7 3 8 2 6 4 (= Huddersfield Surprise Major)

Q. 24.19. Work out other symmetric leads with five and six bells.

Q. 24.20. Invent a continuation of the opening shown, by resuming from row 32 with $(BC)(DE)(FG)$.

Q. 24.21. Analyse the set of permutations shown in table 24.20, the beginning of a method known as 'Stedman Triples', in which the tenor (No. 8) is 'resting behind', i.e. does not take part in the changes, but is struck as the last note of each change. The method ranks as Triples (changes on seven bells), not Major.

Table 24.20

1	1 2 3 4 5 6 7 8	29	4 7 1 2 5 6 3 8	57	6 5 1 4 3 2 7 8
2	2 1 3 5 4 7 6 8	30	4 1 7 5 2 3 6 8	58	5 6 4 1 2 3 7 8
3	2 3 1 4 5 6 7 8	31	1 4 7 2 5 6 3 8	59	5 4 6 2 1 7 3 8
4	3 2 4 1 6 5 7 8	32	1 7 4 5 2 3 6 8	60	4 5 6 1 2 3 7 8
5	2 3 4 6 1 7 5 8	33	7 1 4 2 5 6 3 8	61	4 6 5 2 1 7 3 8
6	2 4 3 1 6 5 7 8	34	1 7 2 4 6 5 3 8	62	6 4 5 1 2 3 7 8
7	4 2 3 6 1 7 3 8	35	1 2 7 6 4 3 5 8	63	6 5 4 2 1 7 3 8
8	4 3 2 1 6 5 7 8	36	2 1 7 4 6 5 3 8	64	5 6 2 4 7 1 3 8
9	3 4 2 6 1 7 5 8	37	2 7 1 6 4 3 5 8	65	6 5 2 7 4 3 1 8
10	4 3 6 2 7 1 5 8	38	7 2 1 4 6 5 3 8	66	6 2 5 4 7 1 3 8
11	4 6 3 7 2 5 1 8	39	7 1 2 6 4 3 5 8	67	2 6 5 7 4 3 1 8
12	6 4 3 2 7 1 5 8	40	1 7 6 2 3 4 5 8	68	2 5 6 4 7 1 3 8
13	6 3 4 7 2 5 1 8	41	7 1 6 3 2 5 4 8	69	5 2 6 7 4 3 1 8
14	3 6 4 2 7 1 5 8	42	7 6 1 2 3 4 5 8	70	2 5 7 6 3 4 1 8
15	3 4 6 7 2 5 1 8	43	6 7 1 3 2 5 4 8	71	2 7 5 3 6 1 4 8
16	4 3 7 6 5 2 1 8	44	6 1 7 2 3 4 5 8	72	7 2 5 6 3 4 1 8
17	3 4 7 5 6 1 2 8	45	1 6 7 3 2 5 4 8	73	7 5 2 3 6 1 4 8
18	3 7 4 6 5 2 1 8	46	6 1 3 7 5 2 4 8	74	5 7 2 6 3 4 1 8
19	7 3 4 5 6 1 2 8	47	6 3 1 5 7 4 2 8	75	5 2 7 3 6 1 4 8
20	7 4 3 6 5 2 1 8	48	3 6 1 7 5 2 4 8	76	2 5 3 7 1 6 4 8
21	4 7 3 5 6 1 2 8	49	3 1 6 5 7 4 2 8	77	5 2 3 1 7 4 6 8
22	7 4 5 3 1 6 2 8	50	1 3 6 7 5 2 4 8	78	5 3 2 7 1 6 4 8
23	7 5 4 1 3 2 6 8	51	1 6 3 5 7 4 2 8	79	3 5 2 1 7 4 6 8
24	5 7 4 3 1 6 2 8	52	6 1 5 3 4 7 2 8	80	3 2 5 7 1 6 4 8
25	5 4 7 1 3 2 6 8	53	1 6 5 4 3 2 7 8	81	2 3 5 1 7 4 6 8
26	4 5 7 3 1 6 2 8	54	1 5 6 3 4 7 2 8	82	3 2 1 5 4 7 6 8
27	4 7 5 1 3 2 6 8	55	5 1 6 4 3 2 7 8	83	3 1 2 4 5 6 7 8
28	7 4 1 5 2 3 6 8	56	5 6 1 3 4 7 2 8	84	1 3 2 5 4 7 6 8
				85	1 2 3 4 5 6 7 8

Q. 24.22. A 'double' method has two 'planes of symmetry'. The place notation using letters, not figures for Double Norwich Court Bob Major is

Write out these rows in full, and study the symmetry.

Q. 24.23. When two perfectly elastic particles of equal masses moving in a straight line collide, the result of the collision is that they exchange velocities.

Suppose we have three such balls moving in a straight line, what is the maximum number of collisions which can occur? Repeat with four balls, and generalise for n balls. Suppose the velocities are $u_1, u_2, u_3, \ldots, u_n$, with $u_1 > u_2 > u_3 > \ldots > u_n$. Then if the first collision which occurs is that between, say, the 3rd and 4th balls, the velocities after the first collision will become $u_1, u_2, u_4, u_3, u_5, \ldots$ In what way does this problem resemble the ringing of changes of bells? What can we say about the velocities after all possible collisions have occurred?

Stedman Triples, the first eighty-four rows of which were printed above, contains $7! = 5040$ changes and takes about three or four hours to ring. The complete 40,320 changes on eight bells (Major) were first rung by relays of men at Leeds (Kent) in 1761, and the whole operation took twenty-seven hours. A system known as Plain Bob Major (rather similar to Plain Bob Minor) employs the changes $\boldsymbol{b} = (AB)(CD)(EF)(GH)$ and $(BC)(DE)(FG)$ in alternation, with $(CD)(EF)(GH)$ at each treble lead. This produces the permutation 1 3 5 2 7 4 8 6 at the beginning of the second coset. Plain Bob Major was rung by eight men at the Loughborough bell foundry in July 1963, and took 17 hours 58½ minutes.

Q. 24.24. Prove that the plain course contains seven leads, each of sixteen changes. Devise continuations using the bob $(BC)(EF)(GH)$ and the single $(EF)(GH)$.

Historical Note

As long ago as the eighteenth century, well before group theory had been discovered, methods were being devised of constructing sets of bell changes. A notable contributor to these efforts in the nineteenth century was a certain W. H. Thompson (1840–1934) who was a paper-and-pencil bell ringer. Being a Cambridge wrangler, it is not surprising that he showed a remarkable grasp of the problems of handling permutations. For this, he devised his own notation, and those blocks of changes which we are familiar with as cosets, he referred to with considerable insight as Q-sets. One of the problems to which he turned his attention was the proof of the impossibility of getting all $n!$ changes by the use of a limited number of devices. A brief account of his methods and achievements is given by T. J. Fletcher in a paper 'Campanological Groups', reprinted from the *American Mathematical Monthly*, Vol. 63, No. 9, Part I, Nov. 1956. Another notable contribution to the applications of group theory to change-ringing is by Prof. Rankin, mathematician but non-ringer. His mathematical survey of the method known as 'Grandsire Triples' may be found in the *Proceedings of the Cambridge Philosophical Society* (1948).

An alternative approach to the application of group theory to change ringing is summarised in an article in the *Mathematical Gazette*, May 1969, pp. 129–33, 'Mathematical Groups in Campanology', by B. D. Price. This contains some new ideas not mentioned in the previous pages, and it is well worth applying those methods to the material of the present chapter.

Unsolved problem in change-ringing: to show that Stedman triples requires at least two singles to complete the 5040 changes. (Solutions to the editor of the *Ringing World!*)

25 *Groups in geometrical situations*

On page 206, Ch. 13, we saw that the group generated by two elements a, b, each of period 2 is bound to be dihedral, and that it may be a finite or an infinite group according to the circumstances in which it occurs. These circumstances are dependent entirely on the *period of the element ab*: if ab has period n, then the group is D_n; if ab is not of finite period, then the group is D_∞.

Suppose for example the group is generated by the functions $f(t) = 1/t$; $g(t) = 1 - t$. Then $fg(t) = 1/(1 - t)$, and this is of period 3, so we shall undoubtedly get the group D_3. (This is the group of the cross-ratios, see pp. 327 ff.) On the other hand, if $f(t) = -1/t$ and $g(t) = 1 - t$, this would give an infinite group with structure D_∞. Each of these examples is capable of a geometric interpretation. If P $(at^2, 2at)$ is a given point on the parabola $y^2 = 4ax$, and if the chord PFQ passes through the focus F $(a, 0)$, then it is well-known that Q has parameter $t' = -1/t = f(t)$. Moreover, the gradient of the line joining two points on the parabola whose parameters are t and t' is $2/(t + t')$, so that if this gradient were 2, we should have

$$\frac{2}{t + t'} = 2 \Rightarrow t' = 1 - t = g(t).$$

Figure 25.01 shows the effect of successive transformations on a given point (labelled '1'). Those pairs of points which are linked by the function f are focal chords; those which are linked by the function g are parallel chords (of gradient 2). Note first, that since we may regard the parallel chords as all passing through a point G at infinity, we have here precisely the same situation as we had in Ch. 4 (see p. 32), and again in Ch. 11 (p. 134). Second, we have here a Cayley diagram for the group generated by f and g, but it is metrical Cayley diagram in its present form, instead of the usual topological one. The group in the present example is D_∞.

On p. 134, we mentioned that we should get $fg = gf$, leading to the group D_2, only when F and G are *conjugate points*, i.e. when G lies on the polar of F, in this case, the directrix of the parabola. The example above showed this in the case when G was the point at infinity on the directrix.

Q. 25.01. Draw the diagram when G is the point $(-a, a)$, as well as in other cases.

Q. 25.02. Show how the group D_2 may be set up on the rectangular hyperbola $x = ct$, $y = c/t$, by taking $f(t) = 1/t$, $g(t) = -t$. The sets of four points will be found to form rectangles. Alter one of the functions so as to obtain a more general example, with the sets of four points forming parallelograms.

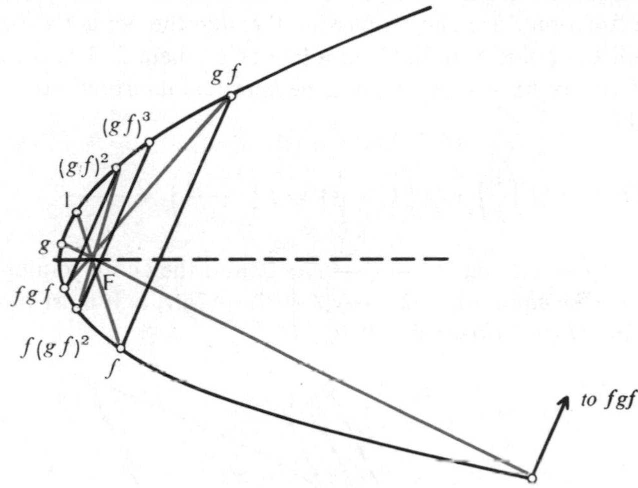

Fig. 25.01. Product of transformations of period 2 (involutions) on a parabola.

Dihedral groups on a circle and on a parabola

A particularly simple special case of the above is when the conic is a circle, with the points F and G at infinity. The case when F and G are conjugate points occurs when their directions are perpendicular, and here our sets of four points, related by the transformations 1, f, g, and $fg = g$ form the vertices of rectangles (see fig. 25.02).

Q. 25.03. Interpret subgroups and their cosets in this diagram.

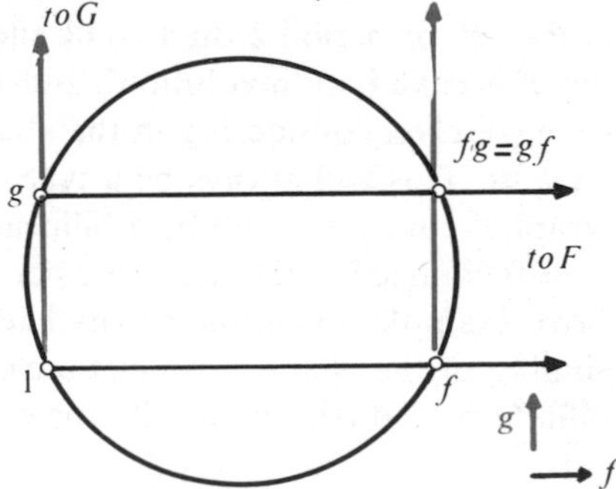

Fig. 25.02. The group D_2 on a circle.

By using the functions $f(t) = 1/t$, and $g(t) = 1 - t$, we may generate the group D_3. We shall illustrate this on the parabola $y^2 = 4ax$ (see fig. 25.03). You may care to consider the same transformations applied to points on the rectangular hyperbola $xy = c^2$. This time, those pairs of points linked by the function f are chords passing through the point $F\,(-a, 0)$, while G is still the point at infinity on a line of gradient 2. The transformations fg and gf do not have such a simple geometrical interpretation, but $fgf = gfg$ does. For:

$$fgf(t) = fg\left(\frac{1}{t}\right) = f\left(1 - \frac{1}{t}\right) = f\left(\frac{t-1}{t}\right) = \frac{t}{t-1}.$$

If $t' = t/(t - 1)$, then $tt' - t' - t = 0$, and the chord joining the points t and t' whose equation is $2x - y(t + t') + 2att' = 0$ must pass through the point $H\,(0, 2a)$ (see fig. 25.03).

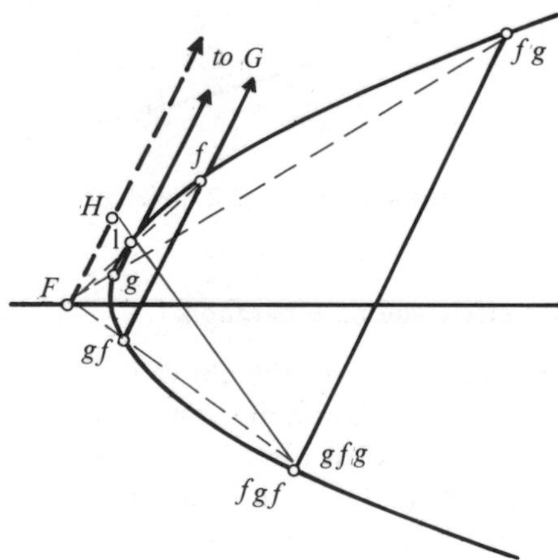

Fig. 25.03. The group D_3 on a parabola.

A transformation $P \leftrightarrow P'$ of period 2 on a conic such that PP' passes through a fixed point F is called an 'involution', and F is the 'pole' of the involution. What we have been considering in the above paragraphs is the *product of two involutions*. This will always be a type of transformation known as a 'homography', and there will be a bilinear relation of the form $att' + bt + ct' + d = 0$ connecting the algebraic parameters of the pairs of points. In the above example, the homography had period 3; in the example at the beginning of the chapter, we met a case when the homography had infinite period (this is usually the case).

Q. 25.04. Show that, if FG meets the conic at M and N, then M and N are *self-corresponding points* of fg and of gf.

It is always possible to transform the parameter of points on the conic in such a way that two particular points have parameters zero and infinity. Selecting M and N, the points of intersection of FG with the conic above for this purpose, we may show that the equations of the two involutions will take the forms:

Chord passing through F: $t' = \dfrac{k_1}{t} = f(t)$,

chord passing through G: $t' = \dfrac{k_2}{t} = g(t)$.

So we have

$$gf(t) = k_2/(k_1/t) = \frac{k_2}{k_1}\,t = \mathrm{K}t\,(\text{say}),$$

$$fg(t) = (k_1/k_2)\,t = \frac{1}{\mathrm{K}}\,t,$$

$$fgf(t) = f(\mathrm{K}t) = k_1/\mathrm{K}t = \frac{k_1^2}{k_2 t},$$

and this is another involution whose pole H also lies on FG, since the equation is of like form to those of the two original involutions. In the example on the parabola above, the point H was found to be $(0, 2a)$, and does indeed lie on the line joining $F\,(-a, 0)$ to G which has gradient 2.

Now the cross-ratio of four points on a conic is that of their algebraic parameters:

$$(t_1\; t_2\; t_3\; t_4) = \frac{(t_1 - t_2)(t_3 - t_4)}{(t_1 - t_4)(t_3 - t_2)}.$$

For the homography $t' = \mathrm{K}t$ which has the points $t = 0$ and $t = \infty$ as self-corresponding points (or 'double points'), we find that the cross-ratio of a pair of corresponding points separating the double points is:

$$(0\; t'\; \infty\; t) = \frac{(0 - t')(\infty - t)\dagger}{(0 - t)(\infty - t')} = \frac{t'}{t} = \mathrm{K}.$$

Thus any pair of the homography make a constant cross-ratio with the double points, its value being K.

For a homography to have period 2, we must have $\mathrm{K}^2 = 1 \Rightarrow \mathrm{K} = -1$ $(\mathrm{K} \neq 1)$, so that a pair of such a homography will separate the double points harmonically – the homography is in this case an involution. For a homography to have period 3, we must have $\mathrm{K}^3 = 1 \Rightarrow \mathrm{K} = \omega = \operatorname{cis} \frac{2}{3}\pi$, or $\mathrm{K} = \omega^2$. In the same way in general, when $\mathrm{K} = \operatorname{cis} 2\pi/n$, we find fg to be of period n, and the group generated is D_n.

† A deliberate loose notation, adopted for brevity.

We have already illustrated simple examples of the product of two involutions, but the case which may be most readily understood is that when the conic is a circle and the points F and G are at infinity. In order to obtain the group D_3, fg must be of period 3, and this means that the directions of the points F and G at infinity must be separated by 60° (see fig. 25.04). Starting from any selected point '1' on the circle, we have

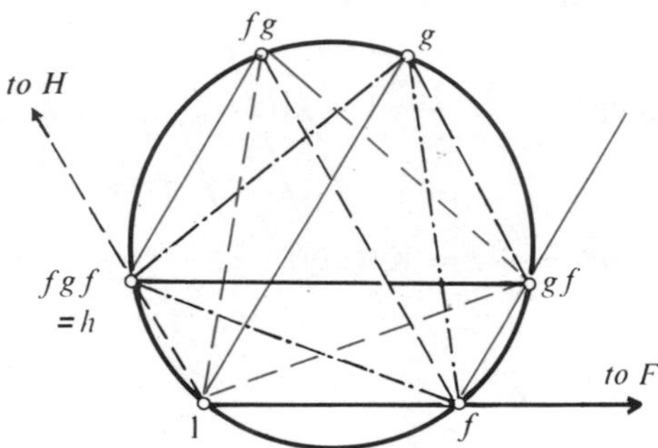

Fig. 25.04. The group D_3 on a circle.

labelled the remaining points of the set of six according to the generators f and g, and this gives the figure the character of a Cayley diagram. However, $fgf = gfg = h$ is another element of period 2, and this gives rise to three chords of the hexagon in the direction passing through the point H as shown by the broken lines. The subgroup $\{1, gf, fg\}$ makes an equilateral triangle, as does the coset $\{f, g, h\}$. If the angle between the directions of F and G had been 45°, we should have obtained the group D_4.

Q. 25.05. Draw the diagram in this case, marking the points in terms of f and g. Pick out subgroups and cosets, showing that the subgroup Gp (fg) and its cosets are squares. Interpret the rectangles and the diameters in group-theoretical terms.

If the angle θ between the directions f and g had been incommensurate with 360°, we should have obtained an infinite group D_∞. In this case, gf represents a rotation about the centre of the circle through 2θ and if θ is an irrational multiple of π, then this rotation has infinite period (see fig. 25.05).

For an interesting extension of the above ideas, see the article 'Illustrations of Simple Group Theory', by Joan Holland, *Mathematical Gazette*, Feb. 1964, especially the section dealing with the representation of groups on the rectangular hyperbola and in three dimensions on a twisted cubic (pp. 50 ff.).

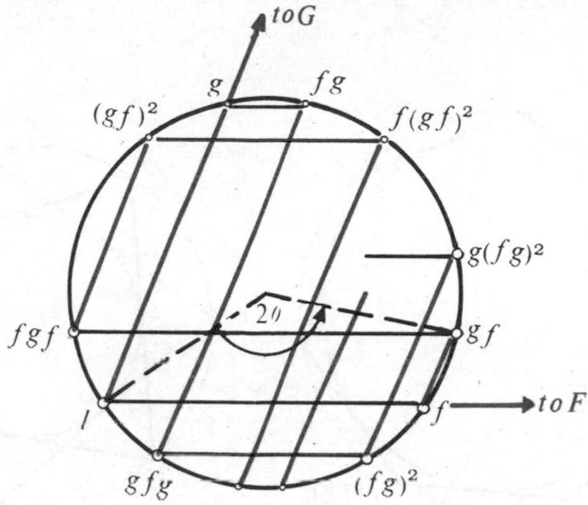

Fig. 25.05. The group D_∞ on a circle.

★Poncelet's porism

We saw that, if fg or gf is of period n, then Gp (f, g) is of the form D_n, while fg itself generates the subgroup C_n. When the conic is a circle and F and G are at infinity as in the last paragraph, the subgroup C_n appeared as an equilateral triangle in the case $n = 3$, and this equilateral triangle itself circumscribes a circle as shown in fig. 25.06. In the general case, we shall have a regular n-gon inscribed in the given circle and

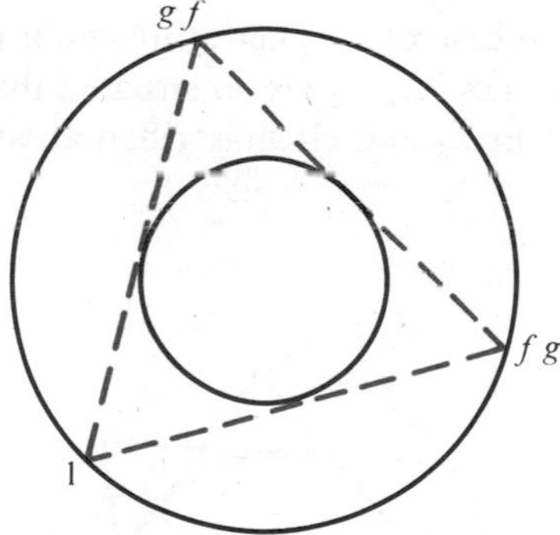

Fig. 25.06. Poncelet's porism (concentric circles, $n = 3$).

circumscribing a smaller (concentric) circle as in figs. 25.071, 25.072. Evidently, in view of the rotational symmetry of the figure, this polygon will be closed whichever point on the given circle is selected as our starting point 1 – there are an infinity of n-gons which may be inscribed in the given circle and circumscribed about the smaller one. If fg were

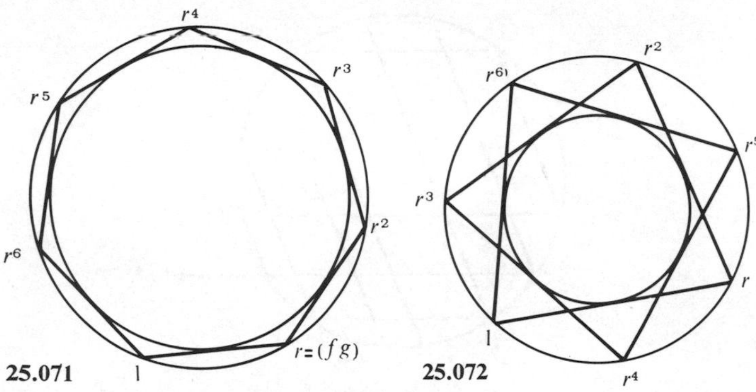

Fig. 25.07. $r = fg$ of period 7.

25.071–2. The circles are so related that regular 7-gons may be inscribed to the larger and circumscribed about the smaller.

not of finite period, then the two circles would not be so related (fig. 25.073).

This is a special case of an interesting result relating to conics in general: if two conics Σ_1, Σ_2 are so situated that an n-sided polygon may be inscribed in Σ_1 and circumscribed about Σ_2 (see fig. 25.08), then

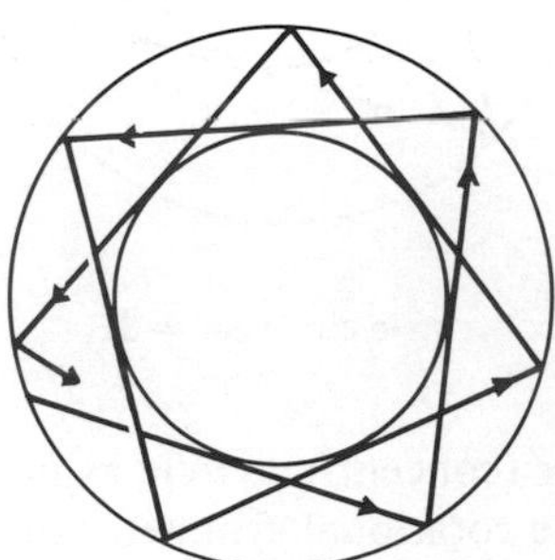

25.073. The circles are not so related.

an infinity of such polygons exist, so that wherever one selected the point A_0 on Σ_1, the polygon $A_0A_1A_2A_3A_4$ (in the case $n = 5$) obtained by drawing tangents to Σ_2 would be closed. The phenomenon is known as Poncelet's porism.

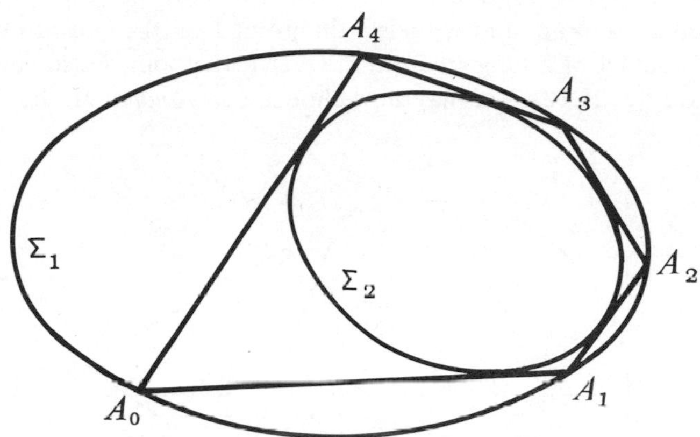

Fig. 25.08. Poncelet's Porism ($n = 5$). An infinity of pentagons may be inscribed to Σ_1 and circumscribed about Σ_2.

In fig. 25.09 we give an illustration of the group D_3 on a circle in the case when F and G are not at infinity, and the points are labelled $1, f, g, fg, gf$, and fgf. The operation $h = fgf$ being of period 2 is an involution, and the join of any two points labelled 1 and $fgf (= gfg)$ must also pass through H, which is off the paper in fig. 25.09. But remember that *any* of the six points could have been used as the starting point and labelled 1, and such a re-lettering has been shown in the diagram, where we have replaced $1, f, g, fg, gf, fgf$ respectively by $fg, g, fgf, gf, 1, f$. Two distinct polygons have been shown: the re-lettering refers to the red one. (It is tempting to believe that this re-lettering corresponds to an automorphism of this group, but this is not so: why?) The line joining 1 and fgf, i.e. gf and g, will therefore also pass through H, and there are thus three such lines: $1, fgf$; gf, g and f, fg, and these are the right cosets $\{1, h\}\,1$; $\{1, h\}\,g$ and $\{1, h\}\,f$.

★**Q. 25.06.** Why the *right* cosets, why not the *left* cosets?

Compare fig. 25.04 where F and G are at infinity. The operations of period 3, i.e. fg and gf give chords which circumscribe a fixed ellipse (not a circle, and not concentric with the given circle this time), and the triangles formed by the coset $\{f, g, h\}$ also circumscribe this same ellipse (why?).

It may be asked, how did we succeed in drawing the figure? How did we manage to select points F and G in such a way that fg came to have period 3, i.e. so that fgf and gfg coincided? One of the points, F, may be chosen at random, and we select the point 1 on the circle to be a point of contact of a tangent from F, so that the point f also coincides with 1 (see fig. 25.10). We may now choose g *at random* on the circle,

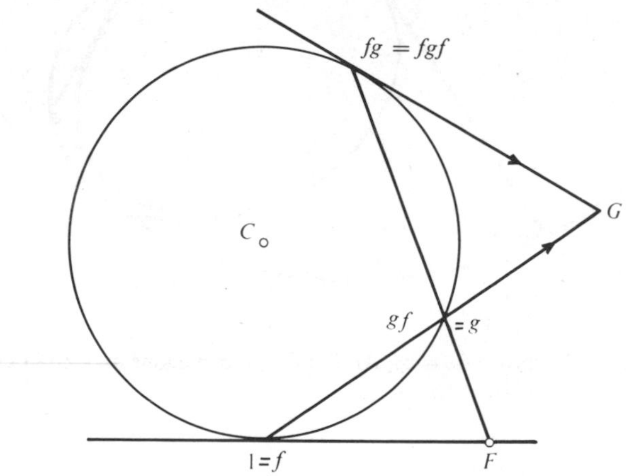

Fig. 25.10. Construction of points F and G so that fg may be of period 3.

thereby establishing that G must lie on the line joining it to 1, and also noting that the point g is also the point gf. This determines the point fgf, which coincides with fg. G must lie on the tangent to the circle at this latter point, and its position is now known. But note that we had freedom to choose the position of g, and so there is an infinity of possible positions for G once F has been selected.

★**Q. 25.07.** Experiment with other positions of g, and with other positions of F on the tangent at 1.

★**Q. 25.08.** Consider the special configuration resulting when C 1 F g is taken to be a *square*.

★**Q. 25.09.** Show that, if FG meets the circle in M and N (not real points), then the cross-ratio $(MFNG)$ is equal to a cube root of 1. Use this to calculate a position of G when the circle is $x^2 + y^2 = 1$, and F is $(0, 2)$.

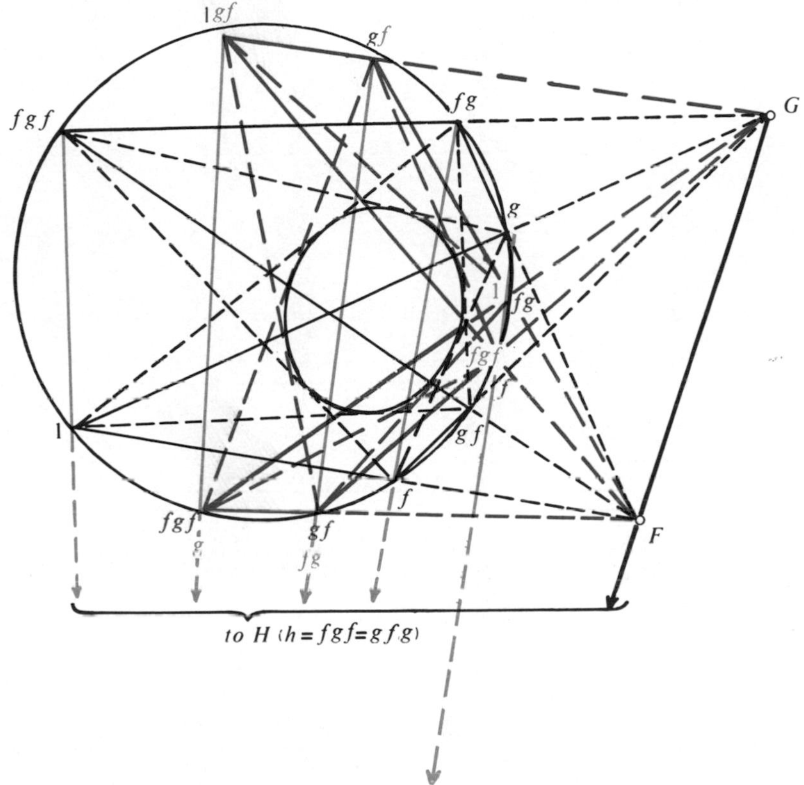

Fig. 25.09. The group D_3 on a circle – general case.

The re-lettering of geometrical configurations

In fig. 25.11 it is given that $AQ \parallel BR$, $BP \parallel CQ$, and we are required to prove that $AP \parallel CR$. A large number of methods are available here, but we mention a proof using areas, which runs as follows:

$$\begin{aligned} AQ \parallel BR \Rightarrow \triangle AQR &= \triangle AQB \Rightarrow \triangle AQR + \triangle APQ \\ &= \triangle AQB + \triangle APQ \\ \Rightarrow \triangle APR &= \text{quad. } ABQP \\ \text{Similarly,} \quad \triangle PAC &= \text{quad. } PQBA \end{aligned} \Bigg\} \Rightarrow \triangle APR = \triangle PAC \Rightarrow PA \parallel CR.$$

A pupil of mine once wrote 'by symmetry' where I have written 'similarly'. I crossed it out and told him the correct word was 'similarly'. It was not until some time later, after studying some group theory, that it occurred

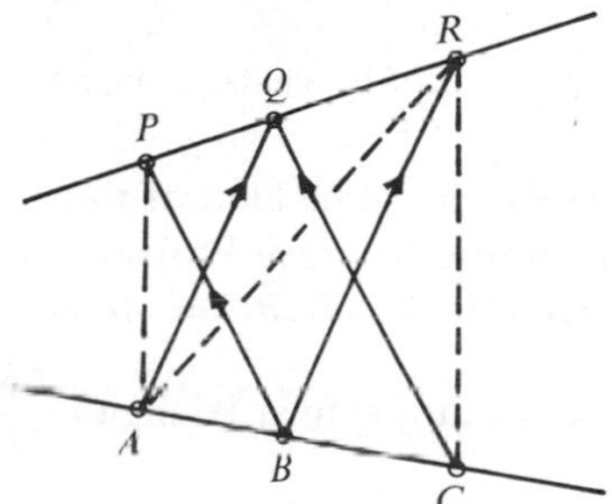

Fig. 25.11. Special case of Pappus' theorem.

to me that I should have raised no objection to the use of the words 'by symmetry'. For although this figure is not symmetrical in the *metrical* sense, yet there is a sense in which it may be described as symmetrical inasmuch as, if we turn the figure over so that the line ABC is at the top and PQR at the bottom, it is just as good a figure as the one we started with. This 'turning upside-down' process is in effect the same as the process of re-lettering the figure by making the interchanges (AP), (BQ), (CR). Indeed, the first part of the proof is still quite valid when these interchanges are made:

the step $AQ \parallel BR \Rightarrow \triangle AQR = \triangle AQB$, is replaced by
$PB \parallel QC \Rightarrow \triangle PBC = \triangle PBQ$, and this is perfectly correct.

When the pupil said 'by symmetry', he was implying the *algebraic* symmetry of the notation. One is doing the same sort of thing when one deduces from the formula $a^2 = b^2 + c^2 - 2bc \cos A$, the two other like formulae:

$$b^2 = c^2 + a^2 - 2ca \cos B \quad \text{and} \quad c^2 = a^2 + b^2 - 2ab \cos C.$$

Note that in fig. 25.11 another lettering is possible: (PR), (AC), corresponding to the figure being turned over so that PA is on the right and RC on the left.

Q. 25.10. Verify the validity of the statements in the proof under the transpositions (PR), (AC).

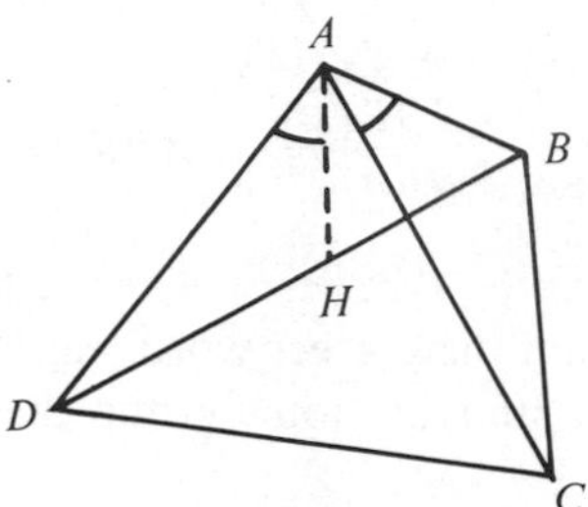

Fig. 25.12. Ptolemy's theorem. Proof by similar triangles.

Another proof in which the same kind of thing happens is illustrated in fig. 25.12. We are attempting to prove Ptolemy's theorem, that in any cyclic quadrilateral, $AB.CD + AD.BC = AC.BD$. Making

$\angle DAH = \angle BAC$, we obtain similar triangles $\dfrac{ABC}{AHD}$, so that

$$\frac{BC}{HD} = \frac{AC}{AD} \Rightarrow AD.BC = AC.HD \qquad \ldots \text{(1)}$$

Similarly because triangles $\dfrac{ADC}{AHB}$ are similar,

$$\frac{DC}{HB} = \frac{AC}{AB} \Rightarrow AB.DC = AC.HB \qquad \ldots \text{(2).}$$

The result (2) can be obtained from (1) by the transposition (BD); that is, if the letters B and D were interchanged in the fig. 25.12, the whole proof would still be valid, but each half of the proof would apply to the other part of the figure.

We might say that the proof of this theorem is invariant under the transposition (BD), and so has the group C_2. Note that the final *result*: $AB.CD + AD.BC = AC.BD$ is invariant under the transpositions $(AB)(CD)$; (AC); (BD); $(AC)(BD)$; $(AD)(BC)$; as well as the cycles of period 4: $(ABCD)$, $(DCBA)$. These are the elements h, f, p, r, y, j and t of S_4 (notation as on p. 219) which together with the identity form the group D_4. They represent, in fact, those permutations of the vertices of $ABCD$ which keep AC and BD as diagonals (compare pp. 132, 222, 311, 326).

The complete quadrilateral

In fig. 25.13 we have a quadrilateral *ABCD* with sides *AB*, *BC*, *CD*, *DA* labelled *a*, *b*, *c*, *d* and the points of intersection of *a* and *c* and of *b* and *d* respectively denoted *E* and *F*. We now consider the situation arising from a relettering of this figure. Since there are six points which have been named, there are $6! = 720$ possible ways of relettering the figure, but we shall restrict our attention to methods of relettering according to particular schemes. The first scheme requires that the three lines *p*, *q*, *r* (diagonals) shall always be named *AC*, *BD* and *EF*, though not in that order. This cuts out 672 of the possible reletterings as it will be found that the number of ways of permuting the names of the six points so that diagonals are

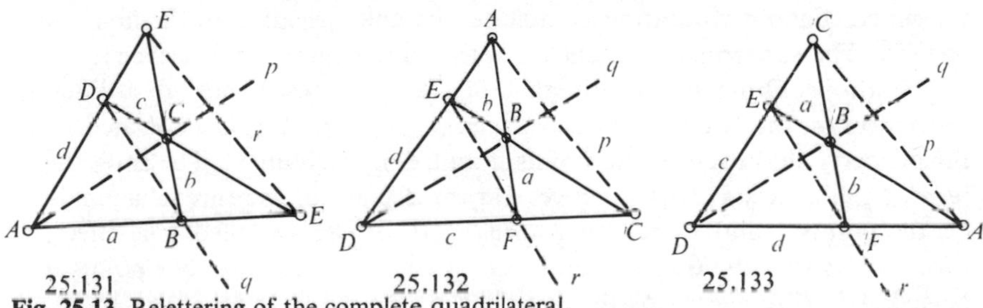

Fig. 25.13. Relettering of the complete quadrilateral.

preserved is only forty-eight. For *p*, *q* and *r* may be permuted in six ways, for each of which the names of the ends of the diagonals *AC*, *BD*, *EF* may be interchanged. Thus in fig. 25.132, *p*, *q*, *r* have been replaced by *q*, *r*, *p* respectively, and this diagram shows only one of the eight ways in which the diagonals could be lettered, any combination of the transpositions (*AC*), (*BD*) and (*EF*) being available. In the original version (1), these three transpositions generate the group $C_2 \times C_2 \times C_2$ (see p. 280). This group is therefore a subgroup of the group of order 48, while the reletterings of (2) form one of the six cosets of this subgroup (one for every permutation of *p*, *q* and *r*).

In our second scheme, we shall further restrict the lettering by requiring all collinearities to be preserved. This will exclude (2), for example, since in (1) *ADF* was a straight line, whereas in (2) it was not. The relettering shown in (3) is acceptable by this scheme, for *E* and *F* are once again the intersections of *a* and *c* and of *b* and *d* respectively. Indeed, once one has chosen *A*, *B*, *C* and *D*, then there is only one possible position for each of *E* and *F*. Evidently there are twenty-four possible permutations of the letters which satisfy the requirements stipulated. The relettering of fig. 25.133 could be represented by the permutation $\begin{pmatrix} A & B & C & D & E & F \\ D & F & B & E & A & C \end{pmatrix}$

of the vertices, or by the permutation $\begin{pmatrix} a & b & c & d \\ d & b & a & c \end{pmatrix}$ of the four sides. The latter is, in fact, the most convenient representation, for one may set up a 1, 1 correspondence between the permissible reletterings and the permutations of the four sides. Thus we must have the group S_4, and table 25.01 shows a complete list of the reletterings, each as a permutation of vertices and of sides. For convenience the vertices are shown in the order which keeps the three diagonals separate. In the final column we show the permutations of the diagonals, the prime being added when the letters of the diagonal have been interchanged, e.g. $\begin{pmatrix} A & C & B & D & E & F \\ D & B & F & E & A & C \end{pmatrix}$. Fig. 25.133 is given as $(q'\ r'\ p)$, the q' and r' indicating that (BD) and (EF) have been interchanged, whereas (AC) retains its original order, so p is unprimed. Each permutation is indicated by the operation in the first column. The subgroup A_4 is shown in the final column by the even permutations. Other subgroups stand out, each corresponding to a different restriction on the lettering. For example, $\{e, s^2, f, g\}$ (D_2) is associated with the property that each of the points A and C is unchanged. The subgroup $\{e, s^2, t^2, u, u^2, u^3, k, l\}$ (D_4) leaves r in third place of the three diagonals, i.e. the points E and F are either unchanged, or else switched. The effect, therefore, is that the quadrilateral formed by the remaining four points is named $ABCD$ in every possible way so that the cyclic order is preserved (compare pp. 189 ff., 221–2, 311, 499, Ex. 13.01). Another subgroup is $\{e, f, h, l, x, x^2\}$ (D_3), and its elements correspond to those cases where, in the column headed 'diagonals', none of the diagonals is primed, i.e. each of the pairs AC, BD and EF retains its original order.

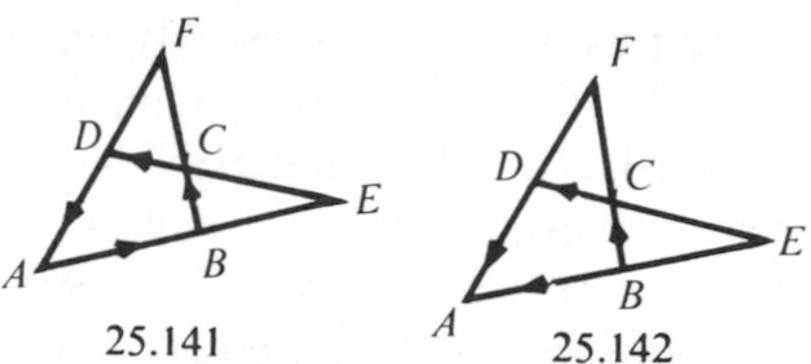

Fig. 25.14. Equal forces along sides of quadrilateral.

Q. 25.11. Find other subgroups, and the invariant property associated with each. Which subgroup arises when the figure is re-lettered so that the points A, B and E are always in the same line, but are permuted between themselves? Discover and interpret cosets of the several subgroups.

Q. 25.12. A convex hexagon $ABCDEF$ is relettered so that the cyclic order of the vertices is retained, either clockwise or anticlockwise. Show that the group of these permutations is D_6, and generalise.

Q. 25.13. Which group is associated with the relettering of the vertices of a tetrahedron? a bitetrahedron?

Q. 25.14. Four equal forces act along the sides of a quadrilateral $ABCD$, the positive direction of each being as indicated in fig. 25.14. The two equal forces

Table 25.01

operation	perm. of vertices	perm. of sides	perm. of diagonals	period	odd or even
e	$A\ C\ \ B\ D\ \ E\ F$	$a\ b\ c\ d$	$p\ q\ r$	1	even
s	$C\ A\ \ F\ E\ \ B\ D$	$b\ d\ a\ c$	$p'\ r'\ q$	4	odd
s^2	$A\ C\ \ D\ B\ \ F\ E$	$d\ c\ b\ a$	$p\ q'\ r'$	2	even
s^3	$C\ A\ \ E\ F\ \ D\ B$	$c\ a\ d\ b$	$p'\ r\ q'$	4	odd
t	$E\ F\ \ D\ B\ \ C\ A$	$c\ d\ b\ a$	$r\ q'\ p'$	4	odd
t^2	$C\ A\ \ B\ D\ \ F\ E$	$b\ a\ d\ c$	$p'\ q\ r'$	2	even
t^3	$F\ E\ \ D\ B\ \ A\ C$	$d\ c\ a\ b$	$r'\ q'\ p$	4	odd
u	$B\ D\ \ C\ A\ \ F\ E$	$b\ c\ d\ a$	$q\ p'\ r'$	4	odd
u^2	$C\ A\ \ D\ B\ \ E\ F$	$c\ d\ a\ b$	$p'\ q'\ r$	2	even
u^3	$D\ B\ \ A\ C\ \ F\ E$	$d\ a\ b\ c$	$q'\ p\ r'$	4	odd
f	$A\ C\ \ E\ F\ \ B\ D$	$a\ c\ b\ d$	$p\ r\ q$	2	odd
g	$A\ C\ \ F\ E\ \ D\ B$	$d\ b\ c\ a$	$p\ r'\ q'$	2	odd
h	$E\ F\ \ B\ D\ \ A\ C$	$a\ b\ d\ c$	$r\ q\ p$	2	odd
j	$F\ E\ \ B\ D\ \ C\ A$	$b\ a\ c\ d$	$r'\ q\ p'$	2	odd
k	$D\ B\ \ C\ A\ \ E\ F$	$c\ b\ a\ d$	$q'\ p'\ r$	2	odd
l	$B\ D\ \ A\ C\ \ E\ F$	$a\ d\ c\ b$	$q\ p\ r$	2	odd
x	$B\ D\ \ E\ F\ \ A\ C$	$a\ c\ d\ b$	$q\ r\ p$	3	even
x^2	$E\ F\ \ A\ C\ \ B\ D$	$a\ d\ b\ c$	$r\ p\ q$	3	even
y	$D\ B\ \ F\ E\ \ A\ C$	$d\ b\ a\ c$	$q'\ r'\ p$	3	even
y^2	$E\ F\ \ C\ A\ \ D\ B$	$c\ b\ d\ a$	$r\ p'\ q'$	3	even
z	$B\ D\ \ F\ E\ \ C\ A$	$b\ d\ c\ a$	$q\ r'\ p'$	3	even
z^2	$F\ E\ \ A\ C\ \ D\ B$	$d\ a\ c\ b$	$r'\ p\ q'$	3	even
w	$D\ B\ \ E\ F\ \ C\ A$	$c\ a\ b\ d$	$q'\ r\ p'$	3	even
w^2	$F\ E\ \ C\ A\ \ B\ D$	$b\ c\ a\ d$	$r'\ p'\ q$	3	even

at B have resultant along the external bisector of $\angle ABC$; similarly the two equal forces at D have resultant along the external bisector of $\angle CDA$. These external bisectors meet at P. Then the resultant of the system acts through P. Similarly the resultant acts through Q, the intersection of the external bisectors of $\angle$s DAB, BCD. A similar argument shows that the resultant also passes through R, the intersection of the external bisectors of the angles at E and F. Therefore PQR is a straight line. Now suppose the force in AB is reversed (fig. 25.142), and denote the operations of reversing the forces in AB, BC, CD and DA by a, b, c and d. The effect of a above is that we shall now get the *internal* bisectors of $\angle$s ABC, DAB and E and three different collinear points P', Q', R'. The same three points would, of course, arise if all four forces were reversed. How many possible sets of collinear points are there? Complete the table showing the effect of the operations, the signs in the six columns denoting whether the internal (+) or external (−) bisector of the angle at that vertex is to be taken; for example, in row bc, we have a '+' under B. This means that when the forces in BC and CD are reversed, it will be the internal bisector of $\angle ABC$ which will be drawn.

Consider the group aspect of this problem. Show that the eight lines, each containing three points, form two distinct complete quadrilaterals.

	A	*C*	*B*	*D*	*E*	*F*
e	−	−	−	−	−	−
a	+	−	+	−	+	−
b	−	+	+	−	−	+
c						
d						
bc			+			
.						
.						
.						

Orthocentric quadrilateral: nine-point circle

Another configuration for which the process of relettering provides interesting group implications is the figure of the triangle and its orthocentre together with the nine-points circle. In fig. 25.151, D is the

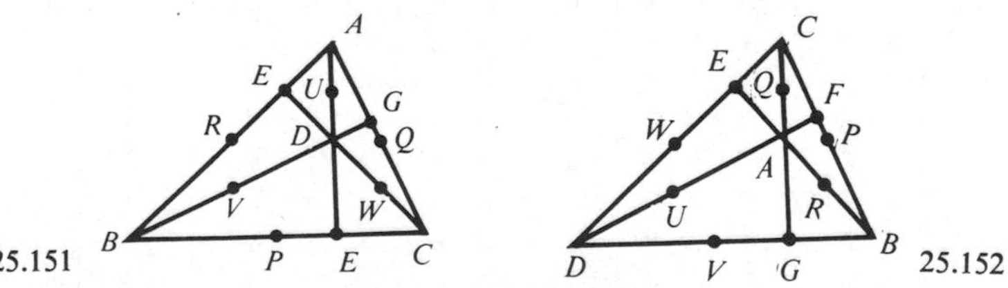

Fig. 25.15. Relettering of nine-point circle configuration.

orthocentre of $\triangle ABC$, with F, G, E the feet of the altitudes from A, B, C. The nine-points circle passes through the mid-points of the six line segments BC, CA, AB, AD, BD, CD, which we call P, Q, R, U, V, W. Now each of the vertices of the (orthocentric) quadrilateral $ABCD$ is the orthocentre of the triangle formed by the other three, so we may reletter this figure in twenty-four ways, one of which, $\begin{pmatrix} A & B & C & D \\ C & D & B & A \end{pmatrix}$, is shown in fig. 25.152. This particular permutation of the four vertices induces changes in the names of the nine points as indicated by the permutation $\begin{pmatrix} E & F & G & P & Q & R & U & V & W \\ E & G & F & V & P & W & Q & U & R \end{pmatrix}$, for in our relettered figure we are to retain the convention that collinearity in the original figure will be invariant, so that F, which was the intersection of AD and BC in the original figure, is also the intersection of AD and BC in the relettered figure. Further, Q is the mid-point of AC in both original and relettered figures. The permutations of the nine points which result from the permutations of A, B, C and D evidently have the effect of keeping the sets $\{E, F, G\}$ and $\{P, Q, R, U, V, W\}$ in separate compartments.

Considering, first, only those permutations induced in the six mid-points P, Q, R, U, V, W, we see that the relettering gives rise to a set of twenty-four permutations of these six mid-points which have the group structure S_4. Now the group S_4 may be realised as the symmetry group of the cube, which has six faces. Could these twenty-four permutations of the six mid-points be the same as the permutations of the six faces of the cube? (See Ch. 17, p. 297.) For example, in fig. 25.152 we showed the permutation $\begin{pmatrix} P & Q & R & U & V & W \\ V & P & W & Q & U & R \end{pmatrix}$ of the six mid-points. The answer to the question posed is 'Yes'! For just as PU, QV and RW are *diameters* of the nine-points circle, so we may use these pairs of letters to denote *opposite faces* of the cube. To make the correspondence more vivid, think of $ABCD$ as being a tetrahedron, and draw parallel planes through each pair of opposite edges to form a parallelepiped. Then the six diagonals of its faces and their mid-points are precisely the mid-points of opposite faces of the parallelepiped (see fig. 25.16). They form, in fact, the vertices of an

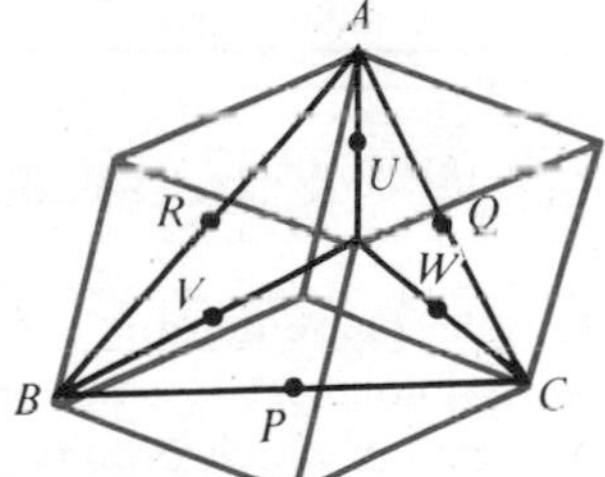

Fig. 25.16. Cube and inscribed tetrahedron.

octahedron, the *dual* (see pp. 297, 303) solid of the parallelepiped. When the tetrahedron is orthocentric (opposite edges perpendicular), the parallelepiped is a rhomboid, and the lines PU, QV, RW are mutually perpendicular.

Q. 25.15. Catalogue various subgroups as we did for fig. 25.13, each being characterised by some property, such as the points U and P remain undisturbed.

Next, we shall forget about the points P, Q, R, U, V, W, and concentrate on the seven points A, B, C, D, E, F, G (see fig. 25.17). We shall also forget about D being the orthocentre and take the case when A, B, C, D are four general points of the plane. The configuration may now remind you of that on p. 275, fig. 16.07 (with a completely fresh set of letters). There we discussed briefly a 'seven-point geometry' in which one thinks of the three points EFG in the present diagram as being collinear, so that there are in this figure seven points lying by threes on the seven lines,

$$ABE, \quad ADF, \quad AGC, \quad BDG, \quad CDE, \quad BFC, \quad EFG.$$

Each line gives rise to a complementary quadrilateral; for example, *BFC* leaves the quadrilateral *AEDG*, and *EFG* leaves the quadrilateral *ABCD*, and so on.

We now reletter this figure so that collinearity is preserved. One of the possible reletterings is shown in fig. 25.172, and this corresponds to the permutation $\begin{pmatrix} A & B & C & D & E & F & G \\ F & B & G & A & C & D & E \end{pmatrix}$. How many such permutations can there be? We may choose A to be any one of the seven points, and B to be any of the remaining six. Thus the letters A and B may be assigned in forty-two ways. Having done this, there are only four possible positions for C, since ABC must not be in line. However, once the points A, B, C are all allocated, all the others fit into place, there being no further

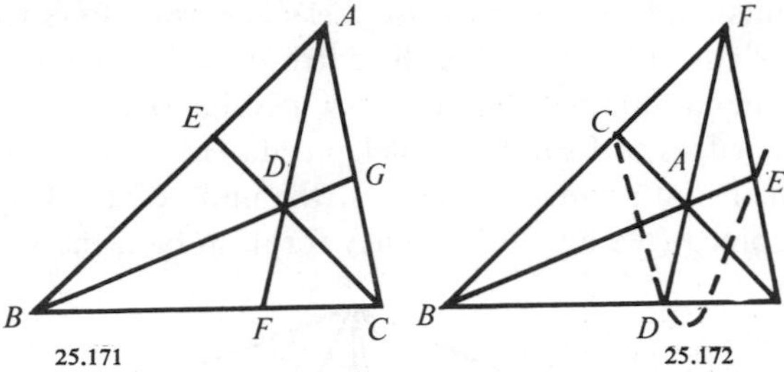

Fig. 25.17. Relettering of configuration of the finite geometry of seven points.

freedom of choice. For example, the remaining point on the line *CA* must be *G*, since we have agreed that collinearity shall be preserved. Thus there are 168 permutations of the vertices which preserve collinearity (always regarding the three points labelled *E*, *F*, *G* in the original figure 25.171 as being collinear), and they form the famous 'simple' group of order 168 alluded to on p. 272, Ch. 16 as the group of automorphisms of

Table 25.02

	perm.							period
e	A	B	C	D	E	F	G	1
a	A	B	D	C	E	G	F	2
$p^2 = b$	A	B	G	F	E	D	C	2
c	A	B	F	G	E	C	D	2
p	A	E	F	C	B	G	D	4
d	A	E	G	D	B	F	C	2
p^3	A	E	D	G	B	C	F	4
f	A	E	C	F	B	D	G	2

$C_2 \times C_2 \times C_2$. Now we consider a few of the subgroups. First, those permutations which leave A invariant and keep B and E in the same two positions, either leaving them alone or switching them. These permutations are as shown in table 25.02, and they form the group D_4.

Q. 25.16. Is this group normal in the group of order 168?

Q. 25.17. If A is left invariant and the only remaining restriction is on the collinearity of sets of three points, what is the order of the group, and what is its structure?

Q. 25.18. Here is a set of permutations which do not preserve collinearity. Do they form a group, and what property of fig. 25.17 are they associated with?

A	B	C	D	E	F	G
A	E	G	F	B	D	C
E	B	F	G	A	C	D
G	F	C	E	D	B	A
D	C	B	A	G	F	E
C	D	A	B	F	E	G
B	A	D	C	E	G	F
F	G	E	D	C	A	B.

Q. 25.19. Discover which groups, other than D_4 above, are subgroups of the group of order 168 of the permutations of the letters which preserve collinearity.

Q. 25.20. What is the connection between the relettering of fig. 25.171 and the relettering of fig. 25.13?

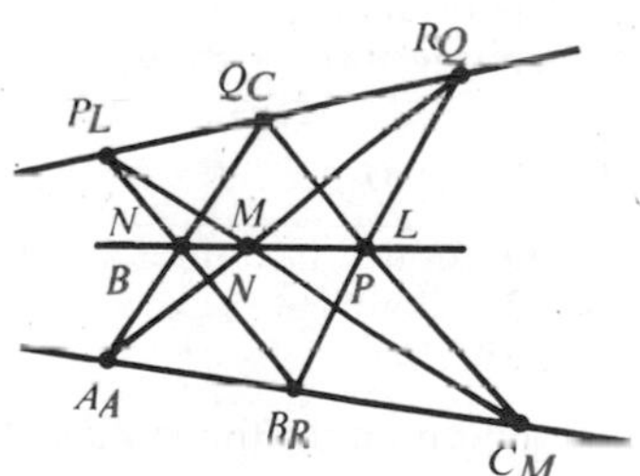

Fig. 25.18. Pappus' configuration – relettering.

Pappus' theorem

In fig. 25.18 we reproduce the configuration for Pappus' theorem (see Ch. 16, p. 273). The red letters show one of the 108 possible reletterings which preserve the nine sets of three collinear points, and this corresponds to the permutation $\begin{pmatrix} A & B & C & P & Q & R & L & M & N \\ A & R & M & L & C & Q & P & N & B \end{pmatrix}$. These 108 permutations are a group, and we consider the subgroup (of order $108/9 = 12$) which leaves A invariant (the 'stabiliser' of A). The permutations are shown in table 25.03. The periods of the permutations leave no doubt that the group is D_6, and is a subgroup of S_8, the group of all possible permutations which leave A invariant.

Table 25.03

	perm.									period
e	A	B	C	P	Q	R	L	M	N	1
a	A	B	C	L	M	N	P	Q	R	2
b	A	C	B	P	R	Q	L	N	M	2
c	A	C	B	L	N	M	P	R	Q	2
d	A	R	M	P	N	B	L	C	Q	2
f	A	R	M	L	C	Q	P	N	B	6
g	A	M	R	P	B	N	L	Q	C	3
h	A	M	R	L	Q	C	P	B	N	2
j	A	N	Q	P	C	M	L	R	B	2
k	A	N	Q	L	R	B	P	C	M	6
l	A	Q	N	P	M	C	L	B	R	3
m	A	Q	N	L	B	R	P	M	C	2

Q. 25.21. Consider the geometrical meaning of various subgroups of the above group, e.g. $\{e, a, b, c\}$ (D_2), which retains B and C in their original positions. Find subgroups with structure C_6 and D_3, and interpret.

Q. 25.22. Is the subgroup D_6 of the table above normal in the full group of order 108?

Q. 25.23. Find the group of permutations which retains the line ABC but permutes these three letters among themselves, e.g.

$$\begin{pmatrix} A & B & C & P & Q & R & L & M & N \\ B & C & A & Q & R & P & M & N & L \end{pmatrix}.$$

Q. 25.24. Consider the permutations which interchange the pairs (AP), (BQ) and (CR), and compare with p. 489. Combine these with the permutations which interchange the pairs (AC) and (PR).

Q. 25.25. Note that in this subgroup, the triangles APL, BQM, CRN are all taken into each other. Consider the permutations of period 3. There are the cycles $y^{-1}z$ $(= g)$ and yz^{-1} $(= l)$ where $x = (A\ P\ L)$, $y = (B\ Q\ M)$, $z = (C\ R\ N)$: find Gp (x, y, z). Show that $p = (A\ B\ C)(P\ Q\ R)(L\ M\ N)$ belongs to the set of 108 permutations, and that $pg \neq gp$. (It may be of interest to note that of the 108 permutations, there are 26 of period 3 which, with the identity, form a subgroup isomorphic with that described on pp. 100–103.)

Pascal's hexagon

In fig. 25.191 we have a hexagon $AQCPBR$ inscribed in a conic, illustrating the well-known result – Pascal's theorem – that the points of intersection of the pairs of opposite sides L $(= BR \cap CQ)$, M $(= CP \cap AR)$, $(N = AQ \cap BP)$ are collinear. The reason for the peculiar lettering becomes apparent when we look at the version of the configuration which has been relettered as in fig. 25.192. Here we have a 'crossed' hexagon, and the configuration is more compact since L, M and N are no longer off the page. But the most striking thing about fig. 25.192 is that it bears a strong resemblance to Pappus' configuration, fig. 25.18. The latter is, in fact, merely a special case of Pascal's hexagon when the conic happens to be a line-pair.

Now the six vertices of the hexagon may be relettered in 720 ways, and we consider subgroups of this group S_6 of permutations of order 720. One interesting one is that which *leaves the Pascal line* (LMN) *invariant.* (In fig. 25.192, the Pascal line is in a different position from that in fig. 25.191.)

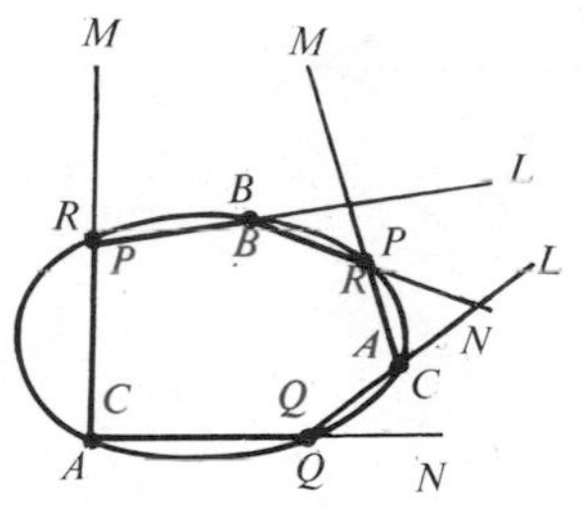

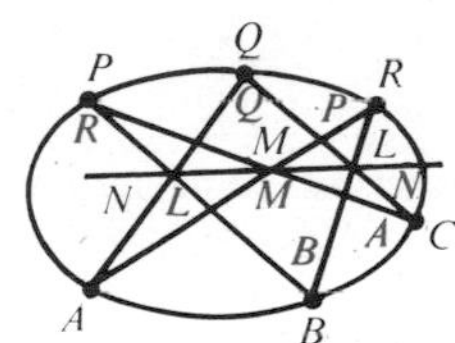

Fig. 25.19. Pascal's Hexagon – relettering.

One relettering which does not move the Pascal line, i.e. which merely reshuffles the points L, M and N, is shown in red, and corresponds to the permutation $\begin{pmatrix} A & Q & C & P & B & R \\ C & Q & A & R & B & P \end{pmatrix}$. There are twelve such permutations, and evidently they are obtained by cyclic permutations of the vertices, and their 'reverses' (compare pp. 189 ff., 221–2, 311, 492, Ex. 13.01) – rather like six beads on a necklace being moved round, with the possibility that the necklace may also be turned over. These form the group D_6 (compare Q. 21.23), and clearly there will be $720/12 = 60$ discrete Pascal lines for any particular set of six points on the conic.

Q. 25.26. Consider other subgroups of S_6, such as that which leaves A invariant, that which leaves L invariant, that which shuffles the sets $\{A, B, C\}$ $\{P, Q, R\}$ among themselves, etc. Consider also cosets, e.g. those permutations which move A to B, etc. Is it possible to find a group of order 60, every permutation of which leads to a distinct position of the Pascal line?

Desargues' perspective triangle theorem

Finally, in fig. 26.20 is shown the configuration for Desargues' perspective triangle theorem. An alternative lettering, shown in red, corresponds to the permutation $\begin{pmatrix} O & A & B & C & P & Q & R & L & M & N \\ R & M & C & L & P & O & Q & B & N & A \end{pmatrix}$. There are 120 possible reletterings which preserve the ten sets of three collinear points, leading to a group of order 120. You should verify the order of this group.

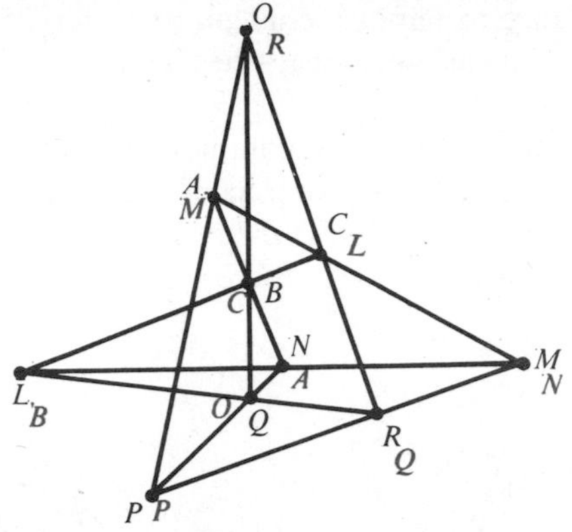

Fig. 25.20. Desargues' configuration – relettering.

Q. 25.27. Find the following subgroups:

(a) O fixed, and the set A, B, C shuffled among themselves, e.g.

$$\begin{pmatrix} O & A & B & C & P & Q & R & L & M & N \\ O & C & B & A & R & Q & P & N & M & L \end{pmatrix}.$$

(b) O fixed, and triangles ABC and PQR interchanged as well as having their vertices reshuffled, e.g. $\begin{pmatrix} O & A & B & C & P & Q & R & L & M & N \\ O & Q & R & P & B & C & A & M & N & L \end{pmatrix}$.

(c) O fixed, and the other nine letters freely moved round, always preserving collinearity.

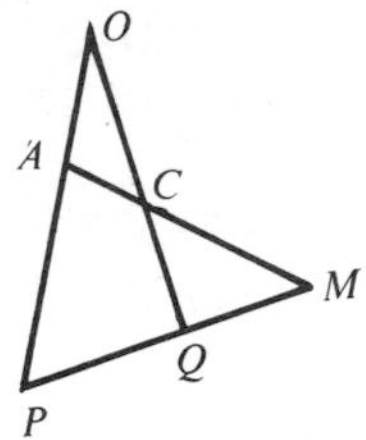

Fig. 25.21.

(d) Note that the figure contains a number of complete quadrilaterals, such as the one shown in fig. 25.21. Consider the subgroup of those re-letterings which permute the six points, as for example described by the permutation $\begin{pmatrix} O & A & C & P & R & M \\ R & P & O & M & C & A \end{pmatrix}$, which induces the permutation $\begin{pmatrix} B & Q & L & N \\ Q & L & B & N \end{pmatrix}$ on the remaining vertices.

(e) Can you discover the structure of the group of order 120?

In connection with (d) and (e) above, note that Desargues' configuration may be obtained as follows: Let 1, 2, 3, 4, 5 be any five points in space (no three collinear), and let Π be any plane preferably not containing any of the five points, and not parallel to the join of any pair of them. Let the line joining 1 and 2 meet Π in a point which we label 12. Similarly there will be ten (= ${}_5C_2$) such points: 12, 13, 23, 14, 24, 34, 15, 25, 34, 45, and these form Desargues' configuration (see fig. 25.22). For the ten lines (such as 235) are the common lines with Π of the ten planes which connect the five points in ten sets of three. Each complete quadrilateral is obtained from *four* points in space, for example, the selection 1, 2, 3 and 4 gives

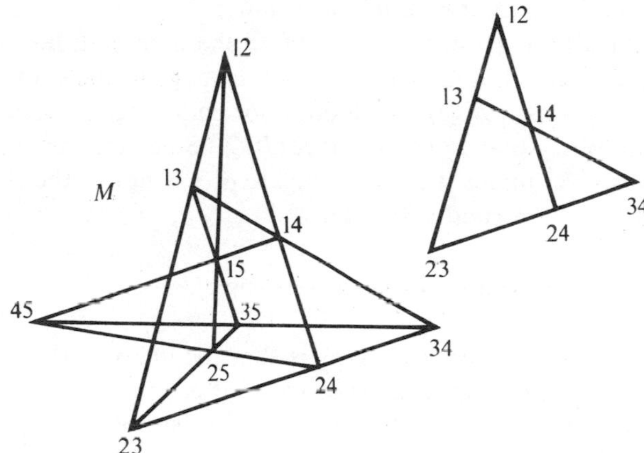

Fig. 25.22. Desargues' configuration by projection from five points in space.

rise to the complete quadrilateral of fig. 25.21. This consideration enables us easily to see that there are five complete quadrilaterals corresponding to the ${}_5C_4$ selections of four from the five points.

Stabiliser subgroups

On p. 498 we referred to the permutations which retain the line *ABC* in Pappus, configuration, but permutes these three letters among themselves, e.g. $(A\ B\ C)(P\ Q\ R)(L\ M\ N)$. There are twelve such possible reletterings which preserve the collinearities of the diagram, forming a subgroup of the group of order 108. This is an example of what is called a *stabiliser* because the subset $\{A, B, C\}$ is unchanged (or 'stable') as a set by those permutations, these forming the 'stabiliser' of the subset $\{A, B, C\}$. The twelve permutations listed on p. 498 are the stabiliser of the single element A.

In general, given a set $S = \{A_1, A_2, A_3, \ldots, A_n, B_1, B_2, \ldots, B_m\}$ the stabiliser of the subset $X = \{B_1, B_2, \ldots, B_m\}$ is the set of all those permutations from the group under consideration which leave the subset X invariant (the B's may, of course, be permuted among themselves).

Q. 25.28. Prove that the stabiliser is a subgroup.

Q. 25.29. Find other instances of stabilisers in this chapter, in Ch. 24 and elsewhere.

Q. 25.30. What is the order of the stabiliser subgroup for $\triangle APL$ in the group of re-letterings for Pappus Theorem (pp. 497–8).

Note, however, that the stabiliser of a subset is not necessarily a normal subgroup. A single counter-example will suffice. Suppose we consider the stabiliser of the set $\{A, B\}$ in the group of all the permutations of $\{A, B, C, D, E\}$. Let $y = (A\ B)(P\ R)$, which belongs to the stabiliser of $\{A, B\}$, and $x = (A\ B\ R\ P\ Q)$ which does not belong to the stabiliser. Then you may verify that $xyx^{-1} = (B\ R)(P\ Q)$. Since this replaces B by R, (though A is 'stable' under it), the conjugate of y is not in the stabiliser, so we do not have a normal subgroup.

Q. 25.31. Construct a much easier counter-example.

★An important theorem on stabilisers is that the index in the main group of the stabiliser subgroup of a single element is equal to the number of distinct elements into which that element may be transformed by all the elements of the group. For example, in the group of order 108 discussed above, the order of the stabiliser group of A is 12, and its index is therefore 9. This is the number of points into which A may be transformed by the 108 permutations of the main group. Taking a simpler example, the twelve permutations in table 16.03 represent the group D_6. The stabiliser of A contains four permutations. $(B\ C)$, $(P\ Q)$, $(B\ C)(P\ Q)$, and the identity, so its index in D_6 is 3, which is exactly the number of distinct positions to which A is moved by the twelve permutations of the group. Note that D_2 is not normal in D_6.

★**Q. 25.32.** Check the truth of the theorem as it applies to the stabiliser of P.

★**Q. 25.33.** Repeat for the twelve sets of permutations for the groups C_{12} and $C_2 \times C_6$ in tables 16.01 and 16.02.

We may prove this result by showing that the transforms of the given element X under the permutations of a group G may be placed into 1,1 correspondence with the cosets of the stabiliser S of X in the group G. This may be seen from the list of the twenty-four permutations of four letters on p. 318. The first six of these are the stabiliser subgroup of S_4 for the letter D; the next six are the coset which replaces C by D, the two remaining cosets replacing B and A respectively by D. Thus the number of cosets is seen to be the same as the number of elements which are replaced by D. Resuming the general proof, suppose p and q

are two permutations of G which change X into the same element, i.e. $p(X) = q(X)$. Then $q^{-1}p(X) = X$, so that $q^{-1}p$ belongs to S, and thus p and q belong to the same left coset of S, and conversely. Therefore for every element into which X may be transformed, there is a unique left coset of S, and vice versa, and so the result follows.

26 *Patterns*

Patterns are everywhere: the panes of glass in a window, tiles on a roof, the wooden slats in a fence, bricks in a wall; railway sleepers, piles of logs, trees in a forest; wallpaper, fabrics, china. Nor do we find them in man-made objects alone; the snow crystal under a microscope shows a beautiful hexagonal pattern; the honeycomb, the flower-head, pineapples, fir cones, even the humble crystals of common salt. Patterns ... owe their regularity and beauty to an underlying mathematical structure of which even a deliberate designer may be only partly conscious. A knowledge of this structure is, however, essential to one who wishes to analyse and interpret patterns; to the crystallographer, and those researchers who unravel the structures of molecules from X-ray diffraction patterns.

S.M.P., *Additional Mathematics, Book I* (C.U.P.), p. 168.

It has been said that the first examples of higher mathematics in the history of the world were symmetric repeating patterns designed by the ancient Egyptians. ... Weyl's fascinating book or the designs of the Dutch artist Escher will give a good start to those who may like to collect ornaments and patterns.

Some Lessons in Mathematics, ed. T. J. Fletcher (C.U.P.), p. 299.

... If the geometry of the Greeks was a disinterested pursuit of beauty rather than of utility, it is astonishing that it was not based on decorative ornament rather than surveying, and that group theory was not discovered long before Euclidean geometry. Perhaps this was because they did not make the crucial move from perceiving symmetries in isolation, to combining them.

T. J. Fletcher

We have already studied the symmetry of *finite* geometrical configurations and how the symmetries of such shapes may combine to give a group. In general these are finite groups, but you should be able to give at least one example of a plane figure which has an infinite symmetry group. In this chapter we shall broaden our ideas of symmetry, and deal with the question: 'How does one describe the symmetry group of a geometrical

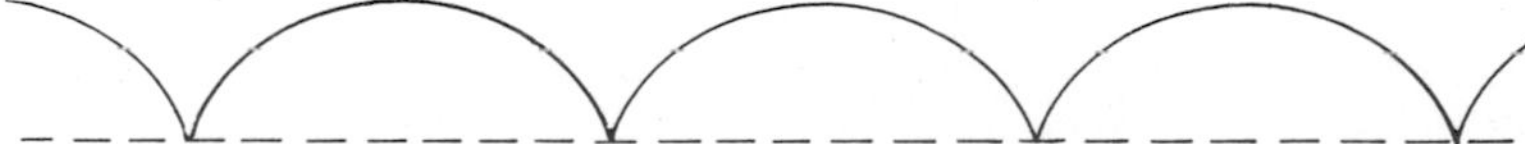

Fig. 26.01. The cycloid.

configuration which *extends to infinity*?' We can, for example, say of a regular pentagon that its symmetry group is D_5; or a parabola that its group is D_1; but how do we give the symmetry group of the cycloid (see fig. 26.01) which extends to infinity in one dimension? Again, we may describe the symmetry of a turbine with twenty blades as C_{20}, but what is the symmetry group of the network of squares on an ordinary piece of graph paper which is regarded as extending to infinity in two dimensions?

Before attempting to answer these questions, we will say a little about what is meant by a 'pattern'. The essence of a pattern is that a basic pattern-unit, or 'motif', or germ-idea, is taken, and to it are applied certain geometric transformations which preserve distance (rotations, translations, reflections and glide reflections), in such a way that the aggregate of the figures so generated possesses form and symmetry. If, for example, the motif is R and it is given reflections in two perpendicular mirrors, as in fig. 13.17, we obtain an extremely simple pattern in which the motif appears four times, and whose group is D_2 (with four elements). Again, in fig. 13.141 the group of the pattern was D_9, with the motif appearing eighteen times. If the mirrors had been inclined at an angle which was incommensurate with a revolution (say $60\sqrt{2}$ degrees), the group would contain a rotation through $120\sqrt{2}°$, and this would generate an infinite number of images which would completely blacken the area between the two concentric circles as in fig. 26.02. This may not be regarded as a pattern.†

Fig. 26.02. Non-pattern generated by rotations through irrational multiples of π.

However, it is perfectly possible for the motif to appear an infinite number of times, and the group to be an infinite one, though any finite region of the plane would contain only a finite number of images of the motif. The simplest example of this is the pattern generated by a single translation, and this is shown in fig. 10.07. Other examples of infinite patterns may be found in figs. 13.143, 13.15, 13.16, 14.06. We may summarise this paragraph by saying that **pattern is the result of systematic repetitions of a motif.**

We shall dismiss finite patterns in one paragraph. It can be proved (see J. H. Cadwell, *Topics in Recreational Mathematics* (C.U.P), p. 118; Coxeter, *Introduction to Geometry* (Wiley), p. 35; S.M.P., *Additional Mathematics*, Book I (C.U.P.), Ch. 8) that those plane patterns in which the motif appears only a finite number of times must have a fixed point, and that their groups are either cyclic or dihedral.‡ Examples of these may

† This is discussed more fully in S.M.P. Book I (C.U.P.), p. 177.

‡ The result is attributed to Leonardo da Vinci, and is sometimes known as Leonardo's Theorem.

be seen in figs. 8.01, 8.03, 11.04, 11.09, 12.01–7, 13.01, 13.04, 13.10, 13.11, 13.14, and these sort of patterns are familiar in everyday life on plates and saucers, tablecloths, carpets, ornamental wrought-iron work, and so on, if we confine our attention to examples which are effectively two dimensional.

The analytical approach

Now there are two possible approaches to the study of infinite repeating patterns: the analytic and the synthetic. Let us apply the analytic approach to the pattern of fig. 26.03 – that is, we take the completed 'work of art'

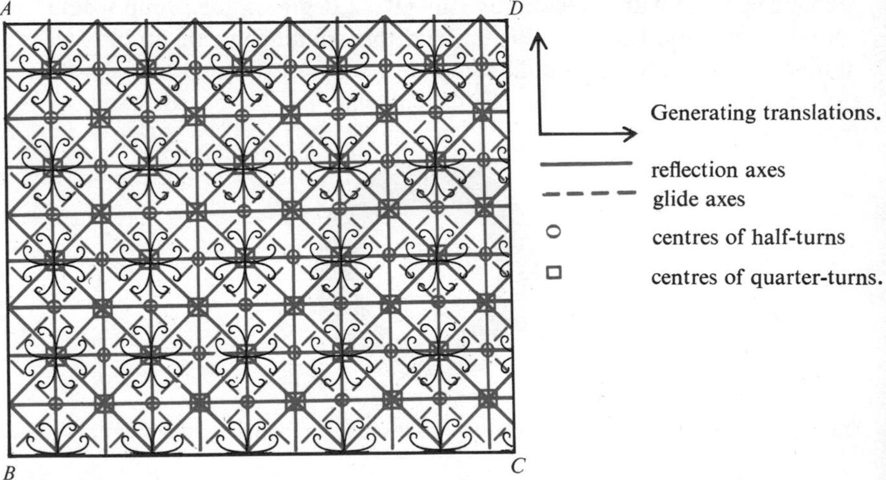

Fig. 26.03. The pattern **p4m**.

and scrutinise it to discover its symmetries. The figure is supposed to extend to infinity in both directions, though, for convenience, the portion of the pattern we are studying has been 'boxed in' to the rectangle $ABCD$. Note, however, that, as it happens, that rectangle and the configuration it contains have *no symmetry whatever!*

Perhaps the most obvious symmetries that strike one are the mirror axes shown by the thick red lines. Those which run at 45° to the sides of rectangle may not have been quite so evident. More likely to be missed are the glide reflections in the broken lines. This is easier to see if we isolate a small fragment of the pattern as in fig. 26.04 or fig. 26.042. Another basic symmetry ingredient is the rotational symmetry. Those points marked □ are centres of rotational symmetry of order 4; that is to say, if the *whole pattern* is rotated through 90°, 180° or 270° about any one of these points, it will be brought into coincidence with itself – it would look just the same as before. Less obvious are the centres of half-turn symmetry, which are marked ○. At first sight, you may think that these points too are centres of quarter-turn symmetry, but if you actually carry out the operation of rotating through a right-angle about one of these points, you may easily convince yourself that this is not so.

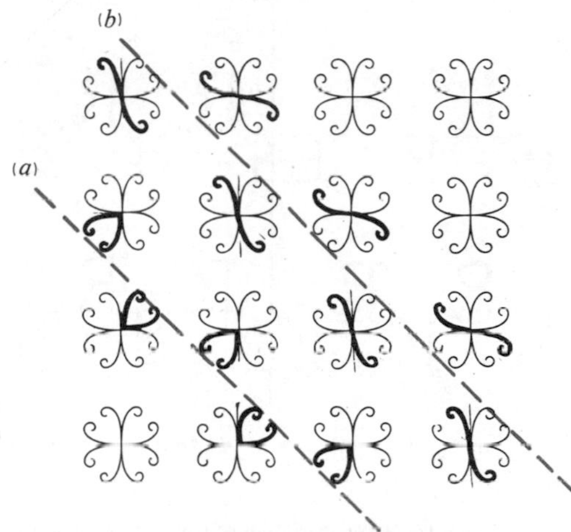

Fig. 26.04. Glide reflections abstracted from the pattern of fig. 26.03.

It may be thought that we have now exhausted the symmetries of this pattern. What remains besides the mirror symmetries about an infinity of axes in four different directions, the glide reflections about an infinity of axes in each of two perpendicular directions, and the infinity of rotational symmetries of periods 2 and 4? The answer is so simple that it may not have occurred to you: we have the translations. If the distance between a pair of thick lines (reflection axes) is 1 cm, then if the whole pattern is shifted (translated) by 2 cm to the right it will be brought into coincidence with itself; similarly if it is moved 2 cm to the left, or vertically up or down the page. But this is not the full story: evidently we may move the pattern 2, 4, 6, 8, . . . cm to the right, left, up or down, and we may combine these translations simultaneously. For example, we may shift the whole pattern 8 cm to the left and 6 cm upwards, and it will still look the same.

This example of a plane pattern was selected because it contained every possible type of symmetry – reflection, glide reflection, rotation and translation. Some of the infinite patterns we shall study will include only some of these symmetry ingredients. For example, the two-dimensional pattern of fig. 26.051 contains translations and glide reflections, but no rotations, because all the R's are 'facing to the right'. Fig. 26.052 contains half-turns as well as translations, but no mirrors or glides, because all the R's are 'the right way round'.

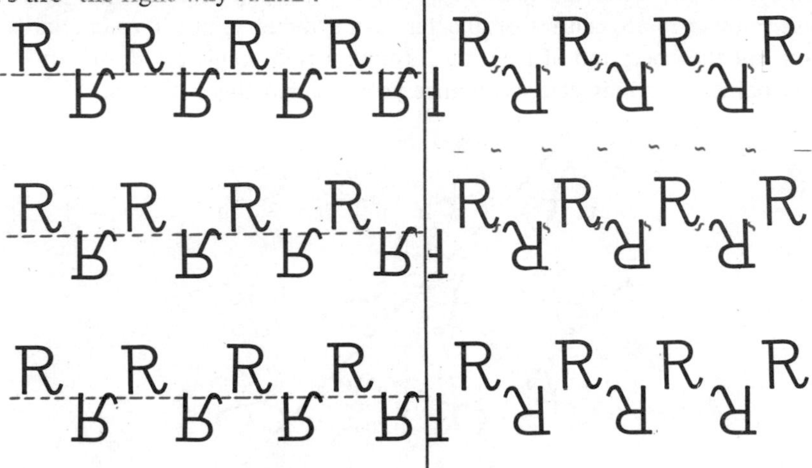

Fig. 26.051. Pattern containing glide reflections. **26.052.** Pattern containing half-turns.

Q. 26.01. Every infinite pattern must contain . . . (which type of isometry?)

Q. 26.02. Draw a strip pattern (one dimensional) which contains neither rotations nor glide reflections but does not contain mirror reflections.

Q. 26.03. Draw a plane pattern (two dimensional) which contains centres of rotation of order 3 (through 120° and 240°), but no mirrors or glides.

Q. 26.04. Analyse the patterns of the figures in the text in previous chapters and the present chapter in the same way as has been done in the above discussion.

Q. 26.05. Comment on the symmetries of the strip patterns of the first two rows, and the plane patterns suggested in fig. 26.06.

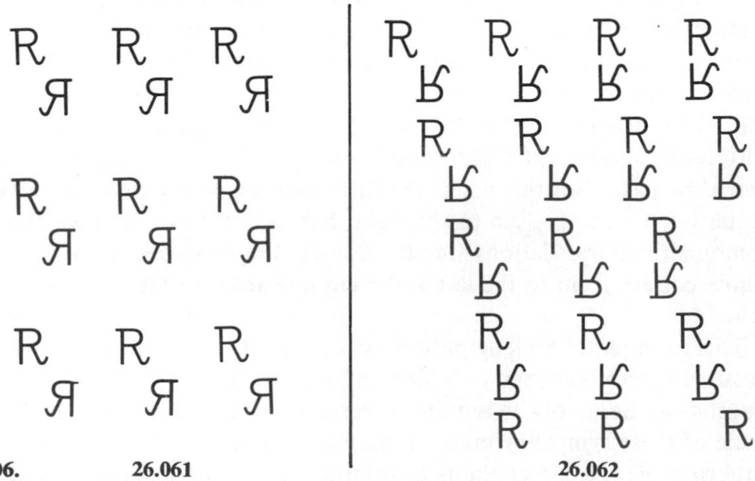

Fig. 26.06. **26.061** **26.062**

The synthetic approach

By this we mean *building up* the pattern from the basic motif – starting with a single 'crochet', how can one manoeuvre it to arrive at the pattern of fig. 26.03? By what sequence of operations may one *generate* the whole pattern?

For example, we might get the pattern (fig. 26.07),

26.071 t^{-2} t^{-1} 1 t t^2

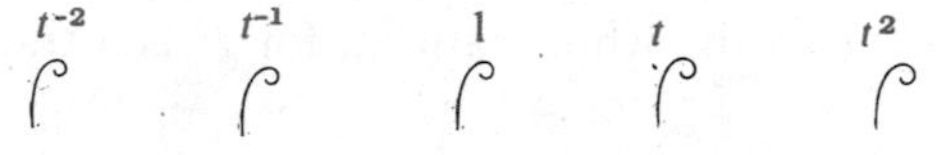

by using a single translation (t). Combining this with a reflection ($= a$) in one of the vertical axes, we should get:

26.072 t^{-2} at^2 t^{-1} at 1 | a t ta t^2

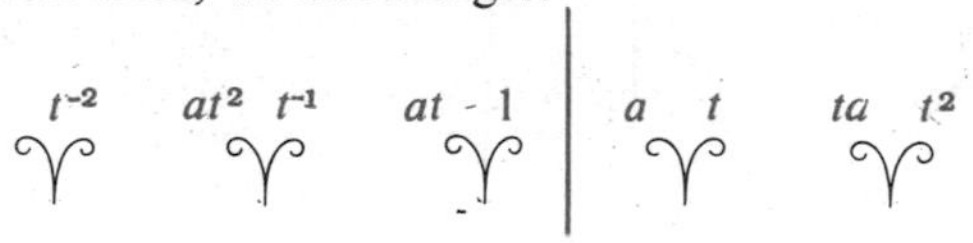

This pattern has generators t and a, and should be compared with fig. 13.143. If we now include a third generator, a reflection in a horizontal mirror b:

at^2 t^{-1} at 1 | a t ta t^2

26.073

bat^2 bt^{-1} bat b | ba bt bta bt^2

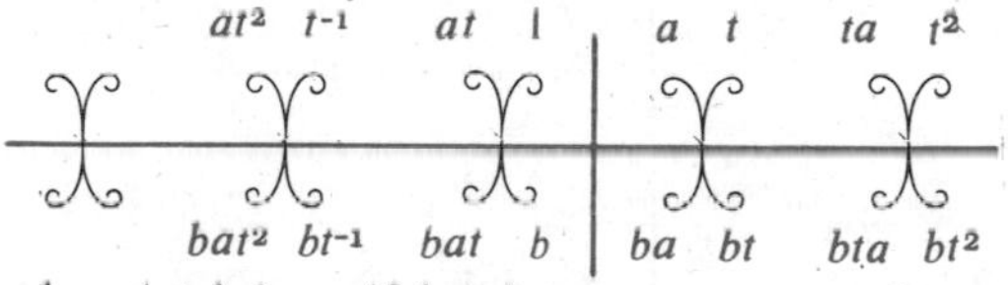

we still have only a 'strip', or 'frieze' pattern – a one-dimensional form. If we wish to extend to infinity in the vertical direction, we shall have to adjoin a fourth generator. This may be done in a large number of ways, the most obvious of which is to use a translation of 2 cm vertically (denoted s). Reflection in another horizontal line 2 cm above the first would do equally well, and we should get:

s

1 a t

26.074

b

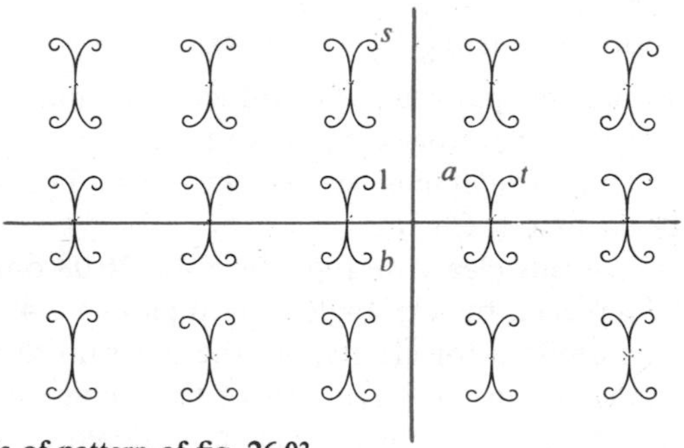

Fig. 26.071–4. Synthesis of pattern of fig. 26.03.

and the pattern is not yet complete, though it is growing!

Q. 26.06. What further generators are needed to complete the pattern? Label the remaining positions of the motifs in terms of the generators.
Q. 26.07. Draw the same pattern using a different motif (e.g. R instead of).

A more subtle way of extending fig. 26.073 into two dimensions would be to use a glide reflection, and fig. 26.075 shows the effect of a single glide reflection g, the glide axis being shown by a broken line. Repeated application of g will give us the whole pattern, for g^2 is a translation which

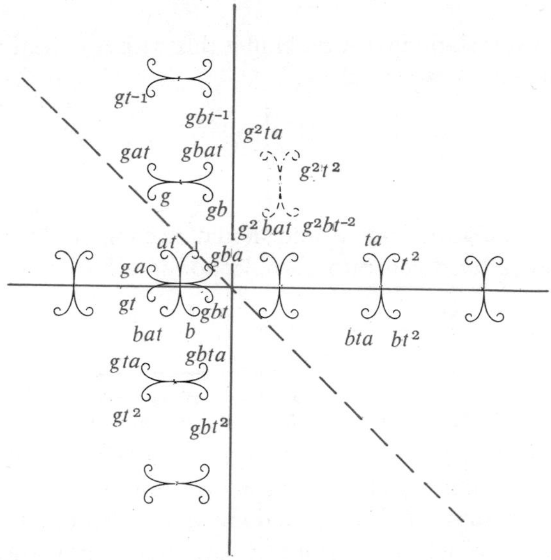

26.075 Glide reflection added to fig. 26.073.

has components 2 cm to the left and 2 cm upwards. Even powers of g give the portion of the pattern in which is upright, and odd powers of g give the other portion where is horizontal.

We have built up this pattern – the synthetic method – using a number of generators. These generators could have been selected in a large variety of ways. Can we find a *minimum* set of generators? The answer is that we have already done so – in Ch. 14!

In order to make this clear, we reproduce fig. 26.08 only with the crochet motif replaced by a triangle with angles 45°, 45°, 90°. Most of the 's have been omitted for clarity, but the diagram shows how the operations of *reflection* in the three sides of the triangle will generate the pattern exactly as they did in the example considered in Ch. 14, pp. 233 ff. This figure should be compared with the whole of the discussion on those pages.

Thus we have succeeded in forming this pattern from only three basic generators, three reflections. You should have no difficulty in seeing why the final pattern contains translations, rotations, and glide reflections as well as reflections; for we know – indeed we may verify from fig. 14.06 that – *cb* and *bc* are quarter-turns; *ab* and $(bc)^2$ are half-turns; *abc* is a glide reflection (cf. pp. 285, 368, fig. 26.25), and $(abc)^2$ is a translation.

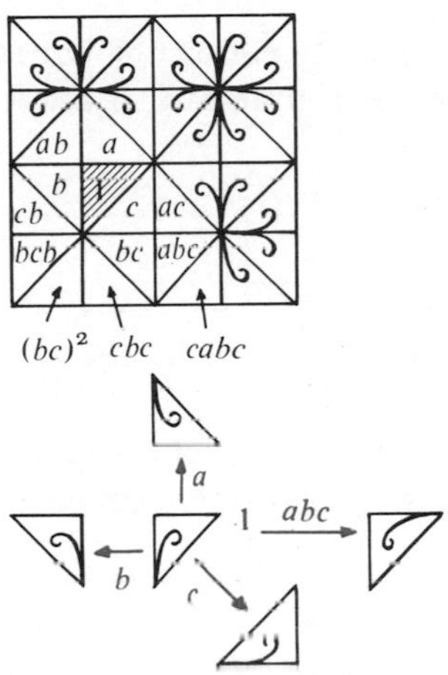

Fig. 26.08. Generators of **p4m** (compare fig. 26.03).

Q. 26.08. Repeat the work of building up a pattern, only using, instead of the 45°, 45°, 90° triangle, a set square with angles 30°, 60°, 90° the triangle itself bearing a suitable motif.

Fundamental regions

Now while the *pattern* is built up from a basic *motif*, the *plane* itself is built up from the right-angled isosceles *triangle* which is known as a 'fundamental region' for the pattern.

In the example we have taken our basic motif to be the crochet. There is no reason why we should not take a larger motif, such as a double crochet, or , or even ; indeed we might even take as a motif and apply to it translations through 2 cm horizontally and vertically. But we deliberately chose the *smallest possible* motif, and in this case our motif had no symmetry of its own. The reason for taking the smallest possible motif is that this reveals more about the structure of the whole pattern. Moreover, even the smallest possible motif is not necessarily unique. The motif for the pattern could just as well be as shown in fig. 26.09 in place of R. For the pattern of fig. 26.03, the motif could be as in fig. 26.10. The fact that the motif may itself possess some *internal* symmetry of its own

R R R R or

Fig. 26.09. Alternative motifs.

does not, however, necessarily endow the pattern as a whole with any extra symmetry. To take a simple case, if the R's in fig. 26.09 above were

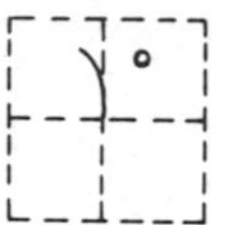

Fig. 26.10. Alternative motif for fig. 26.03.

replaced by ◺ (fig. 26.11), the pattern would have no further symmetry even though each triangle is isosceles – reflection in the axis m is not an element of the group.

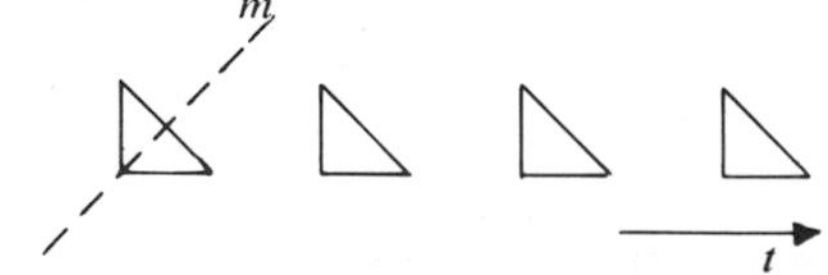

Fig. 26.11. Strip pattern.

Q. 26.09. Draw the pattern generated by m and t. (You must show every possible 'word' such as mt^4mtmt^{-2}; the pattern will fill the plane.)

Q. 26.10. Give various examples of a minimum motif for the curve $y = \sin x$.

Again, the two-dimensional pattern of fig. 26.12 has R as its motif. This pattern would have no more symmetry if the motif had been M or E, or even an *equilateral* triangle, as shown.

Q. 26.11. Which of the capital letters of the alphabet *would* give the pattern extra symmetry if it were used in place of R?

The above pattern is the simplest plane pattern generated by two translations. Every other possible plane pattern will contain this basic type as a subgroup, and if the R is replaced by  , we do indeed obtain the richer pattern of fig. 26.03 provided the generating translations are equal in distance and in the direction of two perpendicular axes of symmetry of this motif.

If the generators were equal and perpendicular and the motif were R, the pattern, fig. 26.123 would possess no additional symmetry, even though the R's are now arranged over a network, or lattice, of squares. There can be no glides or reflections since all the R's are the right way round, and

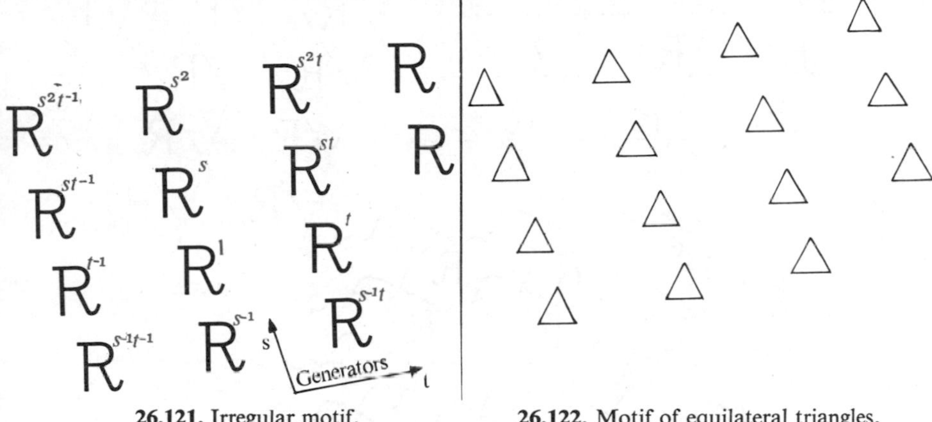

26.121. Irregular motif.

26.122. Motif of equilateral triangles.

Fig. 26.12. Pattern generated by two translations.

there can be no rotations because they are all standing up straight. So there can only be translations – we have the *simplest* type of plane pattern again.

26.123. The pattern is only **p1**, in spite of the square lattice.

It is tempting to suppose that the fundamental regions for this simple plane pattern are necessarily parallelograms. Fig. 26.132 shows that that choice of parallelogram is by no means unique, but fig. 26.133 shows the the fundamental region does not have to be a parallelogram at all, but may be extremely irregular in shape! Generally speaking, the richer the pattern is in symmetry, the less choice one has for a fundamental region. The plane pattern of fig. 26.03 has the 45°, 45°, 90° triangle as a fundamental region, and this is unalterable.

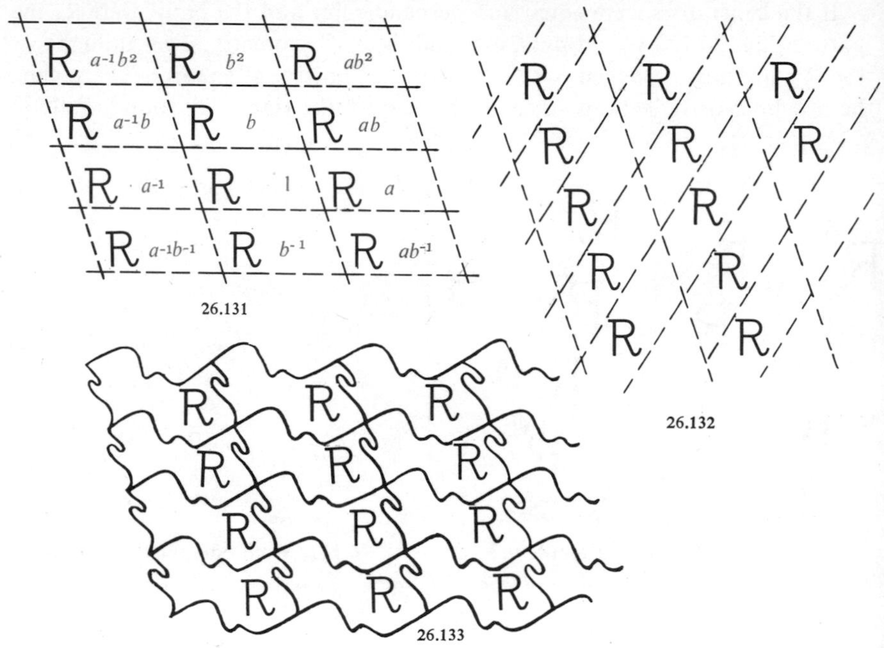

Fig. 26.13. Alternative fundamental region for **p1**.

A plane pattern with half-turn symmetry, such as could be produced by applying the generating translations to a letter such as S, N or Z will have fundamental regions which could be triangles, as in fig. 26.141, but could also be the curvilinear 'triangles' of fig. 26.15. A striking choice of a

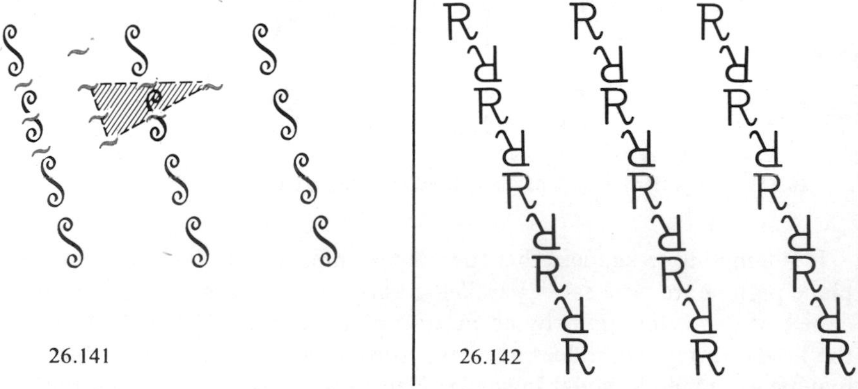

Fig. 26.14. The pattern **p2** generated by a half-turn and two translations, or by three half-turns.

26.141. Motif.

26.142. Motif R.

fundamental region for the plane pattern **pg** is to be found in Coxeter op. cit., p. 57, where an Escher drawing is reproduced in which the fundamental region is a knight on horseback! In this type of pattern the centres of half-turn symmetry are more numerous than one might expect – they are situated at the vertices of the fundamental triangles, and *also* at the mid-points of their sides (some of these are marked in fig. 26.141).

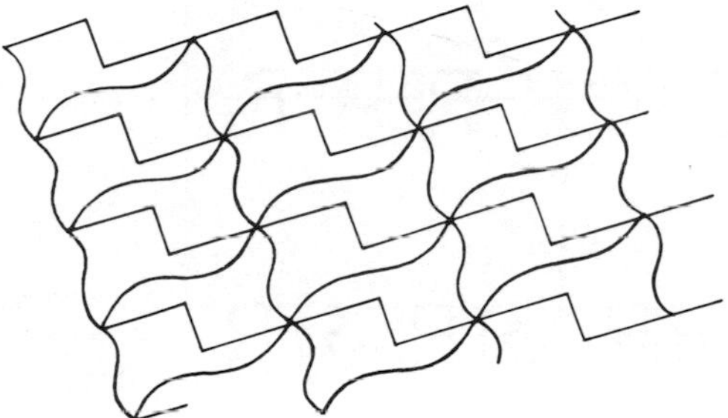

Fig. 26.15. Possible fundamental region for the pattern of fig. 26.14.

Limitations on possible patterns: the seven frieze patterns

Now there is evidently no limit to our freedom of choice of motif – it may be R or ꞁ, or a flower, or a horse, or a beetle, a triangle, or a sausage. However, it is perhaps surprising to learn that once the motif has been selected the number of ways of developing a pattern from it by combinations of isometries is strictly limited. In fact there are exactly seven distinct 'frieze', or 'strip' patterns, that is, repeating patterns extending to infinity in one† dimension. The two dimensional repeating patterns are limited to 17, while in three dimensions there are as many as 230 'space groups'.

Classification of patterns

Having examined the analytic and the synthetic methods of approaching the study in the case of a typical pattern, we now consider the question of the *classification of patterns*, beginning with the simple case of one† dimension.

During the course of the previous page, examples of all seven types of frieze pattern have appeared. We give these below (fig. 26.16), the motif in each case being R, and the generators are also shown, together with each image of the motif expressed in terms of the generators selected. The patterns have been numbered from 1 to 7, and in the final column we give instances of other figures in this book where the patterns may be found.

† Strictly, 1½ dimensions, according to Coxeter, op. cit., p. 49.

	Pattern labels	Generators	Structure	Se als Fi
26.161	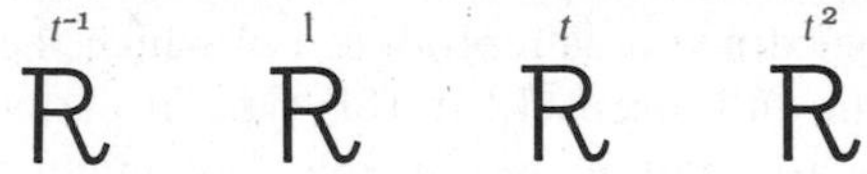 t^{-1} 1 t t^2	→ t 1 translation	C_∞	10. 26.
26.162	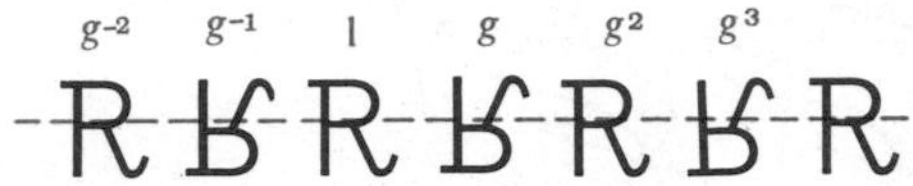 g^{-2} g^{-1} 1 g g^2 g^3	g (glide reflection)	C_∞	26.
26.163	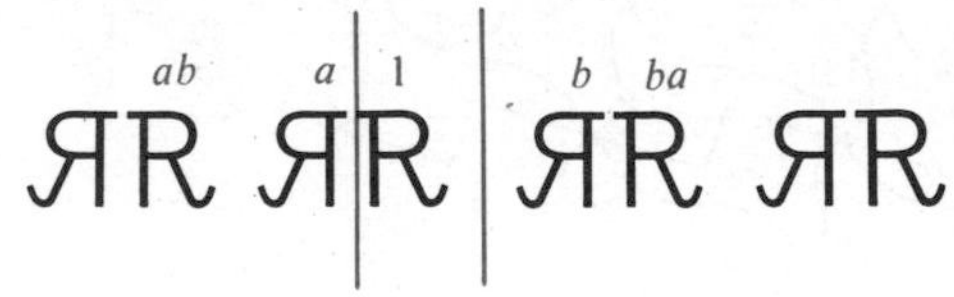 ab a 1 b ba	2 reflections, a and b or reflection and translation ($t = ba$)	D_∞	13. 26. 26.
26.164	 ab aba 1 a ba b	2 half-turns (a and b) or half-turn and translation ($t = ba$)	D_∞	13.
26.165	 bab 1 a / ba b ab aba	1 reflection (a) and 1 half-turn (b) ($g = ab$; $t = (ab)^2$)	D_∞	13.
26.166	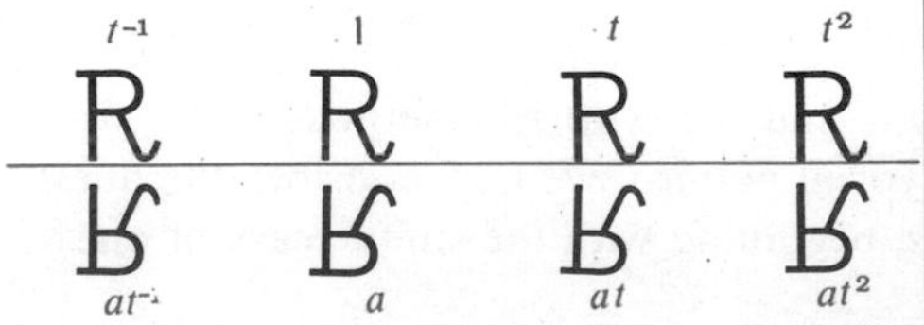 t^{-1} 1 t t^2 / at^{-1} a at at^2	1 reflection (a) and 1 translation (t) or reflection and glide reflection ($g = at$)	$C_\infty \times D_1$	
26.167	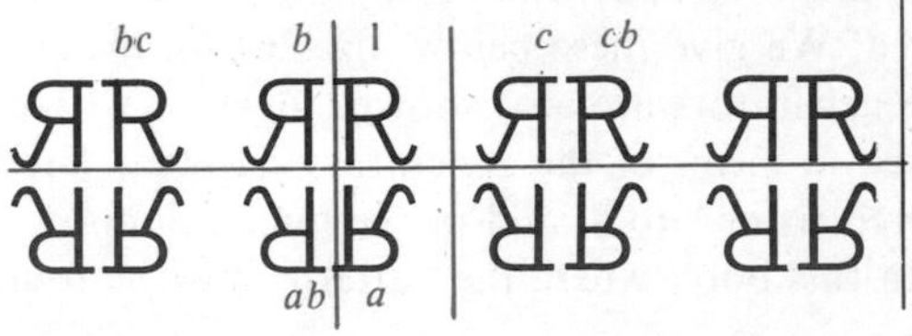 bc b 1 c cb / ab a	3 reflections a, b and c Other sets of generators may include $t = cb$; $g = acb$, etc,	$D_\infty \times D_1$	26.

Fig. 26.16. The seven frieze (strip) patterns.

These seven strip patterns may be easily remembered by means of a simple mnemonic based upon the alternative forms shown in fig. 26.168. We get the word SHEAR or SHARE from numbers 1, 3, 4, 6 and 7, while No. 2 may be remembered as FOOTSTEPS, and No. 5 is the sine curve, or corrugated iron, or battlements, whichever you prefer.

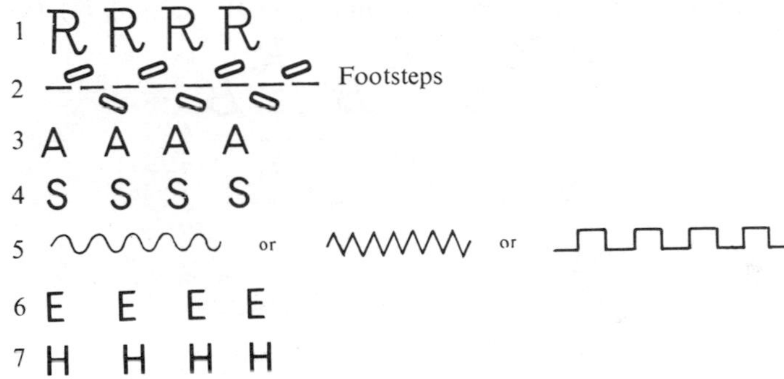

Fig. 26.168.

Q. 26.12. Make a cardboard template as shown in fig. 26.17 with a motif cut out so that its outline may be drawn round with a pencil. a^* is a reflection axis which moves with the template, and B^* is a small hole which is used for the purpose of performing half-turns about a pin passed through it. The centre of half-turns also, of course, moves with the template. Now generate pattern No. 5 by performing the operations a^* and b^* in all possible combinations. Label each

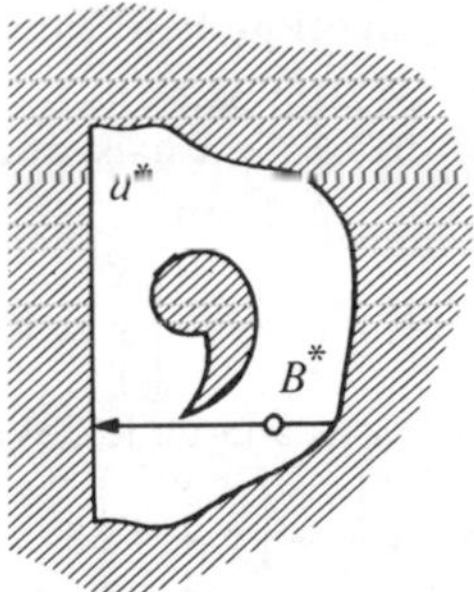

Fig. 26.17. Template for generating pattern by reflection and half-turn.

position of the motif: 1, a^*, b^*a^*, $a^*b^*a^*$, etc. (Compare notation with p. 367.) Note the 'guide line' through B should be drawn on both sides of the template: its purpose is to enable half-turns to be gauged more accurately. When reflections are being performed, it is best to rule a pencil line along the straight edge, then turn the template over, matching the positions of the arrow.

Q. 26.13. Devise a template, or modify the one in the previous question, so that each of the remaining six strip patterns may be constructed.

Q. 26.14. To which types of frieze pattern do each of the following belong? In each case give a motif for the pattern:

(a) T⊥T⊥T . . .; (b) MUMUMUM . . .; (c) DD · DD · DD . . .;
(d) the cycloid; (e) the curve $y = 3 \sin x + \sin 3x$;
(f) KꓘKꓘKꓘK . . .; (g) FĿ · FĿ · FĿ . . .; (h) podpodpod . . .;
(i) CHECKCHECKCHECK . . .; (j) pobopobop . . ./(k), (l), (m), (n), (o), (p) (see fig. 26.18).

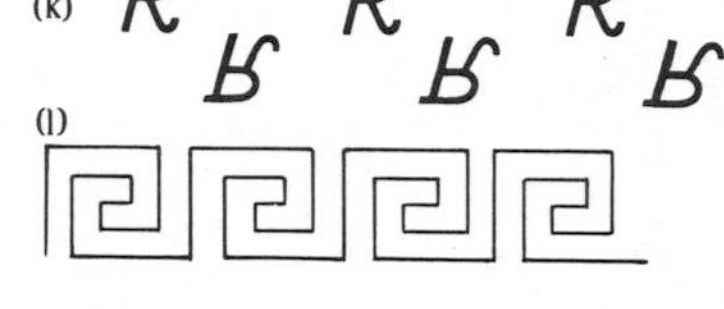

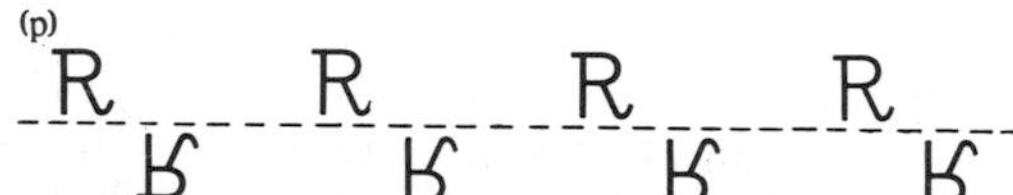

Fig. 26.18.

Q. 26.15. Find other mnemonic words besides SHARE and SHEAR.

Q. 26.16. Evidently all of types 1 to 6 are subgroups of type 7. Give a complete account of which patterns are subgroups of which others.

Q. 26.17. Work out the theory of strip patterns by using matrices as follows: We take the axis of the pattern to be the x-axis, and a translation matrix may be taken to be

$$t = \begin{bmatrix} 1 & 0 & 1 \\ 0 & 1 & 0 \\ 0 & 0 & 1 \end{bmatrix} \text{ operating on the point } \begin{bmatrix} x \\ y \\ 1 \end{bmatrix}; \text{ a half-turn, } h = \begin{bmatrix} -1 & 0 & 0 \\ 0 & -1 & 0 \\ 0 & 0 & 1 \end{bmatrix}.$$

Work out the matrices for reflections and glides. What is the matrix which corresponds to the translation $\begin{bmatrix} x \\ y \\ 1 \end{bmatrix} \to \begin{bmatrix} x \\ y+1 \\ 1 \end{bmatrix}$?

(See *Some Lessons in Mathematics*, ed. T. J. Fletcher (C.U.P.), p. 341.)

The two-dimensional (Wall-paper) patterns

There are, as we stated, seventeen distinct patterns for a given choice of motif. One of these (code-name **p4m**) has been dealt with in detail. As was noted during that study, these plane patterns *must* contain translations in two different directions, but may also contain reflections, glide reflections and rotations. There are only five of the seventeen patterns in which the motifs are not 'turned over', i.e. which contain only translations and rotations (see figs. 26.19–23). This is a consequence of a theorem, known as Barlow's theorem, which states that the only possible periods of rotation in such patterns are 2, 3, 4 and 6.†

In analysing a given plane pattern, the easiest way to detect whether it contains rotations is to actually carry out the rotation and observe the effect; for example, if the pattern looks the same as a result of a quarter turn of the whole sheet, then we know that the symmetries include rotations of period 4 (and the code-name for the pattern will contain the figure 4). If a rotation through 120° makes no difference to the aspect of the pattern, then we may have rotations of period 3, or possibly of period 6. Having detected rotational symmetry, it remains to discover the positions of the centres of these rotations.

Q. 26.18. If the pattern looks the same when turned upside-down, what may we infer? What may we infer if it looks different?

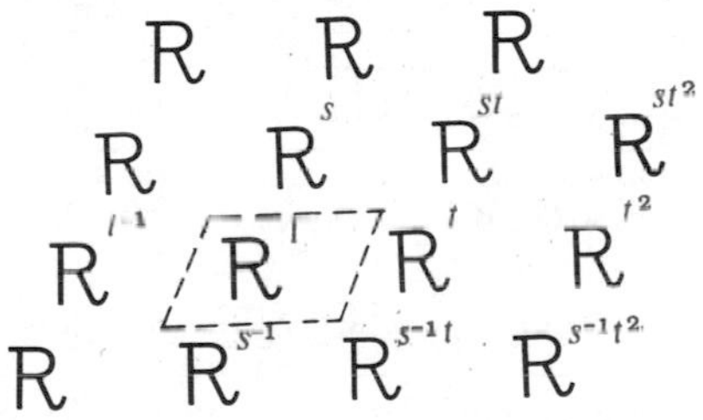

Fig. 26.19. The five direct plane patterns; **p1**, generated by two translations.

In the first five figures we show these five 'direct' patterns, and indicate possible generators. A possible fundamental region is shown by broken lines in each case. Centres of rotational symmetry are shown, where appropriate, as follows:

- ~ half-turn centre
- △ rotations through $\pm 120°$
- ⊡ rotations through $90°n$
- ⬡ rotations through $60°n$.

Q. 26.19. Make a template of cardboard as shown in fig. 26.24. Pierce holes A^* and P^* to be centres of half-turns are third-turns respectively. Mark in the guide-lines shown to enable the rotations to be gauged with reasonable accuracy. Now trace the pattern obtained by successive applications of the operators p^* and a^*.

† For a proof, see Coxeter, op. cit., p. 60; S.M.P. op. cit., p. 179; Cadwell, p. 122; Weyl, *Symmetry* (Princeton U.P.), p. 101.

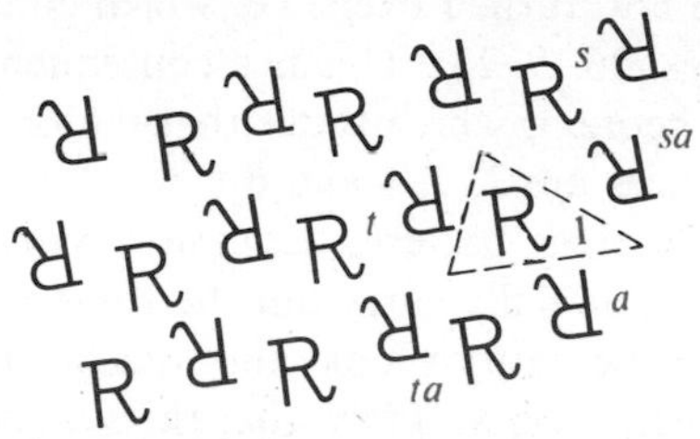

Fig. 26.20. p2; half-turn and two translations, or three half-turns (*a*, *ta*, *sa*).
Q. Is the group Abelian?

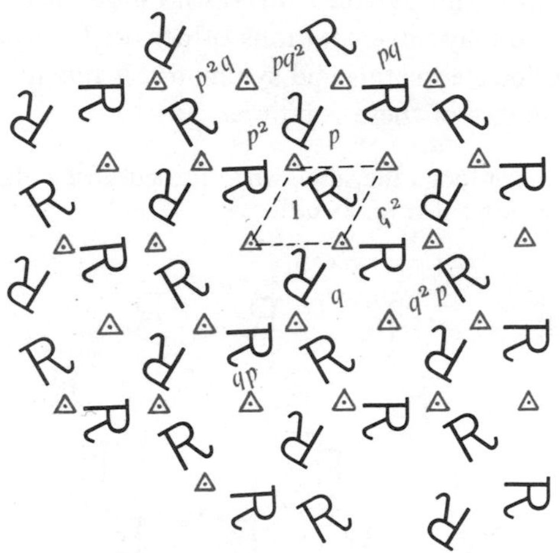

Fig. 26.21. p3; rotations through 120° about two points (p q); or one 120° rotation and a translation (e.g. $t = pqp$).

Q. Find alternative fundamental regions.

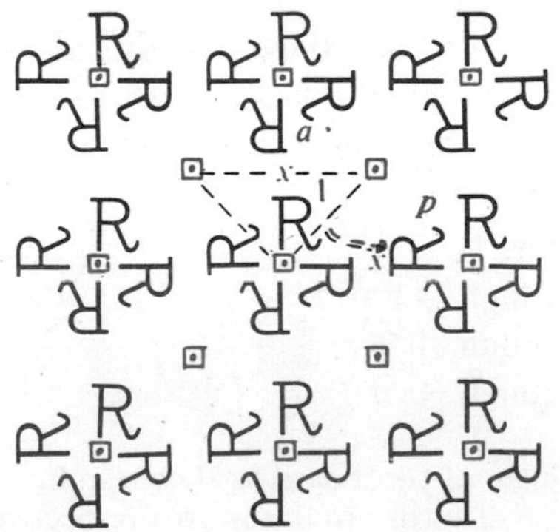

Fig. 26.22. p4; half-turn and a quarter-turn p.

Q. Label the remaining positions of the motif in terms of a and p.
Can you find an alternative fundamental region?
Find an alternative set of generators.
If motifs each with C_4 symmetry (such as swastikas) were arranged on a parallelogram (not square) lattice, would the pattern be **p4**?

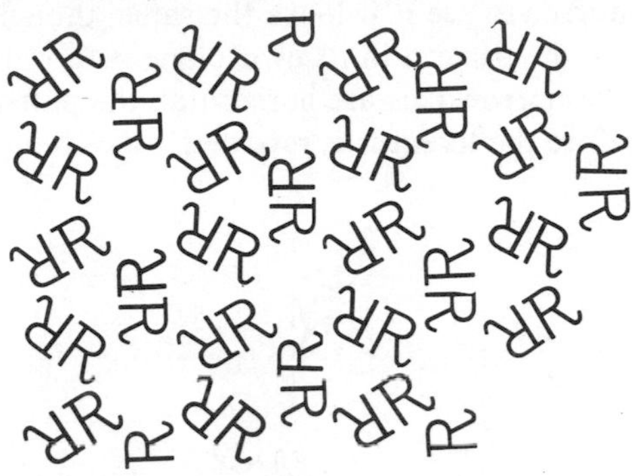

Fig. 26.23. p6; a half-turn and a rotation through 120°.

Q. Give some alternative generators.
Copy the diagram and mark in the various centres of rotation indicating their type.
Show that C_6 is a subgroup.
Select two generators, and starting with any motif as '1', label the remaining images of the motif in terms of these generators.
Mark in a fundamental region for **p6**.

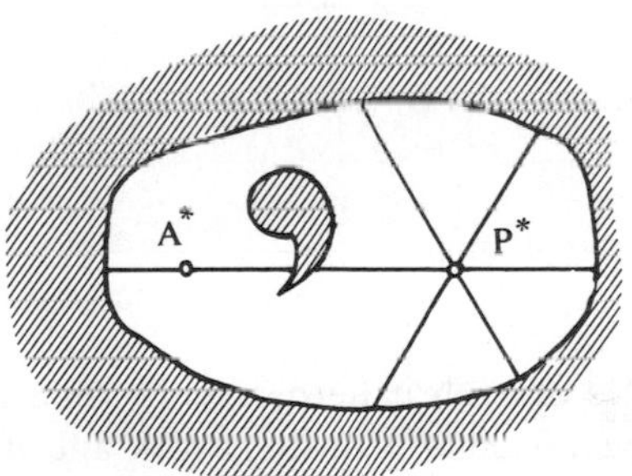

Fig. 26.24. Template for constructing **p4**.

The twelve plane repeating patterns which contain opposite isometries

Now it is common for wall-paper patterns to contain opposite isometries – reflections and glides. Fig. 26.25 shows the sort of way in which a glide reflection may appear where the motif is a flower. All the remaining twelve plane patterns contain reflections and glides, and some contain rotations as well. The code lettering is suggestive of the types of group operations which the pattern described contains. For example, fig. 26.03 is called **p4m** because it contains quarter-turns and mirror reflections. (It also contains glides as well.) If one is analysing a given pattern, then the simplest way of discovering whether it contains reflections is to hold the pattern up to a mirror to see if it looks the same, though it may be necessary to rotate the pattern in its own plane before this is apparent. For example, if the mirror axes are horizontal, the pattern should be given a quarter-turn before the likeness is revealed.

Fig. 26.25. Glide reflection (flower motif).

We now give examples of each pattern, using R as a motif in each case, with comments, and suggestions for further work and investigation. See figs. 26.26–37.

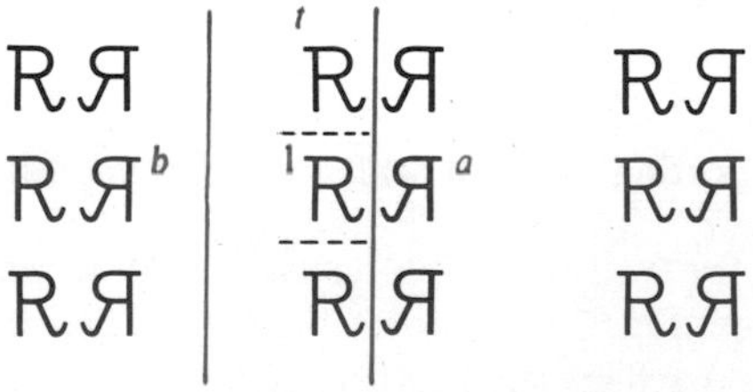

Fig. 26.26. pm; two reflections in parallel axes and a translation.

Q. Label the other motifs in terms of a, b, t. Does the translation have to be parallel to the reflection axes? What is at?
Hold the pattern up to a mirror; then turn it on its side.
Show that this pattern may be described by a motif with D_1 symmetry situated at the vertices of a rectangular lattice. Draw it, using the letter M as motif.

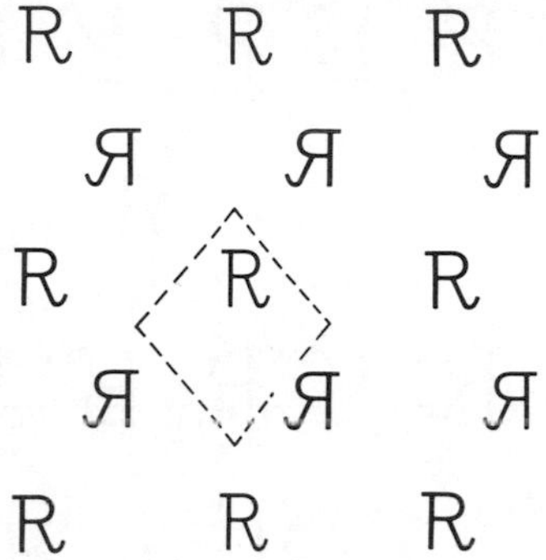

Fig. 26.27. pg; two parallel glide reflections.

Q. Find alternative generators.
Label motifs in terms of g_1 and g_2.
What are g_1g_2 and g_2g_1?
Hold the pattern up to a mirror, and note that it does not look the same.
Find a fundamental region which is a *kite*.
Explain why it is essential that the subgroup which contains those R's which are the right way round shall be at the vertices of a *rectangular* lattice.

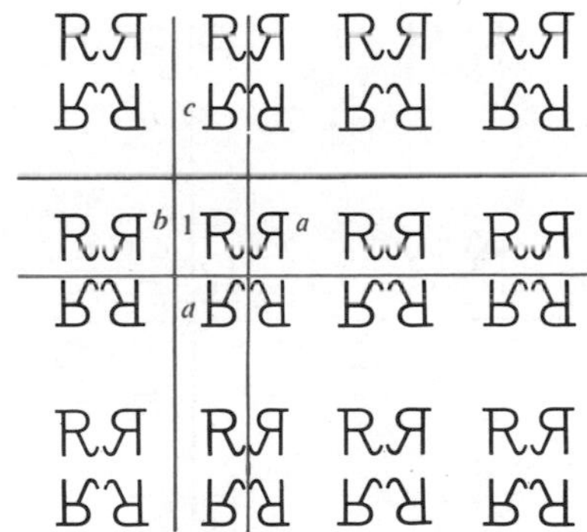

Fig. 26.28. pmm; reflections in four sides of a rectangle.

Q. Show that the pattern may be generated by two translations applied to a motif which itself has D_2 symmetry, as for example the subgroup $\{1, a, d, ad\}$ in the accompanying diagram. Draw this, using the letter H as a motif.
Find alternative fundamental regions.
View the pattern in a mirror, and also turned on its side in a mirror.
Turn the pattern upside-down. Is it unchanged? What does this indicate?
Show the positions of the various rotation centres.
Find some relations between the generators *a*, *b*, *c* and *d*.

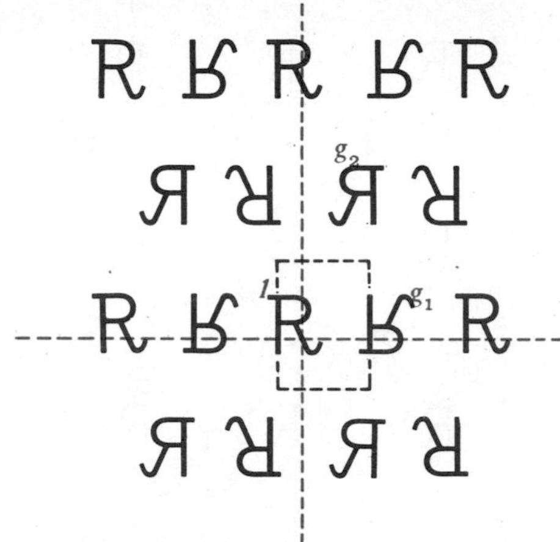

Fig. 26.29. **pgg**; two perpendicular glide reflections.

Q. Why do the glide reflections have to be perpendicular?
Locate the centres of half-turns.
Label the motifs in terms of g_1 and g_2.
What are g_1g_2 and g_2g_1? Express a vertical and a horizontal translation in terms of these generators.
Which pattern is generated by two non-perpendicular glides?
Identify the subgroups: Gp(g_1^2); Gp(g_1^2, g_2^2); Gp(g_1, g_2^2); Gp (g_1^2, g_2); Gp(g_1, g_1g_2); etc., and describe the patterns in each case.

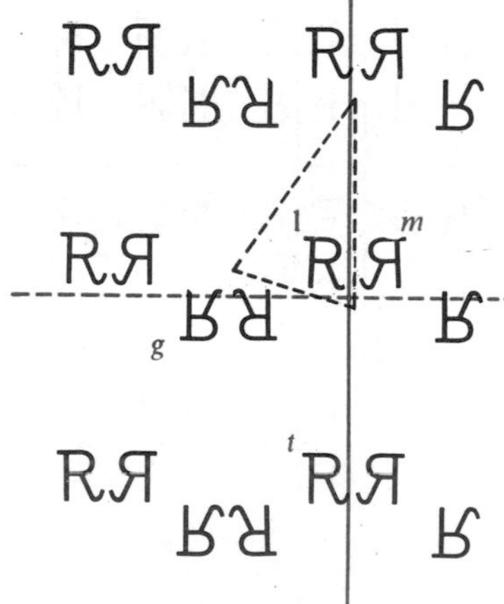

Fig. 26.30. **pmg**; reflection m and glide g in perpendicular axes, and a translation t not parallel to g.

Q. Show that the pattern may be generated by m and two half-turns. Express these half-turns in terms of m, g and t.
Find glide reflections other than the odd powers of g.
Prove that, if a fundamental region is a triangle, then one side lies along a reflection axis, and the other two pass through half-turn centres.
Find subgroups of Gp(m, g, t).

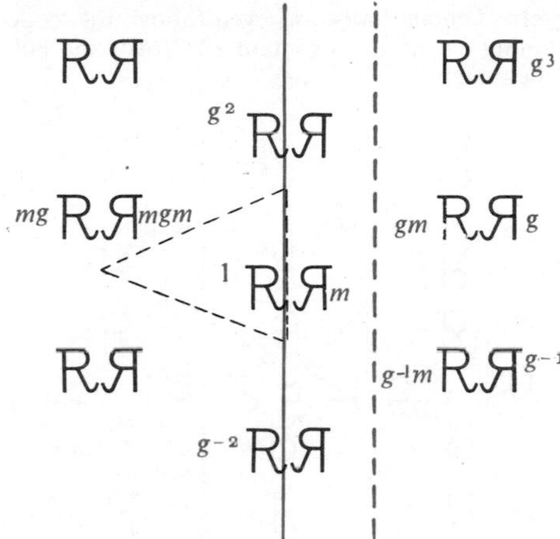

Fig. 26.31. cm; a reflection and a parallel glide reflection (sometimes known in the wall-paper trade as a 'half-drop pattern').

Q. Give a quick reason why there are no half-turns.
Express remaining motif images in terms of m and g. Show that the pattern may be obtained from a motif with D_1 symmetry (such as the letter A) at the vertices of a lattice of rhombuses. How must the A's be orientated in relation to the lattice? Would there be any further symmetry if the rhombuses were squares?
What freedom of choice of fundamental region have we?
Find subgroups of Gp(m, g).
Show that the group is defined by the relations $m^2 = 1$, $mg^2 = g^2m$.

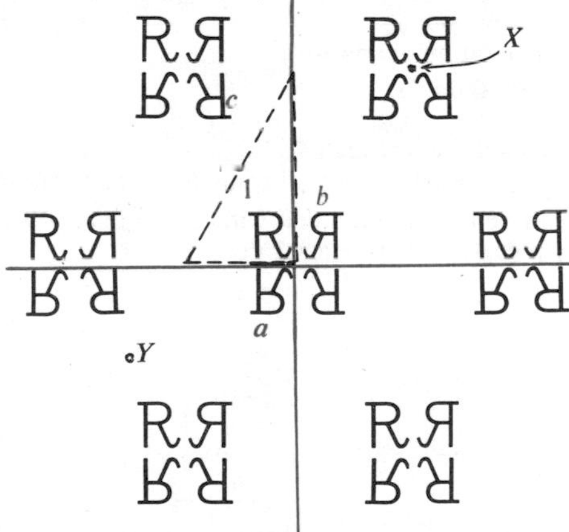

Fig. 26.32. cmm; two perpendicular reflections (a and b) and a half-turn (c).
Q. Show $g = ac = (ca)^{-1}$ is a glide reflection.
Gp(a, b) is a finite subgroup D_2.
Gp(b, c) is an infinite strip pattern – of which type? $(bc)^2$ is a translation to the right.
Find other glides, translations, half-turns, and subgroups containing them.
Describe the pattern in terms of a symmetrical motif unit and a lattice.
Express the following operations on the motif labelled 1 in terms of a, b and c:

(1) a half-turn about X, (2) a reflection in the horizontal axis through X,
(3) a half-turn about Y, (4) a glide reflection in the vertical line through Y.

Some of the above patterns contain half-turns even though the 'code-name' does not contain a '2'. The remaining five patterns contain 120°, 90°, and 60° turns. We omit **p4m** since it has been adequately shown in fig. 26.03.

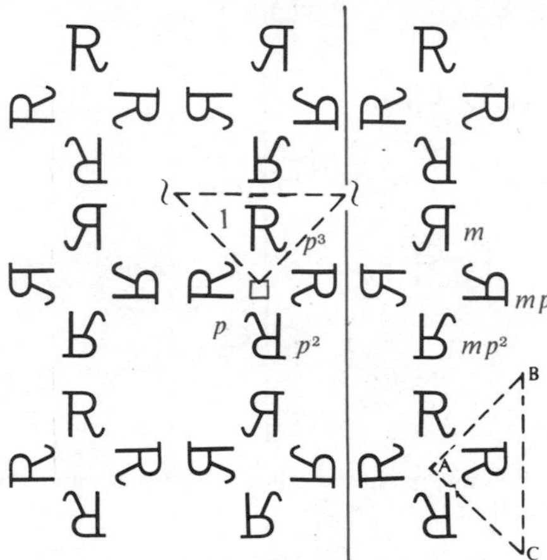

Fig. 26.33. p4g; reflection and quarter-turn (*m* and *p*).

Q. Label the motifs in terms of m and p. mp^2 is a glide reflection in a horizontal axis through P. What is a glide reflection in the vertical axis through P?
Find the axes of the glide reflections mp and pm.
Show that the vertices of the triangular fundamental region are a quarter-turn centre and two-half-turn centres.
Find a square fundamental region.
Express in terms of the generators m and p:

(1) a clockwise quarter-turn about A,
(2) a half-turn about B,
(3) a reflection in a horizontal line through C,
(4) the operation which takes the fundamental triangle to the position ABC.

Show that the pattern can be built from a unit with C_4 symmetry (e.g. swastika) using two perpendicular glide reflections with equal pitches.

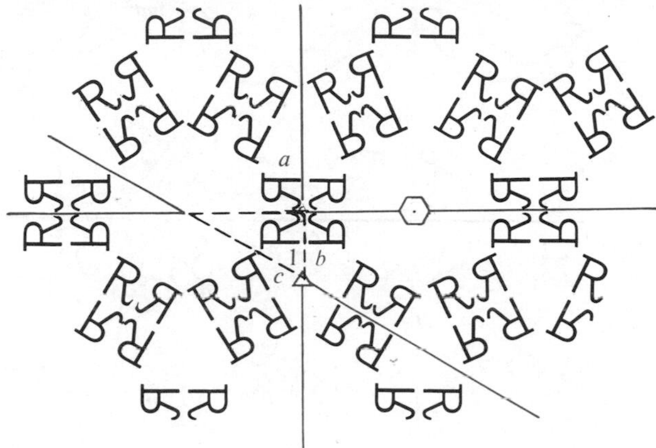

Fig. 26.34. p6m; three reflections in the sides of a fundamental triangle with angles 30° 60°, 90°.

Q. Can you find a system of generators which include a 60° rotation, e.g. about the centre marked ⬡?
Compare with **p4m**. Make a template for generating **p6m** in a similar way to that which was used for **p4m**.
Discover the mirror axes, glide axes and rotation centres, and indicate the periods of the latter.
Find some subgroups.
Show that the pattern may be built up by a suitable arrangement of equilateral triangles.

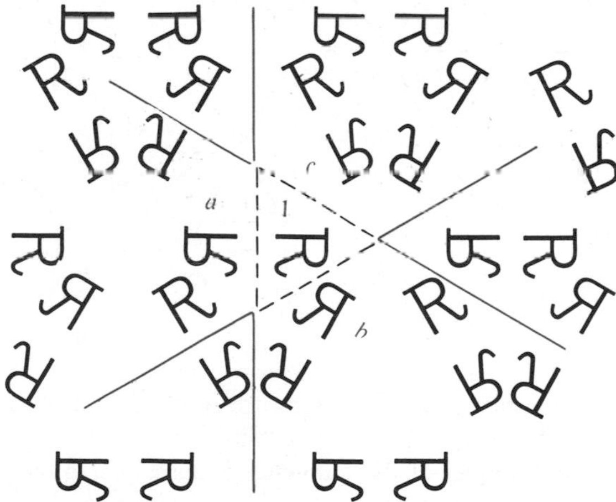

Fig. 26.35. p31m; reflections in three sides of equilateral triangle.

Q. Express some glide reflections in terms of *a*, *b* and *c*, and identify the glide axis corresponding to each.

Note: the difference between **p31m** and **p3m1** is most readily understood by studying figs. 26.371 and 26.372. The subgroup $\{1.\ a, b, ab, aba\}$ (D_3) of **p31m,** and one of the D_3 subgroups of **p3m1** are replaced by an equilateral triangle motif in each case.

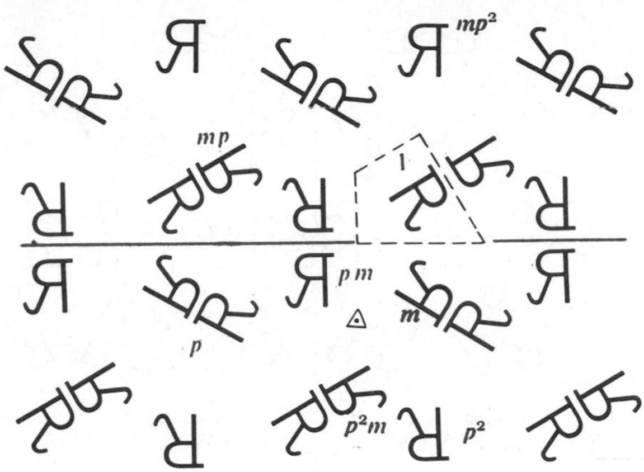

Fig. 26.36. p3m1; a reflection (m) and a third-turn (p).

Q. Find a fundamental region which is a triangle instead of the kite shown.

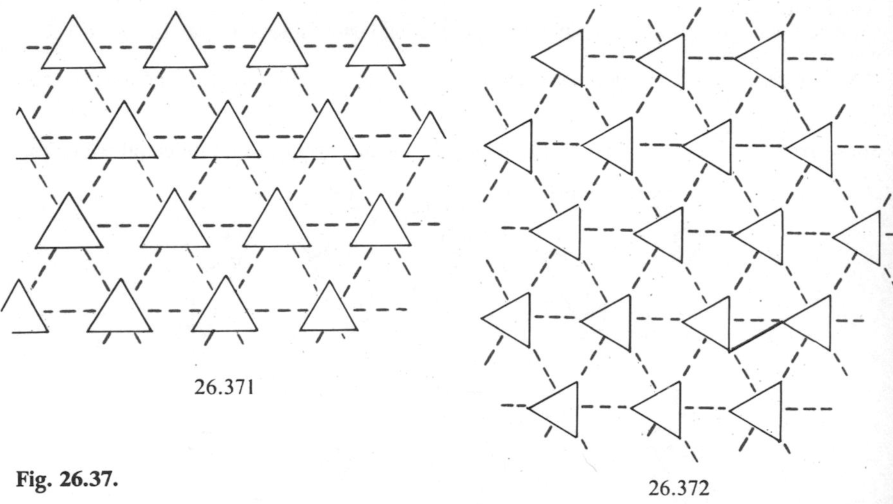

Fig. 26.37.

Q. 26.20. Which of the plane pattern groups are Abelian? Which of the strip pattern groups are Abelian?

Q. 26.21. Construct examples of the following patterns, using the motif indicated:
(a) **pm** and **pgg** using Z; (b) **pm, cm, pmg, pmm** using A; (c) **pmm, p2, cmm, pmg**, using H; (d) **p1, cmm, pmg** using E; (e) **p3** using .

Q. 26.22. Which plane pattern is generated by a translation and a reflection – consider all possible cases.

Q. 26.23. Find the type of pattern in each of the portions of pattern shown in fig. 26.38 (28 parts).

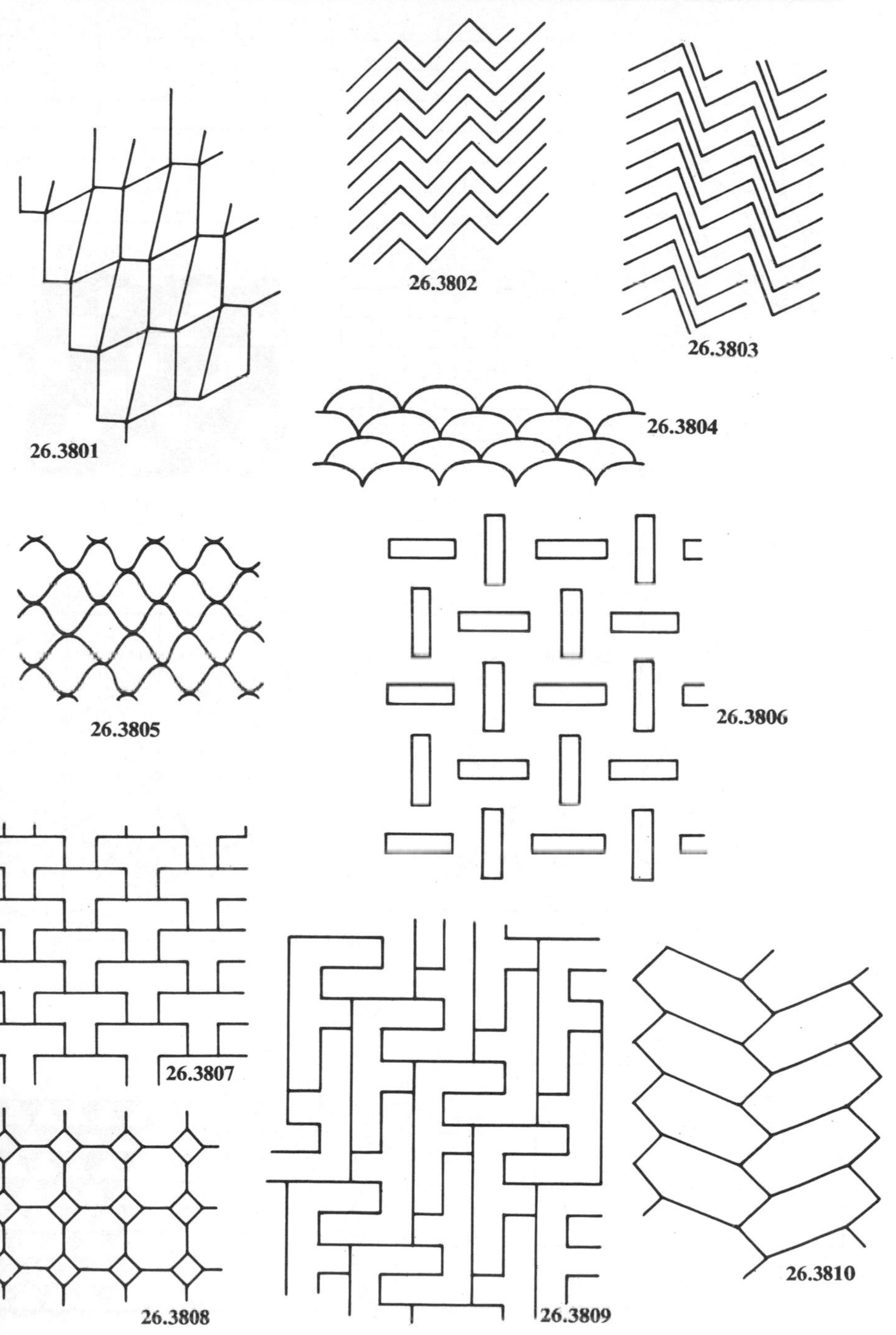
26.3801
26.3802
26.3803
26.3804
26.3805
26.3806
26.3807
26.3808
26.3809
26.3810

Fig. 26.38.

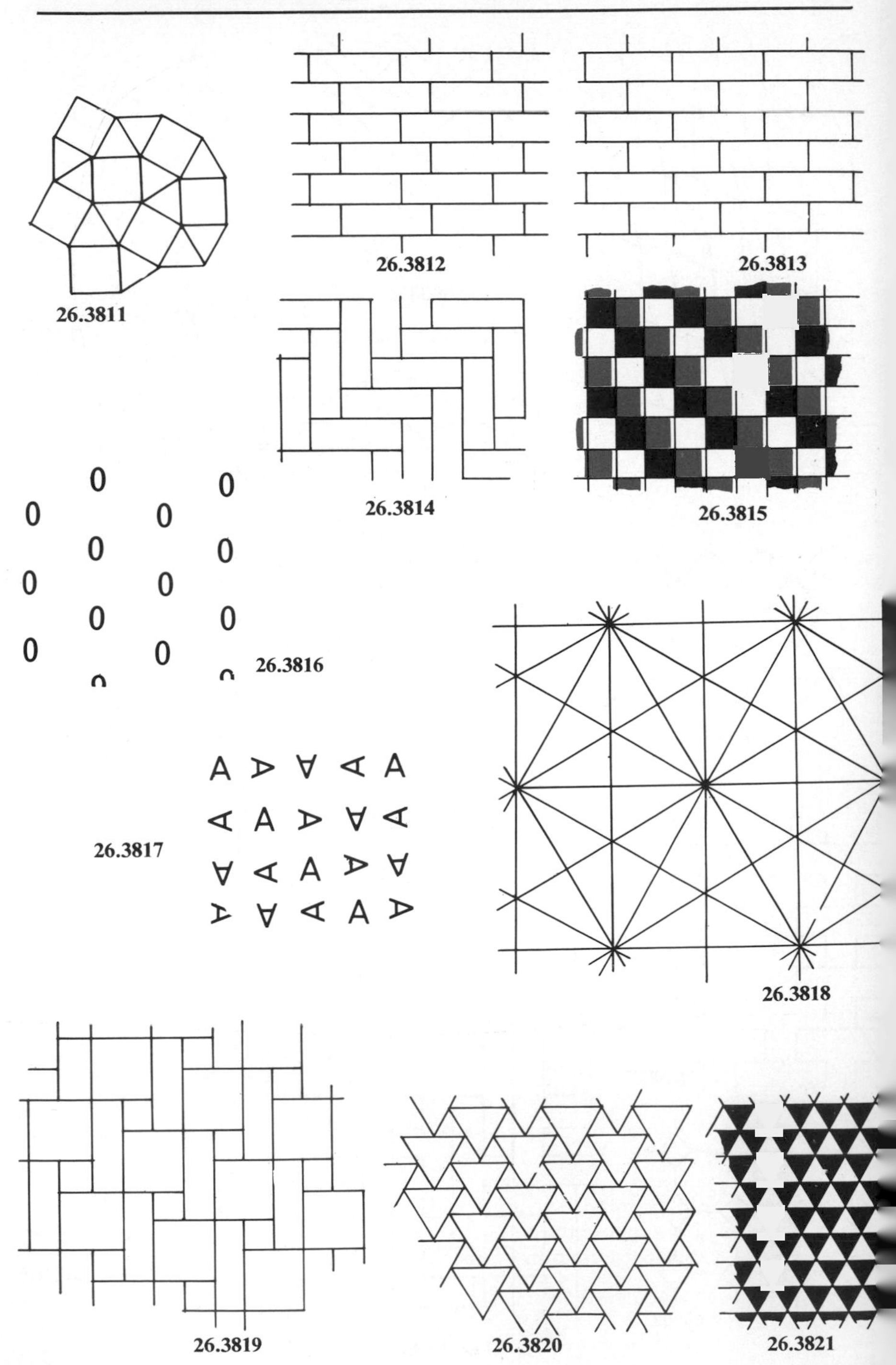

26.3811 26.3812 26.3813 26.3814 26.3815 26.3816 26.3817 26.3818 26.3819 26.3820 26.3821

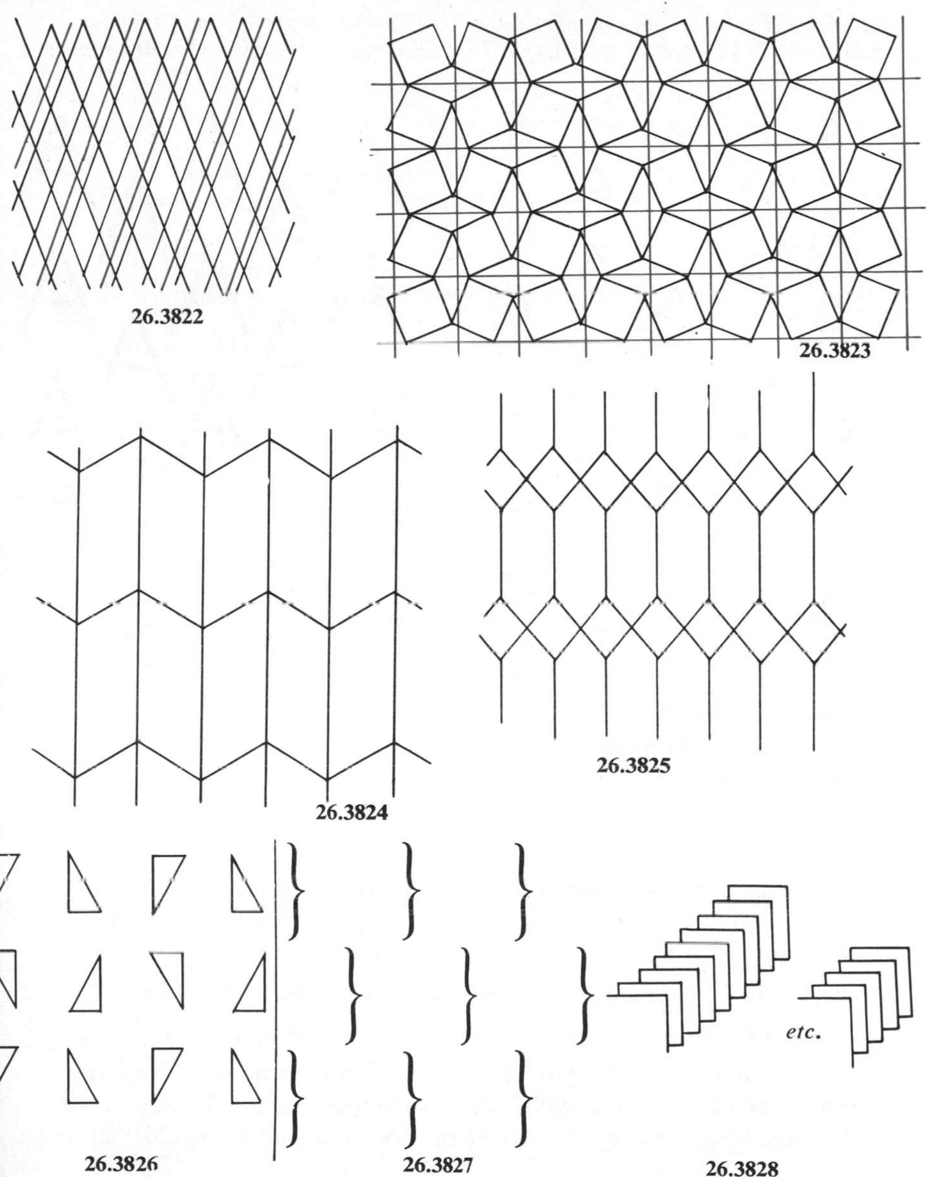

26.3822

26.3823

26.3824

26.3825

26.3826

26.3827

26.3828

Q. 26.24. Consider the reduced groups in the above cases when the regions are coloured according to certain schemes (e.g. the pattern of an infinite network of squares is **p4m**; but suppose the squares are coloured alternately black and white as an infinite chess-board, then the pattern would lose symmetry ...

Q. 26.25. Which plane patterns may be built up from strip pattern Nos. 2, 4, 5 and 6?

Q. 26.26. Find whether plane patterns are unaltered in group structure when each motif is enlarged, e.g. as in fig. 26.39. Repeat for the case when the motifs are rotated.

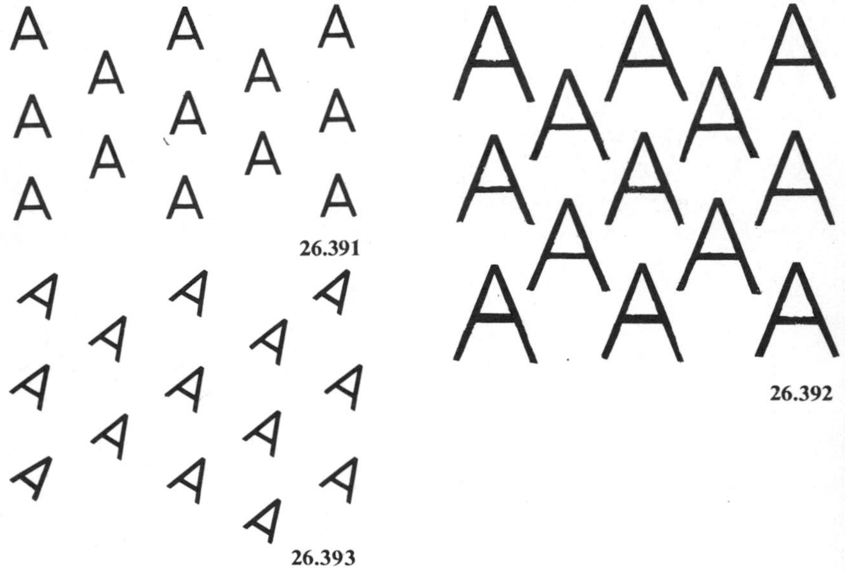

Fig. 26.391. The pattern **cm**.
26.392. Effect of enlargement of motif.
26.393. Effect of rotation of motif.

Sub-patterns

We have seen that a pattern may be generated by an augmented version of the basic motif. For example, **p4m** (fig. 26.03) was generated from the motif by the operations of reflection in the sides of a 45°, 45°, 90° triangle. Had one taken as motif, the pattern could have been generated as **p4** by a quarter-turn and a half-turn. Larger motifs, such as and might also have been selected, as we have already mentioned. Now the selection of a larger motif will always† correspond to a subgroup of the full group of symmetries of the pattern. For example (see fig. 26.072), the

† Provided the symmetries of the basic unit are symmetries of the pattern as a whole. For example, in fig. 26.03 (**p4m**), each motif of the form has D_4 symmetry, and the pattern is the same, in effect, as a pattern of square motifs arranged with their centres forming a square lattice as in fig. 26.123. Now if these motifs were all given a rotation through 45°, we should still have **p4m**; but if they were rotated through, say, 30°, we should merely have **p2**; each motif would still possess its individual D_4 symmetry, but of these symmetries only the half-turns would belong to the symmetries of the *whole pattern*, so the squares might just as well be S's.

subgroup D_1 $\{1, at\}$ corresponds to the choice of ϒ; the choice of the subgroup $\{1, a, b, ab\}$ corresponds to (), or, in effect to X, with symmetry group D_2. Again, in **pmm** (fig. 26.28), the subgroup $\{1, a, d, ad\}$ corresponds to the motif RЯ/ꓤᴚ, i.e. a motif which itself possesses D_2 symmetry.

Now it is quite possible to generate a plane pattern by an '*infinite motif*', i.e. to have a sub-pattern containing an infinite number of R's, and such a sub-pattern would correspond to an infinite subgroup. For example, in **pmm** (fig. 26.28), the subgroup generated by a vertical and a horizontal translation (*ab* and *cd*) would look like fig. 26.123, **p1**. This pattern needs only to be given a reflection about two perpendicular axes correctly placed to yield the plane pattern **pmm**. Again, referring to fig. 26.28, **pmm**, Gp (a, b, d) is a strip pattern (type 7), and by using this infinite strip as an augmented motif, it needs only to be subjected to the vertical translations c to give the pattern **pmm**. Let us now look at **pgg** (fig. 26.29). We may imagine this pattern divided into horizontal (or vertical) strip patterns, each of type 2 ('footsteps'). Starting with a single one of these strip patterns, giving it a half-turn about a suitable point will move it to the position of the adjacent strip, and the whole pattern may be generated by two half-turns of the elemental strip pattern of type 2.

You should look at the remaining plane patterns to see how each may be analysed into (and also synthetised from) the various frieze patterns, and also how many different types of frieze pattern are embedded in each plane pattern. Evidently types **p1**, **p2**, **p3**, **p4** and **p6** cannot possibly contain strips of type 2, 3, 5, 6 and 7. (Why?) Which plane patterns contain all seven strip patterns, each corresponding to an infinite subgroup? (there are five of them), and it is an excellent test to see if you can pick them out. Finally, **pmg** (fig. 26.30) contains all but one of the frieze patterns. (Which one?)

Subgroups of the plane patterns

Examples have already appeared of sub-patterns which themselves are groups being contained within the plane patterns, and also within the frieze patterns. In many cases they may be recognised visually, or else they may be discovered by constructing a group with a suitable choice of generators. Consider for example **p6**, a version of which, using a 'hook' as motif, is given in fig. 26.40. Those motifs which have been blacked in form a sub-pattern whose group structure corresponds to **p2**. This subgroup may be generated, for example, by three half-turns, or by a half-turn and two translations which themselves are generators of the original group. All those hooks lying horizontally in the figure belong to this subgroup, and since they are just one third of the whole population of hook motifs, it follows that the subgroup of the pattern **p2** has index 3 in the group of the pattern **p6**.

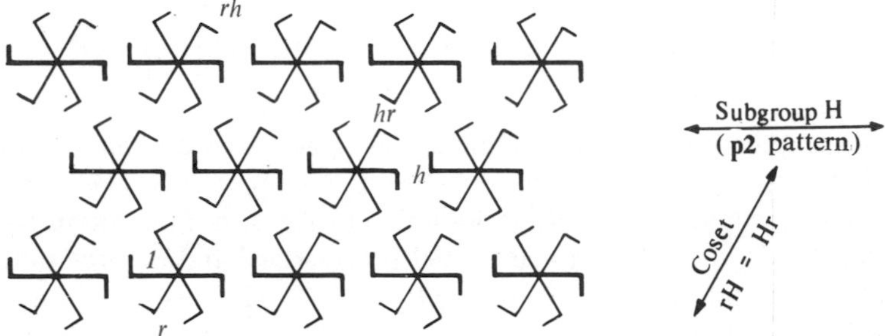

Fig. 26.40. p2 as a subgroup of **p6**.

Q. 26.27. Draw the same diagram and mark in those motifs which form the subgroup whose pattern is **p3**. What is the index of this subgroup?

In the latter case, the subgroup is normal, since its index is 2, but we may wonder whether **p2** is a normal subgroup of **p6**. To answer this, we consider the left and right cosets of the subgroup **p2** by an element not in **p2**, e.g. an anticlockwise rotation r through $\pi/3$ about a selected centre of hexagonal symmetry, such as the point marked O. Let h be any element of the **p2** subgroup. Then r takes the motif marked h to the position marked rh, and this is in the left coset. It is evident that all elements of this left coset, which are obtained from elements of the subgroup by a rotation through $+\pi/3$ will lie in the direction parallel to this. Now let us find the right coset. We must now identify hr, so that the rotation r is first applied and followed by the element of the subgroup, *which is either a translation or a half-turn*. Therefore hr also lies in the same direction as rh, and the right coset consists of all those hooks which point in this direction. Thus the left and right cosets coincide, so we know that **p2** is a normal subgroup in **p6**.

Q. 26.28. Show that **p1** is normal in **p6**, with index 6.

Q. 26.29. Show that **p1** is normal in **pmg** with index 4, and that **cm** is also a subgroup of **pmg** of index 4, but is not a normal subgroup.

★ As a further example we consider the pattern **pgg**, which is reproduced in fig. 26.41. The reason for the selection of this as an example is that when we printed the pattern **pgg** (see fig. 26.29), we asked a number of questions which may have led you to discover some subgroups already. Visually, one subgroup immediately stands out as the pattern of all those R's which are *the right way round*, and since just one quarter of all the R's belong to this subgroup, it follows that it has index 4. The pattern of the subgroup is of course **p1**, and it is generated by two translations, g_1^2 and g_2^2, so we may describe it as Gp (g_1^2, g_2^2).

★**Q. 26.30.** Discover, by premultiplying and postmultiplying the elements of the subgroup by, say, g_1 whether it is normal in **pgg**.

★**Q. 26.31.** What is Gp (g_1)? Gp (g_1^2)? Gp $(g_1 g_2)$?

If we now take all those R's which are not turned over, i.e. R and ꓤ, we have the pattern **p2** as a subgroup of index 2, and this may be generated by two translations g_1^2 and g_2^2 and a half-turn such as $g_2 g_1$.

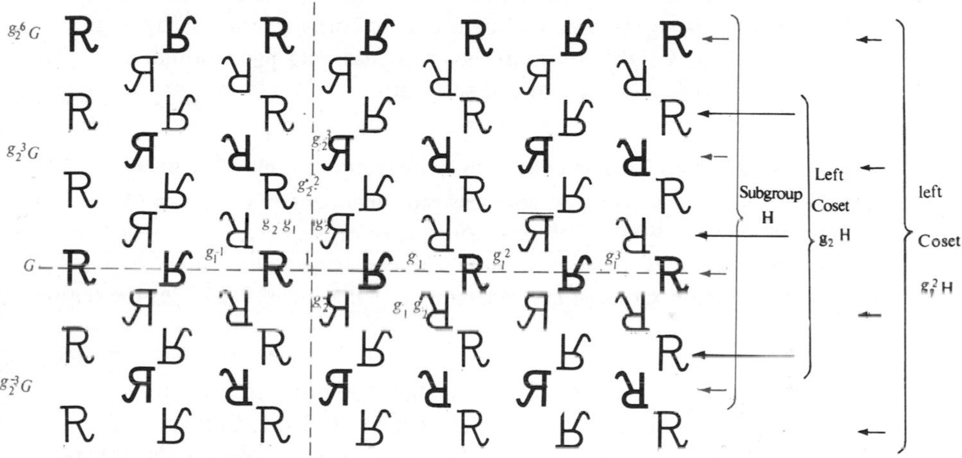

Fig. 26.41. **pgg** as a sub-group of itself.

★**Q. 26.32.** Show that Gp (g_1, g_2^2) is the pattern **pg**, and is also a subgroup of index 2. Draw the motifs which belong to it. Find another way of obtaining the sub-pattern **pg**.

The above subgroups of **pgg** – **p1**, **p2** and **pg** exhaust the subgroups which are *different* patterns from the parent pattern. However, we now proceed to show that **pgg** is also a proper subgroup of itself, with the remarkable fact that it is of index 3, and is a non-normal subgroup of itself!

The fact that there is an infinite group with a proper subgroup having the same structure is nothing new: we have seen it in cases of subgroups of (**Z**, +), and other realisations of C_∞. For example, the even integers under addition are a subgroup of (**Z**, +) with index 2 and also with structure C_∞.

★**Q. 26.33.** Find a proper subgroup of (**Z**, +) of index 3, and structure C_∞.

Those of you who pursued the questions posed when **pgg** was first discussed may have succeeded in discovering the sub-pattern with **pgg** symmetry. One of the simplest ways of obtaining it is as Gp (g_1, g_2^3), for g_2^3 is a glide reflection of three times the pitch of g_2, and those motifs which belong to this subgroup are arrayed along every third row of the pattern in fig. 26.41. These rows are, in fact, left cosets of $G_1 = \text{Gp}(g_1)$ by the elements g_2^{3k} ($k \in \mathbf{Z}$). They form a decomposition of the sub-pattern Gp (g_1, g_2^3). Calling the latter group H, we now form left and right cosets of H by an element not in H, such as g_2.

Now g_2H contains g_2g_1, which is in the row above G_1. It may be verified that *all* elements of this coset lie in the rows immediately above the rows of H. Whereas Hg_2 contains g_1g_2, as well as g_2, and the elements of this right coset will be found to form a series of zig-zags. Thus the left and right cosets do not coincide, and **pgg**, though a subgroup of **pgg**, is not a normal subgroup.

★**Q. 26.34.** Show how **pgg** may arise as a subgroup of itself of index 9.

★**Q. 26.35.** Show that **p4m** and **p6m** between them contain every one of the seventeen plane patterns as subgroups.

★**Q. 26.36.** Show that the least index of **p4g** as a proper subgroup of itself is 9.

★**Q. 26.37.** Find, in terms of the generators *a*, *b* and *t* (see fig. 26.26) the centre of **pm**.

For the reader who wishes to pursue this study of patterns more thoroughly, it is given exhaustive treatment in H. S. M. Coxeter and W. O. J. Moser, *Generators and Relations for Discrete Groups* (Springer-Verlag, Berlin). The twenty-four patterns are also considered in detail with discussion of fundamental regions and of Cayley diagrams in each case. Moreover, defining relations are given for each group. Note that the group C_∞, and the corresponding frieze patterns require *no* defining relations. Apart from this, the simplest is **p1**, and using the notation of fig. 26.19 the group is defined by $st = ts$. Another simple one is **pg**, defined in our notation by $g_1^2g_2^2 = 1$; while **p6** has defining relations $a^2 = p^3 = (ap)^6 = 1$, *a* and *p* being the generators we quoted, being rotations of period 2 and 3. Some of the generators given in the Coxeter and Moser book are, however, different from those suggested with the descriptions of the patterns in the present chapter.

EXERCISES 26

1 Show that every one of the plane patterns may be generated by the use of a specially constructed template, as was done for **p4** (see fig. 26.04). Design templates for other selected patterns.

2 Draw a sketch of the graph $y = \cos x + \cos 2x$, and describe its strip pattern. Repeat for $y = \sin x \pm \sin mx$ $(m = 2, 3, \ldots)$; $\cos x \pm \cos mx$ $(m = 2, 3, \ldots)$; $\sin x \pm \cos mx$; $\cos x \pm \sin mx$. Try to invent trigonometrical functions to give as many different patterns as possible.

3 A frieze pattern is defined by the abstract relations $a^2 = b^2 = 1$. Say which pattern is produced when a and b are half-turns; a and b are reflections; a is a reflection and b is a half-turn, considering the several cases when the mirror axes are in various directions.

4 A square card is placed in the middle of a large table. The card is now repeatedly turned over on the table, at each turn keeping one of the edges fixed. Each of the four possible operations is labelled a, b, c, d, so that a, for example, refers to a half-turn about a particular edge of the card wherever the card may have been moved to. If the operation d is forbidden, show that we get an infinite frieze pattern. Label positions of the card in terms of a, b and c, and identify the associated group.
Make the same investigation in the case of the two-dimensional pattern obtained when d is permitted.

5 A network of equal squares covers the whole plane. Taking a particular square as origin, any other square may be specified by a pair of rectangular 'coordinates'. If the following squares are blackened, what pattern do we obtain: those whose coordinates are $(4m, 4n)$ and those with coordinates $(4p + 2, 4q + 1)$ where $m, n, p, q \in \mathbf{Z}$.

6 Consider the infinity of solutions in integers of the equation $7x + 4y = N$ where N is a multiple of 52. If each point in the Cartesian plane whose coordinates are such a pair (x, y) is marked, consider the pattern over the whole plane which is thereby obtained.

7 Consider the effect in each of the patterns printed in this chapter having R as motif, of replacing R by A, E, H, S; when does the replacement cause the pattern to be richer?

8 Describe the patterns of fig. 26.38.

9 Show how the plane patterns may be 'synthetised' from the frieze patterns by adjoining operations not included in the frieze patterns.

10 Show that if a **p4** pattern has a single axis of bilateral symmetry, then it must be **p4m**.

11 Show that the fundamental region for **p1**, if we insist on its being convex, must be either a parallelogram or a certain type of hexagon.

12 Show that the 'symmetry chart' (see p. 390) for each of the seventeen plane patterns is different. What do you conclude?

13 Note that **pmm** contains glide reflections even though there is no 'g' in the code-name. In fig. 26.28, express some of the glide reflections in terms of the generators a, b, c, d.

14 Draw Cayley diagrams for **p4m** and other plane patterns groups.

15 Show that **cm** may be generated by a reflection and a translation, and express these operations in terms of the generators selected for fig. 26.31.

16 Show by means of a drawing how it is possible that **cmm** is a subgroup of **pmm**, while **pmm** is also a subgroup of **cmm**.

17 Draw an example of the pattern **cmm** which would become **p31m** or **p3m1** if the angles of the rhombuses of the lattice were 60°.

18 Show both from the visual aspect and by using generators that **pmm** is a normal subgroup of **cmm**.

19 Is **p1** a normal subgroup of **pgg**?

20 Show that **pm** and **pmm** have structures $D_\infty \times C_\infty$ and $D_\infty \times D_\infty$ respectively.

21 Show that **p2** may be defined $a^2 = b^2 = c^2 = (abc)^2$ where a, b and c are half-turns (compare the work on p. 369).

22 Find out all you can about a group defined $a^2 = b^2 = 1$; $ac = ca$, $bc = cb$, $ab \neq ba$ in cases where c has various periods. Which pattern does it correspond to when c has infinite period?

Appendix I

A binary operation on the reals

The binary operation ↘ suggested on p. 8 by the example:

$$0{\cdot}234 \mathbin{↘} 0{\cdot}181 = 0{\cdot}213841$$

though apparently artificial, has an interesting application in the following way: it may be applied to set up a 1,1 correspondence between the points of a square and the points of a line segment. For any point of the square may be represented by an ordered pair of real numbers (x, y) where $0 \leqslant x \leqslant 1$ and $0 \leqslant y \leqslant 1$,† and these may then be combined by the above operation to produce a single number z, also lying between 0 and 1,† which may be used to define the position of a point on the line segment. Thus the point (0·234, 0·181) of the square will correspond to the point 0·213841 of the line segment; while $(\sqrt{2} - 1, \frac{2}{7})$, i.e. (0·41421356 . . ., 0·28571428 . . .) maps into the point 0·4218452711345268 . . . on the line segment. Contrariwise, the point 0·8373030303 . . . on the line segment would correspond to the unique point (0·87, 0·3333 . . .) of the square. Thus the 1,1 relationship is established, and of course the *sizes* of the square and of the line segment are of no importance – the sides of the square could be 1,000 miles, and the length of the line segment one millionth of an inch. The argument shows that there are 'as many points' in the minute line segment as there are in the enormous square. Here we touch the fringe of the subject of *transfinite numbers*.

Q. 01. Show how the points of a cuboid may be placed into 1,1 correspondence with the points of a line segment.

Q. 02. Is it possible to set up a 1,1 correspondence between the points of any finite plane region and those of a finite line segment?

Q. 03. Is it possible to set up a 1,1 correspondence between all the points of a plane and the points of a line?

† For example, if the square is taken to be *OUWV*, where *U* is (1,0) and *V* is (0,1), then (x,y) would be the coordinates of the point; while for a point *P* of a line segment *AB*, we could take z to be the ratio AP/AB.

Appendix II

Some proofs relating to identities and inverses

Suppose we have a set S with an operation $*$ which we shall denote for brevity by juxtaposition, which has the following properties: closure, associativity, left identity and left inverse. By the last two we mean that there exists (at least one) element e such that $ex = x$ for *all* x in S (left identity), and that for each and every x in S, there exists a y in the set such that $yx = e$ (left inverse). We proceed to show first, that e is also a right identity, i.e. that $xe = x$ for all x in S; and that a left inverse is also a right inverse, i.e. that $yx = e \Rightarrow xy = e$.

To prove that a left-identity is also a right-identity

Let y be a left inverse of x, so that $yx = e$ and let z be a left inverse of y, so that $zy = e$.

Now	$e = e\,e$	(left-identity)
$\Rightarrow$	$yx = (yx)e$	(y is left inverse of x, so $yx = e$)
$\Rightarrow$	$zyx = zyxe$	(premultiplying by z)
$\Rightarrow$	$(zy)x = (zy)xe$	(associative)
$\Rightarrow$	$e\,x = e\,x\,e$	(z is left inverse of y, so $zy = e$)
$\Rightarrow$	$x = x\,e$	(e is left-identity). Q.E.D.

Note: the above proof is entirely dependent upon (a) associativity and (b) the provision of inverses.

To prove that a left inverse is also a right inverse

i.e. given $yx = e$, to prove that $xy = e$.
Since e is now known to be both a right- and left-identity, we have

	$e\,y = y\,e$	
$\Rightarrow$	$(yx)y = y\,e$	(as $yx = e$, given)
$\Rightarrow$	$zyxy = z\,y\,e$	(premultiplying by z, where $zy = e$)
$\Rightarrow$	$(zy)xy = (zy)e$	(associative)
$\Rightarrow$	$e\,x\,y = e\,e$	($zy = e$)
$\Rightarrow$	$x\,y = e$	(e is the identity). Q.E.D.

To prove the identity is unique

Suppose that e and e' are two distinct identities, so that $xe = ex = xe' = e'x = x$ for all x in the set, as proved above.

Then putting $x = e$ in $xe' = x$ gives $ee' = e$
and putting $x = e'$ in $ex = x$ gives $ee' = e'$.

It follows that $e = e'$, a contradiction.

To prove the uniqueness of inverses

Suppose that x possesses two right inverses y and z, with $xy = e = xz$, and let w be a left inverse of x. Then $w(xy) = w(xz) \Rightarrow (wx)y = (wx)z$ (associative). But $wx = e$, so $ey = ez$, i.e. $y = z$, and x has a unique inverse. Q.E.D.

Appendix III

Where are groups to be found?

The importance of groups is not so much that they solve problems by processes of calculation, as that they provide a general framework of reference against which many problems which previously seemed quite distinct are seen to have a common structure – to be in fact the same problem.

1. Papy's book (13), written for training college students in Belgium, provides very many exercises in which group ideas are used to explore the number system and ordinary two- or three-dimensional space. The insight which this method gives is far from trivial, and after experiencing it these familiar things are never the same again.

2. Klein's *Erlanger Programme* provided a theoretical background for the classification of geometry by group principles. This type of thinking is worked out in terms of a school course on geometrical constructions by Jeger (9).

3. Wherever symmetry occurs groups describe it. Repeating patterns and the great variety of symmetrical objects which children may find can be classified according to the groups to which they belong, just as children learn to classify plants and animals in biology. This approach has important technological applications in crystallography. These groups are very well described by matrices, and the interplay of groups and matrices is a most important theme.

4. The Platonic solids and other related solids can be studied from the point of view of their symmetry groups.

5. Practical problems on symmetry occur in molecular structure – for example, when a chemist wishes to know how many organic compounds there may be within a particular chemical formula. There are applications of group theory in quantum mechanics and in the theory of infra-red spectra. For an account of the applications to chemistry at school level see Wells (14).

6. Bell-ringing may scarcely be considered practical, but bell ringers used the idea of the decomposition of a group into cosets a hundred years before Lagrange (4).

7. Traditional questions on permutations and combinations could now be seen as part of the school study of groups. Presumably when these were part of every sixth-former's mathematical education the questions were thought to be practical. The old-fashioned questions usually asked what was the order of a group, or what was the index of one group in another

8. The application of group theory to the theory of equations was a critical step in the history of mathematics, and the demonstration of the insolubility of the quintic changed the whole course of algebra. School work on the symmetric functions of the roots of equations might look more in this direction (9, 10).

9. The Hamming Code (7) is an example of systems of coding which are used in computers and communication systems for the automatic correction of errors. These systems are frequently vector spaces. Every vector space is a commutative group, so in all of the many practical applications of vector spaces group ideas are present. The Hamming Code works by decomposing its group into cosets. Some solutions of the famous 'Twelve Penny' problem depend on similar ideas, as do other processes of automatic sorting. Do some sets of window reader cards work this way?

10. Finite arithmetic, finite geometrics, and the decomposition of a group into cosets are of practical application in the design of experiments and in industrial quality control. In this connection see an expository article by the children of Form 2A of Shelley School, Crediton, Devon (5), and also the excellent article by Finney (1).

11. Group structures can be found in music. See examples in Bach, Beethoven, Chopin, Debussy, and Schoenberg. Of course, these composers did not know group theory, but they were frequently working out a regular pattern systematically just as a group theorist does.

12. Groups are associated with braids (11), plaits, and knots.

13. When one group is mapped homorphically on to another, the set of elements of the domain which map on to the identity element in the range is called the *kernel*. Examples of this idea are provided by the measurement of angles, projection in geometry, differentiation, etc. (10). The idea, when established, will assist the solution of systems of linear equations, difference and differential equations, describing them all in the same conceptual framework.

14. Many properties of recurring decimals and many tests of divisibility all turn round group ideas. These are connected with other properties which arise when a pack of cards is shuffled; and I have described elsewhere (2) how all these phenomena are interrelated. Similar properties arise again in chain codes (8) which have applications in computing and automatic machinery.

15. There are applications of groups to film animation (3).

16. Pell's equation occurs in a number of circumstances, certainly in a number of puzzle books, and one very neat solution is to use a group of matrices. This method applies also to certain iterative methods of obtaining approximations to square roots.

17. Topology deals extensively with homology groups. These first arose in the study of electrical networks, and the boy at school who solves problems by means of Kirchhoff's Laws is expected to develop his own methods of solving these problems, with very little of the underlying theory being explained. He is doing homology at an intuitive level.

18. Gardner (6) gives many pleasant examples from Scottish Country Dancing.

February 1968 T. J. FLETCHER

References

1 Finney, D. J. 'Statistical Science and Agricultural Research', *Mathematical Gazette*, **31**, No. 293 (Feb. 1947), pp. 21–30.

2 Fletcher, B. & T. J. 'Algebraic Structures in a pack of Cards', *Ideas actuales de la Matematica y su Didactica.* Direccion General de Ensenanza Media, Madrid, 1964.

3 Fletcher, T. J. 'Film Groups', *Mathematical Gazette.*

4 Fletcher, T. J. 'Campanological Groups', *American Mathematical Monthly.*

5 Form 2A, Shelley School, Crediton. Groups. Mathematics Teaching, No. 32, Autumn 1965, pp. 15–20.

6 Gardner, K. L. *Discovering Modern Algebra.* Oxford University Press, 1966.

7 Hamming, R. W. Error Detecting and Error Correcting Codes. *Bell Lyst. Tech. J.*, **16**, No. 2 (April 1950), pp. 147–60.

8 Heath, F. W. & Gribble, M. W. *Chain Codes and the Electronic Applications.* Proc. I.E.E. Monograph, **392**, 1961.

9 Jeger, M. *Transformation Geometry*, trans. A. W. Deicke and A. G. Howson. Allen and Unwin, 1964.

10 Littlewood, D. E. *Skeleton Key of Mathematics.* Hutchinson, 1949.

11 Littlewood, D. E. *A University Algebra.* Heinemann, 1950.

12 Mansfield, D. E. & Bruckheimer, M. *Background to Set and Group Theory.* Chatto and Windus, 1965.

13 Papy, G. *Groups.* Macmillan, 1964.

14 Wells, A. F. *The Third Dimension in Chemistry.* Oxford University Press, 1966.

T. J. FLETCHER

Appendix IV

Table of values of Euler's function $\phi(n)$, the number of numbers less than n and coprime to n. The table of values is taken from the print-out by an I.B.M. 360 computer of a program composed by L. D. Smith and J. S. Forster while in the VI form at the R.G.S. Newcastle-upon-Tyne. The print-out ran up to $n = 3{,}290$. Note: Italic type is used when n is prime, so that $\phi(n) = n - 1$.

n	$\phi(n)$	n	$\phi(n)$	n	$\phi(n)$	n	$\phi(n)$	n	$\phi(n)$	n	$\phi(n)$	n	$\phi(n)$	n	$\phi(n)$	n	$\phi(n)$	n	$\phi(n)$
1	*0*	*2*	*1*	*3*	*2*	4	2	*5*	*4*	6	2	*7*	*6*	8	4	9	6	10	4
11	*10*	12	4	*13*	*12*	14	6	15	8	16	8	*17*	*16*	18	6	*19*	*18*	20	8
21	12	22	10	*23*	*22*	24	8	25	20	26	12	27	18	28	12	*29*	*28*	30	8
31	*30*	32	16	33	20	34	16	35	24	36	12	*37*	*36*	38	18	39	24	40	16
41	*40*	42	12	*43*	*42*	44	20	45	24	46	22	*47*	*46*	48	16	49	42	50	20
51	32	52	24	*53*	*52*	54	18	55	40	56	24	57	36	58	28	*59*	*58*	60	16
61	*60*	62	30	63	36	64	32	65	48	66	20	*67*	*66*	68	32	69	44	70	24
71	*70*	72	24	*73*	*72*	74	36	75	40	76	36	77	60	78	24	*79*	*78*	80	32
81	54	82	40	*83*	*82*	84	24	85	64	86	42	87	56	88	40	*89*	*88*	90	24
91	72	92	44	93	60	94	46	95	72	96	32	*97*	*96*	98	42	99	60	100	40
101	*100*	102	32	*103*	*102*	104	48	105	48	106	52	*107*	*106*	108	36	*109*	*108*	110	40
111	72	112	48	*113*	*112*	114	36	115	88	116	56	117	72	118	58	119	96	120	32
121	110	122	60	123	80	124	60	125	100	126	36	*127*	*126*	128	64	129	84	130	48
131	*130*	132	40	133	108	134	66	135	72	136	64	*137*	*136*	138	44	*139*	*138*	140	48
141	92	142	70	143	120	144	48	145	112	146	72	147	84	148	72	*149*	*148*	150	40
151	*150*	152	72	153	96	154	60	155	120	156	48	*157*	*156*	158	78	159	104	160	64
161	132	162	54	*163*	*162*	164	80	165	80	166	82	*167*	*166*	168	48	169	156	170	64
171	108	172	84	*173*	*172*	174	56	175	120	176	80	177	116	178	88	*179*	*178*	180	48
181	*180*	182	72	183	120	184	88	185	144	186	60	187	160	188	92	189	108	190	72
191	*190*	192	64	*193*	*192*	194	96	195	96	196	84	*197*	*196*	198	60	*199*	*198*	200	80
201	132	202	100	203	168	204	64	205	160	206	102	207	132	208	96	209	180	210	48
211	*210*	212	104	213	140	214	106	215	168	216	72	217	180	218	108	219	144	220	80
221	192	222	72	*223*	*222*	224	96	225	120	226	112	*227*	*226*	228	72	*229*	*228*	230	88
231	120	232	112	*233*	*232*	234	72	235	184	236	116	237	156	238	96	*239*	*238*	240	64
241	*240*	242	110	243	162	244	120	245	168	246	80	247	216	248	120	249	164	250	100
251	*250*	252	72	253	220	254	126	255	128	256	128	*257*	*256*	258	84	259	216	260	96
261	168	262	130	*263*	*262*	264	80	265	208	266	108	267	176	268	132	*269*	*268*	270	72
271	*270*	272	128	273	144	274	136	275	200	276	88	*277*	*276*	278	138	279	180	280	96
281	*280*	282	92	*283*	*282*	285	140	285	144	286	120	287	240	288	96	289	272	290	112
291	192	292	144	*293*	*292*	294	84	295	232	296	144	297	180	298	148	299	264	300	80
301	252	302	150	303	200	304	144	305	240	306	96	*307*	*306*	308	120	309	204	310	120
311	*310*	312	96	*313*	*312*	314	156	315	144	316	156	*317*	*316*	318	104	319	280	320	128
321	212	322	132	323	288	324	108	325	240	326	162	327	216	328	160	329	276	330	80
331	*330*	332	164	333	216	334	166	335	264	336	96	*337*	*336*	338	156	339	224	340	128
341	300	342	108	343	294	344	168	345	176	346	172	*347*	*346*	348	112	*349*	*348*	350	120
351	216	352	160	*353*	*352*	354	116	355	280	356	176	357	192	358	178	*359*	*358*	360	96
361	342	362	180	363	220	364	144	365	288	366	120	*367*	*366*	368	176	369	240	370	144
371	312	372	120	*373*	*372*	374	160	375	200	376	184	377	336	378	108	*379*	*378*	380	144
381	252	382	190	*383*	*382*	384	128	385	240	386	192	387	252	388	192	*389*	*388*	390	96
391	352	392	168	393	260	394	196	395	312	396	120	*397*	*396*	398	198	399	216	400	160
401	*400*	402	132	403	360	404	200	405	216	406	168	407	360	408	128	*409*	*408*	410	160
411	272	412	204	413	348	414	132	415	328	416	192	417	276	418	180	*419*	*418*	420	96
421	*420*	422	210	423	276	424	208	425	320	426	140	427	360	428	212	429	240	430	168
431	*430*	432	144	*433*	*432*	434	180	435	224	436	216	437	396	438	144	*439*	*438*	440	160
441	252	442	192	*443*	*442*	444	144	445	352	446	222	447	296	448	192	*449*	*448*	450	120
451	400	452	224	453	300	454	226	455	288	456	144	*457*	*456*	458	228	459	288	460	176
461	*460*	462	120	*463*	*462*	464	224	465	240	466	232	*467*	*466*	468	144	469	396	470	184
471	312	472	232	473	420	474	156	475	360	476	192	477	312	478	238	*479*	*478*	480	128
481	432	482	240	483	264	484	220	485	384	486	162	*487*	*486*	488	240	489	324	490	168
491	*490*	492	160	493	448	494	216	495	240	496	240	497	420	498	164	*499*	*498*	500	200
501	332	502	250	*503*	*502*	504	144	505	400	506	220	507	312	508	252	*509*	*508*	510	128
511	432	512	256	513	324	514	256	515	408	516	168	517	460	518	216	519	344	520	192
521	*520*	522	168	*523*	*522*	524	260	525	240	526	262	527	480	528	160	529	506	530	208
531	348	532	216	533	480	534	176	535	424	536	264	537	356	538	268	539	420	540	144
541	*540*	542	270	543	360	544	256	545	432	546	144	*547*	*546*	548	272	549	360	550	200
551	504	552	176	553	468	554	276	555	288	556	276	*557*	*556*	558	180	559	504	560	192
561	320	562	280	*563*	*562*	564	184	565	448	566	282	567	324	568	280	*569*	*568*	570	144
571	*570*	572	240	573	380	574	240	575	440	576	192	*577*	*576*	578	272	579	384	580	224
581	492	582	192	583	520	584	288	585	288	586	292	*587*	*586*	588	168	589	540	590	232
591	392	592	288	*593*	*592*	594	180	595	384	596	296	597	396	598	264	*599*	*598*	600	160

Answers

Lack of space prevents some answers from being given in full. In cases where it has been possible to print only a selection of answers, that question is prefixed (S). As far as possible those answers which have been omitted are those which may most easily be dispensed with.

Chapter 2

Questions

Q. 2.01. Union of sets.
Q. 2.02. See footnote, p. 16.
Q. 2.03. $\Lsh$ multiplies corresponding binary digits, modulo 2.
Q. 2.04. Elements in A or B but not in both.
Q. 2.05. Symmetry between x and y.

Exercises

1. (a) H.C.F. (b) difference (c) exponentiation (d) × modulo 10 (e) e.g. $a * b = 2ab$ (mod. 7).
2. (a) 006 (b) 0·064078 (c) 0·600487 (d) 17.
3. Not if proper meaningful words are considered.
6. May need more precise definition.
8. E.g. $\overrightarrow{AB} + \overrightarrow{CD} = \overrightarrow{AD}$; yes (compare Ex. 7.01).
9. (a) E.g. $x * y = 0$ when $x = 0$ or $y = 0$ or $x = y = 0$
   (b) E.g. $x * 0 = 1$; $0 * x = 2$; $0 * 0 = 3$.

Chapter 3

Questions

Q. 3.02. Yes.

Exercises

2. Include $-i$; include zero.
3. (a) YNYN (b) YYYN (c) YNNN (d) YYYY (e) YNYY (f) YYYY (g) YNYY (h) YYYY (i) NNYY (j) YYYN (k) YYYY.
4. Yes, yes.
5. E.g. given circles centres A, B, radii a, b, take the circle centre mid-point of AB, radius $a + b$.
7. (a) YYYN (b) YYYY.
8. C = closed, M = commutative; C′ = not closed, etc.
   (1) C′M′; CM′; CM′ (2) C′M; CM; CM (3) C′M; CM; CM
   (4) C′M; C′M; CM (5) CM; CM; CM.
   Note: The above applies in cases when the operation is meaningful, e.g. in (2), $x \neq 0$, $y \neq 0$.
9. No – include identity; No – see fig. 8.02.
10. xy, yx, xyx $(= yxy)$.
12. Specify numbers of rows and columns.

13. (a) +, −, (× only if degree specified is zero) (b) +, −, ×.
16. All pieces except knight.
17. No.
19. No.
20. No, no.
21. Yes, no.
22. No.

Chapter 4

Questions

Q. 4.01. $(8 - 2x)/(5 - x)$; $4x/(1 + 3x)$; x.

Q. 4.04. $\begin{pmatrix} 0 & 0 & 0 & 0 & 1 \\ 0 & 0 & 1 & 0 & 0 \\ 0 & 1 & 0 & 0 & 0 \\ 0 & 0 & 0 & 1 & 0 \\ 1 & 0 & 0 & 0 & 0 \end{pmatrix}$ All twelve results different; closed.

Q. 4.05. Not possible.

Exercises

1. No – identity missing.
2. Identity.
4. $120 - 44 = 76$, no.
6. No, e.g. $x = \begin{pmatrix} A & B & C & D \\ A & C & B & D \end{pmatrix}$, $x^2 = \begin{pmatrix} A & B & C & D \\ A & B & C & D \end{pmatrix}$.

   No, e.g. $y = \begin{pmatrix} A & B & C & D \\ C & B & D & A \end{pmatrix}$, $y^2 = \begin{pmatrix} A & B & C & D \\ D & B & A & C \end{pmatrix}$.
7. 48, no.
8. No; no 1's on leading diagonal.
11. 4; 12.
12. Not closed; e.g. $x = \begin{pmatrix} 1 & 2 & 3 & 4 & 5 & 6 \\ 2 & 1 & 4 & 3 & 6 & 5 \end{pmatrix}$, $y = \begin{pmatrix} 1 & 2 & 3 & 4 & 5 & 6 \\ 4 & 6 & 5 & 1 & 3 & 2 \end{pmatrix} \Rightarrow$

    $yx = \begin{pmatrix} 1 & 2 & 3 & 4 & 5 & 6 \\ 6 & 4 & 1 & 5 & 2 & 3 \end{pmatrix}$; all 720 permutations need to be included.
14. $apx + pb + q$.
16. $f_1f_2(x) \equiv (x - 1)/x$; six possible functions (the six cross-ratios) $ff(x) \equiv (x - 1)/x$; $fff(x) \equiv x$.

Chapter 5

Questions

Q. 5.01. Yes.

Q. 5.03. Yes – shaded area.

Q. 5.04. Yes; yes; e.g. 'neither *a* nor *b*'.
Q. 5.05. Yes.
Q. 5.06. Yes.

Q. 5.07. $a = b = 1$; or $a = 1, b = 0$; or $a = 0, b = 1$.
Q. 5.08. No; $A * A = A$; $A * 0 = A = 0 * A$.

Exercises

1. (a) Yes (b) no (c) no (d) yes (e) yes, yes (f) no (g) no.
3. Yes; no – adjoin $(z - 1)/z$, $z/(z - 1)$; z, $1/z$, $(z - a)/(az - 1)$, $(az - 1)/(z - a)$ $(a \neq 1)$.
6. $(0, 0) * (0, 0) * (1, 1) =$ either $(0, 2)$ or $(0, 4)$.
7. $(3p + 3qr, 3q, 3r)$.
9. E.g. $OACB$ is a parallelogram where $C = A * B$.
11. No, no, no.
12. No.
14. No.
15. No, e.g. $5(31) = 5$, $(53)1 = 3$; no, e.g. $(BB)C = C$, $B(BC) = D$.
17. Interpreting abc as $(ab)c$ when not associative: (a) 3, (b) none, necessarily, though for example, for subtraction, $a - b - c = a - c - b$.
19. For $k = -1$, $\tan(\tan^{-1} x + \tan^{-1} y)$; for $k = 1$, $\tanh(\tanh^{-1} x + \tanh^{-1} y)$.
22. $x * y = |x - y|$; $x * y = x$; $x * y = x|y|$; $x * y = x + [y]$; $x * y = x + |y|$ $(x > 0)$, $x - |y|$ $(x < 0)$, etc. see also Ex. 5.21 above and Ex. 8.33; also the following: using the operation of Ex. 5.7 above with addition and multiplication modulo 10, taking the decimal part of each real number as a series of ordered triples drawn from {0, 1, 2, . . ., 9}, we should have, e.g.
24·857,036 . . . * 9·772,814 . . . = 33·429,440 . . ., while
9·772,814 . . . * 24·857,036 . . . = 33·529,040 . . .

23(S). $\oplus$ (a) $xy/(x + y)$ (b) $xy/(xy + x + y)$
(c) $(xy - 1)/(x + y - 2)$ (e) $(x^2 + y^2 + k)^{\frac{1}{2}}$
(g) $(x + y)/(1 - xy)$ (h) $xy + \sqrt{(x^2 - 1)(y^2 - 1)}$
(i) $(xy + 1)/(x + y)$

$\otimes$ (a) xy (b) $(x - y)/(x + y - 1)$
(c) $xy - x - y + 2$ (d) $x + y$.

Chapter 6

Questions

Q. 6.01. (a) Right-identity only, = 1 (b) right-identity only, = 0; H.C.F.: $x * xy = x$; L.C.M. $x * 1 = x$.
Q. 6.02. $0x^n + 0x^{n-1} + \ldots + 0x + 1$.
Q. 6.03. (a) $x * y = y - x$: 0 is a left-identity, no right-identity (b) impossible, see p. 541.
Q. 6.04. $i(x) \equiv x$.
Q. 6.05. Universal set $\mathscr{E}$; null set $\emptyset$; $\emptyset$.
Q. 6.06. Yes; e.g. {0, 7, 8}; for primes, identity always present; for composites, a set not including identity can be found.
Q. 6.07. {6, 8}.
Q. 6.08. {1, 3, 5, 9, 11, 13}; 1; {1, 9, 11}; {1, 13}.
Q. 6.09. {6, 3, 9, 12}; {5, 10}.
Q. 6.11. $n = 18$ {10, 2, 4, 8, 14, 16}.
Q. 6.12. No identity, e.g. $0 \sim (-1) = 1$; $(-1) \sim 0 = 1$.

Exercises

1. (a) None (b) 0.
2. (a) (0, 1) (b) (1, 0).
4. The unison.
5. No.
6. (a) **0** (b) none (c) none.
7. 10.
9. No.
10. No left-identity; not associative.
11. Not well-defined when $A \equiv U$ or $B \equiv V$. Associative but no identity.
12. Not commutative; not defined in first two cases.
13. Not associative; no identity.
14. $A * A$ may be selected arbitrarily; not associative; not well defined when $A = 0$ or $B = 0$.
16. No identity; not associative.
17. E.g. $43 \circ 83 = 43$; note that $(10x + y) \circ (10x + y) = 10x + y$.
18. No identity.
19. However, {4, 8, 12, 16} are a group, with identity 16.

Chapter 7

Questions

Q. 7.01. $\begin{pmatrix} 1 & 2 & 3 & 4 \\ 4 & 1 & 3 & 2 \end{pmatrix}$; $\begin{pmatrix} 1 & 2 & 3 & 4 & 5 & 6 \\ 3 & 6 & 4 & 1 & 2 & 5 \end{pmatrix}$; $\begin{pmatrix} 1 & 2 & 3 & 4 & 5 \\ 5 & 3 & 1 & 4 & 2 \end{pmatrix}$.

Q. 7.02. $x \circ y = \begin{pmatrix} 1 & 2 & 3 & 4 & 5 \\ 5 & 3 & 2 & 1 & 4 \end{pmatrix}$; $(x \circ y)^{-1} = \begin{pmatrix} 1 & 2 & 3 & 4 & 5 \\ 4 & 3 & 2 & 5 & 1 \end{pmatrix} = y^{-1} \circ x^{-1}$;

$$x^{-1} = \begin{pmatrix} 1 & 2 & 3 & 4 & 5 \\ 1 & 5 & 4 & 2 & 3 \end{pmatrix};\ y^{-1} = \begin{pmatrix} 1 & 2 & 3 & 4 & 5 \\ 4 & 5 & 1 & 2 & 3 \end{pmatrix}.$$

Q. 7.04. Y^{-1} could mean 'empty C into B'; $Z^{-1} = \begin{pmatrix} A & B & C \\ B & C & A \end{pmatrix}$.

Q. 7.05. (S) (a) $1 + 1/x$ (b) $x/(1 + x)$
(c) $(5x - 7)/(3x - 4)$ (d) $(3x - 6)/(2x - 3)$
(e) $(x - 1)/(2x - 1)$ (f) e^x
(g) $\sin^{-1} x$ (h) e^{e^x}
(i) $x^{\frac{1}{3}}$ (j) $\log_a x$
(k) $(dx - b)/(-cx + a)$, provided $ad \neq bc$.

ff(x) (a) $(x - 1)/(2 - x)$ (b) $x/(1 - 2x)$
(c) $(5x - 7)/(3x - 4)$ (d) x
(e) x (f) $\log \log x$
(i) x^9 (j) a^{a^x}
(k) $[(a^2 + bc)x + (ab + bd)]/[(ca + dc)x + (bc + d^2)]$

(ff)$^{-1}$(x) (a) $(2x + 1)/(x + 1)$ (b) $x/(2x + 1)$
(c) $(4x - 7)/(3x - 5)$ (d) x
(e) x (f) e^{e^x}
(i) $x^{1/9}$
(k) $[(bc + d^2)x - (ab + bd)]/[(-ca - dc)x + (a^2 + bc)]$, $(a + d = 0)$.

Q. 7.06. $0^{-1} = 0$; $U^{-1} = U$.
Q. 7.07. Right-identity and right inverses only.

Q. 7.09. $A \cap B'$; $B \cap A'$; $A \triangle B \triangle C$ = set of elements in exactly 1 or 3 sets; $A \cup B$; $A \cap B'$.

Q. 7.11. (1) No vector may be so transformed
(2) lines parallel (or coincident)
(3) No possible combination of $\binom{a}{c}$ and $\binom{b}{d}$ can give $\binom{p}{q}$ in general, since these two vectors are in the same line.

Q. 7.12. Its determinant shall be zero.

Q. 7.13. $z^{-1}y^{-1}x^{-1}$, etc.

Exercises

1. Associative; no identity or inverses. Vector addition (free vectors).
2. No.
3. $(-3 - 2\sqrt{5})/11$; $(2\sqrt{3} + 1)/23$; $(-2 - 3i)/13$; $(2 - i)/5$; $(a\sqrt{2} - b\sqrt{3})/(2a^2 - 3b^2)$.
5. $-3x^3 + 5x^2 - 2$.
6. x.
7. a; $bd = a$, $cb = a$, $dc = a$, $e^2 = a$.
8. 0; $1 \times 3 = 3 \times 1 = 0$; $2 \times 4 = 4 \times 2 = 0$; $5 \times 5 = 0$.
9. Inverse of **a** is $-$**a**, i.e. the given vector reversed in sense.
11. 0; each numeral is its own inverse.
12. YX and $X^{-1}Y^{-1}$. *Note:* XX^{-1} operates on a passage in French, while $X^{-1}X$ operates on a passage in English; $(YX)^{-1} = X^{-1}Y^{-1}$.
13. $-x/y(x + y)$, $1/y$; $y \neq 0$, $x + y \neq 0$.
14. $(1, 0)$; $(1/a, -b/a)$.
15. $y^{-q}x^{-p}$; $z^{-r}y^{-q}x^{-p}$; $y^{-r}x^{-p}y^{-q}$.
16. The most general form for a *rational algebraic* function.
17. 1; $(-2x - 1)/5$; x; $x(x + 3)/20$; generally inverse of $ax + b$ is $(b - ax)/(a^2 + b^2)$.
18. $1 + 0x$; no inverses when $a^2 = b^2k$; inverses exist when k is not the square of a rational.
19. $$A = \begin{pmatrix} 0 & 1 & 0 \\ 0 & 0 & 1 \\ 1 & 0 & 0 \end{pmatrix} \quad B = \begin{pmatrix} 0 & 0 & 1 \\ 1 & 0 & 0 \\ 0 & 1 & 0 \end{pmatrix}; \begin{pmatrix} 0 & 1 & 0 \\ 0 & 0 & 1 \\ 1 & 0 & 0 \end{pmatrix}; \begin{pmatrix} 0 & 0 & 1 \\ 0 & 1 & 0 \\ 1 & 0 & 0 \end{pmatrix}; \begin{pmatrix} 0 & 0 & 1 & 0 \\ 1 & 0 & 0 & 0 \\ 0 & 1 & 0 & 0 \\ 0 & 0 & 0 & 1 \end{pmatrix}.$$
20. $$\begin{pmatrix} 0 & 0 & 0 & 1 & 0 \\ 0 & 0 & 0 & 0 & 1 \\ 1 & 0 & 0 & 0 & 0 \\ 0 & 0 & 1 & 0 & 0 \\ 0 & 1 & 0 & 0 & 0 \end{pmatrix}; \text{ its transpose.}$$
21. $MX = P \Rightarrow X = M^{-1}P$.
22. Yes.
23. $R^{-1}(z) = iz$; $T^{-1}(z) = \frac{1}{2}(z - i)$; $TRT^{-1}(z) = iz + i + 1$; $RTR^{-1}(z) = 2z - 1$ $T^{-1}RT(z) = iz - \frac{1}{2} - \frac{1}{2}i$; $R^{-1}TR(z) = 2z + 1$.
25. This is effectively multiplication on the Argand diagram. Identity U. If inverse of A is B, make $\triangle BOU$ similar to $\triangle UOA$.
27. Associative (proof by Pascal's hexagon). Identity A. Tangent at O meets UV at X; AX meets conic again at A^{-1}.
28. The construction corresponds to $(a, b) * (c, d) = (ac, b + ad)$ if O is taken as $(0, 0)$ and U is $(1, 0)$ in the Cartesian plane. U is the identity; inverse of

(a, b) is $(1/a, -b/a)$. AN perpendicular to OU; take L on OU such that $OL \,.\, ON = OU^2$. Then LA^{-1} is perpendicular to OU, and OU bisects $\angle AOA^{-1}$.

29. (S) 803: $r^8m^2r^3 = r^{-2}m^{-1}rm^3r^3$; 996: $r^{-1}m^{-2}rm^3r^6 = mr^{-4}m^{-3}rm^3$; 27; $1\bar{2}6$: $rmr^{-2}mr^6$.

Chapter 8

Questions

Q. 8.01. Must have 1 in top left-hand corner and symmetry about leading diagonal. Compare p. 148.

Q. 8.03. (a) kite, isosceles trapezium (one axis); rectangle, rhombus (2 axes) square (4 axes) (b) parallelograms.

Q. 8.04. $+$ mod. 2; $\times$ mod. 3; composition of signs under $\times$; $\times$ mod. 4; $\times$ mod. 8; $\times$; E = even, O = odd.

Q. 8.06. No – identity absent; fails for closure.

Q. 8.07. Fig. 17.01, No. 11; figs. 12.041, 12.042.

Q. 8.08. Symmetric about leading diagonal.

Q. 8.09. (i) No (ii) yes (iii) no (iv) no.

Q. 8.10. **R**, $-$ has right-identity and inverses; $x * y = 2x + y$ has left-identity and inverses.

Q. 8.11. 1110, 1110, 1110, 1110, 1111, 1110, 0000, 0100, 0100, 0100, 1111, 0000, 1111, 0100, 0111.

Q. 8.12. 0111, 0100, 0111, 0100.

Exercises

1. Yes (b), (e), (g), (i), (k), (n), (p), (q), (s), (v), (x).
2. Like Table 11.04.
3. $\{e, p, q, pq = qp\}$.

4.

1	2	4	5	7	8	
2	4	8	1	5	7	
4	8	7	2	1	5	4; 7; $n = 6r$ $(r \in \mathbf{Z})$;
5	1	2	7	8	4	$n = 3r$ $(r \in \mathbf{Z})$; $\{1, 8\}$; $\{1, 4, 7\}$.
7	5	1	8	4	2	
8	7	5	4	2	1	

5.

1	3	4	5	9	
3	9	1	4	5	
4	1	5	9	3	$9x = 5 \Rightarrow x = 3$.
5	4	9	3	1	
9	5	3	1	4	

6. Table as in Q. 8.02.
7. $\begin{pmatrix} 1 & 0 \\ 0 & 1 \end{pmatrix}$; $\begin{pmatrix} -1 & 0 \\ 0 & -1 \end{pmatrix}$.
9. $\{\omega^0, \omega^1, \omega^2, \omega^3, \omega^4, \omega^5\}$ addition of indices mod. 6.
10. See Table 16.10.
11. (c) Table 8.03.
12. $C_3 \times C_3$ (see Table 10.07).
13. C_6, as Table 12.02.
14. All 2-groups.

15. (S) E.g. the roundabout sign has three rotational symmetries, the square board has eight symmetries, but there is no symmetry in the combined figure.
16. c; a; q; a or b or c; b; a or b or c.
18. $\begin{pmatrix} \text{ch u} & -\text{sh u} \\ -\text{sh u} & \text{ch u} \end{pmatrix}$, yes.
20. Yes, yes (zero being excluded).
21. (a) Not closed e.g. $(2 - \sqrt{5}) + (3 + \sqrt{5}) = 5$
    (b) not closed e.g. $\sqrt{5} \times \sqrt{5} = 5$.
22. C_∞, compare $\{10^x, x \in \mathbf{Z}\}$ under $+$.
23. $-1/x$; similar to Table 8.09.
24. $f^2 = g$, $fg = i$; (C_3).
25. $f^2(x) = 1/(2 - 4x)$; $f^3(x) = (1 - x)/(1 - 4x)$; $f_4(x) = (2x - 1)/4x$; $f^{-1}(x) = (4x - 1)/(4x + 2)$; C_6, as in Table 12.02.
26. C_4, similar to Table 8.08.
27. $C_2 \times C_2 \times C_2$, similar to Table 16.10.
28. $C_3 \times C_3$, as in Table 10.07.
30. Identity $(1, 0)$; inverse $(1/a, -b/a)$ ($a \neq 0$, so elements of the form $(0, b)$ must be excluded, for group structure).
31. No inverses; $\mathbf{Q}$ or $\mathbf{R}$, so long as we exclude the rational functions having zero polynomials in numerator or denominator.
32. No identity, e.g. $53 * 0 = 3$, not 53.
33. Identity $0.1010\, e_5e_6e_7 \ldots$ not unique. Extend to have $c = \gamma \,.\, c_1c_2c_3c_4c_5c_6 \ldots$ Then if $A_r = \begin{pmatrix} a_{4r+1} & a_{4r+2} \\ a_{4r+3} & a_{4r+4} \end{pmatrix}$, and $C_r = A_rB_r$, we shall have identity $0{\cdot}10101010\ldots$ (= 10/99!), but no inverses, as they would have to include negative elements. The method may be adapted as for example in Ex. 5.23 which has identity $0{\cdot}0000\ldots$, but fails for inverses.

Chapter 9

Questions

Q. 9.01. $ax = ay \Rightarrow a^{-1}ax = a^{-1}ay \Rightarrow x = y$.
Q. 9.02. Abelian.
Q. 9.03. Table 5.01 not associative; (a), (b) group tables; (c) not a group: $ac = e \neq ca$; (d) not a group: $c(ab) \neq (ca)b$.
Q. 9.07. Yes, Q_4 (cf. Table 9.13).
Q. 9.08. Transpose rows and columns.
Q. 9.09. (s) E.g. C_4: $\begin{pmatrix} 1&0&0&0\\0&1&0&0\\0&0&1&0\\0&0&0&1 \end{pmatrix}, \begin{pmatrix} 0&1&0&0\\0&0&1&0\\0&0&0&1\\1&0&0&0 \end{pmatrix}, \begin{pmatrix} 0&0&1&0\\0&0&0&1\\1&0&0&0\\0&1&0&0 \end{pmatrix}, \begin{pmatrix} 0&0&0&1\\1&0&0&0\\0&1&0&0\\0&0&1&0 \end{pmatrix}$
Q. 9.11. $(yz - x, -y, -z)$.
Q. 9.12. They are the subgroup Gp. (a, b) with structure $C_3 \times C_3$; no – see the brief reference to this group in the Preface.
Q. 9.13. They form D_4.
Q. 9.14. No longer associative.
Q. 9.15. (a) $bp = pab \Rightarrow ab = p^2bp \Rightarrow (ab)^2 = p^2bpp^2bp = 1 \Rightarrow aba = b \Rightarrow ab = ba$
(b) elements are p^2, ap, ap^2, p^2a, apa, ap^2a.
Q. 9.16. $a^2b^2c^2$; a^2bc; a^2; a^2c^2; abc^2.

Q. 9.17. $pr^2 = rp^2 \Rightarrow (pr^2)^2 = rp^2pr^2 = r^3 \Rightarrow pr^2pr^2 = r^3 \Rightarrow pr^2p = r$
$\Rightarrow p^2r^2p^2 = r^3 \Rightarrow r^2p^2 = pr^3 = (pr^2)r = rp^2r$. But $r^2p^2 = r(rp)^2 = rpr^2$, so $rpr^2 = rp^2r \Rightarrow r = p$.
Q. 9.18. Use induction.

Exercises

3. (a) $a^{-1}cb^{-1}$ (b) $cb^{-1}a^{-1}$ (c) $b^{-1}a$ (d) $c^{-1}b^{-1}a^{-1}$ (e) a^2 (f) $a^{-1}bca^{-1}b^{-1}$.
4. $xy = yx$ and $xy^2 = y^3x \Rightarrow yxy = y^3x \Rightarrow xy = y^2x = yx \Rightarrow y = 1$.
5. $p = q^{-1}r^{-1}q = rq^{-1}r^{-1} \Rightarrow r^{-1}q = qrq^{-1}r^{-1}$, etc.
7. Yes.
8. $a(bc) = 1$, $(ab)c = a$; $ad = 1$ but $da \neq 1$; $c^2 = 1$, but 2 is not a factor of 5 (Lagrange).
10. No; 10; D_5 (see Ch. 13).
11. $r^3a = arar \Rightarrow a = rarar \Rightarrow 1 = (ra)^3$.
12. $C_3 \times C_3 \times C_3$ (see Ch. 16).
13. Generally, $a^2 = 1$ and $axa = x^n \Rightarrow x^{n^2-1} = 1$.
14. $pxp^{-1} = x^4 \Rightarrow x = p^{-1}x^4p$ and $x^{16} = (pxp^{-1})^4 = px^4p^{-1}$. Also $x^{16} = (p^{-1}x^4p)^{11} = p^{-1}x^{64}p$; thus $px^4p^{-1} = p^{-1}x^{64}p \Rightarrow x^{64} = p^2x^4p^{-2} = p^{-1}x^4p$ (since $p^3 = 1$) $= x$. Hence $x^{63} = 1$, and period of x is at most 63.
16. $a^2 = 1 = r^6$, $ar = r^4a \Rightarrow (ar)^2 = r^4aar = r^5 \Rightarrow (ar)^{12} = 1$, while $(ar)^n \neq 1$ for $n < 12$. $(ar)^{-1} = r^{-1}a = r^5a = ar^2$.
17. pq, qp, pq^2, all of period 6.
18. $pr^3 = rp \Rightarrow pr = rpr^3 = r^2p$, etc.

Chapter 10

Questions

Q. 10.01. $x^r = x^s \Rightarrow x^{r-s} = 1$, etc.
Q. 10.02. When $xy = yx$.
Q. 10.03. Positions of e in table symmetrically situated about leading diagonal.
Q. 10.04.

Element	e	a	b	c	d	f	g	h	j	k	l	m.
Period	1	6	2	4	4	4	4	4	4	3	3	6

Q. 10.05. $\{e, a^2, a^4\}$, $\{e, a^3\}$.
Q. 10.06.

Element	e	a	a^2	a^3	a^4	a^5	b	b^3	ab	ba	a^2b	ba^2.
Period	1	6	3	2	3	6	4	4	4	4	4	4

Q. 10.07. E.g. $\begin{pmatrix}1 & 2 & 3 & 4 & 5\\ 2 & 1 & 3 & 4 & 5\end{pmatrix}$ (one pair transposed), and $\begin{pmatrix}1 & 2 & 3 & 4 & 5\\ 2 & 1 & 4 & 3 & 5\end{pmatrix}$ (two pairs transposed).
Q. 10.08. PARTICLE; LUNATIC.
Q. 10.09. $x^3 = (1\ 4\ 2\ 5\ 3)$; $x^4 = (1\ 5\ 4\ 3\ 2)$; $x^5 =$ identity.
$y = (1\ 2\ 3\ 4\ 5\ 6)$; $y^2 = (1\ 3\ 5)(2\ 4\ 6)$; $y^3 = (1\ 4)(2\ 5)(3\ 6)$, etc.
Q. 10.10. $pq = qp$ when the cycles are disjoint.
Q. 10.11. (a) 20 (b) 9 (c) 18 (d) 12 (e) 72 (f) 120.
Q. 10.12. $C_{12} \cong$ Gp. (x) where $x = (1\ 2\ 3)(4\ 5\ 6\ 7)$.
Q. 10.13. INCH, PHANTOM, PYRAMID, EQUALITY, FIDGET.
Q. 10.14. $(p\ s\ e)(r\ u)(c\ t\ o\ i\ n)$, period $3 \times 2 \times 5$; (I E T N C A U O), period 8.
Q. 10.16. (T R E G I L N); inverse $\begin{pmatrix}\text{T R I A N G L E}\\ \text{N T G A L E I R}\end{pmatrix}$, period 7.
(R C E A)(P L I); inverse $\begin{pmatrix}\text{R E P L I C A}\\ \text{A C I P L R E}\end{pmatrix}$, period 12.
(O F)(S R M T E A), period 6; (H L M O T A I N); (H T O A)(M I N)(L).

Q. 10.17. (G R N D); (R N); (G D)(R N); (T C E R A); (T C), etc.
Q. 10.18. $ba = (1\ 2\ 4)$ period 3.
Q. 10.19. CLAIMED, DECIMAL, TUESDAY.
Q. 10.21. (a) 2 (b) 2 (c) 4 (d) 2 (e) 4 (f) 4 (g) 4 (h) 5 (i) 2 (j) 3 (k) 6 (l) 9 (m) 9 (n) 6 (o) infinite (p) infinite.
Q. 10.23. P reflects in x-axis, Q in y-axis; $PQ = QP = H$, a half-turn. Table like 8.09 or 11.07, 11.01.
Q. 10.24. $a = d = -1$, $b = c = 0$, or $a + d = 0$, $bc = 1 - a^2$ for period 2. For period 4, $a^2 + d^2 + 2bc = 0$, and $ad - bc = +1$ ($\Rightarrow a + d = \pm\sqrt{2}$).
Q. 10.25. $\begin{pmatrix} \cos\theta & \sin\theta \\ -\sin\theta & \cos\theta \end{pmatrix}$; $\begin{pmatrix} \cos r\pi/6 & -\sin r\pi/6 \\ \sin r\pi/6 & \cos r\pi/6 \end{pmatrix}$ $r = 5, 7, 11$.
Q. 10.26. $\theta = (p/q)\pi$ $(p, q \in \mathbf{Z})$.
Q. 10.27. SR is a quarter-turn then a shear.
Q. 10.29. $\begin{pmatrix} \omega^2 & 0 \\ 0 & \omega \end{pmatrix}, \begin{pmatrix} -\omega^2 & 0 \\ 0 & -\omega \end{pmatrix}, \begin{pmatrix} \omega & 0 \\ 0 & \omega^2 \end{pmatrix}, \begin{pmatrix} -1 & 0 \\ 0 & -1 \end{pmatrix}, \begin{pmatrix} 1 & 0 \\ 0 & 1 \end{pmatrix}.$
Q. 10.30. A has period 2, P period 3; $AP^2 = PA$, like Table 8.07.
Q. 10.31. Only $a = 1$, $b = 0$.
Q. 10.32. Only -1 besides $+1$.

Exercises

1. Period 2.
2. (a) 4 (b) 9 (c) 20 (d) 1800 (e) 9 (f) 5 (g) 360 (h) infinite.
3. 120°, 72°, etc.
4. (a) $(xy)^n = 1 \Rightarrow x(yx)^{n-1}y = 1 \Rightarrow (yx)^{n-1} = x^{-1}y^{-1} = (yx)^{-1}$, etc., *or* $xy = a \Rightarrow y = x^{-1}a$, so $yx = x^{-1}ax$, etc.
   (b) Periods of xyz, yzx, zxy equal.
   Periods of xyz, yzx unequal counter example: $x = (1\ 2\ 3)$, $y = (1\ 4)$, $z = (2\ 4)$, then $xyz = (1\ 4\ 3)$, period 3, whereas $yxz = (1\ 2)(3\ 4)$, period 2.
   (c) Counter-example easily found as in (b).
5. $(ab)^2 = 1 \Rightarrow abab = 1 \Rightarrow aba = b \Rightarrow ab = ba$.
6. Elements of periods 3, 4, . . . inverse pairs, leaving identity and an odd number of elements of period 2.
7. $x^{2n} = 1 \Rightarrow x^n$ of period 2 and no other element of period 2.
8. As in No. 6.
9. As in No. 6; no.
10. See Chapter 20, p. 364.
11. Table 13.05, periods see Table 13.01; Table 19.04, periods see Table 14.02.
12. S_4: 9 of period 2, 8 period 3, 6 period 4 (see Table 14.02).
    S_5: 10 type (1 2), 15 type (1 2)(3 4), 20 type (1 2 3), 30 type (1 2 3 4), 24 type (1 2 3 4 5), 20 type (1 2 3)(4 5), i.e. 25 period 2, 20 period 3, 30 period 4, 24 period 5, 20 period 6.
14. D_2; $t' = -t$ reflects in axis; $t' = -1/t$ gives focal chord; on $xy = c^2$ we have inscribed rectangles with sides of gradient ± 1.
15. Either a multiple of 12, or infinite.
16. $r = qp^{-1}q^{-1}$ – see No. 10 above.
17. xax^{-1} of period 2 by No. 10 $\Rightarrow$ contradiction.
18. m odd; $m \in 3\mathbf{Z}$.

19. It is the same. If period of x^p = period of x^q, $= m$, then period of $x = mk$, where k = H.C.F. (p, q).
20. xy cannot be located.
22. x period 2, y period 4, z period 21, w period 8.
23. $ar^2 = r^2a$ also of period 6. Group is C_6 ($\cong C_3 \times C_2$, see Ch. 16).
24. Six; (1 9)(1 4)(1 11)(2 7)(2 10)(2 14)(2 5)(2 3)(8 12)(8 15) for example.
25. Always even or always odd (see Ch. 18).
27. The group is $C_3 \times C_3$ (see Ch. 16).
28. $a + d = 0$. *Note:* Throughout the Q matrices must be assumed to be non-singular, i.e. $ad - bc \neq 0$, otherwise f(x) degenerates to a mere 'constant function'.
    Homogeneous because of the equivalence relation – only the *ratios* $a:b:c:d$ matter.
29. $\begin{pmatrix} 0 & 1 \\ -1 & 1 \end{pmatrix}^3 = \begin{pmatrix} 1 & 0 \\ 0 & -1 \end{pmatrix}$, and latter is equivalent (cf. No. 28) to $\begin{pmatrix} 1 & 0 \\ 0 & 1 \end{pmatrix}$.
30. fg of period 3 requires $a^2 - ab + b^2 + 1 = 0$, not true for any real a, b.
31. Cf. quaternions, Ch. 15, especially Ex. 15.27.
32. Each of period 2.
33. Interpretation similar to the example pp. 120–2 in text.
34. $(a \quad y)\begin{pmatrix} 1 & 1 \\ 1 & 0 \end{pmatrix} = (x + y \quad x)$; $(x \quad y)\begin{pmatrix} 1 & 1 \\ 1 & 0 \end{pmatrix}^2 = (2x + y \quad x)$, etc.

    $$M^2 = \begin{pmatrix} 2 & 1 \\ 1 & 1 \end{pmatrix} \quad M^3 = \begin{pmatrix} 3 & 2 \\ 2 & 1 \end{pmatrix} \quad M^4 = \begin{pmatrix} 5 & 3 \\ 3 & 2 \end{pmatrix} \quad M^n = \begin{pmatrix} u_{r+2} & u_{r+1} \\ u_{r+1} & u_r \end{pmatrix}$$

    where u_r is general term of Fibonacci series.
37. Period 3; $\begin{pmatrix} e & f & d \\ h & i & g \\ b & c & a \end{pmatrix}$; period 6; $\begin{pmatrix} h & f & a \\ e & c & g \\ b & i & f \end{pmatrix}$.
38. Independent cycles of periods 3, 4, 5, 7, 9, 11, 13 give period equal to their L.C.M. (= 180,180).
39. E.g. $\begin{pmatrix} 0 & -1 \\ 1 & -1 \end{pmatrix}$ (M^2 of No. 33, or of M in the text, p. 120–2).
40. $(xy^2)^3 = 1 \Rightarrow yx^2 = (xy^2)^2 \Rightarrow xyx^{-1}y = x(yx^2)y = x(xy^2)^2y = x^2y^2x$
    $(yx)^3 = 1 \Rightarrow (yx)^2 = x^2y^2 \Rightarrow yxyx^{-1} = (yxyx)x = x^2y^2x$.

Chapter 11

Questions

Q. 11.01. 4.

Q. 11.02. {1, 9, 11, 19}; {1, 5, 7, 11}.

Q. 11.03.

ABCD	*e*
BADC	*a*
DCBA	*b*
CDAB	*c*

Q. 11.04. C is the centre and the chords are perpendicular.

Q. 11.05. Reflection in real axis and half-turn about $\frac{1}{2}a$. If a not real, transformations not closed as $\overline{a - z} \neq a - \bar{z}$.

Q. 11.06. $(2z - 2)/(z - 2)$.

Q. 11.07. $\begin{pmatrix} 1 & 0 \\ 0 & 1 \end{pmatrix}$, $\begin{pmatrix} -1 & 0 \\ 0 & -1 \end{pmatrix}$, $\begin{pmatrix} 0 & 1 \\ 1 & 0 \end{pmatrix}$, $\begin{pmatrix} 0 & -1 \\ -1 & 0 \end{pmatrix}$.

Q. 11.08. Impossible to change an odd number of coins.
Eight operations, $C_2 \times C_2 \times C_2$ (see Ch. 16).
Q. 11.09. E.g. $\{1, i, -1, -i\}$, $\times$; rotations through multiples of 90° (see Table 11.13).
Q. 11.11. E.g. period 5: $4\pi/5$; addition neglecting the integral part.
Q. 11.12. $(\mathbf{Z}, +) \cong (n\mathbf{Z}, +) \quad (n \in \mathbf{N})$.
Q. 11.16. $a = 3$.
Q. 11.17. 6, 7, 11.
Q. 11.18. Mod. 5 scale: 1 3 4 2 } clockwise or anticlockwise, but not
Mod. 7 scale: 1 3 2 6 4 5 } essentially different.
1, 3, 9, 7; 1, 5, 7, 17, 13, 11.
Q. 11.19. n prime.
Q. 11.20. $p - 1$.
Q. 11.22. C_{12} (see Ch. 22).
Q. 11.23. $D_3 \cong S_3$ (see above).
Q. 11.24. $D_3 \cong S_3$.

Exercises

1. E.g. 'All men are equal, but some are more equal than others.'
2. $i^4 = 1$; $2^4 = 6$ (mod. 10).
3. The fifth row in each is of period 6 and generates C_6.
4. (a) 1, 1 correspondence impossible
(b) non-Abelian and Abelian respectively
(c) rotation of period 4 in second case only
(d) $z^n = 1$ for many z, whereas $n(x, y) = (0, 0)$ only in the case $x = y = 0$.
5. $(\mathbf{Z}, +) \cong (a^z, \times)$ $a \in \mathbf{R}^+$; $(\mathbf{R}, +) \cong (a\mathbf{R}, +)$ $a \in R\backslash 0$.
6. $z = x * y \Rightarrow \tan^{-1} z = \tan^{-1} x + \tan^{-1} y \pmod{\pi}$,
$2 \tan^{-1} z = 2 \tan^{-1} x + 2 \tan^{-1} y \pmod{2\pi}$ for rotations.
7.

9.11	e	a	g	d	b	f	c	h
9.13	1	r	r^2	r^3	a	ar	ar^2	ar^3

one of twenty-four solutions, see Ch. 22.

9. Addition of binary numerals followed by subtraction of 11 (= 3), no carrying. In (a) the group is $C_2 \times C_2 \times C_2$ (b) $C_3 \times C_3$ (see Ch. 16).
11. (a) $C_3 \times C_2$ (b) $C_3 \times C_3$ (c) $C_2 \times C_2 \times C_2$ (colours and styles being changed in cyclic order).
13. *Note:* The line $x = 0$ must be excluded. Not isomorphic – given group does not contain elements, of finite period, whereas the rotations do.
14. $\{1, a, a^2, a^3, a^4\}$, $\times$, where $a^5 = 1$; e.g. $3^5 = 1$ (mod. 11) giving $\{1, 3, 4, 5, 9\}$, $\times$ mod. 11.
15. 2; 8.
16. 168 (see Ch. 16, pp. 272–7).
18. $x \to -x$ is not an automorphism of $(\mathbf{R}, \times)$, as $xy = z, \neq (-x)(-y) = (-z)$
$z \to \bar{z}$ is an automorphism of $(\mathbf{C}, \times)$, since $z_1 z_2 = z_a \Rightarrow \bar{z}_1 \bar{z}_2 = \bar{z}_3$.
19. Aut. $C_8 \cong D_2$, etc. (see Table 17.01, and Ch. 22).
20. Yes, e.g. the only automorphisms of $(\mathbf{Z}, +)$ are $x \to -x$ and $x \to x$.
21. $p - 1$; $(p-1)(q-1)$ $(p \neq q, p \neq 1, q \neq 1)$, $p(p-1)$ $(p = q \neq 1)$.
23. See *Mathematical Gazette*, Oct. 1969, p. 293, D. F. Robinson 'Permutations in a Group Table.'
24. a^{-1}; $a^{-1}x^{-1}a^{-1}$; $x \to xa^{-1}$ (but not an automorphism).

26. (a) $A\ (x_1, y_1)$, $B\ (x_2, y_2)$ in Cartesian plane; $C = A * B$ is point (x_1, y_2). AC passes through U, point at infinity on y-axis; BC passes through V, point at infinity on x-axis. Still associative but not commutative. No identities, so no inverses (though $(x, y) * (z, y) = (x. y)\ \forall z$).

Chapter 12

Questions

Q. 12.01. (1 2 3 4 5 6), etc.
Q. 12.02. Reg. hexagonal pyramid; prism of fig. 12.04 (full group); prisms with sections as in fig. 12.03, etc.
Q. 12.03. $f^3(x) = (2 - x)/(1 - 2x)$; $f^5(x) = (2x + 1)/(x + 1)$; subgroups $\{i, f^2, f^4\}$; $\{i, f^3\}$; e.g. $g(x) = (2x + 1)/(4 - 4x)$.
Q. 12.04. $\pm\frac{2}{3}\pi$ rotations combined with reflection ($E\ D$).
Q. 12.05. $\pm\frac{2}{3}\pi$ rotations; $f^3(x) = (2 - x)/(1 - 2x)$.
Q. 12.07. $\text{cis}(\pm\frac{1}{2}\pi)$; $\text{cis}(\pm k\pi/4)$ ($k = 1, 3, 5, 7$).
Q. 12.09. Generated by 3 or 12 (identity 15); $\{1, 3, 5, 9, 11, 13\}$; $\{8, 2, 4, 6, 10, 12\} \times$ mod. 14.
Q. 12.11. Compare Ch. 16, p. 256.
Q. 12.12. 'Full period primes': 7, 17, 19, 23, 29, 47, 59, 61, 97, 109, 113, 131, 149, 167, 179, 181, . . .
Q. 12.14. Because if kq is a multiple of n, where n is composite, one cannot conclude that n divides either k or q.
Q. 12.16. Compare with table for $\{1, 2, 3, \ldots, 12\} \times$ mod. 13, as described in text.
Q. 12.17. 9.
Q. 12.18. No (identity absent).
Q. 12.19. $C_1 \subset C_2 \subset C_4 \subset C_{12}$.
Q. 12.20. C_8: C_1, C_2, C_4, C_8; C_9: C_1, C_3; C_{10}: C_2, C_5.
Q. 12.21. $C_3 \times C_3$ (as in Table 10.07).
Q. 12.22. $1 + 2j$, $2 + 2j$, $2 + j$.
Q. 12.23. $x^2 + x + 1 = 0$; $x^4 + x^3 + x^2 + x + 1 = 0$; $x^6 + x^5 + \ldots + 1 = 0$.
Q. 12.24. $x^2 - x + 1 = 0$.
Q. 12.26. (c) $\{2, 5, 6\}$ $\{7, 8, 11\}$.
Q. 12.28. E.g. Gp. (f, g), where $f(x) = 1/x$, $g(x) = x + 1$.

Exercises

1. C_{24}; C_5; C_{360}; C_{72}; C_{180}; C_{18}.
3. $a^2 = 1$ (mod. p) $\Rightarrow (a + 1)(a - 1) = 0$ mod. $p \Rightarrow a = 1$ or $p - 1$, as p is prime.
4. In $\{0, 1, 2, \ldots, 2n - 1\}$, + mod. $2n$, only n has period 2.
5. No (identity absent).
6. $f(x) = 3/(3 - x)$, period 6: Gp. (f) $\cong C_6$; $f(x) = \frac{1}{2} - x$, period 2: Gp. (f) $\cong C_2$; $f^2(x) = \frac{1}{2}x - \frac{1}{4}$, infinite period: Gp. (f) $\cong C_\infty$; $f(x) = 1/(2 - 2x)$, period 4: Gp. (f) $\cong C_4$; $f(x) = -ix$, period 4: Gp. (f) $\cong C_4$.
7. See last two parts of No. 6 above; or, e.g. $(x - 1)/(x + 1)$, $f(x) = \omega x$ ($\omega^8 = 1$), $\{\omega^r x\}$, $r = 0, 1, 2, \ldots, 7$; Subgroups C_4 ($r = 0, 2, 4, 6$), C_2 ($r = 0, 4$).
8. Clockwise rotation through $\frac{1}{3}\pi$ $\begin{pmatrix} \cos\theta & \sin\theta \\ -\sin\theta & \cos\theta \end{pmatrix}$ where $\theta = k\pi/3$ ($k = 0, 1, 2, 3, 4, 5$).

9. Square of given matrix $= \begin{pmatrix} 0 & 1 \\ -1 & 1 \end{pmatrix}$.
10. Period 3; (1 3 5)(2 6 4) and identity.
11. No.
12. (a) $\begin{pmatrix} i & 0 \\ 0 & \omega \end{pmatrix}$ $(i^4 = 1, \omega^3 = 1)$ (b) (x, y); $x = 0, 1, 2, 3$; $y = 0, 1, 2$ with $x_1 + x_2$ mod. 4 and $y_1 + y_2$ mod. 3. We show here that $C_{12} \cong C_4 \times C_3$, compare Ch. 16.
13. Yes.
17.

Period	1	2	3	4	5	6	8	9	10	12	15	18	20	24	30	36	40
C_{15}	1	–	2	–	4	–	–	–	–	–	8	–	–	–	–	–	–
C_{18}	1	1	2	–	–	2	–	6	–	–	–	6	–	–	–	–	–
C_{24}	1	1	2	2	–	2	4	–	–	4	–	–	–	8	–	–	–
C_{36}	1	1	2	2	–	2	–	6	–	4	–	6	–	–	–	12	–
C_{40}	1	1	–	2	4	–	4	–	4	–	–	–	8	–	–	–	16
C_{60}	1	1	2	2	4	2	–	–	4	4	8	–	8	–	8	–	16

18. (a) Order 40
(b) 20, for $73^5 = 97 = -7$; $73^{10} = 49 \Rightarrow 73^{20} = 1$; alternately by using $73 = (-3)^3$
(c) $19 \times 79 = 1$ (mod. 100)
(d) $3^{20} = (10 - 1)^{10} = 1$ (mod. 100), so 3 is of period 20 in the group. If group were C_{40}, then there would have to be a number of period 40 whose square was 3, and there is no such number. So the group is not C_{40}, yet contains elements of period 20 – structure is $C_{20} \times C_2$.
Let $2n = 2^k(2m + 1)$; $(2n)^{20} = (2^{20})^k(2m + 1)^{20} = 76^k \,.\, 1 = 76$ (mod. 100). *Note:* 76 is the identity in the group $\{76, 4, 16, 64, 56, 24, 96, 84, 36, 44\}$, $\times$ mod. 100.
19. If $x^{mr} = 1$, then $(x^r)^m = 1$, and Gp. $(x^r) \cong C_r$.
21. $[\text{cis } 2\pi k/m]^{mr} = 1$ for $r = 1, 2, 3, \ldots$, but is only a primitive root when $r = 1$.
22. $C_3 \times C_3$; (1, 1), (1, 2), (2, 1), (2, 2) form the group D_2 under $\otimes$.
23. See Table 17.01; also Ch. 22; either 1, 3, 7 or 9 may be taken as a (primitive) generator of $\{0, 1, 2, \ldots, 9\}$, + mod. 10.
24. These are the members of the set which are coprime with n. See No. 26 below.
26. Closure established in first part. Also if r has period m, then r^{m-1} is inverse of r.
28. Generator of C_∞ may be mapped on to ± 1 in $(\mathbf{Z}, +)$.
29. See fig. 25.062.
31. The angles are rational multiples of π and are added mod. 2π.
32. The composition of rotations corresponds to addition of reals mod. 2π; so the angle θ should be mapped on to the real number $\theta k/2\pi$ in the group $(k\mathrm{S}, + \text{ mod. } k)$.
33. A step up or down.
34. $a^k b^k = 1 \,.\, b^k = b^k$, mod. $(ab - 1)$. But $a^k b^k = (ab)^k = 1$ mod. $(ab - 1)$, so $b^k = 1$. a and b are inverses in the group of residues coprime to $ab - 1$ under multiplication mod. $(ab - 1)$.
35. 1 replaced by 3 induces the cycle (1 3 2 6 4 5) of period 6.

Chapter 13

Questions

Q. 13.01. $bf = r^3$.
Q. 13.02. $\{1, a, r^3, d\}$
Q. 13.04. 24, 12, 24.
Q. 13.05. D_5: $r^5 = a^2 = 1$, $ar = r^4a$.
Q. 13.08. D_4 (e.g.) z, $2 - z$, $2/(2 - z)$, $(2z - 2)/(z - 2)$, $(z - 2)/(z - 1)$, $z/(z - 1)$, $2 - 2/z$, $2/z$
D_3 (e.g.) z, $1/(1 - z)$, $(z - 1)/z$, $(2z - 1)/(z - 2)$, $(2 - z)/(1 + z)$, $(z + 1)/(2z - 1)$.
Q. 13.09. $\begin{pmatrix} \pm 1 & 0 \\ 0 & \pm 1 \end{pmatrix}$, $\begin{pmatrix} 0 & \pm 1 \\ \pm 1 & 0 \end{pmatrix}$.
Q. 13.10. D_3, D_6.
Q. 13.12. If only one generator, then cyclic.
Q. 13.16. $\{1, r^2, r^4, b, g, d\}$, $\{1, r^3, f, b\}$.
Q. 13.17. C_2 (five times), C_4 $\{1, r, r^2, r^3\}$, D_2 $\{1, r^2, a, ar^2\}$ and $\{1, r^2, ar, ar^3\}$
Q. 13.18. D_n is the group of symmetries of one of the two regular n-gons.
Q. 13.19. The table shows that $D_6/C_3 \cong D_2$ (quotient group, see Ch. 21).
Q. 13.20. $ar = r^{n-1}a \Rightarrow rar = r^na = a$; rotate, reflect, then rotate again, same as the reflection.
Q. 13.21. $a^2 = b^2 = 1$, $(ab)^n = 1$.
Q. 13.22. Like fig. 26.32.
Q. 13.23. Like fig. 13.15.
Q. 13.24. $ri = ir$, so not D_n.
Q. 13.25. ab of period 6. Group is D_6; e.g. $(A\ C)(B\ E)(D\ F)$ and $(A\ E)(B\ F)$; D_5; D_8.
Q. 13.26. $a^2 = b^2 = 1 \Rightarrow (aba)^2 = 1$, and $aba \neq a$, $aba \neq b$.
Q. 13.27. D_5; the functions of D_3 (Q. 13.08), together with $(z + 1)/(2 - z)$, $(2 - z)/(1 - 2z)$, $(2z - 1)/(z + 1)$, $1 - z$, $z/(z - 1)$ and $1/z$.

Exercises:

1. *Note:* (a) and (c): plane regular pentagon may be regarded as a regular pentagonal prism of very small thickness; (c) and (d): pentagonal bipyramid and prism are dual solids; (b) requires opposite symmetries (five reflections).
2. $n = 1$ or 2.
3. Think of a very thin rectangle; a very thin isosceles trapezium.
4. + gives cyclic groups only; × gives Abelian groups, so only D_2.
5. A decagon consisting of a regular pentagon with five isosceles triangles based on each side.
6. In D_3, $(ap)^2 = 1$.
7. $(z - 1)/z$, $1/(1 - z)$, z, $z/(z - 1)$. See F. J. Budden, *Complex Numbers* (Longmans), Ch. 4, pp. 118–28.
8. E.g. $g(z) = 2 - z$.
9. fg has period 3, so we have D_3. Other functions are $1/(1 - x)$, $(x - 1)/x$, x and $(3x - 2)/(x - 3)$.
10. $\mathrm{gf}(x) = 2 - 2/x$ of period 4, so group is D_4.
11. $\mathrm{f}^2(z) = 1/z$; $\mathrm{f}^3(z) = (z - i)/(1 - iz)$; 8; D_4 since $\mathrm{fg} = \mathrm{gf}^3$.
12. The six functions of D_3 in Q. 13.08 together with the six of Q. 13.26.

13. D_3 generated by $(A\ B)(X\ Y)$ and $(A\ C)(X\ Y)$
D_6 generated by $(B\ C)$ and $(A\ C)(X\ Y)$: *Note:* We see D_6 as a subgroup of S_5.
14. D_4; D_8.
15. Closed set of permutations of six objects are a subgroup of S_6.
17. $Hh \to tH$ and $Hh \to hT$ each of period 4.
18. $a = \begin{pmatrix} 1 & 0 \\ 0 & -1 \end{pmatrix}$ $p = \frac{1}{2}\begin{pmatrix} -1 & -\sqrt{3} \\ \sqrt{3} & -1 \end{pmatrix}$ $p^2 = \frac{1}{2}\begin{pmatrix} -1 & \sqrt{3} \\ -\sqrt{3} & -1 \end{pmatrix}$;
$b = pa = \frac{1}{2}\begin{pmatrix} -1 & \sqrt{3} \\ \sqrt{3} & 1 \end{pmatrix}$; $c = p^2a = \frac{1}{2}\begin{pmatrix} -1 & -\sqrt{3} \\ -\sqrt{3} & 1 \end{pmatrix}$.
19. Compare Nos. 8, 9 above and 27 below.
20. $aba = bab \Rightarrow (ab)^3 = 1$.
21. $a^2 = b^2 = (ab)^4 = 1$, or $a^2 = r^4 = 1$, $ar^3 = ra$.
22. $1 \subset C_2 \subset D_2 \subset D_6$; $1 \subset C_3 \subset D_3 \subset D_6$.
23. See Ch. 22 and Table 17.01.
24. (a) One element of period n and any independent element of period 2.
(b) Two elements of period 2 such that their product is of period n; e.g. for the regular hexagon they may be reflections in lines at an angle 30° but *not* 60°.
26. Yes.
27. $\begin{pmatrix} 1 & 0 \\ 0 & 1 \end{pmatrix}, \begin{pmatrix} 0 & -1 \\ 1 & 1 \end{pmatrix}, \begin{pmatrix} -1 & -1 \\ 1 & 0 \end{pmatrix}, \begin{pmatrix} -1 & -1 \\ 0 & 1 \end{pmatrix}, \begin{pmatrix} -1 & 0 \\ 0 & -1 \end{pmatrix}, \begin{pmatrix} 0 & -1 \\ -1 & 0 \end{pmatrix},$
$\begin{pmatrix} 0 & 1 \\ -1 & -1 \end{pmatrix}, \begin{pmatrix} 1 & 0 \\ -1 & -1 \end{pmatrix}, \begin{pmatrix} 1 & 1 \\ -1 & 0 \end{pmatrix}, \begin{pmatrix} 1 & 1 \\ 0 & -1 \end{pmatrix}.$
28. As in Table 9.06.
30. Yes – all six perms. of the three carbon copies occur with equal frequency.
31. E.g. *aba* or *bab* of period 2.
32. (a) $f(x) = (6x - 3x^2)/(x^2 - x + 1)$ (b) $f(x) = 1 + x + x/(1 - x)$.
33. (a) E.g. $g(x) = -x$ (b) as long as fg is of infinite period we shall get D_∞. If f and g are chosen at random, this will highly probably happen, e.g. $g(x) = -x$, $h(x) = 1 - x$.
34. (a) No, not closed (b) no, see Q. 13.23 (c) no – only one element of period 2.
35. Invert w.r. to a point of intersection so that circles → lines; angle of cut must be 60°.
36. (a) Compare No. 35 (b) invert w.r. to a limiting point to obtain concentric circles; then *ab* is an *enlargement*, and so of infinite period; D_∞.
37. E.g. let -5 correspond to *ababa* in D_∞, where $a^2 = b^2 = 1$
$+3$ correspond to *bab*
-5×3 correspond to $(ababa)(bab) = abababab$, etc.
Use these labels in fig. 13.143. See also *Mathematical Gazette*, Dec. 1970, p. 368.
38. $AB = \begin{pmatrix} 1 & 1 \\ 0 & 1 \end{pmatrix}$, a shear, of infinite period.
39. Group is C_n. Reflections in the tangents would generate an infinite pattern, so they do not belong to the group.

Chapter 14

Questions

Q. 14.01. C_3, C_2.
Q. 14.02. When $x \neq e$, and period of x is not the order of the whole group.
Q. 14.04. Whole group.
Q. 14.05. In an infinite group, h does not necessarily have finite period.
Q. 14.07. Centre of D_4 $\{e, r^2\}$; of D_5 is $\{e\}$; of D_6 is $\{e, r^3\}$.
Q. 14.09. $\{1, f, p, r\}$; $\{1, b, w, y\}$; $\{1, h, r, y\}$.
Q. 14.10. $D_3\{1, a, b, c, d, f\}$; $\{1, i, m, b, p, g\}$; $\{1, l, u, f, g, w\}$.
Q. 14.11. D_4 $\{1, h, r, y, n, k, b, w\}$; $\{1, h, r, y, s, x, g, a\}$.
Q. 14.12. Those generated by (1 2) and (3 4); by (1 3) and (2 4); by (1 4) and (2 3).
Q. 14.13. Consider those perms of N objects in which $N - n$ are not moved. ${}_5C_2 = 10$; ${}_5C_4 = 5$.
Q. 14.15. The whole group.
Q. 14.17. $a = dc^3, b = c^3d, f = cdc, g = c^3, h = d^3, j = dcd, k = c^2$ $(= d^2)$, $l = cd, m = dc$.
Q. 14.18. Any two elements not in the same subgroup for any of the subgroups listed on p. 216.
Q. 14.19. Gp.$(a, d) = \{1, a, d, f, b, c\}D_3$; Gp.$(k, i) \cong S_4$; Gp.$(w, y) = \{1, w, y, b\}D_2$; Gp.$(l, m) = \{1, c, h, l, m, d, r, y, q, i, u, v\} =$ Gp.(d, p) A_4; Gp.$(j, r) = \{1, j, r, t\}C_4$; Gp.$(d, p) =$ Gp.$(a, b, g) =$ Gp.$(s, p) =$ Gp.$(j, k) \cong S_4$; Gp.$(j, y) = \{1, j, y, r, t, p, h, f\}D_4$.
Q. 14.20. Any two coprime integers. Gp. (10) = integers divisible by 10 $(= Z_{10})$. Gp. (-4) = Gp. (4).
Q. 14.21. In $(\mathbf{R}, \times)$, Gp. $(\alpha) = \{\alpha^n : n \in \mathbf{Z}\}$; Gp. $(-1) = \{-1, 1\}$.
Q. 14.23. $C_2 \times C_2 \times C_2$, like Table 16.09; stabiliser of {A, B} = {(C D), (E F), (C D)(E F), identity}.
Q. 14.24. (a) $\begin{pmatrix}1 & 0\\0 & 1\end{pmatrix}, \begin{pmatrix}-1 & 0\\0 & -1\end{pmatrix}, \begin{pmatrix}0 & 1\\1 & 0\end{pmatrix}, \begin{pmatrix}0 & -1\\-1 & 0\end{pmatrix}$, D_2; (b) not closed, no identity.
Q. 14.25. S_4; the permutation matrices form S_3; those whose determinant is +1 form C_3.
Q. 14.26. $n \to -10n$.
Q. 14.27. ± 10; or a pair $10m$, $10n$ where n, m coprime; No (identity absent). Subgroups of $(\mathbf{Z}, +)$ are $(n\mathbf{Z}, +)$ $(n \in \mathbf{Z})$.
Q. 14.28. E.g. $(\mathbf{R}, +)$.
Q. 14.29. (a) No (b) No (c) Yes (d) Yes (e) e.g. $\{x^n : n \in \mathbf{Z}\}$, $\times$ where $x \in \mathbf{R}$, $|x| \neq 1$, $x \neq 0$.
Q. 14.30. $\{1, -1\}$.
Q. 14.32. On the line $2x + y = 0$.
Lines drawn through origin only (must contain identity, 0).
Q. 14.33. E.g. $\{2^n \text{cis } n\theta : n \in \mathbf{Z}\}$ for given θ.
Q. 14.37. An 'opposite' transformation, containing a reflection (see pp. 187–8). The 2×2 matrices with determinant ± 1 preserve area.
Q. 14.38. D_9.
Q. 14.39. They form a group, the centraliser of A.
Q. 14.40. See fig. 13.15 (D_∞).
Q. 14.41. E.g. translations in a given direction; e.g. translations in three dimensions.

Q. 14.42. No.
Q. 14.44. $RT(z) = (z + a)\text{cis}\,\alpha$, but $TR(z) = z\,\text{cis}\,\alpha + a$.
Q. 14.45. T generates C_∞ (see fig. 10.07).
Q. 14.46. Rotation about A; the identity, cf. Q. 14.53.
Q. 14.47. Even powers of abc, etc.
Q. 14.48. $a = cbc$.
Q. 14.50. (a) D_3 (b) D_6 are the finite subgroups.
Q. 14.52. Not a group, e.g. $ba = b$!
Q. 14.54. 3.
Q. 14.55. (a) Compatible with matrix multiplication (b) compatible with matrix addition.
Q. 14.56. No; see p. 66.

Exercises

1. E.g. any cyclic subgroup generated by an element.
2. $\{1, a, g, h\}D_2$; h: $\{1, a, g, h, r, s, x, y\}D_4$; i: $\{1, i, m\}C_3$; n: $\{1, k, n, y\}C_4$; r: $\{1, f, h, j, p, r, t, y\}D_4$; none have centraliser $\{1\}$; none have C_2; eight elements of period 3 have centraliser C_3.
4. S_3; no, fails for closure.
6. If $a \in A \cap B$, and if $a^m = 1$ $(m \neq 1)$, then m divides $|A|$ and $|B|$, contradiction.
7. None except $\{\mathbf{0}\}$.
8. (one of twelve possible automorphisms, see Ch. 22).

10.3	e	a	a^2	a^3	a^4	a^5	c	d	f	g	h	i
14.3	1	a	l	k	m	b	c	d	f	g	h	i

13. (a) nx $(n \in \mathbf{Z})$; (b)(c)(d) ordered quadruples of integers (a, b, c, d) under $+$. (e) $ax^2 + b(3x + 1)$, $a, b \in \mathbf{Z}$.
15. Centre of S_n is identity $(n \neq 2)$.
20. $b = j^2aja$; $g = j^2aj^2$.
21. Bilinear transformations include inversions (see *Complex Numbers*, F. J. Budden, Longmans, Ch. 4). Invariants – circles preserved, conformal.
(a) $c = 0$, $d = 1$ direct similarities.
(b) $a, b, c, d \in \mathbf{R}$ gives a subgroup not easily recognised geometrically.
22. See *Mathematical Gazette*, Feb. 1970, 'On Functions which form a group', F. J. Budden, p. 9.
23. Any subset of r numbers can be taken as a cycle; six subgroups of S_5 generated by (1 2 3 4 5), (1 2 3 5 4), (1 2 4 3 5), (1 2 4 5 3), (1 2 5 4 3) and (1 2 5 3 4).
24. Group is infinite.
25. (a) Yes (b) yes (c) no, but there is an isomorphism $\begin{pmatrix} 0 & 0 & 0 \\ 0 & p & q \\ 0 & r & s \end{pmatrix} \to \begin{pmatrix} p & q \\ r & s \end{pmatrix}$, and the group of 3×3 matrices of the first form is a subgroup of $(\mathcal{M}_3, +)$ (d) yes.
26. (a) E.g. polynomials with integer coefficients; real coefficients.
(b) E.g. rational functions with rational coefficients; complex coefficients.
(c) E.g. $(a, 0) \oplus (c, 0) = (ac, 0)$.
29. Finite, e.g. $\{1, a, c, ac\}$, $\{1, bcb, a, abcd\}D_2$; infinite, e.g. Gp. (abc) generated by a glide reflection, or by a translation.
30. E.g. $\begin{pmatrix} a & b \\ 0 & 1 \end{pmatrix}$ where $a \in \mathbf{Q}$, $a \neq 0$, $b \in \mathbf{R}$.

31. Yes; compare Q.14.37 and p. 301, direct and opposite symmetries.
32. And so the centre of the affine group is the group of direct similarities from the given origin 0.
33. The identity only.
34. Lines parallel to given line, and those perpendicular to it have directions preserved.
35. E.g. the translations in a particular direction.
36. Group is $C_p \times C_p$ (see Ch. 16). Use Lagrange.

Chapter 15

Questions

Q. 15.02. $a^2 = b^2 = 1$, ab of period 6.

Q. 15.04. E.g. j, k (see Q. 14.19, also Ch. 18); S_n requires two generators (see Q. 18.29, ff.).

Q. 15.05. In every case, yes.

Q. 15.07. E.g. $k^3 = 1$, $(gk)^2 = (kg)^2$, $(gk)^4 = 1$, etc.

Q. 15.08. $C_4\{1, x, x^2, x^3\}$ letting $y = x^2$.

Q. 15.09. $xy = yx$, $yz = zy$, $xz = zx$.

Q. 15.10. Group is C_4.

Q. 15.11. e, period 1; a, period 2; the rest, period 4 (Q_4); e.g. f, g.

Q. 15.12. E.g. $e = \begin{pmatrix}1 & 0\\ 0 & 1\end{pmatrix}$; $m = \begin{pmatrix}\omega & 0\\ 0 & \omega^2\end{pmatrix}$; $c = \begin{pmatrix}0 & \omega\\ -\omega^2 & 0\end{pmatrix}$; $l = \begin{pmatrix}\omega^2 & 0\\ 0 & \omega\end{pmatrix}$;

$h = \begin{pmatrix}0 & \omega^2\\ -\omega & 0\end{pmatrix}$; $f = \begin{pmatrix}0 & 1\\ -1 & 0\end{pmatrix}$; $k = \begin{pmatrix}-1 & 0\\ 0 & -1\end{pmatrix}$; $j = \begin{pmatrix}0 & -1\\ 1 & 0\end{pmatrix}$;

$a = \begin{pmatrix}-\omega & 0\\ 0 & -\omega^2\end{pmatrix}$; $g = \begin{pmatrix}0 & -\omega\\ \omega^2 & 0\end{pmatrix}$; $b = \begin{pmatrix}-\omega^2 & 0\\ 0 & -\omega\end{pmatrix}$; $d = \begin{pmatrix}0 & -\omega^2\\ \omega & 0\end{pmatrix}$.

(One of twelve possible answers, see Ch. 22.)

Q. 15.14. $\{e, a, b, c\}$, $\{e, d, a, f\}$, $\{e, g, a, h\}$.

Q. 15.15. In Q_n, there are n elements of period 4 together with those in the subgroup C_n; thus $4 + 2$ in Q_4; $6 + 0$ in Q_6; $8 + 2$ in Q_8, etc. One subgroup C_4 for each pair of inverse elements of period 4.

Q. 15.16. $\begin{pmatrix}-1 & 0\\ 0 & -1\end{pmatrix}$ is the only quaternion of period 2.

Q. 15.17. Q_4 is a subgroup of S_8 but not of S_7 nor of $S_4 \times C_2$.

Q. 15.18. $x \leftrightarrow b$, $y \leftrightarrow d$. To see this, note the isomorphism $a + ib \leftrightarrow \begin{pmatrix}a & b\\ -b & a\end{pmatrix}$, whereby $i \leftrightarrow \begin{pmatrix}0 & 1\\ -1 & 0\end{pmatrix}$ and $1 \leftrightarrow \begin{pmatrix}1 & 0\\ 0 & 1\end{pmatrix}$. Compare Ex. 15.17.

Q. 15.19.

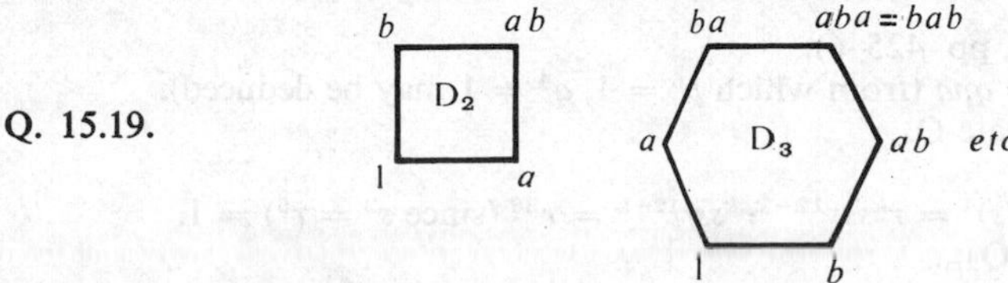

Q. 15.20.

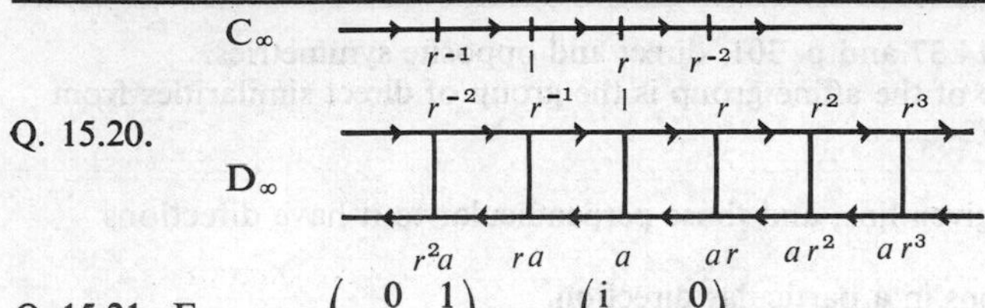

Q. 15.21. E.g. $p = \begin{pmatrix} 0 & 1 \\ -1 & 0 \end{pmatrix}, q = \begin{pmatrix} i & 0 \\ 0 & -i \end{pmatrix}.$

$1 \leftrightarrow e, p \leftrightarrow d, p^2 \leftrightarrow a, p^3 \leftrightarrow f, q \leftrightarrow b, q^{-1} \leftrightarrow c, pq \leftrightarrow h, qp \leftrightarrow g.$

Q. 15.23. See pp. 425–6; p and q must be independent generators, i.e. not inverses.

Q. 15.24. E.g. $p = qpq, q = pqp.$

Exercises

1. $q = pqp \Rightarrow q^2 = pqpq = p(qpq) = p^2;$
$q = p(pqp)p = p^2qp^2 = q^2qq^2 = q^5 \Rightarrow q^4 = 1$, etc.
2. Not true, e.g. in $C_6 = \text{Gp.}(p)$, $x = p$, $y = p^4$, then $x^2 = p^2 = y^2$, yet y has period 3.
3. True, provided it is a *proper* subgroup.
4. $x = y^4$ and $y^6 = 1$.
5. $x = y^6$, $y^{10} = 1$.
6. (a) $x = y^6$; $y^9 = y^6y^3 = x^4x^2 = 1$, C_9 (b) we may have C_5 with $xy = 1$.
7. $pq = qp^3$, $p^2 = q^2 \Rightarrow p^6 = 1$.
8. C_6.
9. D_3.
10. By agreeing $m^9 = 1$.
11. 8; like Table 16.11.
12. (a) $ap = pa$ (b) $ap = p^2a$ (c) $(ap^6 = 1$ for Q_6; $(ap)^3 = 1$ for A_4, (Table 17.02).
14. Q_4 and D_4 both satisfied.
17. No, $a = p^2$ necessarily here.
18.

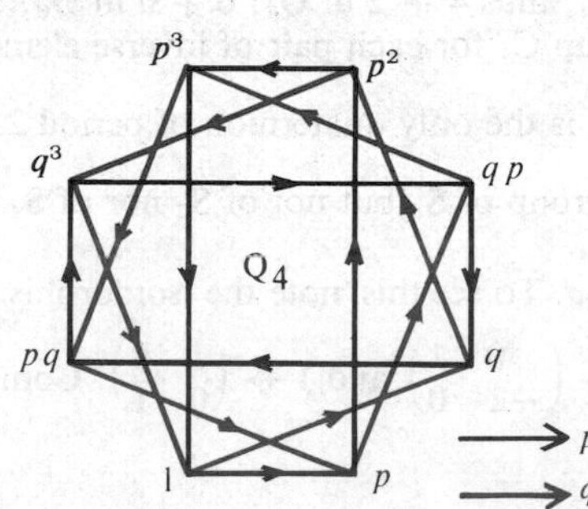

19. 24; S_4 (see Ch. 22, pp. 425–6).
20. Q_6; $q = pqp$, $p^2 = qpq$ (from which $p^6 = 1$, $q^4 = 1$ may be deduced).
21. 10; p^2 and q generate Q_4.
23. Compare No. 21.
24. $r^ks = sr^{12-k} \Rightarrow (r^ks)^4 = r^kssr^{12-k}r^kssr^{12-k} = r^{36}$ (since $s^2 = r^6$) $= 1$.
25. $p^2 = q^m = (pq)^2$; Q_{2m}.

26. $\begin{pmatrix} 1 & 0 \\ 0 & -1 \end{pmatrix}$; elements of period 4 when $z^n = 1$.

27. $a^2 = r^3 = 1$, $ara = r^2ar^2$, or $(ar)^3 = 1$.

28. Order divisible by 3, and ≥ 6; $C_3 \times C_3$; A_4 (Table 17.02).

29. (a) As in No. 28 (b) $p^3 = q^3 = 1$, $p^2q = q^2p$.

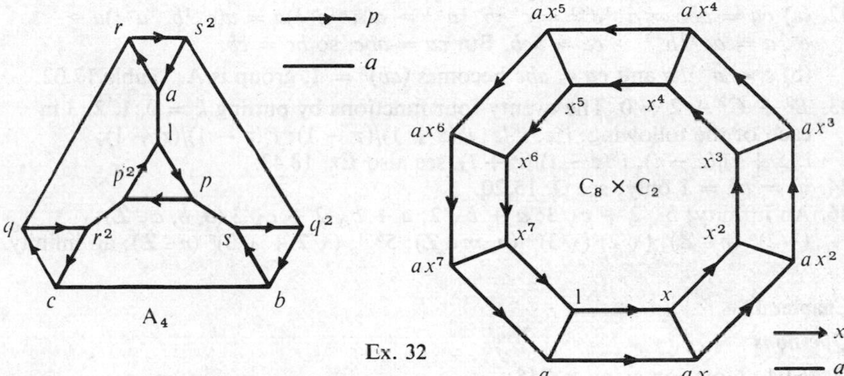

Ex. 32

30. E.g. in S_4, Table 14.02, taking a, j and q.

33. Each of period 6; group of order 24.

34. (a) $ar = r^2a \Rightarrow r = ar^2a \Rightarrow ar^2 = ra$. But $ar^2 = arr = r^2ar = r^2r^2a = r^4a$. Hence $ra = r^4a \Rightarrow r^3 = 1$; also $r^5 = 1$, so finally $r = 1$. (b) $rqr = qrq$.

35.

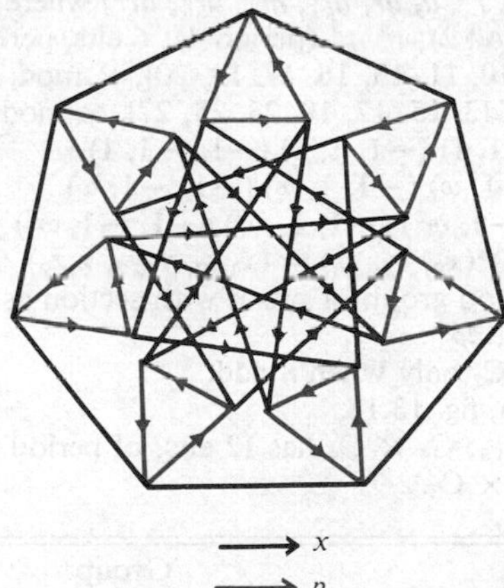

36. (a) D_4 (b) S_4.

37. $M = \begin{pmatrix} 0 & 1 \\ 1 & 0 \end{pmatrix}$, $S = \begin{pmatrix} 1 & 1 \\ 0 & 1 \end{pmatrix}$ Any matrix $\begin{pmatrix} a & b \\ c & d \end{pmatrix}$ may be built up by repeated multiplications of M and S with $A = \begin{pmatrix} a & 0 \\ 0 & 1 \end{pmatrix}$, $B = \begin{pmatrix} b & 0 \\ 0 & 1 \end{pmatrix}$, $C = \begin{pmatrix} c & 0 \\ 0 & 1 \end{pmatrix}$, $D = \begin{pmatrix} d & 0 \\ 0 & 1 \end{pmatrix}$. See *Maths Gazette*, Oct. 1970 p. 284 (D. A. T. Soffe).

The affine group defined by an x-stretch, a reflection in $x = y$, and an x-shear.

39. Some of the edges may be superfluous, e.g. in fig. 15.032 the sides of the convex hexagon do not belong to the Cayley graph showing Gp. (a, b).
40. Half-turn $\begin{pmatrix} 0 & 0 & 1 \\ 0 & -1 & 0 \\ 1 & 0 & 0 \end{pmatrix}$ about line $x = z$, $y = 0$; $(pr)^3 = 1$.
41. E.g. $R = c$, $S = g$, $RS = j$.
42. (a) $ca = abc \Rightarrow a^{-1}c^{-1} = c^{-1}b^{-1}a^{-1} \Rightarrow a(a^{-1}c^{-1})a = a(c^{-1}b^{-1}a^{-1})a \Rightarrow c^{-1}a = ac^{-1}b^{-1} \Rightarrow ca = acb$. But $ca = abc$, so $bc = cb$.
(b) $c = a^{-1}ba$ and $ca = abc$ becomes $(ab)^3 = 1$, group is A_4. Table 17.02.
43. $B^2 + C^2 + 2 = 0$. The twenty-four functions by putting $k = 0, 1, 2, 3$ in each of the following: i^kz; i^k/z; $i^k(z + 1)/(z - 1)$; $i^k(z - 1)/(z + 1)$; $i^k(z + i)/(z - i)$; $i^k(z - i)/(z + i)$, see also Ex. 18.43.
44. $a^2 = b^2 = 1$ only; see Q. 15.20.
46. An infinity; $b\sqrt{2} + c\sqrt{3}$; $a + b\sqrt{2}$; $a + b\sqrt{2} + c\sqrt{3}$ $(a, b, c \in \mathbf{Z})$; $(\sqrt{2})^n$ $(n \in \mathbf{Z})$; $(\sqrt{2})^n(\sqrt{3})^m$ $(n, m \in \mathbf{Z})$; $5^{n/3}$; $(\sqrt{2} + \sqrt{3})^n$ $(n \in \mathbf{Z})$; an infinity.

Chapter 16

Questions

Q. 16.04. Notation as on p. 255:
$$\left.\begin{aligned}(g_p, g_q')(g_r, g_s') &= (g_p * g_r, g_q' *' g_s')\\ (g_r, g_s')(g_p, g_q') &= (g_r * g_p, g_s' *' g_q')\end{aligned}\right\}\text{equal, as } *, *' \text{ commutative.}$$
Q. 16.06. E.g. $C_2 \times C_4$ may be generated by perms of 1, 2, 3, 4, 5, 6 generated by (1 2 3 4) and (5 6).
Q. 16.07. $\{1, r, r^2, r^3, r^4, r^5, a, ar, ar^2, ar^3, ar^4, ar^5\}$ where $r^6 = a^2 = 1$, $ar = ra$; r^3, a, ar^3 (period 2); r^2, r^4 (period 3); 6 elts. period 6.
Q. 16.09. {1, 2, 4, 5, 8, 10, 11, 13, 16, 17, 19, 20}, × mod. 21.
{1, 3, 5, 9, 11, 13, 15, 17, 19, 23, 25, 27}, × mod. 28.
Q. 16.10.
$$\left.\begin{matrix}(1, 1, 1)\ (1, -1, 1)\ (-1, 1, 1)\ (-1, -1, 1)\\ (1, 1, \omega)\ (1, -1, \omega)\ (-1, 1, \omega)\ (-1, -1, \omega)\\ (1, 1, \omega^2)\ (1, -1, \omega^2)\ (-1, 1, \omega^2)\ (-1, -1, \omega^2)\end{matrix}\right\}\begin{matrix}\text{where}\\ \omega = \cos\tfrac{2}{3}\pi + i\sin\tfrac{2}{3}\pi\end{matrix}$$
and $(x_1, y_1, z_1)*(x_2, y_2, z_2) = (x_1x_2, y_1y_2, z_1z_2)$.
Q. 16.11. $C_2 \times C_6$ rotation group of prism with section as in fig. 12.03.
Q. 16.12. $\{1, d, p, dp, p^2, dp^2\}$.
Q. 16.13. $D_{2n} \cong D_n \times C_2$ only when n odd.
Q. 16.15. Bi-tetrahedron, fig. 13.18.
Q. 16.17. $C_3 \times C_5 \cong C_{15}$; $C_4 \times C_4$ has 12 elts. of period 4 (see p. 265).
Q. 16.18. $C_3 \times C_6$; $D_3 \times C_3$.

Q. 16.19.

	Group:					
Period:	C_{24}	$C_{12} \times C_2$	$D_2 \times C_6$	D_{12}	$D_2 \times D_3$	$D_4 \times C_3$
2	1	3	7	13	15	5
3	2	2	2	2	2	2
4	2	4	–	2	–	2
6	2	6	14	2	6	10
8	4	–	–	–	–	–
12	4	8	–	4	–	4

contd.

Period:	Group: $D_3 \times C_4$	$Q_4 \times C_3$	$Q_6 \times C_2$	Q_{12}	S_4	$A_4 \times C_2$
2	7	1	3	1	9	7
3	2	2	2	2	8	8
4	8	6	12	14	6	—
6	2	2	6	2	—	8
8	—	—	—	4	—	—
12	4	12	—	—	—	—

Q. 16.21. $C_4 \times C_2$: $r^4 = a^2 = 1$, $ra = ar$; $C_3 \times C_3$: $p^3 = q^3 = 1$, $pq = qp$; $C_6 \times C_2$: $r^6 = a^2 = 1$, $ra = ar$; $D_3 \times C_3$: $p^3 = a^2 = q^3 = 1$, $pq = qp$, $aq = qa$, $apap = 1$.
Q. 16.22. See also p. 268, text.
Q. 16.23. b = half-turn; a, c reflections about symmetry axes inclined at $\pi/4$.
Q. 16.24. $C_2 \times C_2 \times C_2 \times C_2 \cong D_2 \times D_2$.
Q. 16.25. $C_{10} \times C_{10}$.
Q. 16.26. 32; $(C_2)^5$.
Q. 16.27. $\dfrac{16.09 \quad e \quad a \quad b \quad c \quad d \quad f \quad g \quad h}{16.10 \quad e \quad x \quad y \quad p \quad r \quad i \quad z \quad q}$ for example.
Q. 16.28. Make BMQ collinear (fig. 16.06).
Q. 16.29.

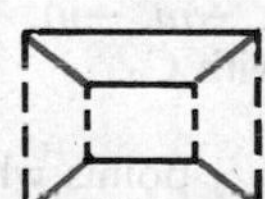

—— a
– – – b
—— c

Q. 16.30. ab, ac, bc behind intersections of appropriate mirrors; abc behind common point of three mirrors.
Q. 16.31. 30°: $C_2 \times D_6$; 72°: $C_2 \times D_5 \cong D_{10}$; 5°: $C_2 \times D_{36}$; parallel: $C_2 \times D_\infty$.
Q. 16.32. Yes.
Q. 16.33. $a = (1\ 2)$, $b = (3\ 4)$, $c = (5\ 6)$, then others are e, ab, bc, ca, abc.
Q. 16.34. $\{e, r, r^2, r^3\}$, $\{e, r^3, ar, ar^3\} \cong C_4$; $\{e, r^2, a, ar^2\} \cong D_2$; eight automorphisms, form D_4.
Q. 16.37. $p^4 = a^2 = 1$, $ap = pa$

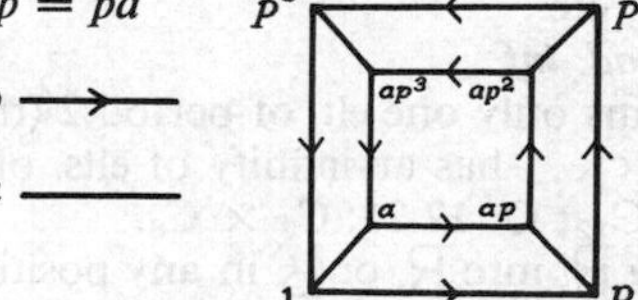

Q. 16.38. Cube is a special type of prism of type fig. 16.13.
Q. 16.39. r^2 unchanged in any automorphism, whereas a and ar^2 may be interchanged.
Q. 16.40. {1, 3, 5, 7, 9, 11, 13, 15}.
Q. 16.41. {1, 3, 7, 9, 11, 13, 17, 19}; {1, 7, 11, 13, 17, 19, 23, 29} × mod. 30.

Q. 16.42. $\begin{pmatrix}0&0\\0&0\end{pmatrix}\begin{pmatrix}1&0\\0&1\end{pmatrix}\begin{pmatrix}2&0\\0&2\end{pmatrix}\begin{pmatrix}3&0\\0&3\end{pmatrix}\begin{pmatrix}2&2\\2&2\end{pmatrix}\begin{pmatrix}3&2\\2&3\end{pmatrix}\begin{pmatrix}0&2\\2&0\end{pmatrix}\begin{pmatrix}1&2\\2&1\end{pmatrix}$,
$+$ mod. 4,
$\begin{pmatrix}1&0\\0&1\end{pmatrix}\begin{pmatrix}i&0\\0&i\end{pmatrix}\begin{pmatrix}-1&0\\0&-1\end{pmatrix}\begin{pmatrix}-i&0\\0&-i\end{pmatrix}\begin{pmatrix}1&0\\0&-1\end{pmatrix}\begin{pmatrix}i&0\\0&-i\end{pmatrix}\begin{pmatrix}-1&0\\0&1\end{pmatrix}$
$\begin{pmatrix}-i&0\\0&i\end{pmatrix}$, $\times$.

Q. 16.44. (g) $\{1, 9, 16, 22, 29, 53, 74, 79, 81\}$, $\times$ mod. $91 \cong C_3 \times C_3$,
$\{1, 4, 7, 10, 13, 16, 19, 22, 25\}$, $+$ mod. $27 \cong C_9$.
Q. 16.45. $p^3 = q^3 = r^3 = 1$, $pq = qp$, $pr = rp$, $qr = rq$; $C_3 \times C_9$.
Q. 16.48. $xy = yx$, $x^n \neq 1$ $y^n \neq 1$ $\forall n$.
Q. 16.49. E.g. (x, y, z), $+$, where $x, y, z \in \mathbf{Z}$.
Q. 16.52. $(V_3, +) \cong (\mathbf{R}, +) \times (\mathbf{R}, +) \times (\mathbf{R}, +)$.
Q. 16.54. D_∞.
Q. 16.55. E.g. Gp (r^2), Gp (ar^2), Gp (a, r^2), etc.
Q. 16.56. D_∞, fig. 13.15.
Q. 16.57. Fig. 26.073.

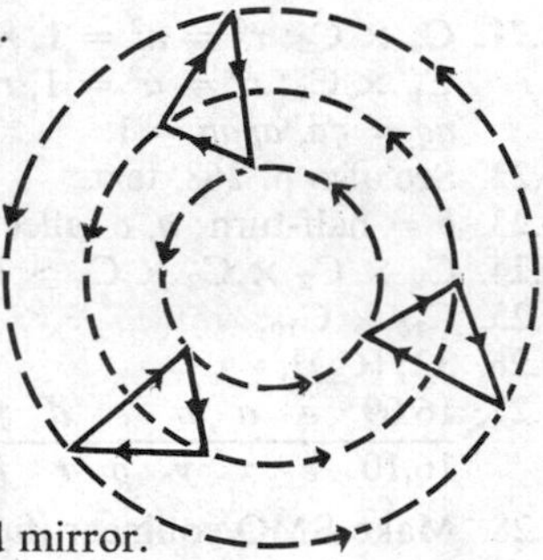

Ex. 6

$C_3 \times C_3$

Exercises

3. $C_n \times C_2$: $r^n = a^2 = 1$, $ar = ra$.
4. $C_4 \times C_3 \cong C_{12}$.
6. $p^3 = q^3 = 1$, $pq = qp$.
7. 26.
8. Mirrors at 60°, 45°, 30°, each perp. to third mirror.
9. Line $(l, m, n) \to (-l, m, n) \to (-l, -m, n) \to (-l, -m, -n)$.
12. The direct product of each of the groups quoted with C_2.
13. $C_3 \times C_9$.
16. (i) 35 (ii) 145. No (finite geometry has $N^2 + N + 1$ points where $N = p^m$, p prime).
17. 9, 21, 25, 27.
18. C_{18}, $C_6 \times C_3$; C_{20}, $C_{10} \times C_2$; C_{36}, $C_{18} \times C_2$, $C_{12} \times C_3$, $C_6 \times C_6$; C_{40}, $C_{20} \times C_2$, $C_{10} \times C_4$, $C_{10} \times C_2 \times C_2$.
19. C_{16}, $C_8 \times C_2$, $C_4 \times C_4$, $C_4 \times C_2 \times C_2$, $C_2 \times C_2 \times C_2 \times C_2$; C_{18}, $C_6 \times C_3$.
23. True for Abelian gps. when $p \neq q$. Note $p = 2$, $q = 3$ gives C_6 and D_3.
26. $C_{10} \times C_4$ (e.g. 11 has period 10, 7 has period 4).
28. $D_4 \times C_2$.
29. $D_6 \times C_2$ hexagonal prism; $D_4 \times C_2$ square bi-pyramid.
30. Identify g_1 with $(g_1, 1)$, etc.
31. $C_2 \times C_2 \times C_2 \times \ldots$ *ad. inf.*
32. Gp of Ex. 12.31 contains only one elt. of period 2 (the half-turn), whereas $C_2 \times C_3 \times C_4 \times C_5 \times \ldots$ has an infinity of elts. of period 2.
33. Ex. 3.15: $C_4 \times C_3 \cong C_{12}$; Q. 12.21: $C_3 \times C_3$.
36. (a) isometries changing R into R or R in any position,
(b) isometries changing R into R or R in any position,
(c) direct isometries.
37. $x^n \neq 1$, $y^n \neq 1$ $\forall n$, $xy = yx$.
38. $C_3 \times C_5$ ($\cong C_{15}$).
39. $C_2 \times C_2 \times C_\infty$.

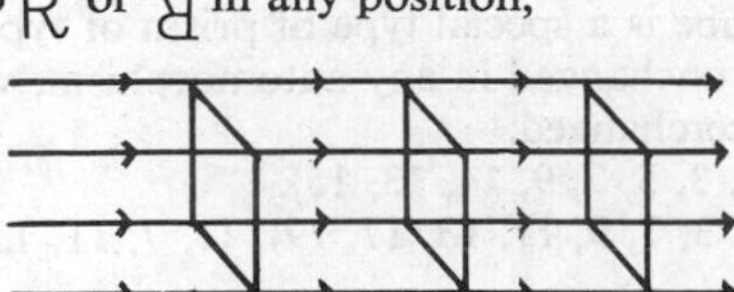

Ex. 39

40. $C_\infty \times C_\infty \times C_2$: two linked parallel plane lattices, $D_2 \times C_\infty$ as in Ex. 39; $C_\infty \times C_\infty \times C_\infty$ the lattice $\{(x, y, z), x, y, z \in \mathbf{Z}\}$.
41. $C_2 \times C_4 \times C_\infty$.

Chapter 17

Questions

Q. 17.01.

Period:	Group: C_{16}	$C_8 \times C_2$	$C_4 \times C_4$	$C_2 \times C_2 \times C_2$
2	1	3	3	7
4	2	4	12	8
8	4	8	–	–

Period:	Group: $(C_2)^4$	D_8	Q_8	$D_4 \times C_2$	$Q_4 \times C_2$
2	15	9	1	11	3
4	–	2	10	4	12
8	–	4	4	–	–

Q. 17.02. C_1: F, G, J, L, P, Q, R; C_2: N, S, Z; D_1: A, B, C, D, E, K, M, T, U, V, W, Y; D_2: H, I, O, X (D_4).
Q. 17.03. D_1: (*a*), (*b*), (*c*), (*d*), (*f*), (*i*), (*k*); C_1 (*e*); C_2 (*g*), (*h*); C_4 (*j*).
Q. 17.04. Must be subgroups of D_4; approx. one eighth; must contain a quarter-turn, i.e. C_4 or D_4.
Q. 17.06. D_1: kite, isos. trapezium; C_2: parallelogram; D_2: rectangle, rhombus; C_4: none; D_4: square. Fig. 12.03; fig. 12.041; fig. 25.04; in fig. 26.3825; fig. 17.10 (*b*); in fig. 26.3810.
Q. 17.09. Bi-pyramid with base a regular *n*-gon.
Q. 17.11. Prisms with cross-sections as in, e.g. fig. 12.034, 12.04.
Q. 17.13.

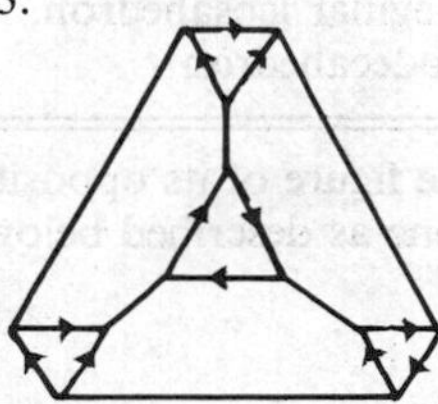

$\longrightarrow p$ $b = p^2ap$ $r = pa$
$c = pap$ $r^2 = ap^2$
$—— a$ $q = apa$ $s = ap$
$q^2 = ap^2a$ $s^2 = p^2a$
A_4

Q. 17.14. No; the whole group in each case; $p^3 = q^3 = (pq^2)^2 = 1$.
Q. 17.15. No.
Q. 17.16. Two given matrices could be *a* and *p*.
Q. 17.18. $\{1, p, p^2\}$ leaves altitude from *A* invariant; $\{1, a, b, c\}$ leaves a half-turn axis unchanged.

Q. 17.20. E.g. $\begin{pmatrix} 1 & 0 & 0 \\ 0 & 0 & -1 \\ 0 & 1 & 0 \end{pmatrix}$, take product of this with the twelve matrices on p. 295. Each contains either three or one minus signs.

Q. 17.21. $\begin{pmatrix} 0 & -1 & 0 \\ 0 & 0 & -1 \\ 1 & 0 & 0 \end{pmatrix}$, $\begin{pmatrix} 0 & 0 & 1 \\ -1 & 0 & 0 \\ 0 & -1 & 0 \end{pmatrix}$; $\begin{pmatrix} -1 & 0 & 0 \\ 0 & 0 & \pm 1 \\ 0 & \pm 1 & 0 \end{pmatrix}$ for axes lying in plane $x = 0$;

$\begin{pmatrix} 0 & 0 & -1 \\ 0 & 1 & 0 \\ 1 & 0 & 0 \end{pmatrix}$, $\begin{pmatrix} -1 & 0 & 0 \\ 0 & 1 & 0 \\ 0 & 0 & -1 \end{pmatrix}$, $\begin{pmatrix} 0 & 0 & 1 \\ 0 & 1 & 0 \\ -1 & 0 & 0 \end{pmatrix}$; $\begin{pmatrix} 1 & 0 & 0 \\ 0 & 1 & 0 \\ 0 & 0 & -1 \end{pmatrix}$.

Half-turn about Oz; third-turn combined with central inversion.

Q. 17.22. Cube is special case of cuboid and has reflections in perp. planes, see fig. 16.04. $C_4 \times C_2$ is subgroup of $S_4 \times C_2$ (cf. fig. 16.13). Not Q_4 (see p. 306).

Q. 17.24. Full group $A_5 \times C_2$, order 120; C_2, C_3, C_5 subgroups of A_5 (not C_4).

Q. 17.26. No rotations through $\pi/3$.

Q. 17.27. Rotations about altitudes give A_3 ($\cong C_3$), so four subgroups.

Q. 17.28. Fig. 17.08. Face *ABCDE* invariant under rotations through $2k\pi/5$ and is exchanged with opposite face by five half-turns; these ten rotations give D_5.

Q. 17.29.

Group	Rotations	Full group
C_n	Prisms, e.g. figs. 12.034, 12.04, 16.13; bi-pyramids	Pyramids on base like fig. 12.03, 12.04, 16.13
D_n	Regular n-gon prisms and bi-pyramids, e.g. fig. 13.09	Regular n-gon pyramid.
A_4	Regular tetrahedron	Defaced* regular tetrahedron
S_4	Cube, reg. octahedron	Regular tetrahedron
A_5	Regular icosahedron and dodecahedron	Defaced* regular 12- or 20-hedron
$C_n \times C_2$	–	Prisms and bi-pyramids, e.g. figs. 12.034, 12.04, 16.13
$D_n \times C_2$ (n even)	–	Regular n-gon prism or bi-pyramid
$A_4 \times C_2$	–	Defaced* cube (octahedron)
$S_4 \times C_2$	–	Cube, regular octahedron
$A_5 \times C_2$	–	Regular icosahedron, dodecahedron

* The object of the 'defacing' is to rob the figure of its opposite symmetries, e.g. for the cube this may be done as described below under Ex. 17.35.

Exercises

1. Ellipse, etc.
2. D_{48}, $D_{48} \times C_2$; D_{15}, $D_{15} \times C_2 \cong D_{30}$; D_3, D_3 when each thread makes an integral number of turns.
3. (a) D_1 (b) D_2; intersection makes no difference.

4. Line through fourth vertex to give square in middle, C_4.
5. (a) C_4 (b) D_4.
6.

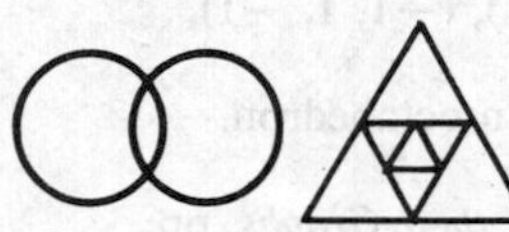
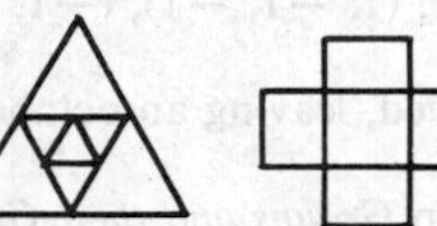
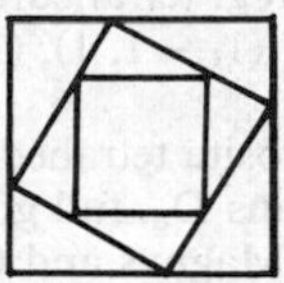
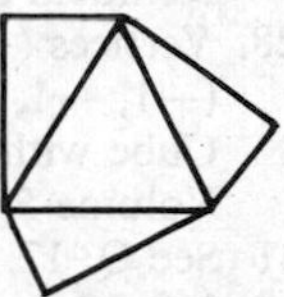

7. (a) D_1 (b) D_2 (c) D_4 (d) D_2 (e) D_4 (f) D_4 (g) D_n (n odd), D_{2n} (n even) (h) D_2 (i) D_3 (j) D_5 (k) D_1 (about $\theta = \pi/2$).
9. C_2; D_2.

10.	(a)	(b)	(c)	(d)	(e)
Rotation group	C_1	C_2	C_2	C_1	D_3
Full group	C_2	D_2	$C_2 \times C_2$	C_2	D_6

12. No symmetry other than the identity.
13. C_2, includes the half-turn $(A\ D)(B\ C)$.
14. (a) C_3 (b) D_3; No.
15. C_2 $(A\ B)(C\ D)$ only, basically same as Ex. 13; net contains paralellograms in each case.
16. Same as cube, S_4.
17. No, e.g. D_3; $D_3 \times C_{2n+1}$.
18. Half-turn about line joining $(\frac{1}{2}, 0, 0)$ and $(\frac{1}{2}, \frac{1}{2}, \frac{1}{2})$ only
    (a) two pairs opp. edges equal, and as in Exs. 17.13, 17.15
    (b) as in Ex. 17.14 (c) three pairs opp. edges equal.
19. C_2 Parallelogram prism with four rectangular faces.
    C_3 Bi-tetrahedron, ABC equilateral, $AD = BD = CD = a$, $AE = BE = CE = b \neq a$.
    C_4 Frustrum of square pyramid.
    D_2 Cuboid, D_3 all faces congruent rhombi, D_4 square prism.
    A_4 Cube with faces suitably marked, e.g. by joining consecutive face diagonals to form inscribed regular tetrahedron, or as in Ex. 35.
20. D_2; D_4.
21. $(a\ b\ c\ d)$ quarter-turn about diagonal,
    $(a\ b\ c)$ $2\pi/3$ rotation about joins of centres of opp. faces,
    $(a\ b)$ reflection in rhombic plane of symmetry,
    $(a\ b)(c\ d)$ reflection in square plane of symmetry.
22. Fig. 17.05 (a) $2\pi/3$ rotation about CC' (b) half-turn about join of centres of $ABCD$, $A'B'C'D'$ (c) $\pi/2$ rotation about join of centres of $AC'D'B$, $A'CDB'$ (d) half-turn about joins of mid-points of BC, $B'C'$ (e) $(c\ d)$ (f) $(a\ c\ b\ d)$.
23. (a) Cyclic order of letters unchanged (or reversed)
    (b) A pair of opp. faces of cube remain parallel to their initial positions.

24. Gp $(a, r) \cong C_4 \times C_2$; central inversion (1 7)(2 8)(3 5)(4 6) $= ar^2$.
26. Yes.
27. Removed tetrahedron must be regular, otherwise group is reduced. Removal of single reg. tetrahedron reduces symmetry to C_3.
28. Vertices $(-1, 1, 1), (1, -1, 1), (1, 1, -1), (1, -1, -1), (-1, 1, -1), (-1, -1, 1)$.
Cube with two opposite tetrahedra removed, leaving an octahedron.
Volume $5\frac{1}{3}$. Rotations D_3, full group D_6.
31 See Q. 17.13. Also Magnus and Grossman *Groups and their Graphs*, pp. 118–19.
32. S_4 to A_4: remove 4 from alternate vertices (e.g. A, B', C, D', fig. 17.04).
S_4 to C_4: remove 4 from corners of one face (e.g. A, B, C, D).
35. Discrepancy derives from fact that the full group of cube and of icosahedron each contain central inversion, whereas tetrahedron does not.
Solid with full group $A_4 \times C_2$ – a cube whose quarter-turns have been eliminated, e.g. each face divided into two congruent rectangles as shown; or else to form a polyhedron, each face surmounted by congruent 'roofs' (like in fig. 17.08, but avoiding regular dodecahedron). *Note*: one form of this solid is called a 'pyritohedron' and occurs in nature as the crystal form of iron pyrites.
36. Identity only if A, B coincide.

Chapter 18

Questions

Q. 18.01. O, O, E, E, E, E, E.
Q. 18.02. (a) reversed (b) unchanged.
Q. 18.05. (2 6)(1 5)(6 8)(3 8)(7 8)(4 10)(9 10).
Q. 18.06. O, E, O, E.
Q. 18.08. Parities O, O, E, O, E, E, E; *IKCMHRDOPJBNAFEGQL*; odd; period 12.
Q. 18.09. Parity of $x, x^2, x^3, \ldots$ alternates when length of cycle even; remains even when length of cycle odd.
Q. 18.10. Even if $n = 0$ or 1 (mod. 4), odd if $n = 2$ or 3 (mod. 4).
Q. 18.12. 15 even of type (1 2)(3 4); 10 odd of type (1 2); 20 even of type (1 2 3), 20 odd of type (1 2 3)(4 5); 30 odd of type (1 2 3 4); 24 even of type (1 2 3 4 5).
Q. 18.16. $\{1, h, r, y\}$ is a *normal* subgroup (see Chs. 20, 21).
C_4: rotations about join of mid-points of opp. faces.
Q. 18.17. Cycles (A B) and (C D) disjoint.
Q. 18.18. $a^2 = b^2 = c^2 = 1, (ab)^3 = 1, (gab)^4 = 1$.
(S) e.g. $i = bg$; $t = gba$; $u = agba$; $s = bagbg$; $y = agbagb$.

Q. 18.19. $\{1, r, h, y, c, d, i, l, m, q, u, v\}$.
Q. 18.20. E.g. $S_4 = \text{Gp}(a, j)$, see p. 320.
Q. 18.22. E.g. $S_3 \subset A_6$: (1 2)(3 4), (1 5)(3 6), (2 5)(4 6), (1 5 2)(3 6 4) and (1 2 5)(3 4 6).
Also $S_3 \subset A_5$: (1 2)(3 4), (1 5)(3 4), (2 5)(3 4), (1 2 5), (1 5 2).
Q. 18.23. No (cf. p. 316).
Q. 18.25. $a^2 = j^4 = (aj)^3 = 1$.
Q. 18.27. (1 2 6 5 3 7 4); (1 4 7 3 5 6 2).
Q. 18.28. S_5, 6; S_6, 6; S_7, 12; S_8, 15; S_9, 20.
Q. 18.29. S_n generated by $(1\ 2\ 3\ \ldots\ n-1)$ and $(n-1\ n)$.
Q. 18.30. See Q. 18.12. In S_6: *Odd* 120 type (1 2 3 4 5 6), 120 type (1 2 3)(4 5), 90 type (1 2 3 4), 15 type (1 2), 15 type (1 2)(3 4)(5 6).
Even: 144 type (1 2 3 4 5), 40 type (1 2 3), 90 type (1 2 3 4)(5 6) 40 type (1 2 3)(4 5 6), 45 type (1 2)(3 4).
Q. 18.32. A_n generated by ternary cycles.
Q. 18.33.

m	2	3	4	5	6	7	8	9	10	11	12	13	14	15	16	17	18
n	2	3	4	5	5	7	8	9	7	11	7	13	9	8	16	17	11.

Q. 18.34. $(1\ 2)(2\ 3)(3\ 4)\ \ldots\ (n-1\ n)$.
Q. 18.35. Gp (a, r) where $a = (B\ C)(P\ Q)$, $r = (A\ B\ C)$.
Q. 18.37. $x = (1\ 5\ 2\ 6\ 3\ 4)$, $x^2 = (1\ 2\ 3)(4\ 5\ 6)$, $x^3 = (1\ 6)(3\ 5)(2\ 4)$, etc.
Q. 18.38. $(A\ B)$.
Q. 18.40. $yx = (2\ 7\ 5\ 6\ 8)(3\ 4)$.
Q. 18.41. $f(x_1, x_2, x_4, x_3)$ and 7 others equal to it; $AB^2 + BD^2 + DC^2 + CA^2$.
Q. 18.42. D_4, D_2, S_3, C_3, D_4, D_4, S_3, C_2, D_6, C_3, S_3, S_3, S_3, D_2, C_2, S_2, D_2, D_2, C_2, D_4, C_2, D_2.

Exercises

2. Odd, odd, even.
4. $pqr = \begin{pmatrix} \text{A B C D E} \\ \text{D B C E A} \end{pmatrix} = (\text{A D E})$; $r^2p = \begin{pmatrix} \text{A B C D E} \\ \text{A D C E B} \end{pmatrix} = (\text{B D E})$; both even; p even q,r both odd.
5. (a) (E A T S L); (E L)(T A); (P S E T A); (E A)(T L S)
(b) (E R S G A T N I M); (E S G)(M T N I R A); (E M A R)(I S G T N)
(c) (S D)(A E N)(I T); (A T N I); (S I)(A N T E)
(d) (G S O U)(R E N); (e) (S P A R)(E L); (S P A R L Y); (S P L R Y)
(f) (A L); (A T L); (S A L E T R); (S A L T E R)
(g) (N A G O S E D); (G S A D)(R I)(N E); (E G O I S N A)(R D); (G E N A O I S).
6. E.g. $x = (1\ 2\ 4)$ (see pp. 364–5).
7. True in general – no. of intersections = no. of inversions of order.
8. $(n-1)!$ cycles of length n.
10. $x^{-3}y^{-1}$.
11. $x = (1\ 5\ 2)(3\ 4\ 7\ 6)$; $y = (1\ 7\ 2)(3\ 4\ 5\ 6)$.
12. $|G| \in 2Z + 1 \Rightarrow$ all perms. even.
14. $r = qp^{-1}q^{-1} = (1\ 2\ 4\ 3)$.
15. E.g. (1 2 3)(3 4) = (1 2 3 4).
17. $C_4 \times C_2$ Abelian, so r commutes with a. Thus $a = (5\ X)$, or $(5\ X)(Y\ Z)$.
18. For overlapping transpositions, (1 2)(2 3) = (1 2 3), otherwise (1 2)(3 4) = (1 2)(1 3)(3 1)(3 4) = (1 3 2)(1 3 4).
21. $(1\ 2\ X)(1\ 2\ Y)(1\ 2\ Z)(1\ 2\ X)(1\ 2\ Y) = (X\ Y\ Z)$, so any ternary cycle is guaranteed.

22. No, as $60 \div 8 = 7\frac{1}{2}$.
23. 3-cycles: 20 in S_5, 40 in S_6,
    4-cycles: 30 in S_5, 90 in S_6 (cf. Q. 18.12 and 18.30).
26. (1 3), and similarly for (1 r).
27. Two generators necessary for S_n.
29. $b \to c$, $c \to b$, $p \to q \Rightarrow bp \to cq$, i.e. $q \to s$. Also $pa \to qa$, so $r \to s$.
30. $pq = (1\ 2\ 4)(3\ 5)$, so Gp (p, q) contains elts. of periods 4, 5 and 6 and its order is a multiple of 60. But q is an odd perm, so Gp $(p, q) \not\cong A_5$. Hence Gp $(p, q) \cong S_5$ (using Q. 21.26).
32. A_5: When p is a transposition, Gp $(p, q) \cong S_5$.
33. $S_3 \times C_2 \cong D_6$.
34. $p^4 = 1, p^2 = q^2 = (pq)^2$, so Q_4.
35. E.g. if $p = (1\ 2\ 3\ 4)$, p^2 is not a 4-cycle.
36. $D_2 \subset D_4$.
38. $p_1^2 - 2p_2$; $p_1p_2 - 3p_3$; $p_1^3 - 3p_1p_2 + 3p_3$.
40. $p = (1\ 2\ 3)$, $a = (1\ 2)(3\ 4)$, Gp $(p, a) \cong A_4$.
41. (a) $a = (1\ 2)$, $p = (3\ 4\ 5)$ Gp $(a, p) \cong C_2 \times C_3 \cong C_6$
    (b) $a = (1\ 2)$, $p = (2\ 3\ 4)$, Gp $(a, p) \cong S_4$.
42. 13057.
43. A_4: $\pm z$; $\pm\frac{1}{z}$; $\pm\left(\frac{z \pm i}{z \mp i}\right)$; $\pm i\left(\frac{z \pm 1}{z \mp 1}\right)$.

Chapter 19

Questions

Q. 19.01. E.g. $Hp^2 = \{p^2, p^2x, b, bx\}$ sends A to third place.
Q. 19.02. Subgroup is $\{1, p, p^2, a, b, c\} \cong D_3$.
Q. 19.03. L. and R. cosets by x or by cx are same, others different.
Q. 19.04. $\{ax, bx, cx\}$ neither a subgroup nor a coset; $\{1, p, p^2, a, b, c\}$.
Q. 19.06. Subgroup is normal (see Chs. 20, 21).
Q. 19.07. $\{1, p, p^2\}$ $\{a, r, s^2\}$ $\{b, s, q^2\}$ $\{c, q, r^2\}$.
Q. 19.09. $\{g, h, m, n, t, u\}$ $\{i, k, p, r, v, x\}$ $\{j, l, q, s, w, y\}$.
Q. 19.10. $\{1, r, j, t\}$ $\{a, i, s, u\}$ $\{p, g, n, v\}$ $\{c, h, m, w\}$ $\{d, l, g, x\}$ $\{f, k, p, y\}$
$H = \{1, f, h, j, p, t, y, r\}$ $aH = \{a, d, g, i, q, u, x, s\}$
$Ha = \{a, c, g, l, v, m, s, x\}$, $H' = \{1, b, w, y\}$, $aH' = \{a, c, v, x\}$,
$H'a = \{a, d, q, s\}$, etc.
Q. 19.11. Both L. and R. cosets must contain all the remaining elts.
Q. 19.13. S_4 $\{1, h, r, y\}$; A_4 $\{1, a, b, c\}$; Q_4 see Q. 19.14; D_6: C_6, C_3 and $\{1, r^3\}$; Q_6: C_6, C_3, $\{e, k\}$.
Q. 19.14. In Q_4 only subgroup of order 2 is $\{1, r^2\}$ (table 9.13).
Not true in Q_6. No normal subgroups in A_n ($n \geqslant 5$) and 'simple' groups (p. 410).
Q. 19.16. (a) $2\mathbf{Z}$, $2\mathbf{Z} + 1$ (b) $3\mathbf{Z}$, $3\mathbf{Z} \pm 1$ (c) $6\mathbf{Z}$, $6\mathbf{Z} + 1$, etc.
Q. 19.17. Coset of $\overrightarrow{OA}$ is the set of vectors $\overrightarrow{OP}$ where P lies in plane through A parallel to plane of V_2.
Q. 19.18. The same.
Q. 19.19. Coset by $a + ib$ ($a, b \in \mathbf{R}$) is $\{(x + a) + i(y + b) : x, y \in \mathbf{Z}\}$.
Q. 19.20. Parallel lines; yes, any line is a coset of the parallel line through origin.
Q. 19.23. All cubics with coefficient of x^3 equal to 1.
Q. 19.25. Reciprocals of linear functions, etc.
Q. 19.27. The set $\{ai : a \in \mathbf{R}\}$ of 'purely imaginary' numbers.
Q. 19.28. E.g. the bilinear functions (Q. 19.25).

Q. 19.30. A set of rotations about points on the perp. bisector of OA' where O is mid-point of AA'.

Q. 19.31. $r_{A,\alpha} * r_{O,\theta} = r_{C,\alpha+\theta}$, and left coset consists of such rotations for varying θ.

Q. 19.33. No, since reflection in real axis does not in general commute with a rotation.

Q. 19.35. Set of rotations through α about any point of plane. Yes.

Q. 19.36. The set $\begin{pmatrix} 0 & 0 & p \\ 0 & q & 0 \\ r & 0 & 0 \end{pmatrix}$ $(p \neq 0, q \neq 0, r \neq 0)$.

Q. 19.37. Take A as origin, AB as z-axis, perp. axis as y-axis, and consider
$$\begin{pmatrix} -1 & 0 & 0 \\ 0 & 1 & 0 \\ 0 & 0 & -1 \end{pmatrix} \begin{pmatrix} \cos\theta & -\sin\theta & 0 \\ \sin\theta & \cos\theta & 0 \\ 0 & 0 & 1 \end{pmatrix}.$$

Q. 19.39. That set of matrices $\begin{pmatrix} p & q \\ r & s \end{pmatrix}$ such that $ps - qr = 2, -1, 4$ respectively.

Q. 19.40. See table 21.03. C_3 is not normal, so impossible.

Q. 19.41, 42.

	1	r	r^2	r^3	a	b	c	d	
a	a	d	c	b	1	r^3	r^2	r	
b	b	a	d	c	r	1	r^3	r^2	
c	c	b	a	d	r^2	r	1	r^3	
d	d	c	b	a	r^3	r^2	r	1	(D_4)
a	a	b	c	d	1	r	r^2	r^3	
b	b	c	d	a	r	r^2	r^3	1	
c	c	d	a	b	r^2	r^3	1	r	
d	d	a	b	c	r^3	1	r	r^2	$(C_4 \times C_2)$
a	a	d	c	b	r^0	r	1	r^0	
b	b	a	d	c	r^3	r^2	r	1	
c	c	b	a	d	1	r^3	r^2	r	
d	d	c	b	a	r	1	r^3	r^2	(Q_4)
a	a	b	c	d	r	r^2	r^3	1	
b	b	c	d	a	r^2	r^3	1	r	
c	c	d	a	b	r^3	1	r	r^2	
d	d	a	b	c	1	r	r^2	r^3	(C_8)

Q. 19.43. The set of reflections (or flip-overs).

Q. 19.44. E.g. coset of C_4 is rotations through $\pm 45°$, $\pm 135°$.

Q. 19.46. Those which are not equivalence relations, following show where they fail (R, S or T): (e) T (j) friends, T, R(?), S(?) (o) R, T.
In (k) and (n), assume k non-zero. *Note*: (e) needs defining more precisely.

Q. 19.48. E.g. elts. of a group having the same period, as in footnote p. 353.

Q. 19.52. $A + B = H$; $\mathbf{2} + \mathbf{4} = \mathbf{1}$, etc.

Q. 19.54. Follows trivially from Q. 19.50.
Q. 19.55. In latter case, $(A \cap B)C \subset AC \cap BC$.
Q. 19.56. E.g. in D_3 (table 8.03) $A = \{b, c\}$, $B = \{p, q\}$, $C = \{e, p\}$.

Exercises

2. 4.
4. Centre H. $\forall h \in H$, $ghg^{-1} = h$.
5. L. cosets, see Q. 19.10. R. cosets different.
9. E.G. in D_2 (table 3.10), $H_1 = \{I, X\}$; $H_2 = \{I, Y\}$.
11. $Hx^{-1} = Hy^{-1} \Rightarrow \exists\ h$ s.t. $hx^{-1} = y^{-1} \Rightarrow y = xh \Rightarrow y \in xH \Rightarrow yH = xH$.
12. Elts. of same coset differ by multiples of 10.
13. $xG \cup G$ not necessarily closed (e.g. contains x but not x^2) so is not a group. Case quoted is a counter-example.
16. (a) $a = 1$ (translations) gives a normal subgroup,
    (b) $b = 0$ (spiral similarities about origin) gives a non-normal subgroup.
17. In D_3 (table 8.03), $p\{b, c\} = \{c, a\}$.
18. Coset by $\begin{pmatrix} p & q \\ r & s \end{pmatrix}$ (left or right) is set of matrices whose determinant $= ps - qr$.
19. (a) 8 cosets (b) coset by $\begin{pmatrix} p & q \\ r & s \end{pmatrix}$ is the set $\left\{\begin{pmatrix} p + 2a & q + 2b \\ r + 2c & s + 2d \end{pmatrix} (a, b, c, d \in \mathbf{Z})\right\}$.
21. (a), (b) yes; (c), (d) fail for transitivity.
22. (a) $k = m$ when A, B are cosets of a normal subgroup,
    (b) If A is a subgroup, and B is a left (or right) transversal (see Ex. 19.26 below), then AB is the whole group; e.g. in S_4 (p. 219), if $A = \{1, a, b, c, d, f\}$ and $B = \{a, k, n, s\}$ then aA, kA, nA, sA are four distinct cosets, so AB is the whole group.
23. $h_1, h_2 \in H$. $h_1 \to xh_1x^{-1}$, $h_2 \to xh_2x^{-1}$, then $h_1h_2 \to x(h_1h_2)x^{-1}$ establishes isomorphism; $xHx^{-1} = H \Rightarrow H$ is normal in G.
24. G is $D_4 \times C_2$; HK is $C_2 \times C_2 \times C_2$.
25. E.g. $r = (1\ 5\ 4)$; index of Gp (p, q) in Gp (p, q, r) is 5.
26. m^r transversals. To prove Hx_1^{-1} and Hx_2^{-1} disjoint, suppose $h_1x_1^{-1} = h_2x_2^{-1}$, then $x_2h_2^{-1} = x_1h_1^{-1}$ so $x_2 \in x_1H$, contrary to data.
28. $A_5 \times C_2$ contains no elts. of period 4, so S_4 not a subgroup.

Chapter 20

Questions

Q. 20.01. E.g. $s^{-1}ys = (1\ 8\ 6\ 7\ 3)(2\ 5\ 4)(9\ 10)$.
Q. 20.02. $rar^{-1} = (2\ 3)$, $r^2ar^{-2} = (3\ 4)$, $r^{-1}ar = (4\ 1)$; $S_5 \cong$ Gp (a, s) where $s = (1\ 2\ 3\ 4\ 5)$.
Q. 20.03. $(a^*b^*c^*)^{-1} = c^*b^*a^*$; $(a^*b^*c^*)^2$ is a translation, translations commute.
Q. 20.04. Rotation about P' where P is mid-point of $P'Q$.
Q. 20.06. c is a reflection in the reflection of a in b.
Q. 20.07. If r is rotation through θ about P, then ara^{-1} is a rotation through $-\theta$ about reflection of P in a.
Q. 20.08. $r_2r_1r_2^{-1}$ is rotation through α about $r_2(A)$.
$r_1^{-1}r_2r_1r_2^{-1}$ is translation through $4AB \sin \frac{1}{2}\alpha \sin \frac{1}{2}\beta$ in direction $\frac{1}{2}(\alpha + \beta)$ to BA

Q. 20.09. $gfg^{-1}(x) \equiv x/(a+bx)$; $ghg^{-1}(x) \equiv (c+dx)/(a+bx)$;
$jfj^{-1}(x) \equiv ax - b$; $jhj^{-1}(x) \equiv (ax-b)/(d-cx)$;
$hgh^{-1}(x) \equiv [(ac-bd)x + b^2 - a^2]/[(c^2-d^2)x + bd - ac]$.

Q. 20.10. $g^{-1}fg(x) \equiv (2-x)x$; $gfg^{-1}(x) \equiv x^2/(2x-1)$.

Q. 20.11. $RMR^{-1} = \begin{pmatrix} 1 & 0 \\ 0 & -1 \end{pmatrix}$; reflection in $y = 0$.

$R^{-1}MR = \begin{pmatrix} \cos 4\theta & \sin 4\theta \\ \sin 4\theta & -\cos 4\theta \end{pmatrix}$; reflection in $y = x \,.\, \tan 2\theta$.

Q. 20.12. $MAM^{-1} = \begin{pmatrix} d & c \\ b & a \end{pmatrix}$.

Q. 20.15. $\{1, d, f, g, h, j\}$; $\{a, d, h, l\}$; $\{m, f, j, k\}$; $\{m, c, g, k\}$; $\{a, c, g, l\}$; $\{c, g, d, h, f, j\}$; $\{a, f, j, l\}$.

Q. 20.16. D_3: $\{e\}\ \{p, q\}\ \{a, b, c\}$ D_4: $\{e\}\ \{g\}\ \{a, c\}\ \{b, d\}\ \{f, h\}$
D_6: $\{1\}\ \{r^3\}\ \{r, r^5\}\ \{r^2, r^4\}\ \{a, c, f\}\ \{b, d, g\}$
Q_4: $\{e\}\ \{r^2\}\ \{r, r^3\}\ \{a, d\}\ \{b, c\}$.

Q. 20.17. Identical when $x \in$ centre.

Q. 20.19. Classes represent perms. of types $(A\ B)$; $(A\ B)(C\ D)$; $(A\ B\ C)$ and $(A\ B\ C\ D)$ respectively.

Q. 20.20. $\{b, q, s^2\} = q\{1, p, p^2\}$.

Q. 20.21. E.g. normaliser of a is $\{1, a, r^3, d\} = \mathrm{H}$. $b\mathrm{H} = \{b, r, f, r^4\}$, whereas $\mathrm{H}b = \{b, r^5, f, r^2\}$.

Q. 20.22. See Q. 20.21 $c\mathrm{H} = \{c, r^2, g, r^5\}$, $\mathrm{H}c = \{c, r, g, r^4\}$, etc.

Q. 20.23. If x and y transform a alike, $xax^{-1} = yay^{-1} \Rightarrow (y^{-1}x)a = a(y^{-1}x)$ $\Rightarrow y^{-1}x \notin N$. Thence we deduce x, y lie in different cosets.

Q. 20.24. Isomorphism established by $\phi(h) = aha^{-1}$, so if $\phi(h_1) = ah_1a^{-1}$, $\phi(h_2) = ah_2a^{-1}$, then $\phi(h_1)\phi(h_2) = ah_1a^{-1}ah_2a^{-1} = ah_1h_2a^{-1} = \phi(h_1h_2)$.

Q. 20.25. (a) $1, a, u, x$ (b) b, y, v, p transform H into $\{1, c, w, x\}$,
c, z, w, p^2 transform H into $\{1, b, v, x\}$.

Q. 20.27. No, C_4 contains only two elts. of period 4, whereas a normal subgroup would have at least three.

Q. 20.28. $x\mathrm{H} = \mathrm{H}x \Rightarrow px = xp$ or else $p^2x = xp \Rightarrow x = pxp$.

Q. 20.29. D_6: $\{1, r^3\}$, C_3, D_3, C_6. A_4: $\{1, a, b, c\}$. S_4: A_4 and D_2 $\{1, h, r, y\}$.

Q. 20.30. Q_6: $1, p, p^2, r, r^2, r^3, pr\ (= rp^2), p^2r\ (= rp), pr^2\ (= r^2p), p^2r^2\ (= r^2p^2)$ $pr^3\ (= r^3p^2), p^2r^3\ (= r^3p)$.

Q. 20.31. $G \cong A_4$.

Q. 20.32. If det. $M = \pm 1$, det. $A = \Delta \neq 1$, then det. $(AMA^{-1}) =$ det. A det. M det. $A^{-1} = \Delta \times (\pm 1) \times (1/\Delta) = \pm 1$, so A transforms subgroup into itself.

Q. 20.33. No, e.g. if

$$Y = \begin{pmatrix} 0 & \omega \\ -\omega^2 & 0 \end{pmatrix} \quad (\omega = \operatorname{cis} \tfrac{2}{3}\pi), \qquad X = \begin{pmatrix} x_1 & x_2 \\ -x_2 & x_1 \end{pmatrix},$$

then

$$YXY^{-1} = \begin{pmatrix} x_1 & x_2\omega^2 \\ -x_2\omega & x_1 \end{pmatrix},$$

which is not in the subgroup.

Q. 20.34. C_3, C_4 not normal in S_4. In S_5, if $a = (4\ 5)$, $r = (1\ 2\ 3\ 4)$, then $ara^{-1} = (1\ 2\ 3\ 5)$ which is not in S_4, so S_4 is not normal in S_5.

Q. 20.35. Yes.

Q. 20.36. Yes.

Q. 20.37. No, e.g. when $s(z) \equiv 1/z$, $srs^{-1}(z) \equiv z/(p+qz)$, which is not in the subgroup.

Q. 20.38. Those listed in Q. 20.29, e.g. $\{1, r^2, r^4\}$.

Q. 20.39. Rotations about it, reflections in a plane containing it, central inversion, or combinations of these.
Q. 20.41. 1, h, r, y.
Q. 20.42. The half-turn axes in this case are moved to *new* positions by some of the other rotations.
Q. 20.43. E.g. $C_n \lhd D_n$.
Q. 20.45. Note, this Q. provides an alternative definition for 'centre'.
Q. 20.47. E.g. in D_3, $c\{a, b\} = \{q, p\}$; $\{a, b\}c = \{p, q\}$, but $ca \neq ac$.
Q. 20.48. S is a subset of the centre, and S generates a normal subgroup of G.
Q. 20.51. Sufficient to show (1 2 3), (1 3 2), (1 2 4), (1 4 5) in same class,
e.g. (4 5)(1 2)(1 2 3)(1 2)(4 5) = (1 3 2)
(3 4)(1 2)(1 3 2)(1 2)(3 4) = (1 2 4), etc.
True that all ternary cycles in A_n are conjugate.
Q. 20.52. $z(x^{-1}y^{-1}xy)z^{-1} = (zx^{-1}z^{-1})(zy^{-1}z^{-1})(zxz^{-1})(zyz^{-1}) = x'^{-1}y'^{-1}x'y'$, where $x' = zxz^{-1}$, $y' = zyz^{-1}$, so subgroup is normal.
Q. 20.54. Converse not true: counter-example – any Abelian group.

Exercises

1. xax^{-1} has period 2 for all x, so $xax^{-1} = a$.
2. $x = (2\ 4)(1\ 4) = (1\ 2\ 4)$.
3. $y(xy)y^{-1} = yx$.
4. A quarter-turn about z-axis anticlockwise (b) the same clockwise.
6. $gxg^{-1} = g^{-1}xg \Rightarrow g^2x = xg^2$.
7. E.g. a and b in D_3 (table 8.03); or in Q_6 (table 14.01), centraliser of c is $\{1, c, g, k\}$, of f is $\{1, f, j, k\}$ and $c = bfb^{-1}$.
9. $ar = r^{n-1}a \Rightarrow ara^{-1} = r^{n-1}$.
10. (a) $y^2 = 1 \Rightarrow y = y^{-1} = xyx^{-1} \Rightarrow xy = yx$, so an elt. x must exist which commutes with y,
(b) there must exist an x s.t. $(xy)^2 = x^2$; y cannot belong to the centre,
(c) In Ex. 20.05, c and d are inverses but not conjugate.
15. From (4), $q = pr^{-1}p^{-1}$, so q and r^{-1} have same period, hence $r^4 = 1$.
In Q_6, j, d, c satisfy (1), (2), (3), (4), (5) but not (6).
16. (5) $p = q^{-1}r^{-1}q$ (6) $p = rq^{-1}r^{-1} \Rightarrow q^{-1}r^{-1}q = rq^{-1}r^{-1} \Rightarrow qrq = rqr$ (7).
Also (5) $q = rqp$ (4) $qp = pr^{-1} \Rightarrow q = rpr^{-1} \Rightarrow qr = rp$ (8).
$1 = q^4 = q^3(rqp)$ (by (5)) $= q^2(qrq)p = q^2(rqr)p$ (by (7)) $= q(qrq)rp = q(rqr)rp$ $= (qr)^3$ by (8).
17. No, in both cases, centre of rotation is moved by TRT^{-1}.
18. No, e.g. if $Q = \begin{pmatrix} i & 2i \\ 2i & -i \end{pmatrix} \notin Q_4$, $P = \begin{pmatrix} i & 0 \\ 0 & -i \end{pmatrix} \in Q_4$, then QPQ^{-1} $= 0{\cdot}2 \begin{pmatrix} -3i & 4i \\ 4i & 3i \end{pmatrix} \notin Q_4$.
19. $xHx^{-1} = H$, $yHy^{-1} = H \Rightarrow xyH(xy)^{-1} = xyHy^{-1}x^{-1} = xHx^{-1} = H$, establishes closure, etc. H does not need to be normal.
20. The set is a coset of the centraliser of a. Note also that $xRy \Leftrightarrow x^{-1}yay^{-1}x = a \Leftrightarrow (x^{-1}y)a(x^{-1}y)^{-1} = a$, so that if $x^{-1}y = z$, we have $zaz^{-1} = a$, i.e. z commutes with a. Thus z belongs to the centraliser of a, which is a subgroup N. This means that $x^{-1}y \in N$, from which it follows (see p. 381) that x and y belong to the same left coset of N.
21. $rsr^{-1}s^{-1}$ is a translation.
22. G is of order 21.

23. $\mathbf{BA} = \mathbf{B(AB)B}^{-1}$ = rotation ϕ about axis U moved by $\mathbf{B}$ to position V. Take U as z-axis, V as rotation axis for $\mathbf{C}$, then $\mathbf{AB} = \mathbf{M}$ and $\mathbf{BA} = \mathbf{C}$, so tr. $(\mathbf{C})$ = tr. $(\mathbf{M}) = 2\cos\phi + 1 \Rightarrow \phi = \cos^{-1}\frac{1}{2}(\text{tr. }(\mathbf{C}) - 1)$.

Chapter 21

Questions

Q. 21.01. $\{10^x, x \in \mathbf{R}\}, \times$.
Q. 21.02. No.
Q. 21.03. The set of constants.
Q. 21.04. Yes, each coset of K maps into a distinct elt. of the image group.
Q. 21.07. $(a\mathrm{K}b\mathrm{K})c\mathrm{K} = ab\mathrm{K}c\mathrm{K} = ab(\mathrm{K}c)\mathrm{K} = abc\mathrm{KK} = (abc)\mathrm{K}^2 = abc\mathrm{K}$. Similarly $a\mathrm{K}(b\mathrm{K}c\mathrm{K}) = abc\mathrm{K}$.
Q. 21.08. No, see p. 409; also $C_2 \lhd D_2 \lhd A_4$, yet $C_2 \ntriangleleft A_4$.
Q. 21.09. C_2.
Q. 21.11. $G \cong D_2 \times C_3 \Rightarrow G/C_2 \cong C_6$; $G/C_3 \cong D_2$; $G/C_6 \cong C_2$.
Q. 21.13. $G/C_4 \cong C_2 \Rightarrow G \cong C_2 \times C_4$, C_8, D_4 or Q_4
$G/C_5 \cong C_2 \Rightarrow G \cong C_{10}$ or D_5
$G/C_6 \cong C_2 \Rightarrow G \cong C_2 \times C_6$, C_{12}, D_6 or Q_6
$G/D_2 \cong C_2 \Rightarrow G \cong D_4$, $C_2 \times C_2 \times C_2$ or $C_2 \times C_4$
$G/C_2 \cong C_3 \Rightarrow G \cong C_6$
$G/D_2 \cong C_3 \Rightarrow G \cong A_4$ or $C_2 \times C_6$
$G/C_2 \cong D_2 \Rightarrow G \cong D_4$, Q_4, or $C_2 \times C_2 \times C_2$.
Q. 21.14. E.g. $D_4/C_2 \cong D_2$; $D_6/C_3 \cong D_2$.
Q. 21.16. mn elts. of G map into the n left cosets, m elts. of G into each L. coset.
Q. 21.17. $C_2 \lhd D_2 \lhd A_4$ is only counter-example for groups of order up to 15.
Q. 21.18. S_4 not normal in S_5, e.g. (1 5) (1 2 3 4)(1 5) = (1 5 3 4), taking S_4 to be the stabiliser of 5.
Q. 21.19. For $H_1H_2 \lhd G$, use $x(h_1h_2)x^{-1} = (xh_1x^{-1})(xh_2x^{-1})$.
Q. 21.20. $1 \lhd C_n \lhd C_{2n} \lhd D_{2n}$ and $1 \lhd C_2 \lhd C_{2n} \lhd D_{2n}$.
Q. 21.21. None is normal (see p. 386).
Q. 21.22. (a) centre is $C_3 \times C_2$, so $G/H \cong C_2 \times C_2$,
(b) $G/H \cong C_6$
(c) $1 \lhd C_3 \lhd C_6 \lhd C_{12} \lhd C_3 \times D_4$; $1 \lhd C_2 \lhd C_4 \lhd D_4 \lhd C_3 \times D_4$.
Q. 21.23. Centre of $D_6 \times C_2 \cong C_2 \times C_2$; $G/H \cong D_3$.
Q. 21.24. A_6 (order 360).
Q. 21.26. True since a subgroup of S_5 of order 60 has to be normal.
Q. 21.27. Consider the map $z\mathrm{H} \rightarrow z^3$.
Q. 21.28. The finite subgroups of $(\mathbf{C}, \times)$ are sets of nth roots of 1.
Q. 21.29. $D_\infty/C_\infty \cong C_2$.
Q. 21.30. $\mathrm{K/T} \cong (\mathbf{S}, + \text{ mod. } 1) \cong$ group of rotations about a fixed point.
Q. 21.31. Consider $z = \rho \operatorname{cis} \phi$; $z_1z_2 = \rho_1\rho_2 \operatorname{cis}(\phi_1 + \phi_2)$, and result follows.
Q. 21.32. (a) $(\mathbf{V}_2, +)/(\mathbf{V}_1, +) \cong (\mathbf{V}_1, +)$ (b) $(\mathbf{V}_3, +)/(\mathbf{V}_2, +) \cong (\mathbf{V}_1, +)$, e.g. in $(\mathbf{V}_3, +)$, cosets of $(\mathbf{V}_2, +)$ are a set of parallel planes: consider the map from each coset to the length of the perp. from O to the corresponding plane (or line in (a)).
Q. 21.33. $(\mathbf{C}, +)$/(Gaussian integers) $\cong$ addition of complex numbers within the square $OUWV$ (fig. 10.04) $\cong (\mathbf{S}, + \text{ mod. } 1) \times (\mathbf{S}, + \text{ mod. } 1)$.
Q. 21.35. Normal: (1), (7); not normal (2), (3), (4), (5), (6).
In (7), $G/H = (\mathbf{R}, \times)$, for matrices with equal determinants are mapped into the same coset, and multiplication of cosets corresponds to multiplication of the appropriate determinants.

Exercises

1, 2. H is centre, $G/H = D_2$.
4. (b) isomorphism $r = \pm 1, 3$; homomorphism $r = 2, 4$.
(c), (d), (e), (f), (g), (j), No (h) homoporphism $x \to |x|$.
6. $xy = z$ and $x^2y^2 = z^2 = xyxy \Rightarrow xy = yx$.
If x of period $2m$, then x^r and x^{m+r} both map into x^{2r}.
7. $x^{-1}y \in H$; $(xz)^{-1}(yz) = z^{-1}(x^{-1}y)z \in H$ if $H \lhd G$ (see p. 355).
8. E.g. $G \cong S_4$, $A \cong A_4$; $B \cong C_2$.
9. $\{0, 4\} \to 0$, $\{1, 5\} \to 1$, $\{2, 6\} \to 2$, $\{3, 7\} \to 3$; $\{0, 2, 4, 6\} \to 0$, $\{1, 3, 5, 7\} \to 1$.
10. C_4.
12. Kernel is the normal subgroup D_2, whose operations leave the three perp. axes invariant but change the cube's diagonals. For $S_6 \to S_4$, consider permutations of the six faces.
13. Consider the map $\mathbf{R} \to \mathbf{S}$ where $x \to x - [x]$, e.g. $7{\cdot}41 \to 0{\cdot}41$ (as in Ex. 21.14).
14. $x \to (x - [x])2\pi$, e.g. $-4{\cdot}25 \to (-4{\cdot}25 - (-5))\ 2\pi = 3\pi/2$. Kernel is $(\mathbf{S}, + \text{ mod. } 1)$.
15. Kernel is $\{1, \omega, \omega^2, \omega^3, \omega^4, \omega^5\}$, where $\omega = \text{cis}\ \frac{1}{3}\pi$.
16. $z \to z^8$ homomorphism; $z \to \bar{z}$ and $z \to 1/z$ isomorphisms; $z \to 2z + 1$ neither.
17. E.g. $z \to |z|$.
18. arg. $(z_1z_2) =$ arg. $z_1 +$ arg. z_2, so $z \to$ arg. z is a homomorphism from $(\mathbf{C}, \times) \to (\mathbf{S}, + \text{ mod. } 1)$, since arg. z may be expressed in revolutions $(0 \leq \text{arg. } z < 1)$.
19. Kernel is polynomials having $x^2 + 1$ as a factor.
20. E.g. $\begin{pmatrix} a & b \\ c & d \end{pmatrix} \to (ad - bc)^n$ where $n = -1, 3, \ldots$ For kernel, $ad - bc = 1$.
21. $\phi(xy^{-1}) = \phi(x)\phi(y^{-1}) = \phi(x)[\phi(y)]^{-1} = \phi(x)[\phi(x)]^{-1} = 1$.
22. Yes, since $(\mathbf{x} + \mathbf{y}) . \mathbf{a} = \mathbf{x} . \mathbf{a} + \mathbf{y} . \mathbf{a}$.
23. $G \cong C_\infty \times C_\infty \times C_\infty$; $H \cong C_\infty$; $G/H \cong C_\infty \times C_\infty$.
24. $(\mathbf{C}, \times)/(\mathbf{R}, \times) \cong (\mathbf{S}, + \text{ mod. } 1)$.
$(\mathbf{C}, \times)/(\mathbf{Q}, \times) \cong (\mathbf{R}, \times)/(\mathbf{Q}, \times) \times (\mathbf{S}, + \text{ mod. } 1)$.
25. Map the positive rationals on to 0, the negatives on to 1, then $(\mathbf{Q}, \times)$ is mapped on to $\{0, 1\}$, + mod. 2.
26. (a) the rotations about a fixed point $\cong (\mathbf{S}, + \text{ mod. } 1)$,
(b) the rotations about a fixed point and reflection in a fixed line.
27. E = even; O = odd; $C_\infty/C_\infty \cong C_n$.
28. $G/H \cong D_2$.
29. E.g. $\begin{pmatrix} a & b \\ 0 & d \end{pmatrix} \to \log(ad)$.
30. Represent D_2: $\begin{pmatrix} 0 & 0 \\ 0 & 0 \end{pmatrix}, \begin{pmatrix} 1 & 0 \\ 0 & 0 \end{pmatrix}, \begin{pmatrix} 0 & 0 \\ 0 & 1 \end{pmatrix}, \begin{pmatrix} 1 & 0 \\ 0 & 1 \end{pmatrix}$, + mod. 2. a, b, c, d replaced by 0 or 1 according to whether even or odd.
31. Transformation represented by matrix $\begin{pmatrix} a & b \\ c & d \end{pmatrix} \to$ transformation represented by complex bilinear function $(az + b)/(cz + d)$ $(a, b, c, d \in \mathbf{R})$. Kernel is set of enlargements, represented by $\begin{pmatrix} k & 0 \\ 0 & k \end{pmatrix}$.
32. Kernels are the lines $x + y = 0$, $x = 0$, $2x + y = 0$, $y = 0$.

33. Kernel is the line $x:y:z = -4:1:3$. Any matrix $\begin{pmatrix} a & b & c \\ p & q & r \end{pmatrix}$ of rank 2 provides for $(\mathbf{V}_3, +) \to (\mathbf{V}_2, +)$.
34. Subgroup not normal.
36. See *Mathematical Gazette* quoted p. 403 above.

Chapter 22

Questions

Q. 22.01. D_2.
Q. 22.03. $z(ygy^{-1})z^{-1} = (zy)g(zy)^{-1}$ establishes closure of inner automorphisms, etc.
Q. 22.04. If C_8 is represented $\{0, 1, 2, 3, 4, 5, 6, 7\}$, + mod. 8, the automorphisms are (1 3)(5 7)(2 6), (1 5)(3 7)(2 6) and (1 7)(2 6)(3 5).
Q. 22.05. Compare Q. 22.04 above.
Q. 22.06. D_4.
Q. 22.07. Suppose two generators a, b in Abelian group. Any elt. may be expressed $a^p b^q$ $(= b^q a^p)$.
But $a^p b^q \to a^{kp} b^{kq}$; $a^r b^s \to a^{kr} b^{ks}$;
So $(a^p b^q)(a^r b^s) = a^{p+r} b^{q+s} \to a^{k(p+r)} b^{k(q+s)} = (a^{kp} b^{kq})(a^{kr} b^{ks})$.
k prime to n necessary for correspondence to be 1 to 1. Row 3 is the only one; $r \to r^5$ (a) not an automorphism (products not preserved) (b) a homomorphism, e.g. for $C_2 \times C_6$, $r \to r^3$ maps the group on to $\{1, a, r^3, ar^3\}$.
Q. 22.08. The subgroup stabilising $\{1, p, p^2\}$ is D_6. Each coset contains 2 of each period 2, 3, 4 and 6, and 4 of period 8. Group has 13 of period 2, 8 of period 3, 6 of period 4, 8 of period 6, and 12 of period 8.
Q. 22.09. Half-turns about three perp. axes.
Q. 22.11. See table 17.01.
Q. 22.13. The outer automorphisms of A_4 are transforms of elts. of A_4 by elts. in S_4 but not in A_4. Compare also Q. 22.15.
Q. 22.14. Aut. $(S_n) \cong S_n$ (all inner).
Q. 22.15. They are changed; e.g. in A_4 (table 17.02), $\{p, q, r, s\} \leftrightarrow \{p^2, q^2, r^2, s^2\}$.

Chapter 23

Exercises

3. Pitch raised in ratio 1·35, i.e. just over a fourth.
4. (a) 6 in. (b) 3 in.
5. $(x + 50)/(x + 78) = 5/6 \Rightarrow x = 90$ in.; just over 11 in.
6. 5·9%; 260, 275, 292, 309, 328, 347, 368, 390, 413, 437, 463, 491, 520.

Chapter 24

Questions

Q. 24.01. Cannot be done unless bells rest in same place for more than 3 consecutive blows, e.g. **a, b, g, b, g, b, a, b, a, b, g, b, g, b, a, b, a, b, g, b, g, b, a.**
Q. 24.05. $r = gfg^{-1}$, $s = p$, $t = q$.
Q. 24.07. Almost identical; identical over a plain course of sixty changes.
Q. 24.09. Grandsire doubles. Plain course of thirty changes: **c, a, e,** . . ., with **a, e** alternating, and **c** following **a** at each treble lead.

Q. 24.11. See p. 471.
Q. 24.12. Yes. With notation of table 24.13, rows 5, 17, 29, 41, 53 would be t, pt, p^2t, p^3t, p^4t, so we have *right* cosets.
Q. 24.13. A group must contain equal numbers of even and odd permutations (if not all even).
Q. 24.15. Row 345.
Q. 24.18. $(\mathbf{a\ z\ b\ z\ b\ z})^6 = 1$, i.e. rounds return after thirty-six changes.
Q. 24.21. At rows $4 + 12n$, bells in places A, B, C start a quick six, at rows $10 + 12n$, a slow six, meanwhile dodging in places (D, E) and (F, G). Seventh place is made to link the sixes.
Q. 24.23. Adjacent balls exchange velocities, just as adjacent bells are exchanged. Only one pair at a time, unless two or more impacts are simultaneous.

A	*B*	*C*	*D*
1	2	3	4
1	2	4	3
1	4	2	3
4	1	2	3
4	2	1	3
4	2	3	1
4	3	2	1

No further collisions with velocities ending in reversed order. In the above, it is assumed that the first collision is between C and D, the second between B and C, and so on. Other orders are of course possible.

Chapter 25

Questions

Q. 25.02. E.g., t, $-t$, $2/t$, $-2/t$.
Q. 25.05. 1, fg, $gfgf$, gf and f, g, gfg, fgf are squares.
1, f, $gfgf$, gfg is a D_2 subgroup, and the other rectangle is its coset.
Q. 25.06. Because the operation g, written on the right, is applied first, and then h in the case of the chord joining g to H.
Q. 25.09. G (α, β) needs to satisfy $4(2\alpha - 1)^2 = 3(\alpha^2 + \beta^2 - 1)$.
Q. 25.11. Stabiliser of $\{A, B, E\}$ is $\{e, x, x^2, f, h, l\}$, D_3.
Q. 25.12. D_n for convex n-gon.
Q. 25.13. S_4; S_5. See fig. 13.18. If the sets $\{P, Q\}$ and $\{A, B, C\}$ are stable, the group is D_6 ($\cong D_3 \times C_2$).
Q. 25.15. Stabiliser of points A and U is D_2, generated by $(A\ D)$ and $(B\ C)$; Stabiliser of $\{P, Q, R\}$ is D_3, generated by $(B\ C)$ and $(A\ B)$.
Q. 25.16. No (the group of order 168 is simple).
Q. 25.17. 24; S_4.
Q. 25.18. Each point in turn is fixed and pairs of points on the three lines through it interchanged. Not a group. Can be made into $C_2 \times C_2 \times C_2$ by introducing X and interchanging each invariant point in turn with it, e.g. 2nd perm becomes $(A\ X)(B\ E)(C\ G)(D\ F)$.
Q. 25.19. S_4 and subgroups. Also C_7, generated by $(A\ F\ C\ G\ E\ D\ B)$.
Q. 25.20. $ABCD$ is a quadrangle in both figures, with EFG the diagonal point triangle.
Q. 25.21. C_6, generated by $f = (B\ R\ Q\ C\ M\ N)$; D_3 $\{e, a, m, h, g, 1\}$ preserves triangles APL, BQM, CRN.
Q. 25.22. No; for if $x = (A\ B)(P\ Q)(L\ M)$, then left coset replaces A by B in all twelve cases, whereas right coset contains bx, which replaces A by C.
Q. 25.23. D_6.
Q. 25.24. D_2.

Q. 25.24. D_2.
Q. 25.25. $C_3 \times C_3 \times C_3$ (not in the group of order 108, since x, y, z do not themselves belong to it).
Q. 25.27. (a) D_3 (b) D_6 (c) same as (b); (d) S_4 (e) S_5, by considering fig. 25.22.
Q. 25.30. 36.
Q. 25.31. In the six perms of A, B, C, the stabiliser of A is of order 2, and this is not normal in S_3.
Q. 25.32. Stabiliser of P has index 2.
Q. 25.33. In $C_2 \times C_6$, stabiliser of P is of index 2, stabiliser of A is of index 6.

Chapter 26

Questions

Q. 26.01. A translation.
Q. 26.02. Fig. 26.016.
Q. 26.03. Fig. 26.21.
Q. 26.05. Glide reflections do not belong; group contains only translations, the motif being R Я.
Q. 26.06. Quarter-turn, or reflection in mirror at 45°, or a glide reflection.
Q. 26.08. Fig. 26.3818.
Q. 26.09. Like fig. 26.31.
Q. 26.10. E.g. the portion $0 \leqslant x \leqslant \pi/2$.
Q. 26.11. Those with half-turn symmetry: H, I, N, O, S, X, Z.

Q. 26.14. (a) 26.165, (b) 26.163, (c) 26.166, (d) 26.163,
(e) 26.165, $0 \leqslant x \leqslant \pi/2$ (f) 26.167, (g) 26.161, FE
(h) 26.164, P (i) 26.166, (j) 26.162, P
(k) 26.162, R (l) 26.164, (m) 26.164,
(n) 26.163, (o) 26.161, R△ (p) 26.161, R

Q. 26.18. Contains, or does not contain, half-turns.
Q. 26.20. Abelian frieze patterns: 26.161, 2 and 6.
Abelian plane patterns: **p1** only.
Q. 26.21.

(S) (*b*) **pm** A A / A A **cm** A A A / A **pmg** A ∀ A ∀ / A A **pmm** A A / ∀ ∀ etc.

(*c*) **pmm** H H / H H **cmm** H H H / H (*d*) **cmm** ⊟ / ⊟ ⊟ / ⊟ **pmg** E Ǝ E Ǝ / E E

Q. 26.22. **cm**, unless translation is parallel or perp. to reflection, in which case 26.166 and 26.163 respectively.
Q. 26.23. 26.3801 **p2**; −2, **p2**; −3, **p2**, −4, **cm**; −5, **cmm**; −6 **p4g**; −7, **cm**; −8, **p4m**; −9, **p2**; 26.3810, **pgg**; −11, **p4**; −12, **pmg**; −13, **cmm**; −14, **pgg**; −15, cm; −16, **cmm**; −17, **p1** (motif is a 4 × 4 square of characters); −18, **p6m**; −19, **p2**; −20, **p31m**; −21, **p31m**; −22, **p2**; −23, **p4g**; −24, **pmg**; −25, **pmm**; −26, **pgg**; −27, **cm**; −28, **pm**.

Q. 26.25. No. 2: all twelve indirect patterns.
No. 4: **p2, p4, p6, pmm, pgg, pmg, cmm, p4g, p4m, p6m.**
No. 5: **pmm, pmg, cmm, p4m, p4g, p6m.**
No. 6: **pm, pmm, pmg, cm, cmm, p4g, p4m, p6m, p31m, p3m1.**
Q. 26.26. All preserved under enlargement; only the five direct patterns preserved under rotation of motif.
Q. 26.27. 2.
Q. 26.30. Yes.
Q. 26.31. Fig. 26.162; Fig. 26.161; of order 2.
Q. 26.32. Gp (g_1^{2n}, g_2^{2m+1}).
Q. 26.33. The set 3**Z**.
Q. 26.34. E.g. Gp (g_1^3, g_2^3), or Gp (g_1, g_2^9).
Q. 26.36. Not normal.
Q. 26.37. Gp (t).

Exercises

2. As fig. 26.163; $\sin x + \sin mx$: $m = 2$, 26.164; $m = 3$, 26.165, etc.
3. Figs. 26.164, 26.163, 26.165.
4. As in fig. 26.167, with a and b interchanged; **pmm**.
5. **pmg.**
6. A lattice of rhombi, **cmm**.
8. See Q. 26.23.
9. E.g. **pmm** from 26.165 by two perp. reflections,
p6m from 26.167 by half-turn and 120° turn about suitable points.
11. Hexagons have three pairs of opp. sides equal and parallel.
12. Pattern may be identified by the symmetry chart.
13. E.g. *dbacb*.
15. E.g. *m* and *gm*.
16. Each of index 2 in the other.
19. Yes, of index 4.
22. *a* and *b* reflections in parallel mirrors, *c* a translation parallel to the mirrors, or a glide in one of them, gives **pm**.

Bibliography

1. Books on group theory, beyond the standard of the present work, i.e. more suitable for students in mathematical courses at universities and colleges.

W. Burnside, *Theory of Groups of finite order*. Dover, 1911.

H. S. M. Coxeter and W. Moser, *Generators and relations for discrete groups*. Springer Verlag, 1957.

M. Hall, *Theory of Groups*. Macmillan, 1959.

M. Hamermesh, *Group theory and its applications to physical problems*. Addison Wesley, 1962.

B. Higman, *Applied group theoretical and matrix methods*. Dover, 1954.

A. G. Kurosh, *The Theory of Groups* (two volumes). Chelsea, 1955.

W. Lederman, *Introduction to the theory of finite groups*. Oliver and Boyd, 1949.

I. D. Macdonald, *The theory of groups*. Oxford University Press, 1968.

M. Newman, *Matrix representation of groups*. National Bureau No. 60, 1968.

G. Papys, *Groups*. Collier Macmillan, 1964.

E. M. Patterson and D. E. Rutherford, *Elementary Abstract Algebra*. Oliver and Boyd, 1965.

E. Schenkman, *Group theory*. Van Nostrand, 1965.

O. U. Schmidt, *Abstract theory of groups*. W. H. Freeman, 1966.

W. R. Scott, *Group theory*. Prentice-Hall, 1964.

H. Wielandt, *Finite permutation groups*. Academic Press, 1964.

J. J. Rotman, *The Theory of Groups: An Introduction*. Allyn and Bacon, 1965.

2. Books on algebra of university standard, with chapters on groups, or with extended treatment of groups.

I. T. Adamson, *Introduction to field theory*. Oliver and Boyd, 1964.

G. Birkhoff and S. Maclane, *A survey of modern algebra*. Collier Macmillan, 1953.

J. B. Fraleigh, *A first course in abstract algebra*. Addison Wesley, 1966.

F. M. Hall, *Abstract algebra* (two volumes). Cambridge University Press, 1969.

D. E. Littlewood, *Skeleton key of mathematics*. University Library, Hutchinson, 1949.

E. A. Maxwell, *Algebraic structure and matrices*. Cambridge University Press, 1965.

H. Meschkowski, *Introduction to modern mathematics*. Harrap, 1968.

G. D. Mostow *et al.*, *Fundamental structures of algebra*. McGraw-Hill, 1963.

B. L. Van der Waerden, *Modern algebra*. Ungar, N.Y., 1949.

3. Books of standard approximately equivalent to, or just lower than, the present work.

On groups:

P. Alexandroff, *Introduction to the theory of groups*. Blackie/Hafner, 1959.

A. W. Bell, *Algebraic structure*. Allen and Unwin, 1966.

A. W. Bell and T. J. Fletcher, *Symmetry groups*. Assoc. Teachers of Mathematics, 1964.

J. A. Green, *Sets and groups*. Routledge and Kegan Paul, 1965.
I. Grossman and W. Magnus, *Groups and their graphs*. Random Ho., 1964.
F. L. Hardy, *Exercises in group theory*. Prentice Hall, 1970.
D. E. Mansfield and M. Bruckheimer, *Background to set and group theory*. Chatto and Windus, 1966.

Containing treatment of groups:
C. P. Benner *et al.*, *Topics in modern algebra*. Harper and Row, N.Y., 1962.
J. H. Cadwell, *Topics in recreational mathematics*. Cambridge University Press, 1966.
ed. N. J. Hardman, *Exploring university mathematics* (2). Pergamon, 1966.
R. North, *The art of algebra*. Pergamon, 1965.
W. W. Sawyer, *A concrete approach to abstract algebra*. W. H. Freeman, 1959.

4. Elementary and popular books containing introductory work on groups.
I. Adler, *The new mathematics*. Signet Science Lib., 1959.
A. Balfour, *Introduction to sets, groups and matrices*. Heinemann, 1965.
G. Beaumont and P. Caldwell, *Introduction to groups*. Assoc. of Teachers of Mathematics.
B.B.C., Booklets on the T.V. series 'Maths today'. B.B.C. Publications, 1969.
Z. P. Dienes and E. W. Golding, *Groups and coordinates*. E. S. A. Hutchinson.
ed. T. J. Fletcher, *Some lessons in mathematics*. Cambridge University Press, 1964.
J. S. Friis, *Groups* ('Expanding Mathematics' series). Basil Blackwell, 1967.
A.T.M. team, *Mathematical reflections*. Cambridge University Press, 1970.
K. L. Gardner, *Discovering modern algebra*. Oxford University Press, 1966.
C. Lanczos, *Numbers without end*. Oliver and Boyd, 1968.
D. T. E. Marjoram, *Modern mathematics* (three books). Pergamon, 1964.
E. A. Maxwell, *Gateway to abstract algebra*. Cambridge University Press, 1965.
J. Moakes, *The core of mathematics*. Macmillan, 1964.
M. S. Norton, *Finite mathematical systems*. John Murray, 1966.
D. Pedoe, *The gentle art of mathematics*. English Universities Press, 1958.
D. A. Quadling, *The same but different*. Bell, 1970.
W. W. Sawyer, *Prelude to mathematics*. Pelican, 1955.
S.M.P. books, (e.g.) *Additional mathematics book 1*. Cambridge University Press, 1966.
D. F. Taylor, *Amateur modern mathematics*. Basil Blackwell, 1967.
L. F. Taylor, *Numbers*. Faber and Faber, 1970.

5. References from the *Mathematical Gazette*.
Feb. 1956, p. 15, T. J. Fletcher, Film Groups.
Oct. 1961, p. 181, F. M. Hall, Group theory in the sixth form.
Oct. 1961, p. 207, W. W. Sawyer and R. C. Lyness, Periodic sequence and cycles.
Feb. 1962, p. 1, R. North, On functions which form a group.
Feb. 1963, p. 25, R. E. Green, Primes and recurring decimals.
Oct. 1963, p. 209, A. W. Fuller, Rotation groups and permutation groups.
Feb. 1964, p. 47, J. Holland, Illustrations of simple group theory.
Dec. 1964, p. 384, S. N. Collings, Maximum length decimals. (See also May 1956, p. 137, C. L. Wiseman.)
Feb. 1965, p. 26, R. J. Wilson, Modern mathematics in secondary schools.
Feb. 1966, p. 21, E. J. F. Primrose, On functions which form a group.
May 1966, p. 105, D. S. Macnab, The cubic curve and an associated structure.

May 1966, p. 132, E. D. Bender, Logical mappings.
Oct. 1966, p. 262, D. A. Quadling *et al.*, The use of axiomatic method in secondary teaching.
Oct. 1966, p. 290, D. B. Hunter, Permutations and rearrangements.
Dec. 1966, p. 398, T. J. Fletcher, Applications of groups. (See App. 3.)
Feb. 1967, p. 11, J. J. Malone, Uses of Sylov theory.
Dec. 1967, p. 309, E. J. F. Primrose, Cyclic projectivities.
Oct. 1968, p. 263, H. G. Forder, Groups from one axiom.
Dec. 1968, p. 342, R. L. Goodstein, Free variable axioms for groups.
Dec. 1968, p. 354, F. H. Francis, Patterns in group tables.
Feb. 1969, p. 19, F. J. Budden, Transformation geometry in the plane by complex number methods.
May 1969, p. 129, B. D. Price, Mathematical groups in campanology.
May 1969, p. 162, E. H. Thompson, Note on the finite rotations of rigid bodies.
Oct. 1969, p. 293, D. F. Robinson, Permutations on a group table.
Oct. 1969, p. 295, H. M. Cundy, A review reviewed.
Feb. 1970, p. 9, F. J. Budden, On functions which form a group.
May 1970, p. 125, M. Holt, Mathematics for tomorrow's children.
Oct. 1970, p. 237, A. L. Davies, Rotating the fifteen puzzle.
Oct. 1970, p. 284, D. A. T. Soffe, The reduction of a general matrix.
Oct. 1970, p. 299, J. Huckaba and A. Kirch, Additive subgroups of the rationals.
Dec. 1970, p. 346, M. Holt, Group on the Mobius ring.
Dec. 1970, p. 368, F. J. Budden, A non-commutative associative operation on the reals.
Dec. 1970, p. 372, S. K. Abdali, Verification of associativity of a binary operation.
Feb. 1971, p. 38, V. Bryant, Reducing classical axioms.
Feb. 1971, p. 59, B. Spielman, A test for associativity.

6. Books on special topics related to group theory, as described in the text.

D. Hilbert and C. Cohn Vossen, *Geometry and the imagination*. Chelsea, 1952.
H. S. M. Coxeter, *Introduction to geometry*. Wiley, 1961.
H. W. Guggenheimer, *Plane geometry and its groups*. Holden-Day, 1966.
P. H. Yale, *Geometry and symmetry*. Holden-Day, 1968.
H. Weyl, *Symmetry*. Princeton University Press, 1952.
M. C. Escher, *The graphic work of M. C. Escher*. Oldbourne Press, 1961.
I. M. Yaglom, *Geometric Transformations*. Random Ho., 1962
M. Jeger, *Transformation geometry*. Allen and Unwin, 1964.

H. M. Cundy and A. P. Rollett, *Mathematical Models*. Oxford University Press, 1952.
L. Fejes Toth, *Regular figures*. Pergamon, 1964.
H. S. M. Coxeter, *Regular polytopes*. Methuen, 1947.
A. F. Wells, *The third dimension in chemistry*. Oxford University Press, 1956.

F. J. Budden, *Complex numbers and their applications*. Longmans Green, 1968.
W. R. Hamilton, *Elements of quaternions* (two volumes). Chelsea, N.Y.

A. C. Aitken, *Determinants and matrices*. Oliver and Boyd, 1939.
J. R. Branfiend and H. W. Bell, *Matrices and their applications*. Macmillan, 1970.
G. Matthews, *Matrices* (two volumes). Edward Arnold, 1964.
T. Brand and A. Sherlock, *Matrices pure and applied*. Edward Arnold, 1970.

E. Artin, *Galois theory*. Notre Dame, 1948.
M. Postnikov, *Foundations of Galois theory*. Oxford University Press, 1962.

Sir James Jeans, *Science and Music*. Cambridge University Press, 1937.
Sir Walford Davies, *The pursuit of music*. Nelson, 1935.
C. A. Taylor, *The Physics of Musical Sounds*. English University Press, 1965.
A. H. Benade, *Horns, strings and harmony*. New York, Anchor, 1960.
F. Winchel, *Music, sound and sensation*. New York, Dover, 1967.
J. Backus: *The acoustical foundation of music*. Murray, 1970.
*W. G. Wilson, *The art of change ringing*. Faber and Faber, 1965.
*W. H. Thompson, *On Grandsire triples*. Macmillan and Bowes, Cambridge, 1886.
*D. A. Bayles, *An introduction to method structure*.
*J. Snowdon, *Grandsire*
*J. Segar, *A blue-line proof*.
A. W. T. Cleaver, *The theory of change – ringing*. John Hilton, 1965.
*A. Y. Bramble, *Method structure in change – ringing*.

*The above booklets may be available from the Central Council of Church Bell Ringers.

Index

More important references are given in **bold type**

Abel, N. H. 411
Abelian group 30, **82**, 96, 141, 238, 244, 259, 280, 282, 287, 291, 338, 363
  non-Abelian 152, 217, 374, 377
  Abelian groups in patterns 528
  automorphisms 422–3
  homomorphic images 171–3, 404–5
abstract group **138,** 206, 221, 307
addition
  in **Z** 74, 125, 144, 181, 227, 251, 341, 396, 413, 416, 419
  in **Q** 87, 227
  in **R** 144, 155, 341, 396
  in **C** 155, 228–9, 238, 341, 416
  of circuits 12
  of matrices **9,** 64, 74
  of ordered pairs 10
  of polynomials 16, 88, 238
adjoin 264, 360, 417, 509
affine group 231, *see also* transformations, groups of
alternating group 220, 291, **316–7,** 330, 393, *see also* groups, A_n
anagram 13
Archimedes semi-regular solids 307, 309
area-preserving transformations **229, 398**
areal coordinates 275–7
Argand diagram 11, 70, 134, 177, **228, 341, 388,** 414
associativity **Ch. 5**
  definition **37**
  failure of 36, 38, 84, 90
  of $\cap$ and $\cup$ 37
  of mappings **39, 40**
  test for 37–42, 96
  use in proofs 65, 90, 100, 102–4, 106, 111, 115, 206–7, 222, 355, 364, 368–9, 380
automorphisms **147–153, 290, Ch. 22**
  inner **152–3,** 382–4, **421–3,** 424, **428**
  of C_n 147, 150, 169, 185–6, 331, 423
  of D_n 149, 151, 211
  of $C_2 \times C_2 \times C_2$ **272–3,** 411
  of $C_n \times C_2$ 281, 423–4
  of S_4 426–7
axes of rotation and reflection (fixed or moving with body) 56, **198–201,** 234, 293, **365–71,** 389, 517
axioms for groups 73, 104

bell-ringing 207, 363, 371, **Ch. 24,** 542
  bob 471–2, 474

bell-ringing—(*contd.*)
  coursing order 470
  dodge 454, 466, 469
  doubles 454, 459–66
  Erin 461
  Grandsire 466, 478
  hunting 454, 457, 469, 474
  lead 459; treble lead 457, 467, 469–70, 475, 478
  linkage 471
  Major 454; Glasgow Surprise 453, 475; Huddersfield Surprise 476; Plain Bob 478
  Maximus 454, 474
  method 451, 474
  Minor 454, 467–75; College 475; Plain Bob 467–74; St Clements 475
  notation 452–3, 460
  'path' (of bell) 454
  peal 456, 470, 474
  place 453, 456, 460
  plain course 467, 469, 473
  principle 461
  Rankin (Professor) 478
  rounds 451, 457
  row 451
  single 463, 467, 474
  six (quick and slow) 455, 466
  Stedman 463–6, 471, 476–9
  Thompson, W. H. 478
  treble, tenor 451
bilinear function
  closure 20, 35
  period of **117–18, 128–9**
  groups of 254, 287, 332, 342
bilinear transformation 237, 239, 372, 389, 482–3
binary operation **Ch. 2**
  on arrows 13
  on cosets **359, 405–6,** 417
  on functions **22,** 35, 42, 74, *see also* functions, groups of
  on matrices **9, 10,** 54
  on musical intervals **429,** 434, 436
  on ordered pairs 10–1, 20, 38, 42, 51, 61, 74, *see also* ordered pairs
  on ordered triples 10, 42, 100, *see also* ordered triples
  on points of circle 20, 40, 49, 52, 71, 481, **484, 487–8**
  on points of conic 53, **480–8**

binary operation—(*contd.*)
on points of cubic 53
on points of plane 4, 11, 43, 52–3, 70–1
on points of sphere 13
on planes 11, 14
on sets **6, 9, 37,** 46
on subsets of groups **356–9,** 361
on symmetries 30, 78, 81
on **R** App. I
binary scale 8, 13, 68, 84, 86, 133, 269, 287
bi-pyramid 193, 310
bi-tetrahedron 210, 225
Buffon 257

campanology, *see* bell-ringing
cancellation 89–90, 103, 106, 111, 346
Cartesian product 266
Cayley's theorem 78, **92,** 143, 154, 157, 184, 194, **201–2,** 211, 214, 271, 279, 311
Cayley diagram (graph) 137, 203, 247–50, 253, 281–2, 285, 288, 294, 310, 343, 419, 480, 537
in three dimensions 250, 275–6, 289, 307
centraliser 222–3, 294, 345, 361, 379–81, 383, 392, 394
centre 217–18, 240, 294, 360, 374, 383, 393, 409, 421, 428
Chain 176, 211, 408–9, 411, 413
circle, *see* binary operation
closure 5, 11, **Ch. 3,** 45
condition for subgroup **217**
failure for **16–18,** 20, **83–4,** 96, 231, **436,** 461
in music 16, 436
column vector 9–10, 31, 45, 65, 296, 301
commutative (operations) **10,** 17
commuting elements 13, 42, 218, **222–3,** 230, 363, 374, 383
failure for commutativity 10, 13, 42, 83, 136, 233
commutator 310, 371, **393,** 395
complement 6
complex numbers 11, 134, 246, 251, **342–5, 388,** 396, *see also* Argand diagram; roots of unity
composition factors 409
composition, law of, *see* binary operation
computers 5, 245
Fortran 8
congruence 5, **59, 340**
conjugate elements **152, 200,** 320, **Ch. 20, 363** (defn), 412, 461, 464, 470, 502
conjugacy classes **374–81,** 384–5, 393–4
of S_4 **380, 384**
self conjugate 374, 393
see also transform
conjugate points w.r. to conic 134, 480–1
conjugate subgroups **381–2,** 385, **390**
contrapositive 138
converse 138, 153
cosets 81, 175, 245, 280, 319, **Ch. 19,** 438, 487–8
binary operation on **359**
in campanology 457–8, 461, 463, 469–71, 478
conjugacy classes and 380
disjoint property **339, 347,** 461, 469, 484
as equivalence classes **355**
groups of **401–8**
in infinite group **339–46, 413–7**
left and right: same **382–3,** 401, 534; different 334, 390, 392, 536
from tables **336**
counter-example 16, 37, 39, 41, 143–4
cross-ratio 11, 43, 221, **327–8,** 480, 483, 488
crystallography 504, 542
cube 221, **297–303,** 495, *see also* rotation groups; symmetry groups
cycles **111–17,** 160, 179, 211, **321–5**
inverse of 324
notation **113, 322–4**
overlapping **117, 324,** 329, 394
parity of 315, 318
period of 112, **114–15,** 320–1, 412–3
product of disjoint **113–16,** 184, 260, 281–2, 286, 315, **321–4,** 364
ternary 322, 330–1, 393
cyclic groups **109,** 115, **145–8,** 150, **Ch. 12,** *see also* groups C_n
composite order **170–5**
definition **157**
infinite 181
prime order **167–70, 176,** 215
subgroups 174–5
cycloid 504
cyclotomic equation **177–81**

decimal (repeating) **164–7**
defining relations (abstract definition) xiii, 99–100, 102–4, 110–11, 193, 197, **Ch. 15, 242–5,** 268, **290–1,** 349, **385,** 525, 536
Desargues' theorem 138, 274, **499–501**
determinant 7, 66, 184, 226, 229–30, 240, 301, 335, 346, **386–7, 398**
dicyclic groups **245–7**
defining relations **245,** *see also* groups Q_n
dihedral group **Ch. 13,** *see also* groups D_n ($n = 1, 2, 3 \ldots$), D_∞
in art 209
in campanology 456, 460, 469

dihedral group—(*contd.*)
defining relations 193, **203,** 242–3
generated by 2 elements period 2 197, **206–9,** 320, 460, 469, 480
dilatation, *see* similarity group
direct (isometry, transformation, symmetry) **187,** 191, 198, 225
in three dimensions **292–3**
direct product groups **Ch. 16,** *see also* groups
definition **255**
of finite with infinite groups **284–6**
of two infinite groups **283–4,** 398, 416
quotient groups of **407–8**
dodecahedron, *see* symmetry groups; rotation groups, icosahedron
dual 138, 297, 303, 307, 495

edges, perms of 297
eigenvector 374
empty set 63–4, 91, 136
enantiomorph, *see also* symmetry groups **188–9,** 225, 262, 270, 288, **292–3,** 297, **301–2,** 306, **335**
enlargement 532, *see also* symmetry groups
equations, theory of 178–81, 326–7, 543
cyclotomic equation **177–81**
Quintic equation 305, **410–13**
equivalence class 233, 236–7, **352, 354–5,** 397, **430,** 437, *see also* residue classes
disjoint property **354–5**
equivalence relation **352,** 374, 394, 429, 437, 561
properties **353**
equivalent 234, 256, 277, 283, 429
Escher 504, 515
Euclidean geometry 274–5, 504
Euler's function **185 6, 545**
even permutation, *see* permutations
exponentiation 3, 16, 46
extension 264

faces, permutations of 278, 288, 297, 311
factor group, *see* quotient groups
Fermat's theorem **166–7**
field 3, 5, 154, 266
finite arithmetic **46–8,** 276, 543, *see also* residue classes
addition mod. *n* 16, 18, 74, 93, 125, **145, 162, 170–5,** 338, 413–14, 437–9
multiplication mod. *n* 8, **47–8,** 52, 54, 74, 91, **132, 145–7, 162–4, 173,** 183–4, 260, 280, 282, 287–8
finite geometry **274–7,** 495–7, 543
Fortran 8, 36

free group 243, 245
frieze, *see strip pattern*
Frobenius 384
full group, *see* symmetry groups
function, rational 88, 239, 343, *see also* bilinear function
functions 5, 74, *see also* bilinear function; rational function
composition of **22,** 35, 42
groups of 85–8, 134–5, 160, 184, **195–6,** 207, 210, 212, 239, 332, 360, 480–2
period of **117–18,** 129
transform of 373
fundamental region **511–14, 523–8,** 537

Galois 411
games 25
Gaussian integers 228, 238, 283, 341, 416
generators xii, 110, 193, **206–8, 223–5, Ch. 15,** 256, 268–72, 281, 283, **290,** 455, 484, *see also* Cayley diagram; defining relations
independence of 244–5, 251
in infinite groups 125, 182, 207, 213, 284
minimum set xii, **241–3,** 252, 333, 510
of patterns **509–15, 519–28,** 533–8
of S_n 318–22, 330–1, 365, 426
single generator 123, 125, **161–3,** 176, 182, 185, 437
geometry 11, 25–6, 40–2, 49–50, 52–3, 61, 70–1, 121, **230–6, Ch. 25,** *see also* finite geometry; pattern; reflections; rotations; symmetry; transformation
glide reflection 185, 208, 285, 368, 390, **505–8, 510–11, 521–7, 534–6**
graph, *see* Cayley diagram
Greeks 504
groups, *see also* cyclic; dihedral; dicyclic; functions; quotient; rotations groups
definition **73**
examples of **74ff**
'opposite' 369
order: of order 4 **139–40**; of order 6 **141–2**; of order 8 266–7, 348–50; of order 12 291, 294; of order 16 291; of order 24 264
of a polynomial, etc. 325–9
regular representation by matrices, **97–99**
C_2 **80–1,** 291–2, 309, 399, 401, 415
C_3 28, 47, 118, 255, 284, 309, 327, 342, 389, 441
C_4 48, 60, 81–2, 89, 110, 140, 143, 147–8, 255, 292, 421, 442, 526
C_5 93, 96, 331
C_6 141, 150, **158–64,** 247 (Cayley graph), 259 ($\cong C_2 \times C_3$), 292
C_7 168

groups—(*contd.*)
C_8 177, 184, 350
C_9 182
C_{10} 60, 185
C_{12} 146, **170–3,** 179–81, 258 ($\cong C_4 \times C_3$), 339, 349, 404, 437–8
C_p **167–170,** 215
C_∞ 125, 144, **181–3,** 185–6, 227, 233, 340–1, 396, 413, 516, 535
D_1 **78–80,** 291–2, 504
D_2 17, 75, 89, **131–40,** 149, 193, 209, 259 ($\cong C_2 \times C_2$), 292, 307–9, 481, 505
D_3 **76–8, 97–9,** 151–2, **196–203,** 220, 328, 379, 389, 399, 472, 482, 484, 487, 492
D_4 94, 196, 220, 226, 236, 241–2, 292, 309, 326, 349, 366, 418, 421, 456–7, 490, 492, 496
D_5 243, 311, 426, 504
D_6 **188–95,** 203, 225, 247 (Cayley graph), 261–2 ($\cong D_3 \times C_2$), 333–4 (cosets in), 379, 381–2, 389, 469, 492, 497–9, 502
D_9 204
D_n **293,** 394, 401, 480
D_∞ 183, **204–9,** 213, 254, 480, 484–5, 504, 516
Q_4 101, **245–7,** 249 (Cayley graph), 252, 331, 350, 394, 418, **424–6**
Q_6 108, **216–17,** 244, 357–9, 379, 390–3, 403, 409, 426, 428
Q_n **245** (defn)
$C_3 \times C_3$ 123, 176, 185, 259, **282–3,** 286–7, 312, 424
$C_2 \times C_4$ 259, **280–2,** 311, 330, 350, 406
$C_2 \times C_6$ **260,** 309, 423–4
$C_4 \times C_4$ 265
$C_2 \times C_2 \times C_2$ **266–80,** 287, 348–9, 491, 496
$C_\infty \times C_2$ 183, **284–6,** 343, 516
$C_\infty \times C_\infty$ 283, 341, 398, 513–14, 519
$D_\infty \times D_1$ 240, 286, 343, 516
A_4 220, 252, 291, **294,** 305–6, 309–10, 322, 336–7, 349, **375–84** *passim*, 390–4 *passim*, **402–3,** 406–9
A_5 272, **303–11,** 331, 393, **410–13,** 463
A_n 330, 332
S_4 254, 332, 335; automorphisms of 426–7; conjugacy classes of 380, 384; cosets in 337–8, 351–2; defining relations of 253, 319; generators of 243, 253, 320, 365; representation by permutations 317–21, 492–4; rotations of cube **297–301,** 389; subgroups of **218–22,** 360, 409, 502
S_5 243, 293, 304, 322, 365, 387, 409, 500
S_6 330, 467, 473, 497, 499
S_6 330, 467, 473, 497, 499

groups—(*contd.*)
S_n 226, 237, 239, 243, 381–2
$A_4 \times C_2$ 306, 310
$A_5 \times C_2$ 300, 306, 310, 331, 362
$S_4 \times C_2$ 226, 279, 301–3, 306

half-turn 82, 121–2, 135, 206–7, 209, 231, 234, 285, **369–70, 507–8,** 511, **515–21,** 532 ff
half-turn symmetry 17–18, 79, **187,** 279, 288, 291, 294, 297–8, 304–5, 318, 336, 389–90, 514
hand calculating machine 71, 251
harmonic range 43, *see also* cross-ratio
H.C.F. 2, 7, 37, 51
hexagon, *see* polygon; regular polygon
homography 482–4
homomorphism **Ch. 21**
homomorphic images: failure when $H \ntriangleleft G$ 403–5; of finite groups 80–1, 171–3, 203–4, 399, 401–5, 410; of infinite groups 396–8, 413–16
homomorphic mapping 81, 396–400, 437–9
kernel 397–8, **400–1,** 419–20, 437, 543
natural homomorphism **401–9,** 414–20; *see also* cosets; quotient groups
homothety, *see* similarity group
hyperbola 482, 484

icosahedron, *see* rotation group; symmetry group
identity 5, 18, 25, **Ch. 8**
definition of **46**
left and right **50–1, 540–1**
notation **30**
uniqueness of 541
identity transformation 17–18
image, *see* homomorphism; homomorphic images
index of subgroup **339, 347,** 360, **384,** 394, 402, 406, 408, 473, 502, 535, 542
infinite groups, *see* groups C_∞, D_∞, etc.
inner products, *see* vectors
inner automorphism, *see* automorphism
integers, *see* addition; multiplication in **Z**
intersection (of sets) 6, 14, **37**
interval, *see* music
invariant subgroup 304, 338 *see* normal subgroup
inverse 5, 51, **Ch. 7**
in campanology 456
definition **61**
failure for, 63
in finite arithmetic 59ff

inverse—(*contd.*)
  function **57**, **57–9**, 67
  left and right 62, **540–1**
  matrix **64–6**
  period of **106–7**
  permutation **55**, 324
  of product **66**, 238
inversion
  central 271, 273, 300, 302–3, 306, 335, 345, 390
  in circle 5, 206, 212, 232
inversions of order **312** ff
involution 5, 81, 481–3, 487
irrationals 15–16, 87, 125, 505, *see also* addition in **Q**; multiplication in **Q**
isometry 74, 84–5, **231**, 343–5, **388**, 505
  direct/opposite 84, 344–5, 388, 415, 419, 421; *see also* rotation groups, symmetry groups, transformations
isomorphism 93, 95, 99, 103, **Ch. 11**, 227, 271, 280–3, 299, 303–4, 382, 396, 420–1, 438
  definition **142**
  of infinite groups 144

juxtaposition **30**, 57, 255

Klein four-group **139**, **149**, 318, *see also* group D_2

Lagrange's theorem **215**
  corollaries 108, 126, 141, 170
  proof **346–7**, 355
  uses of 218, 224, 267, 282, 291, 305, 335, 412
Latin square **90–6**, 104, 141, 402
lattice 513, 523, 525, 537
L.C.M. 2, 7, 37, 51, 68
  for period of cycles **115–16**, **321**
  for period of elements in direct product group **263**
leading diagonal 97, 133, 139
left, *see* identity; inverse; coset
linear group 231
logarithm **144 ff**, 429, 436–7
logic 37, **138**

mapping 39, 40, 58
matrices, *see also* determinant, permutation, transformation, etc.
  summary **9–10**
  addition and multiplication **9**, 64, 74
  groups of 74, 86, **97–9**, **119–24**, 129, **135**, 184, 196, 211–12, 226, 229–30, 240, 258, 268, 282, **295–7**, **301–3**, 335, **386–7**, 398, 416
  identity, null **45**
  inverse **64**
  for rotation 119–21, 230, 395
  scalar 303
  similar **373–4**, 387, 395
  (non)-singular **65**, 229, 386
  square 10, 229
  for strip pattern 518
  trace 395
Mercator 236
mirror, *see* reflection
modulus
  of complex number 342
  of congruences, *see* finite arithmetic
motif **505**, 509–12, 519–22, 532
motivation, for groups 89, 398–401
multiplication tables for groups, *see* tables
multiplication, *see also* product
  in **Z** 5
  in **Q** 74, 87, 227, 416, 419
  in **R** 144, 155, 227, 341–3
  in **C** 155, **228–30**, **342–3**, 387, 414–16, 419
  in $\mathbf{Z}_n$, *see* finite arithmetic
  of matrices **9**, *see also* matrices
  of polynomials 16, 69
music 16, 62, 363, 372, **Ch. 23**, 543
  augmentation/diminution 446
  Bach 436, 443, 448–9
  canon 443
  Chopin 446
  common chord 431
  counterpoint 443, 448
  equal temperament 17, 433, **436–7**, 440
  fifth 431
  form **440–8**
  fugue 443–4
  Greek modes 434
  ground 443
  Handel 443, 449
  harmonic series 430–1
  harmony 434
  imitation 443
  interval 429, 432; closure 436; form a group 436; ratio of frequencies 431–4; tempered 429, 432, 438
  inversion 431
  just intonation 434
  leading note 435
  major/minor 434
  modulate 435
  Ockeghem 446
  octave 354, 429
  pentatonic scale 432
  pitch 429
  Pythagoras 433, 439

music—(*contd.*)
round 440–3
Schoenberg 448
semitone 429, 434, 436
sequence 444–5
strette 443
transpose 444, 62, 372
turning of instruments 433, 440
unison 431

network 11–12, 137, 543, *see also* Cayley diagram
neutral element 30, *see also* identity
nine-point circle 494
non-commutative, *see* commutative operations, Abelian group
normal subgroup 272, 306, 345, 359, 362, **382–3, 502**
definition of 338
chains of 408–9
how to check for **384–90**
in infinite groups 413–6
inner automorphisms 290, 422, 426, **428**
kernel in groups of cosets **400–8**
of patterns 534–8
of S_4, A_4, Q_6, D_6 384–5, 389
normaliser 222, *see also* centraliser
of subset **391–3**
notation **xvii, 28–9,** 57, 404
for cycles **113,** 322–4
for group tables 75, 109
multiplicative **30–1,** 59, 147
null element 45, 83, *see also* identity
null set, *see* empty set

odd permutations, *see* permutations
operand 29, 32, 40
operation 2, *see* binary operation
composition of 18, **Ch. 4**
opposite (isometry, transformation, symmetry) **187–93,** 198, 225, **292–3,** 297, **301–2,** 306, 335, **521–8,** *see also* glide reflection; enantiomorph; reflection
order 5, 116, *see also* period
of group 74, 391
group of order p^m 239, 287
ordered pairs 7, 10, 20, 38, 42, 51, 69, 123–4, 133, 153, 163, 184–5, 255–7, 262, 266, 284, 398
ordered triples 42, 100–2, 260–1, 268–9, 275, 282–3
outer automorphism 421, 423

Pappus' theorem 138, 273–4, 489, **497–9,** 502
parabola 31, 133, 480–2
parity, *see* permutation
partition 178, 181, **354,** 397, 417, 437
Pascal hexagon 498–9
pattern 233, 283, **Ch. 26,** *see also* strip pattern
classification 515
sub-pattern and sub-group 532–6
two-dimensional 505–15, **518–38**
three-dimensional 515
period 100, **Ch. 10,** 166, 170, 519
of conjugate elements **364,** 377
definition of 106
in direct product groups **262–7**
finite period in infinite groups 183
of function **117–18**
infinite period 106, **124**
of matrix **118–24,** 129, 184
number of elements of period, 107, 126, 143, 184, 212, 252, 260, 280, 282, 288, 290, 299, 304, 377, 412, 426
of permutations **111,** 127, 184, 215, **321,** *see also* cycles, period of
of quaternions 246
theorems on, 108, 126, **364**
permutation(s) **Ch. 18,** *see also* cycles
analysis into cycles **114–17, 321**
arrangements 56, 199–201
composition of **22–4,** 33–5
cyclic 26, 94, **111–13,** 151, **157,** 159
of diagonals **298–300,** 309, 335, 491–2
even/odd, parity of 294, **312–17,** 329, 332, 335, 399, 463, 467, 472–4
of five cubes 305–6
groups of **76–8,** 85–6, 132, 153–4, 160–1, 194, 210, 221, 259–62, 333
matrices **31–2,** 69–70, **97–9,** 230, 237, 279, 295, 300
notation 24, 56, **113, 322–4**
similar **364**
polar (triangle) 134, 480, 482
polygon 14, 485–7, *see also* regular polygon
polynomial 46, 68, 74, 88, 238–9, 270, 342
mod. $x^2 + 1$ 69, 419
symmetry of **325–7,** 332
Poncelet's porism 485–7
primitive element of group 150, **162,** 166–7, 169, **176–9,** 185, 423
prism 160, 193–4, 281, 288, 293
probability 257
product, *see also* multiplication
of cosets 359, **401–8,** 415
inverse of **66**
products preserved 94–6, **142–3,** 300, **396–8,** 422, 428
product set 356
projective geometry 273

proper subgroup 214
  and improper 187
Ptolemy 327, 490
punctuation 36
pyramid 209
Pythagoras 92, 433, 439

quadrilateral
  complete 491–3
  orthocentric 494
quaternion **245–7**, 251, 284, 387, 394, 420
quintic equation 305, **410**, 413
quotient group 204, **406–9**, 414–9, *see also* Homomorphism

rational functions 88, 239, 343
rationals 66
realisation (representation) of group **139**, 158, 194, 221, 277, 285
reciprocal 57, 66, 68, 228
recurring (repeating) decimal **164–6**
reflections
  in two dimensions 135, **187–9**, 222, 226, 505, 517–8, 521–8 *passim*
  in three dimensions 79, 161, 188, 269–70, 277, 345
  successive 196, 204–8, **233–6**, 277, 367–8
  in moved axis 200–1, **363–8**
reflexive **353**, 374
regular polygon **168–9, 174–5, 188–94**, 389
regular polyhedron **293–307**, 542, *see also* rotation groups, symmetry groups
relation, *see* equivalence relation
re-lettering 272, 274, **489–501**, *see also* vertices
representative, of equivalence class 352, 430, 437
residue classes 46, 52, **340, 352**, 356–7, 359, 397, **413–14**
right, *see* identity; inverse; coset
ring, 266
roots of unity 70, 86, 147, 150, 158, 162, **167–9, 177–81**, 185, 229, 255, 342, 396, 414
rotation, *see also* half-turn
  about moved axis **369–70**
  composition of 83, 344; about fixed point 19, 20, 125, 144, 289, 343, 418, 505
  centre of 506–7, 519–21
  finite plane groups 81–2, 147, 159, 188–92, 196–201
  matrix for 119–21, 230, 295, 301–2
  and translations 21, 231, 343–4, 370–1, 387
in three dimensions 345

rotation groups: of cube 221, **297–302**, 318–19, 389–90, 426
  of bi-tetrahedron 210
  of cuboid 16–17, 131, 149–50
  of isosahedron **303–6, 412–3**
  of prism 193–4, 201–3
  of tetrahedron **293–7**, 336, 378
row vector 9, 31, 45

self-conjugate subgroup 338, 383, *see* normal subgroup
self-corresponding points 483
semi-group 40
sets
  union of 6, 12, 37
  intersection of 6, 14, 37
  symmetric difference of 9, 37, **63–4**, 91, 136, 279
shear 120–1, 230, 237
similarity group 231, 388
simple group 272, 304, 306, **410**, 496
singular matrix **65**, 229, 386
slide-rule for finite arithmetic **145–7**, 164, 211
Spirograph 14
soluble group 409, 411
stabiliser 333, **497–503**
  theorem on, 502
strip pattern 123, 181–3, **204–9**, 283–6, 343, 415, 508–9, **515–18**, 525, 533, 537
structure 1, 6, 397
subgroup 151, **Ch. 14**, *see also* centraliser; centre; normaliser; normal subgroup
  conjugate 382, 390
  definition **214**
  generated by single element **108–9, 216**, 218, 229
  generated by $\geq$ two elements 183, **223–5**
  of infinite groups 227, 240
  invariant properties **220–1**, 240, 297, 492
  of patterns **532–3**
  theorem **217**
  of C_{12} 174
  of cyclic groups 183
  of Q_6 **216–18**, 223
  of S_4 **218–20**, 221–3
  of D_∞ 208–9
  of D_n 203, 209
  of **C**, + ; of **C**. × EM 228–9
  of $C_2 \times C_2 \times C_2$ 273
  of $C_\infty \times C_2$ 286
  of $S_4 \times C_2$ 303
subtraction 14, 18, 45
Sylov **391**
symmetric groups **92–3**, 214, 311, **326**, *see also* groups S_n
  centre of 218

symmetric relation 353, 363
symmetry chart 390, 537
symmetry groups **78–9, 291–310,** *see also* rotation groups
  in plane 292, 505–6
  bi-pyramid 193, 310
  cube 279, 301–3, 335
  cuboid 270–1, 278–9
  full group in three dimensions **292–3,** 306–7, 542
  icosahedron 300, 310
  prism 160–1, 193, 203, 209, 281, 288, 293
  pyramid 209
  reg. hexagon 188, 389
  tetrahedron 297
symmetry operation **78–9,** 187–8

tables for groups **75,** 90, 97
  construction of 92, 141, 190–3, 348–51
  use to find period 107
ternary cycle, *see* cycle
tetrahedron **293–7,** 336, 378, 495, *see also* rotation group, symmetry group
tetrahedral group, *see* groups, A_4
theorems 143, 166, 218, 222, 238, 263, 273–4, 287, 306, 315, 321, 355, 360, 382, 391, 393, 411, 428, 502
topology 11, 20, 74, 543
torsion 284
transfinite number 143, 539
transform 152, 200, 320, **Ch. 20,** 421, 425, 561, 470, 502, *see also* conjugate elements
  definition **363**
  of geometric operations **365–72**
  of matrix 387
  of subgroup **381,** 386
transformation(s), geometric, *see also* glide reflections, rotation, translation, etc.
  associativity 40
  composition of 21, 30
transformation(s)—(*contd.*)
  groups of 83–4, **230–6, 343–5, 387–9,** 394, 415, 419
  successive 17
transitive **353,** 374, 409
translation 21, 83, 125, 181–3, 206–8, 231–6, 343, 368–71, 387–90, 415, **507–18** *passim*, 534–5
transpose of matrix 32
transposition 111, 221, 271, 279, 289, 306, 316–18, 384, 412, 452–78 *passim*, 490
  change of parity **313,** 463, 467
  product of transpositions 127, **313–15,** 330–1
transversal, left or right 361–2

unary operation 6, 16, 32, 62
union 6, 12, 37
unit 45

vector
  addition 20, 30, 37, 44, 52, 68, 256; $\mathbf{V}_2$, + 74, 283, **340,** 357, 398, 416; $\mathbf{V}_3$, + 238, 341, 416, 420
  column, *see* column vector
  product: scalar 14, 52, 419; vector 14, 44, 52, 84
  row, *see* row vector
  space 176, 283
Venn diagram 37, 63, 91
vertices, permutations of 190–2, 197–9, 222, 225–6, 270–1, 288, 293–4, 296–7, *see also* re-lettering

wallpaper, *see* patterns
Wheatstone Bridge 12
word 25–6, 42, 102, 206, 234, 250, 461, 467

zero 18, 45, *see also* identity
  removal of xvii, 66, 176, 227, 229

1974

The Fascination of [partly covered]